Cell and Molecular Biology of the Testis

CELL AND MOLECULAR BIOLOGY OF THE TESTIS

Edited by

CLAUDE DESJARDINS, Ph.D.

Professor
Department of Physiology
University of Virginia
School of Medicine
Charlottesville, Virginia

LARRY L. EWING, Ph.D.

Late Professor
Department of Population Dynamics
Johns Hopkins University
School of Hygiene and Public Health
Baltimore, Maryland

New York Oxford
OXFORD UNIVERSITY PRESS
1993

Oxford University Press

Oxford New York Toronto
Delhi Bombay Calcutta Madras Karachi
Kuala Lumpur Singapore Hong Kong Tokyo
Nairobi Dar es Salaam Cape Town
Melbourne Auckland Madrid

and associated companies in
Berlin Ibadan

Published by Oxford University Press, Inc.,
200 Madison Avenue, New York, New York 10016

Library of Congress Cataloging-in-Publication Data
Cell and molecular biology of the testis /
editors, Claude Desjardins, Larry L. Ewing.
p. cm. Includes bibliographical references.
Includes index. ISBN 978-0-19-506269-4
1. Testis—Physiology. 2. Molecular biology.
I. Desjardins, Claude, 1938–
II. Ewing, Larry L.
[DNLM: 1. Molecular Biology.
2. Testis—cytology. 3. Testis—physiology.
WJ 830 C3924] QP255.C45 1991 612.6'1—dc20
DNLM/DLC for Library of Congress 90-7755

Printed in the United States of America
on acid-free paper

Preface

Advances in understanding complex biological processes ultimately depend on the development of new experimental strategies. This has been readily apparent in the advances that have been made in uncovering the regulatory activities of multiple cell types in the testis. Cells supporting the endocrine and exocrine functions of the testis have temporally different requirements for hormones and growth factors; the kinetics of cell populations that make up the germ, somatic, and endocrine cell lines are remarkably different; separate mechanisms govern the distribution of solutes among anatomically distinct tissue compartments; and it has become increasingly clear that multiple autocrine and paracrine factors have critical roles in directing germ cell development and Leydig cell secretion. Knowledge of these diverse facets of the testis has evolved in the past decade and has come to include more molecular aspects of cell regulation.

There are thus several reasons for writing this book. The major justification is our feeling that the testis is important, and it can serve as a model system for dissecting molecular events involved in cell proliferation and differentiation, as well as for those involved in the synthesis and secretion of regulatory peptides and steroids. While there is no shortage of monographs on some of these topics, this is the first work to integrate the cellular and molecular basis of regulatory processes in the testis. The book is also directed at readers interested in the mechanisms of Leydig, germ, or Sertoli cell responses to induced change, be it physiological, pathological, or toxicological.

The task of preparing this book was shared with Larry L. Ewing, who died after the chapters were submitted for publication. The thematic organization of the book reflects his complementary collaborative effort. Without his early effort this text would not exist. We shared the goal of producing a book that would be a source of advice for beginners and aficionados alike, and would promote investigators to apply new methodologies to study regulatory mechanisms in testicular cells, provide an easily assimilated synthesis of current thinking, and point to the direction of important new areas of research. Nothing would be more satisfying if this effort in any way engendered one or more of these outcomes.

Sincere thanks go to the individuals who have been willing to contribute to this book. I must thank Ellen Barber, Susan Hobbs, and Andrew Stauffer for their advice and assistance in preparing the proofs, and the staff at Oxford University Press for their encouragement and advice.

Charlottesville, VA C. D.
May 1993

Contents

Contributors

C. WAYNE BARDIN, M.D.
The Population Council
Center for Biomedical Research
1230 York Avenue
New York, New York 10021

PHILIPPE BERTA
Biochimie CNRS - INSERM
BP 5051
34044 Monpellier
Cedex, France

C. YAN CHENG, Ph.D.
The Population Council
Center for Biomedical Research
1230 York Avenue
New York, New York 10021

CURTIS CHUBB
Department of Cell Biology and Neuroscience
The University of Texas
Southwestern Medical Center
Dallas, Texas 75235

YVES CLERMONT, Ph.D.
Department of Anatomy
McGill University
Faculty of Medicine
Montreal PQ, Canada H3A 2B

CLAUDE DESJARDINS, Ph.D.
Department of Physiology
University of Virginia
School of Medicine
Charlottesville, Virginia 22908

PATRICIA K. DONAHOE, M.D.
Division of Pediatric Surgery
Massachusetts General Hospital
Boston, Massachusetts 02114

NATHAN A. ELLIS
The New York Blood Center
310 East 67TH Street
New York, New York 10021

LARRY L. EWING, Ph.D.
Division of Reproductive Biology
Department of Population Dynamics
Johns Hopkins University
School of Hygiene and Public Health
Baltimore, Maryland 21205

PETER N. GOODFELLOW
Laboratory of Human and Molecular Genetics
Imperial Cancer Research Fund
London WC2A 3PX, England

CAROL M. GRUPPI
Department of Genetics and Development
Columbia University College of Physicians
and Surgeons
630 West 168TH Street
New York, New York 10032

GLEN L. GUNSALUS, Ph.D.
The Population Council
Center for Biomedical Research
1230 York Avenue
New York, New York 10021

NORMAN B. HECHT, Ph.D.
Department of Biology
Tufts University
Medford, Massachusetts 02155

LOUIS HERMO
Department of Anatomy
McGill University
Faculty of Medicine
Montreal PQ, Canada H3A 2B

BARRY T. HINTON
Department of Anatomy and Cell Biology
University of Virginia
School of Medicine
Charlottesville, Virginia 22908

DAVID M. DE KRETSER
Department of Anatomy
Monash University
Clayton, Victoria 3168
Australia

DIANE S. KEENEY, Ph.D.
Department of Obstetrics & Gynecology
University of Texas
Southwestern Medical School
Dallas, Texas 75235

TATSUO KURODA, M.D.
Department of Pediatric Surgery
Massachusetts General Hospital
Boston, Massachusetts 02114

MARY M. LEE, M.D.
Department of Pediatric Surgery
Massachusetts General Hospital
Boston, Massachusetts 02114

JOHN R. MCCARREY, Ph.D.
Department of Population Dynamics
Division of Reproductive Biology
Johns Hopkins University
School of Hygiene and Public Health
Baltimore, Maryland 21205

MARVIN L. MEISTRICH, Ph.D.
Department of Experimental Radiotherapy
M.D. Anderson Hospital and Tumor Institute
The University of Texas Medical School
Houston, Texas 77030

DIANA G. MYLES, Ph.D.
Department of Physiology
University of Connecticut Health Center
Farmington, Connecticut 06032

RICHARD OKO
Department of Anatomy
McGill University
Faculty of Medicine
Montreal PQ, Canada H3A 2B

JOANNE M. ORTH
Department of Anatomy and Cell Biology
Temple University
School of Medicine
Philadelphia, Pennsylvania 19140

MARK S. PALMER
The Prion Disease Group
Department of Biochemistry and Molecular Genetics
St. Mary's Hospital Medical School
Norfolk Place
London W2, England

GAIL P. RISBRIDGER
Department of Anatomy
Monash University
Clayton, Victoria 3168
Australia

DAVID M. ROBERTSON
Department of Anatomy
Monash University
Clayton, Victoria 3168
Australia

ANDREW H. SINCLAIR
The University of Sydney
School of Biological Sciences
Macleay Building A12
Sydney, NSW 2006
Australia

HERBERT STERN
Department of Biology
The University of California, San Diego
La Jolla, California 92093

TERRY T. TURNER
Department of Urology
University of Virginia
School of Medicine
Charlottesville, Virginia 22908

KENNETH S. K. TUNG, M.D.
Departments of Pathology and Internal Medicine
Washington University Medical Center
660 South Euclid Avenue
St. Louis, Missouri 63110

VIDA VAMBUTAS
The Center for Reproductive Sciences
Columbia University College of Physicians and Surgeons
630 West 168TH Street
New York, New York 10032

MARIA E. A. B. VAN BEEK
Department of Experimental Radiotherapy
M.D. Anderson Hospital and Tumor Institute
The University of Texas Medical School
Houston, Texas 77030

AXEL G. WADEWITZ
The Center for Reproductive Sciences
Columbia University College of Physicians and Surgeons
630 West 168TH Street
New York, New York 10032

DEBRA J. WOLGEMUTH
Department of Genetics and Development
The Center for Reproductive Science
Columbia University College of Physicians & Surgeons
New York, New York 10032

WILLIAM W. WRIGHT, Ph.D.
Division of Reproductive Biology
Department of Population Dynamics
Johns Hopkins University
School of Hygiene and Public Health
Baltimore, Maryland 21205

BARRY R. ZIRKIN
Division of Reproductive Biology
Department of Population Dynamics
Johns Hopkins University
School of Hygiene and Public Health
Baltimore, Maryland 21205

Cell and Molecular Biology of the Testis

CHAPTER ONE

Cell Biology of Testicular Development in the Fetus and Neonate

Joanne M. Orth

The mammalian testis originates in the young fetus and continues to develop throughout the remainder of fetal life and some portion of the postnatal period. After puberty, when its development is completed, the testis will play a dual role in insuring fertility by producing sperm, its exocrine product, and androgen, its endocrine product. Although both functions are vital in the adult, the bases for these processes are established long before puberty, during perinatal development. Interestingly, in young fetuses of both sexes, the gonads evolve along identical lines through the indifferent stage, and it is only after this point that development diverges in males and females. Thus, although genetic sex is determined in an individual at fertilization, "maleness" of the gonad is a characteristic that is acquired through a series of specific developmental changes which give rise to a recognizable testis.

In general, the main feature that distinguishes a fetal testis from an ovary is the presence of seminiferous cords. These contain the seminiferous epithelium that is separated from the surrounding peritubular tissue by a basement membrane. The seminiferous epithelium is comprised of somatic elements, or Sertoli cells, and germ cells or gonocytes; Leydig cells, the endocrine component of the testis, are located in the highly cellular interstitial tissue between the testis cords. Hence, the definitive fetal testis is made up of a number of different types of cells that are maintained in specific physical and functional relationships with each other. The general aims of this chapter are to (1) review the current state of our knowledge concerning the origin of these cells, (2) discuss how they interact with each other and with the extracellular environment during organogenesis and further testicular development, (3) describe the factors, of both extratesticular and intratesticular origin, that regulate testicular growth and development, and (4) detail, as far as is known, the mechanisms by which these factors act on the developing testis.

I. Organogenesis of the Fetal Testis

A. Origin of Testicular Cells

The source of the somatic and germinal cells of the fetal gonad has been a matter of controversy and confusion for many years. Although it is now generally accepted that the primordial germ cells, or gonocytes, originate extratesticularly and migrate to the gonadal ridge, as described later, the origin of the nongerminal or somatic elements remains partially unresolved. The gonad develops in proximity to the mesonephric kidney and is first apparent as a thickening of the surface epithelium on its ventromedial aspect (Fig. 1–1). Whether this epithelium alone gives rise to the underlying medulla or blastema of the indifferent gonad or

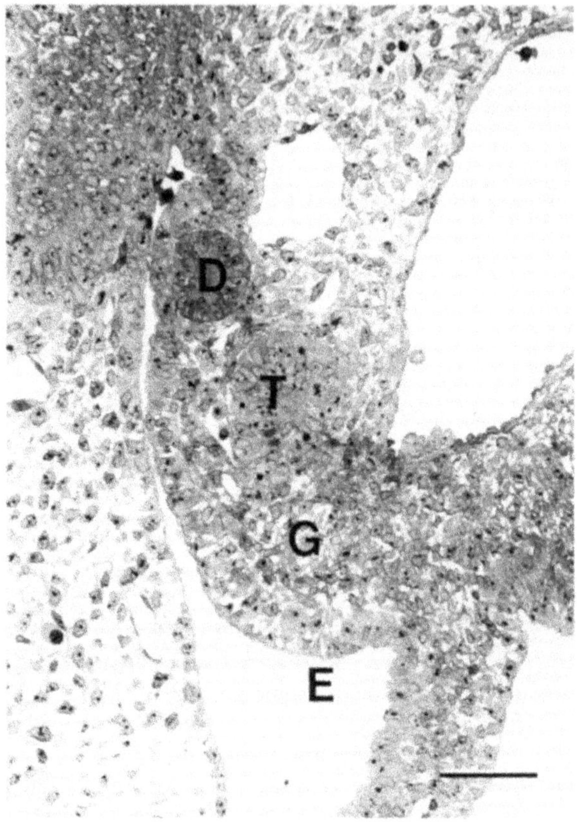

Figure 1–1. An indifferent gonad of a rat fetus aged 12.5 days pc. The gonadal ridge (G) consists of a surface epithelium (E) and underlying loose mesenchyme. The mesonephric duct (D) and a mesonephric tubule (T) are nearby. Bar = 50 μm. (Reprinted with permission from Paranko et al., 1983.)

whether there is a contribution by the nearby mesonephros is less clear (for reviews see Byskov, 1986, and Wartenberg, 1989).

The classic belief was the proliferation of the coelomic epithelium that covers the gonadal ridge is solely responsible for the further development of the testicular medulla and thus gives rise to the seminiferous cords and interstitial tissue of the testis. This concept, however, has proven to be less than satisfactory. Although differentiation of the male gonad appears to differ somewhat among species, it is now generally accepted that the mesonephros contributes at least some somatic cells to the differentiating testis. Indeed, Byskov (1986) proposed that mesonephric cells migrate into the gonadal ridge early in organogenesis and that the involvement of coelomic epithelial cells is minor at best. While this view may hold for some species, controversy still surrounds the issue. For example, elegant experiments with chicken–quail chimeras (Rodemer-Lenz, 1989) provide strong support for a dual origin (i.e., mesonephric and epithelial) of the testis in birds, whereas in some primates, including humans (Wartenberg, 1978), available evidence suggests that the coelomic epithelium provides many of the somatic cells for the fetal gonad. Thus, while the relative importance of a mesonephric or coelomic epithelial source for somatic cells may vary, both elements apparently contribute precursors to the developing testis in most, if not all, species.

The accepted view of germ cell origin in mammals has also changed dramatically from that first proposed. It was believed for decades that the primitive germ cells originated in the covering epithelium of the gonadal ridge, which was called the *germinal epithelium* for that reason. Although diverging views were proposed as early as the mid 1800s (Pflüger, 1863), this theory of gonocyte origin was generally accepted until the 1940s. At that time, Everett (1943) demonstrated that, if mouse gonadal ridges of differing ages were transplanted into adults and left for several weeks, gonocytes did not develop in the transplant unless they had already been recognizable within the gonadal ridge when it was placed in the host. His conclusion that gonocytes develop extragonadally and then relocate to the gonad was later confirmed for many species, including humans (Witschi, 1948) in which the migratory pathway of the germ cells was traced from the yolk sac through the gut mesentery to the gonads.

Today, the extragonadal origin of primitive germ cells in both sexes is universally accepted and the mechanism used by these cells to reach the gonad has been the subject of intensive study both in vivo and in vitro. In general, gonocytes seem to migrate actively in their journey to the genital ridge and display most of the attributes of motile cells at this time (for review, see Eddy et al., 1981). Although the identity of the stimulus for this movement is still unknown, some studies in vitro suggest that the nature of the extracellular matrix on which these cells migrate is important in controlling or directing their movement. For example, fetal gonocytes maintained in culture have been shown to interact with a number of matrix components, including fibronectin (Alvarez-Buylla and Merchant-Larios, 1986) and laminin (DeFelici and Dolci, 1989), which are found along their migratory routes in vivo. A detailed description of the origin of germ cells and of their migration to the developing gonad can be found elsewhere in this volume.

B. Differentiation of the Seminiferous Cords

The way in which Sertoli and spermatogenic cells are organized in seminiferous tubules of the adult testis is vitally important for normal testicular function, as discussed in detail in subsequent chapters. It is in the fetus, however, when the seminiferous cords are established that the groundwork for further changes leading to formation of seminiferous tubules is laid. The results of extensive studies by Jost and coworkers on several species, including rats and rabbits, have contributed greatly to our understanding of the precise events leading to the formation of seminiferous cords (e.g., Jost and Magre, 1984; Jost et al., 1985; Jost and Magre, 1988). As described in rat fetuses, the first recognizable event in an indifferent gonad destined to become a testis is the emergence at 13.5 days postconception (pc) of large clear cells that aggregate together and eventually enclose the germ cells within solid cords (Fig. 1–2). These clear cells are identifiable as Sertoli

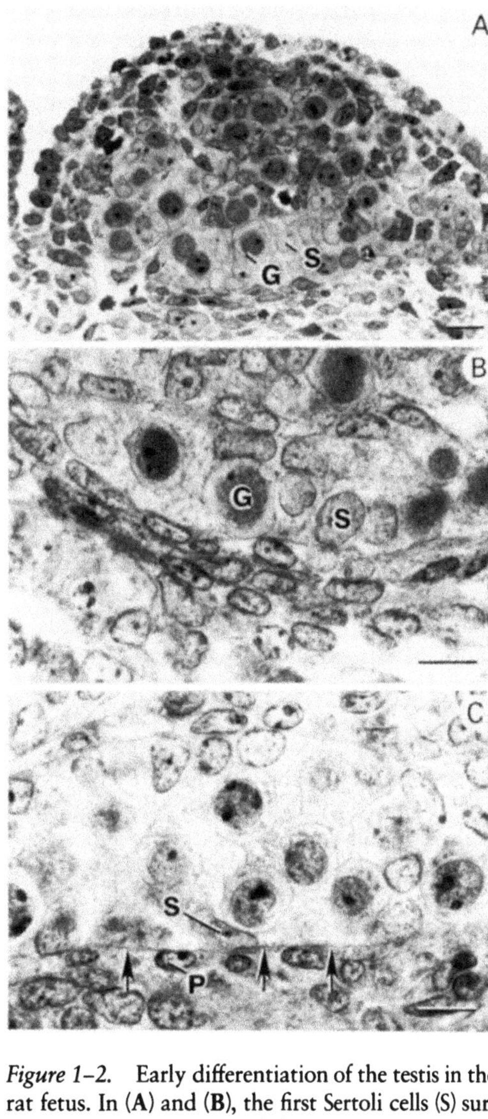

Figure 1–2. Early differentiation of the testis in the rat fetus. In (**A**) and (**B**), the first Sertoli cells (S) surrounding some gonocytes (G) appear in the depth of the gonad on day 13 [13 day 9 hours in (**B**) and 13 day 15 hours in (**A**)]. (**C**) Differentiated seminiferous cord in the testis on day 14 (14 day 14 hours fetus). A basement membrane *(arrows)* is present between the Sertoli (S) and peritubular (P) cells. G, gonocyte. (**A**) Bar = 15 μm. (**B**), (**C**) Bar = 10 μm. (Reprinted with permission from Jost and Magre, 1988.)

cells by their capacity to produce mullerian-inhibiting hormone (MIH; Magre and Jost, 1984). The forming seminiferous cord is rapidly surrounded by a distinct basal lamina, a process that takes less than 24 hours in the fetal rat. At or about the same time, mesenchymal cells flatten and become closely associated with the basal portions of the Sertoli cells to comprise the boundary tissue of the cords (Figs. 1–2 and 1–3). A similar sequence of events is reported to occur in human fetuses, except that segregation of platelike structures containing somatic cells apparently precedes their development into cords (Wartenberg, 1989). In humans, cords become obvious by the end of week 9; however, well-organized placement of peritubular elements around the cords is not apparent until the 18th week.

Thus, the morphological changes that give rise in fetuses to recognizable seminiferous cords have been described in some detail; however, our understanding of the cellular mechanisms involved and of how these events are controlled remains incomplete. It is thought that initiation of testicular organogenesis is either spontaneous or regulated by local factors produced within the urogenital ridge because differentiation of seminiferous cords occurs in vitro, providing the mesonephroi and gonadal primordia are explanted together. Although the precise factor(s) stimulating or regulating aggregation of Sertoli cells into cords are still unknown, the inclusion of serum with cultured gonads (Fig. 1–4; Agelopoulou et al., 1984) or the addition of cyclic adenosine monophosphate (cAMP) (Taketo et al., 1984) is reported to prevent cord formation. Interestingly, gonocytes appear to continue normal development in cAMP-treated testes in which cord formation has been prevented (Taketo et al., 1984). Moreover, the presence of germ cells is apparently not required for cord formation, as evidenced by normal Sertoli cell aggregation in several mouse models lacking germ cells. Thus, although Sertoli–germ cell interaction is critically important later in development (discussed later) as well as in adults, it is not a prerequisite for seminiferous cord formation in fetuses. It should also be noted that cytodifferentiation of Sertoli cells is apparently not dependent on cord formation because MIH-positive Sertoli

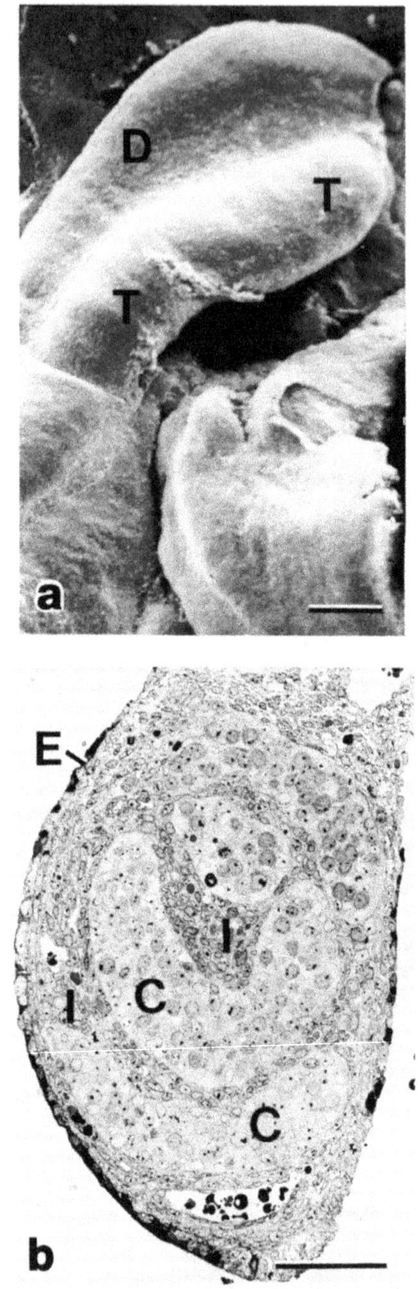

Figure 1–3. (**a**) Scanning electron micrograph of a testis (T) from a rat fetus aged 14.5 days pc. D, mesonephric duct. Bar = 200 μm. (**b**) Thick section of fetal rat testis aged 15.5 days pc. Seminiferous cords (C) and interstitium (I) have differentiated and the surface epithelium (E) has become distinct and started to flatten. Bar = 100 μm. (Reprinted with permission from Paranko et al., 1983.)

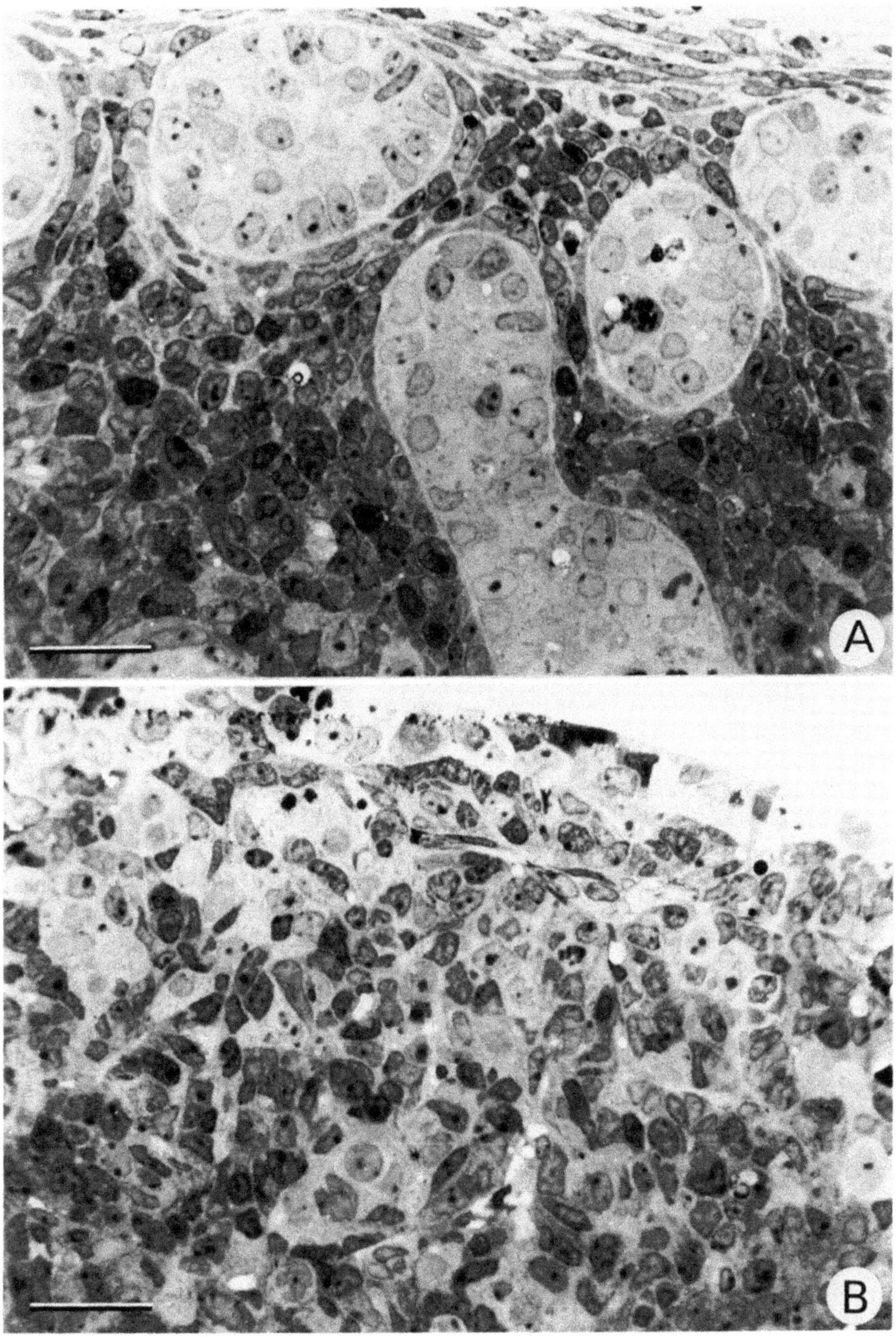

Figure 1–4. Semithick sections through male gonads taken from 13 day 9 hours rat fetuses and cultured in vitro for 4 days. (**A**) Control gonad grown in synthetic medium (CMRL, 1066 medium). It contains well-differentiated seminiferous cords. (**B**) Gonad cultured in the same medium supplemented with fetal calf serum (15 percent). The differentiation of seminiferous cords did not occur under these conditions. Bar = 25 μm. (Reprinted with permission from Jost and Magre, 1988.)

cells appear in cultured gonads even if their aggregation into cords is prevented.

On the other hand, interaction between Sertoli cells and the surrounding extracellular matrix seems to be critical for seminiferous cord formation in fetal testes. Sertoli cells appear to cooperate with a subpopulation of mesenchymal cells, presumably the future peritubular cells, to produce the extracellular material surrounding both cell types in young fetuses. This is suggested by the observation that the basal lamina underlying the seminiferous epithelium in postnatal testes is a joint product of Sertoli and peritubular myoid cells (Skinner et al., 1985). In the fetal testis, the nature and location of this matrix may be critical in promoting aggregation of Sertoli cells into cords. This idea is supported by the finding that cords fail to form in the presence of cAMP (Taketo et al., 1984), an effect that may be related to inhibition of matrix synthesis by the nucleotide. Indeed, when older fetal testes already containing cords are incubated with cAMP, the basement membrane eventually disintegrates and disappearance of the cords themselves occurs shortly thereafter.

Immunocytochemical observations on the normal sequence of events in developing testes have also provided support for the idea that the extracellular matrix is critical for cord formation. For example, before the appearance of cords, fibronectin and laminin are distributed uniformly between the somatic cells of the fetal rat testis (Fig. 1–5a; Paranko et al., 1983; Pelliniemi et al., 1984; Agelopoulou and Magre, 1987). As cells aggregate to form cords by about 13.5 days pc, however, these two matrix factors become restricted to the newly formed basement membrane that also contains collagen types IV and V and heparan sulfate proteoglycan (Fig. 1–5b; Paranko et al., 1983; Pelliniemi et al., 1984). In addition, at least some of these components are essential for aggregation of Sertoli cells into seminiferous cords in vitro. This was demonstrated by Jost et al. (1985), who cultured indifferent gonadal primordia from male rat fetuses in media containing L-azetidine 2-carboxylic acid (LACA), a proline competitor known to interfere with synthesis and secretion of collagen. In these incubations, Sertoli cells aggregared but failed to form well-defined cords; moreover, the expression of fibronectin and laminin in the LACA-treated primordia was suppressed. Unlike controls, no basement membrane formed in these cultures and the Sertoli cells did not become polarized or establish a structural relationship with presumptive peritubular cells. Thus, Sertoli-myoid cell interactions leading to production of extracellular matrix of a specific composition are likely to have important implications for seminiferous cord formation in vitro and perhaps in vivo as well. However, whether the synthesis and organization of these matrix components into a basement membrane causes the appearance of cords or is merely an obligate adjunct of some other essential process(es) remains to be determined, as does the identity of the precise regulatory mechanism(s) underlying cord development.

C. Development of Fetal Leydig Cells

In virtually all species examined, recognizable Leydig cells appear in the interstitium of the fetal testis sometime after development of the seminiferous cords. In rats, 1 or 2 days separate these events while in humans the appearance of Leydig cells follows that of cords by about 1 week. Although not proven directly, the source of fetal Leydig cells is likely to be the mesenchyme of the gonadal blastema, which is thought to contain a subpopulation of undifferentiated cells from which Leydig cells originate. This concept is supported by the observation of a decrease in the number of mesenchymal cells in gonads of fetal pigs that corresponds temporally with the appearance of Leydig cells and is inversely related to the increasing numbers of these cells. Although first appearing as rounded precursor cells with a moderately dense cytoplasm (Fig. 1–6), Leydig cells gain the typical ultrastructural characteristics of steroidogenic cells as development proceeds. Because the appearance of Leydig cells follows that of seminiferous cords, it has been suggested that cord formation may play a role in stimulating differentiation of these cells. This idea is supported by the observation that, when fetal testes containing cords but no Leydig cells are minced to disrupt cords and placed in vitro, Leydig cells do not differentiate as they do in whole testes that retain well-defined cords. However, in other experiments for

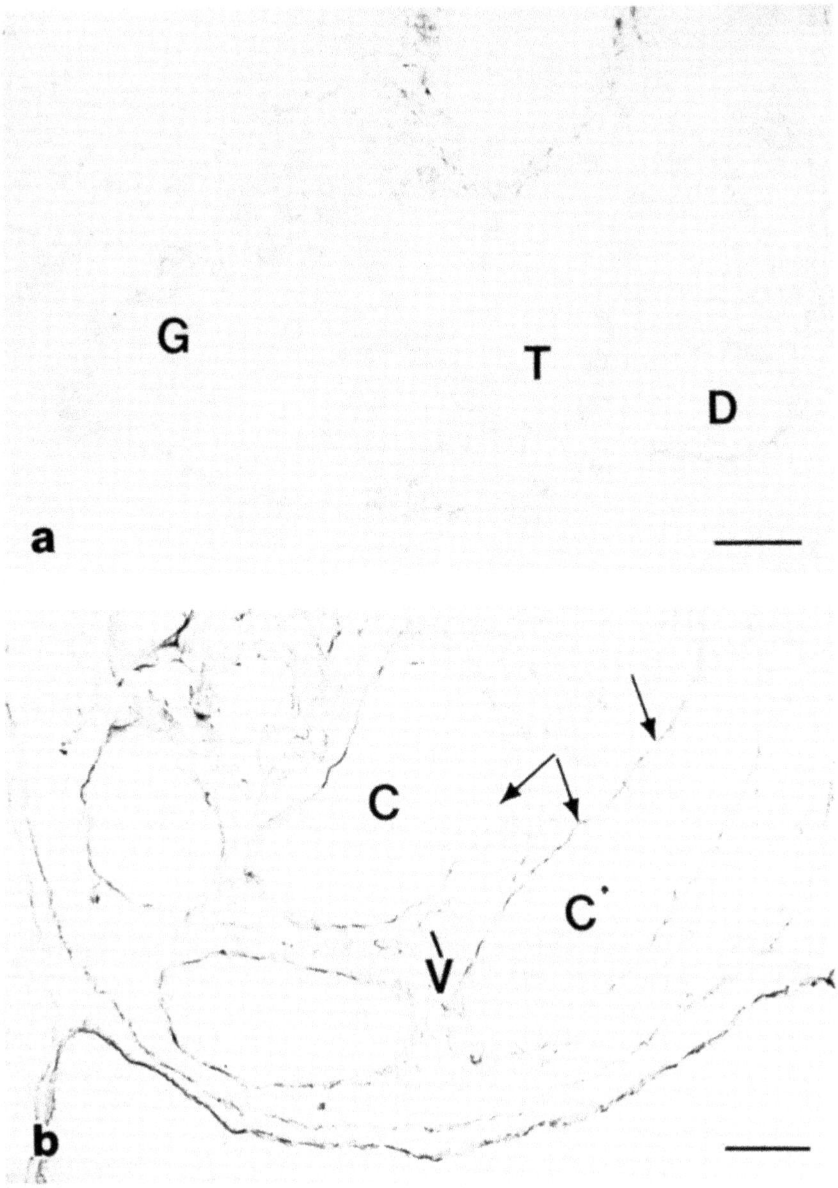

Figure 1–5. (**a**) Indifferent gonad of a rat fetus at 12.5 days pc. Fibronectin, localized immunocytochemically by the peroxidase-antiperoxidase technique, is found in small amounts in the mesenchyme of the gonad (G) and mesonephros. T, mesonephric tubule; D, mesonephric duct. Bar = 50 μm. (**b**) Testis of a rat fetus at 15.5 days pc, treated to localize fibronectin as in (**a**). A continuous fibronectin layer *(arrows)* is found at the basement membrane surrounding the seminiferous cords (C). Some fibronectin is also seen within the interstitial tissue, especially surrounding vessels (V), as well as at or near the surface of the testis. Bar = 50 μm. (Reprinted with permission from Paranko et al., 1983.)

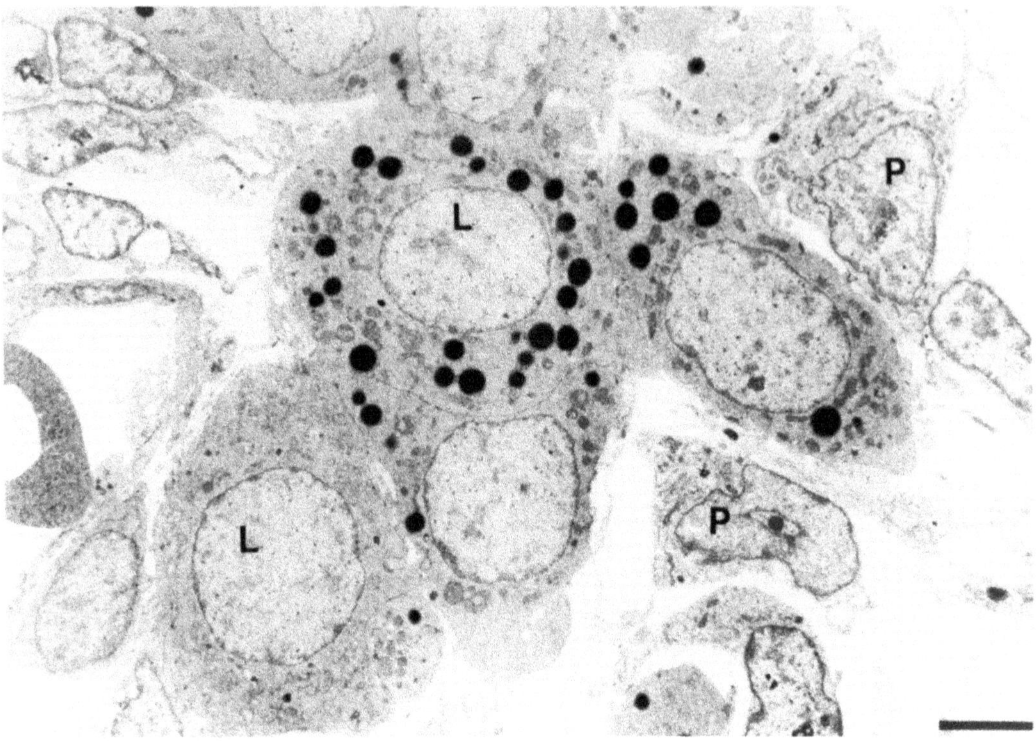

Figure 1–6. Electron micrograph of fetal-type Leydig cells (L) and apparent Leydig cell precursors (P) in the testis of a 19-day-old fetal rat. The mature cells are often in contact with each other, as seen here, and will later form distinct clusters. Bar = 5 μm. (Reprinted with permission from Kuopio et al., 1989a.)

which cord formation in vitro was inhibited by inclusion of fetal calf serum, some cells that were positive for steroidogenic enzymes and capable of producing androgen did appear after several days (Patsavoudi et al., 1985). Although steroidogenesis was at an abnormally low level in these cord-negative gonads, this observation suggests that at least some functional Leydig cells can form in the absence of organized seminiferous cords.

Recent observations also suggest that development of fetal Leydig cells may be similar to that of seminiferous cords in requiring a particular extracellular matrix. Partial or complete basement membranelike structures containing laminin (Fig. 1–7) and collagen type IV (Fig. 1–8) have been detected surrounding groups of fetal rat Leydig cells (Kuopio et al., 1989, 1989a), whereas some patchy remnants of this matrix were found adjacent to Leydig cells of adults (Kuopio and Pelliniemi, 1989). Similar observations of matrix material in association with mouse and human Leydig cells have also been reported. More direct evidence pointing to the importance of the extracellular matrix was provided by Jost et al. (1988) who found that Leydig cells failed to appear when collagen synthesis was impaired by adding LACA, a proline competitor (Fig. 1–9). These interesting findings have led to the suggestion that fetal Leydig cells might comprise a population of epithelial-like cells that are regulated at least to some extent by their association with an extracellular matrix. For example, Leydig cell–matrix interaction during development may promote aggregation of these cells into clusters or islands (see Fig. 1–6) within which the cells themselves could interact and become functionally differentiated. In summary, regardless of the precise events involved, a significant body of current evidence points to involvement of the extracellular matrix in promoting development of fetal Leydig cells.

It is not entirely clear whether Leydig cell

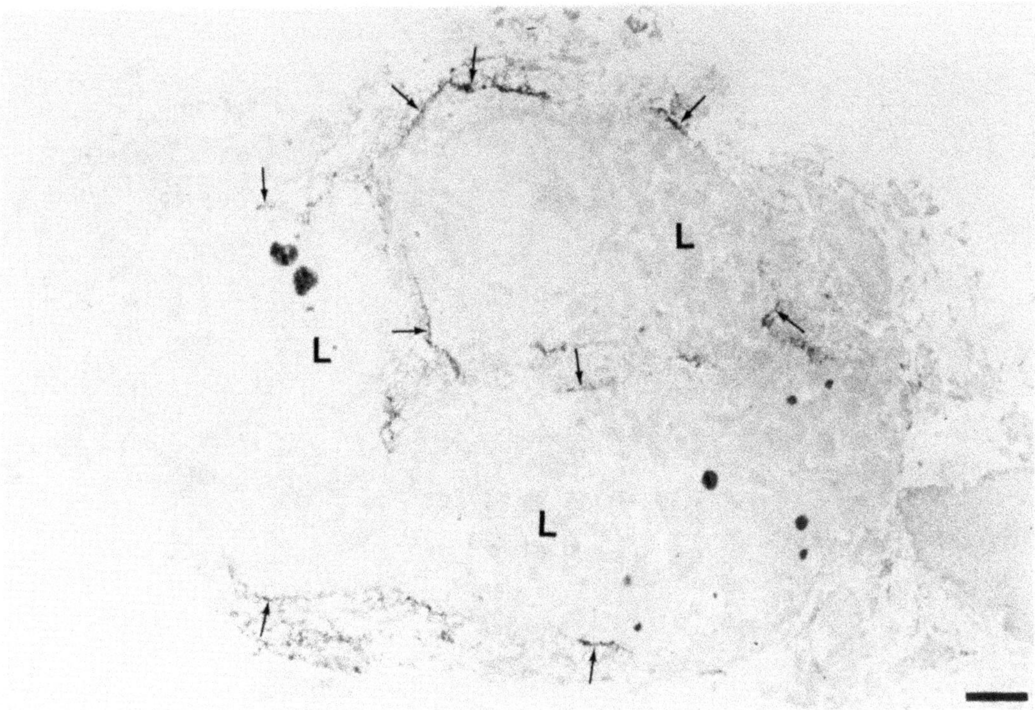

Figure 1–7. Immunocytochemical localization of laminin in a testis of a 3-day-old rat pup. Immunolabeling *(arrows)* is found on the surface of a cluster of Leydig cells as well as between individual cells. L, Leydig cell. Bar = 2 μm. (Reprinted with permission from Kuopio et al., 1989a.)

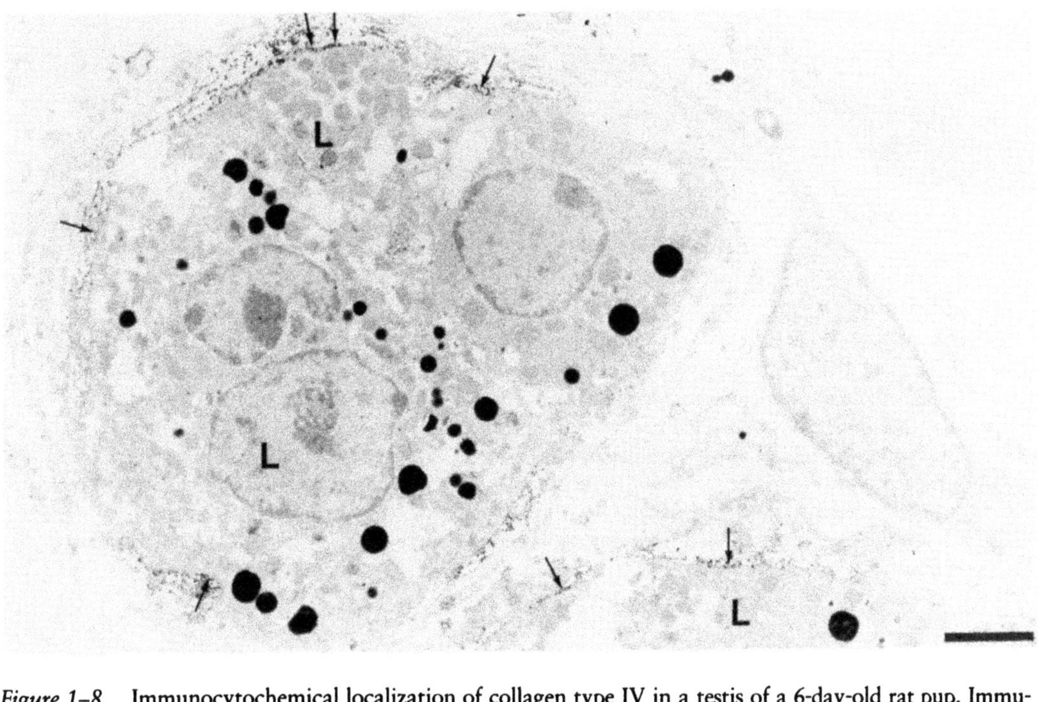

Figure 1–8. Immunocytochemical localization of collagen type IV in a testis of a 6-day-old rat pup. Immunolabeling *(arrows)* is seen primarily on the surface of a cluster of Leydig cells. Bar = 2 μm. (Reprinted with permission from Kuopio et al., 1989a.)

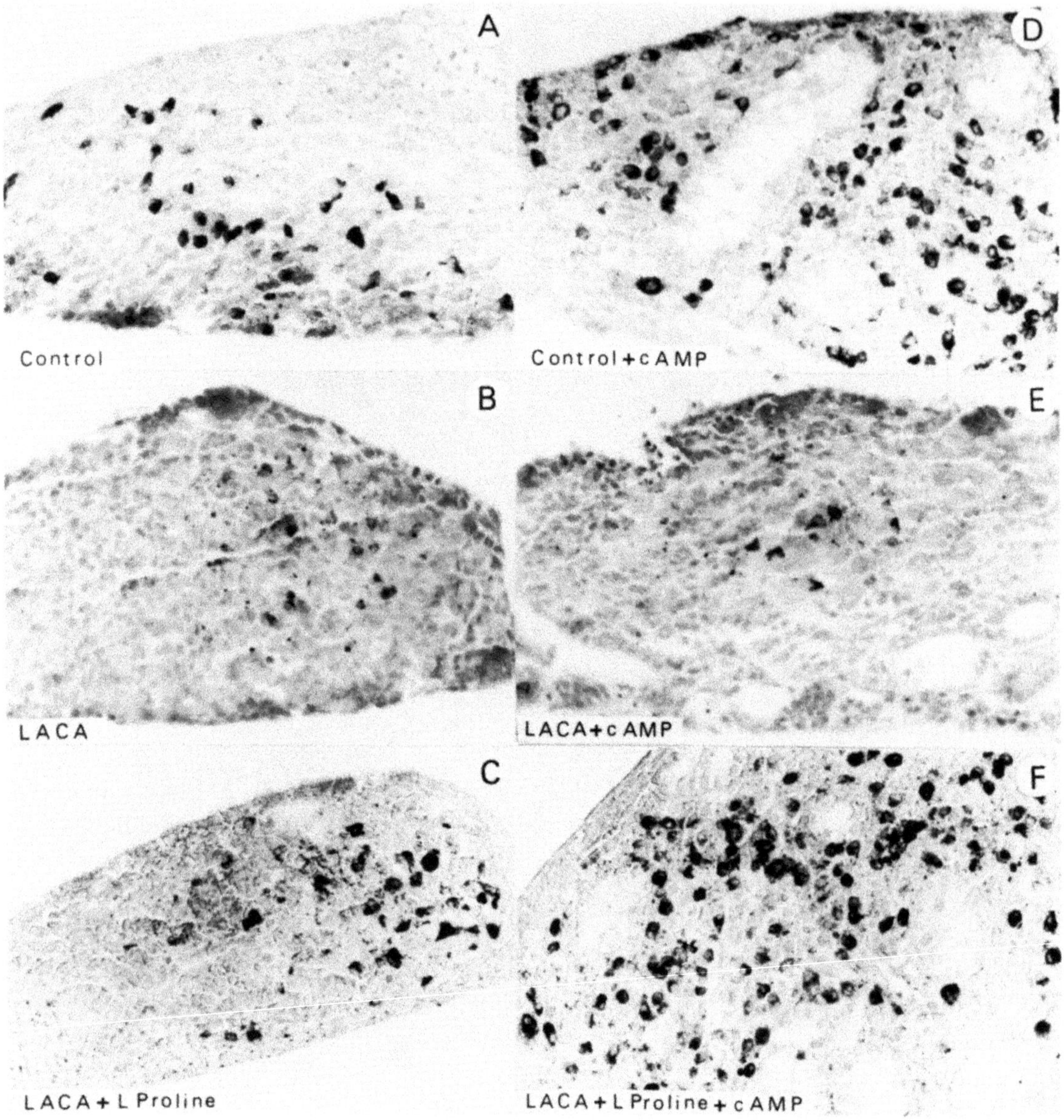

Figure 1–9. Histochemistry for 3β-hydroxysteroid dehydrogenase activity, a functional indicator of the presence of Leydig cells, in male gonads explanted from 13 day 9 hour rat fetuses and cultured for 4 days in vitro. (**A**) Control medium for 4 days. (**B**) Medium containing LACA (100 μg/mL), a proline competitor, for 4 days. (**C**) Medium containing LACA plus L-proline (250 μg/mL). (**D–F**) The same medium, respectively, as **A–C**, but supplemented with db cAMP (1 mM) for the final 24 hours. More Leydig cells are evident in cAMP-treated gonads, whereas LACA prevented their development almost totally in both controls and cAMP-treated gonads. Addition of exogenous proline, however, abolished the effect of LACA in both cases. (Reprinted with permission from Jost et al., 1988.)

differentiation requires extragonadal factors. Some evidence points to a role for pituitary gonadotropins in development of these cells, at least in several species. For example, removal of the fetal pituitary by decapitating fetuses in utero was found to cause some depletion of Leydig cell numbers in rabbits and monkeys. In addition, abnormally few Leydig cells appear in testes of human male anencephalic fetuses. On the other hand, when rat fetal testes were placed in culture at 12 to 14 days pc, before differentiation of Leydig cells, cells positive for steroidogenic enzymes appeared after culture for several additional days in serum-free medium (Patsavoudi et al., 1985). This suggests that at least the initial appearance of Leydig cells may not be under extragonadal control. It is noteworthy that, in those incubations, spontaneous differentiation of Leydig cells was apparently delayed by 1 to 2 days if the cultures were initiated on day 12 or 13 rather than at 14 days pc. Taken together, these observations on the development of Leydig cells in several species under various conditions suggest that initial differentiation of these cells may not require extratesticular factors but that subsequent expansion of this population is likely to be under pituitary control.

II. Testicular Growth During Fetal and Early Postnatal Development

The testis is unusual in its mode of growth in a number of ways. First, it enlarges relatively rapidly during both the fetal and neonatal period even though puberty occurs considerably later in postnatal life. Second, each population of cells within the testis expands during a particular period of development and, at least in the case of Sertoli and spermatogenic cells, these periods are distinct from each other and inversely correlated. Third, the factors that regulate growth of testicular cells, as far as they have been identified and studied, are apparently specific to each population and originate both intratesticularly and extratesticularly. Although our understanding of testicular development and its control is far from complete, progress has been made in determining when and how the testis grows and in clarifying the nature of some of the factors regulating its growth.

A. Cellular Proliferation in the Perinatal Testis

1. SERTOLI CELLS

In adults, Sertoli cells are fully differentiated, nondividing cells that apparently comprise a stable population. Their interactions with spermatogenic cells, as discussed in detail in subsequent chapters, are crucial for maintenance of normal sperm production. Because of this important functional relationship between the germ cells and Sertoli cells of adults and because mature Sertoli cells are nonrenewable, the period of development when these cells divide and the size of the population is established is especially important in laying the foundation for normal function of the adult testis and thus insuring fertility.

Little was known about how and when the size of the Sertoli cell population is determined until observations were made by Clermont and Perey (1957) on rat Sertoli cells in vivo and subsequently by Steinberger and Steinberger (1971) on developing rat testis in vitro. In both studies, mitotic figures believed to represent dividing Sertoli cells were detected in newborn rat testes both in vivo (Clermont and Perey, 1957) and in organ culture (Steinberger and Steinberger, 1971). Additional findings from these two studies suggested that Sertoli cell replication declines with increasing postnatal age.

More recently, the detailed relationship between testicular development and Sertoli cell proliferation was studied with a quantitative autoradiographic approach (Orth, 1982). For this, ^{3}H-thymidine was administered subcutaneously to rat fetuses and pups and the testes were processed for light microscope autoradiography 2 hours later (see Figs. 1–10 and 1–11 for examples of autoradiographs at several ages). The resulting quantitative data are detailed in Figure 1–12. Since the youngest fetuses examined were 16 days old, about 2 days after discrete seminiferous cords appear, it is still not clear whether Sertoli cells divide while becoming organized into cords. As numerous Sertoli cells proliferate at 16 days pc (Fig. 1–10A), however, it seems likely that some of

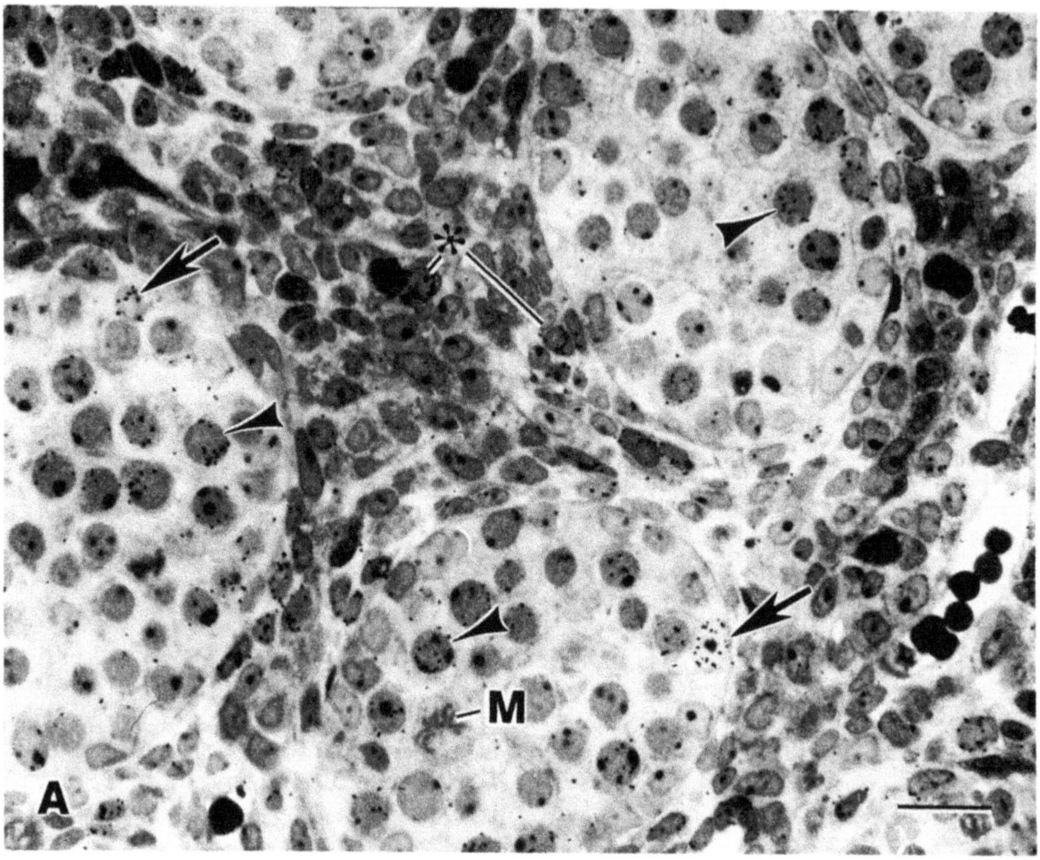

Figure 1–10. Autoradiographs of testes from rat fetuses aged 16 days pc in (**A**) and 17 days pc in (**B**). Both fetuses were given ^{3}H-thymidine subcutaneously while in utero and killed 2 hours later. *Arrows,* labeled Sertoli cells; *arrowheads,* gonocytes. In (**A**), many Sertoli cells and gonocytes are proliferating, whereas in (**B**) numerous Sertoli cells are still labeled but gonocytes have ceased dividing. From day 17 onward, gonocytes enter a "cap" phase, during which organelles become clustered to one side of the nucleus as seen in (**B**). Labeled peritubular cells (P in [**B**]) and dividing undifferentiated interstitial cells (*asterisk* in [**A**]) are common throughout fetal life. Bars = 25 μm. (Reprinted with permission from Orth, 1982.)

these cells divide before this age. When the percentage of Sertoli cells that had incorporated ^{3}H-thymidine was quantified from day 16 pc onward, the greatest expansion of the Sertoli cell population was detected during the latter part of fetal life, between days 18 and 21, a period that coincides temporally with increasing production of androgen by fetal Leydig cells. Postnatally, Sertoli cells divide at a steadily decreasing rate, until division ceases sometime between 10 and 16 days of age (Steinberger and Steinberger, 1971; Orth, 1982). Thereafter, all Sertoli cells apparently become terminally differentiated and mitotically quiescent, entering an extended G_1 phase of the cell cycle sometimes termed "G_0."

The close functional relationship that is believed to exist between Sertoli and spermatogenic cells of adults is apparently established during development, while Sertoli cells are still dividing. This was suggested by the recent development of a "Sertoli cell-depleted" adult rat (Orth et al., 1988). In these animals, Sertoli cell proliferation was impeded in neonates, when germ cells are nondividing (see below), by short-term intratesticular administration of an antimitotic drug. There was no residual effect of this treatment on the onset of germ cell di-

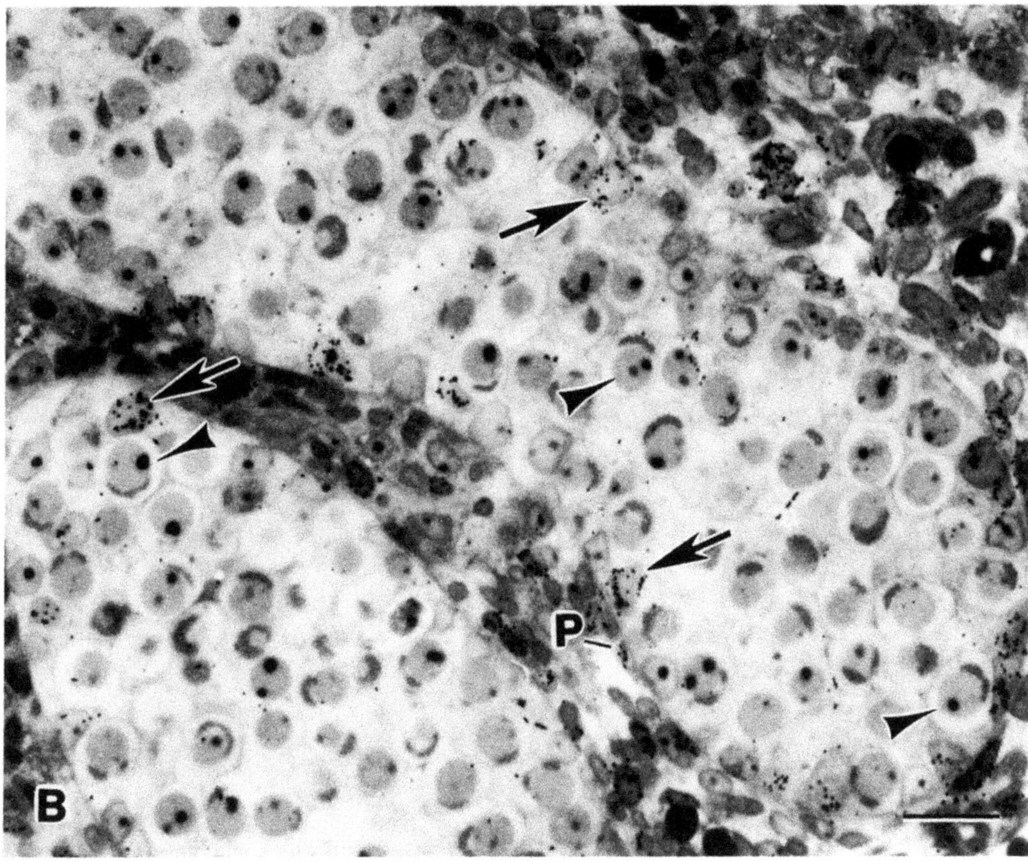

Figure 1–10. (**B**).

vision several days later and it seemed to leave Leydig cells unaffected. Because Sertoli cells were dividing at the time of treatment, however, a significant number of mitotic Sertoli cells were destroyed before the drug was cleared from the testis. As a result, the Sertoli cell population in adults that had been drug-treated as neonates was decreased in size by more than 50 percent compared with controls (Table 1–1). While spermatogenesis in Sertoli cell-depleted adults was qualitatively normal (Fig. 1–13), a comparison of the numbers of immature round spermatids in control and Sertoli cell-depleted adults indicated that these germ cells were decreased by 50 percent in the latter (Table 1–1), a decrease that paralleled the diminution in Sertoli cell population size in these rats. Thus, the number of round spermatids per Sertoli cell was essentially identical in both groups, suggesting that a fixed numerical relationship exists between these cells, and that the number of germ cells supported by each Sertoli cell under normal conditions is maximal and incapable of increasing. These observations in an experimental animal model underscore the long-term impact of interactions between Sertoli and spermatogenic cells during development and point to the size of the Sertoli cell population as an important factor in determining fertility of the adult. Available data concerning the precise mechanism(s) through which immature Sertoli cells and gonocytes may interact in perinatal life are discussed later in this chapter.

2. GERM CELLS

During fetal development of mammalian testes, the primitive spermatogenic cells or gonocytes

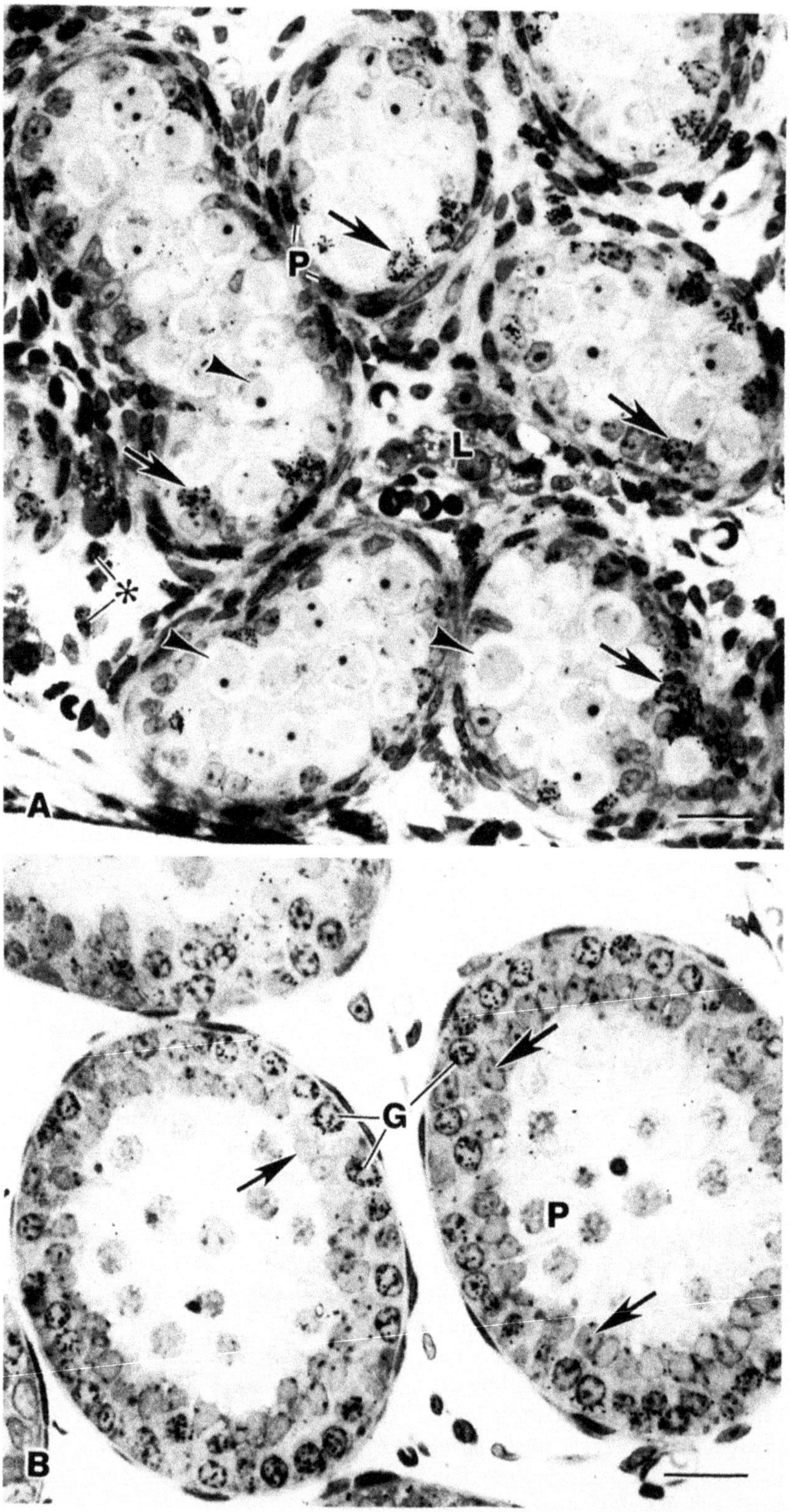

Figure 1–11. Autoradiographs of testis from a rat fetus aged 20 days pc (**A**) or from a pup 21 days after birth (**B**). ^{3}H-thymidine was administered to the fetus subcutaneously and the animal killed 2 hours later. In (**A**), many Sertoli cells *(arrows)* are dividing, while gonocytes *(arrowheads)* are still quiescent. L, Leydig cells; *asterisk,* labeled undifferentiated interstitial cells; P, labeled peritubular cells. In (**B**), all Sertoli cells *(arrows)* are unlabeled and have apparently become terminally differentiated. G, proliferating spermatogonia; P, primary spermatocyte. Bars = 25 μm. (Reprinted with permission from Orth, 1982.)

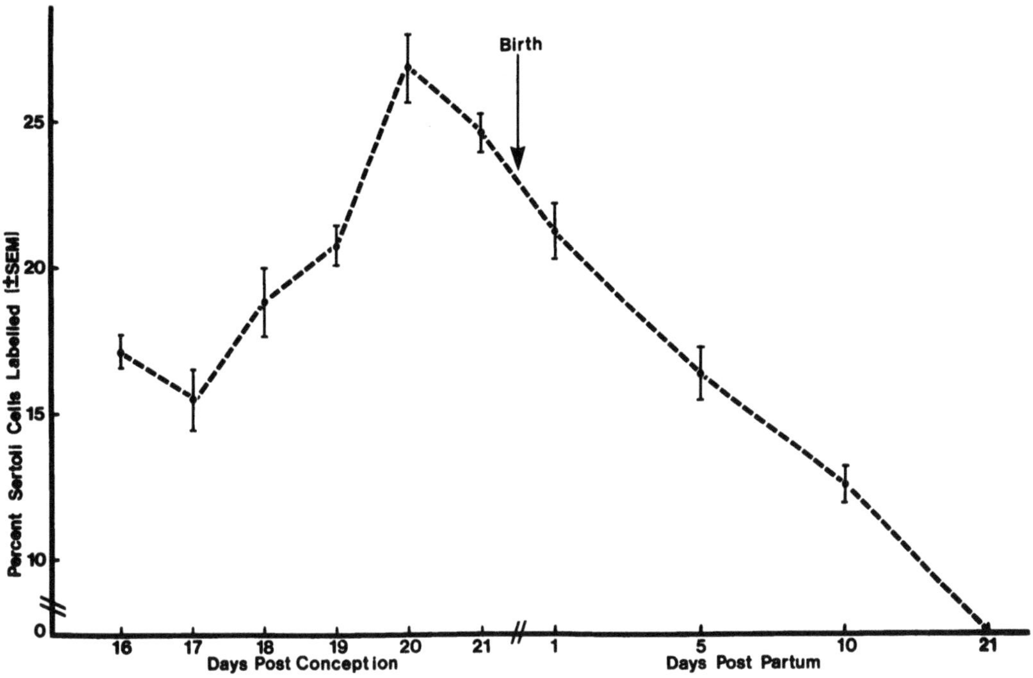

Figure 1–12. The percentage of Sertoli cell nuclei that had incorporated ^{3}H-thymidine and were labeled in autoradiographs of testes from fetal and postnatal rats. Each point is a mean of six animals from two litters (± SEM). In fetuses, significant increases occurred between days 17 and 18 pc ($P < 0.05$) and between days 19 and 20 pc ($P < 0.01$). Between days 20 and 21 pc this percentage fell ($P < 0.05$) and, during the subsequent postnatal period, additional decreases ($P < 0.01$) were detected at each age studied. No labeled Sertoli cells were found on day 21 after birth. (Reprinted with permission from Orth, 1982.)

display several waves of proliferative activity whose duration and extent seem to vary substantially from species to species. The periods of germ cell division that occur before puberty have been termed *prespermatogenesis* (Hilscher et al., 1974) and germ cells participating in these events are described by various terminologies relating to the presence or absence of mitosis at each stage. Since studies of rats have provided the most detailed information, this discussion focuses on the rat, with additional comments added concerning humans.

Table 1–1. Number of Sertoli Cells Per Testis and of Round Spermatids Per Testis and Per Sertoli Cell in Normal and Sertoli-cell-depleted (SCD) Adult Rats.

	n	No. of Sertoli cells/testis	No. of round spermatids/testis	No. of round spermatids/Sertoli cell
Control	4	15.5×10^{6a}	92.2×10^{6b}	5.7
		(0.7×10^{6})	(8.1×10^{6})	(0.2)
SCD	4	7.1×10^{6a}	41.1×10^{6b}	6.0
		(1.1×10^{6})	(8.0×10^{6})	(0.4)

[a,b] $P<0.01$; ± SEM is in parentheses.

Reprinted with permission from Orth et al., 1988.

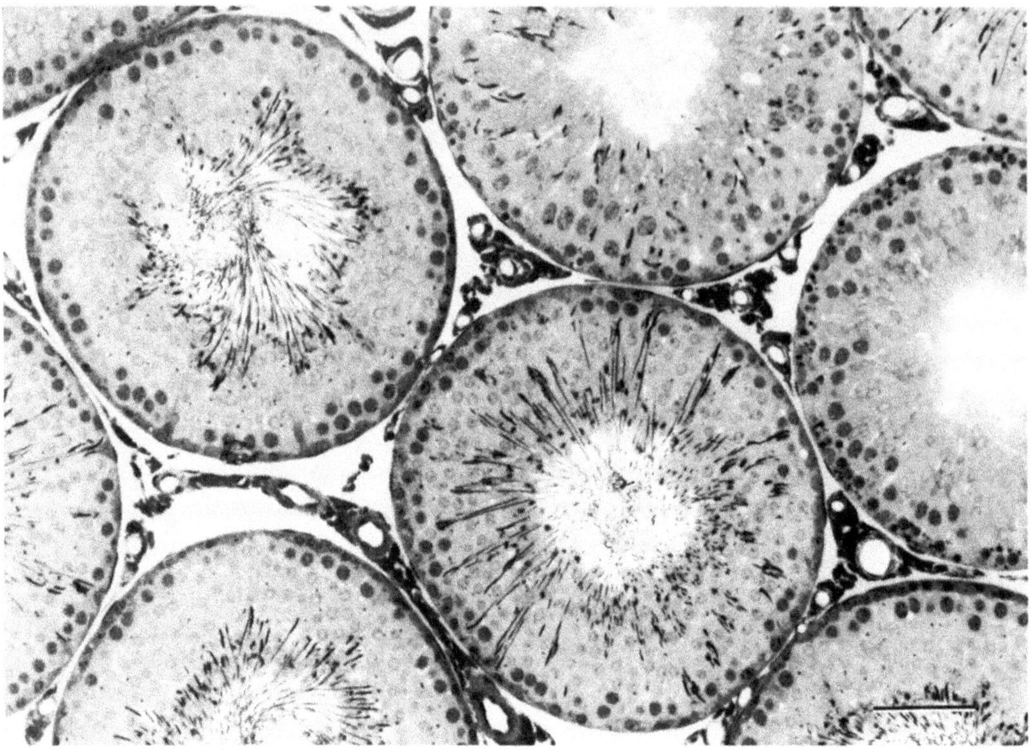

Figure 1–13. A light micrograph of a portion of a testis from a 60-day-old Sertoli cell-depleted rat. This animal had been injected intratesticularly with cytosine arabinoside on day 2 after birth. Seminiferous tubules appear morphologically normal, with mature sperm present in many areas. Bar = 100 μm. (Reprinted with permission from Orth et al., 1988.)

It is generally believed that gonocytes proliferate to at least a limited extent during their early migration into the gonadal ridge. Once enveloped by Sertoli cells during formation of the cords, however, they cease dividing for a period of time. It is interesting that inclusion of gonocytes within the cords is apparently vital for their survival since Byskov (1986) reports that any of these cells remaining in the interstitial areas degenerate. All gonocytes within the testis are mitotically quiescent until fetal day 12 (Hilscher et al., 1974), at which time they reinitiate mitosis and continue to divide for several additional days. In Sprague-Dawley rats, this second wave of proliferation ceases abruptly between days 16 and 17 postconception (Orth, 1982; see Fig. 1–10) and is followed by an obvious, but poorly understood, change in gonocyte morphology. On day 17, these quiescent cells enter a "cap phase" characterized by clustering of organelles, primarily mitochondria (Orth, unpublished observations), to one side of the nucleus (see Fig. 1–10b). During the remainder of fetal life and the first few days after birth, gonocytes remain nonproliferative and enlarge noticeably. It is interesting to note that the period of late fetal development when gonocytes in the male are quiescent corresponds temporally with the time of meiotic prophase in the female, a coincidence that is discussed more fully later.

Gonocytes in the postnatal rat have been the object of relatively few studies and hence their development is not well understood. Hilscher et al. (1974) suggested that rat gonocytes remain nonmitotic until several days after birth. It is during this period of quiescence that substantial numbers of degenerating gonocytes (Byskov, 1986) have been observed in many species, including the rabbit in which fully 50 percent of the gonocytes disappear during the first few postnatal days. Thus, degeneration of gonocytes probably accounts in large measure for the decrease in seminiferous cord diameter and in numbers of germ cells observed by Clermont and Huckins (1968) when they compared neonatal rats with late fetuses. In a recent study aimed at pinpointing the precise time when gonocyte mitosis is reinitiated in neonates, rat pups were injected subcutaneously with tritiated thymidine and their testes subjected to quantitative autoradiography (McGuinness and Orth, 1989a). The results indicated that approximately 7 percent of these cells reinitiate division on day 3 after birth and that proliferation is maintained at this level for the next 2 days (Fig. 1–14). Mitotic postnatal gonocytes were termed *T_2-prospermatogonia* by Hilscher and are believed to be the germ cells that eventually transform into the type A spermatogonia of the prepubertal testis. It is likely that interaction of neonatal gonocytes with neighboring Sertoli (and possible myoid) cells and with the extracellular environment in general is important in triggering events critical for normal onset of spermatogenesis during later testicular development, as discussed in a subsequent section of this chapter.

Because prenatal and postnatal development of the human testis spans a substantially longer period than does that of most experimental animals, the pattern of human gonocyte proliferation and differentiation has been more difficult to define. Although it is clear that the level of germ cell division varies during development of humans, there is some disagreement concerning the exact timing and extent of changes in this population. In a recent study, however, Codesal and co-workers (1989) used cytophotometric techniques to quantify the

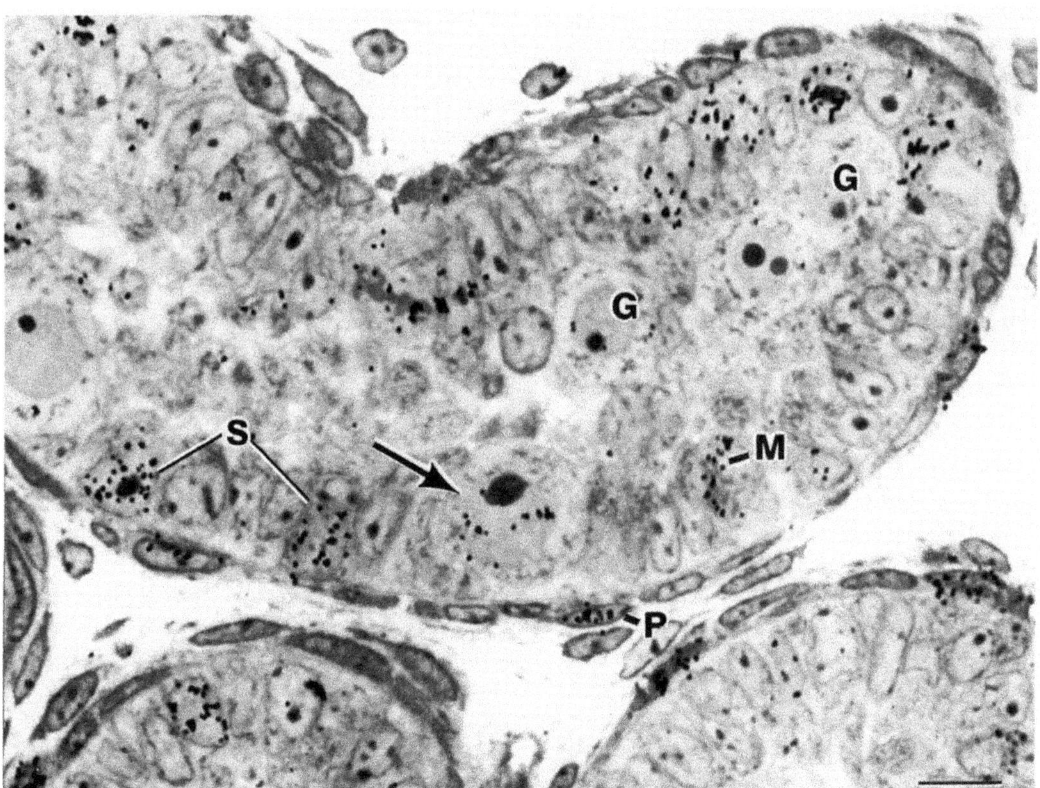

Figure 1–14. An autoradiograph of testis from a 4-day-old rat pup, 1 day after proliferating gonocytes were first detected postnatally. A ^{3}H-thymidine-labeled gonocyte *(arrow)* is seen in this view, adjacent to an apparent gonocyte mitotic figure (M) that is also labeled. G, gonocyte; P, labeled peritubular cells; S, proliferating Sertoli cells. Bar = 10 μm.

percentages of gonocytes with DNA contents consistent with either mitotic quiescence (i.e., cells in $G_{1/0}$) or with partial or completed DNA replication (i.e., S-phase or G_2) in developing human testes. The findings of this study provide a more detailed and precise evaluation of gonocyte proliferation in humans than any previously available study. During the second trimester of fetal life, a high level of germ cell division was detected with this approach, suggesting that a reservoir of gonocytes is produced during that time. Later in gestation, proliferation falls substantially to a low level that is maintained during the immediate neonatal period and throughout infancy. Thus, gonocytes in humans apparently divide continuously during later fetal life and throughout postnatal development, albeit at a lower rate after birth. This is in contrast to the situation for most experimental animals in which division of these cells occurs during two developmental periods, one fetal and one postnatal, that are separated by a time of total quiescence.

It is noteworthy that, in humans as in rats, a decrease in germ cell numbers and in tubular diameters during infancy has been detected even though gonocyte proliferation continues during neonatal life. Thus, it is likely that significant numbers of gonocytes also degenerate during early postnatal life in humans. This supposition is consistent with the general designation of postnatal development of the human testis, before approximately 4 years of age, as a "resting phase of the testis" (Codesal et al., 1989).

3. PERITUBULAR CELLS

Beyond recognition of their role as boundary cells for the seminiferous epithelium, peritubular myoid cells have been largely ignored until recently. Lately it has become clear that, at least in adults, myoid cells can modify testicular function by interacting with Sertoli cells, and possibly with germ cells and/or Leydig cells as well. For example, myoid cells produce a factor, *P-mod-S,* that seems to alter Sertoli cell function in vitro (Skinner and Fritz, 1985), a possibility suggested previously by the observation that Sertoli cells enhance their production of androgen binding protein when cocultured with peritubular cells. Interactions between peritubular myoid cells and other testicular cells of adults are considered in detail in subsequent chapters.

Although our understanding of these important intercellular relationships in the mature testis has grown in recent years, little detailed information is available concerning the peritubular cell population in the developing animal. It is clear, however, that many of these cells incorporate ^{3}H-thymidine in testes of fetal rats (Orth, 1982; see Fig. 1–11A), suggesting that growth of this population accompanies that of fetal Sertoli cells. Although proliferation of fetal and neonatal peritubular cells has not been studied in detail, qualitative observations suggest that their rate of proliferation falls immediately after birth in parallel with that of Sertoli cells (Orth, unpublished observations). In addition, when an antimitotic drug was administered intratesticularly to rats 36 to 48 hours after birth with the aim of producing a Sertoli cell-depleted rat, a limited quantitative analysis of myoid cell proliferation was also carried out (Orth, 1982). The results indicate that the level of myoid cell division is very low in rats 3 days after birth, suggesting that fewer of these cells proliferate just after birth than during fetal life. This is not an unexpected finding considering reports of a neonatal decrease in seminiferous tubular diameter in many species, as discussed previously. Although it is likely that further testicular development and onset of spermatogenesis is accompanied by corresponding growth of the peritubular cell population, information in this area is currently unavailable.

4. LEYDIG CELLS

Although a morphologically distinct population of Leydig cells appears shortly after development of seminiferous cords in the fetus, little is known about the precise origin of these cells or the way in which the size of the fetal population is determined under normal circumstances. Roosen-Runge and Anderson (1959) reported the presence of mitotic figures in the interstitium of very young rat fetuses that they presumed represented dividing Leydig cells. Mitoses (Byskov, 1986) or ^{3}H-thymidine-labeled nuclei (Orth, 1982), however, are rarely if ever seen among morphologically recognizable Leydig cells in older fetuses or neonates. For this reason, it has been suggested by many investigators that most, if not all, fetal Leydig cells originate by differentiation from mesenchyme-like precursors and, once differentiated, no longer divide under normal conditions. Because specific Leydig cell markers that would positively identify such precursors are lacking, this likely possibility remains to be confirmed. It is clear, however, that proliferating mesenchyme-like cells are common within the interstitial tissue in fetal testes at various ages (see Figs. 1–10B and 1–11A). Thus, it is possible that continued division of Leydig cell precursors might occur throughout fetal development.

The classical view of Leydig cells in the postnatal period is that the population shrinks progressively until the approach of puberty. After a detailed stereological study of Leydig cells in the maturing testis, however, Zirkin and Ewing (1987) concluded that testes of 18 to 20-day-old rat fetuses and those of 2 to 3-day-old pups contained the same number of Leydig cells. This suggests that the postnatal loss of Leydig cells reported by others may only be relative to their decreasing density in an enlarging testis. Thus, Zirkin and Ewing's observation that the size of this population in older fetuses is similar to that in neonates, along with the failure of others to detect degenerating Leydig cells at this stage of development, suggests that this population is essentially stable during perinatal life, with little addition of new cells or loss of previously existing ones.

The fate of the fetal generation of Leydig cells and how these cells relate to those of the adult is an area of current study. It has been suggested that fetal Leydig cells are replaced during postnatal development by a new, "adult" generation of Leydig cells that may arise de novo from undifferentiated precursors rather than from the fetal Leydig cells. This view is supported by observations of morphological differences between Leydig cells in fetuses and those in adults. These differences include the presence of abundant lipid droplets in fetal rat cells (see Fig. 1–11B) compared with the virtual absence of these inclusions in cells of mature adults. The common belief that Leydig cell numbers fall postnatally and then rise dramatically around the time of puberty has also tended to lend credence to this idea. The quantitative observations of Zirkin and Ewing (1987), however, indicate that fetal Leydig cells may not disappear during postnatal development of rats but rather could remain stable in number until the population begins to expand prepuberally. Thus, it is also possible that fetal Leydig cells persist into adulthood; a loss of lipid droplets in these cells might reflect a change in the way in which they procure or store precursors required for steroidogenesis. Regardless of the fate of fetal Leydig cells, however, it is likely that most of the rise in Leydig cell numbers that occurs by adulthood reflects differentiation of new cells from precursors since, as detailed previously, recognizable Leydig cells rarely divide under normal circumstances.

III. Control of Fetal and Neonatal Growth of the Testis

The expansion of each cell population in the perinatal testis is apparently controlled by a particular set of factors that interact specifically with that population during the appropriate stage of development. The nature of these interactions is complex, with some regulatory factors originating extratesticularly, primarily from the pituitary, and others being produced within the testis itself and acting in a paracrine fashion. A great deal remains to be learned about the source, nature, and control of each of these factors and about how they act on their specific target. Considerable information is now available, however, regarding proliferation and differentiation of at least some testicular cells, as detailed below.

A. Extratesticular Factors

1. FOLLICLE STIMULATING HORMONE AND SERTOLI CELLS

Although it has been recognized for some time that follicle stimulating hormone (FSH) acts in males by specific receptors on Sertoli cells, the physiological role of this hormone in the adult testis is considerably less clear. For example, as they mature Sertoli cells become generally unresponsive to FSH although they bear FSH receptors, a phenomenon that cannot be explained solely by the appearance of Sertoli cell phosphodiesterase, an enzyme that breaks down cAMP. Moreover, when the action of FSH is specifically blocked by various means in adults, spermatogenesis remains qualitatively normal and testis weight drops only slightly. Thus, the role of FSH in the mature testis in vivo still requires study.

These observations have led many investigators to suggest that FSH is most important during development, a supposition that has been confirmed by several studies. Two approaches, one in vivo and one in vitro, were used to demonstrate that FSH is a major factor regulating proliferation of Sertoli cells in fetal and neonatal rats (Orth, 1984). To interfere with the action of FSH during prenatal development in vivo, 18-day-old fetuses were either decapitated in utero or injected subcutaneously with anti-FSH and the subsequent level of Sertoli cell proliferation in testes of these fetuses was compared with that in testes of appropriate controls. In both cases, a drop of about 50 percent in the portion of the Sertoli cell population labeled with ^{3}H-thymidine was detected 1 day after either treatment (Table 1–2). In addition, when testes from intact or decapitated fetuses were organ-cultured with or without FSH, stimulation of Sertoli cell proliferation by the hormone was detected in both situations, an effect that was mimicked by cAMP (Fig. 1–15). It has also been shown that elevated levels of FSH are associated with abnormally high rates of Sertoli cell proliferation in vivo (Orth et al., 1984). For this, newborn rats were hemicastrated, an approach shown previously to produce compensatory hypertrophy of the remaining testis. In these animals, removal of one testis was followed by markedly elevated rates of Sertoli cell proliferation during the next 4 days (Fig. 1–16). Moreover, in hemicastrated rats FSH levels were more than two times higher than control levels, while circulating luteinizing hormone (LH) was unchanged. Taken together, these findings indicate that FSH is physiologically important in stimulating Sertoli cells to divide during fetal and neonatal development of the testis, and that this effect is cAMP mediated.

Table 1–2. Uptake of ^{3}H-thymidine by Sertoli Cell Nuclei of 19-day-old Rat Fetuses That Were Either Decapitated in Utero or Injected with Anti-FSH the Previous Day Compared with That in Intact and Saline-injected Littermates, Respectively.

Treatment	*n*	% Sertoli cells labeled
Decapitation in utero	16	15.3 ± 0.4^a
Intact littermate controls	12	25.6 ± 1.0^a
Anti-FSH injection	8	14.7 ± 3.4^b
Saline injection	6	28.8 ± 1.0^b

For each fetus, percentages of Sertoli cells that had incorporated label (± SEM) were determined by scoring 1500 or more Sertoli cells in autoradiographs as labeled or unlabeled.

$^a P < 0.001$; $^b P < 0.005$.

Beyond the involvement of cAMP, little is known about the events that cause the onset of Sertoli cell division in response to FSH. It is possible that FSH-mediated production of estrogen by the Sertoli cell might act to stimulate Sertoli cell division in a manner paralleling the way in which FSH-stimulated estrogen synthesis increases granulosa cell division in growing ovarian follicles. In fetal males, the activity of testicular aromatase is markedly enhanced after FSH treatment (Weniger and Zeis, 1988), and it is likely that the Sertoli cell is the site of FSH-stimulated aromatization during development (Dorrington and Armstrong, 1975). In spite of this circumstantial evidence, however, data supporting a direct role for aromatization in Sertoli cell division are lacking.

It is also possible that FSH-mediated estrogen production by immature Sertoli cells may be involved in feedback control of FSH itself by a paracrine effect on androgen output by Leydig cells. Numerous reports suggest that, in addition to inhibin, androgen is capable of regulating FSH output by the pituitary. These include the observation that testosterone treatment prevents the enhancement of Sertoli cell proliferation resulting from hemicastration of neonates (Orth et al., 1984), implying that the elevation in FSH that usually occurs after this surgery is prevented by androgen. Furthermore, estrogen produced by perinatal Sertoli cells may act via a "short-loop" system to suppress androgen secretion by Leydig cells and thus indirectly regulate FSH levels. While the precise relationships among FSH, Sertoli cell aromatase, and production of androgen by fetal Leydig cells are complex and not yet clear, these findings suggest one possible mechanism whereby Leydig cells could be involved in con-

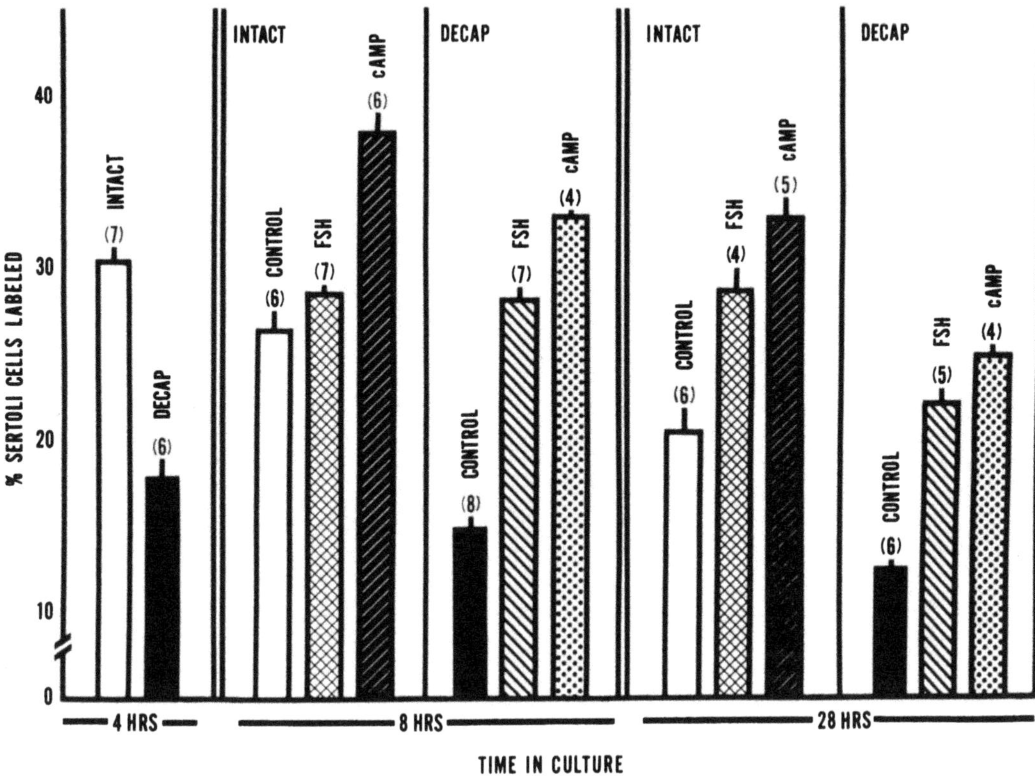

Figure 1–15. The effect of FSH or cAMP on incorporation of ^{3}H-thymidine by Sertoli cells in organ-cultured testes from normal fetuses or fetuses decapitated in utero 1 day earlier. In all cases, the isotope was present for the final 4 hours of incubation. When immediately exposed to label for 4 hours, the percentage of cells dividing in testes from decapitated fetuses was about half of that in controls. After 8 hours of exposure to FSH, the percentage of cells dividing in testes of decapitated rats increased to twice that seen in testes incubated without hormone; similar treatment with cAMP produced an even larger increase in the percentage of cells dividing. Sertoli cell proliferation was also elevated in normal testes incubated under similar conditions with cAMP, whereas FSH had no effect; however, after other control testes were exposed to the hormone or cAMP for a total of 28 hours, Sertoli cell division was significantly and similarly elevated in both cases when compared with untreated testes. (Reprinted with permission from Orth, 1984.)

trolling FSH levels and hence could have an indirect role in determining the final size of the Sertoli cell population.

2. LUTEINIZING HORMONE AND LEYDIG CELLS

As discussed above, the appearance of the first differentiated Leydig cells in the fetal testis seems to be a spontaneous event since it occurs in organ-cultured testes in the absence of extratesticular factors (Picon, 1976) and apparently even in the absence of seminiferous cords (Patsavoudi et al., 1985). Evidence obtained from several species, however, suggests that further development of the Leydig cell population in the fetal testis is regulated by pituitary LH. For example, decapitation of fetal rats in utero on day 17 resulted, by the next day, in a reduction in the collective volume of Leydig cells in these fetuses compared with controls (Eguchi et al., 1978). Moreover, in anencephalic fetuses with low LH levels, Leydig cells are scarce, whereas in human infants with testicular feminization syndrome, in which LH levels are abnormally high, there is an increase in the size of the interstitial compartment occupied by Leydig cells (Byskov, 1986). Direct stereological measurement of Leydig cell numbers was not carried out in any of these studies of fetal Leydig cells. For this reason, it is difficult to determine if the

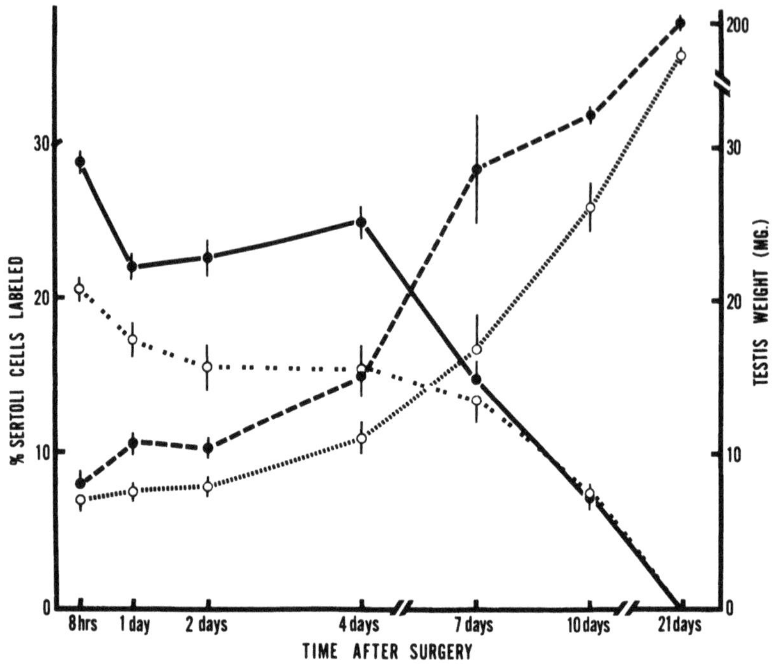

Figure 1–16. The pattern of increased proliferation of Sertoli cells after hemicastration of neonatal rats, compared with the appearance of compensatory testicular hypertrophy in these animals. The percentage of cells incorporating ^{3}H-thymidine was greater in hemicastrates (●———●) than in controls (O---O) at 8 hours, 1, 2, and 4 days after surgery and returned to normal thereafter. The weight of the remaining testis in hemicastrates (●---●) was significantly increased over that in controls (O · · · O) by 1 day postsurgery and at each later interval studied. (Reprinted with permission from Orth et al., 1984.)

reported changes in the total volume of the Leydig cell population reflect altered size of the individual cells or changes in the total numbers of cells.

Additional light may be shed on this issue by data from a number of recent studies of Leydig cells in normal and experimentally manipulated postnatal rats. Kuopio and colleagues (1989b) injected 2-day-old rats with human chorionic gonadotropin (hCG) and used morphometric methods to measure Leydig cell numbers 1 to 5 days later. They detected an almost twofold increase in the number of Leydig cells in treated rats by 1 day and also reported the appearance of occasional mitotic metaphases in recognizable Leydig cells in hCG-injected rats. In another study, Kuopio et al. (1989c) used a different approach to study the same question. For this, testes from newborn rats were transplanted under the kidney capsule of intact or castrated adults and Leydig cell development in the transplants compared in the two hosts. In both cases, after 2 weeks the transplanted testes had grown and vascularized. In addition, qualitative observations suggested that many more Leydig cells were present in those testes transplanted into castrates, in which blood levels of gonadotropins were greatly elevated, than in those placed in intact hosts. Although both of these situations were nonphysiological and encompassed only neonatal life when Leydig cells do not normally increase in number, the findings tend to support the idea that LH may have a role in determining the size of the fetal population of Leydig cells. This question, however, is not yet resolved.

3. EXTRATESTICULAR INFLUENCES ON PERITUBULAR (MYOID) CELLS

Recent findings support the idea that myoid cells surrounding the seminiferous epithelium play more than a passive role in testicular function in adults. However, little is known about when these cells proliferate and differentiate during growth of the testis and about how these events are controlled. It has been reported that the innermost layer of peritubular cells (i.e., those closest to the seminiferous epithelium) undergoes postnatal modification to give rise to "myoid" cells in rats by 2 to 3 weeks of age (Leeson and Leeson, 1963). These cells lose the fibroblastlike morphology that characterizes perinatal cells, becoming flattened and developing abundant intracellular actin filaments.

The postnatal transformation of a portion of the peritubular cell population into apparently contractile cells was studied in vivo by Bressler and Ross (1972). They evaluated the influence of pituitary hormones on peritubular cell maturation by implanting newborn testes beneath the tunica albuginea of adult rats that were intact, hypophysectomized, or hypophysectomized and androgen-treated. In the untreated hypophysectomized hosts, peritubular cells with a myoid appearance failed to arise in the implanted testes. In normal hosts, however, the innermost layer of peritubular cells developed normally in the implants as they reached about 3 weeks of age. These observations provide strong evidence indicating that normal pituitary function is important, at least indirectly, in stimulating differentiation of a myoid subpopulation of peritubular cells during postnatal development. Bressler and Ross also reported that testosterone treatment of hypophysectomized hosts resulted in partial maintenance of myoid cell development in the implants. Because of this observation, some investigators have suggested that androgen is the prime agent supporting myoid cell development. However, myoid cells of testes implanted into androgen-treated hypophysectomized hosts only developed partially, and precocious development of myoid cells did not occur in testes implanted into normal adults even though androgen levels in those hosts were at mature levels. In addition, since both gonadotropins are likely to have been suppressed after androgen treatment of the hypophysectomized hosts in these experiments, it is unclear whether LH, FSH, or both are also required for complete myoid cell development. Thus, these observations imply that control of myoid cell differentiation is complex and may involve several factors.

It is also not clear whether peritubular cells are the direct targets of testosterone, pituitary hormones, or both, or whether interaction between these and other testicular cells is involved in the transformation of some peritubular cells into myoid cells. It should be noted that only the innermost layer of peritubular cells differentiates into myoid cells even though all peritubular cells in newborns are morphologically identical. Thus, interaction between the inner layer of peritubular cells and Sertoli cells or the matrix that separates these two cellular layers may also have a role in the postnatal development of myoid cells. Unfortunately, few additional studies have been carried out on myoid cells in vivo, leaving their differentiation poorly understood.

B. Paracrine Interactions Influencing Testicular Development

1. SERTOLI–LEYDIG CELL COMMUNICATION

Since FSH stimulates perinatal Sertoli cells to divide, circulating levels of FSH during development are critical in determining the size of the Sertoli cell population. It has also become increasingly clear that paracrine factors produced within the testis itself are likely to affect the mitotic response of Sertoli cells to FSH. Furthermore, the fact that one or more of these paracrine agents originates from Leydig cells suggests that communication takes place between the seminiferous and interstitial compartments of the testis during its maturation.

One way in which Leydig cells influence the function of Sertoli cells in both immature animals and adults is by producing proopiomelanocortin (POMC)-derived peptides such as β-endorphin and MSH, peptides that act directly on Sertoli cells (for review, see Bardin et al., 1988). The discovery that these peptides were localized to the testis was followed by conclusive evidence that Leydig cells express the POMC gene and hence are a source of POMC-derived peptides. While their mode of action and physiological effects in adults and immature animals are still subjects of study, several lines of evidence indicate that at least one of these factors, β-endorphin, functions during perinatal life to modify the proliferative response of Sertoli cells to FSH. First, circumstantial evidence for an involvement of β-endorphin in Sertoli cell development was provided by the observation that the pattern of Leydig cell production of POMC-derived peptides corresponds closely to that of growth of the Sertoli cell population. Leydig cells immunostainable for β-endorphin and MSH appear in fetal mice on day 14 and increase in number until birth, after which their numbers decline steadily so that, by 5 days postpartum, cells positive for these peptides are rarely seen (Shaha et al., 1984). Although Sertoli cell division has not been studied in perinatal mice, developmental events in this species generally occur within 1 day of the same event in growing rats. Thus, taken together these data suggest that the portion of the fetal Leydig cell population that produces β-endorphin is maximal just before birth, when the rate of Sertoli cell proliferation is likely to be greatest, and then declines in parallel with a decreasing level of Sertoli cell division.

In later studies, the possibility of a link between Sertoli cell division and Leydig cell production of β-endorphin was confirmed in vivo (Gerendai et al., 1983; Orth, 1986) and its mechanism of action probed with an in vitro approach (Orth and Boehm, 1990a). When the effect of intratesticular injection of antibodies to endorphin on proliferation of neonatal Sertoli cells was studied in vivo, a large increase in the percentage of cells that were dividing occurred by 8 hours postinjection, a change that was sustained for at least several more hours (Table 1–3). The effect of an opiate blocker, naloxone, on FSH-stimulated proliferation of Sertoli cells was also studied in organ-cultured fetal testes (Fig. 1–17). After a day of exposure to both naloxone and FSH, the percentage of cells labeled with ^{3}H-thymidine was substantially increased over that seen in testes incubated with FSH alone. Taken together, these findings strongly suggest that β-endorphin of intratesticular (presumably Leydig cell) origin depresses the proliferative response of perinatal Sertoli cells to FSH in vivo.

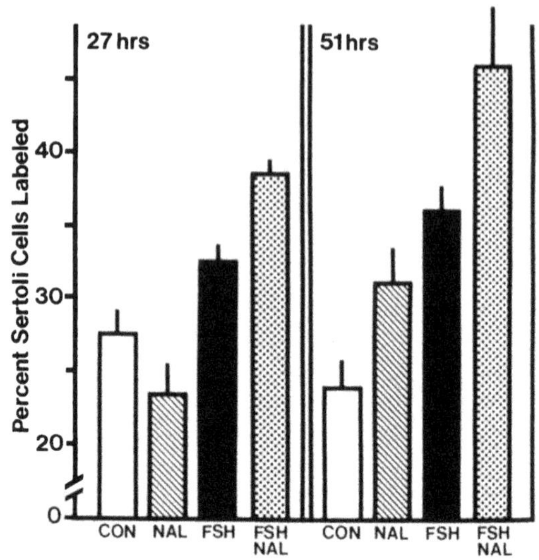

Figure 1–17. The effect of the opiate blocker naloxone, with or without FSH, on Sertoli cell division in organ cultured fetal rat testes. When added 3 hours before FSH, the drug significantly enhanced the response to the hormone after 27 or 51 hours of total incubation. Exposure to the drug alone for 51 hours also increased proliferation over control levels. (Reprinted with permission from Orth, 1986.)

Table 1–3. Uptake of ^{3}H-thymidine by Sertoli Cell Nuclei of 5-day-old Rat Pups After Intratesticular Injection of Either Saline or Antiserum to β-endorphin.

Treatment	*n*	%Sertoli cells labeled (± SEM)
Saline, 8 hours	3	16.6 ± 0.9[a]
Anti-β-endorphin	5	25.2 ± 0.7[a]
Saline, 19 hours	3	15.9 ± 0.7[b]
Anti-β-endorphin	6	24.2 ± 0.5[b]

[a] $p < 0.005$; [b] $p < 0.001$.

The mechanism through which endorphin modifies the proliferative response of perinatal Sertoli cells to FSH has been studied in cells isolated and maintained in vitro (Orth and Boehm, 1990a). In that system, direct treatment of cultures with β-endorphin before the addition of FSH produced a dramatic decrease in Sertoli cell proliferation compared to that seen with FSH alone (Fig. 1–18). Moreover, quantitative analysis of similar cultures and of others exposed to cAMP, with or without β-endorphin, confirmed that the latter depresses the proliferative response of these cells to FSH but not to cAMP (Fig. 1–19). Thus, β-endorphin modifies the effect of FSH on Sertoli cells by acting at a point before generation of the cyclic nucleotide in the cells. Finally, inclusion of pertussis toxin, a blocker of inhibitory gua-

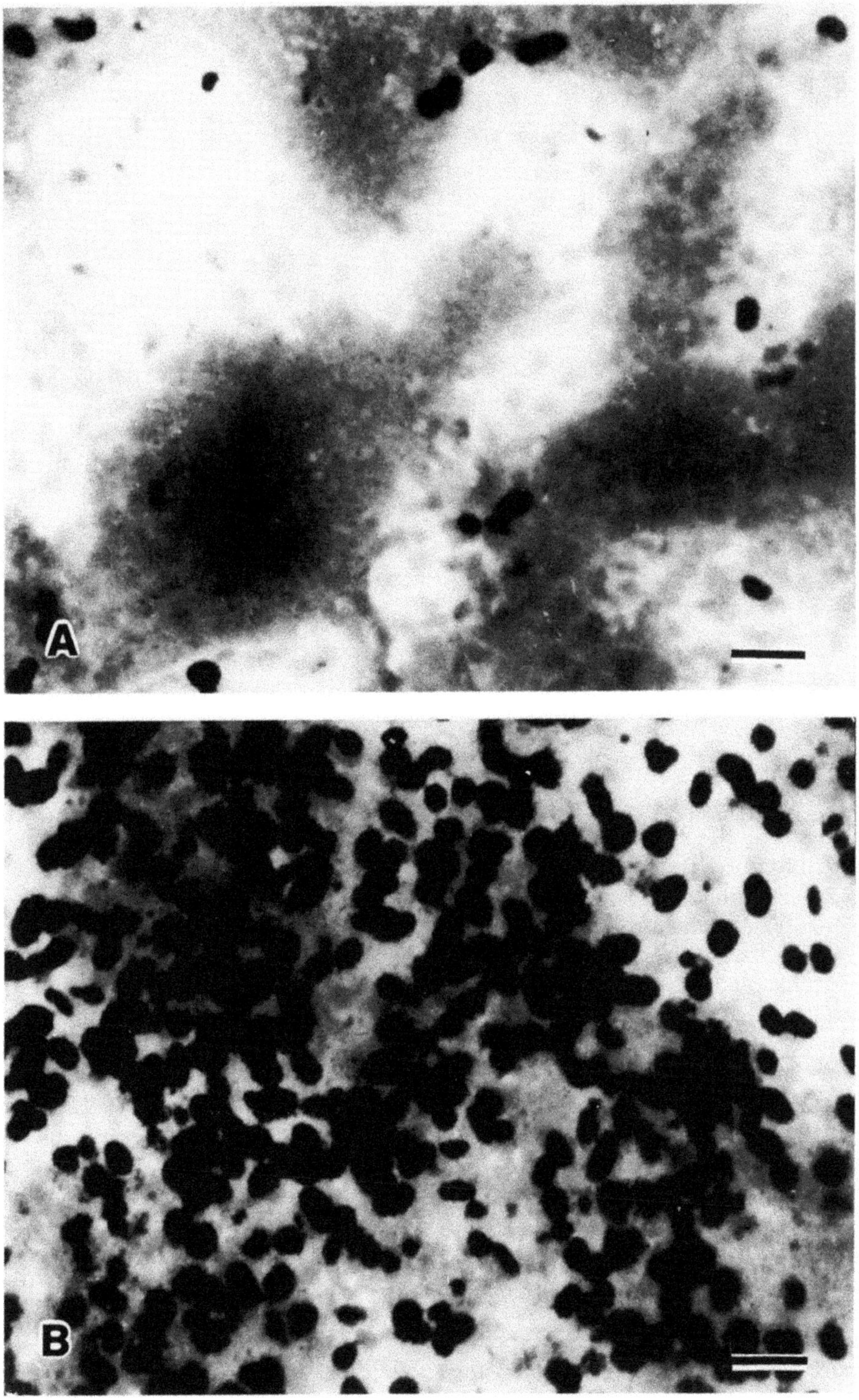

Figure 1–18. Low power views of autoradiographs showing representative areas of Sertoli cells isolated from 6-day-old rat pups and maintained in culture for 4 days, with ^{3}H-thymidine present for the final 24 hours. Individual nuclei of proliferating cells are heavily labeled and appear black. Treatments were (**a**) vehicle alone; (**b**) FSH from days 2 through 4; (**c**) β-endorphin and FSH from days 2 through 4, with the peptide added 3 hours before FSH. Bar = 20 μm. (Reprinted with permission from Orth and Boehm, 1990.)

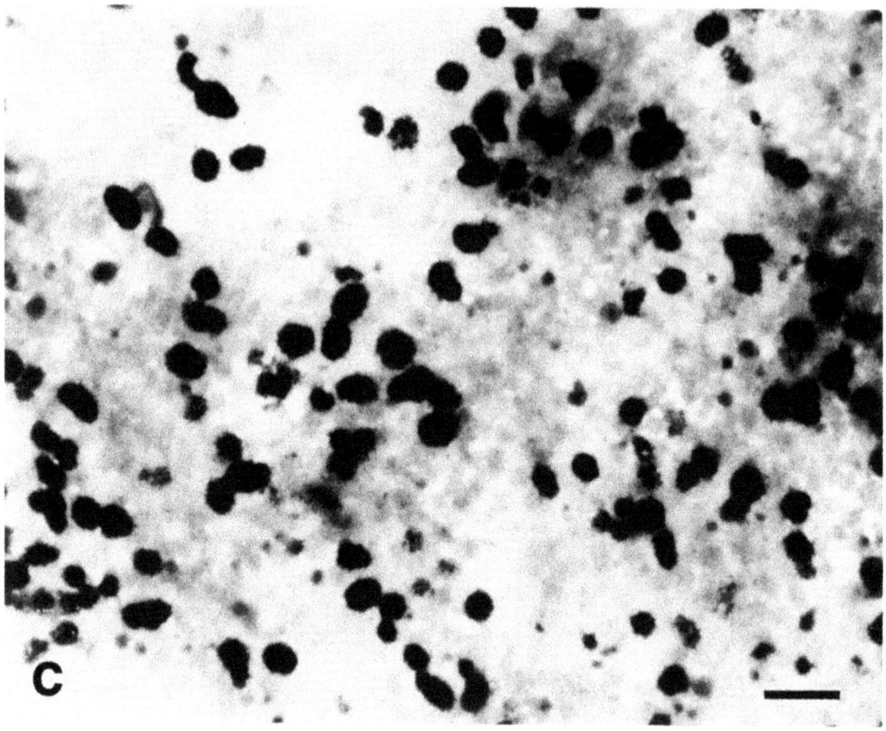

Figure 1–18. (C).

nosine triphosphate (GTP)-binding proteins, in Sertoli cell cultures before sequential addition of FSH and β-endorphin preserved the response of the cells to FSH even though β-endorphin was present (Fig. 1–19). These findings suggest that β-endorphin blunts the increase in adenyl cyclase activity, which normally follows exposure of Sertoli cells to FSH, by interacting with an inhibitory guanine nucleotide-binding protein such as G_i.

It has also been suggested that other POMC-derived peptides, notably adenocorticotropic hormone (ACTH) and α-MSH (melanocyte stimulating hormone), are involved in Leydig cell–Sertoli cell interaction during testicular development. Not only has production of small amounts of these factors by Leydig cells of rodents been measured but, under certain conditions in vitro, peptides with MSH-like activity have been shown to stimulate adenyl cyclase in Sertoli cells. It has, therefore, been postulated that peptides other than β-endorphin may act to make Sertoli cells more sensitive to FSH (Bardin et al., 1988). Such a mechanism might provide a counterbalance to the negative effect of β-endorphin on FSH responsiveness, a possibility not yet confirmed in vivo.

Recently, there has been substantial interest in other factors that might act as paracrine modifiers of Sertoli cell proliferation in developing animals. For example, it has been reported that somatomedin-C, epidermal growth factor, fibroblast growth factor, and retinoic acid are capable of stimulating DNA synthesis by Sertoli cells isolated from immature pigs (Jaillard et al., 1987). In some cases, the observed effects seemed to involve altered sensitivity of the cells to FSH and concomitant changes in cAMP production. In others, enhanced proliferation was produced without involvement of the FSH/cAMP system. The relevance of these interesting findings, however, and that of data from similar studies to the situation in vivo is not clear. This is of particular concern for some factors, such as fibroblast growth factor. This peptide seems capable of stimulating mature Sertoli cells to proliferate, an event that is beyond the normal capacity of

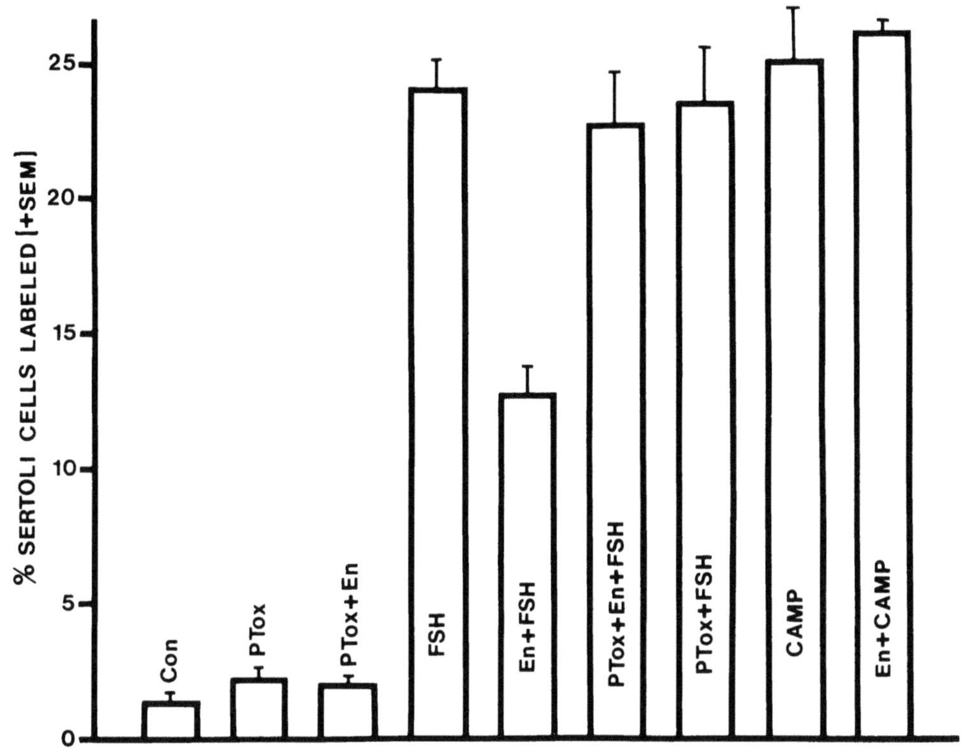

Figure 1–19. The percentage of Sertoli cells labeled by ^{3}H-thymidine in cultures prepared from 6-day-old rats and exposed to FSH, cAMP, or vehicle from days 2 through 4 in vitro, with or without previous inclusion of β-endorphin. In some cultures, pertussis toxin, a blocker of G_i, was added 3 hours before β-endorphin and/or FSH. Both FSH and cAMP increased the percentage of cells dividing compared with controls; previous addition of β-endorphin left the response to cAMP unaffected, but greatly reduced that of the cells to FSH. In the presence of pertussis toxin, however, the response to FSH was undiminished by addition of β-endorphin. Thus, β-endorphin apparently interferes with FSH-induced Sertoli cell proliferation by interacting with an inhibitory GTP-binding protein and blunting production of cAMP within the cells. (Reprinted with permission from Orth and Boehm, 1990.)

these cells. Moreover, it is not known whether these agents are produced in perinatal testes, in which Sertoli cells are proliferative. Thus, although there may be other ways in which paracrine factors may influence Sertoli cell development, this area clearly requires further study.

Additional data exist indicating that perinatal Sertoli cells are able to affect Leydig cells, thus pointing to the existence of two-way communication between the functional compartments of the developing testis. Kerr and Sharpe (1985) showed that effects of FSH on Leydig cell maturation, previously assumed to reflect contamination of FSH preparations with LH, were due to the action of FSH itself. They found that purified FSH was effective in promoting recovery of Leydig cell number and function after hypophysectomy of immature rats. Since FSH receptors are restricted to cells of the seminiferous epithelium, it was concluded that the effects of FSH on Leydig cells were mediated by action of this hormone on Sertoli cells. Further support for this suggestion was provided by Benahmed and co-workers (1986) who assessed the effect of including FSH along with LH on testosterone output by cultured Leydig cells isolated from immature pigs. The resulting data indicated that more androgen was produced in the presence of FSH and LH than in the presence of LH alone, provided Sertoli cells were included in the cultures. Moreover, the effect of FSH on Leydig cell steroidogenesis could be mimicked with medium conditioned by FSH-stimulated Sertoli cells. These findings strongly suggest that Sertoli cells secrete factor(s) in response to FSH that enhance the steroidogenic response of Leydig cells to LH.

Although the results described above imply that Sertoli cells can modify Leydig cell function, the data were obtained using maturing testes that are likely to contain some or all adult-type Leydig cells. As already discussed, the relationship between the fetal and adult generation of these cells is still not well understood. Thus, it is not clear if similar FSH-mediated effects on Leydig cell steroidogenesis also occur in younger (i.e., perinatal) testes that contain exclusively fetal Leydig cells.

Sertoli cells of fetuses, however, do produce at least one likely modifier of Leydig cell function, a gonadotropin releasing hormone (GnRH)-like peptide, in response to FSH (for review, see Sharpe, 1984). Binding sites for GnRH have been detected in testes of rats on the day of birth (Dalkin et al., 1981) and a culture system for fetal Leydig cells has been used to study the interactions among exogenous GnRH, LH, and Leydig cells in vitro (Dufau and Knox, 1985). In addition, exposure of fetal Leydig cells to a low level of this peptide produced a dose-dependent inhibition of androgen output in the presence of LH, leading to the suggestion that Sertoli cells may exert a local effect on nearby Leydig cells during development via secretion of a GnRH-like peptide. The existence of a Sertoli cell–Leydig cell communication system in vivo, based on production of a GnRH-like molecule by Sertoli cells, is an intriguing possibility that awaits confirmation.

In summary, findings from a number of studies in which various approaches were used point to the existence of complex interactions between the interstitial and seminiferous compartments of the testis during perinatal development. At least one form of direct Leydig cell–Sertoli cell interaction, which involves β-endorphin and which acts to regulate the proliferative response of Sertoli cells to FSH, is operational during perinatal development of the testis in vivo. In addition, evidence obtained in vitro makes it likely that mechanisms exist whereby Sertoli cells can, in turn, affect the differentiation and steroidogenic function of Leydig cells in developing testes. These mechanisms include the possible production by Sertoli cells of a GnRH-like peptide that modifies LH-stimulated testosterone synthesis by Leydig cells. As already noted, maintenance of fertility in adults depends on at least two key factors: (1) the size of the Sertoli cell population, which is crucial in determining numbers of sperm produced, and (2) the presence of adequate testosterone, which is necessary to support spermatogenesis. Thus, Sertoli cell paracrine activities are critical to the balance that develops between the exocrine and endocrine functions of the testis.

2. SERTOLI CELL–GONOCYTE INTERACTION DURING PERINATAL DEVELOPMENT

After the initial wave of cell division that occurs early in fetal life, gonocytes of males remain mitotically inactive until after birth. This period of quiescence may be actively imposed on the gonocytes by one or more factors that originate in the testis. Indeed, it has been suggested that an autonomous "meiotic clock" (Byskov, 1986) is present in germ cells of both males and females that, if unopposed, spontaneously moves fetal gonocytes into meiosis at a preprogrammed time in development. This view is supported by observations made on ectopic germ cells in mice of both sexes. During their migration, some germ cells may be retained in the fetal adrenal where they survive for a time. It has been reported that these ectopic cells enter meiosis in both sexes at the same time as normal oogonia that are located in the ovary (Zamboni and Upadhyay, 1983). These observations have led to the suggestion that the onset of meiosis is spontaneous unless prevented, and that a meiosis-preventing substance is produced by the fetal testis to inhibit entry of male gonocytes into meiosis during development.

If a meiosis-preventing factor exists in male fetuses, it is likely to originate from within the seminiferous cords because gonocytes that remain outside the cords but within the fetal testis also enter meiosis before eventually degenerating (Byskov, 1986). Sertoli cells of perinatal rats produce several peptides that are likely candidates for a role in regulating the function of gonocytes. One of these, mullerian inhibiting hormone (MIH), is apparently the first protein secreted by fetal Sertoli cells and, in all species studied including rat, its production is associated with differentiation of the seminiferous cords (for bibliography, see Bardin et al., 1988). Besides its well-recognized role in promoting loss of mullerian derivatives, it has been suggested that MIH may act to prevent entry of gonocytes into meiosis. This proposal gains support from the finding that MIH production parallels the appearance of quiescence of the male germ cells; MIH is high during later fetal life and totally disappears after birth, at least in rats. Moreover, substantial evidence exists supporting the idea that MIH acts as a meiosis inhibitor in females (for review, see Hutson et al., 1989). For example, in contrast to the situation in males, MIH production is undetectable in fetal life when oogonia are meiotic and high at the time of meiotic arrest just before birth. Finally, results of studies in vitro indicate that inclusion of MIH with oocytes prevents the spontaneous onset of meiosis normally displayed when these cells are placed into culture. Although these observations are consistent with the possibility that MIH prevents movement of male gonocytes into meiosis, direct evidence supporting such a role for this peptide is not yet available.

Several other peptides related to MIH, such as inhibin and activin (see Chapter 10 for further details) are produced within the testis. Although first recognized for their effects on pituitary secretion of FSH, it is also thought that these peptides act as paracrine agents in adults to promote Sertoli cell–Leydig cell interaction. Immunostaining with antisera generated against various subunits of these dimeric molecules, however, was high in seminiferous cords of fetuses (Bardin et al., 1989), implying that inhibin or activin, or both, may be important during testicular development as well as at maturity. The limited data available suggest that inhibin reduces ^{3}H-thymidine incorporation into testicular DNA in vitro and that this effect is specifically on germ cells and primarily on type B spermatogonia (Thumann and Bustos-Obregon, 1978). Although this observation in males remains to be confirmed with highly purified inhibin preparations, which have recently become available, some direct evidence has accumulated to suggest that inhibin prevents meiosis in oogonia of females (e.g., see O et al., 1989). Thus, inhibin remains a possible candidate for a meiosis-inhibitor in males as well. In addition, other Sertoli cell-specific proteins not related to MIH may be secreted in response to the status of germ cells in developing rats and thus may have a role in regulating their entry into meiosis. For example, Shabanowitz and colleagues (1986) found a low molecular weight protein, designated T35, whose production by Sertoli cells was greatest in fetal and early postnatal testes, implying that its function may be specific to events occurring only during testicular development. Hence, one or more paracrine messages produced by Sertoli cells and acting on germ cells may be important in modifying or supporting development of gonocytes in perinatal testes.

The findings outlined above raise the intriguing possibility that peptide factors such as MIH or inhibin may act to regulate entry of germ cells into meiosis. Most direct evidence supporting this idea, however, has been gathered in females. In considering these data, it should be noted that significant differences exist between the onset of meiosis in females and in males. Importantly, postnatal gonocytes of males do not pass directly from a period of quiescence into meiosis as do oogonia in females. Rather, they cease dividing for a time before birth, reenter the cell cycle shortly after birth, and then divide *mitotically* at least once before maturing into spermatogonia. Some of these spermatogonia will be recruited to enter meiosis later in postnatal development. The agent(s) stimulating postnatal resumption of *mitosis* by gonocytes of males remain an enigma. It has been shown recently, however, that gonocytes in testes organ-cultured in defined media (McGuinness and Orth, 1992) or co-cultured with Sertoli cells (Orth and McGuinness, 1991) reinitiate mitosis without added factors, indicating that this event is not under extratesticular control. This suggests that stimuli causing germ cells in neonatal males to resume mitosis must originate in the testis itself, perhaps from Sertoli cells.

It is possible that contact between Sertoli cells and gonocytes may mediate changes occurring in these germ cells after birth, including their resumption of mitosis, change in position within the seminiferous cord (see below), and maturation into spermatogonia. This is suggested by the likelihood that contact-mediated recognition between gonocytes and Sertoli cells is important at an earlier time in development when the seminiferous cords are formed. Moreover, from that time until just after birth, these cells are completely surrounded by Sertoli cells (see Figs. 1–10 and 1–11). This raises the possibility that postnatal changes in gonocytes may be stimulated by factors passing directly to them from Sertoli cells.

Recently, a co-culture system was established in which the possibility of communication between Sertoli cells and gonocytes could be studied directly (Orth and Boehm, 1990). In these cultures, gonocytes from 4 to 6-day-old rats adhere to an underlying monolayer of Sertoli cells (Fig. 1–20) plated on a basement membranelike material (MATRIGEL) and survive for 7 or more days. By the day after co-culture, numerous desmosome-like attachments develop between Sertoli cells and adjacent gonocytes (Fig. 1–21). Moreover, when two fluorescent probes, Lucifer yellow, which is permeant to gap junctions, and Rhodamine-dextran, which is not, were introduced into Sertoli cells in these co-cultures with a glass-bead-loading technique (McNeil and Warder, 1987), Lucifer yellow passed within minutes from Sertoli cells into adjacent gonocytes that had not been bead-loaded with either probe (Fig. 1–22; Orth and Boehm, 1989). This indicates that, at least under these conditions in vitro, gap junctions are present between the two cell types. These observations, along with additional ultrastructural evidence for the existence of Sertoli cell–gonocyte gap junctions in vitro (see Fig. 1–21), suggest that gonocytes not only adhere to Sertoli cells but also become metabolically coupled

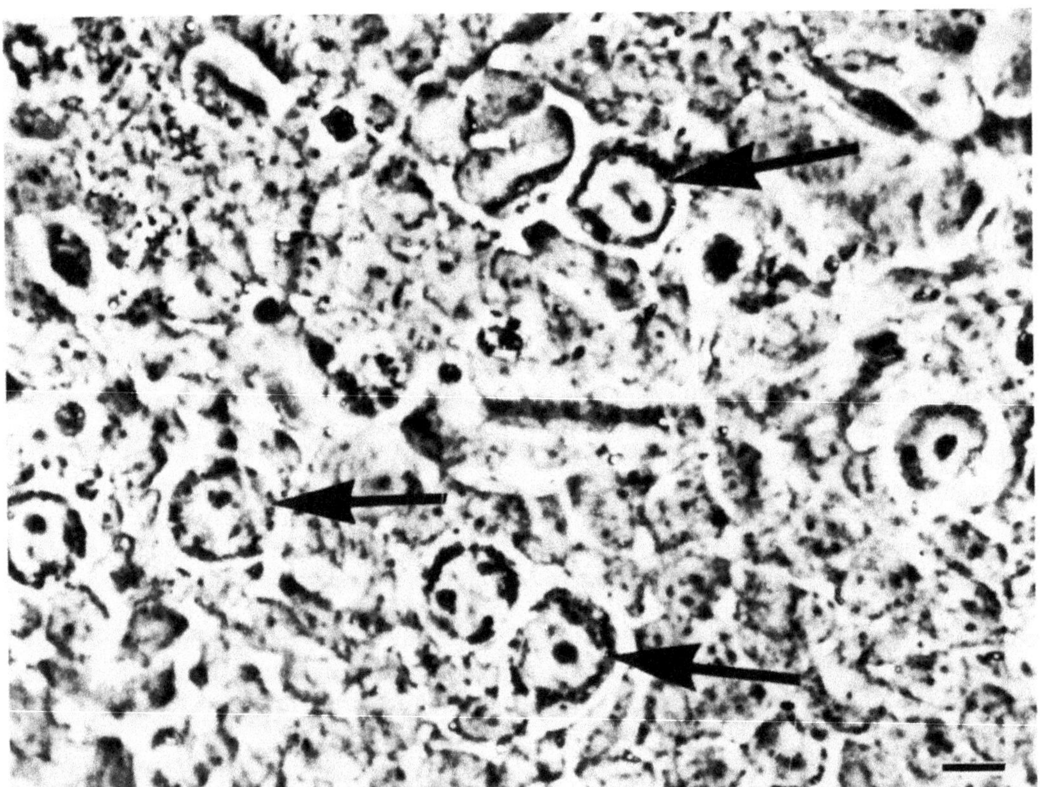

Figure 1–20. A view of a Sertoli cell–gonocyte co-culture, several hours after the cells were isolated from 6-day-old rat pups and plated on a substrate of MATRIGEL. *Arrows* indicate large, round gonocytes that have attached to the underlying monolayer of Sertoli cells. Bar = 10 μm.

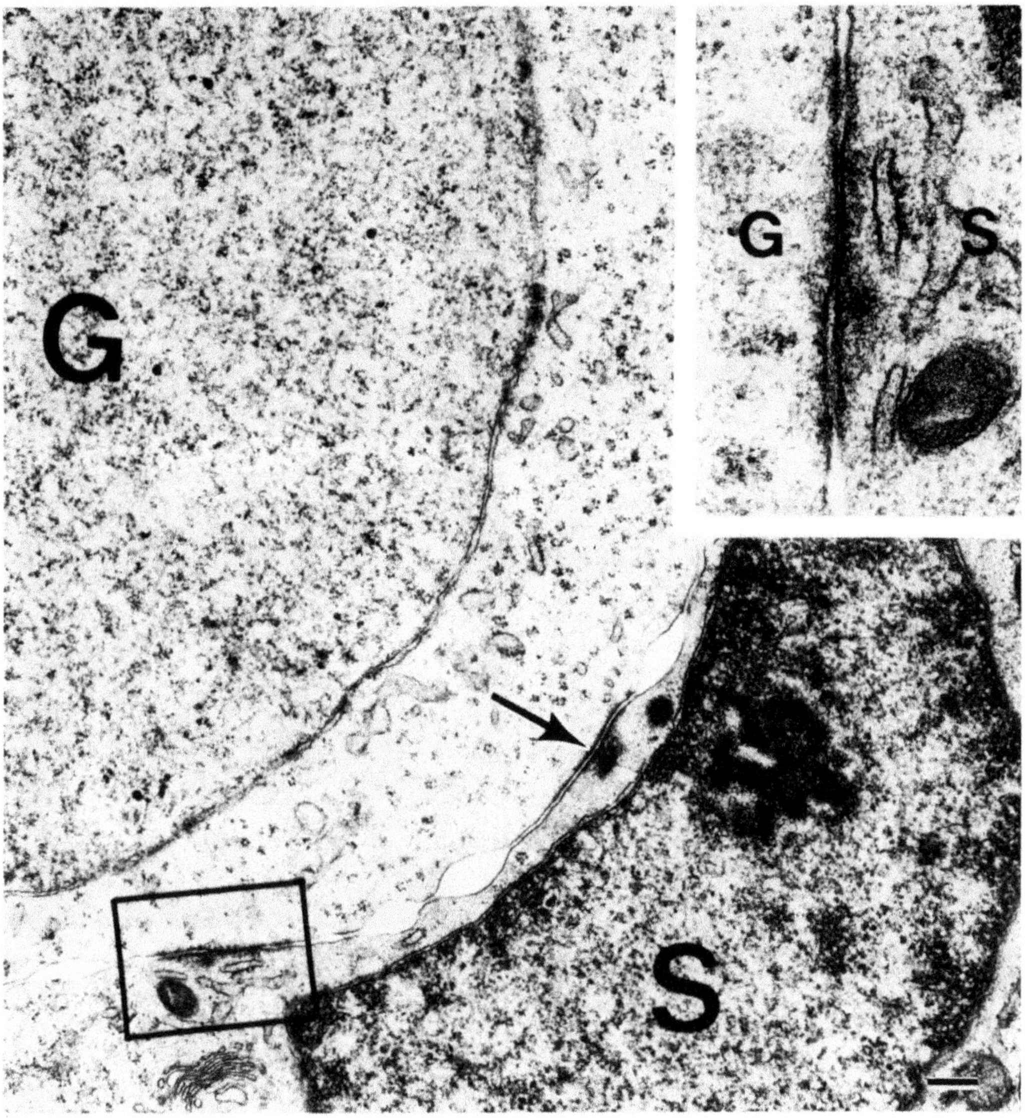

Figure 1–21. An electron micrograph of portions of a Sertoli cell (S) and a gonocyte (G) isolated as single cells from a 6-day-old rat pup and co-cultured as in Figure 1.20. Desmosome-like attachments *(arrow)* are common between the two cell types in vitro. In addition, by the day after plating these cells become coupled by gap junctions, as evidenced by the appearance of areas where the intercellular space between adjacent cells is narrowed *(inset).* Bar = 0.1 μm.

to them. Although the existence of similar contact-mediated interaction has not been confirmed yet in vivo, it seems likely that gonocytes and Sertoli cells recognize and communicate with each other via junctional contacts during perinatal life, an interaction that may have a role in promoting germ cell development.

3. SERTOLI CELL-MYOID CELL INTERACTION IN DEVELOPING TESTES

Since the basement membrane underlying the seminiferous epithelium is a joint product of Sertoli cells and myoid cells, at least in postnatal rats (Skinner et al., 1985), and since components of the extracellular matrix are crucial for seminiferous cord formation as already de-

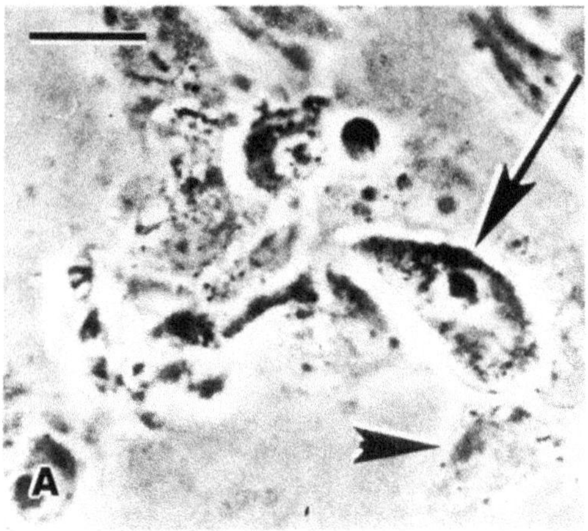

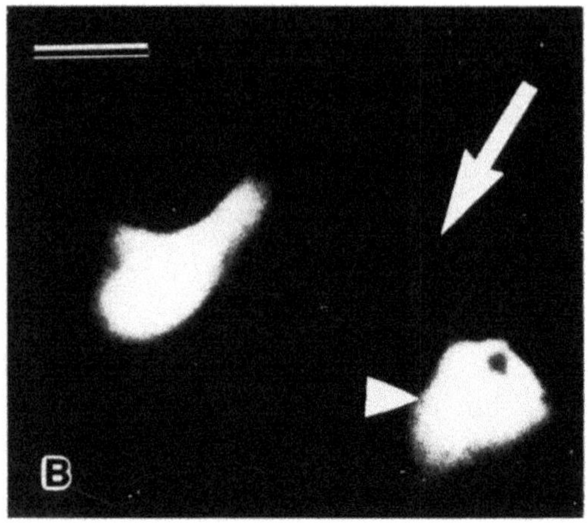

Figure 1–22. Three views of an elongated gonocyte *(arrow)* in co-culture with Sertoli cells *(arrowhead)*. To determine if these cells are metabolically coupled via gap junctions, they were glass-bead loaded simultaneously with two fluorescent probes, Rhodamine-dextran, which is too large to pass through these junctions, and Lucifer yellow, which is gap junction-permeant. Panel **A** is the phase contrast image. In panel **B**, rhodamine fluorescence in the Sertoli cell *(arrowhead)* indicates that it was glass-bead loaded with both probes. The absence of rhodamine fluorescence at the location of the gonocyte *(arrow)* indicates that it was not. Panel **C** shows the distribution of Lucifer yellow. The occurrence of fluorescence in the gonocyte *(arrow)* indicates that it is coupled via gap junctions with the adjacent Sertoli cell from which it received Lucifer yellow. Bar = 10 μm. (See Orth and Boehm, 1989, for details of method.)

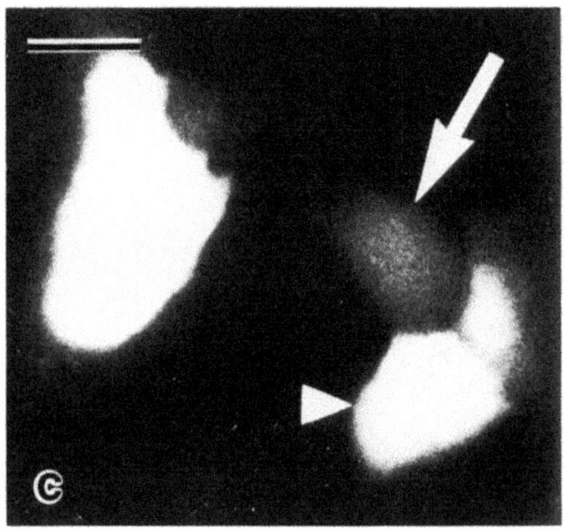

scribed, it is likely that Sertoli cell–myoid cell interactions are important during testicular development. Although studies aimed directly at assessing such interactions in vivo would be extremely difficult to perform, some indirect evidence in this area has been gained from work with Sertoli cell–peritubular cell co-cultures. Data gathered by Tung and Fritz (1986, 1987) suggest that contact-mediated cooperation between these two cell types may be essential for normal formation of testicular cords during fetal development. Utilizing an in vitro approach in which immature Sertoli cells were plated on a preexisting monolayer of peritubular cells, they found that apparent recognition between these two cell types led to the sequential formation of Sertoli cell "islands, ridges, and mounds" separated from peritubular cells by a highly organized basement membrane. When close apposition between the two cell types was blocked by a filter, however, basement membrane material failed to appear between them, suggesting that its formation is mediated by contact between Sertoli cells and peritubular cells. Moreover, when cAMP was added to co-cultures in which the cells were allowed to make contact, the formation of Sertoli cell mounds and deposition of basement membrane-like structures were also blocked. This suggests that inhibition of cord formation in cultured fetal testes incubated with cAMP (Taketo et al., 1984) results from interference by cAMP with Sertoli cell–peritubular cell cooperation, leading to abnormal formation of extracellular matrix in these testes.

Deposition of individual matrix components

by Sertoli cells and myoid cells seems to be specific to each cell type, at least in vitro. For example, Sertoli cells in culture secrete collagen type IV, laminin, and proteoglycans, but not fibronectin. On the other hand, peritubular cells produce various collagens and fibronectin, but not laminin. Although fibronectin is a likely component of the basement membrane of fetal seminiferous cords in vivo, it is lacking in the basal lamina between Sertoli cells and myoid cells of adults (Hadley and Dym, 1987). Thus, fibronectin synthesis by peritubular cells apparently ceases as the testis matures, a change that may relate to differentiation of the innermost layer of peritubular cells into "myoid" cells, as already described. In light of the findings with Sertoli cell–peritubular cell co-cultures described above, these apparent changes in peritubular cell function during testicular development may reflect a change in the way these cells interact with Sertoli cells.

IV. Postnatal Remodeling of the Seminiferous Cords

During the first week or two after birth, seminiferous cords undergo several morphological changes that are extremely important for later function of the mature testis. Two of these, relocation of gonocytes to the basement membrane of the cord and development of Sertoli–Sertoli junctional complexes leading to compartmentalization of the epithelium, apparently involve major alterations in the way cells of the seminiferous epithelium interact with each other. Each of these events is thought to insure survival of the germ cells at a distinct stage of their development into sperm and is thus vital for the fertility of the individual. The following description outlines our present understanding of the cellular and subcellular events underlying these occurrences in seminiferous cords of developing rats.

A. Relocation of Postnatal Gonocytes to the Basement Membrane

1. SEQUENCE OF EVENTS

Before birth, gonocytes of rats are large, round cells that are either located toward the center of the seminiferous cord or, if eccentric in position, are invariably separated from the underlying basement membrane by some portion of the surrounding Sertoli cells (see Fig. 1–10). Shortly after birth, however, some gonocytes have been observed in vivo with cytoplasmic processes directed toward (Novi and Saba, 1968) or in immediate contact with the basement membrane (Beaumont and Mandl, 1963). During this neonatal period many gonocytes also degenerate. The observation that most degenerating cells lie centrally in the cords, with few if any disappearing after reaching the basement membrane (Roosen-Runge and Leik, 1968), suggests that relocation of these cells to the periphery of the seminiferous cord where they contact the underlying extracellular matrix is essential for their survival and further development.

Recent studies on relocation of postnatal rat gonocytes in vivo and in vitro have provided more detailed information regarding when and how these cells change their position. In rats, all gonocytes remain completely surrounded by Sertoli cells until day 4 after birth (day of birth = day 1), when some come into direct contact with the basement membrane (McGuinness and Orth, 1992). At or just before this time, some gonocytes display cellular processes that are directed toward the basement membrane (Fig. 1–23) and that resemble pseudopods in that they lack larger organelles (Novi and Saba, 1968) and may contain bands of microtubules. These observations, along with the fact that gonocytes of fetuses are migratory during their journey to the gonadal ridge, have led to the suggestion that postnatal gonocytes may regain the ability to migrate in order to move to the periphery of the cord where they interact with the basement membrane. Although direct evidence is still lacking, data have recently been gathered that lend some support to this idea. For example, when motile cells, such as macrophages, receive a signal to migrate, the Golgi complex reorients toward the anticipated direction of movement and is thought to facilitate directed insertion of new membrane into the growing pseudopod. In seminiferous cords of rat pups, the Golgi complex was found to enlarge dramatically in some gonocytes a few days after birth (McGuinness and Orth, 1989), at the time when relocation of some of these cells is initiated. In addition, in postnatal gonocytes that were apparently en route to the periphery of the seminiferous cord, the Golgi

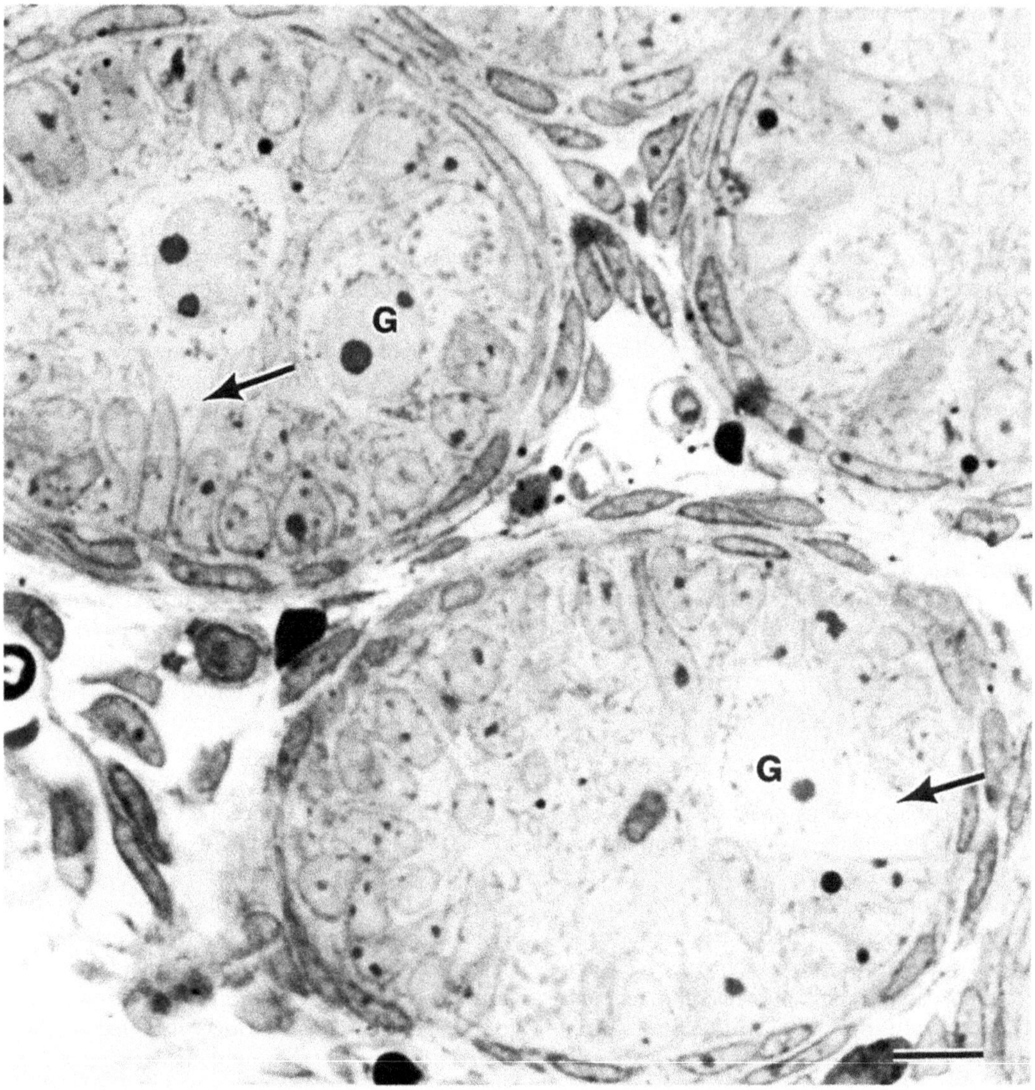

Figure 1–23. A light micrograph of testis from a 4-day-old rat pup. From day 3 onward, gonocytes (G) are commonly seen that have extended a pseudopodlike process *(arrows)* toward the basement membrane underlying the seminiferous epithelium. From day 4 onward, occasional gonocytes can be found that have relocated and are in direct contact with the basement membrane. Bar = 10 μm.

complex was found to be preferentially oriented toward the basement membrane. Both of these observations in vivo are consistent with the possibility that some gonocytes become motile in rats shortly after birth, presumably in response to a signal directing them to change their position within the cord.

With the recent development of co-cultures containing neonatal Sertoli cells, gonocytes and some peritubular cells plated on an artificial basement membrane material (MATRIGEL), it may prove possible to test the above hypothesis more directly. Gonocytes from 5 day old pups cultured under these conditions adhere to and form desmosome-like attachments and gap junctions with adjacent Sertoli cells (see Fig. 1–21; Orth and Boehm, 1990). After a few hours of culture, some gonocytes

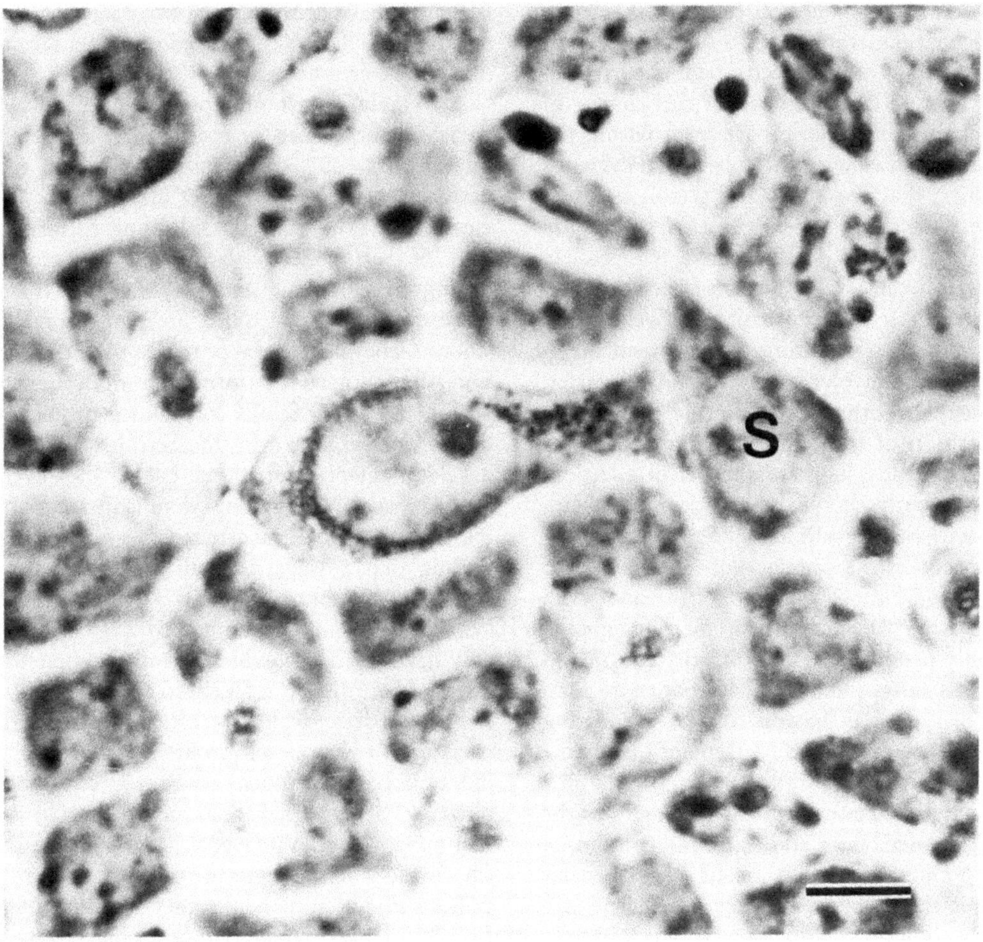

Figure 1–24. A phase contrast micrograph of a Sertoli cell–gonocyte co-culture 2 hours after plating on MATRIGEL. Shortly after plating, all gonocytes that have attached to underlying Sertoli cells are round, as in Figure 1.20. However, within a few hours some gonocytes, such as the one shown here (G), become elongated and develop prominent cellular process, raising the possibility that postnatal gonocytes may regain the ability to migrate, at least in vitro. S, Sertoli cell. Bar = 10 μm.

become elongated (Fig. 1–24) and form cellular extensions similar in morphology to those seen in some of these cells in vivo. Moreover, *newborn* gonocytes cultured under identical conditions acquire the ability to develop cellular processes during the initial days of culture (Orth and McGuinness, 1991). Thus, these findings imply that gonocytes isolated from seminiferous cords of neonates and co-cultured with Sertoli cells resemble those apparently en route to the basement membrane in vivo and may thus be useful in testing the notion that some of these cells resume motility shortly after birth.

2. POSSIBLE FACTORS INVOLVED IN RELOCATION OF POSTNATAL GONOCYTES

Although the precise cellular mechanism(s) used by gonocytes in reaching the basement membrane shortly after birth have not been identified, some preliminary information is available on the impetus for this change in position. When testes of newborn rats were organ-cultured in serum-free defined medium for several days, some gonocytes were observed in direct contact with the basement membrane by the end of the experiment (McGuinness and Orth, 1992). Since all gonocytes were found toward the center of the seminiferous cords and thus surrounded by Sertoli cells at the start of the incubation, translocation of these cells to the periphery of the cords had occurred in vitro. Therefore, a stimulus for this change in position is likely to originate from within the testis itself. In view of the substantial evidence indicating that gonocytes and Sertoli cells recognize each other and interact in potentially important ways, as detailed above, it is possible that Sertoli cells are a source of such stimulatory factors. It is also plausible, however, that a chemotactic signal from the basement membrane may be involved in initiating relocation of gonocytes, perhaps reflecting a change in the composition of the extracellular matrix after birth, as already suggested above. In addition, it was recently shown that gonocytes co-cultured with Sertoli cells are able to develop processes only in the presence of an underlying, laminin-containing matrix (Orth and McGuinness, 1991). Since the basement membrane is jointly produced by Sertoli cells and myoid cells, this raises the possibility that more than one cell type may have a role in stimulating gonocytes to move to the basement membrane shortly after birth.

Finally, since movement of gonocytes to the basement membrane and resumption of gonocyte proliferation both occur shortly after birth, it is possible that these two events are causally related. However, recent findings argue against this (McGuinness and Orth, 1992). When the first newborn gonocytes to divide in organ culture were labeled with ^{3}H-thymidine and their position in the cords followed thereafter, it was found that, although some cells divide before relocating, others do not. These observations suggest that (1) contact with the basement membrane does not by itself stimulate reinitiation of gonocyte division postnatally and (2) cell division is not the underlying impetus for relocation. Thus, proliferation and relocation of gonocytes seem to be separable events that are unlikely to share a common regulatory mechanism.

B. Development of Sertoli–Sertoli Junctional Complexes

1. SEQUENCE OF EVENTS

As is well known, unique junctional complexes between adjacent Sertoli cells constitute the structural basis of the blood testis barrier in the mature testis. The resulting compartmentalization of the seminiferous epithelium appears to be extremely important in providing the microenvironment needed by spermatogenic cells, a subject that has been intensely studied in the past few years (see Bardin et al., 1988 for review). However, the events leading to formation of Sertoli–Sertoli junctional complexes during development and the way in which these events are controlled are poorly understood. Generally, the appearance of the junctions is followed by formation of a lumen within the seminiferous cord and secretion of fluid into the lumen by the Sertoli cells. The development of a functional barrier, however, is apparently more gradual than is its morphological appearance, as attested to by Setchell's finding that passage of small molecules is incompletely blocked even at 30 days of age (Setchell et al., 1988), an age that is considerably beyond that at which terminal differentiation of Sertoli cells is likely to occur.

In rats, barrier formation also correlates with cessation of Sertoli cell proliferation since the junctions begin to appear at about 15 days after birth and achieve their final morphology by 16 to 19 days of age (Vitale et al., 1973). However, the relationship between the capacity of Sertoli cells to divide and their ability to form junctional complexes has not been investigated. The appearance of Sertoli–Sertoli junctional complexes also occurs in most mammals at about the same time as the start of meiosis by the germ cells. This has led to the suggestion that barrier formation may have a role in the onset of spermatogenesis. It is likely that meiosis begins before effective compartmentalization of the epithelium, however, because at least some pachytene spermatocytes are present in testes before the Sertoli–Sertoli junctional complexes are established (Vitale et al., 1973). Thus, a functional correlation between appearance of the blood testis barrier and either differentiation of the Sertoli cells or development of spermatogenesis has not been verified.

2. CONTROL OF BARRIER FORMATION

The factors responsible for stimulating formation of junctional complexes between adjacent Sertoli cells have not been conclusively identified. Although the possibility that gonadotropins may have a role in barrier formation has been probed in several studies, the results have been equivocal. For example, treatment of developing rats with estrogenic compounds delays appearance of the barrier, but does not prevent it (Vitale et al., 1973); this delay may have resulted from a direct effect of estrogen itself on the testis rather than from altered gonadotropin levels caused by estrogen treatment. The results of one study, however, tend to implicate gonadotropins in barrier formation. In that study, newborn mouse testes were implanted into testes of normal or hypophysectomized adults and development of the implants monitored over a number of weeks. The barrier failed to form in implants placed in hypophysectomized hosts while testes placed into normal hosts developed Sertoli–Sertoli junctions at the appropriate time (Bressler, 1976). Although neither direct monitoring of gonadotropin levels nor quantitative analysis of the implanted tissue was carried out for this study, the findings lend additional support to the idea that gonadrotropic hormones are necessary for the blood–testis barrier to develop.

It seems unlikely that the presence of germ cells is a requirement for formation of Sertoli–Sertoli junctional complexes. This conclusion is supported by the finding that morphologically and functionally normal barriers form in several different animal models in which germ cells are wholly or partially depleted. Moreover, when germ cells are essentially lacking from the developing testis, as in the prenatally irradiated, Sertoli cell-enriched rat (Tindall et al., 1975), an apparently normal blood–testis barrier forms, albeit after a delay. Other possibilities, such as involvement of Leydig cell–Sertoli cell communication in promoting barrier formation, have not yet been explored. In addition, Sertoli cell–extracellular matrix interaction may be important in barrier formation as immature Sertoli cells plated in vitro on an artificial basement membrane mimic to a remarkable extent the morphology of these cells in vivo (Byers et al., 1986). This includes formation of inter-Sertoli junctions between cultured cells that are virtually identical to those seen in situ. Thus, although development of these junctions is often viewed as a "preprogrammed" event that occurs more or less spontaneously, it is nevertheless possible that intratesticular and/or extratesticular factors are critical for their formation.

V. Summary

In recent years, substantial progress has been made in identifying events critical for the establishment of a morphologically and functionally differentiated testis in the fetus and neonate. During testicular organogenesis, formation of seminiferous cords seems to occur without the influence of extratesticular factors. It does, however, involve interaction between Sertoli cells and the surrounding extracellular matrix, whose composition is likely to be controlled both by nearby peritubular cell precursors and by the Sertoli cells themselves. Leydig cells make their appearance after definitive cords are present through processes that are, as yet, unexplored and that may be partly or totally independent of cord formation. Initial differentiation of Leydig cells may also involve their interaction with the surrounding extracellular matrix, and may be spontaneous or require only intratesticular factors.

Once the testis is fully formed, further growth and differentiation of each of its constituent cell types, that is, Sertoli cells, gonocytes, and Leydig cells, are regulated by specific factors in a characteristic sequence of events. FSH from the fetal or neonatal pituitary stimulates Sertoli cells to divide at levels that, in rats, are maximal before birth and decline thereafter. The responsiveness of Sertoli cells to this gonadotropin is apparently modulated by paracrine factors originating from Leydig cells and possibly other testicular cells. One such factor, β-endorphin, has been studied in some detail and depresses the cAMP-response of Sertoli cells to FSH by interacting with an intracellular inhibitory GTP-binding protein. This and other paracrine agents are no doubt important in creating a balance between the seminiferous and interstitial compartments during development that will be perpetuated in the mature testis. The size of the Sertoli cell population, which is established during fetal and neonatal life through the mechanisms just described, is crucial in determining the number of sperm produced in the adult, as demonstrated by studies of Sertoli cell-depleted rats. Thus, fertility of the adult appears to be at least partly a consequence of key cellular events occurring in the developing testis during the perinatal period.

Growth of other cell populations in the testis has been studied in less detail. It is clear, however, that rat gonocytes enter a period of quiescence late in fetal life and remain nonmitotic until just after birth. Resumption of mitosis by these cells is independent of extratesticular influence and may involve their interaction with either Sertoli cells, with whom they maintain gap junctions, or with myoid cells, which differentiate from the peritubular cell population shortly after birth. Around the time that some of these germ cells resume mitosis, gonocytes extend cellular processes toward the periphery of the seminiferous cord and eventually translocate to the basement membrane, contacting it. The precise temporal relationship between these events and how they are controlled is likely to be crucial for function of the mature testis and will doubtless be the subject of detailed future studies.

Development of the fetal Leydig cell population apparently relies primarily on addition of new cells via differentiation from mesenchymelike precursors, as Leydig cells rarely divide under normal circumstances. Once established, the size of this fetal population may well be stable not only during neonatal life but possibly persisting into adulthood. Although the initial appearance of Leydig cells in young fetuses seems to be a spontaneous event, that is, not requiring extratesticular influences, further development of this population depends at least in part on LH. Responsiveness of Leydig cells to LH may also be regulated by Sertoli cell-secreted factors, suggesting that two-way interaction operates between these cells during testicular development.

Thus, although many questions remain unanswered, considerable progress has been made in recent years in understanding how testicular cells interact with extratesticular factors, with each other, and with the nearby extracellular environment during fetal and neonatal development. Future studies will undoubtedly continue to focus on these exciting issues.

ACKNOWLEDGMENTS

The author wishes to thank Drs. Alfred Jost and Lauri Pelliniemi for kindly providing some of the original illustrations used in this chapter. The studies carried out in the author's laboratory were supported by grants HD00591 and HD15563 from the National Institutes of Health.

REFERENCES

Agelopoulou, R., and Magre, S. (1987). Expression of fibronectin and laminin in fetal male gonads in vivo and in vitro with and without testicular morphogenesis. Cell Differentiation 21:31–36.

Agelopoulou, R., Magre, S., Patsavoudi, E., and Jost, A. (1984). Initial phases of the rat testis differentiation in vitro. J. Embryol. Exp. Morph. 83:15–31.

Alvarez-Buylla, A., and Merchant-Larios, H. (1986). Mouse primordial germ cells use fibronectin as a substrate for migration. Exp. Cell Res. 165:362–368.

Bardin, C. W., Morris, P., Shaha, C., Feng, Z., Rossi, V. Vaughan, J., Vale, W., Voglymayr, J., and Chen, C. (1989). Inhibin structure and function in the testis. Ann. N. Y. Acad. Sci. 564:10–23.

Bardin, C. W., Cheng, C., Musto, N., and Gunsalus, G. (1988). The Sertoli cell, in *The Physiology of Reproduction* (Knobil, E., et al., eds.) pp. 933–974, Raven Press, New York.

Beaumont, H., and Mandl, A. (1963). A quantitative study of primordial germ cells in the male rat. J. Embryol. Exp. Morph. 11:715–740.

Benahmed, M., Tabone, E., Grenot, C., Sanchez, P., Chauvin, M., and Morera, A. (1986). Paracrine control of Leydig cell activity by FSH dependent proteins from Sertoli cells: An in vitro study. J. Steroid Biochem. 24:311–315.

Bressler, R. (1976). Dependence of Sertoli cell maturation on the pituitary gland of the mouse. Am. J. Anat. 147:447–456.

Bressler, R., and Ross, M. (1972). Differentiation of peritubular myoid cells of the testis: Effects of intratesticular implantation of newborn mouse testes into normal and hypophysectomized adults. Biol. Reprod. 6:148–159.

Byers, S., Hadley, M., Djakiew, D., and Dym, M. (1986). Growth and characterization of polarized monolayers of epididymal epithelial cells and Sertoli cells in dual environment culture chambers. J. Androl. 7:59–68.

Byskov, A. G. (1986). Differentiation of mammalian embryonic gonad. Physiological Reviews 66:71–117.

Clermont, I., and Huckins, C. (1968). Evolution of gonocytes in the rat testis during late embryonic and early post-natal life. Arch. Anat. Hist. Embryol. 51:343–354.

Clermont, I., and Perey, B. (1957). Quantitative study of the cell population of the seminiferous tubules in immature rats. Am. J. Anat. 100:241–260.

Codesal, J., Santamaria, R., Paniagua, R., and Nistal, M. (1989). Proliferative activity of human spermatogonia from fetal period to senility measured by cytophotometric DNA quantification. Arch. Andrology 22:209–215.

Dalkin, A., Bourne, G., Pieper, D., Regiani, S., and Marshall, J. (1981). Pituitary and gonadal gonadotropin-releasing hormone receptors during sexual maturation in the rat. Endocrinology 108:1658–1664.

DeFelici, M., and Dolci, S. (1989). In vitro adhesion of mouse fetal germ cells to extracellular matrix components. Cell Diff. Develop. 26:87–96.

Dorrington, J., and Armstrong, D. (1975). Follicle stimulating hormone stimulates estradio-17β synthesis in cultured Sertoli cells. Proc. Nat. Acad. Sci. (USA) 72:2677–2681.

Dufau, M., and Knox, G. (1985). Fetal Leydig cell culture—An in vitro system for the study of trophic hormone and GnRH receptors and actions. J. Ster. Biochem. 23:743–755.

Eddy, E. M., Clark, J. M., Gong, D., and Fenderson, B. A. (1981). Origin and migration of primordial germ cells in mammals. Gamete Res. 4:333–362.

Eguchi, Y., Arishima, K., Nasu, T., Toda, M., Morikawa, Y., and Hashimoto, Y. (1978). Development of the fetal pituitary–testicular system based on observation of Leydig cells in encephalectomized, hypophysectomized and control fetal rats. Anat. Rec. 190:679–686.

Everett, N. B. (1943). Observational and experimental evidences relating to the origin and differentiation of the definitive germ cells in mice. J. Exp. Zool. 92:49–91.

Gerendai, I., Nemeskeri, A., and Csernus, V. (1983). Naloxone has a local effect on the testis of immature rats. Andrologia 15:398–403.

Hadley, M., and Dym, M. (1987). Immunocytochemistry of extracellular matrix in the lamina propria of the rat testis: Electron microscopic localization. Biol. Reprod. 37:1283–1289.

Hilscher, B., Hilscher, W., Bülthoff-Ohnolz, B., Krämer, U., Birke, A., Pelzer, H., and Gauss, G. (1974). Kinetics of gametogenesis. I. Comparative histological and autoradiographic studies of oocytes and transitional prospermatogonia during oogenesis and prespermatogensis. Exp. Cell Res. 154:443–470.

Hutson, J., Metcalfe, S., MacLauglin, D., Cate, R., Cigarroa, F., Ueno, S., and Donahoe, P. (1989). Mullerian inhibiting substance, in *The Testis* (Burger, H., and DeKretser, D., eds.) pp. 143–180, Raven Press, New York.

Jaillard, C., Chatelain, P., and Saez, J. (1987). In vitro regulation of pig Sertoli cell growth and function: Effects of fibroblast growth factor and somatomedin-C. Biol. Reprod. 37:665–674.

Jost, A., and Magre, S. (1988). Control mechanisms of testicular differentiation. Phil. Trans. R. Soc. (Lond) B 322:55–61.

Jost, A., Perlman, S., Valentino, O., Castanier, M., Scholler, R., and Magre, S. (1988). Experimental control of the differentiation of Leydig cells in the rat fetal testis. Proc. Natl. Acad. Sci. 85:8094–8097.

Jost, A., Valentino, O., Agelopoulou, R., and Magre, S. (1985). Action d'un analogue de la proline sur la différenciation in vitro du testicule fœtal du rat. C. R. Acad. Sc. Paris 301:225–232.

Jost, A., and Magre, S. (1984). Testicular development phases and dual hormonal control of sexual organogenesis, in *Sexual Differentiation: Basic and Clinical Aspects* (Serio, M., et al., eds.) pp. 1–15, Raven Press, New York.

Kerr, J., and Sharpe, R. (1985). Follicle-stimulating hormone induction of Leydig cell maturation. Endocrinology 116:2592–2604.

Kuopoi, T., and Pelliniemi, L. (1989). Patchy basement membrane of rat Leydig cells shown by ultrastructural immunolabeling. Cell Tiss. Res. 256:45–51.

Kuopio, T., Paranko, J., and Pelliniemi, L. (1989a). Basement membrane epithelial features of fetal-type Leydig cells in rat and human testis. Differentiation 40:198–206.

Kuopio, T., Pelliniemi, L., and Huhtaniemi, I. (1989b). Rapid Leydig cell proliferation and luteinizing hormone receptor replacement in the neonatal rat testis after a single injection of human choroinic gonadotropin. Biol. Reprod. 40:135–143.

Kuopio, T., Savouras, P., Pelliniemi, L., and Huhtamiemi, I. (1989c). Transplantation of newborn rat testis under the kidney capsule of adult host as a model to study the structure and function of Leydig cells. J. Androl. 10:335–345.

Leeson, C., and Leeson, T. (1963). The postnatal development and differentiation of the boundary tissue of the seminiferous tubule of the rat. Anat. Rec. 147:243–259.

Magre, S., and Jost, A. (1984). Dissociation between testicular organogenesis and endocrine cytodifferentiation of Sertoli cells. Proc. Natl. Acad. Sci. 81:7831–7834.

McGuinness, M., and Orth, J. (1992). Reinitiation of gonocyte mitosis and movement to the basement membrane in testes of newborn rats in vivo and in vitro. *Anat. Rec.* in press.

McGuinness, M., and Orth, J. (1989). Relocation of postnatal gonocytes to the basement membrane of seminiferous cords may involve re-orientation of the Golgi complex. J. Cell Biol. 109:68a.

McNeil, P., and Warder, E. (1987). Glass beads load macromolecules into living cells. J. Cell Sci. 88:669–678.

Novi, A., and Saba, P. (1968). An electron microscopic study of the development of rat testis in the first ten postnatal days. Zeit. Zellforsch. 86:313–326.

O, W., Robertson, D., and DeKretser, D. (1989). Inhibin as an oocyte meiotic inhibitor. Mol. Cell. Endocrinol. 62:307–311.

Orth, J. (1986). FSH-induced Sertoli cell proliferation in the developing rat is modified by β-endorphin produced in the testis. Endocrinology 119:1876–1878.

Orth, J. (1984). The role of follicle stimulating hormone in controlling Sertoli cell proliferation in testes of fetal rats. Endocrinology 115:1248–1255.

Orth, J. (1982). Proliferation of Sertoli cells in fetal and postnatal rats: A quantitative autoradiographic study. Anat. Rec. 203:485–492.

Orth, J., and McGuinness, M. (1991). Neonatal gonocytes co-cultured with Sertoli cells on a laminin-containing matrix resume mitosis and elongate. Endocrinology 129:1119–1121.

Orth, J., and Boehm, R. (1990a). Endorphin suppresses FSH-stimulated proliferation of isolated neonatal Sertoli cells by a pertussis toxin-sensitive mechanism. Anat. Rec. 226:320–327.

Orth, J. and Boehm, R. (1990). Functional coupling of neonatal rat Sertoli cells and gonocytes in coculture. J Endocrinology 127:2812–2820.

Orth, J., Gunsalus, G., and Lamperti, A. (1988). Evidence from Sertoli cell-depleted rats indicates that spermatid number in adults depends on numbers of Sertoli cells produced during perinatal development. Endocrinology 122:787–794.

Orth, J., Higginbotham, C., and Salisbury, R. (1984). Hemicastration causes and testosterone prevents enhanced uptake of ^{3}H-thymidine by Sertoli cells in testes of immature rats. Biol. Reprod. 30:263–270.

Paranko, J., Pelliniemi, L., Vaheri, A., Foidart, J-M., and Lakkala-Paranko, T. (1983). Morphogenesis and fibronectin in sexual differentiation of rat embryonic gonads. Differentiation 23 (Suppl):S72–S81.

Patsavoudi, E., Magre, S., Castanier, M., Scholler, R., and Jost, A. (1985). Dissociation between testicular morphogenesis and functional differentiation of Leydig cells. J. Endocrinology 105:235–238.

Pelliniemi, L. J., Paranko, J., Grund, S., Fröjdman, K., Foidart, J., and Lakkala-Paranko, T. (1984). Extracellular matrix in testicular differentiation. Ann. N. Y. Acad. Sci. 438:405–416.

Pflüger, E. F. (1863). Über die Eierstöcke der Säugetiere und des Menschen. Leipzig:Engelmann.

Picon, R. (1976). Testosterone secretion by foetal rat testes in vitro. J. Endocrinol. 71:231–238.

Rodemer-Lenz, E. (1989). On cell contribution of gonadal soma formation in quail-chick chimeras during the indifferent stage of gonadal development. Anat. Embryol. 179:237–242.

Roosen-Runge, E., and Leik, J. (1968). Gonocyte degeneration in the postnatal male rat. Am. J. Anat. 122:275–300.

Roosen-Runge, E., and Anderson, D. (1959). The development of the interstitial cells in the testis of the albino rat. Acta Anat. 37:125–136.

Setchell, B., Pöllänen, P., and Zupp, J. (1988). Development of the blood–testis barrier and changes in vascular permeability and puberty in rats. Int. J. Androl. 11:225–233.

Shabanowitz, R., DePhilip, R., Crowell, J., Tres, L., and Kierszenbaum, A. (1986). Temporal appearance and cyclic behavior of Sertoli cell-specific secretory proteins during development of the rat seminiferous tubule. Biol. Reprod. 35:745–760.

Shaha, C., Liotta, A., Krieger, D., and Bardin, C. (1984). The ontogeny of immunoreactive β-endorphin in fetal, neonatal and pubertal testes from mouse and hamster. Endocrinology 114:1584–1591.

Sharpe, R. M. (1984). Intratesticular factors controlling testicular function. Biol. Reprod. 30:29–49.

Skinner, M., and Fritz, I. (1985). Identification of a non-mitogenic paracrine factor involved in mesenchymal–epithelial cell interactions between testicular peritubular cells and Sertoli cells. Mol. Cell. Endocrinol. 44:85–87.

Skinner, M., Tung, P., and Fritz, I. (1985). Cooperativity between Sertoli cells and testicular peritubular cells in the production and deposition of extracellular matrix components. J. Cell Biol. 100:1941–1947.

Steinberger, A., and Steinberger, E. (1971). Replication pattern of Sertoli cells in maturing rat testis in vivo and in organ culture. Biol. Reprod. 4:84–87.

Taketo, T., Thau, R., Adeyemo, O., and Koide, S. (1984). Influence of adenosine 3′5′-cyclic monophosphate analogues on testicular organization of fetal mouse gonads in vitro. Biol. Reprod. 30:189–198.

Thumann, A., and Bustos-Obregon, E. (1978). An "in vitro" system for the study of rat spermatogonial proliferative control. Andrologia 10:22–25.

Tindall, D., Vitale, R., and Means, A. (1975). Androgen binding protein as a biochemical marker of formation of the blood testis barrier. Endocrinology 97:636–648.

Tung, P., and Fritz, I. (1986). Cell–substratum and cell–cell interactions promote testicular peritubular myoid cell histotypic expression in vitro. Dev. Biol. 115:155–170.

Tung, P., and Fritz, I. (1987). Morphogenetic restructuring and formation of basement membranes by Sertoli cells and testis peritubular cells in co-culture: Inhibition of the morphogenetic cascade by cyclic AMP derivatives and by blocking direct cell contact. Dev. Biol. 120:139–153.

Vitale, R., Fawcett, D., and Dym, M. (1973). The normal development of the blood–testis barrier and the effects of clomiphene and estrogen treatment. Anat. Rec. 176:333–344.

Wartenberg, H. (1989). Differentiation and development of the testes, in *The Testis* (Burger, H., DeKretser, D., eds.) 2nd ed, pp. 67–118, Raven Press, New York.

Wartenberg, H. (1978). Human testicular development and the role of the mesonephros in the origin of a dual Sertoli cell system. Andrologia 10:1–21.

Weniger, J.-P., and Zeis, A. (1988). Stimulation of aromatase activity in the fetal rat gonads by cAMP and FSH. Acta Endocrinol. 119:381–385.

Witschi, E. (1948). Migration of the germ cells of human embryos from the yolksac to the primitive gonadal fold. Contrib. Embryol. 32:67–80.

Zamboni, L., and Upadhyay, S. (1983). Germ cell differentiation in mouse adrenal glands. J. Exp. Zool. 228:173–193.

Zirkin, B., and Ewing, L. (1987). Leydig cell differentiation during maturation of the rat testis: A stereological study of cell number and ultrastructure. Anat. Rec. 219:157–163.

CHAPTER TWO

Molecular Genetics of the Human and Mouse Y-Chromosome

Andrew H. Sinclair, Mark S. Palmer, Philippe Berta, Nathan A. Ellis, and Peter N. Goodfellow

I. Introduction

In mammals, sex is determined by chromosomal constitution: males have one X chromosome and one Y chromosome; females have two X chromosomes (reviewed by Goodfellow and Darling, 1988). Analysis of individuals with unusual sex chromosome combinations has shown that the presence of the Y chromosome is crucial for male sex determination and absence of the Y chromosome results in female development (Ford et al., 1959; Welshons and Russell, 1959; Jacobs and Strong, 1959). The Y chromosome acts by inducing the undifferentiated fetal gonads to become testes and subsequent male sexual differentiation is controlled by testicular hormones (Jost et al., 1973; Josso et al., 1977). The Y chromosome gene, or genes, responsible for inducing testis development have been named testis determining factor *(TDF)* in humans and testis determining-Y chromosome *(Tdy)* in mouse.

The Y chromosome is highly adapted for its role in sex determination (Koller and Darlington, 1934). It is composed of two parts with very different properties: the pseudoautosomal region (PAR), shared with the X chromosome, and the Y-specific region. The PAR is thought to be required to correct pairing in male meiosis. Homology between the PAR on the X and Y chromosomes is maintained by recombination within this region. The Y-specific region encodes *TDF* and, perhaps, fertility and growth factors. The fact that the Y-specific region is only found in males makes it unlikely that this region codes for any genes that are not specific for the male phenotype. The reverse is not the case, however, as many genes needed for male differentiation are known to be autosomal.

In this chapter we describe the genetic mapping of the Y chromosome in humans and mice and indicate how this information has been used to identify candidate sequences for *TDF.*

II. Mapping the Human Y Chromosome

A. Cytogenetic Map

The human Y chromosome is among the smallest human chromosomes and is only one third the size of the X chromosome. Cytogenetic analysis of the Y chromosome is difficult because of its small size; nevertheless, several general regions can be identified (Goodfellow et al., 1985, Fig. 2–1).

The pairing region. The terminal part of the Y chromosome short arm pairs with the terminal part of the X chromosome short arm in male meiosis (Chandley et al., 1984). The extent of pairing varies extensively according to the stage of meiosis from 11 to 56 percent of the Y chromosome. Even the smallest pairing region observed is considerably larger than the

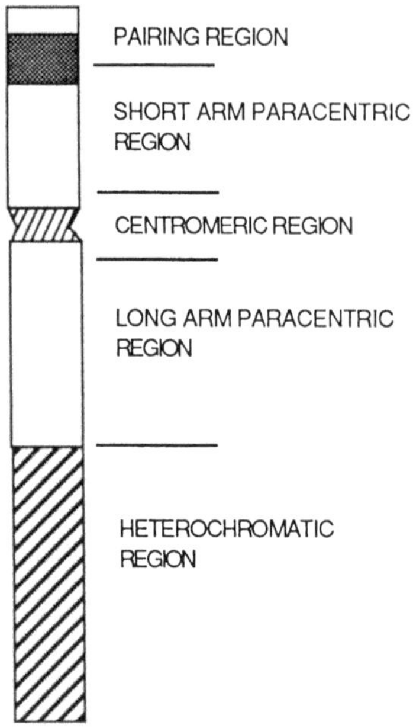

Figure 2–1. A classic map of the human Y chromosome based on cytogenetic analysis. This map was taken from Goodfellow et al., 1985.

known size of the pseudoautosomal region and must extend into the sex determining region.

The short arm paracentric region. Cytogenic analysis of translocated and deleted Y chromosomes suggested that this was the most likely location for *TDF* (Davis, 1981).

The centromeric region. All human chromosomes contain, in the centromeric region, alphoid sequences arranged as blocks of tandem repeats. Alphoid sequences, *DYZ3,* from the human Y chromosome have been cloned, sequenced, and analyzed by short-range and long-range restriction mapping (Wolfe et al., 1985; Tyler-Smith and Brown, 1987). These studies have defined a diagnostic repeat of 5.5 kb, revealed by cleavage with *Eco*RI. The Y chromosome contains multiple copies of this repeat, but, the precise number is highly variable in the population. The function of the alphoid sequences is unknown.

The paracentric long arm euchromatin. Translocations and deletions in this region have been correlated with infertility and effects on growth (Tiepolo and Zuffardi, 1976; Alvesalo and de la Chapelle, 1981).

The heterochromatic region. Approximately half of the Y chromosome is heterochromatic and stains intensely with quinacrine mustard. Two repeat sequences, *DYZ1* and *DYZ2,* comprise the majority of the sequences in the heterochromatic region (Cooke, 1976). These repeats are essentially Y specific; however, related repeats are found on other chromosomes (Smith et al., 1987). The size of the heterochromatic region varies extensively between different males and is present in 1 in 3000 normal females (McKay et al., 1978; Cooke and Noel, 1979). These observations suggest that the heterochromatic region is of limited functional importance.

B. Meiotic and Physical Maps of the Pseudoautosomal Region

Based on cytogenetic studies of meiosis in male rats, Koller and Darlington (1934) proposed that meiotic pairing of mammalian sex chromosomes depended on shared sequences and that recombination would occur between these sequences. Genes that exchange between the sex chromosomes would fail to show complete sex linkage and this inheritance pattern was termed pseudoautosomal (Burgoyne, 1982). Subsequently, the region shared by the sex chromosomes has been widely referred to as the pseudoautosomal region.

The first direct evidence for the existence of the PAR in humans came from studying the gene *MIC2* that encodes a cell surface antigen expressed on many different cell types (Goodfellow, 1983). *MIC2* was mapped to the telomeric region of the short arms of both the human X and Y chromosomes (Buckle et al., 1985). The chromosomal localization was within the known pairing region and was consistent with *MIC2* being a pseudoautosomal gene.

Formal proof of the existence of the PAR was facilitated by the molecular cloning of sequences derived from the Y chromosome short arm. Several DNA probes were found that recognized restriction fragment length polymorphisms (RFLPs) and in family studies these RFLPs were found to recombine between the X and Y chromosomes (Cooke et al., 1985; Simmler et al., 1985). Using these markers and

DNA probes for *MIC2* (Goodfellow et al., 1986a), it was possible to construct a meiotic map of the PAR from *DYS14,* at the telomere to *MIC2,* which is located adjacent to the sex specific region (Rouyer et al., 1986; Goodfellow et al., 1986a; Weissenbach et al., 1987, Fig. 2–2).

Recombination in the PAR has several unique features. The rate of recombination in male meiosis is much higher than for the same region in female meiosis: the distance between *DXYS14* and *MIC2* is about 47 cM in male meiosis and less than 5.0 cM in female meiosis. Despite the large genetic distance between *DXYS14* and *MIC2* double recombinants in this region have not been observed. The lack of double recombinants and the close to 50 percent recombination at the telomere imply that one recombination occurs in the PAR in every male meiotic event.

Recently, three groups have produced long-range restriction maps of the entire PAR in humans (Brown, 1988; Petit et al., 1988; Rappold and Lehrach, 1988). These maps agree that the PAR is approximately 2.6 kb in length and they show a direct correspondence between physical position on the long-range map and genetic localization on meiotic map (Fig. 2–2). The physical maps also identify five CpG-rich islands (also known as HTF islands) within the PAR. CpG-rich islands are often associated with genes and the most proximal island is part of the *MIC2* gene (Pritchard et al., 1987).

Besides *MIC2,* no other genes have been directly demonstrated to be pseudoautosomal, although we have described a pseudoautosomal locus *XGR* that regulates expression of the XG and MIC2 antigens on red blood cells (Goodfellow et al., 1986b). However, it has not been possible to show formally that *XGR* is distinct from *MIC2.* Other candidates for pseudoautosomal loci have been suggested, but, in none of these cases has convincing evidence been presented.

C. A Map of the Y-Specific Region

A large number of sequences have been isolated from the Y-specific region and these have been used, in combination with DNA from patients with chromosomal abnormalities, to produce a detailed "deletion" map of the Y chromosome. The short arm of the Y chromosome has been mapped by reference to the genomes of XX males and XY females. The long arm has been mapped using translocation and deletion chromosomes.

The majority of XX males have been created by an abnormal exchange between the short arms of the X and Y chromosomes that transfers *TDF* from the Y chromosome to the X chromosome (Ferguson-Smith, 1966; Guellaen et al., 1984; Petit et al., 1987). Along with *TDF* a variable amount of Y-derived material is also transferred depending on the position of the Y chromosome breakpoint. The X chromosome breakpoint can be in either the PAR or the X-specific region (Weissenbach et al., 1987). Some XY females are created by the inheritance of a Y chromosome deleted by a terminal exchange.

The original map of Vergnaud and colleagues (1986) contained seven intervals and *TDF* was located in the most distal Y-specific segment proximal to the PAR. In subsequent studies, it became clear that *TDF* was located very close to the pseudoautosomal region boundary (Page et al., 1987). The precise chromosomal localization of *TDF* and of *ZFY,* a candidate gene for *TDF,* is discussed in the section on *ZFY.* In the proximal short arm region Arnemann and associates (1987, 1989) have identified a gene *TSPY* (testis-specific protein Y-linked) of unknown function, which is highly expressed in testis.

In most maps the centromeric region is defined by alphoid sequences and the heterochromatic region of Yq by *DYZ1* and *DYZ2.* No genes, other than *MIC2, TDF, TSPY,* and *ZFY* have been mapped to the Y chromosome short arm. Genes that have been mapped to the centromeric region and long arm of the Y chromosome include *H-Y, AZF, GCY,* and *GBY* (Weissenbach et al., 1989).

H-Y controls the expression of an antigen defined by transplantation studies and by MHC-restricted T-cell killing. As it has been proposed that H-Y antigen is a candidate for the testis determining factor, it is discussed in detail below.

Individuals with deletions of the long arm of the Y chromosome, that include part of paracentric euchromatin, are usually sterile. This

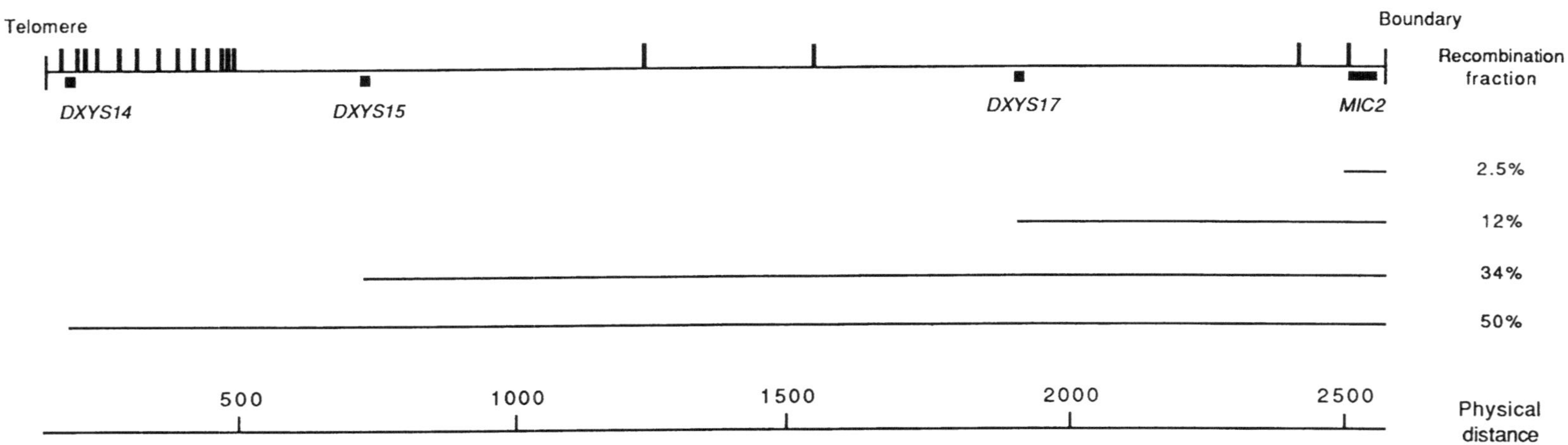

Figure 2–2. A comparison of the genetic and physical maps of the human pseudoautosomal region. The *upper lines* show the location of loci used in genetic and physical mapping and of CpG-rich regions *(bars)*. Probes are shown below the map *(boxes)*. Recombination fractions with respect to sex-specific sequences are at right and physical distance is at bottom in kb. This figure and legend have been reproduced from Ellis and Goodfellow (1989) with the kind permission of the Editor of *Trends in Genetics*.

has been used as an argument for the existence of a spermatogenesis factor, *AZF* (azoospermia factor; Tiepolo and Zuffardi, 1976). Similarly, reduced tooth size in patients with deletions of the Y chromosome long arm has been used to postulate the existence of growth control gene *GCY* (growth control-Y; Alvesalo and de la Chapelle, 1981).

Gonadoblastoma is a characteristic tumor of dysgenic gonads involving germ cells and accessory cells. The occurrence of gonadoblastoma appears to be promoted by the presence of the Y chromosome and this has led to the proposal that a protooncogene, *GBY* (gonadoblastoma-Y), may reside on the Y chromosome (Page, 1987).

H-Y, AZF, GCY, and *GBY* have not been separated genetically and it is possible that they are different manifestations of the same gene.

A major problem with current maps of the Y chromosome is that each map has been constructed using different patient material and different probes. This has two consequences: it is difficult, or impossible, to compare maps from different groups and, to quote Professor E. B. Robson, "Private maps are like private armies—they encourage people to fight." In an attempt to improve this situation, J. Weissenbach has agreed to make available DNA from a selected group of patients that can act as cross-reference points between the different maps.

D. The Pseudoautosomal Region Boundary

The boundary between the sex-specific region and the pseudoautosomal region must function to separate the two regions and to prevent recombination progressing into the sex-specific region.

The boundary was initially located by a combination of long-range restriction mapping and meiotic mapping (Goodfellow et al., 1986a; Pritchard et al., 1987). On a long-range restriction map the boundary was positioned within 150 kb of *MIC2* and this was consistent with the 100-kb indirectly deduced from the meiotic map. The long-range restriction map also identified *DYS104* as the Y-specific locus closest to the boundary. The region between *MIC2* and *DYS104* was cloned by chromosomal walking and the position of the boundary was sought by testing for the transition from X-Y shared to Y-specific sequences. The transition region was identified and sequenced (Ellis et al., 1989). In the immediate vicinity of the boundary four different regions could be identified:

1. The pseudoautosomal region
2. An insertion of an *Alu* repeat on the Y chromosome
3. A 225-bp region of partial (77 percent) sequence homology between the X and Y chromosomes
4. A sex chromosome specific region with no sequence homology between the X and Y chromosomes

The first base pair of abrupt sequence divergence between the X and Y chromosomes is the first base pair of the *Alu* repeat. This suggests that the *Alu* repeat is at the boundary and should be present on the Y chromosome of all normal males and absent in normal females. We have tested 150 individuals: all males have the *Alu* sequence, all females lack this sequence at the boundary. In addition, great apes have a similar boundary structure and have *Alu* inserts on the Y chromosome (N. Ellis, personal communication).

One explanation for these results is that before the formation of the great ape lineage, an *Alu* retroposon inserted 225 bp away from the boundary. The insertion disrupted pairing at the old boundary and prevented recombination in the 225-bp region proximal to the *Alu* repeat. Consequently, the 225-bp region began to diverge between the X and Y chromosomes. The alternative explanation is that the *Alu* inserted precisely at the old boundary and that the 225-bp region was already diverging between the X and Y chromosomes. The latter hypothesis appears to be unlikely, but, can be tested by studying evolution of the boundary region in primates.

At face value, these results are consistent with the boundary being formed by any sequence nonidentity.

III. Mapping the Mouse Y Chromosome

By comparison with humans, genetic mapping of the mouse Y chromosome is relatively crude.

A large number of cloned probes are available to define different Y-specific loci; however, there are few Y chromosome deletions and translocations that can be used for mapping.

Like the human Y chromosome much of the murine Y chromosome is repeated sequences. Eicher and colleagues have estimated that up to one-third of the Y chromosome is composed of the retrovirus-related sequences (M720 and MuRVY) and associated sequences (Phillips et al., 1982; Eicher and Washburn, 1986; Eicher et al., 1989). These have proved to be valuable tools for tracing relationships between different species of mice (Phillips et al., 1982; Tucker et al., 1989). Similarly, other Y chromosome-deprived repeat sequences have been used for distinguishing the Y chromosomes of laboratory mice (Bishop et al., 1985).

Based on two unusual Y chromosome anomalies Y^{sxr} and Y*, Eicher and Washburn (1986) suggested that the mouse Y chromosome could be derived into three functional units (Fig. 2–3), in addition to the centromere and telomeres. Recent work has led to a subdivision of one of these units and a revision of the order. Central to the mapping of the mouse Y chromosome has been the *Sxr* mutation.

A. Y-Specific Region

Sxr was first described by Cattanach and colleagues as an autosomal dominant mutation that caused sex reversal in females (Cattanach et al., 1971). Males carrying the *Sxr* mutation are phenotypically normal and fertile. *Sxr* carrier males mated to normal females produce equal numbers of four types of offspring: normal males, normal females, carrier males, and sex-reversed females. Despite the deceptively simple genetics, it proved impossible to map the *Sxr* mutation to any known linkage group.

Singh and Jones (1982) found that Y-specific Bkm repeat sequences were inherited with the *Sxr* mutation (see below for a description of Bkm sequences). In addition, it was shown that whereas Bkm probes hybridized to the centro-

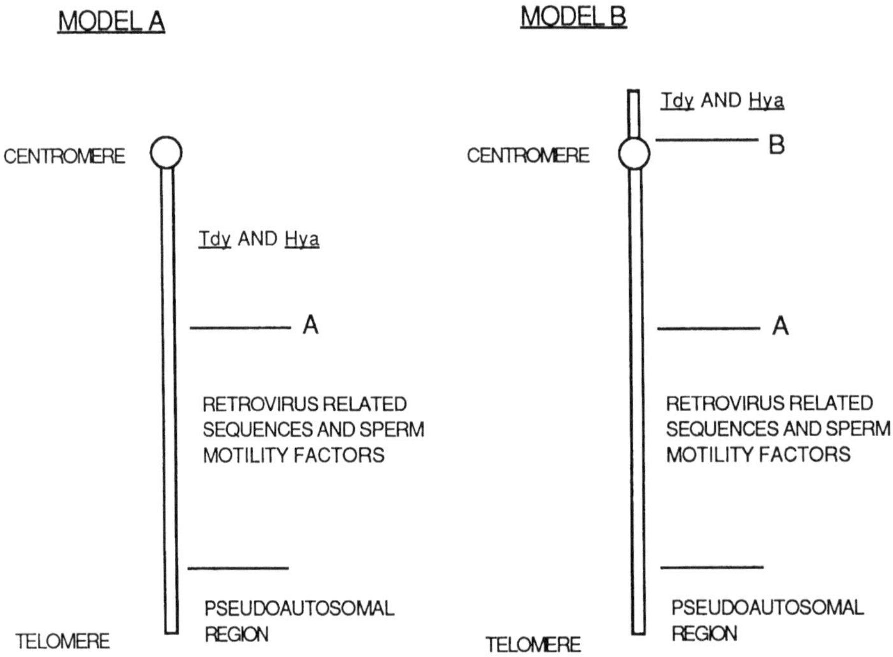

Figure 2–3. Two models for the mouse Y chromosome. Model A is adapted from Eicher and Washburn (1986); Model B is a modification of model A that incorporates new information about Sxr^{a} and Sxr^{b}. The short arm in model B has been divided into two major subregions based on Sxr^{a} (*Tdy, Spy* and *Hy-a* positive) and Sxr^{b} (*Tdy* positive; *Spy* and *Hy-a* negative). Position 1 is the breakpoint in Y*, position 2 is the presumed breakpoint in Sxr^{a}.

meric region of the normal Y chromosome, in carrier males there was an extra site of hybridization at the end of the Y chromosome. In sex-reversed females a site of hybridization to Bkm probes was present on the distal end of the X chromosome. These observations were simultaneously interpreted by several authors in the following way: *Sxr* is a duplication of the region of Y chromosome that contains *Tdy* and this region has been translocated to the distal end of the Y chromosome telomeric to the mouse pseudoautosomal region (Hansmann, 1982; Burgoyne, 1982; Eicher, 1982). In male meiosis a recombination transfers the *Sxr* region from the Y to the X chromosome creating an X chromosome carrying *Tdy*. Inheritance of this chromosome results in sex reversal. Results from high resolution cytogenetics analysis supported this interpretation (Evans et al., 1982). Final proof was achieved by demonstrating transmission of *Sxr* through females due to X-inactivation of the X^{sxr} chromosome (McLaren and Monk, 1982).

Sex-reversed female mice are H-Y antigen positive suggesting that *Hya* (the gene controlling expression of H-Y antigen) and *Tdy* map to the same region (Bennett et al., 1977; Simpson et al., 1981). The presence of two XX chromosomes is incompatible with germ cell development in males, and XX *Sxr* mice fail to produce sperm. In the germ line of XO *Sxr* mice, spermatogenesis occurs, however, although the later stages are depleted and only abnormal nonmotile sperm are produced (Cattanach et al., 1971). This implies the existence of one or more genes required for spermatogenesis outside the region defined by *Sxr*.

In 1984, McLaren and colleagues found a variant of *Sxr*, which they named *Sxr'* (McLaren et al., 1984). To simplify nomenclature when describing new variants of *Sxr*, the original *Sxr* mutation has been renamed Sxr^a and *Sxr'* is now called Sxr^b. Female carriers of Sxr^b are sex reversed, but H-Y antigen negative. This excludes *Hya*, as defined by transplantation and T cell killing, as being *Tdy*. Molecular analysis of Sxr^b has clearly demonstrated that it has arisen by deletion of sequences present in Sxr^a (Bishop et al., 1988). It is assumed that the deletion is due to an unequal crossover at meiosis and using repeat probes it can be demonstrated that small rearrangements occur frequently in the *Sxr* region (McLaren et al., 1988).

XO Sxr^b mice have a more severe block in spermatogenesis than XO Sxr^a mice and this has been used to define a spermatogenesis factor, *Spy*, present in Sxr^a but deleted from Sxr^b (Burgoyne et al., 1986). Similarly, it has been suggested that a growth gene, *Gdy*, may be located in the Sxr^a region and absent in Sxr^b. The precise relationship, or even identity, between *Hya, Spy,* and *Gdy* remains to be determined (Sutcliffe and Burgoyne, 1989).

The original in situ hybridization experiments with *Bkm* sequences had been interpreted as suggesting that *Tdy* was located close to the centromere on the long arm of the Y chromosome. However, the high density of *Bkm* sequences resulted in a massive signal with a general location in the centromeric region. Bishop and colleagues (Roberts et al., 1988) repeated the in situ hybridization with low copy sequence derived from the Sxr^a region. These experiments localize the *Sxr* region to the minuscule short arm of the Y chromosome ("the ears").

The Y* chromosome appears to have been created by a chromosomal rearrangement without an obvious duplication (Eicher and Washburn, 1986). The rearrangement has resulted in the repositioning of the PAR such that a large part of the Y chromosome behaves like *Sxr* and is transferred to the X chromosome during male meiois. The Y sequences transferred include *Hya, Tdy, Bkm,* and some of the retroviral-related sequences. Unlike XO *Sxr* animals, XO derY males are fertile, which maps spermatogenesis factors to this Y* defined region.

In summary, Sxr^a, Sxr^b, and Y* can be used to define five regions of the Y chromosome (Fig. 2–3). Assuming *Sxr* is not caused by a complex mechanism, *Tdy* and *Hya* map to the short arm of the mouse Y chromosome.

B. The Pseudoautosomal Region

There are only two pseudoautosomal loci known in the mouse: *Sts* and the retroviral insertion site *Mov15* (Harbers et al., 1986).

Sts encodes the enzyme steroid sulphatase. The equivalent human *STS* locus is X-linked (Shapiro et al., 1978), although a pseudogene is present on the long arm of the Y chromosome (Yen et al., 1988). X-linkage for murine *Sts* was also suggested by the finding that XX oocytes contain twice the enzyme levels of XO oocytes (Gartler and Rivest, 1983). Close linkage between *Sts* and *Sxr*, however, provided proof that *Sts* is pseudoautosomal (Keitges et al., 1987).

Sts exchanges between the sex chromosomes in male meiosis with a frequency close to 50 percent (Keitges et al., 1987; Nagamine et al., 1987). The retroviral insertion *Mov15* exchanges with a lower frequency. This has allowed a skeleton map of the mouse PAR to be constructed: tel *(Sxr)—Sts—(Mov15)*. Unfortunately, it has not been possible to clone the normal sequences at the *Mov15* insertion site to exclude the possibility that the viral insertion has changed the genetic behavior of the PAR. This control is important because, unlike the human case, several double recombinants have been seen in the PAR of mouse (Soriano et al., 1987).

IV. Candidate Genes for Controlling Sex Determination

The mechanism of sex determination has exercised the minds of people for many centuries. The history of this topic is replete with quasi-scientific explanations that range from coital position and passion through to conjunctions of an astrological nature. Since the establishment of the chromosomal basis of sex determination, the responsible genetic elements have been sought assiduously and in the last decade three major candidates emerged: H-Y antigen, Bkm sequences, and *ZFY*. At the time of writing this review, none of these candidates appears to be the long sought for testis-determining gene.

A. H-Y Antigen

H-Y antigen was first described by Eichwald and Silmzer (1955). Female mice grafted with skin from male animals of the same inbred strains could mount a rejection response. The graft rejection was typical of the reaction generated by a minor histocompatibility antigen and like other minor histocompatibility loci it proved to be very difficult to raise antibodies to the H-Y antigen. (With hindsight, we believe that the difficulty in producing antibodies is due to the biology of the MHC and T cell killing. The class 1 and 2 MHC antigens are able to present processed fragments of internal molecules on the cell surface and these are recognized by T cells. Most attempts to raise antibodies assumed that H-Y antigen was a cell surface expressed molecule.) Eventually, Goldberg and colleagues (1971) were able to produce male-specific antibodies that could kill sperm in a complement-mediated cytotoxicity assay. Using this assay, and minor modifications, the groups associated with Goldberg, Wachtel, and Wolf provided a large body of data that correlated the presence of H-Y antigen with the presence of the Y chromosome and/or testes. Unfortunately, these studies were hindered by the poor quality of the antisera and the reproducibility of the assays. This largely prevented the successful exchange of reagents and made it very difficult to compare results from different laboratories. The current status of H-Y antigen is confused and there is no consensus answer to the following questions:

1. Is there more than one serologically detectable H-Y antigen?
2. Is H-Y serologically the same as H-Y antigen defined by T cell killing and transplantation?

This confusion has resulted in the suggestion that serologically detected H-Y antigen should be renamed serologically determined male antigen (sdma; Silvers et al., 1982) or serological sex specific antigen (SXS; Wiberg, 1987). Until there is a consensus about the molecular nature of the sdma/SXS it is not possible to address the question of its relationship to *TDF*. Readers wishing to pursue this topic further should read the reviews by Goodfellow and Andrews (1982), Silvers and associates (1982), Wiberg (1987), and Goldberg (1988).

Fortunately, the relationship between H-Y antigens defined by transplantation and T cell killing (hereafter referred to as H-Y antigen) and *TDF* is simple: H-Y antigen is not related to *TDF* or *Tdy*. Mice carrying Sxr^a are H-Y antigen positive, but mice carrying Sxr^b are H-Y antigen negative (McLaren et al., 1984); as the latter mice have testes, H-Y antigen cannot be a product of *Tdy*. Similarly, XX human males generated by abnormal exchange between the sex chromosomes are frequently H-Y antigen negative (Simpson et al., 1987).

B. Bkm Sequences

Jones and colleagues isolated a satellite sequence from the Banded Krait. Although sequences hybridizing to this satellite were present at lower levels throughout the genome, there was a large concentration of this sequence on the sex determining W chromosome of snakes (Singh et al., 1976). The same concentration of sequences were also present on the Y chromosome of some mammals and were shown to be associated with *Sxr* in the mouse (Singh and Jones, 1982). Sequence analysis showed the Bkm satellite to be largely composed of the tetranucleotides GACA and GATA (Epplen et al., 1982). The apparent conservation of *Bkm* on the sex-determining chromosome in snakes and mammals led to the hypothesis that *Bkm* might play a crucial role in sex chromosome structure or function. *Bkm* might even be equated with *TDF* (Chandra, 1985). More detailed analysis of mammalian sex chromosomes failed to find high concentrations of *Bkm* sequences on the Y chromosome of all species (K. Reed, personal communication). It is not possible to exclude a role for Bkm sequences in sex determination; however, there is no evidence to support directly this contention. The Y chromosome is a home for many repetitive elements and expansion of different families of repeats appears to have occurred in different species.

C. ZFY

The position of *TDF* on the human Y chromosome was precisely localized using sex-reversed individuals and by making the following assumptions:

1. Exchange XX males have inherited *TDF* along with the Y-derived sequences.
2. XY females, with Y chromosome deletions, have lost *TDF*.

By chromosome walking between the breakpoints present in an XX male (LGL203) and an XY female (WHT1013), Page and colleagues (1987) cloned an interval of 150 kb that should have contained *TDF* (Fig. 2–4). This interval mapped 140 kb from the pseudoautosomal boundary and contained an HTF island previously identified as a possible candidate gene (Pritchard et al., 1987). Within the cloned sequences and adjacent to the HTF island, several blocks of conserved sequences were found and one of these included an open reading frame that encoded a polypeptide with several zinc-finger motifs. Sequences homologous to this exon were found on the Y chromosome of all eutherian mammals tested. A related, but different, sequence was also found on the X chromosome in humans and other eutherians. The Y-located gene was named *ZFY* (zinc-finger gene Y chromosome) and the X-located gene was named *ZFX*. The situation in the mouse is even more complex: two *XFY* related genes are present on the Y chromosome, *Zfy-1* and *Zfy-2;* there is one X-located gene, *Zfx;* and an autosomal pseudogene, *Zfa*. Both *Zfy-1* and *Zfy-2* map within the *Sxr*ᵃ interval, but only *Zfy-1*

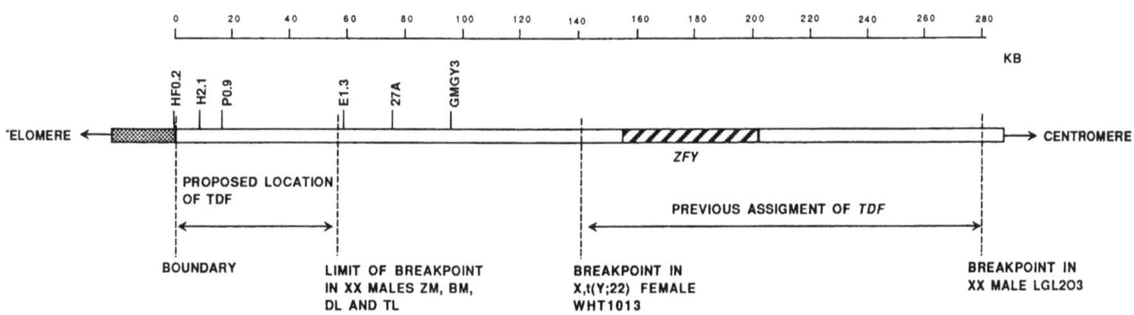

Figure 2–4. Map of the distal short arm of the human Y chromosome. This Figure has been reproduced from Palmer et al., 1989, with the kind permission of the Editor of *Nature*.

maps within the *Sxr*^{b} interval (Mardon et al., 1989; Nagamine et al., 1989).

Sequencing of transcripts has demonstrated a close relationship between *ZFY* and *ZFX;* both genes can encode proteins containing a highly acidic domain as well as a putative nuclear localization signal and a zinc-finger domain with 13 contiguous zinc fingers (Palmer et al., 1990). These features are consistent with *ZFY* and *ZFX* being DNA binding transcription factors and would imply a genetic regulatory role for these proteins. Transcripts from mouse *Zfy*-related genes have a similar structure (Mardon and Page, 1989; Ashworth et al., 1989).

The similarity between *ZFY* and *ZFX* encouraged the idea that these genes might be functionally identical and that sex determination might be controlled by dosage—with the assumption that *ZFX* was subject to X-inactivation (Page et al., 1987). Surprisingly, *ZFX* has been shown to escape X-inactivation, even though it is located outside the region on the X chromosome previously known to escape inactivation (Schneider-Gadicke et al., 1989). In addition, several of the sequence differences between *ZFX* and *ZFY* in the finger region occur at residues in the protein that might predict changes in the DNA binding specificity.

Despite the genetic evidence that *ZFY* could be equated with *TDF,* several lines of evidence were less supportive of this hypothesis. Although *ZFY*-related sequences are Y-located in all eutherians, they are autosomal in metatherian mammals (marsupials) (Sinclair et al., 1988). This suggests either that a different Y-located gene is sex determining in marsupials or that *ZFY* is not *TDF.* Also slightly puzzling was the tissue distribution of expression of *ZFY* and *Zfy-1: ZFY* is transcribed in all male tissues, whereas *Zfy-1* expression is restricted to the gonads.

Recently, further evidence has been presented that argues strongly against equating *TDF* and *ZFY.* Palmer and colleagues have described four exchange XX males with Y-derived sequences that do not include *ZFY* (Palmer et al., 1989; Fig. 2–4). This implies that one of the two assumptions used in the genetic mapping of *ZFY* is wrong. The apparent inconsistency with Page and colleagues' previously published results (Page et al., 1987) was resolved when a second deletion was found on the Y chromosome of the XY female previously used to locate *TDF* (Page et al., 1990). Palmer and co-workers were then able to predict that *TDF* should be located within 60 kb of the pseudoautosomal boundary. Finally, Koopman and colleagues (1989) demonstrated that *Zfy-1* expression is restricted to germ cells in the male mouse. As germ cell sex is irrelevant for sex determination (McLaren, 1988), this also augurs that *Zfy-1* and *Tdy* cannot be equated.

D. SRY, a New Candidate for TDF

The portion of the Y chromosome inherited by the rare XX males, described by Palmer et al. (1989) was narrowed down still further. One probe detected the breakpoints in these individuals that clustered at 35 kb immediately adjacent to the pseudoautosomal boundary (Sinclair et al., 1990; Fig. 2–5).

D.1. SEARCH OF THE SEX-DETERMINING REGION OF THE Y CHROMOSOME

A search of 35 kb of the Y chromosome for a Y-specific conserved sequence revealed several Y-specific sequences in the human genome but only one that was Y-specific in human, bovine, and murine genomes (Fig. 2–5). This sequence of 2.1 kb defines the gene *SRY* (sex determining region Y gene) (Sinclair et al., 1990). In the mouse, the homologous gene *Sry* is located on the Sxr′ fragment, which is the smallest region of the Y chromosome known to be sex determining (Gubbay et al., 1990). Furthermore, Gubbay and colleagues (1990) were able to show that *Sry* is absent from a mutant Y chromosome that has deleted the testis determining gene(s). The *SRY* gene also detects homologous Y-specific sequences on a wide range of eutherian mammals (Sinclair et al., 1990). At a low stringency *SRY* cross-reacts with autosomal loci in a number of species. Sequencing of the *SRY* gene in human, rabbit, and mouse, revealed a large open rading frame of 669 bp. A search of the DNA data base with the nucleotide sequence of *SRY* failed to find any other related sequence. However, when the predicted protein product of *SRY* was used to search the

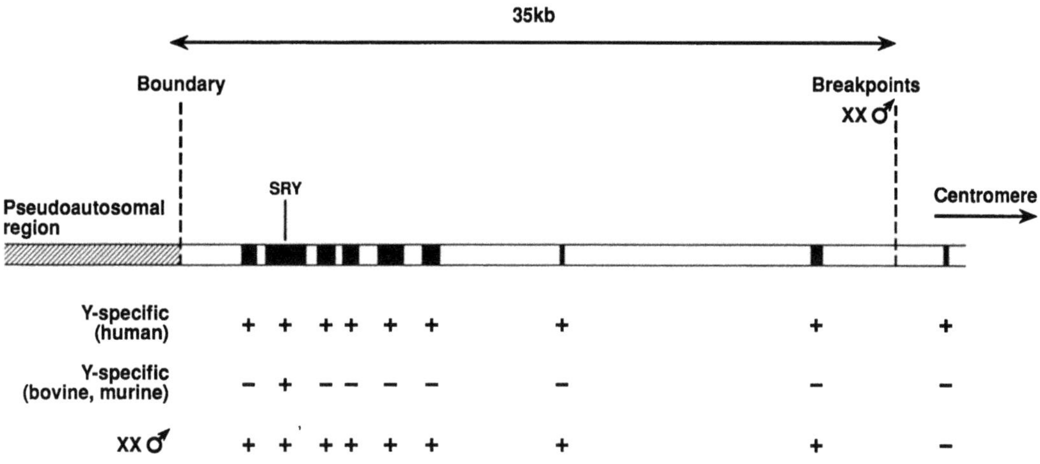

Figure 2–5. Refined map of the distal short arm of the human Y chromosome: shaded region (left), pseudoautosomal region; broken line, boundary between pseudoautosomal and Y-specific regions. Breakpoints of the XX males, broken lines at 35 kb. Black boxes, probes detecting single-copy Y-specific human DNA fragments (+). When these probes were hybridized to male and female DNA from bovine and murine genomes, only *SRY* detected Y-specific fragments (+). All of the probes between the boundary and the breakpoints hybridized to sequences in the XX males.

protein data base, several similar proteins were identified, all having a conserved HMG box motif in common (Sinclair et al., 1990). This motif was originally found in high-mobility group proteins, but has since been found in several other proteins with predicted or known DNA binding capacity. The ability of the *SRY* protein to bind DNA would be consistent with a role in regulating the expression of other genes in a developmental pathway.

D.2. Tissue Distribution and Expression of SRY

Using northern blot analysis, *SRY* expression was shown to be restricted to the testis in humans, where it produces a transcript of 1.1 kb (Sinclair et al., 1990). Koopman and colleagues (1990) were able to demonstrate that in the fetal mouse, *Sry* expression precedes and coincides with the first signs of testicular development. Other experiments showed that *Sry* is expressed in the testes of mutant W/W mouse embryos, which form testes lacking in germ cells. However, in adult mouse testes, *Sry* is also expressed but by contrast its expression requires the presence of germ cells (Koopman et al., 1990).

D.3. Genetic Evidence Equating SRY and TDF

If *SRY* is the testis-determining gene, then mutations in *SRY* could result in the failure of testes formation and hence the female phenotype would result. Consequently, it would be predicted that some XY females with gonadal dysgenesis who possess *SRY* should have mutations in the gene. *SRY* sequences were amplified by PCR from 11 XY females and 50 normal males. The amplified products were analysed using the SSCP assay (single strand conformational polymorphism) to look for sequence variants. All the normal male sequences gave the same pattern as those of 9 of the XY females. However, two XY females (patients AA and JN) showed sequence variation (Berta et al., 1990). DNA from these two individuals and their fathers were amplified, cloned, and sequenced. In the XY female AA there was a base change from G to A that causes a change at the amino acid level from methionine to isoleucine in the conserved DNA binding motif. The father of AA has a normal *SRY* sequence. However, the other XY female JN shares a base change in *SRY* with her father (Berta et al., 1990). There are several possible explanations for the sequence variant found in JN. This variant could cause conditional sex reversal and is dependent upon other genetic or environmental factors; the father is mosiac for wild type and variant sequences, or the variant could be a low frequency polymorphism and sex reversal is due to an autosomal or X-linked sex reversing gene.

The *de novo* mutation in the *SRY* gene of AA is associated with sex reversal and implies *SRY* is required for testis formation and male sex determination. Support for this contention also comes from Jager and colleagues (1990), who discovered an XY female with a *de novo* frameshift mutation in the DNA binding region of the *SRY* gene.

Conclusion

SRY has been shown to meet many of the criteria expected for the testis-determining gene. *SRY* maps to the sex-determining region of the human and mouse Y chromosome, is Y-specific and conserved in a wide range of eutherian mammals and encodes a protein with a potential DNA binding motif. Its expression in mouse embryos is restricted to the somatic portion of the genital ridge, and its spatial and temporal profile of expression correlates with the onset of testis development. Finally, the presence of *de novo* mutations in the *SRY* gene of XY females implies that *SRY* gene is necessary for testis development. With the completion of this review a gene encoding the GM-CSF receptor has been shown to be pseudoautosomal in man (Gough et al., 1990), and *Sry* has been equated with *Tdy* by the construction of transgenic mice (Koopman et al., 1991).

REFERENCES

Alvesalo, L., and de la Chappelle, A. (1981). Tooth sizes in two males with deletions of the long arm of the Y-chromosome. Ann. Hum. Genet. 45:49–54.

Arnemann, J., Epplen, J. T., Cooke, H. J., Sauermann, U., Engel, W., and Schmidtke, J. (1987). A human Y-chromosome DNA sequence expressed in testicular tissue. Nucl. Acids Res. 15:8713–8724.

Arnemann, J., Thuering, S., Kahler, S., Jakubiezka, S., Berg, L. P., Burfeind, P., and Schmidtke, J. (1989). Molecular organisation and expression of the human Y specific DNA sequence pJA36B2. Cytogenet. Cell. Genet. 51:953.

Ashworth, A., Swift, S., and Affara, N. (1989). Sequence of cDNA for murine *Zfy-1,* a candidate for Tdy. Nucl. Acids Res. 17:2864.

Bennett, D., Mathieson, B. J., Scheid, M., Yanigasawa, K., Boyse, E. A., Wachtel, S., and Cattanach, B. M. (1977). Serological evidence for H-Y antigen in Sxr, XX sex-reversed phenotypic males. Nature 265:255–257.

Berta, P., Hawkins, J. R., Sinclair, A. H., Taylor, A., Griffiths, B. L., Goodfellow, P. N., and Fellous, M. (1990). Genetic evidence equating *SRY* and the testis-determining factor. Nature 348:448–450.

Bishop, C. E., Boursot, P., Baron, B., Bonhomme, F., and Hatat, D. (1985). Most classical *Mus musculus domesticus* laboratory mouse strains carry a *Mus musculus musculus* Y chromosome. Nature 315:70–72.

Bishop, C. E., Weith, A., Mattei, M. G., and Roberts, C. (1988). Molecular aspects of sex determination in mice: An alternative model for the origin of the *Sxr* region. Phil. Trans. R. Soc. (Lond.) B322:119–124.

Brown, W.R.A. (1988). A physical map of the human pseudoautosomal region. Embo. J. 7:2377–2385.

Buckle, V., Mondello, C., Darling, S., Craig, I. W., and Goodfellow, P. N. (1985). Homologous expressed genes in the human sex chromosome pairing region. Nature 317:739–741.

Burgoyne, P. S. (1982). Genetic homology and crossing over in the X and Y chromosomes of mammals. Hum. Genet. 61:85–90.

Burgoyne, P. S., Levy, E. R., and McLarn, A. (1986). Spermatogenic failure in male mice lacking H-Y antigen. Nature 320:170–172.

Burgoyne, P. S. (1989). Thumbs down for zinc finger? Nature 342:860–862.

Cattanach, B. M., Pollard, C. E., and Hawkes, S. G. (1971). Sex-reversed mice: XX and XO males. Cytogenetics 10:318–337.

Chandley, A. C., Goetz, P., Harbreave, T. B., Joseph, A. M., and Speed, R. M. (1984). On the nature and extent of XY pairing at meiotic prophase in man. Cytogenet. Cell. Genet. 38:241–247.

Chandra, H. S. (1985). Sex determination: A hypothesis based on noncoding DNA. Proc. Natl. Acad. Sci. (USA). 82:1165–1169.

Cooke, H. (1976). Repeated sequence specific to human males. Nature 262:182–186.

Cooke, H. J., and Noel, B. (1979). Confirmation of Y/autosome translocations using recombinant DNA. Hum. Genet. 50:39–44.

Cooke, H. J., Brown, W.R.A., and Rappold, G. (1985). Hypervariable telomeric sequences from the human sex chromosome are pseudoautosomal. Nature 317:688–692.

Davis, R. M. (1981). Localization of male determining factors in man: A thorough review of structural anomalies of the Y chromosome. J. Med. Genet. 18:161–195.

Eicher, E. M. (1982). Primary sex determining genes in mice: A brief review, in *Prospects for Sexing Mammalian Sperm* (Amman, R. P., and Seidel, G. E., eds.) pp. 121–135, Colorado: Colorado Associated University Press.

Eicher, E. M., and Washburn, L. K. (1986). Genetic control of primary sex determination in mice. Am. Rev. Genet. 20:327–360.

Eicher, E. M., Hutchison, K. W., Phillips, S. J., Tucker, P. K., and Lee, B. K. (1989). A repeated segment on the mouse Y chromosome is composed of retroviral-related, Y-enriched and Y-specific sequences. Genetics 122:181–192.

Eichwald, E. J., and Silmzer, C. R. (1955). Communication Transplants. Bull. 2:148–149.

Ellis, N., and Goodfellow, P. N. (1989). The mammalian pseudoautosomal region. Trends Genet. 5:406–410.

Ellis, N. A., Goodfellow, P. J., Pym, B., Smith, M., Palmer, M., Frischauf, A-M., and Goodfellow, P. N. (1989). The pseudoautosomal boundary in man is defined by an *Alu* repeat sequence inserted on the Y chromosome. Nature 337:81–84.

Epplen, J. T., McCarrey, J. R., Suton, S., and Ohno, S. (1982). Base sequence of a cloned snake W-chromosome DNA fragment and identification of a male specific putative mRNA in the mouse. Proc. Natl. Acad. Sci. (USA). 79:3798–3802.

Evans, E. P., Burtenshaw, M., and Cattanach, B. M. (1982). Meiotic crossing over between X and Y chromosomes of male mice carrying the sex reversing *(Sxr)* factor. Nature 300:443–445.

Ferguson-Smith, M. A. (1966). X-Y chromosomal interchange in the aetiology of true hermaphroditism and of XX Klinefleter's syndrome. Lancet ii:475–476.

Ford, C. E., Jones, K. W., Polani, P., De Almeida, J. C., and Brigg, J. H. (1959). A sex chromosome anomaly in a case of gonadal sex dysgenesis (Turner's syndrome). Lancet i:711.

Gartler, S. M., and Rivest, M. (1983). Evidence for an X-linkage of steroid sulfatase in the mouse: Steroid sulfatase levels in oocytes of XX and XO mice. Genetics 103:137–141.

Goldberg, E. H., Boyse, E. A., Bennett, D., Scheid, M., and Carswell, E. A. (1971). Serological demonstration of H-Y (male) antigen on mouse sperm. Nature 232:478–480.

Goldberg, E. H. (1988). H-Y antigen and sex determination. Phil. Trans. R. Soc. (Lond. B) 322:73–81.

Goodfellow, P. N., and Andrews, P. W. (1982). Sexual differentiation and H-Y antigens. Nature 295:11–13.

Goodfellow, P. (1983). Expression of the 12E7 antigen is controlled independently by genes on the human X and Y chromosomes. Differentiation 23:25–39.

Goodfellow, P. N., Darling, S. M., and Wolfe, J. (1985). The human Y chromosome. J. Med. Genet. 22:332–344.

Goodfellow, P. J., Darling, S. M., Thomas, N. S., and Goodfellow, P. N. (1986a). A pseudoautosomal gene in man. Science 234:740–743.

Goodfellow, P. J., Pritchard, C., Tippett, P., and Goodfellow, P. N. (1986b). Recombination between the X and Y chromosomes at the YG locus; implications for the relationship between *MIC2, XG* and *YG*. Ann. Hum. Genet. 51:161–167.

Goodfellow, P. N., and Darling, S. M. (1988). Genetics of sex determination in man and mouse. Develop. 102:251–258.

Gough, N. M., Gearing, D. P., Nicola, N. A., Baker, E., Pritchard, M., Callen, D. F., and Sutherland, G. R. (1990). Localization of the human GM-CSF receptor gene to the X-Y pseudoautosomal region. Nature 345:734–736.

Gubbay, J., Collignon, J., Koopman, P., Capel, B., Economou, A., Munsterberg, A., Vivian, N., Goodfellow, P. N., and Lovell-Badge, R. (1990). A gene mapping to the sex-determining region of the mouse Y chromosome is a member of a novel family of embryonically expressed genes. Nature 346:245–250.

Guellaen, G., Casanova, M., Bishop, C., Geldwerth, D., Audre, G., Fellous, M., and Weissenbach, J. (1984). Human XX males with Y single copy DNA fragments. Nature 307:172–173.

Hansmann, I. (1982). Sex reversal in the mouse. Cell 30:221–332.

Harbers, K., Soriano, P., Muller, U., and Jaenisch, R. (1986). High frequency of unequal recombination in the pseudoautosomal region shown by proviral insertion in transgenic mice. Nature 324:682–685.

Jacobs, P. A., and Strong, J. A. (1959). A case of human intersexuality having a possible XXY sex determining mechanism. Nature 183:302–303.

Jager, R. J., Anvert, M., Hall, K. and Scherer, G. (1990). A human XY female with a frame shift mutation in the candidate testis-determining gene *SRY*. Nature 348:452–454.

Josso, N., Picard, J.-Y., and Tran, D. (1977). The antimullerian hormone. Rec. Progr. Horm. Res. 33:117–167.

Jost, A., Vigier, B., Prepin, J., and Perchellet, J. P. (1973). Studies on sex differentiation in mammals. Recent Prog. Horm. Res. 29:1–41.

Keitges, E. A., Schorderet, D. F., and Gartler, S. M. (1987). Linkage of the steroid sulfatase gene to the sex reversed mutation in the mouse. Genetics 116:465–468.

Koller, P. C., and Darlington, C. D. (1934). The genetical and mechanical properties of the sex chromosomes. 1. *Rattus novegicus*. J. Genet. 29:159–173.

Koopman, P., Gubbay, J., Collignon, J., and Lovell-Badge, R. (1989). *Zfy* gene expression patterns are not compatible with a primary role in mouse sex determination. Nature 342:941–943.

Koopman, P., Munsterberg, A., Capel, B., Vivian, N. and Lovell-Badge, R. (1990). Expression of a candidate sex-determining gene during mouse testis differentiation. Nature 348:450–452.

Koopman, P., Gubbay, J., Vivian, N., Goodfellow, P., and

Lovell-Badge, R. (1991). Male development of chromosomally female mice transgenic for *Sry* [see comments]. Nature 351:117–121.

Mardon, G., and Page, D. C. (1989). The sex determining region of the mouse Y chromosome encodes a protein with a highly acidic domain and 13 zinc fingers. Cell 56:765–777.

Mardon, G., Mosher, R., Disteche, C. M., Nishioka, Y., McLaren, A., and Page, D. C. (1989). Duplication, deletion and polymorphism in the sex determining region of the mouse Y chromosome. Science 243:78–80.

McKay, R.D.G., Bobrow, M., and Cooke, H. J. (1978). The identification of a repeated DNA sequence involved in the karyotype polymorphism of the human Y chromosome. Cytogenet. Cell. Genet. 21:19–32.

McLaren, A., and Monk, M. (1982). Fertile females produced by inactivation of an X chromosome of 'sex-reversed' mice. Nature 300:446–448.

McLaren, A., Simpson, E., Tomonari, K., Chandler, P., and Hogg, H. (1984). Male sexual differentiation in mice lacking H-Y antigen. Nature 312:552–555.

McLaren, A. (1988). Somatic and germ-cell sex in mammals. Phil. Trans. R. Soc. (Lond. B) 322:3–9.

McLaren, A., Simpson, E., Epplen, J. T., Studer, R., Koopman, P., Evans, E. P., and Burgoyne, P. S. (1988). Location of genes controlling H-Y antigen expression and testis determination on the mouse Y chromosome. Proc. Natl. Acad. Sci. (USA) 85:6642–6645.

Nagamine, C. M., Michot, J.-L., Roberts, C., Guenet, J.-L., and Bishop, C. E. (1987). Linkage of the murine steroid sulphatase locus, *Sts*, to sex reversed *Sxr:* A genetic and molecular analysis. Nucl. Acids Res. 15:9227–9238.

Nagamine, C. M., Chan, K., Kozak, C. A., and Lau, Y. F. (1989). Chromosome mapping and expression of a putative testis determining gene in the mouse. Science 243:80–83.

Page, D. C. (1987). Hypothesis: A Y-chromosomal gene causes gonadoblastoma in dysgenic gonads. Develop. 101S:151–155.

Page, D. C., Mosher, R., Simpson, E. M., Fisher, E.M.C., Mardon, G., Pollack, J., McGillivray, B., de la Chapelle, A., and Brown, L. G. (1987). The sex-determining region of the human Y chromosome encodes a finger protein. Cell 51:1091–1104.

Page, D. C., Fisher, E.M.C., McGillivray, B. and Brown, L. G. (1990). Additional deletion in sex-determining region of human Y chromosome resolves paradox of X,t (Y;22) female. Nature 346:279–281.

Palmer, M. S., Sinclair, A. H., Berta, P., Ellis, N. A., Goodfellow, P. N., Abbas, N. E., and Fellous, M. (1989). Genetic evidence that *ZFY* is not the testis determining factor. Nature 342:937–939.

Palmer, M. S., Berta, P., Sinclair, A. H., Pym, B., and Goodfellow, P. N. (1990). Characterization and comparison of human *ZFY* and *ZFX* transcripts. Proc. Natl. Acad. Sci., in press.

Petit, C., de la Chapelle, A., Levilliers, J., Castillo, S., Noel, B., and Weissenbach, J. (1987). An abnormal terminal exchange accounts for most but not all cases of human XX maleness. Cell 49:595–602.

Petit, C., Levilliers, J., and Weissenbach, J. (1988). Physical mapping of the human pseudoautosomal region. Comparison with the genetic linkage map. EMBO J. 7:2369–2376.

Phillips, S. J., Birkenmeier, E. H., Callahan, R., and Eicher, E. M. (1982). Male and female mouse DNAs can be discriminated using retroviral probes. Nature 297:241–243.

Pritchard, C. A., Goodfellow, P. J., and Goodfellow, P. N. (1987). Mapping the limits of the human pseudoautosomal region and a candidate sequence for the male-determining gene. Nature. 328:273–275.

Rappold, G. A., and Lehrach, H. (1988). A long range restriction map of the pseudoautosomal region by partial digest PFGE analysis from the telomere. Nucl. Acids Res. 16:5361–5377.

Roberts, C., Weith, A., Michot, J. L., Mattei, M. G., and Bishop, C. E. (1988). Molecular and cytogenetic evidence for the location of *Tdy* and H-Y on the mouse Y chromosome short arm. Proc. Natl. Acad. Sci. (USA) 85:6646–6649.

Rouyer, F., Simmler, M. C., Johnsson, C., Vergnaud, G., Cooke, H., and Weissenbach, J. (1986). A gradient of sex linkage in the pseudoautosomal region of the human sex chromosomes. Nature 319:291–295.

Schneider-Gadicke, A., Beer-Romero, P., Brown, L. G., Nussbaum, R., and Page, D. C. (1989). *ZFX* has a gene structure similar to *ZFY,* the putative human sex determinant, and escapes X-inactivation. Cell 57:1247–1258.

Shapiro, L. J., Weiss, R., Buxman, M. M., et al. (1978). Enzymic basis of typical X-linked ichthyosis. Lancet ii:756–757.

Silvers, W. K., Gasser, D. L., and Eicher, E. M. (1982). H-Y antigen, serologically detectable male antigen and sex determination. Cell 28:439–440.

Simmler, M.-C., Rouyer, F., Vergnaud, G., Nystrom-Lahti, M., Ngo, K. Y., de la Chapelle, A., and Weissenbach, J. (1985). Pseudoautosomal DNA sequences in the pairing region of the human sex chromosomes. Nature 317:692–697.

Simpson, E., Edwards, P., Wachtel, S., McLaren, A., and Chandler, E. (1981). H-Y antigen in *Sxr* mice detected by H-2 restricted cytotoxic T cells. Immunogenet. 13:335–358.

Simpson, E., Chandler, P., Goulny, E., Disteche, C. M., Ferguson-Smith, M. A., and Page, D. C. (1987). Separation of the genetic loci for the H-Y antigen and for testis determination on the human Y chromosome. Nature 326:876–878.

Sinclair, A. H., Foster, J. W., Spencer, J. A., Page, D. C., Palmer, M., Goodfellow, P. N., and Graves, J.A.M. (1988). Sequences homologous to *ZFY,* a candidate for the sex determining gene, are autosomal in marsupials. Nature 336:780–783.

Sinclair, A. H., Berta, P., Palmer, M. S., Hawkins, J. R., Griffiths, B. L., Smith, M. J., Foster, J. W., Frischauf, A. M., Lovell-Badge, R. and Goodfellow, P. N. (1990). A gene from the human sex determining region encodes a protein with homology to a conserved DNA-binding motif. Nature 346:240–244.

Singh, L., Purdon, I. F., and Jones, K. W. (1976). Satellite DNA and evolution of sex chromosomes. Chromosoma 59:43–51.

Singh, L., and Jones, K. W. (1982). Sex reversal in the

mouse *(Mus musculus)* is caused by a recurrent non-reciprocal crossover involving the X and an aberrant Y chromosome. Cell 28:205–216.

Smith, K. D., Young, K. E., Talbot, C. C. Jr., and Schmeckpeper, B. J. (1987). Repeated DNA of the human Y chromosome. Develop. 101S:77–92.

Soriano, P., Keitges, K. A., Schorderet, D. F., Habers, K., Gartler, S. M., and Janisch, R. (1987). High rate of recombination and double cross-overs in the mouse pseudoautosomal region during male meiosis. Proc. Natl. Acad. Sci. (USA) 84:7218–7220.

Sutcliffe, M. J., and Burgoyne, P. S. (1989). Analysis of the testis of H-Y negative XO *Sxr*^{b} mice suggests that the spermatogenesis gene *(Spy)* acts during the differentiation of the A spermatogonia. Develop. 107:373–380.

Tiepolo, L., and Zuffardi, O. (1976). Localization of factors controlling spermatogenesis in the nonfluorescent portion of the Y chromosome long arm. Hum. Genet. 34:119–124.

Tucker, P. K., Lee, B. K., and Eicher, E. M. (1989). Y chromosome evolution in the subgenus *Mus* (Genus *Mus*). Genetics 122:169–179.

Tyler-Smith, C., and Brown, W.R.A. (1987). Structure of the major block of alphoid satellite DNA on the human Y chromosome. J. Mol. Biol. 195:457–470.

Vergnaud, G., Page, D. C., Simmler, M.-C., Brown, L., Rouyer, F., Noel, B., Botstein, D., de la Chapelle, A., and Weissenbach, J. (1986). A deletion map of the human Y chromosome based on DAN hybridisation. Am. J. Hum. Genet. 38:109–124.

Weissenbach, J., Levilliers, J., Pettit, C., Rouyer, F., and Simmler, M.-C. (1987). Normal and abnormal interchanges between the human X and Y chromosomes. Develop. 101S:67–74.

Weissenbach, J., Goodfellow, P. N., and Smith, K. D. (1989). Report of the committee on the genetic constitution of the Y chromosome, Human Gene Mapping 10. Tenth International Workshop on Human Gene Mapping. Cytogenet. Cell Genet. 51:438–449.

Welshons, W. J., and Russell, L. B. (1959). The Y chromosome as the bearer of the male determining factors in the mouse. Proc. Natl. Acad. Sci. (USA) 45:560–566.

Wiberg, U. H. (1987). Facts and consideration about sex-specific antigens. Hum. Genet. 76:207–219.

Wolfe, J., Darling, S. M., Erickson, R. P., Craig, I. W., Buckle, V. J., Rigby, P.W.J., Willard, H. F., and Goodfellow, P. N. (1985). Isolation and characterization of alphoid centromeric repeat family from the human Y chromosome. J. Mol. Biol. 182:447–485.

Yen, P. H., Marsh, B., Allen, E., Tsai, S. P., Ellison, J., Connolly, L., Neiswanger, K., and Shapiro, L. J. (1988). The human X-linked steroid sulfatase gene and a Y-encoded pseudogene: Evidence for an inversion of the Y chromosome during primate evolution. Cell 55:1123–1135.

CHAPTER THREE

Development of the Germ Cell

John R. McCarrey

I. Introduction

A. Background and Questions

From an evolutionary view the primary purpose of an individual in a population is to reproduce and test the relative fitness of his or her phenotype. If sufficient, the corresponding genotype will then be passed on to the next generation. The vehicle for this process is the germ cell. Unlike all other cell lineages in the body (collectively known as the soma or somatic cells), the germ cell line does not face death as a mandatory endpoint. Rather, by contributing genetic information to the next generation, including that required for the development of new germ cells, the germ-cell line can be viewed as a potential continuum from generation to generation.

The ontogeny of the germ cell line can be divided into two parts: (1) the developmental phase and (2) the differentiation phase. The developmental phase includes the period from initial embryonic development, through the time when primordial germ cells (PGCs) are first recognized in the early embryo, to the time they subsequently migrate to and colonize the developing gonads in the fetus and undergo sexual differentiation. The differentiation phase in the male is the period of spermatogenesis that occurs within the testis and includes the differentiation of prospermatogonia into spermatogonia, spermatocytes, spermatids, and finally spermatozoa. This chapter deals with regulatory events that occur during the developmental phase of the germ cell line in the male.

A number of questions exist regarding the development of germ cells. During gametogenesis, germ cells undergo substantial differentiation, including specific changes in gene expression, biochemical functions, cell surface characteristics, cell movements, cell shape and cell function. Few, if any, cell types are any more differentiated than the mature spermatozoan. Yet throughout this differentiation process the germ cell genome must retain, or at some point reestablish, genetic totipotency in order to contribute to the development of a complete individual in the next generation. Does this require a special mechanism to protect genetic totipotency that must be in place when the germ cells are first formed, and is this related to the extragonadal origin of germ cells? Shortly after the germ cells colonize the developing genital ridges, the fetal gonads initiate sexual differentiation either into testes or ovaries. This triggers the sexual differentiation of the germ cells toward oogenesis in the female or spermatogenesis in the male. Although an intriguing relationship between the sexual differentiation of germ cells and the time of onset of meiosis has been noted, the mechanisms by which these processes are initiated in the developing testis and mediated in the developing male germ cell are yet to be defined.

Thus, a number of questions exist regarding the development of germ cells. These include:

1. Where and when do germ cells arise during embryogenesis?
2. How are germ cells initially set aside from somatic cells?

3. Is genetic totipotency maintained in germ cells, and, if so, how?
4. What advantage does extragonadal origin confer on germ cell development?
5. How is the migration of germ cells to the developing gonads regulated?
6. How is the development of a sexual phenotype (sexual differentiation) regulated in germ cells?

It is the overall goal of this chapter to address these questions and to provide a summary of the processes that give rise to the stem spermatogonia that subsequently contribute to spermatogenesis in the adult testis. Although a great deal is known about the differentiation phase of the germ cell line that occurs within the testis, much less is known about the preceding developmental phase. The major areas of

Table 3-1. The Male Germ Cell Line

		Time of first appearance*	
	Cell type	Mouse	Rat
	Zygote	1d pc[1]	1d pc[7]
	Morula	2d pc[1]	4d pc[7]
Pre-germ cells	Inner cell mass	3d pc[1]	4.5d pc[7]
	Primitive ectoderm	4d pc[1]	5d pc[8]
	Epiblast	6d pc[1]	7d pc[8]
Primordial germ cells	PGCs	7d pc[2]	10d pc[9]
	M-Prospermatogonia	12d pc[3]	14d pc[10]
Prespermatogenesis	T_1-Prospermatogonia	14d pc[3]	17d pc[10]
	T_2-Prospermatogonia	1d pp[3]	4d pp[10]
Undifferentiated	A_s-Spermatogonia	6d pp[4]	4–6d pp[11]
spermatogonia	A_{pr}-Spermatogonia	6d pp[4]	4–6d pp[11]
	A_{al}-Spermatogonia	6d pp[4]	4–6d pp[11]
Differentiating	A_{1-4}-Spermatogonia	8d pp[4]	4–6d pp[11]
spermatogonia	In-Spermatogonia	8d pp[4]	6d pp[11]
	B-Spermatogonia	8d pp[4]	6d pp[11]
	Pre-Leptotene	10d pp[4]	9–12d pp[11]
	Leptotene	10d pp[4]	9–12d pp[11]
Meiosis I	Zygotene	12d pp[4]	15–18d pp[11]
Primary spermatocytes	Pachytene	14d pp[4]	15–18d pp[11]
	Diplotene	17–18d pp[5]	23–26d pp[11]
	Dictyate	17–18d pp[5]	23–26d pp[11]
Meiosis II	Secondary spermatocytes	18d pp[4]	26d pp[11]
	Round spermatids	20d pp[4]	26d pp[11]
Spermiogenesis	Condensing spermatids	30d pp[6]	38d pp[11]
	Spermatozoa	35d pp[6]	45d pp[11]

* First reported appearance, times may vary slightly by strain.
d pc = days post coitum.
d pp = days post partum.
[1]Eddy and Hahnel, 1983.
[2]Ginsburg et al., 1990.
[3]Burgoyne, 1987.
[4]Bellve et al., 1977.
[5]Nebel et al., 1961.
[6]Kramer and Erickson, 1981.
[7]Beaudoin, 1980.
[8]Beddington, 1983.
[9]Kemper and Peters, 1987.
[10]Hilscher et al., 1974.
[11]Clermont and Perey, 1957.

current and past investigations of germ cell development will, therefore, be presented in an attempt to define the key questions that remain and the most promising directions for future research.

B. Terminology

The germ cell line is that lineage that by descent gives rise to the gametes. By this definition, the zygote is considered the first cell in the germ line of an individual, just as it is the first cell in all other lineages within the body. The germ cell line, however, is distinguished from all other cell lineages by virtue of its production of the reproductive (germ) cells. A germ cell is a cell whose surviving descendants *all* become gametes (McLaren, 1985a).

The germ cell line is one of the best characterized lineages in the mammal. Several different stages, time points, and/or cell types within this lineage have been described. A list of the various cell types as they occur chronologically in the developing male embryo, fetus, and newborn are provided in Table 3–1. For completeness, the cell types that subsequently occur during spermatogenesis in the adult are also included.

II. Overview of Germ Cell Development and Differentiation

A. Life Cycle of the Germ Cell Line

By virtue of the potential continuity of the germ cell line from one generation to the next, the development and differentiation of this lineage can be viewed as a cycle repeated in each generation (Fig. 3–1). The union of parental gametes at fertilization marks the beginning of development of a new individual. Early in embryonic development the PGCs are set aside from the somatic cells. In all animal species studied, the PGCs are first recognized in an ex-

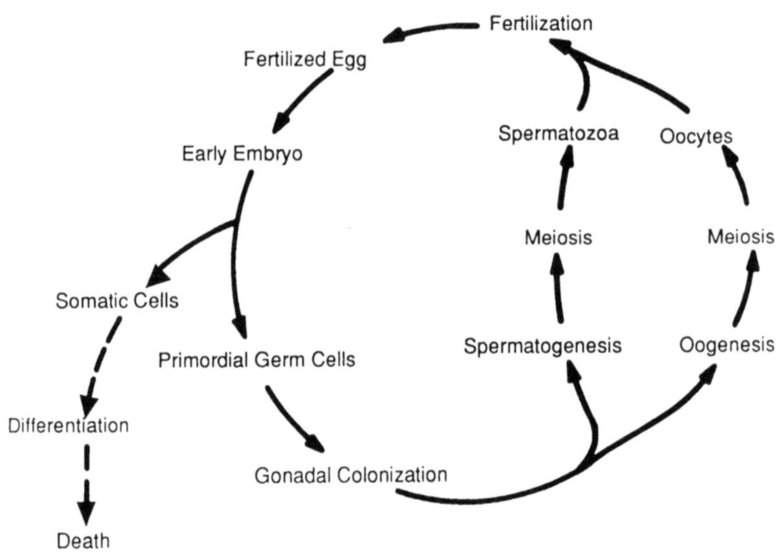

Figure 3–1. The life cycle of the germ cell line in the mammal is schematically represented. The union of the sperm (spermatozoa) and the egg (oocytes) at fertilization gives rise to a new individual fertilized egg (zygote). After cleavage and early embryogenesis, the germ cell line is set aside in the form of primordial germ cells as distinct from the somatic cells. The somatic cells undergo various types of differentiation depending on the particular somatic lineage, all of which are lost when the individual dies. The primordial germ cells arise outside the gonads in the embryo, migrate to and colonize the gonads, and then give rise to the stem cells for gametogenesis in each sex. The process of spermatogenesis, including meiosis, in the male gives rise to the haploid male gamete, the spermatozoan. The process of oogenesis, including meiosis, in the female gives rise to the haploid female gamete, the oocyte. The gametes then participate in fertilization to begin the cycle anew.

tragonadal location and subsequently migrate to the developing genital ridges (reviewed by McCarrey and Abbott, 1979). Then, as a result of primary sex differentiation, the somatic elements of the gonads begin to develop into either testes or ovaries and the germ cells initiate spermatogenesis in the former. The period of gametogenesis extends into the adult and includes the entry of stem spermatogonia into meiosis to produce haploid cells that subsequently differentiate into mature gametes. These gametes then become competent to participate in fertilization, thus beginning the cycle anew.

B. Recognition of Germ Cells

During the developmental phase germ cells can be recognized by (1) morphology, (2) position within the embryo or fetus, (3) specific histochemical staining, and/or (4) detection of specific cell surface markers (reviewed by Eddy et al., 1981). The ultimate validity of each of these recognition methods has been established by following the identified cell line into the differentiation phase to confirm that the cells thus identified do indeed give rise to gametes. The most commonly used marker for PGCs in mammals has been specific staining for alkaline phosphatase activity (Chiquoine, 1954). By this method, PGCs are first detectable in the extraembryonic mesoderm just posterior to the definitive primitive streak at 7 days postcoitum (pc) in the mouse embryo (Ginsburg et al., 1990) and then in the primary endoderm and early hindgut at about 8.5 days pc, and can be followed throughout their subsequent migration to the gonads, which is complete by 12 days pc (Tam and Snow, 1981). Thereafter, the germ cells continue to demonstrate detectable alkaline phosphatase activity until at least 14 days pc. Once the germ cells have colonized the gonads they are identifiable by their large size, morphology, and characteristic location within the gonad (Hilscher et al., 1974). In recent years detection of specific cell surface markers has become an effective and increasingly common method to identify and follow PGCs. These markers include antigens for which specific antibodies are available (Fox et al., 1981; Wylie et al., 1985; Hahnel and Eddy, 1986) and carbohydrate moieties that can be detected by binding of specific lectins (De Felici, 1984). With one possible exception that will be discussed later, the use of cell surface markers has not provided additional direct information about the origin or development of PGCs; however, these markers do facilitate certain experimental manipulations that cannot be carried out using the alkaline phosphatase marker (see below and McCarrey et al., 1987).

C. Kinetics of Germ Cell Development

Estimates of the number of PGCs present at various stages of development have been made for the mouse (Tam and Snow, 1981) and rat (Beaumont and Mandl, 1963). Estimates of PGC numbers before, during, and shortly after PGC migration have relied on detection of alkaline phosphatase activity and assume that this activity is detectable in all PGCs at any particular stage. During the remainder of fetal development germ cell numbers have been estimated on the basis of direct observations using morphology as the means of identification in association with stereology and autoradiography. Studies of germ cell numbers in the late embryonic–early fetal period have been made primarily on the mouse (Tam and Snow, 1981; Ginsburg et al., 1990), whereas studies of the mid–late fetal period have been carried out more commonly on the rat (Hilscher et al., 1974; Beaumont and Mandl, 1963) (Table 3–2). From these kinds of data a generalized scheme for the kinetics of germ cell development in mammals has been devised. The differentiation of an initial population of cells into "mature" (alkaline phosphatase-containing) PGCs appears to occur between 7 and 8 days pc in the absence of any mitotic activity (Ginsburg et al., 1990). From 8.5 to 13.5 days pc in the mouse (late embryonic–early fetal period) the primordial germ cells replicate by mitosis at a uniform rate, with a doubling time of about 16 hours (Tam and Snow, 1981). Throughout this period the germ cells develop in a similar manner in males and females. However, once the germ cells initiate sexual differentiation (at about 13.5 days pc in the mouse and 14.5 days pc in the rat), the subsequent kinetics of germ cell development show a dramatic sexual dimor-

Table 3–2. Representative Numbers of Developing Germ Cells in the Male

Species	Age	No. Germ Cells*
	7.5d pc	50
	8.5d pc	145
Mouse[1]	9.5d pc	364
	10.5d pc	1,012
	11.5d pc	2,999
	13.5d pc	25,791
	14.5d pc	19,897
	15.5d pc	43,799
	16.5d pc	70,140
	17.5d pc	95,905
	18.5d pc	118,001
	19.5d pc	129,966
	20.5d pc	122,442
Rat[2]	21.5d pc	129,159
	Birth	104,073
	1.5d pp	120,637
	2.5d pp	129,873
	3.5d pp	126,444
	4.5d pp	71,706
	5.5d pp	118,180
	6.5d pp	174,540

*Mean numbers; see original reports for variation, times may vary slightly by strain.

d pc = days post coitum; d pp = days post partum.

[1]Data from Tam and Snow (1981) and Ginsburg et al. (1990).

[2]Data from Beaumont and Mandl (1963).

phism: oogonia enter meiotic prophase (McLaren, 1988), while prospermatogonia continue to divide mitotically then enter a phase of mitotic arrest.

The development of prospermatogonia in male fetuses has been described by Hilscher and colleagues (1974). The M-prospermatogonia continue to divide by mitosis until about 14 days pc in the mouse or 17 days pc in the rat when the T_1-prospermatogonia begin to form. These are "postmitotic" spermatogonia (Hilscher et al., 1974) that remain quiescent until after birth. The T_1-prospermatogonia show very little, if any, mitotic activity, and there is some degeneration. Shortly after birth in the mouse, and at about 6 days postpartum (pp) in the rat the germ cells undergo a burst of mitotic activity to form the T_2-prospermatogonia. These cells then give rise to the primitive type A or type A_s spermatogonia that act as stem cells for spermatogenesis in the adult testis.

III. Embryonic Origin of the Germ Cell Line

A. *Descriptive, Experimental, and Comparative Embryology*

1. OBSERVATIONS IN VIVO

The PGCs are first distinguishable from somatic cells after 7 to 8 days pc in the mouse embryo on the basis of either alkaline phosphatase activity (Chiquoine, 1954; Ginsburg et al., 1990) or binding of specific antibodies (Hahnel and Eddy, 1986). At this time the PGCs are seen posterior to, and then at the caudal end of the primitive streak, at the root of the allantoic mesoderm, and in the underlying yolk sac splanchnopleure (Chiquoine, 1954; Ginsburg et al., 1990). It was originally suggested that the PGCs originate in the endodermal cell layer (Mintz and Russell, 1957; Chiquoine, 1954) or in the mesoderm (Spiegelman and Bennett, 1973; Ozdzenski, 1967) in this region; however, this now appears to be a secondary location rather than the sight of origin of PGCs. The observation of PGCs in locations intermediate between the endoderm and splanchnic mesoderm, along with observations of morphological characteristics indicating a resemblance between PGCs and splanchnic mesoderm cells, leads to the suggestion that PGCs originate in either the mesoderm or ectoderm and then become located in the endoderm (Clark and Eddy, 1975; Spiegelman and Bennett, 1973). At this stage, the EMA-1 and EMA-6 antibodies recognize the same cells as those identified as PGCs by alkaline phosphatase staining. These antibodies, however, also react with the free surfaces of ectodermal (epiblast) and some endodermal cells, but not with mesodermal cells (Hahnel and Eddy, 1986). Taken together the alkaline phosphatase/morphology studies seem to mitigate the argument for an endodermal origin of PGCs and the antibody studies fail to support a mesodermal origin. This leaves the primary ectoderm/epiblast as the most likely site of origin of PGCs.

2. IMPLICATIONS FROM EXPERIMENTAL MANIPULATIONS

The lack of any specific marker for PGCs before 7 days pc in the mouse may reflect the actual lack of distinct germ cells before this stage. However, a number of investigations using techniques of experimental embryology have indicated that the potential to form germ cells is limited to specific tissues at earlier stages. Snow (1981) performed an elegant series of experiments to test the potential of isolated pieces of mouse embryos excised at 7 days pc and maintained in culture. He found that PGCs developed only from pieces of the egg cylinder isolated from the caudal end of the primitive streak at the base of the allantois, and not from any other part of the embryo. This demonstrated that significant regionalization of potential to form PGCs is already established within the tissues of the mouse embryo by 7 days pc. The cuts made by Snow to produce these pieces of the egg cylinder transected all tissues in the embryo, so that each piece included primary endoderm, mesoderm, and epiblast/ectoderm. Thus, this experiment did not distinguish which of these layers contains the precursors of the PGCs. Hahnel and Eddy (1986), however, subsequently demonstrated that 7-day pc egg cylinders from which the primary endoderm had been removed by immunosurgery retained the ability to develop PGCs when cultured for another 24 hours. This eliminated the endoderm as a source of germ cells. Ginsburg and co-workers (1990) recently identified PGCs directly in the mesoderm at 7 days pc. They suggested that the progenitors of PGCs form part of the developing extraembryonic mesoderm, and subsequently PGCs move back to the mesoderm of the primitive streak and then into the visceral and hindgut endoderm.

Gardner and co-workers (1985) demonstrated that genetically marked primitive ectoderm/epiblast cells isolated from mouse embryos at 5 days pc (prior to gastrulation and the formation of mesodermal cells) could be injected back into host blastocysts at 4 days pc and would give rise to both somatic and germ cells in the resulting chimera. Similarly single inner cell mass cells isolated from one blastocyst at 4 days pc and injected into another 4 days pc blastocyst also gave rise to both somatic and germ cells (Gardner, 1975). However, primitive endoderm cells from embryos at 5 days pc injected back into 4 day pc blastocysts contributed only to the extraembryonic endoderm of the resultant chimera, and not to any lineages, including the germ cell line, in the embryo proper (Gardner, 1982; Gardner and Rossant, 1979). This established that cells deriving from the inner cell mass in the blastocyst and subsequently from the primitive ectoderm/epiblast in the 5-day pc embryo are the precursors of the PGCs. However, this did not formally rule out the possibility that the PGCs come from mesodermal cells that have themselves derived from primitive ectoderm cells during gastrulation.

3. COMPARATIVE STUDIES

A number of studies of germ cell development in nonmammalian species have provided insight into the possible mechanisms of germ cell origin in mammals (for review see Eddy et al., 1981). With regard to the issue of ectoderm/epiblast versus mesoderm origin of PGCs, recent studies of germ cell development in the chick are informative. Eyal-Giladi and colleagues (1981) performed experiments to analyze germ cell origin in the avian embryo in a manner analogous to that used by Gardner and co-workers for the mouse. they generated chick–quail chimeras by transplanting the epiblast from one species onto the hypoblast of the other before gastrulation. They observed that nearly all of the germ cells in the resultant chimeras displayed nuclear staining characteristic of the species from which the epiblast was derived, indicating that, in birds, the epiblast is the source of PGCs.

Urven and colleagues (1988) showed that the EMA-1 antibody that recognizes PGCs in the mouse fetus also recognizes PGCs in the chick embryo. In a study of germ cell origin in the early chick embryo, they found that just before gastrulation the epiblast layer becomes mosaic for EMA-1 binding, with some EMA-1-positive cells seen apparently budding off the ventral surface into the space between the epiblast and the hypoblast (Fig. 3–2a). These authors used the periodic acid–Schiff (PAS) histochemical marker to demonstrate that the EMA-1-positive cells were indeed PGCs. Subsequently, during gastrulation the epiblast appeared to bind the EMA-1 antibody uniformly and PGCs could be seen in the space between the epiblast and the hypoblast, including in the newly formed mesodermal layer (Fig. 3–2b). Thus, this work demonstrated that PGCs derive *directly* from the epiblast in the chick embryo *be-*

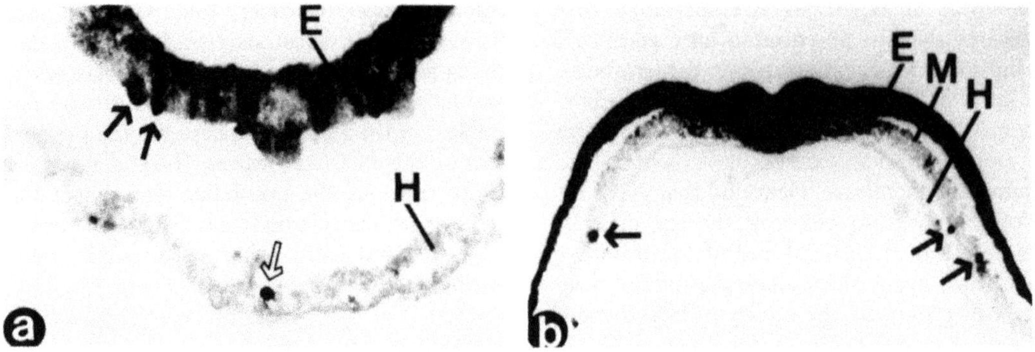

Figure 3–2. The origin of primordial germ cells (PGCs) in the chick embryo. (**A**) A chick embryo after approximately 12 hours of incubation (stage XII) is labeled with the EMA-1 antibody to reveal PGCs. The epiblast (E) and hypoblast (H) layers are visible before gastrulation. The epiblast shows a mosaic pattern of labeling with EMA-1, including ingressing cells emerging from the ventral aspect to become PGCs *(solid arrows)*. A PGC is also visible in the hypoblast layer *(open arrow)*. (From Urven et al., 1988, with permission.) (**B**) After gastrulation the mesodermal layer (M) is visible between the epiblast (E) and the hypoblast (H) in a slightly later chick embryo (24 hours of incubation). Labeling with the EMA-1 antibody reveals PGCs in the mesodermal layer *(arrows)*. (Micrograph provided by L. Urven.)

fore and during early gastrulation, and do not come from mesodermal cells, nor from any other germ layer after gastrulation.

Urven and colleagues (1988) noted similarities between their findings and those reported by Hahnel and Eddy (1986), who used the EMA-1 antibody to characterize germ cell development in the mouse. The latter authors observed periods during the development of the mouse embryo at 6 to 7 days pc when the epiblast bound EMA-1 in a uniform manner (Fig. 3–3a). In experiments in which 7-day pc epiblasts immunosurgically denuded of endoderm were cultured for 24 hours, they noted individual EMA-1-positive cells at the periphery (Fig. 3–3b). Interestingly, in the sections shown, the EMA-1-positive cells occurred only on the aspect of the epiblast that would be adjacent to the primitive endoderm in normal intact embryos. This is directly analogous to the emer-

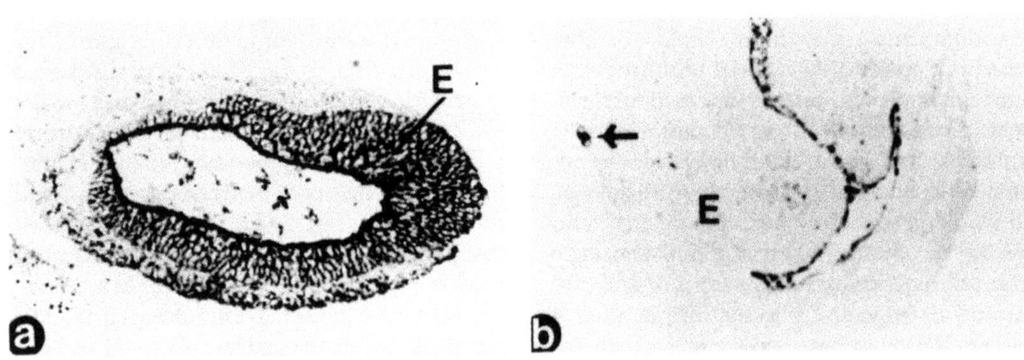

Figure 3–3. The origin of primordial germ cells (PGCs) in the mouse embryo. (**A**) A 7-day pc mouse embryo labeled with the EMA-1 antibody after immunosurgery to remove the primary endoderm. The epiblast (E) shows a uniform binding of the EMA-1 antibody at this stage. (From Hahnel and Eddy, 1986, with permission.) (**B**) A 7-day pc mouse embryo labeled with the EMA-1 antibody after immunosurgery and subsequent culture for 24 hours. The epiblast (E) is generally negative for EMA-1 binding except for specific cells *(arrow)* on the "ventral" aspect of the epiblast. (From Hahnel and Eddy, 1986, with permission.)

gence of PGCs from the ventral aspect of the epiblast in the chick (Fig. 3–2a) (Urven et al., 1988). The proposed origin of PGCs in mammals is represented in Figure 3–4 a–e.

B. Migration of Primordial Germ Cells to the Gonads

Of the many similarities noted among various species regarding the development of the germ cell line, the single most striking is the consistently extragonadal, often extraembryonic, occurrence of PGCs during embryogenesis of every higher animal species studied, including insects, amphibians, birds, and mammals (see McCarrey and Abbott, 1979, for review). In each case the PGCs arise outside the site of the gonadal anlagen and undergo a lengthy migration to the developing gonads where they initiate gametogenesis. This migration of PGCs in the mouse embryo is schematically represented in Figure 3–4 f–j. The mechanisms by which the PGCs carry out this migration, including the method of locomotion, the selection of appropriate substrata, and the directed movement toward the genital ridges are all areas of past and current investigation, and are yet to be fully understood. Even more elusive is an explanation of why the germ cells must apparently arise outside the developing gonadal tissue. At first glance this would seem to be a most inefficient developmental strategy. Yet, the fact that this process persists in every vertebrate species and many other animal species suggests that it is indispensable and must play a critical role in germ cell development.

1. MIGRATION PATHWAY

As noted, the PGCs are first directly identifiable in the extraembryonic mesoderm just posterior to the definitive primitive streak at 7 days pc. However, the PGCs arise from the epiblast. Thus, their location in the extraembryonic mesoderm and then in the yolk sac and hindgut endoderm must be the result of an initial relocation of the PGCs after their emergence from the epiblast. This is presumably analogous to the dropping down of the PGCs from the epiblast into the mesoderm and onto the hypoblast (equivalent to the endoderm) in the chick embryo as observed by Urven and colleagues (1988). The experiments with isolated pieces of 7-day pc mouse embryos performed by Snow (1981) demonstrated that when PGCs formed in a fragment that produced an allantois but lacked the hindgut endoderm, the germ cells were dispersed over a wide region, predominantly in an uncharacteristic location toward the distal end of the allantois. However, when PGCs developed in a fragment that also produced a hindgut endoderm, the germ cells became tightly arrayed exactly at the point of invagination, indicating a preference to locate in this position when possible.

Exactly how the PGCs become relocated from the epiblast to the hindgut epithelium is not clear. However, the advantage of this location for PGC migration would seem to lie in the subsequent relocation of the gut from a relatively external position in the 8-day pc embryo to an internal position in the 9-day pc embryo (Fig. 3–4 f–h). Thus, by associating themselves with the invaginating hindgut, the PGCs obtain rapid "free passage" to the interior of the embryo.

Once internalized by development of the hindgut, the PGCs begin an active interstitial migration. Using apparent amoeboid movement they move cranially in the hindgut epithelium (Fig. 3–4h), then dorsally and cranially up into the mesentery that supports the hindgut (Fig. 3–4i), then around the angle of the coelom and into the genital ridges/developing gonads (Fig. 3–4j) (Clark and Eddy, 1975; Eddy et al., 1981). The PGCs begin to reach the gonads by 11 days pc in the mouse, and complete colonization is achieved by 12.5 days pc when sexual differentiation begins.

2. MECHANISMS OF PRIMORDIAL GERM CELL MIGRATION

The migration of PGCs from their point of origin in the epiblast to their target organ, the developing gonads, involves a combination of passive, directed movements ("hitch-hiking" during major rearrangements of tissues), active, directed migration of individual PGCs, and possibly active, nondirected migration as well. The passive, directed movement is that associated with the invagination of the hindgut described previously. To participate in this movement the PGCs must become located in the hindgut endoderm after they emerge from the epiblast, which they appear to do by active mi-

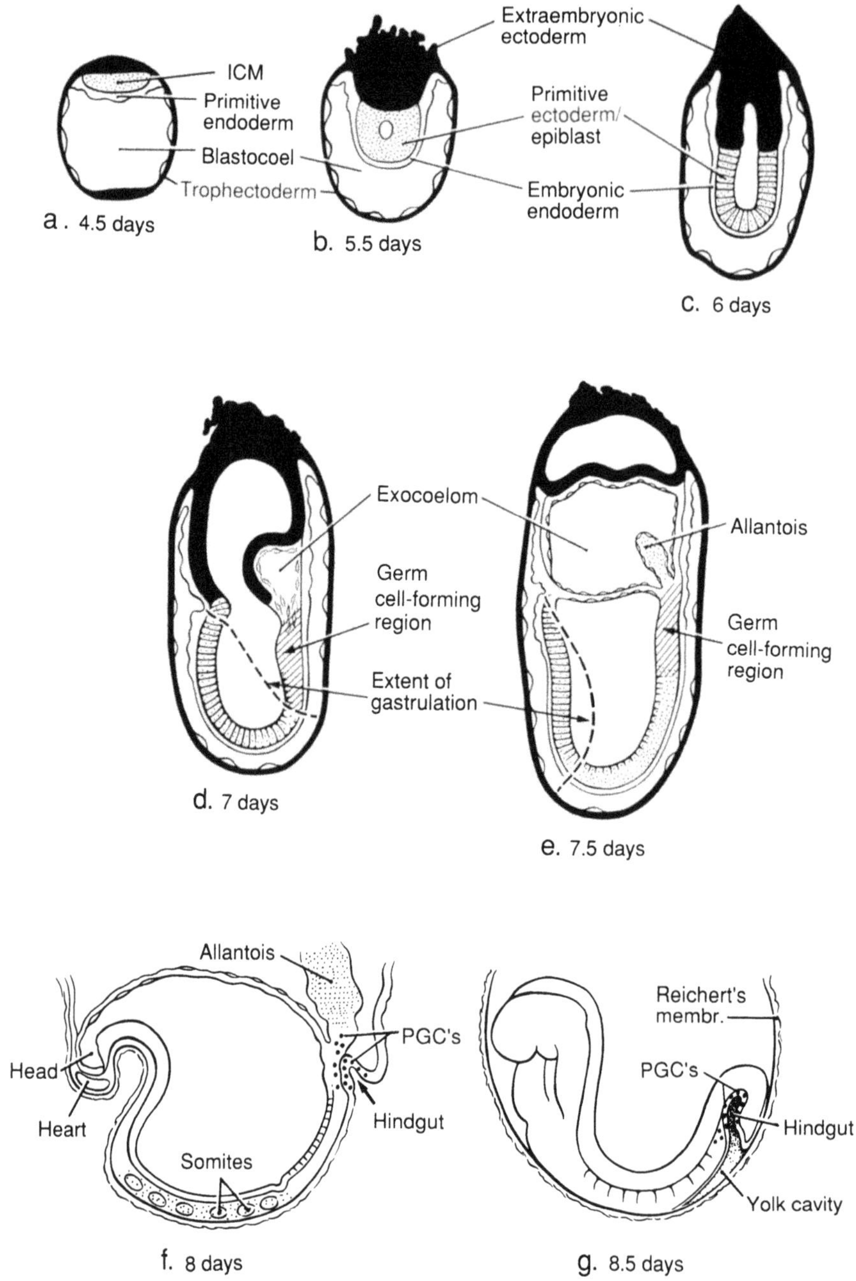

Figure 3–4. The origin and migration of primordial germ cells in the mammalian embryo. (**a–c**) The early development of the mouse embryo, including the tissues that will give rise to the germ cells *(stippled)*. At 4.5 days pc (**a**) the inner cell mass (ICM) cells are located within the blastocyst. By 5.5 days pc (**b**) and 6 days pc (**c**) the primitive ectoderm/epiblast has formed. (**d–e**) Formation of germ cells in the "germ-cell forming region" *(crosshatched)* during gastrulation. At 7 days pc (**d**) and 7.5 days pc (**e**) gastrulation is underway and the germ cell-forming region at the caudal end of the primitive streak includes cells that will become primordial germ cells. (**f–g**) Passive internalization of the germ cells. At 8 days pc (**f**) the primordial germ cells (PGCs) are identifiable in the region of the hindgut invagination. At 8.5 days pc (**g**) the hindgut invagination begins to transport the PGCs to the interior of the embryo. (**h–j**) Active migration of the germ cells. At 9 days pc (**h**) the PGCs are present in the epithelium of the hindgut and are beginning to actively migrate toward the genital ridges. By 10 days pc (**i**) the PGCs are leaving the hindgut epithelium and moving into the dorsal mesentery. At 11 days pc (**j**) the PGCs have moved around the angle of the coelom and begin to colonize the gonads. (Drawings **a–f** after Hogan et al., 1986, with germ cell-forming regions added on the basis of findings of Snow, 1981.)

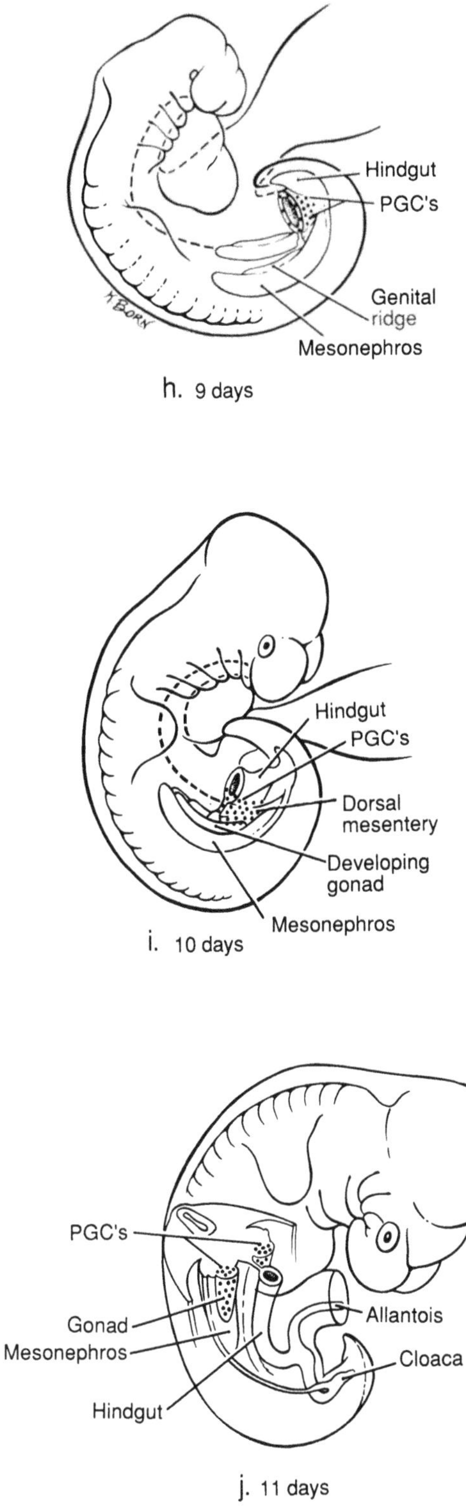

Figure 3–4. **(h–j).**

gration (whether directed or nondirected is not clear). Morphological indications of amoeboid movement have been reported in PGCs at this stage (Spiegelmen and Bennett, 1973; Clark and Eddy, 1975), and the results reported by Snow (1981) suggest an attraction of the PGCs to the hindgut invagination.

The PGCs are not individually motile during the invagination of the hindgut, but then resume their active migration cranially through the endoderm of the newly formed hindgut. This movement is clearly directional, following an interstitial route directly to the developing gonads. At this stage the PGCs are elongated and possess large, blunt pseudopodia and have their organelles clustered on one side of the nucleus, all characteristics of actively migrating cells (Eddy et al., 1981). In addition, PGCs isolated from the mouse at this stage demonstrate active motility over feeder layers of an embryo-derived cell line in vitro, and also exhibit characteristics of invasiveness (Donovan et al., 1986). Similarly active locomotion of PGCs from human fetuses has been seen in vitro (Kuwana and Fujimoto, 1983).

The directionality of PGC movement has been ascribed to one or both of two possible influences: (1) an ordered association with somatic cells along the interstitial migration pathway to achieve a form of contact guidance or directed differential adhesion, or (2) chemotactic attraction by a factor emanating either from the developing gonads themselves or a nearby structure in the embryo. Presumably any significant association or interaction between PGCs and somatic cells would be mediated by extracellular substances such as fibronectin. Fujimoto and colleagues (1985) described the distribution of fibronectin (FN) along the migratory route of PGCs in the mouse embryo. They noted that the doral mesentery is rich in FN in embryos at 9.5 to 11.0 days pc, and that the FN was located predominantly at the points of contact between the surfaces of extended PGC pseudopodia and neighboring somatic mesentery cells. Detection of the abundant FN present in the mesentery at 9.5 to 11.0 days pc was significantly reduced by 11.5 to 12.0 days pc after the PGCs had passed through this tissue. However, it is yet to be established whether fibronectin, or any other extracellular molecule, actually forms a "pathway" to direct PGC migration.

The suggestion that PGCs migrate toward a chemotactic agent has persisted in the literature since early studies (Witschi, 1948), with little supporting evidence until very recently. Early evidence for chemotactic attraction of migrating PGCs came from work in avian systems (reviewed by McCarrey and Abbott, 1979). Dubois (1968) demonstrated that the chick embryo germinal epithelium can exert a chemotactic effect on chick PGCs in culture. Recently, Kuwana and colleagues (1987) used scanning electron microscopy and time-lapse video photomicroscopy to carry out an extensive analysis of the behavior of chick PGCs in response to the influence of genital ridge tissue in vitro. They documented the directed movement of PGCs toward the gonadal tissue over thin and thick layers of collagen substrata, and noted similarities in the appearance of the migrating PGCs in vitro with those observed in vivo.

That a chemotactic effect similar to that observed in the chick embryo might also influence the migration of mammalian PGCs was indicated in an experiment by Rogulska and coworkers (1971). These authors grafted PGC-containing hindguts of early mouse embryos into the coeloms of early chick embryos and observed subsequent relocation of the mouse PGCs toward the developing gonads in the chick. Recently Godin and colleagues (1990) showed that isolated mouse PGCs would migrate toward 10.5 day pc genital ridge tissue in preference to other explanted organs in vitro. This provides direct evidence that the genital ridges exert a long-range tropic effect on migrating PGCs that may play a role in guidance of these migration cells. It remains to determine the nature and source of this chemotactic agent and its mechanism of action over a significant distance. Recent progress in defining specific mutations that disrupt PGC migration discussed below may now shed light on this mechanism.

When the PGCs arrive in the gonads they lose the morphological characteristics associated with migrating cells (Clark and Eddy, 1975). The PGCs, however, seem to retain the potential for some migratory ability for several days thereafter as has been demonstrated by dissociating PGC-colonized gonads and studying the behavior of the liberated PGCs in vitro (reviewed in Eddy et al., 1981). Recently, however, Donovan and colleagues (1986) have argued that PGCs undergo a major alteration in phenotype upon entry into the developing gonads, including a significant decrease in migratory ability. Clearly any residual migratory potential in these cells is not normally used once the PGCs colonize the gonads, as these cells are on the verge of initiating gametogenesis.

3. GENETIC MUTANTS DEFICIENT IN PRIMORDIAL GERM CELL MIGRATION

Two single gene mutations, both pleiotropic in effect, have been known for many years to disrupt the migration of PGCs in the mouse embryo. These are the *White-spotting* (*W*) gene described by Mintz and Russell (1957) and the *Steel (Sl)* gene described by Bennett (1956). Pleiotropy of *Sl* includes effects on erythropoiesis, hair pigment formation, and the development of germ cells. These effects are seen in the *Sl*/+ heterozygote, but are more extreme in the *Sl*/*Sl* homozygote. Bennett (1956) suggested that the action of *Sl* may impact on both the migratory ability and proliferation of PGCs during the migratory period, but *not* on the initial formation of PGCs. As a result these mutants contain a normal number of PGCs at 8.5 days pc, but by 14 days pc they are totally devoid of germ cells. At 9.5 days pc the PGCs in *Sl* mutants are present in reduced numbers and exhibit morphological abnormalities in comparison with normal littermates. The PGCs in this mutant apparently undergo normal formation and passive migration by the invagination of the hindgut, but cannot complete the subsequent active, directed migration to the developing gonads that is seen in normal embryos. Rather, these PGCs show an unusual clumping and tend to remain in the region of the hindgut and allantois, with no further migration or cell division.

Mice carrying a mutant allele at the *W* locus display a pleiotropic phenotype similar to that described above for *Sl* mutants. The *W* or W^v alleles and multiple other mutant alleles at this locus (Geissler et al., 1981) cause severe deficiencies in blood formation, hair pigment and germ cell development (Mintz and Russell, 1957). As with *Sl*/*Sl* mutants, *W*/*W* or W/W^v

mutants show no abnormality in germ-cell formation through 8.5 days pc in the mouse embryo. However, beginning at 9.5 days pc there is a marked deficiency in germ cell numbers in these mutants. By 12.5 days pc the germ-cell count in *W*/*W* mutant embryos is approximately 2 percent of that observed in normal embryos (Mintz and Russell, 1957). Although no morphological abnormalities in PGC locations or associations have been observed in *W*/*W* mutants, Mintz and Russell (1957) noted a distinct retardation of migration of the PGCs.

The pleiotropic nature of the *W* and *Sl* mutations has contributed toward an understanding of the mechanism of action of the normal counterpart of each of these genes. Defects in migrating cells can occur as a result of genetic lesions in the cells themselves (cell-autonomous mutations) or genetically determined changes in the environment or to environmental cues that impinge on the cells during migration (noncell-autonomous mutations) (Morrison-Graham and Weston, 1989). On the basis of analysis of the effects of *W* and *Sl* mutations on development and migration of neural crest cells, it appears that *Sl* mutations represent the noncell-autonomous class, with defects manifest in the embryonic tissue environment, whereas *W* mutations represent the cell-autonomous class, with defects manifest within the migrating cells themselves (reviewed by Morrison-Graham and Weston, 1989). These conclusions were based on experiments in which migrating cells from mutants were recombined with an appropriate tissue environment from normal individuals, or vice versa. This can be accomplished by tissue grafts or by the generation of chimeras in vivo, or through co-culture experiments in vitro. In such experiments, migrating cells from cell-autonomous mutants still express developmental defects in association with a normal tissue environment, whereas migrating cells from noncell-autonomous mutants develop normally when provided with a normal environment. A recombination experiment to test the developmental potential of germ cells from the *Sl* mutant was carried out by Nakayama and colleagues (1988). They generated chimeras between *Sl*/*Sl* and normal, +/+ mouse embryos and documented the production of functional spermatozoa from *Sl*/*Sl* germ cells. This indicated that germ cells carrying the *Sl* mutation can develop normally when placed in a genetically normal environment, supporting the notion that *Sl* is a noncell-autonomous mutation.

On the basis of the results summarized above, Morrison-Graham and Weston (1989) speculated that the noncell-autonomous *Sl* gene produces an altered environmental signal associated with the extracellular matrix, that specifically affects the stem cell populations giving rise to melanocytes, blood cells and germ cells. They further suggested that the cell autonomous *W* mutation represents a lesion in the receptor mechanism for the signal encoded by the normal allele of the *Sl* locus. Recent reports that identify the *W* locus as the *c-kit* protooncogene (Geissler et al., 1988; Chabot et al., 1988), and the *Sl* locus as a gene producing the ligand for the *c-kit* gene product (Zsebo et al., 1990; Copeland et al., 1990), support this hypothesis.

Identification of the *W* locus as one and the same as the *c-kit* proto-oncogene was indicated by mapping the *c-kit* oncogene to mouse chromosome 5 (Yarden et al., 1987) and was confirmed by demonstrating disruption of the *c-kit* gene by three mutant *W* alleles (Geissler et al., 1988; Chabot et al., 1988). The *c-kit* polypeptide shows structural similarities to the receptors for colony-stimulating factor 1 (CSF-1) and platelet-derived growth factor (PDGF), and has been shown to be a transmembrane cell-surface glycoprotein possessing kinase activity (Qiu et al., 1988; Yarden et al., 1987). Yarden and colleagues (1987) demonstrated that the *c-kit* polypeptide is expressed as a 145,000 mol wt cell surface molecule with tyrosine-specific autophosphorylation activity. They suggested that the *c-kit* polypeptide acts as a transmembrane receptor for an extracellular ligand and uses its ligand-induced tyrosine kinase activity to generate an intracellular signal that eventually triggers mitogenesis.

The suggestion is that the normal allele of the *W* locus is the *c-kit* protooncogene encoding a cell-autonomous receptor that functions in a mechanism that induces mitogenic activity. This is consistent with the phenotype produced by mutant alleles at the *W* locus, which includes a lack of mitogenic activity in PGCs. This model requires that the *c-kit* polypeptide bind with a noncell-autonomous ligand. Recently it was demonstrated that the product of a gene at the *Sl* locus termed either "stem cell factor" (SCF) (Zsebo et al., 1990), "mast cell growth factor" (MGF) (Copeland et al., 1990) or "steel factor" (SLF) (Witte, 1990) is the ligand for the *c-kit* gene product, and that this factor is deleted or altered in the various *Sl* mutants that display defective PGC development. A testable prediction of the model of action of the *c-kit* and *Sl* genes during PGC development is that the *c-kit* (*w*) gene should normally be expressed in migrating PGCs, whereas the *Sl* gene should be expressed in surrounding somatic cells. A recent report by Matsui and co-workers (1990) has verified these predictions. These authors found that the *Sl* gene product is expressed during embryogenesis in cells associated with both the migratory pathways and the homing sights of PGCs, melanoblasts and haemotopoietic stem cells. Upon colonization of the genital ridges by PGCs, the *Sl* gene product was detected in gonadal somatic cells, but not in germ cells, whereas the *c-kit* gene product was detected in gonadal germ cells, but not in somatic cells (see Section VI). That the disruption of a single mechanism by mutations at either the *Sl* or *w* loci can account for the breakdown of both mitogenic *and* migratory ability of PGCs is not unexpected since growth factors (e.g., PDGF) are known to promote cellular proliferation *and* differentiation, and in this case migratory capability would represent the latter.

IV. Genetic Determination and Developmental Allocation of the Germ Cell Line

The genetic and developmental determination of the germ cell line refers to the mechanisms by which the germ cell line is set aside and maintained as distinct from the somatic cell lineages. A number of theories about this process have been proposed (see reviews by Beams and Kessel, 1974; McCarrey and Abbott, 1979; Eddy et al., 1981; Snow and Monk, 1983; Eddy and Hahnel, 1983; and McLaren, 1983, for examples). Among many others, the following questions are especially pertinent: (1) Is there evidence for a continuous germ line maintained by the presence of a germ plasm or by other means, or is a new germ cell line established in each individual? (2) Are PGCs predetermined by intrinsic factors or are they induced to differentiate by factors or influences from surrounding somatic cells? (3) Is the determination of the germ cell line the result of an active genetic mechanism operative very early during embryogenesis (before or during gastrulation) or a passive process that yields germ cells as the "left overs" after other germ layers and/or cell lineages have been established? (4) Is genetic totipotency maintained throughout the life-cycle of the germ cell line or must it be reestablished at some point? (5) Does DNA methylation, or the lack thereof, play a role in germ cell determination, and is this related to the extragonadal origin of PGCs?

A. Continuity of the Germ Cell Line: Potential Existence and Function of Germ Plasm versus Induction of PGCs Anew in Each Individual

The existence of a continuous germ plasm throughout the life cycle of the germ line is well established in *Drosophila* (Mahowald and Boswell, 1983). A number of elegant experiments have demonstrated that inclusion of this germ plasm is necessary and sufficient for the determination of germ cells in this species (see Illmensee and Mahowald, 1974, for example). This germ plasm is present at the vegetal pole of the unfertilized *Drosophila* egg, and the cells that subsequently incorporate this plasm in the developing embryo ultimately become the germ cells. Although the mechanism is still not understood, this plasm clearly acts as a germ cell determinant. The recent isolation of a monoclonal antibody that recognizes a component of the *Drosophila* germ plasm indicates the involvement of a 72,000 to 74,000 mol wt protein identifiable on Western blots and opens the door to more extensive molecular analyses of the composition and function of germ plasm (Hay et al., 1988). Because the germ plasm is constantly present throughout the life cycle of the individual, it can be said that there is a continual germ line in *Drosophila*.

Evidence for the existence of a germ plasm

associated with the germ cell lineage in vertebrate species has been reviewed extensively (see Beams and Kessel, 1974; Eddy, 1975; McCarrey and Abbott, 1979; and Eddy et al., 1981, for examples). The largest body of work supporting a functional germ plasm in vertebrates comes from experiments on amphibian species, and includes the identification of a morphologically distinct germ plasm, demonstrations that destruction of the germ plasm inhibits germ cell development, and that transplanted cells containing the germ plasm form germ cells (Smith and Williams, 1975; Wakahara, 1978; Ikenishi, 1987). Eddy and Hahnel (1983) summarized the predicted ontogeny of a germ cell determinant present continuously throughout embryogenesis in the mouse. They argued the material should be present in most or perhaps all cells through at least the eight-cell stage, in at least some of the inner cell mass cells at 3.5 to 4.5 days pc, in at least some of the cells of the epiblast in 6-day pc embryos, and in cells in the caudal portion of the primitive streak of 7-day pc embryos. There has been no report of any germ plasm-like material at any of these stages; however, it is not clear that sufficient studies to detect such material have been performed. A morphologically recognizable material termed nuage (Eddy, 1975) has been observed in PGCs in the hindgut of mouse embryos at 9 days pc (Spiegelman and Bennett, 1973) and in all subsequent stages of the germ cell line (Eddy and Hahnel, 1983). Although this nuage has been suggested as a mammalian germ plasm, no experimental tests of the functional capacity of this material to induce germ cell development have been reported.

An argument (McLaren, 1983) that has been levied against the existence of a predetermined germ plasm in the mouse embryo is that individual cells taken from the inner cell mass at 3.5 days pc, or from the epiblast at 4.5 days pc, and injected into new host blastocysts contributed to both germ and somatic cell lineages in the resultant chimeras (Gardner, 1977). These results suggest that there was no absolute segregation of a germ plasm analogous to that observed in *Drosophila.* As Eddy and Hahnel (1983) point out, however, it has not been feasible to test *all* of the cells from a single inner cell mass or epiblast in this manner; therefore, the possibility that some cells have been determined to form germ cells at these stages cannot be formally excluded. Furthermore, these results certainly do not rule out the existence of a germ plasm that is present in the cytoplasm of inner cell mass cells, but does not become completely segregated so as to determine individual germ cells until later stages.

A unique event that frustrates the analysis of the germ cell (or any other) lineage during the period around gastrulation is the extensive cell mixing that occurs at this stage in mammalian embryos (Rossant, 1985, 1987). Analysis of aggregation chimeras between different species of mice in which the source of each cell can be traced by an in situ marker has revealed that there is extensive cell mixing within the primitive ectoderm (from which the PGCs derive) soon after implantation. This cell mixing continues within the ectoderm layer of the egg cylinder up to the time the primitive streak is formed. Rossant (1985) concluded that the occurrence of such cell mixing at the primitive streak stage suggests that cells are not strictly allocated or committed to definitive structures or lineages at this stage, unless the cell mixing is actually a process of sorting out of precommitted precursors. Indeed it is difficult to envision how a preexisting, localized germ plasm of the sort that exists in *Drosophila* could become incorporated into specific cells in the inner cell mass or epiblast and remain localized in a discrete group of cells while such extensive mixing takes place.

If there is no discrete, identifiable component of the germ cell line during the early stages of embryogenesis (zygote through gastrula), the continuity of the germ cell line from generation to generation is called into question. Indeed, Snow and Monk (1983) argued that the germ cell line is not continuous, but arises anew relatively late during embryonic development, *after* gastrulation, and the formation of the primary germ layers. They suggested that the PGCs are not predetermined, but are formed from those cells that are "left over" after the allocation of somatic tissues has occurred. This view is not consistent with the evidence discussed earlier that the PGCs come directly from the epiblast just *before* and *during* gastrulation. Furthermore, Soriano and Jaenisch (1986) used retroviral infection as a marker to study the formation of the germ cell line and concluded that at least three founder cells contribute to the germ cell line, and that it is set aside *before* somatic tissue allocation.

McLaren (1983) proposed that the germ cell lineage arises from the epiblast at the time of gastrulation in the mouse. According to this view, the epiblast represents a pluripotent tissue that contains cells capable of differentiating into either germ cells or somatic cells, consistent with the findings of Gardner (1977) discussed previously. Thus, it would seem that the concept of continuity of the germ cell line must be modified to include a period during which the germ cell line and somatic cell lines are initially indistinguishable, as is apparently the case in the early epiblast. What remains to be explained is how the regionalization of the epiblast is generated such that PGCs arise only in the area posterior to the caudal end of the primitive streak. The possibility of some sort of region-specific induction could be invoked to explain this; however, it is not clear why this induction would influence only some of the epiblast cells in this region while others would remain uninduced.

B. Genetic Totipotency in the Germ Cell Line: Maintained or Reestablished?

By definition, a characteristic of the genetic information carried by the gametes is that it is totipotent, so that, in conjunction with the genetic information from a gamete of the opposite sex, it can direct the development of a complete individual in the next generation. Although classic experiments with amphibian embryos and tadpoles demonstrated that the genetic information in a nucleus from a differentiated somatic cell could still direct the development of a complete new individual (Gurdon, 1962), it is not clear that all somatic cell genomes in mammals retain functional genetic totipotency throughout their lifetimes, especially in those cells that become "terminally differentiated." Rather, it may be that genes that are not required for the differentiation of a particular cell type become *irreversibly* repressed in that cell type, thus eliminating genetic totipotency. The experiments of Gurdon (1962) and others demonstrated that cellular differentiation is not initiated by the elimination of irrelevant portions of the genome, nor, initially, by elimination of the potential to express those irrelevant genes. However, these experiments did not discount the possibility that such potential (i.e., totipotency) is lost at some point during the lifetime of a terminally differentiated cell as the result of an irreversible regulatory mechanism.

Germ cells, judged by any criterion, undergo significant differentiation but still emerge as genetically totipotent gametes. Thus, the question is raised as to whether the complete genetic potential that is present in the initial zygotic genome is continually maintained in the precursors of the germ cell line and throughout the development and differentiation of the germ cells themselves, or, as has been suggested by Monk (1981), is lost at some point during germ cell differentiation and then reestablished at some later point during gametogenesis. If the former is the case, there must be a unique mechanism operating in the cells of the germ line to protect their totipotency. If the latter is the case, an equally unique mechanism must be activated at some point during gametogenesis to reestablish genetic totipotency.

The existence of genetic totipotency in the germ cell line is difficult to test. The blastocyst injection experiments of Gardner and colleagues discussed earlier provide one means of testing cellular potency. Thus, cells from the primitive ectoderm at 4.5 days pc and from the epiblast at 5.5 days pc appear to retain the ability to give rise to all cell types in the embryo proper, but have already lost the potential to form structures that derive from the primitive endoderm, such as the yolk sac (Gardner and Rossant, 1979; Gardner et al., 1985). However, this approach provides a test for *cellular* potential that may be more restricted than *genetic* potential. The most commonly used test for genetic potential involves transplantation of the nucleus of the cell to be tested back into an enucleated egg in an attempt to provide the necessary cytoplasmic components for complete development under appropriate direction by a totipotent genome. This is the approach that was used by Gurdon (1962) to establish the constancy hypothesis in amphibians. Recently the application of this approach to compare the genetic potential of inner cell mass cells and male primordial germ cells in the mouse was reported by Tsunoda and colleagues (1989). They found that nuclei from PGCs taken from male mouse fetuses at 12.5 to 16.5 days pc were able to direct development to the blastocyst stage more frequently (6 to 20 percent success) than were nuclei taken from inner cell mass cells (3 percent success). At the very least this would seem to indicate that PGCs from male fetuses as late as 16.5 days pc have retained as much genetic potential as exists in inner cell mass cells of the blastocyst.

None of the reconstituted embryos derived from male PGC nuclei developed to postimplantation stages (Tsunoda et al., 1989). The authors suggest that this may be due to the absence of an appropriately "imprinted" maternal genome (discussed later) that is required along with a paternally imprinted genome for complete embryogenesis (Reik et al., 1987; Sapienza et al., 1987). If this is the case, it would imply that reprogramming of the maternal genome that contributed to the fetuses from which the male PGCs were isolated had occurred by 12.5 days pc, and that the characteristic paternal imprint had already been established.

To the degree that *both* a maternally and paternally derived genome are required for normal development of an embryo (McGrath and Solter, 1984), even the gametes are not completely totipotent. This requirement is based on a differential parental imprint in the gametic genomes from each sex and contributes to the first restriction of developmental potential in the mammalian embryo, that distinguishing between embryonic and extraembryonic tissues (see review by Surani, 1986). Experimental embryos derived completely from a maternal genome fail to develop normal extraembryonic tissues, whereas those derived completely from a paternal genome fail to develop normal embryonic structures (Surani et al., 1987).

The extraembryonic tissues that show deficient development in the absence of a paternal genome are derivatives of the trophectoderm, which contribute to the placenta. It is interesting that it is only in placental mammals that this sort of gametic imprinting has been observed. It may be that gametic imprinting is a unique mechanism that has evolved in placental mammals to establish a transient state in which expression of genes for trophectoderm development can be initiated, and that this state is not compatible with that required to express genes for embryonic development. Thus, the paternal imprint would initially prevail to determine the development of the trophectoderm cells while the inner cell mass cells remain uncommitted. Then, *after* the dominant but transient paternal imprint had expired in the early embryo, cells from the inner cell mass would have lost the ability to form trophectoderm, and could then interpret the maternal imprint to form embryonic tissues. Whatever the nature of this imprint, it must be reversible in the germ cell line at each generation (Monk, 1987). Such reprogramming of gametic imprinting in the male to facilitate development of the trophectoderm in the next generation could represent a form of the restoration of totipotency that was suggested by Monk (1981) to occur at each generation in the germ cell line. At the same time, the potential to form all cell types in the embryo proper could be maintained throughout the life cycle of the germ line, at least in the female. Many aspects of gametic imprinting are yet to be understood, including: (1) the mechanism by which the imprint is established in a heritable fashion without changing the DNA sequence itself, (2) at what point during the development and differentiation of the germ cell line in each sex an imprint is established or reestablished, (3) which genes are imprinted in the developing embryo, and (4) how imprinting modulates the expression of those genes.

C. DNA METHYLATION AND GERM CELL DETERMINATION

For many years the observation that the germ cells arise extragonadally and subsequently migrate to the developing gonads, rather than arising de novo in the gonadal anlagen, has suggested the existence of a protective mechanism against influences in the embryo proper that might compromise the genetic potential of the germ cells (reviewed in McCarrey and Abbott, 1979). The evidence that PGCs emerge from the epiblast to become transiently located in the extraembryonic yolk sac tissues suggests that such compromising events might occur transiently throughout the epiblast at this stage and are then completed by the time the PGCs reenter the embryo proper by invagination of the hindgut. One candidate for a de novo event that appears to occur in somatic cells but not in germ cells is DNA methylation.

DNA methylation refers to methylation of the pyrimidine base cytosine in the dinucleotide sequence CpG in the DNA. The presence of methylation can be tested by comparing the ability of a pair of restriction enzymes (isoschizomers), one of which will cut only unmethylated DNA and one which will cut regardless of methylation, to digest the same sequence in DNA that includes a CpG dinucleotide (McClelland, 1983). With appropriate probes the methylation status of repeated DNA sequences or low copy number sequences (the latter includes most structural genes and genes that are likely to be involved in regulation of developmental processes) can be assessed.

The methylation status of a particular gene sequence, or a chromosomal domain, or even an entire chromosome has been correlated with the expression potential/transcriptional activity of the gene sequences involved (see Naveh-Many and Cedar, 1981; Doerfler, 1983; Gartler and Riggs, 1983, for reviews). Typically, genes that are hypomethylated (especially in their 5′-promoter regions) are active and hypermethylated genes are inactive. In most cases, DNA methylation is assumed not to be the primary regulator of gene activity/inactivity, but rather to stabilize a particular state. Thus the DNA methylation associated with the inactive X chromosome in female somatic cells appears to follow and to stabilize inactivation (Migeon et al., 1988; Lock et al., 1987). Methylated DNA appears to be unable or unlikely to form potentially active chromatin structures, whereas the absence of DNA methylation favors the formation of such structures (Keshet et al., 1986). Thus, methylated DNA may be unable to interact with transcription factors that would predispose or initiate gene expression, thereby maintaining a gene in a stably repressed state.

The stability of the methylation effect on gene expression is facilitated by the heritability of DNA methylation patterns. In double stranded DNA every CpG dinucleotide is paired with an antiparallel CpG dinucleotide, and when the DNA is methylated the cytosines on both strands are methylated. When semiconservative replication occurs, hemimethylated DNA is produced. It then becomes fully methylated again by the action of a maintenance methylase, thus maintaining the methylation pattern and the repressed state in a heritable fashion (Razin and Riggs, 1980).

A generally greater level of DNA methylation exists in spermatozoa as compared with oocytes (Sanford et al., 1987). Specifically, sperm DNA is hypermethylated at many low-copy or unique gene sequences (Sanford et al., 1987; Rahe et al., 1983), and hypomethylated at highly repetitive sequences (Sanford et al., 1984; Ponzetto-Zimmerman and Wolgemuth, 1984). Oocyte DNA is hypomethylated at dispersed repetitive sequences (Sanford et al., 1984) and highly repetitive sequences as well as at low-copy or unique gene sequences (Sanford et al., 1987; Singer-Sam et al., 1990). However, these differences are not apparent in PGCs, which appear to be hypomethylated in both sexes (Monk et al., 1987). The increase in DNA methylation observed in spermatogenic cells may represent changes in chromatin structure associated with DNA packaging during spermiogenesis and/or paternal gametic imprinting (Sanford et al., 1987; Monk et al., 1987).

Methylation of DNA sequences in cleavage-stage mouse embryos and in inner cell mass cells was found to be intermediate between that in oocytes and spermatozoa, indicating that the differential methylation levels (imprints) contributed by each gamete persist into embryonic development (Sanford et al., 1987). Monk and colleagues (1987) analyzed the overall DNA methylation levels (with no distinction between repetitive and unique sequences) of early embryonic cells of the blastocyst, primitive epiblast, and primitive endoderm cells of embryos at 6.5 days pc, and cells of the embryo proper and extraembryonic chorion at 7.5 days pc. They found that the blastocyst was relatively under methylated, and that this hypomethylated state persisted in the primitive endoderm at 6.5 days pc and in the chorion at 7.5 days pc (also observed by Chapman et al., 1984). However, the epiblast was already more methylated than the endoderm at 6.5 days pc, and by 7.5 days pc the DNA in the cells of the embryo proper was notably hypermethylated.

This indicates that a significant de novo methylation event occurs throughout the cells in the epiblast beginning at 6.5 to 7.5 days pc.

It is at 6.5 to 7.5 days pc that the PGCs begin to leave the epiblast and become associated with the tissues of the developing yolk sac endoderm. If the purpose of this relocation is to escape the de novo DNA methylation occurring within the cells of the epiblast, then the prediction would be that the PGCs should remain under methylated. Indeed this is the case as noted by Monk and co-workers (1987) who measured DNA methylation in male and female germ cells at 12.5 and 14.5 days pc. In all cases the germ cells demonstrated methylation levels lower than those found in *either* the embryonic or extraembryonic cells at 6.5 to 7.5 days pc. The germ cell genomes remained hypomethylated at a level similar to that in the inner cell mass cells of the blastocyst.

The major limitation of the experiments reported by Monk and colleagues (1987) is that they are based on overall methylation levels of genomic DNA with no knowledge of which specific CpG sites were involved and whether or not these sites were associated with particular genes. Recently Driscoll and Migeon (1990) analyzed the methylation status in DNA from germ cells of human fetuses at 57 specific CpG sights in five X-linked housekeeping genes and two tissue specific genes (one X-linked and one autosomal). They found that these sites were consistently unmethylated in fetal germ cells of both sexes and methylated in somatic cells of both sexes. Furthermore, they observed that the unmethylated state was maintained into meiosis in female germ cells but not in male germ cells. Thus, single-copy genes in sperm DNA and in DNA from adult testis (most of which is meiotic and postmeiotic germ cells) were methylated just as they were in somatic cells.

In summary, it appears that the precursors of the germ cell line arise in the pluripotent epiblast at a time when all cells of that tissue still have the potential to contribute to all structures in the embryo proper. The PGCs escape from the epiblast layer just before a major de novo methylation event that may be required to restrict or differentially program epiblast cells to participate in one of the definitive germ layers after gastrulation (Monk et al., 1987), but might also irreversibly restrict the genetic potential of the germ cells if they remained in this tissue. The PGCs thus find a haven amid the cells of the yolk sac endoderm that do not undergo any significant de novo methylation. They then return to the interior of the embryo proper by the invagination of the hindgut, having maintained their genetic totipotency in the form of a hypomethylated genome. Thus, in mammals an equivalent of a continuous germ plasm may be found in the continuous hypomethylated state of the genome. To the extent that this is maintained throughout the development and early differentiation of the germ cells in both sexes, and throughout oogenesis in females, the germ line potential can be viewed as continuous.

V. Sexual Differentiation in the Germ Cell Line

Shortly after germ cells colonize the developing gonads the process of primary sex differentiation begins in the embryo. This includes *both* sexual differentiation of the bipotential gonads *and* sexual differentiation of the germ cells. The mechanisms by which these two parts of primary sex differentiation are initiated and regulated are distinct but related.

A. The Relationship Between Gonadal and Germ Cell Sex Differentiation

Sexual differentiation of the gonads refers to the differentiation of the somatic elements of the bipotential (indifferent) gonad into either a testis or an ovary. This process has been reviewed extensively elsewhere (see McCarrey and Abbott, 1979; Gordon and Ruddle, 1981; Eicher and Washburn, 1986, for examples). Differentiation of the testis appears to require the action of at least one gene on the Y chromosome (Sinclair et al., 1990; Gubbay et al., 1990) and appears to be manifest initially in the developing Sertoli cells (Burgoyne et al., 1988). Neither the presence of, nor any influence from, the germ cells is required for the initiation of this process (reviewed in McCarrey and Abbott, 1979; McLaren, 1985b). In both *W* and

Sl mutant mice, which are devoid or severely depleted in germ cells at the time of primary sex differentiation, the somatic elements of the gonads still differentiate appropriately (Bennett, 1956; Mintz and Russell, 1957; Handel and Eppig, 1979). Studies of the effects of destruction of PGCs by treatment of rat embryos with busulfan or by other means demonstrated that elimination or depletion of PGCs did not affect the initial sexual differentiation of the somatic elements of the gonads (Merchant, 1975; Merchant-Larios and Coello, 1979; Dierickx and Verhoeven, 1980), and similar findings in the chick embryo have also been reported (McCarrey and Abbott, 1978, 1982).

Sexual differentiation of the gonad can be detected first at about 12.5 days pc in the mouse. Initiation of sexual differentiation of the germ cells occurs about a day later at 13.5 days pc (Monk and McLaren, 1981). After arriving at the gonads at 10 to 11 days pc, the PGCs undergo further mitotic divisions in both sexes. Then, at about 13.5 days pc in female fetuses, germ cells begin to enter meiotic prophase and cease further division, whereas in males germ cells continue to divide mitotically for another day or two. Male germ cells then enter a mitotic arrest at about 14.5 days pc, and do not begin meiosis until well after birth, at puberty. The regulation of the initiation and maintenance of this sexual dimorphism is not understood.

Mammalian germ cells do not appear to have the ability to undergo sexual differentiation in complete autonomy. Isolated PGCs from pregonadal stages or early gonadal stages do not undergo sexual differentiation autonomously in culture (De Felici and McLaren, 1983). This may be due in part to as-yet-unknown requirements of germ cells in culture, as germ cells at this or any other stage do not grow or divide well in culture and can, at best, merely be maintained for periods of a few days. Thus, at the very least, germ cells appear to require a nurturing environment that is normally provided by the gonads (Godin et al., 1990).

Whether the initiation of sexual differentiation of the germ cells is regulated by an intrinsic mechanism (presumably the sex chromosome constitution or "sexual genotype" of the germ cells) or by extrinsic influences of the gonadal environment (the "sexual phenotype" of the gonads) can be studied in cases of natural or experimental sex reversal. Situations in which germ cells of one sexual genotype are surrounded by gonadal cells of the opposite sexual phenotype can be experimentally produced in XX/XY chimeras or observed in mutations that produce XY females or XX males (see McLaren, 1983; Eicher and Washburn, 1986, for reviews). Most XX/XY chimeras develop as males with testes, and in these testes the majority of germ cells, presumed to include a significant number of XX cells, initiate differentiation in the male direction (McLaren, 1983). Female XX/XY chimeras are rare; however, Ford and colleagues (1975) reported two cases of progeny of an XX/XY female that apparently developed from XY germ cells. Eicher and Washburn (1986) developed a breeding scheme to produce XY females on a regular basis, and observed that these animals produced large numbers of functional XY oocytes. The *sex reversed (Sxr)* mutation in the mouse causes karyotypically XX or XO embryos to develop as males (Cattanach et al., 1971). In these animals the germ cells initiate differentiation in the male direction (reviewed in McLaren, 1983). The consistent observation in these examples is that the germ cells initiate sexual differentiation in accord with the sexual phenotype of the surrounding gonad, regardless of their own sexual genotype. Thus, the *initiation* of sexual differentiation appears to be regulated by influences extrinsic to the germ cells.

There are old examples of misguided PGCs that do not reach the gonads during their migratory phase, but rather colonize other tissues (Chiquoine, 1954; Falin, 1969; Jirasek, 1976), and in these cases it was concluded that germ cells could not differentiate in extragonadal locations. However, more recent studies (Upadhyay and Zamboni, 1982; Zamboni and Upadhyay, 1983; McLaren, 1984) have shown that germ cells can differentiate in ectopic locations, especially in the adrenal glands. Of particular interest in these studies was the finding that germ cells that colonize the adrenal undergo differentiation to form oocytes *regardless* of the sex of the individual. Thus, in the absence of a testicular environment, XY germ cells initiate oogenesis on a schedule similar to that observed in females. This suggests that, as is the case with primary gonadal and secondary sex differentiation, the female (oogenic) pathway is predetermined to occur autonomously and does so unless the male (spermatogenic) pathway is induced by the testicular environment.

B. Meiosis and Germ Cell Sex Differentiation

A distinguishing feature of germ cell sex differentiation in the male is that the cells continue to undergo mitotic replication and do *not* enter meiosis during fetal stages when the female germ cells do. Thus, a primary influence of the fetal testicular environment on male germ cells is to inhibit their entry into meiosis. It has been suggested that the fetal testis produces a "meiosis-preventing substance" (MPS) and that the ovary produces a "meiosis-inducing substance" (MIS) (Byskov and Saxen, 1976; Byskov, 1978). The observation that germ cells in the adrenals enter meiosis indicates that an ovarian environment is not a strict requirement for female germ cells to initiate sexual differentiation. However, the adrenals and gonads are both of mesonephric origin (Zamboni and Upadhyay, 1983); therefore, similar environmental influences could emanate from both.

McLaren (1984, 1985b) has reviewed the evidence for the "inhibition" versus "stimulation" hypotheses for regulation of initiation of meiosis in germ cells. The former holds that fetal entry into meiosis is an autonomous property of germ cells in both sexes, but is inhibited in testicular germ cells (as proposed by Upadhyay and Zamboni, 1982). The latter presumes that germ cells are nonmeiotic unless induced to initiate meiosis by factors (MIS) present in the mesonephric region (including in the adrenals), and that this induction is prevented in testicular cells (by MPS) (as proposed by Byskov, 1978). McLaren (1984) concluded that "It is not obvious that one (hypothesis) fits the data better than the other," and that improved culture systems are required to assess the degree of autonomy of germ cells. However, she further states that "The phenotypic sex of a germ cell, i.e., whether it undergoes oogenesis or spermatogenesis, depends crucially on the stage at which it enters meiosis" (McLaren, 1985b). She points out that the sex chromosome constitution of a germ cell does not influence its ability to enter meiosis if appropriately induced by its tissue (usually gonadal) environment (McLaren, 1984). Thus, the entry of germ cells into meiosis at fetal versus postnatal stages appears to be the primary determinant of female versus male germ cell sex differentiation, respectively, and this process is normally dictated by the sexual phenotype of the gonad.

C. Function of the X Chromosome in Germ Cell Sex Differentiation

It has been stated that "the somatic elements of the gonad are responsible for initiating germ cell development in either the male or female direction, but the extent of response to this impetus is dependent upon the genotype of the germ cells themselves" (McCarrey and Abbott, 1979). The first part of this statement is supported by the evidence reviewed above that the gonadal phenotype regulates the time of onset of meiosis, which initiates sexual differentiation in the germ cells. The second part is supported by the observation that in most cases germ cells of one sexual genotype do not give rise to functional gametes of the opposite sexual phenotype (evidence reviewed in McLaren, 1984). Thus, XX germ cells either in XX/XY male chimeras or XX,*Sxr* males degenerate shortly after birth in the mouse and fail to participate in spermatogenesis (Mystkowska and Tarkowski, 1968; Cattanach et al., 1971), and, with the exception of the genetically unusual situation reported by Eicher and Washburn (1986) in which functional XY oocytes were produced in mice carrying a Y chromosome from a mutagenized strain (POS A) on a genetically normal (C57B1/6J) background, reports of functional XY oocytes are extremely rare (Ford et al., 1975). The sexual genotype of a germ cell is dictated by its sex chromosome constitution, and in the male germ cell there is evidence that both the X and Y chromosomes play a role in the sexual differentiation of these cells.

In somatic cells the presence of different numbers of X chromosomes in males and females is compensated by X-chromosome inactivation in female cells so that they express genes from only one X chromosome just as male somatic cells do (Gartler and Riggs, 1983).

This is initially the case in developing germ cells as well. One of the two X chromosomes in developing female germ cells undergoes random inactivation at the same time as this occurs in other cells derived from the epiblast (McMahon et al., 1981); therefore, PGCs of both sexes express genes from only one X chromosome (Monk and McLaren, 1981). However, as the germ cells initiate sexual differentiation in the female, the second X chromosome reactivates at about 12.5 days pc, just before the time these cells enter meiosis (Gartler et al., 1975; McLaren, 1984). Thus, at the time of sexual differentiation, female germ cells contain two active X chromosomes whereas male germ cells have only one. When male germ cells enter meiosis at puberty and thereafter, the single X chromosome in primary spermatocytes appears to be inactivated (evidence reviewed in Handel, 1987). Thus, in addition to a critical difference in the timing of entry of germ cells into meiosis in each sex, there appears to be a dramatic difference in X-chromosome activity, with meiotic oocytes containing two active X chromosomes and meiotic spermatocytes containing none.

The evidence above suggests that specific X-chromosome activity states are correlated with the ability to enter meiosis in germ cells of both sexes—two active X chromosomes and fetal entry into meiosis in the female germ cell; zero active X chromosomes and postnatal entry into meiosis in the male germ cell. However, this correlation breaks down in the XX,*Sxr* mutant male mouse in which the XX germ cells are prevented from entering meiosis by the testicular environment but still reactivate their second X chromosome at a time analogous to that when ovarian XX germ cells reactivate their second X chromosome (Monk and McLaren, 1981). Thus, reactivation of the second X chromosome may be a necessary condition for fetal entry of germ cells into meiosis in females, but it is not a sufficient condition to stimulate fetal entry of XX germ cells into meiosis in males.

That X-chromosome function plays a critical role in the normal differentiation of male germ cells is indicated by the observation that any aberration in the number (more than one X chromosome) or structure (e.g., X-autosome translocations) of the X chromosome results in defective spermatogenesis (reviewed by Handel, 1987). This is true more often than with similar aberrations involving the Y chromosome or the autosomes, and typically affects gametogenesis more severely in males than in females. In the case of supernumerary X chromosomes in male germ cells, the observed effects occur before meiosis. In both XX,*Sxr* and XXY mice the primary defect is a failure of perinatal T-spermatogonia to undergo the burst of mitotic activity that normally occurs *before* the formation of Type A spermatogonia and their subsequent entry into meiosis (Burgoyne, 1978; McLaren, 1981). Whether this defect is related to an abnormality of X-chromosome activity, such as one X chromosome remaining transcriptionally active after the other has undergone the normal inactivation that appears to occur during male germ cell differentiation, cannot be assessed because it is currently not known when the single X chromosome in normal XY male germ cells becomes inactivated. Indeed, as pointed out by Handel (1987), the hypothesis that X-chromosome inactivation occurs in XY spermatogenic cells has come to be accepted as dogma in the absence of truly convincing data. Nonetheless, the presence of a second X chromosome in male germ cells is clearly detrimental since XX,*Sxr* germ cells do not survive while XO,*Sxr* germ cells are able to progress to postmeiotic stages of spermatogenesis (Cattanach et al., 1971). A complete understanding of the function of the X chromosome in the differentiation of the male germ cell will require more information regarding whether or not it does become completely inactivated, and, if so, when this occurs. Then, it may be possible to determine if there are specific X-linked genes whose aberrant expression contributes to the defective spermatogenesis observed in mutants with supernumerary X chromosomes as suggested by Forejt (1982).

D. Function of the Y Chromosome in Germ Cell Sex Differentiation

The initial influence of the Y chromosome on germ cell sex differentiation comes from its role in inducing the bipotential gonad to form a testis. Germ cells in a fetal testis do not enter meiosis and thus initiate male differentiation. There does not appear to be a requirement for a Y chromosome within the germ cells themselves to facilitate formation of T_1-prospermatogonia in a fetal testis, since both XX and XO germ cells can accomplish this (Burgoyne, 1987). The ability of at least some XO,*Sxr* germ cells to progress to postmeiotic stages (Cattanach et al., 1971), and the fact that all sperm-specific structures are assembled in these cells (Kot and Handel, 1990), suggest that there is also not a strict requirement for a complete Y chromosome throughout most of spermatogenesis. However, the *Sxr* mutation actually represents a portion of the Y chromosome that is carried on one of the X chromosomes in XX,*Sxr* individuals (Singh and Jones, 1982), and Burgoyne and co-workers (1986) proposed that this portion of the Y chromosome includes a gene *(Spy)* required for the proper differentiation of male germ cells through spermatogenesis.

The suggestion that the Y-linked *Spy* gene is required for spermatogenesis was based on the discovery of a variant of the *Sxr* mutation (McLaren et al., 1984) (the original mutation is now termed Sxr^a and the variant is Sxr^b, McLaren et al., 1988). As described above, the Sxr^a mutation results in the development of testes in an XX individual, and facilitates the development of at least some postmeiotic spermatids in XO individuals. The Sxr^b variant also causes development of testes in XX or XO individuals, but results in a significant impairment to spermatogenesis in XO individuals that is not seen in the XO,Sxr^a mutant (Burgoyne et al., 1986). This led Burgoyne and colleagues (1986) to propose the existence of the *Spy* gene that is carried by the Sxr^a mutation but is lacking in the Sxr^b mutant. Recently Sutcliffe and Burgoyne (1989) presented evidence that there is a reduction in mitotic activity among T_1-prospermatogonia in XO,Sxr^b mice, resulting in a shortage of T_2-prospermatogonia and undifferentiated Type A spermatogonia. They further noted a spermatogenic block during the early differentiating spermatogonial stages such that progression to intermediate or Type B spermatogonia did not occur. McLaren and colleagues (1984) noted that unlike the Sxr^a mutation, the Sxr^b mutation does not encode the H-Y antigen, and she suggested that the H-Y antigen gene *(Hya)* might have a function in normal spermatogenesis. Sutcliffe and Burgoyne (1989) pointed out that Sxr^b also lacks one of the two zinc-finger-encoding genes *(Zfy-2)* that is present on the mouse Y chromosome and carried by Sxr^a. The relationship between the *Spy, Hya,* and *Zfy-2* genes, whether one and the same, interacting, or independent, is yet to be established, as is the mechanism by which any or all of these genes and/or their products affect the differentiation of male germ cells.

In summary, the hypothesis that sexual differentiation of the germ cells is initiated according to the sexual phenotype of the gonads, but that the ability of the germ cells to respond to this influence is limited by their sexual genotype remains a valid representation of the evidence. However, this concept can now be stated more specifically. The sexual phenotype of the gonads regulates the initiation of sexual differentiation of the germ cells by determining whether the germ cells will enter meiosis during the fetal period and thus initiate oogenesis, or be prevented from entering meiosis at this stage, thus initiating spermatogenesis. The ability of the germ cells to respond to this impetus is initially undaunted, regardless of their genotype. However, the ability of germ cells in the testis to undergo normal spermatogenesis at perinatal, postnatal, and puberal stages is limited by their sexual genotype, requiring appropriate contributions from both the X and Y sex chromosomes.

VI. Differential Gene Expression in Developing Germ Cells

Despite the requirement to retain or reestablish genetic totipotency in the germ cell line, the development and differentiation of germ cells must proceed by a mechanism similar to that operating in all other cell types, namely that cellular differentiation is mediated by differential gene expression. That specific, differential gene expression is associated with the differentiation phase of the germ cell line during spermatogenesis is well established (Hecht, 1986). To the extent that germ cells undergo changes in morphology, cell surface characteristics, biochemical functions, cell movements, and other cell functions, there must be differential gene expression occurring during the developmental phase as well. Indeed, it is most likely that whatever mechanism maintains or reestablishes genetic totipotency in the germ cell line is also mediated by differential gene expression in these cells. Unlike the differentiation phase, however, there is very little known about differential gene expression during the development of germ cells in the mammal. This is an area that is clearly ripe for experimental inquiry.

A. Evidence for Differential Gene Expression in Developing Germ Cells

That differential gene expression does occur in developing germ cells is indicated by the presence of molecules that have been exploited as specific markers for PGCs, including alkaline phosphatase (Chiquoine, 1954) and various cell surface molecules that bind specific antibodies (Hahnel and Eddy, 1986; Wylie et al., 1985; Fox et al., 1981) or lectins (De Felici, 1984). However, no quantitative analysis of these molecules on a per cell basis, or any analysis of precursors or RNA transcripts that contribute to these molecules has been reported, thus precluding any insight into the mechanism(s) regulating the differential appearance of these molecules in germ cells as opposed to somatic cells.

Evans and colleagues (1979) carried out a study in which PGCs isolated from gonads at 12.5 days pc were labeled with ^{35}S-methionine to facilitate analysis of cellular proteins by two-dimensional gel electrophoresis. They compared germ cell proteins with those from ectoderm of 7 and 6-day pc embryos, endoderm of 7 and 6-day pc embryos, epiblast plus endoderm of 5-day pc embryos, and samples of embryonal carcinomas. Of over 1000 polypeptides resolved on the gels, 19 showed consistent variation among the samples studied and 18 of these were consistently absent in germ cells, including four that were present in *all* the other tissues. Unfortunately, this study failed to reveal any proteins that were made in PGCs but *not* in any of the other cell types. Nonetheless, a sensitive comparison of proteins in PGCs with those in other tissues and cell types still seems to represent a promising approach to identify germ cell-specific gene products.

At the biochemical level, Brinster and Harstad (1977) demonstrated differential activity in mouse germ cells at 15 days pc with respect to metabolism of pyruvate and glucose, and lactate dehydrogenase activity. They found that developing germ cells were twice as active as oocytes, but only two-thirds as active as the blastocyst in converting pyruvate to carbon dioxide. They also reported that the fetal germ cells, like the mature oocyte, have unusually high lactate dehydrogenase activity and contain a preponderance of B subunits.

The evidence discussed earlier that the normal allele at the *W* locus in mice acts in a cell-autonomous manner (Morrison-Graham and Weston, 1989) also indicates expression of a particular gene product in germ cells. As also noted previously, the identification of this locus as the source of the *c-kit* polypeptide and the availability of a cDNA clone for this gene (Geissler et al., 1988; Chabot et al., 1988) facilitated an analysis of the expression of the *c-kit* protooncogene at the RNA level in developing germ cells as reported by Matsui and colleagues (1990). Expression of *c-kit* in other cell types affected by the *W* mutation has recently been demonstrated (Nocka et al., 1989). Similarly, the availability of DNA sequence for alkaline phosphatase genes (Millan and Manes, 1988; Hahnel and Schultz, 1989) has allowed an analysis of the germ cell specificity of members of this gene family. Hahnel and coworkers (1990) used the polymerase chain reaction (PCR) to show that of three functional genes encoding alkaline phosphatase in the mouse genome, only one, the so-called tissue-non-specific (TNS) gene, is expressed in PGCs. Although the extent of information regarding differential gene expression in developing germ cells is currently limited, improved methods for isolating germ cells and studying gene expression should facilitate a significant expansion of knowledge in this area.

B. Methods of Isolating Developing Germ Cells

Obviously it is possible to characterize gene expression or any other biochemical or molecular phenomena occurring in germ cells only if sufficient material can be obtained to perform the necessary experiments. This has been a major hindrance in the study of all but morphological phenomena in developing germ cells.

Several methods to isolate or enrich for germ cells from fetal gonads have been reported. These methods all exploit one or more of the following characteristics of developing germ cells: (1) germ cells are frequently larger than surrounding somatic cells; (2) unlike gonadal somatic cells, germ cells tend to float in culture media and do not attach to most substrata; (3) germ cells tend to be less tightly bound to somatic cells than somatic cells are to one another within the gonads; and (4) germ cells display specific markers on their cell surface. Heath (1978) described a method (similar to that originally used by Blandau et al., 1963) involving compression of fetal gonads between a slide and coverslip to rupture the gonad and liberate the loosely bound germ cells that were then collected by pipette. They reported obtaining populations that were 75 to 95 percent pure with viabilities of 70 to 80 percent. De Felici and McLaren (1982) reported methods involving collagenase or EDTA-assisted dissociation of fetal gonads followed by size fractionation on a Percoll gradient. They obtained several thousand viable PGCs per 13.5-day pc gonad with 10 to 20 percent contamination. Gong and colleagues (1978) used a similar approach but relied on a 2 to 4 percent gradient of bovine serum albumin for separation of PGCs from somatic cells. They found 80 to 90 percent of the recovered cells had PGC-like morphology. Seidel and Jaenisch (1988) reported enrichment of PGCs to 20 to 50 percent purity by culturing dissociated fetal gonadal cells for 2 hours and then recovering the nonattached cells in the culture medium.

McCarrey and colleagues (1987) used antibody-directed flow sorting to isolate relatively large and pure populations of mouse PGCs. This method involved labeling dissociated gonadal cells from 12.5 day pc mouse embryos with the PGC-specific antibody, EMA-1 (Hahnel and Eddy, 1986). Subsequently, a secondary, fluorescinated antibody that binds to the EMA-1 primary antibody was added to render the EMA-1-labeled cells fluorescent. This population was then subjected to flow sorting on a fluorescence-activated cell sorter (FACS). The result was the acquisition of 50,000 to 250,000 cells per experiment (25 to 35 embryos dissected) of which approximately 99 percent were PGCs as determined independently by alkaline phosphatase activity (Fig. 3–5). This approach affords a method of obtaining reasonably large and pure populations of PGCs sufficient for biochemical or molecular analysis.

G. Methods of Analyzing Differential Gene Expression in Developing Germ Cells

As noted, the availability of methods to obtain sufficient material in conjunction with new biochemical and molecular methods for the analysis of differential gene expression should now allow a significant and rapid increase in our knowledge about differential gene expression in developing germ cells. Many improved methods are now available to detect specific gene products directly or indirectly at both the protein and RNA levels, as well as on the basis of specific cell functions, morphologies, or states.

The study of Evans and co-workers (1979) still stands as an excellent example of the sort of approach that must be followed. While not new, the analysis of proteins by two-dimensional gel electrophoresis is an effective method to look at differential gene expression at the protein level. It is unfortunate that further studies of this sort have not been reported. There have been other studies at the protein level in the form of characterization of specific cell surface molecules by the production of antibodies (e.g., Fox et al., 1981; Wylie et al., 1985; Hahnel and Eddy, 1986). In cases where such antibodies recognize a native protein molecule, the opportunity now exists to use the antibody as a probe to screen an expression cDNA library to obtain a cDNA clone of the gene responsible for the production of that protein (Young and Davis, 1983). Alternatively if a particular protein can be purified from PGCs and sufficient amino acid sequence information obtained, it is also possible to use this information to generate synthetic oligonucleotide probes that can be used to detect the corresponding cDNA clone in an appropriate library (Suggs et al., 1981).

The potential advantages of applying molecular biology to the study of differential gene expression in developing germ cells are manifold. The level of sensitivity is generally greater than that obtainable at the protein level. Thus,

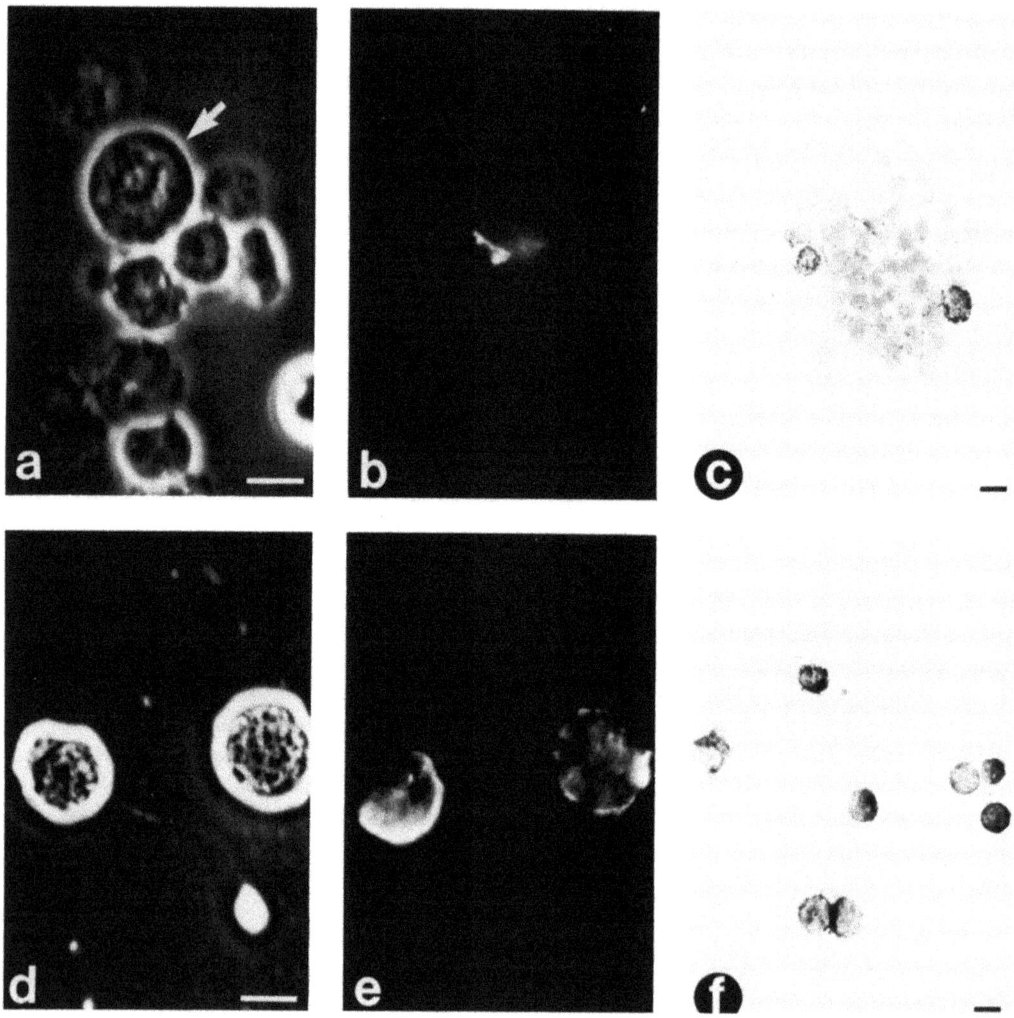

Figure 3–5. Isolation of primordial germ cells (PGCs) by antibody-directed flow sorting. (**a–c**) Before sorting. (**a**) Dispersed gonadal cells from a mouse fetus at 12.5 days pc viewed with phase optics. (**b**) The same cells as those in (**a**) viewed with ultraviolet illumination to reveal EMA-1-antibody-binding to one of the cells (equivalent cell noted with an arrow in (**a**). (**c**) Dispersed 12.5 days pc gonadal cells stained for alkaline phosphatase activity, showing approximately 10 percent positive cells. (**d–f**) After sorting. (**d**) Germ cells recovered from fluorescence-activated cell-sorting (FACS) viewed with phase optics. (**e**) The same cells as those in (**d**) viewed with ultraviolet illumination to reveal EMA-1-antibody-binding to both cells. (**f**) Germ cells recovered from FACS stained for alkaline phosphatase activity showing 95 to 100 percent positive cells. Bars = 50 μm.

for example, the number of gene products represented in a cDNA library is typically much greater than that resolvable by two-dimensional gel electrophoresis. Furthermore, once a cDNA clone representing the RNA transcript for a particular gene product is obtained, significant information about both the corresponding protein product and the gene can be readily pursued. Nucleotide sequencing of the cDNA clone can yield information about the amino acid sequence that can frequently provide identifying information about the protein itself. Alternatively by using the cDNA as a probe to screen a genomic DNA library, a clone of the actual gene for a particular product can be obtained. This then opens the door to studies of how differential expression of the gene, at least at the transcriptional level, is regulated.

Studies of the regulation of differential gene expression can be accomplished by several approaches. The isolation of a genomic clone for a particular gene that is known to be differentially expressed provides access to the 5′-regulatory or promoter sequences, which are typically involved in the regulation of differential transcription of so-called "tissue-specific" or more accurately "cell-specific" genes. A cDNA clone of a particular gene product allows an analysis of the differential presence, processing, and/or turnover of the corresponding RNA transcribed from a particular gene. This facilitates an analysis of whether the control of differential expression of that particular gene is exerted primarily at the transcriptional or posttranscriptional level.

Unfortunately, at present, with the exception of the very short list discussed previously, there are very few genes that are known, or even suspected of being differentially expressed in developing germ cells. How can additional candidates for genes that are differentially expressed in developing germ cells be identified? One particularly powerful approach afforded by molecular biology is differential screening of cDNA libraries (Sargent and Dawid, 1983). Inasmuch as a cDNA library represents a collection of the transcripts from genes expressed in the particular tissue or cell type from which RNA was isolated to generate the cDNA library, clones present in a library from one tissue but not in a library from another will represent genes that are differentially expressed in the former tissue. This comparison can be achieved by what have become routine hybridization techniques (see Sargent, 1987, for example). In the case of developing germ cells, the difficulty comes in generating a specific cDNA library as this requires a significant amount of RNA isolated from a purified population of germ cells. However, the techniques now available for the isolation of large numbers of PGCs (McCarrey et al., 1987), as well as techniques for the production of cDNA libraries from minimal amounts of starting material (Weng et al., 1989) now render this task feasible. This approach has been initiated in the author's laboratory.

Although the paucity of cell numbers during the developmental phase of the germ line is a hindrance to any investigation using molecular and/or biochemical techniques, the relative abundance of different, well-characterized stages in both the developmental and differentiation phases is an asset. From the standpoint of investigating and ultimately understanding the mechanisms that contribute to the development and differentiation of any lineage, the ability to access and study multiple stages in this process is a distinct advantage. This allows changes in the differentiated state of the germ cell, differential gene expression within the germ cell, and any molecular mechanism(s) predisposing differential gene expression within the germ cell to be followed and correlated.

As noted in the introduction to this chapter, the germ cell line is one of the best characterized lineages in the mammal (Table 3–1). Of particular importance for any study of the mechanisms involved in regulating the development and differentiation of the germ cell lineage is the fact that many of these stages are accessible so that reasonably large and pure populations of germ cells can be obtained for molecular studies. A summary of the isolatable stages in the male germ cell line is presented in Table 3–3. Included are stages from both the developmental and differentiation phases. With advances such as the polymerase chain reaction to amplify specific nucleic acid sequences (Saiki et al., 1985), even very small numbers of cells may now be sufficient to facilitate investigation at the molecular level. The opportunity to apply the growing body of knowledge about growth factors, their receptors, transcription factors, and other regulatory genes, along with the tools of molecular biology, has opened a new era in the study of the mechanisms governing PGC development.

VII. Conclusions

A large number of investigations throughout this century have led to what is now a fairly complete description of the embryology of germ cell development. As with most investigations, however, each new piece of information has raised new questions. In this case the questions that remain are predominantly in the area of mechanisms that regulate the origin,

Table 3–3. Isolatable Stages in the Germ Cell Line of the Male Mouse

Age/source	Cell type	Method of isolation
8–10d Embryo/Hindgut	PGC	FACS[6]
11–13d Embryo/Gonads	PGC/M-Prospermatogonia	FACS[6]
14–20d Fetus/Testis	T_1-Prospermatogonia	Dispersal/Gradient[7]
New-born Pup/Testis	T_2-Prospermatogonia	Dispersal/Gradient[7] Staput[8]
6d Pup/Testis	Primitive type A gonia	
8d Pup/Testis	Adult type A gonia	Staput[8]/Elutriation[10]
8d Pup/Testis	Adult type B gonia	Staput[8]/Elutriation[10]
16–18d Pup/Testis	Pre-Lep[1] Spermatocytes	Staput[8]/Elutriation[10]
16–18d Pup/Testis	Lep/Zyg[2] Spermatocytes	Staput[8]/Elutriation[10]
16–18d Pup/Testis	Early Pach[3] Spermatocytes	Staput[8]/Elutriation[10]
Adult/Testis	Late Pach[3] Spermatocytes	Staput[9]/Elutriation[10]
Adult/Testis	Round spermatids	Staput[9]/Elutriation[10]
Adult/Testis	Residual bodies	Staput[9]/Elutriation[10]
Adult/Testis	Condensing spermatids	Elutriation[10]
Adult/Epidiymis	Caput Epididymidal Sperm[4]	Dissection/Extrusion[11]
Adult/Epidiymis	Cauda Epididymidal Sperm[5]	Dissection/Extrusion[11]

[1]Preleptotene of meiosis I.
[2]Leptotene/zygotene of meiosis I.
[3]Pachytene of meiosis I.
[4]Caput sperm—immotile, not fertile, carry cytoplasmic droplet.
[5]Cauda sperm—motile, fertile, no cytoplasmic droplet (equivalent to ejaculated sperm).
[6]McCarrey et al., 1987.
[7]De Felici and McLaren, 1982.
[8]Bellve et al., 1977.
[9]Romrell et al., 1976.
[10]Meistrich et al., 1981.
[11]Cornwall et al., 1986.

development, and differentiation of the germ cell line in both the male and female mammal. The era of descriptive studies and classic experimental embryology studies would seem to be poised to yield to a new era of molecular and biochemical studies that hopefully will provide insight into regulatory mechanisms. However, as this transition is made, it is of critical importance that the knowledge and approaches from embryology not be abandoned or ignored, but rather used as a foundation on which future molecular studies can be built. If this can be done, the male germ cell line provides an opportunity for significant inroads into greater understanding of the mechanisms that regulate development in general, as well as those that are unique to the development of the germ cell line itself.

ACKNOWLEDGMENTS

The author is grateful to L. Urven, B. Migeon, and M. A. Handel for sharing upublished data; L. Urven, M. A. Handel, and E. M. Eddy for critical reading of the manuscript and valuable suggestions; and K. Born for art work. The author is a recipient of a career development award from the NIH (HD 00829), and acknowledges support from research grants from the NIH (HD 23126), the NSF (BMB 8904741), and the USDA (CRCR-1-2301), a program project grant from the NIH (AG 08321), and a population center grant from the NIH (HD 06308).

REFERENCES

Beams, H. W., and Kessel, R. G. (1974). The problem of germ cell determinants. Int. Rev. Cytol. 39:413–479.

Beaudoin, A. R. (1980). Embryology and teratology, in *The Laboratory Rat* (Baker, H. J., Lindsey, J. R., and Weisbroth, S. H., eds.) Vol. II, pp. 75–101, Academic Press, New York.

Beaumont, H. M., and Mandl, A. M. (1963). A quantitative study of primordial germ cells in the male rat. J. Embryol. Exp. Morph. 11:715–740.

Beddington, R. (1983). The origin of the foetal tissues during gastrulation in the rodent, in *Development in Mammals* (Johnson, M. H., ed.) Vol. 5, pp. 1–32, Elsevier, New York.

Bellve, A. R., Cavicchia, J. C., Millette, C. F., O'Brien, D. A., Bhatnagar, Y. M., and Dym, M. (1977). Spermatogenic cells of the prepuberal mouse, isolation and morphological characterization. J. Cell Biol. 74:68–85.

Bennett, D. (1956). Developmental analysis of a mutation with pleiotrophic effects in the mouse. J. Morphol. 98:199–233.

Blandau, R. J., White, B. J., and Rumery, R. E. (1963). Observations on the movements of the living primordial germ cells in the mouse. Fertil. Steril. 14:482–489.

Brinster, R. L., and Harstad, H. (1977). Energy metabolism in primordial germ cells of the mouse. Exp. Cell Res. 109:111–117.

Burgoyne, P. S. (1987). The role of the mammalian Y chromosome in spermatogenesis. Development 101 (Suppl):133–141.

Burgoyne, P. S. (1978). The role of the sex chromosomes in mammalian germ cell differentiation. Ann. Biol. Anim. Bioch. Biophys. 18:317–325.

Burgoyne, P. S., Buehr, M., Koopman, P., Rossant, J., and McLaren, A. (1988). Cell-autonomous action of the testis-determining gene: Sertoli cells are exclusively XY in XX–XY chimaeric mouse testes. Development 102:443–450.

Burgoyne, P. S., Levy, E. R., and McLaren, A. (1986). Spermatogenic failure in male mice lacking H-Y antigen. Nature 320:170–172.

Byskov, A. G. (1978). Regulation of initiation of meiosis in fetal gonads. Int. J. Androl. (Suppl) 2:29–38.

Byskov, A. G., and Saxen, L. (1976). Induction of meiosis in fetal mouse testis in vitro. Devel. Biol. 52:193–200.

Cattanach, B. M., Pollard, C. E., and Hawkes, S. G. (1971). Sex-reversed mice: XX and XO males. Cytogenetics 10:318–337.

Chabot, B., Stephenson, D. A., Chapman, V. M., Besmer, P., and Bernstein, A. (1988). The proto-oncogene c-kit encoding a transmembrane tyrosine kinase receptor maps to the mouse W locus. Nature 335:88–89.

Chapman, V., Forrester, L., Sanford, J., Hastie, N., and Rossant, J. (1984). Cell lineage-specific undermethylation of mouse repetitive DNA. Nature 307:284–286.

Chiquoine, A. D. (1954). The identification, origin and migration of the primordial germ cells in the mouse embryo. Anat. Rec. 118:135–146.

Clark, J. M., and Eddy, E. M. (1975). Fine structural observations on the origin and association of primordial germ cells of the mouse. Dev. Biol. 47:136–155.

Clermont, Y., and Perey, B. (1957). Quantitative study of the cell population of the seminiferous tubules in immature rats. Am. J. Anat. 100:241–267.

Copeland, N. G., Gilbert, D. J., Cho, B. C., Donovan, P. J., Jenkins, N. A., Cosman, D., Anderson, D., Lyman, S. D., and Williams, D. E. (1990). Mast cell growth factor maps near the steel locus on mouse chromosome 10 and is deleted in a number of steel alleles. Cell 63:175–183.

Cornwall, G. A., Smyth, T. B., Vindivich, D., Harter, C., Robinson, J., and Chang, T.S.K. (1986). Induction and enhancement of progressive motility in hamster caput epididymal spermatozoa. Biol. Reprod. 35:1065–1074.

De Felici, M. (1984). Binding of fluorescent lectins to the surface of germ cells from fetal and early postnatal mouse gonads. Gamete Res. 10:423–432.

De Felici, M., and McLaren, A. (1983). In vitro culture of mouse primordial germ cells. Exp. Cell Res. 144:417–427.

De Felici, M., and McLaren, A. (1982). Isolation of mouse primordial germ cells. Exp. Cell Res. 142:476–482.

Dierickx, P., and Verhoeven, G. (1980). Effect of different methods of germinal cell destruction on rat testis. J. Reprod. Fert. 59:5–9.

Doerfler, W. (1983). DNA methylation and gene activity. Am. Rev. Biochem. 52:93–124.

Donovan, P. J., Stott, D., Cairns, L. A., Heasman, J., and Wylie, C. C. (1986). Migratory and postmigratory mouse primordial germ cells behave differently in culture. Cell 44:831–838.

Driscoll, D. J., and Migeon, B. R. (1990). Sex difference in methylation of single-copy genes in human meiotic germ cells: implications for X chromosome inactivation, parental imprinting, and origin of CpG mutations. Somatic Cell Mol. Genet. 16:267–282.

Dubois, R. (1968). La colonisation des ébauches gonadiques par les cellules germinales de l'embryon de poulet, en culture in vitro. J. Embryol. Exp. Morph. 20:189–213.

Eddy, E. M. (1975). Germ plasm and the differentiation of the germ cell line, in *International Review of Cytology* (Bourne, G. H., and Danielli, J. F., eds.) Vol. 43, pp. 229–280, Academic Press, New York.

Eddy, E. M., and Hahnel, A. C. (1983). Establishment of the germ cell line in mammals, in *Current Problems in Germ Cell Differentiation,* Symposium of British Society for Developmental Biology (McLaren, A., and Wylie, C. C., eds.) pp. 41–69, Cambridge University Press, Cambridge.

Eddy, E. M., Clark, J. M., Gong, D., and Fenderson, B. A. (1981). Origin and migration of primordial germ cells in mammals. Gamete Res. 4:333–362.

Eicher, E. M., and Washburn, L. L. (1986). Genetic control of primary sex determination in mice. Ann. Rev. Genet. 20:327–360.

Evans, M. J., Lovell-Badge, R. H., Stern, P. L., and Stinnakre, M. G. (1979). Cell lineages of the mouse embryo and embryonal carcinoma cells; Forssman antigen distribution and patterns of protein synthesis, in *Cell Lineage, Stem Cells* and *Cell Determination,* INSERM Symposium No. 10 (Le Dourarin, N., ed.) pp. 115–129, Elsevier/North Holland Biomed. Press.

Eyal-Giladi, H., Ginsburg, M., and Farbarov, A. (1981). Avian primordial germ cells are of epiblastic origin. J. Embryol. Exp. Morph. 65:139–147.

Falin, L. I. (1969). The development of genital glands and the origin of germ cells in human embryogenesis. Acta Anat. 72:195–232.

Ford, C. E., Evans, E. P., Burtenshaw, M. D., Clegg, H. M., Tuffrey, M., and Barnes, R. D. (1975). A functional "sex-reversed" oocyte in the mouse. Proc. R. Soc. Lond. B. 190:187–197.

Forejt, J. (1982). X-Y involvement in male sterility caused by autosome translocations—A hypothesis, in *Genetic Control of Gamete Production and Function* (Crosignani, P. G., and Rubin, B. L., eds.) pp. 135–151, Academic Press, New York.

Fox, N., Damjanov, I., Martinez-Hernandez, A., Knowles, B. B., and Solter, D. (1981). Immunohistochemical localization of the early embryonic antigen (SSEA-1) in postimplantation mouse embryos and fetal and adult tissues. Dev. Biol. 83:391–398.

Fujimoto, T., Yoshinaga, K., and Kono, I. (1985). Distribution of fibronectin on the migratory pathway of primordial germ cells in mice. Anat. Rec. 211:271–278.

Gardner, R. L. (1982). Investigation of cell lineage and differentiation in the extraembryonic endoderm of the mouse embryo. J. Embryol. Exp. Morph. 68:175–198.

Gardner, R. L. (1977). Developmental potency of normal and neoplastic cells of the early mouse embryo, in *Birth Defects* (Littlefield and de Grouchy, eds.) Exerpta Med. Int. Congr. Ser., No. 432, pp. 154–166.

Gardner, R. L. (1975). Analysis of determination and differentiation in the mammalian embryo using intra- and inter-species chimeras, in *The Developmental Biology of Reproduction,* 33rd Symposium of the Society for Developmental Biology (Markert, C. L., ed.) pp. 207–236, Academic Press, New York.

Gardner, R. L., Lyon, M. F., Evans, E. P., and Burtenshaw, M. D. (1985). Clonal analysis of X-chromosome inactivation and the origin of the germ line in the mouse embryo. J. Embryol. Exp. Morph. 88:349–363.

Gardner, R. L., and Rossant, J. (1979). Investigation of the fate of 4.5 day post-coitum mouse inner cell mass cells by blastocyst injection. J. Embryol. Exp. Morph. 52:141–152.

Gartler, S. M., and Riggs, A. D. (1983). Mammalian X-chromosome inactivation. Ann. Rev. Genet. 17:155–190.

Gartler, S. M., Andina, R., and Gant, N. (1975). Ontogeny of X-chromosome inactivation in the female germ line. Exp. Cell Res. 91:454–457.

Geissler, E. N., McFarland, E. C., and Russell, E. S. (1981). Analysis of pleiotropism at the dominant white-spotting (W) locus of the house mouse: a description of ten new W alleles. Genetics 97:339–361.

Geissler, E. N., Ryan, M. A., and Housman, D. E. (1988). The dominant-white spotting (W) locus of the mouse encodes the c-kit proto-oncogene. Cell 55:185–192.

Ginsburg, M., Snow, M.H.L., and McLaren, A. (1990). Primordial germ cells in the mouse embryo during gastrulation. Development 110:521–528.

Godin, I., Wylie, C., and Heasman, J. (1990). Genital ridges exert long-range effects on mouse primordial germ cell numbers and direction of migration in culture. Development 108:357–363.

Gong, D., Kratzer, P. G., and Eddy, E. M. (1978). Isolation and characterization of germ cells from mouse embryos. J. Cell Biol. 75:171a.

Gordon, J. W., and Ruddle, F. H. (1981). Mammalian gonadal determination and gametogenesis. Science 211:1265–1271.

Gubbay, J., Collignon, J., Koopman, P., Capel, B., Economou, A., Munsterberg, A., Vivian, N., Goodfellow, P., and Lovell-Badge, R. (1990). A gene mapping to the sex-determining region of the mouse Y chromosome is a member of a novel family of embryonically expressed genes. Nature 346:245–250.

Gurdon, J. B. (1962). Adult frogs derived from the nuclei of single somatic cells. Devel. Biol. 4:256–273.

Hahnel, A. C., and Schultz, G. A. (1989). Cloning and characterization of a cDNA encoding alkaline phosphatase in mouse embryonal carcinoma cells. Clinica Chimica Acta, in press.

Hahnel, A. C., and Eddy, E. M. (1986). Cell surface markers of mouse primordial germ cells defined by two monoclonal antibodies. Gamete Res. 15:25–34.

Hahnel, A. C., Rappolee, O. A., Millan, J. L., Manes, T., Ziomek, C. A., Theodosiou, N. G., Werb, Z., Pedersen, R. A., and Schultz, G. A. (1990). Two alkaline phosphatase genes are expressed during early development in the mouse embryos. Development 110:555–564.

Handel, M. A. (1987). Genetic control of spermatogenesis in mice, in *Results and Problems in Cell Differentiation* (W. Hennig, ed.) Vol. 15, pp. 1–62, Springer-Verlag, Berlin.

Handel, M. A., and Eppig, J. J. (1979). Sertoli cell differentiation in the testes of mice genetically deficient in germ cells. Biol. Reprod. 20:1031–1038.

Hay, B., Ackerman, L., Barbel, S., Jan, L. Y., and Jan, U. N. (1988). Identification of a component of *Drosophila* polar granules. Development 103:625–640.

Heath, J. K. (1978). Mammalian primordial germ cells, in *Development in Mammals* (Johnson, M. H., ed.) Vol. 3, pp. 267–298, North Holland, New York.

Hecht, N. B. (1986). Regulation of gene expression during mammalian spermatogenesis, in *Experimental Approaches to Mammalian Embryonic Development* (Rossant, J., and Pedersen, R. A., eds.) pp. 151–193, Cambridge University Press, New York.

Hilscher, B., Hilscher, W., Bulthoff-Ohnolz, B., Kramer, U., Birke, A., Pelzer, H., and Gauss, G. (1974). Kinetics of gametogenesis: I. Comparative histological and autoradiographic studies of oocytes and transitional prospermatogonia during oogenesis and prespermatogenesis. Cell Tiss. Res. 154:443–470.

Hogan, B., Constantini, F., and Lacy, E. (1986). *Manipulating the Mouse Embryo, A Laboratory Manual,* Cold spring Harbor Laboratory, Cold spring Harbor, N.Y.

Ikenshi, K. (1987). Functional gametes derived from explants of single blastomeres containing the "germ plasm" in *Xenopus* laevis: A genetic study. Devel. Biol. 122:35–38.

Illmensee, K., and Mahowald, A. P. (1974). Transplantation of posterior pole plasm in *Drosophila.* Induction of germ cells at the anterior pole of the egg. Proc. Natl. Acad. Sci. (USA) 7:1016–1020.

Jirasek, J. E. (1976). Principles of reproductive embryology, in *Disorders of Sexual Differentiation* (Simpson, J. L., ed.) pp. 51–110, Academic Press, New York.

Kemper, C. H., and Peters, P.W.J. (1987). Migration and proliferation of primordial germ cells in the rat. Teratology 36:117–124.

Keshet, I., Lieman-Hurwitz, J., and Cedar, H. (1986). DNA methylation affects the formation of active chromatin. Cell 44:535–543.

Kot, M. C., and Handel, M. A. (1990). Spermatogenesis in XO, *Sxr* mice: Role of the Y chromosome. J. Exp. Zool. 256:92–105.

Kramer, J. M., and Erickson, R. P. (1981). Developmental program of PGK-1 and PGK-2 isozymes in spermatogenic cells of the mouse: Specific activities and rates of synthesis. Devel. Biol. 87:37–45.

Kuwana, T., Miyayama, Y., Kajiwara, Y., and Fujimoto, T. (1987). Behavior of chick primordial germ cells moving toward gonadal primordium in vitro: Scanning electron microscopic study. Anat. Rec. 219:164–170.

Kuwana, T., and Fujimoto, T. (1983). Active locomotion of human primordial germ cells in vitro. Anat. Rec. 205:21–26.

Lock, L. F., Takagi, N., and Martin, G. R. (1987). Methylation of the Hprt gene on the inactive X occurs after chromosome inactivation. Cell 48:39–46.

Mahowald, A. P., and Boswell, R. E. (1983). Germ plasm and germ cell development in invertebrates, in *Current Problems in Germ Cell Differentiation,* Symposium of British Society for Developmental Biology (McLaren, A., and Wylie, C. C., eds.) pp. 3–17, Cambridge University Press, Cambridge.

Matsui, Y., Zsebo, K. M., and Hogan, B.L.M. (1990). Embryonic expression of a haematopoietic growth factor encoded by the Sl locus and the ligand for c-kit. Nature 347:667–669.

McCarrey, J. R., Hsu, K. C., Eddy, E. M., Klevecz, R. R., and Bolen, J. L. (1987). Isolation of viable mouse primordial germ cells by antibody-directed flow sorting. J. Exp. Zool. 242:107–111.

McCarrey, J. R., and Abbott, U. K. (1982). Functional differentiation of chick gonads following depletion of primordial germ cells. J. Embryol. Exp. Morph. 68:161–174.

McCarrey, J. R., and Abbott, U. K. (1979). Mechanisms of genetic sex determination, gonadal sex differentiation, and germ-cell development in animals. Adv. Genet. 20:217–290.

McCarrey, J. R., and Abbott, U. K. (1978). Chick gonad differentiation following excision of primordial germ cells. Devel. Biol. 66:256–265.

McClelland, M. (1983). The effect of site specific methylation on restriction endonuclease cleavage (update). Nuc. Acids Res. 11:r169–r173.

McGrath, J., and Solter, D. (1984). Completion of mouse embryogenesis requires both the maternal and paternal genomes. Cell 37:179–183.

McLaren, A. (1988). The developmental history of female germ cells in mammals. Oxford Rev. Reprod. Biol. 10:162–179.

McLaren, A. (1985a). Mouse germ cells: Fertilization to birth. Arch. Anat. Micr. Morphol. Exper. 74:5–9.

McLaren, A. (1985b). Relation of germ cell sex to gonadal differentiation, in *The Origin and Evolution of Sex* (Halvorson, H., and Monroy, A., eds.) pp. 289–300, Alan R. Liss, Inc., New York.

McLaren, A. (1984). Meiosis and differentiation of mouse germ cells, in *Controlling Events in Meiosis,* 38th Symposium of the Society for Experimental Biology (Evans, C. W., and Dickinson, H. G., eds.) pp. 7–23, Company of Biologists, Cambridge.

McLaren, A. (1983). Primordial germ cells in mice. Biblthca. Anat. 24:59–66.

McLaren, A. (1981). The fate of germ cells in the testis of fetal sex-reversed mice. J. Reprod. Fert. 61:461–467.

McLaren, A., Simpson, E., Epplen, J. T., Studer, R., Koopman, P., Evans, E. P., and Burgoyne, P. S. (1988). Location of the genes controlling H-Y antigen expression and testis determination on the mouse Y chromosome. Proc. Natl. Acad. Sci. (USA) 85:6642–6645.

McLaren, A., Simpson, E., Tomonari, K., Chandler, P., and Hogg, H. (1984). Male sexual differentiation in mice lacking H-Y antigen. Nature 312:552–555.

McMahon, A., Fosten, M., and Monk, M. (1981). Random X-chromosome inactivation in female primordial germ cells in the mouse. J. Embryol. Exp. Morph. 64:251–258.

Meistrich, M. L., Longtin, J., Brock, W. A., Grimes, S. R., and Mace, M. L. (1981). Purification of rat spermatogenic cells and preliminary biochemical analysis of these cells. Biol. Reprod. 25:1065–1077.

Merchant, H. (1975). Rat gonadal and ovarian organogenesis with and without germ cells. An ultrastructural study. Devel. Biol. 44:1–21.

Merchant-Larios, H., and Coello, J. (1979). The effect of busulfan on rat primordial germ cells at the ultrastructural level. Cell Differ. 8:145–155.

Migeon, B. R., Axelman, J., and Beggs, A. H. (1988). Effect of aging on reactivation of the human X-linked HPRT locus. Nature 335:93–96.

Millan, J. L., and Manes, T. (1988). Seminoma-derived Nagao isozyme is encoded by a germ-cell alkaline phosphatase gene. Proc. Natl. Acad. Sci. (USA) 85:3024–3028.

Mintz, B., and Russell, E. S. (1957). Gene-induced embryological modifications of primordial germ cells in the mouse. J. Exp. Zool. 134:207–237.

Monk, M. (1987). Genomic imprinting: Memories of mother and father. Nature 328:203–204.

Monk, M. (1981). A stem-line model for cellular and chromosomal differentiation in early mouse development. Differentiation 19:71–76.

Monk, M., Boubelik, M., and Lehnert, S. (1987). Temporal and regional changes in DNA methylation in the embryonic, extraembryonic and germ cell lineages during mouse embryo development. Development 99:371–382.

Monk, M., and McLaren, A. (1981). X-chromosome activity in fetal germ cells of the mouse. J. Embryol. Exp. Morph. 63:75–84.

Morrison-Graham, K., and Weston, J. A. (1989). Mouse mutants provide new insights into the role of extracellular matrix in cell migration and differentiation. Trends in Genet. 5:116–121.

Mystkowska, E. T., and Tarkowski, A. K. (1968). Observations on CBA-p/CBA-T6T6 mouse chimaeras. J. Embryol. Exp. Morph. 20:33–52.

Nakayama, H., Kuroda, H., Onoue, H., Fujita, J., Nishimune, Y., Matsumoto, K., Nagano, T., Suzuki, F., and Kitamura, Y. (1988). Studies of Sl/Sl^d—+/+ mouse aggregation chimaeras. II. Effect of the steel locus on spermatogenesis. Development 102:117–126.

Naveh-Many, T., and Cedar, H. (1981). Active gene sequences are undermethylated. Proc. Natl. Acad. Sci. (USA) 78:4246–4250.

Nebel, B. R., Amarose, A. P., and Hackett, E. M. (1961). Calendar of gametogenic development in the prepuberal male mouse. Science 134:832–833.

Nocka, K., Majumder, S., Chabot, B., Ray, P., Cervone, M., Bernstein, A., and Besmer, P. (1989). Expression of *c-kit* gene products in known cellular targets of *W* mutations in normal and *W* mutant mice—Evidence for an impaired *c-kit* kinase in mutant mice. Genes Develop. 3:816–826.

Ozdzenski, W. (1967). Observations on the origin of primordial germ cells in the mouse. Zool. Pol. 17:367–379.

Page, D. C., Mosher, R., Simpson, E. M., Fisher, E.M.C., Mardon, G., Pollack, J., McGillivray, G., de la Chapelle, A., and Brown, L. G. (1987). The sex-determining region of the human Y chromosome encodes a finger protein. Cell 51:1091–1104.

Ponzetto-Zimmerman, C., and Wolgemuth, D. J. (1984). Methylation of satellite sequences in mouse spermatogenic and somatic DNAs. Nuc. Acids Res. 12:2807–2822.

Qiu, F., Ray, P., Brown, K., Barker, P. E., Jhanwar, S., Ruddle, F. H., and Besmer, P. (1988). Primary structure of c-kit: Relationship with the CSF-1/PDGF receptor kinase family—Oncogenic activation of v-kit involves deletion of extracellular domain and c-terminus. EMBO J. 7:1003–1011.

Rahe, B., Erickson, R. P., and Quinto, M. (1983). Methylation of unique sequence DNA during spermatogenesis in mice. Nuc. Acids Res. 11:7947–7961.

Razin, A., and Riggs, A. D. (1980). DNA methylation and gene function. Science 210:604–610.

Reik, W., Collick, A., Norris, M. L., Barton, S. C., and Surani, M. A. (1987). Genomic imprinting determines methylation of parental alleles in transgenic mice. Nature 328:248–251.

Rogulska, T., Ozdzenski, W., and Komer, A. (1971). Behavior of mouse primordial germ cells in the chick embryo. J. Embryol. Exp. Morph. 25:155–164.

Romrell, L. J., Bellve, A. R., and Fawcett, D. W. (1976). Separation of mouse spermatogenic cells by sedimentation velocity. A morphological characterization. Devel. Biol. 49:119–131.

Rossant, J. (1987). Cell lineage analysis in mammalian embryogenesis. Curr. Topics Devel. Biol. 23:115–146.

Rossant, J. (1985). Interspecific cell markers and lineage in mammals. Phil Trans. R. Soc. (Lond) B 312:91–100.

Saiki, R. K., Scharf, S., Faloona, F., Mullis, K. B., Horn, G. T., Erlich, H. A., and Arnheim, N. (1985). Enzymatic amplification of B-globin genomic sequences and restriction site analysis for diagnosis of sickle cell anemia. Science 230:1350–1354.

Sanford, J. P., Clark, H. J., Chapman, V. M., and Rossant, J. (1987). Differences in DNA methylation during oogenesis and spermatogenesis and their persistence during early embryogenesis in the mouse. Genes Devel. 1:1039–1046.

Sanford, J. P., Forrester, L., and Chapman, V. (1984). Methylation patterns of repetitive DNA sequences in germ cells of Mus musculus. Nuc. Acids Res. 12:2823–2836.

Sapienza, C., Peterson, A. C., Rossant, J., and Balling, R. (1987). Degree of methylation of transgenes is dependent on gamete of origin. Nature 328:251–254.

Sargent, T. D. (1987). Isolation of differentially expressed genes, in *Guide to Molecular Cloning Techniques* (Berger, S. L., and Kimmel, A. R., eds.). Methods in Enzymology, Vol. 152, pp. 423–432, Academic Press, New York.

Sargent, T. D., and Dawid, I. B. (1983). Differential gene expression in the gastrula of *Xenopus* laevis. Science 222:135–139.

Seidel, G. E., and Jaenisch, R. (1988). A rapid method for isolating murine primordial germ cells. Biol. Reprod. 38(Suppl. No. 1):129.

Sinclair, A. H., Berta, P., Palmer, M. S., Hawkins, J. R., Griffiths, B. L., Smith, M. J., Foster, J. W., Frischauf, A.-M., Lovell-Badge, R., and Goodfellow, P. N. (1990). A gene from the human sex-determining region encodes a protein with homology to a conserved DNA-binding motif. Nature 346:340–344.

Singer-Sam, J., Grant, M., LeBon, J. M., Okuyama, K., Chapman, V., Monk, M., and Riggs, A. D. (1990). Use of a HpaII-polymerase chain reaction assay to study DNA methylation in the Pgk-1 CpG island of mouse embryos at the time of X-chromosome inactivation. Mol. Cell. Biol. 10:4987–4989.

Singh, L., and Jones, K. W. (1982). Sex reversal in the mouse *(Mus musculus)* is caused by a recurrent nonreciprocal crossover involving the X and an aberrant Y chromosome. Cell 28:205–216.

Smith, L. D., and Williams, M. A. (1975). Germinal plasm and determination of the primordial germ cells, in *The Developmental Biology of Reproduction,* 33rd Symposium of the Society for Developmental Biology (Markert, C. L., and Papaconstantinou, J., eds.) pp. 3–24, Academic Press, New York.

Snow, M.H.L. (1981). Autonomous development of parts isolated from primitive streak-stage mouse embryos. Is development clonal? J. Embryol. Exp. Morph. 65(Suppl):269–287.

Snow, M.H.L., and Monk, M. (1983). Emergence and migration of mouse primordial germ cells, in *Current Problems in Germ Cell Differentiation,* Symposium of British Society for Developmental Biology (McLaren, A., and Wylie, C. C., eds.) pp. 115–135, Cambridge University Press, Great Britain.

Soriano, P., and Jaenisch, R. (1986). Retroviruses as probes for mammalian development: Allocation of cells to the somatic and germ cell lineages. Cell 46:19–29.

Spiegelman, M., and Bennett, D. (1973). A light- and electron-microscopic study of primordial germ cells in the early mouse embryo. J. Embryol. Exp. Morph. 30:97–118.

Suggs, S. V., Wallace, R. B., Hirose, T., Kawashima, E. H., and Itakura, K. (1981). Use of synthetic oligonucleotides as hybridization probes: isolation of cloned cDNA sequences for human B_2-microglobin. Proc. Natl. Acad. Sci. (USA) 78:6613–6617.

Surani, M.A.H. (1986). Evidences and consequences of differences between maternal and paternal genomes during embryogenesis in the mouse, in *Experimental Approaches to Mammalian Embryonic Development* (Rossant, J., and Pedersen, R. A., eds.) pp. 401–435, Cambridge University Press, New York.

Surani, M.A.H., Barton, S. C., Norris, M. L. (1987). Influence of parental chromosomes on spatial specificity in androgenic-parthenogenetic chimaeras in the mouse. Nature 326:395–397.

Sutcliffe, M. J., and Burgoyne, P. S. (1989). Analysis of the testes of H-Y negative XOSxr^{b} mice suggests that the spermatogenesis gene (Spy) acts during the differentiation of the A spermatogonia. Development 107:373–380.

Tam, P.P.L., and Snow, M.H.L. (1981). Proliferation and migration of primordial germ cells during compensatory growth in the mouse embryo. J. Embryol. Exp. Morph. 64:133–147.

Tsunoda, Y., Tokunaga, T., Imai, H., and Uchida, T. (1989). Nuclear transplantation of male primordial germ cells in the mouse. Development 107:407–411.

Upadhyay, S., and Zamboni, L. (1982). Ectopic germ cells: Natural model for the study of germ cell sexual differentiation. Proc. Natl. Acad. Sci. (USA) 79:6584–6588.

Urven, L. E., Erickson, C. A., Abbott, U. K., and McCarrey, J. R. (1988). Analysis of germ line development in the chick embryo using an anti-mouse EC cell antibody. Development 103:299–304.

Wakahara, M. (1978). Induction of supernumery primordial germ cells by injecting vegetal pole cytoplasm into *Xenopus* eggs. J. Exp. Zool. 203:159–164.

Weng, D. E., Morgan, R. A., and Gearheart, J. D. (1989). Estimates of mRNA abundance in the mouse blastocyst based on cDNA library analysis. Molec. Reprod. Develop. 1:233–241.

Witschi, E. (1948). Migration of the germ cells of human embryos from the yolk sac to the primitive gonadal folds. Carnegie Inst. Contr. Embryol. 32:67–80.

Witte, O. N. (1990). Steel locus defines new multipotent growth factor. Cell 63:5–6.

Wylie, C. C., Stott, D., and Donovan, P. J. (1985). Primordial germ cell migration, in *Development Biology: A Comprehensive Synthesis* (Browder, L., ed.) Vol. 2, The cellular basis of morphogenesis, Plenum Press, New York.

Yarden, Y., Kuang, W., Yang-Feng, T., and Ullrich, A. (1987). Human proto-oncogene c-kit: A new cell surface receptor tyrosine kinase for an unidentified ligand. EMBO J. 6:3341d–3351d.

Young, R. A., and Davis, R. W. (1983). Efficient isolation of genes by using antibody probes. Proc. Natl. Acad. Sci. (USA) 80:1194–1198.

Zamboni, L., and Upadhyay, S. (1983). Germ cell differentiation in mouse adrenal glands. J. Exp. Zool. 228:173–193.

Zsebo, K. M., Williams, D. A., Geissler, E. N., Broudy, V. C., Martin, F. H., Atkins, H. L., Hsu, R.-Y., Birkett, N. C., Okino, K. H., Murdock, D. C., Jacobsen, F. W., Langley, K. E., Smith, K. A., Takeishi, T., Cattanach, B. M., Galli, S. J., and Suggs, S. V. (1990). Stem cell factor is encoded at the Sl locus of the mouse and is the ligand for the c-kit tyrosine kinase receptor. Cell 63:213–224.

CHAPTER FOUR

Genetic Control of Spermatogenesis and Steroidogenesis

Curtis Chubb

I. Introduction

The germinal epithelium within seminiferous tubules produces millions of spermatozoa on a daily basis throughout the adult life of mammals. This proliferative process, named spermatogenesis, is even more extraordinary when one considers that the product is a motile cell with a condensed nucleus, a haploid genome, and minimal cytoplasm. It is logical to assume that the genetic control of spermatogenesis must be tightly regulated because failure to produce functional spermatozoa would doom a species to extinction.

The basic features of spermatogenesis are common to all mammals (De Kretser and Kerr, 1988). Spermatogenesis in each species, however, does have unique characteristics. For example, humans require 64 days to convert spermatogonia to spermatozoa, whereas mice only need 35 days to accomplish the same task (De Kretser and Kerr, 1988). Human spermatozoa are 60 μm long; mouse spermatozoa are 120 μm long (Eddy, 1988). The significance of these differences is not known, although unique surface recognition proteins are essential to ensure that the spermatozoon binds to the appropriate ovum.

The primary focus of this review is the genetic control of spermatogenesis in the mouse. A spectrum of gene mutations is discussed that have been reported to perturb mouse spermatogenesis without concomitant effects on the endocrine system. These mutations provide important insight about genes controlling spermatogenesis. At present, information about these genes at the molecular level is minimal. Due to the increasing awareness of the homologies between mouse and human DNA, comparative gene maps are used to identify human chromosomal segments that are homologous to those bearing genes that impact spermatogenesis in the mouse. In addition, a possible genetic linkage between the reproductive and nervous systems is explored.

Leydig cell steroidogenesis represents another critical function of the mammalian testes. Interestingly, the genes encoding steroidogenic enzymes were initially identified in the human genome and are presently being mapped in the mouse. A brief discussion of these genes is included in the review. Comparative gene maps are used to identify homologous chromosomal segments in the mouse genome.

At the end of the review, new and emerging molecular technologies that should assist in defining the genetic control of spermatogenesis and steroidogenesis are discussed.

II. Genes Affecting Spermatogenesis

A. Important Aspects of Spermatogenesis

Of signal importance to the understanding of the present review is an appreciation of the in-

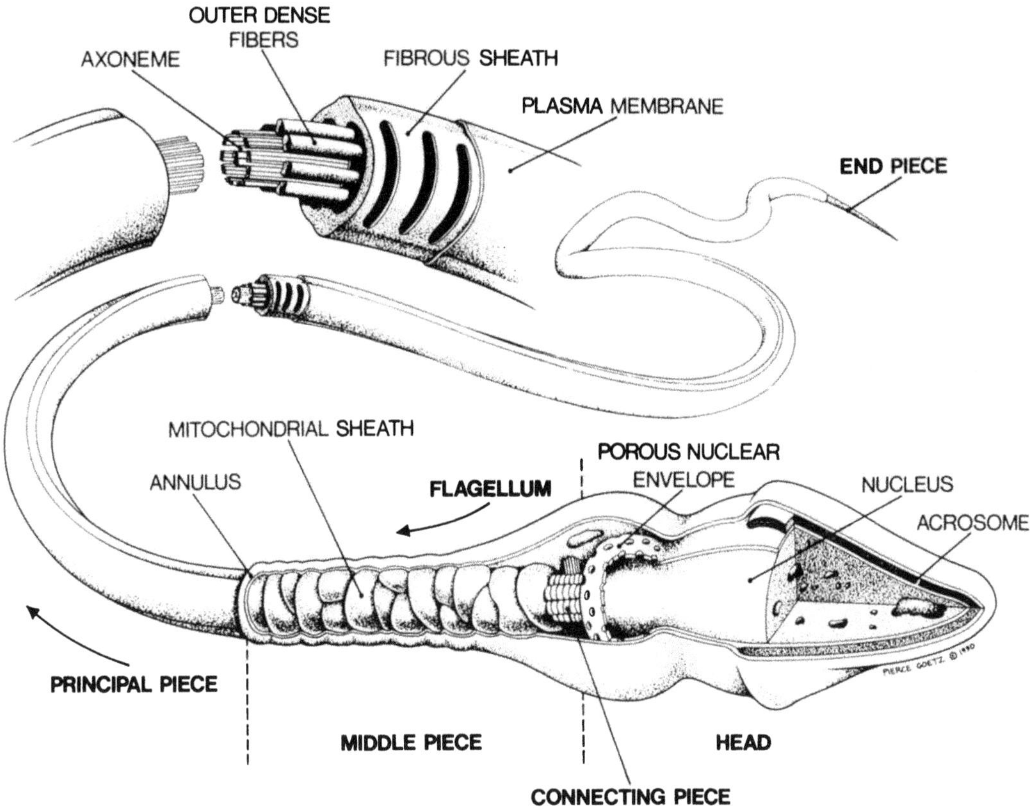

Figure 4–1. An illustration of the mammalian spermatozoon. Terminology based on review by Eddy (1988).

tricacies of spermiogenesis—the phase of spermatogenesis responsible for spermatid differentiation. The principal components of the spermatozoon (Fig. 4–1) are synthesized at specific steps of spermiogenesis (see following summary). Mature spermatids are classified as step 16 spermatids in the mouse.

Axoneme: steps 1–7
Connecting piece: steps 1–5
Acrosome: steps 1–7
Fibrous sheath: steps 2–11
Outer defense fibers: steps 8–16
Manchette: steps 8–16
Nuclear shape: steps 8–11
Chromatin condensation: steps 12–16
Mitochondrial sheath: steps 15–16
Loss of cytoplasm: steps 15–16

A complete description of mammalian spermatogenesis is provided in other chapters of this text (see Chapters 3, 8, 12, and 14) and in comprehensive reviews (Bellvé, 1979; Irons and Clermont, 1982a, b; Bellvé and O'Brien, 1983; Guraya, 1987; DeKretser and Kerr, 1988).

B. Sources of Gene Mutations

Important information concerning genes regulating spermatogenesis can be inferred from the phenotypic effects caused by specific gene mutations. For this review, a key assumption is that mutations of genes regulating spermatogenesis will impair the development of a normal spermatozoon. The described gene mutations were either spontaneous or induced by chemical or radiation mutagenesis. A brief review of these forms of mutagenesis is presented in the following paragraphs.

Spontaneous mutations occur in mice at a rate of 1 in 100,000 per locus (Russell et al., 1979). Dominant gene mutations can be quickly recognized if they induce abnormalities in the reproductive system, since the detrimental action of the mutation would be expressed even in heterozygotes. In contrast, recessive gene mutations altering reproduction would be seen in homozygotes only. Since the continuation of inbred mouse strains requires brother × sister mating (Lyon, 1989), the probability is increased considerably for detecting mice that are homozygous for recessive gene mutations.

Chemical mutagenesis constitutes another avenue for the production of gene mutations. The low mutation rates of most chemical mutagens limit their usefulness (Russell et al., 1979; Bulfield, 1981). However, ethylnitrosourea is of great interest because it produces mutations at a rate 10 to 15 times higher than other mutagenic chemicals and induces single base-pair changes (Bode, 1984; Justice and Bode, 1986).

Irradiation by x-rays or neutrons has been used frequently to create mutations in the mouse (Bulfield, 1981). The disadvantage of this method is that radiation induces chromosomal rearrangements and deletions that may alter the function of several genes.

C. Genes Regulating Spermatogenesis

For the most part, this author has used the reported phenotypic actions of gene mutations to identify genes regulating spermatogenesis. To be included in the review, the phenotypic effects of the gene mutation on spermatogenesis had to be defined morphologically. Inbred mice bearing the gene mutations played an important role in these morphological investigations. Two hundred ninety-four inbred mouse strains are available (Festing, 1989). Because inbred mice are produced after 20 consecutive generations of brother × sister mating, 99 percent of their gene loci should be homozygous (Fitch and Atchley, 1985). Consequently, the actions of a gene mutation can be defined by comparing mice from one inbred strain that either have the gene mutation or have the normal (wild-type) gene. These mice are described as coisogenic or congenic (Lyon, 1989). In addition to standardizing the genetic background, other parameters that may cause variability, such as environment and age, can be controlled. This standardization permits any difference in spermatogenesis between the mutant and normal mice to be assigned to the gene mutation.

Certain gene loci known to impair spermatogenesis are normal alleles. These are included in the review and are considered with the gene mutations. To be included in the following list, the gene mutations had to fulfill three criteria: (1) they had to have a primary effect on germ cells in the absence of a reported endocrine effect, (2) their effect on spermatogenesis had to be morphologically defined, and (3) they had to be reported for the mouse. Several gene mutations/genes that induce sterility in male mice are not included because they did not meet the criteria. For example, distorter *(Tcd)* and responder *(Tcr)* genes within the *t*-complex on mouse chromosome 17 cause the production of dysfunctional spermatozoa by an unknown mechanism and in the absence of detectable morphological changes (Olds-Clarke and Peitz, 1985; Silver, 1985; Lyon, 1986; Handel, 1987). As a result, these genes are not included in this review.

The gene mutations are divided into three categories according to their actions on primordial germ cells/gonocytes, the progression of spermatogenesis, or spermiogenesis. General features of the gene mutation are followed by a discussion designed to highlight their effects on spermatogenesis and, where applicable, the nervous system. The gene mutations are listed alphabetically by their gene symbol. General references about the gene mutations can be found in Green (1989) and the current issues of the *Mouse Genome* (Oxford University Press).

1. GENES REGULATING PRIMORDIAL GERM CELLS AND/OR GONOCYTES

The primordial germ cells must migrate from the extraembryonic yolk sac to the gonadal ridges (Bellvé, 1979). During this migration, they increase in number from 76 at day 8 of gestation to 5711 at day 12 of gestation (Silvers, 1979). Within the seminiferous cords, the primordial germ cells are referred to as gonocytes. The following gene mutations disrupt this developmental pattern and cause neonatal testes to be devoid of spermatogonia. (See Chapter 3 for more discussion of primordial germ cells.)

an (Hertwig's anemia)

General features: recessive gene mutation; probably radiation-induced; mouse chromosome 4; sterile male and female homozygotes; high neonatal mortality; affects erythropoiesis and gametogenesis.
(Russell, 1985; Green, 1989)

Discussion: Mice homozygous for the Hertwig's anemia mutation have testes with significantly decreased numbers of germ cells. In contrast to the actions of the sterility inducing alleles of the *Sl* and *W* gene loci, primordial germ cells reach the embryonic testes of *an/an* mice but then undergo cell death (Russell, 1985).

at (atrichosis)

General features: recessive gene mutation; spontaneous; mouse chromosome 10; sterile male and female homozygotes; affects hair growth and skin texture.
(Chubb and Nolan, 1984; Green, 1989)

Discussion: Atrichosis mouse testes contain seminiferous tubules consisting of Sertoli cells only (Chubb and Nolan, 1984; Chubb, 1989). The primary action of the gene mutation is unknown. This author postulates that the progenitor germ cells are deleteriously affected.

Sl (Steel)

General features: Sl, Sl^{d}, Sl^{m}, and *Sl^{so}* are mutant alleles of the *Sl* locus that cause male and female sterility; semidominant mutations; *Sl, Sl^{d}, Sl^{so}*—spontaneous; *Sl^{m}*—radiation-induced; mouse chromosome 10; affects melanogenesis, erythropoiesis, gametogenesis.
(Silvers, 1979; Green, 1989)

Discussion: Four mutant alleles of the *Sl* locus cause sterility in mice: *Sl* (steel), *Sl^{d}* (steel-Dickie), *Sl^{m}* (steel-Miller), and *Sl^{so}* (sooty). Other mutant alleles of the *Sl* locus do not cause male sterility (Silvers, 1979; Green, 1989). All of the *Sl*-sterile alleles appear to interfere with the proliferation of primordial germ cells during their migration to the genital ridges (Silvers, 1979; Green, 1989). The action of *Sl*-sterile alleles is discussed in Section II.F.

W (Dominant spotting)

General features: W, W^{b}, and *W^{v}* are mutant alleles that cause male and female sterility; semidominant mutations; spontaneous; mouse chromosome 5; affects melanogenesis, erythropoiesis, gametogenesis.
(Silvers, 1979; Green, 1989)

Discussion: Although the original *W* gene mutation causes in utero death when homozygous, mice survive to maturity when homozygous for the other mutant alleles [*W^{b}* (Ballantyne's spotting), *W^{v}* (viable dominant spotting)] (Silvers, 1979; Green, 1989). Other mutations of the *W* locus do not cause sterility. In contrast to the *Sl*-sterile alleles, the *W*-sterile alleles prevent the migration of the primordial germ cells to the genital ridge. The action of these mutant alleles is discussed in Sections II.F and IV.C.

2. GENES REGULATING THE PROGRESSION OF SPERMATOGENESIS

fl (flipper-arm)

General features: recessive mutation; possibly radiation-induced; not mapped; sterile male homozygotes; female homozygotes have reduced fertility; skeletal defect.
(Green, 1989)

Discussion: Mice homozygous for the flipper-arm gene mutation are difficult to study because of high neonatal mortality. Our laboratory reported the only analysis of testicular function in flipper-arm mice—one mouse was studied (Chubb and Nolan, 1985).

Morphometric analyses suggested that the gene mutation affected spermatogenesis after the formation of step 7 spermatids; only 50 percent of the normal number of elongated spermatids were observed. No other morphological studies were performed. Conclusions about the ultrastructure of the spermatids cannot be made. Spermatozoa were not detected in the epididymis.

Hst-1 (hybrid-sterility)

General features: alleles at the *Hst-1* locus are designated as *$Hst\text{-}1^{s}$* and *$Hst\text{-}1^{f}$* for laboratory mice and *$Hst\text{-}1^{wf}$* and *$Hst\text{-}1^{ws}$* for wild mice (allelism is assumed for all four genes); *only* hybrid male mice with the *$Hst\text{-}1^{s}/Hst\text{-}1^{ws}$* genotype are sterile; the alleles are not mutations; mouse chromosome 17; no other known effects.
(Forejt and Iványi, 1975; Green, 1989)

Discussion: The *Hst-1* gene locus is interesting because sterile hybrid males are the progeny of unique crosses between laboratory and wild mice. The PWK inbred strain originated from wild mice *(Mus musculus musculus)* captured in Czechoslovakia and is homozygous for Hst^{ws}. Inbred strains of laboratory mice *(Mus musculus domesticus)* are homozygous for either $Hst\text{-}1^{s}$ or $Hst\text{-}1^{f}$. Only male progeny of pairings between PWK mice and laboratory mice homozygous for $Hst\text{-}1^{s}$ (e.g., BALB/c) will be sterile (Forejt and Iványi, 1975). Testes of sterile hybrid mice are characterized by spermatogenesis blocked at the pachytene spermatocyte stage (Chubb and Nolan, 1987; Chubb, 1989).

jsd (juvenile spermatogonial depletion)

General features: recessive mutation; spontaneous; mouse chromosome 1; sterile male homozygote; fertile female homozygote.

(Beamer et al., 1988)

Discussion: Testes of mice homozygous for the *jsd* mutation appear to experience transient spermatogenesis. In 9-week-old mice, seminiferous tubules contain Sertoli cells and spermatogonia only (Beamer et al., 1988).

sks (skeletal fusion with sterility)

General features: recessive mutation; spontaneous; mouse chromosome 4; sterile male and female homozygotes; skeletal defects.

(Handel et al., 1988)

Discussion: The *sks* gene mutation interferes with the normal progression of spermatogenesis (Handel et al., 1988). The most advanced germ cell types are pachytene spermatocytes or leptotene spermatocytes in 91 and 5 percent of the tubules, respectively. Cells in different stages of degeneration and sloughing are noted.

Spy (spermatogenesis gene on Y chromosome)

Discussion: The *Spy* gene is hypothesized to be required for normal spermatogenesis (Burgoyne, 1987). Although not mapped, Burgoyne (1987) indicated the *Spy* gene to be near the centromere of the mouse Y chromosome.

3. GENES REGULATING SPERMIOGENESIS

azh (abnormal spermatozoon head shape)

General features: recessive mutation; probably spontaneous; mouse chromosome 4; male and female homozygotes are fertile.

(Meistrich, M., personal communication; Cole et al., 1988)

Discussion: The primary defect of the *azh* gene mutation involves the abnormal positioning of the manchette during steps 8–9 of spermiogenesis (Cole et al., 1988; Meistrich et al., 1990). This causes *all* of the sperm heads to be shaped abnormally. The data support the contribution of the manchette to spermatid head shaping during spermiogenesis. The normal fertility of males homozygous for *azh* questions the importance of normal head shape for successful fertilization. Flagella are normal.

bs (blind-sterile)

General features: recessive mutation; spontaneous; mouse chromosome 2; sterile male homozygotes; fertile female homozygotes; bilateral lenticular cataracts.

(Varnum, 1983; Green, 1989)

Discussion: Blind-sterile mouse testes are characterized by a spectrum of spermatogenic defects. Seminiferous tubules within one testis may contain normal germinal epithelium, Sertoli cells only, or hypoplastic germinal epithelium (Sotomayor and Handel, 1986; Chubb, 1989).

The reason for the variability in gene expression is not known. However, a constant effect of the blind-sterile mutation is the failure of acrosomal development (Sotomayor and Handel, 1986). Although nuclear condensation, nuclear elongation, and manchette development proceed normally, the spermatid heads are abnormally shaped. The sperm flagellum appears normal (Sotomayor and Handel, 1986).

c (albino)

General features: c^{3H} and c^{6H} are mutant alleles of the *c* locus; recessive mutations; lethal; radiation-induced; mouse chromosome 7; male mice with genotype c^{3H}/c^{6H} are sterile; female mice with genotype c^{3H}/c^{6H} have apparently normal oogenesis except all progeny die at mid-gestation; runted.

(Lewis et al., 1978; Green, 1989)

Discussion: The c^{3H} and c^{6H} mutations are probably small deletions (Lewis et al., 1978; Green, 1989). Mice with the c^{3H}/c^{6H} genotype are viable but sterile due to defects in spermiogenesis (Lewis et al., 1978). Spermatogenesis is normal until step 13 spermiogenesis when the spermatid heads become vacuolated and abnormal in shape. Nuclear chromatin does not condense correctly. Epididymal spermatozoa have reduced motility—the alterations in flagella have not been investigated.

hop (hop-sterile)
hop^{hpy} (hydrocephalic-polydactyl)

General features: hop-sterile and hydrocephalic-polydactyl mutations are allelic; recessive mutations; *hop*—spontaneous; hop^{hpy}—radiation-induced; mouse chromosome 6; sterile male homozygotes; fertile female homozygotes; hopping gait.

(Johnson and Hunt, 1971; Green, 1989)

Discussion: Mice homozygous for either the *hop* or hop^{hpy} mutations are sterile due to defective spermiogenesis. These mice are characterized by spermatozoa without flagella or with severely abnormal flagella (Johnson and Hunt, 1971; Bryan, 1977b; Handel, 1987; Green, 1989). Flagella are not visible in the lumen of seminiferous tubules (Johnson and Hunt, 1971; Bryan, 1977b). The centrioles are present but appear to be unable to initiate or maintain the axonemal structure (Bryan, 1977b; Searle, 1982). Outer dense fibers are absent. In addition, the acrosome becomes distorted starting with step 7 spermatids in mice homozygous for hop^{hpy} (Bryan, 1977b). Abnormal sperm heads are also observed (Johnson and Hunt, 1971; Searle, 1982).

olt (oligotriche)

General features: recessive mutation; spontaneous; not mapped; sterile male homozygotes; fertile female homozygotes; thinning of ventral coat.

(Moutier, 1976)

Discussion: Oligotriche mouse spermatozoa have condensed nuclei and acrosomes, but their flagella are completely absent (Chubb, 1992). The kinetics of spermatogenesis and testicular steroidogenesis are normal in oligotriche mice (Chubb, 1992).

p (pink-eyed dilution)

General features: p^{bs}, p^{s}, p^{6H}, p^{25H} mutant alleles cause sterility in male mice bearing any combination of the four alleles; recessive mutations; radiation-induced; mouse chromosome 7; fertile female homozygotes—some with reduced fertility; nervous behavior; lack of coordination; abnormal behavior.

(Sidman et al., 1965; Hunt and Johnson, 1971; Wolfe et al., 1977; Handel, 1987; Green, 1989)

Discussion: Four mutant alleles of the *p* locus are known to induce male sterility: p^{bs} (*p*-black-eyed sterile), p^{s} (*p*-sterile), p^{6H} (*p*-6H), p^{25H} (*p*-25H). Other *p* locus alleles do not induce male sterility (Green, 1989). The primary effect of the *p*-sterile alleles is to interfere with acrosome formation during early spermiogenesis (Hunt and Johnson, 1971; Bryan, 1977b; Searle, 1982; Handel et al., 1987). A disorganized Golgi apparatus has been implicated (Bryan, 1977a; Hash and Wolfe, 1979). The abnormal acrosome development is postulated to result in irregular spermatid head shapes (Bryan, 1977a; Handel, 1987). In fact, the percentages of the different head shapes are used as markers for the different combinations of *p*-allele genotypes (Wolfe et al., 1977). The manchette appears to be unaffected (Bryan, 1977a). Chromatin in the posterior region of spermatid nuclei in p^{6H}/p^{6H} and p^{s}/p^{s} mice does not condense normally (Bryan, 1977a; Handel et al., 1987). Although the p^{6H} and p^{25H} mutations induce the formation of spermatids with multitails and various numbers of nuclei (Hunt and Johnson, 1971; Johnson and Hunt, 1971; Silvers, 1979; Handel et al., 1987), p^{s}/p^{s} mouse spermatids have ultrastructurally normal and motile sperm flagella (Hollander et al., 1960; Bryan, 1977a, b; Silvers, 1979). Interestingly, the only *p*-alleles that cause abnormal behavior also induce male sterility (Silvers, 1979). The pleiotropic actions of these *p*-alleles may be due to the mutations being deletions of more than one gene locus (Melvold, 1974). The definition of pleiotropism is the presence of multiple phenotypic effects caused by a single gene mutation.

pcd (Purkinje cell degeneration)

General features: recessive mutation; spontaneous; mouse chromosome 13; sterile male homozygotes; fertile female homozygotes; postnatal degeneration of all cerebellar Purkinje cells; moderate ataxia.

(Mullen et al., 1976; Green, 1989)

Discussion: The *pcd* gene mutation acts on differentiating spermatids. Early stages of spermiogenesis are normal with nuclear elongation, acrosome formation, and manchette development proceeding as expected (Handel and Dawson, 1981). However, defects occur during late spermiogenesis that induce head shape abnormalities and errors in the assembly of the outer dense fibers (Handel and Dawson, 1981). Epididymal spermatozoa are immotile and possess abnormal heads and flagella (Mullen et al., 1976; Handel and Dawson, 1981).

qk (quaking)

General features: recessive mutation; spontaneous; mouse chromosome 17; sterile male homozygotes; fertile female homozygotes; severe myelin deficiency of central nervous system with less severe myelin deficiency of peripheral nervous system; quaking tremors.

(Sidman et al., 1964; Bennett et al., 1971; Green, 1989)

Discussion: Quaking mouse testes exhibit normal spermatogenesis until spermiogenesis when assembly of flagella becomes disorganized. The sperm flagellum virtually disappears by step 9 of spermiogenesis and lumens of the seminiferous tubules are devoid of the characteristic long flagella seen in normal testes (Bennett et al., 1971; Chubb, 1992). Outer dense fibers and the fibrous sheath are not formed but the manchette is not affected (Bennett et al., 1971). In addition, the sperm heads are misshapen although nuclear condensation and elongation are normal (Bennett et al., 1971; Handel et al., 1987).

Ethylnitrosourea mutagenesis was used to create a new mutation at the quaking locus (assigned the symbol qk^k). Interestingly, male mice with the qk/qk^k genotype express the characteristic quaking phenotype but are fertile (Bode, 1984). These results suggest that the original quaking gene mutation spans more than one gene locus. Lipid analyses performed to understand the primary action of the mutation support this conclusion. Lipid analyses of quaking mouse brains revealed a depletion of complex lipids including sphingomyelin (Baumann et al., 1968); however, similar lipid analyses of mutant mouse testes failed to demonstrate a similar defect (Coniglio et al., 1975).

wr (wobbler)

General features: recessive mutation; spontaneous; not mapped; male and female homozygotes have not bred; degeneration of motor neurons in spinal cord/brain stem.

(Sidman et al., 1965; Duchen and Strich, 1968; Handel, 1987; Green, 1989)

Discussion: The wobbler mutation has a direct effect on spermiogenesis. Electron microscopic analyses revealed that Golgi in early spermatids do not produce proacrosomal granules, the acrosome never develops, and head shapes are bizarre (Kelly and Woolley, 1980). Chromatin condensation is not affected. Although flagella of spermatozoa in the cauda epididymidis appear normal, they are immotile and 1 to 4 of the axonemal outer doublets and associated outer dense fibers are missing (Leestma and Sepsenwol, 1980). Interestingly, the flagellum defects were noticed in only 5 percent of spermatozoa in the testis but in 70 percent of spermatozoa following epididymal transit (Leetsma and Sepsenwol, 1980).

D. Common Features of Genes Regulating Spermatogenesis

The information presented in Section II.C reveals several interesting common features about the gene loci involved in regulating spermatogenesis. This author focuses the discussion on the autosomal genes; this restriction only eliminates the *Spy* gene on the Y chromosome. First, the alleles are either recessive or semidominant. Consequently, both homologous chromosomes must bear the allele affecting spermatogenesis before the phenotypic effect is expressed. Second, 78 percent of the alleles disrupting spermatogenic progression and spermiogenesis cause male-specific sterility. This fact suggests that spermatogenesis and oogenesis have unique genetic control systems. Third, 100 percent of the alleles influencing primordial germ cells/gonocytes effect male *and* female sterility. These observations indicate that the sex chromosome composition of primordial germ cells (X/X vs. X/Y) does not modulate the actions of these gene mutations. Fourth, 64 percent of the alleles impeding normal spermiogenesis have pleiotropic effects on the nervous system. This linkage is discussed in more detail in Section II.H.

Another feature common to the reviewed gene alleles affecting spermatogenesis is that the gene locus is defined only according to its phenotype and not in molecular terms. For example, it is not known if the listed gene mutations are single-base changes, deletions, or rearrangements. We also do not know if the mutations are of structural or regulatory genes.

This author has constructed a theoretical diagram illustrating the genes regulating spermiogenesis and the primary division of the spermatozoon that is controlled by their protein products (Fig. 4–2). For those mutations that may perturb the development of both the head and the flagellum, their location in Figure 4.2 is based on what is perceived as their primary action, as discussed in the next two paragraphs. More complete information about the phenotypic effects can be found in Section II.C discussions.

Genes that have an integral role in the development of the acrosome are blind-sterile, *p*-sterile alleles, and wobbler: mutations of these genes cause acrosomal defects and concomitant abnormalities in sperm head shape. In contrast, mutation of the abnormal spermatozoon head shape gene results in abnormally shaped sperm heads due to a specific disturbance of the manchette. The other genes required for the development of normally shaped sperm heads control less-defined processes; these genes are *c*-sterile alleles and Purkinje cell degeneration. Mutations of the *c, pcd,* and *wr* loci also interfere with the normal assembly of sperm flagella. However, sperm flagella are normal in mice homozygous for the *azh, bs,* and *p* gene mutations.

Genes involved in controlling the complex assembly of the sperm flagellum include hop-sterile, oligotriche, and quaking. Mutations of these genes cause spermatids to differentiate without developing a flagellum. In every instance, the sperm head is abnormally shaped. These observations support the conclusion that normal flagellum assembly is necessary for normal sperm head differentiation. But the reverse situation is different—normal flagella *can* evolve in spermatids with impaired head development.

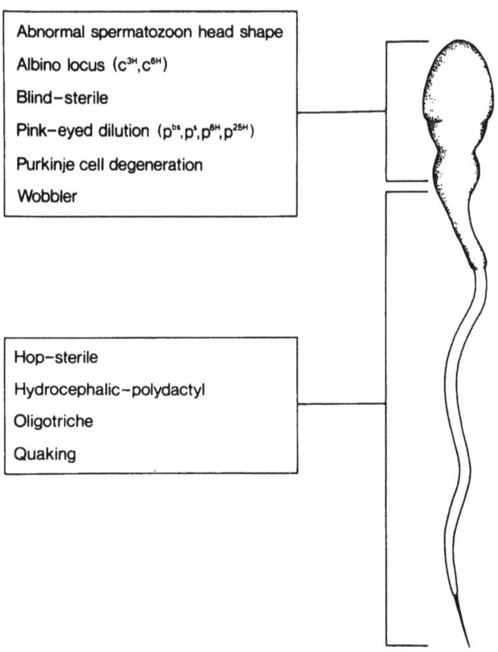

Figure 4–2. A schematic indicating the primary site of action of the putative genes that control spermatogenesis.

E. Gene Expression Analyses

A major argument of this review is that gene mutations provide important experimental tools for elucidating the genetic control of spermatogenesis. Another experimental approach to achieving this goal is to identify initially the protein products that are unique to specific germ cell classes. This approach has been used to demonstrate that proteins for the basal plate, connecting piece, outer dense fibers, fibrous sheath, and outer mitochondrial membranes are synthesized *exclusively* during spermiogenesis (Bellvé, 1979; O'Brien and Bellvé, 1980; Irons and Clermont, 1982a). In contrast, other spermatozoal proteins, including those of the acrosome and axoneme, are synthesized primarily by spermatocytes and stored for later use during spermiogenesis. One must be aware that proteins secreted by Sertoli cells can be integrated into differentiating germ cells (Kierszenbaum et al., 1988).

Other studies of protein synthesis have also demonstrated that unique proteins are produced by the different classes of germ cells

(O'Brien, 1987; Willison and Ashworth, 1987). The appearance of these protein products can be used to monitor the expression of specific genes during spermatogenesis, and to raise antibodies that could be used to screen complementary DNA (cDNA) libraries.

cDNA probes have been used to confirm that gene expression during spermatogenesis is stage-specific (Slaughter et al., 1989; Thomas et al., 1989). Importantly, cDNA probes have provided unequivocal evidence that genes are expressed in haploid spermatids. For example, genes encoding α-tubulin are expressed in round spermatids at developmental times corresponding to the assembly of manchette and axoneme microtubules (Distel et al., 1984). Additional investigations using cDNA probes confirmed that mouse protamine genes are only expressed in haploid spermatids (Hecht et al., 1986). These latter experimental results were applied creatively in a transgenic mouse study, which is discussed in Section II.F.

Together, the protein and messenger RNA (mRNA) data support the conclusion that a significant proportion of the ultrastructural and biochemical changes occurring in spermiogenesis are controlled by genes expressed in the spermatid (see Chapter 16 for more discussion).

F. How Gene Mutations May Affect Germ Cells

The observation that germ cells originating from one spermatogonium are joined by intercellular bridges due to incomplete cytokinesis prompted conjecture about the purpose of this syncytial arrangement. One possible reason for the syncytium is to ensure the coordinated progress of germ cells during spermatogenesis. An alternative explanation is of direct importance to the genetic control of spermatogenesis: the intercellular bridges facilitate the exchange of cellular products between germ cells. As a consequence, the mRNA and/or protein of one gene can be shared with other germ cells of the same clone. Evidence supporting the second hypothesis is discussed in the following paragraph.

The first evidence originated from the observation that spermatids carrying specific *t*-haplotypes inactivate other clonal spermatids not bearing the specific *t*-haplotypes (Silver, 1985). More convincing evidence originated from an investigation reported by Braun and co-workers (1989). They produced transgenic mice carrying a chimeric gene composed of the human growth hormone structural gene and the transcriptional regulatory sequences for the mouse protamine gene. Recall that the mouse protamine gene is expressed *only* in spermatids. Because only one of the two homologous chromosomes carried the chimeric transgene, it was transmitted to only 50 percent of the haploid spermatids. Hybridization with DNA probes, however, detected human growth hormone mRNA in *every* spermatid. These results provide strong evidence that gene expression products are shared between spermatids.

The interchange of mRNA/proteins between germ cells is important to explain the action of gene mutations perturbating spermiogenesis. Gene mutations that disrupt spermiogenesis are all *recessive,* meaning that both homologous chromosomes *must* carry the gene mutation before the mutant phenotype is expressed. In mice heterozygous for the gene mutation, normal spermatozoa are produced even though they bear the gene mutation. The rescue of spermatids with the gene mutation integrated in their haploid genome has two possible explanations. First, mRNA/protein transcribed from the normal gene in spermatocytes could be equally distributed to spermatid daughter cells by the meiotic divisions. The second explanation is that mRNA/proteins originating in spermatids with the normal gene are transported to spermatids with the gene mutation via intercellular bridges. Both of these situations could result in normal spermiogenesis in mice heterozygous for the gene mutation. The explanation that is applicable to each gene mutation has not been determined.

The discussion has centered on a primary effect of the mutations on the germ cells. However, the deleterious effects of the gene mutations could be secondary. For instance, the mutations could alter the function of Sertoli cells and thereby prevent normal germ cell differentiation. One approach to distinguishing between primary and secondary effects of gene mutations is to produce mice chimeric for normal cells and cells homozygous for the specific gene mutation. Chimeric mice result from in vitro mixing normal and mutant cells from two-cell embryos and later transferring blastocysts containing the mixture of cells to pseudopregnant host females. The transplanted blastocysts develop into chimeric mice with testes consisting of mutant and normal cells.

The genotype of spermatozoa from chimeric mouse testes is tested by inbreeding the progeny. The production of functional spermatozoa bearing the gene mutation would suggest that its action on germ cells was secondary as a percentage of spermatogenesis would be supported by Sertoli cells with the normal genotype. If the mutation effect on germ cells is primary, spermatozoa bearing the gene mutation would not be produced. Handel and co-workers (1987) used this experimental approach to demonstrate that quaking and *p*-6H gene mutations had *primary* effects on germ cells.

Transplantation methods have been used to study how mutations of the viable-dominant spotting (*W*) and steel (*Sl*) gene loci cause deficiencies in melanocytes, blood, and germ cells (Russell, 1985). Hematopoietic and melanoblast stem cells from Sl/Sl^d mutant mice function correctly when transplanted to normal hosts. In contrast, the defects in the stem cells from W/W^v mutant mice are not corrected by transplantation. Together these experimental results suggest that *Sl* gene mutations disrupt the microenvironment necessary for differentiation and proliferation of the stem cells, whereas the *W* gene mutations had primary effects on the stem cells.

Although primordial germ cells from Sl/Sl^d mutants could not be transplanted to normal hosts, later experiments used chimeras of Sl/Sl^d and wild-type (+/+) cells to determine if the gene mutations also disrupted the microenvironment of germ cells (Nakayama et al., 1988). The chimeric mice produced functional spermatozoa carrying the gene mutations. The results demonstrate that the *Sl* gene mutations have secondary effects on germ cells since their main action is on the cells creating the specialized microenvironment.

The actions of the *Sl* and *W* gene mutations provide examples of both spurious and genuine pleiotropism. The actions of *Sl* gene mutations on primordial germ cells are examples of *spurious* pleiotropism as their primary effects are not expressed within the germ cells. In contrast, the *W* gene mutation effects on pigment, blood, and germ cells constitute *genuine* pleiotropism as the effects result from primary actions in each of the different cell types.

G. Homologies Between Mouse and Human Genes Affecting Spermatogenesis

Comparative gene mapping between humans and mice is an exciting and rapidly expanding field of study. Increasing amounts of evidence support the concept that DNA from the 20 mouse chromosomes has been shuffled among the 23 human chromosomes. Cytogenetics and the mapping of marker genes have confirmed the homology of several chromosomal segments between mice and humans (Sawyer and Hozier, 1986; Lalley et al., 1989; Nadeau, 1989). At present, 37 percent of the mouse autosomal chromosomes are known to be conserved in human chromosomes.

Comparative gene maps for mouse and human genomes were used to develop Table 4–1. In Table 4–1, the chromosomal locations of human genes regulating spermatogenesis have been identified on the basis of the available information about mouse and human DNA homologies. Of the gene loci listed in section II.C,

Table 4–1. Locations of Human Chromosomal Segments Homologous to Mouse Genes Regulating Spermatogenesis Based on Comparative Gene Mapping Data[a]

	Gene location	
Mouse gene symbol	Mouse chromosome	Human chromosome
an	4	9p
at	10	?[b]
azh	4	?
bs	2	20
c	7	?
fl	?	?
hop	6	7q
Hst-1	17	?
jsd	1	2q
olt	?	?
p	7	11p
pcd	13	?
qk	17	6q
sks	4	1p
Sl	10	12q
Spy	Y	Y
W	5	4q
wr	?	?

[a]Human chromosome location based on comparative mapping information (Lyon, 1988; Lalley et al., 1989; Nadeau, 1989).

[b]Insufficient data.

83 percent have been mapped on mouse chromosomes. Sixty-six percent of the mapped mouse genes are located on mouse chromosomal segments that have homologies in humans. The homologies between mouse and human chromosomes should assist in isolating genes regulating spermatogenesis in both species.

H. Link Between Spermatogenesis and the Nervous System

A provocative aspect of the gene mutations disrupting spermiogenesis is the high frequency of concomitant defects in the nervous system. The effects on the nervous system may reflect either genuine or spurious pleiotropism as discussed previously. Genuine pleiotropic effects could assist in deciphering the primary action of gene mutations because the defect may be common to each cell type. This possibility is the primary motivation for including this section in the review.

Significant events occur in mouse reproductive and nervous systems during the first 25 days of life. In the reproductive system, the testes develop rapidly and the first spermatids appear (Desjardins, 1985). Accompanying these morphological changes is a transient increase in the secretion of 5α-reduced androgens (Desjardins, 1985). In the central nervous system, a similar transient increase of 5α-reductase activity occurs within myelin—this time period corresponds to intense myelinogenesis by oligodendrocytes (Baumann et al., 1968; Melcangi et al., 1988). Pleiotropic actions of specific gene mutations suggest the existence of unrecognized parameters common to the reproductive and nervous systems. The gene mutations selected for discussion are *pcd, qk,* and *p*-sterile alleles.

The *pcd* gene mutation causes the degeneration of all cerebellar Purkinje cells during the neonatal period (Mullen et al., 1976). Concomitantly, significant defects in spermiogenesis result from the action of the *pcd* mutation. The primary action of the *pcd* gene mutation has not been deciphered.

Mutation of the quaking gene disrupts spermiogenesis and nervous system myelination. Hypomyelination, however, does not always cause male sterility. Mice homozygous for the shiverer or myelin deficient gene mutations are fertile although characterized by severe myelin deficiency throughout the central nervous system (Popko et al., 1987; Green, 1989). The ability to be fertile in the presence of defective myelination supports the hypothesis that the quaking gene mutation spans more than one gene locus. This conclusion concurs with the phenotype of mice bearing the quaking gene mutation induced by ethylnitrosourea; these mice quake but are fertile as discussed in section II.C. The primary actions of the quaking gene mutation, however, have not been elucidated.

The *p*-sterile alleles disrupt acrosomal development by modifying the function of the Golgi apparatus (Bryan, 1977a; Hash and Wolfe, 1979). Moreover, the net charge of spermatid plasma membranes in mice homozygous for *p*-sterile alleles is "+" although the normal charge is "−" (Hash and Wolfe, 1979). The abnormal positive charge may be associated with the absence of gangliosides—molecules also involved in the transmission of nervous impulses. Consequently, the altered function of the Golgi may interfere with the normal production of gangliosides in both reproductive and nervous tissues.

The experimental data are too limited to form a conclusion concerning the existence of biochemical pathways common to nervous tissue cells and germ cells. Although numerous other gene mutations with well-defined actions on the nervous system induce male sterility (Green, 1989), the causes of the sterility have not been reported. The gathering of this information may assist in understanding if the pleiotropic actions of gene mutations affecting the nervous and reproductive systems are genuine or spurious.

III. Genes Encoding Steroidogenic Enzymes

A. Overview of Testicular Steroidogenesis

Steroidogenesis is the other important function of the mammalian testes. Testosterone, the primary steroid secretion of testes, is required for the development and maintenance of the accessory reproductive organs, male sexual behavior, differentiation of the male reproductive tract, and spermatogenesis.

Four principal steroidogenic enzymes within

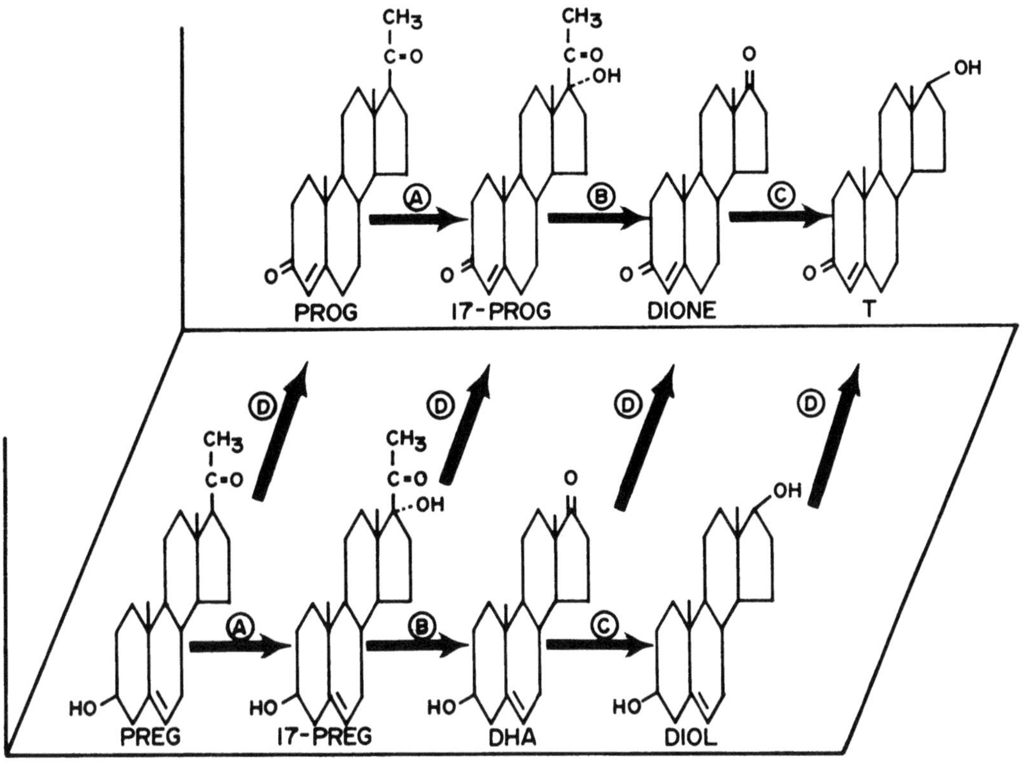

Figure 4–3. A diagram illustrating the location of the principal steroidogenic enzymes in the testosterone biosynthetic pathway. The steroidogenic enzymes are 17α-hydroxylase/C17-20 lyase (A and B), 17β-hydroxysteroid dehydrogenase (C), and 3β-hydroxysteroid dehydrogenase/$\Delta^{5\text{-}4}$ isomerase (D). The cholesterol side-chain cleavage enzyme produces pregnenolone (PREG). Other steroids are abbreviated as DHA (dehydroepiandrosterone), DIOL (androstenediol), DIONE (androstenedione), 17-PREG (17α-hydroxypregnenolone), PROG (progesterone), 17-PROG (17α-hydroxyprogesterone), and T (testosterone).

Leydig cells act in concert to convert cholesterol to testosterone. These are the cholesterol side-chain cleavage enzyme ($P450_{scc}$), 17α-hydroxylase/$C_{17\text{-}20}$ lyase ($P450_{17\alpha}$), 3β-hydroxysteroid dehydrogenase/Δ^{5-4} isomerase (3βHSD), and 17β-hydroxysteroid dehydrogenase (17βHSD) (Fig. 4–3). (See Chapter 7 for more discussion about steroidogenic enzymes.)

B. Genes Encoding Human Steroidogenic Enzymes

The genes encoding for steroidogenic enzymes differ from those controlling spermatogenesis in two important ways. First, the protein products of the steroidogenic enzyme genes have been purified and antibodies raised to them. These antibodies have been used to screen genomic and cDNA libraries to identify the DNA sequences of the genes. In contrast, the primary products of genes controlling spermatogenesis have not been identified. Second, the steroidogenic enzyme genes were identified in the human genome first—only one has been mapped in the mouse chromosome. In contrast, the spermatogenesis-controlling genes have been studied in the mouse—but not in the human. The available information about genes encoding steroidogenic enzymes is summarized in the next two paragraphs.

$P450_{scc}$ is encoded by one gene on human chromosome 15 (Chung et al., 1986; Miller, 1988). The other P450 enzyme in the testosterone biosynthetic pathway, $P450_{17\alpha}$, is encoded by one gene on human chromosome 10 (Chung et al., 1987; Miller, 1988). The finding that there is only one $P450_{17\alpha}$ gene is interesting because $P450_{17\alpha}$ enzymes in testes and adrenals have significant differences in activity. In addition, the proteins responsible for transferring electrons from NADPH to $P450_{scc}$ have been purified and their genes located within the human genome. These two proteins and their gene locations are adrenodoxin reductase (human chromosome 17cen→q25) and adrenodoxin (human chromosome 11q13→qter) (Miller, 1988; Morel et al., 1988).

Recently, the cDNAs for human placental 17βHSD (Peltoketo et al., 1988) and 3βHSD (The et al., 1989) have been cloned and sequenced. The results of a recent 17βHSD study (Tremblay et al., 1989) suggest that the detected enzyme primarily converted estrone to estradiol and that perhaps a distinct 17βHSD catalyzes the interconversion of androstenedione and testosterone. The gene encoding the "estrogenic" 17βHSD was localized to human chromosome 17cen→q25. For the purposes of this review, this author assumes that only one 17βHSD exists. The gene encoding 3βHSD is located on human chromosome 1 (Lorence et al., 1990).

C. *Homologous Genes Affecting Mouse Steroidogenesis*

Similar to that performed for the spermatogenesis genes, comparative gene maps were implemented in determining mouse chromosomal segments that are homologous to human chromosomal segments bearing genes encoding steroidogenic enzymes and associated proteins (Table 4–2). Homologous mouse chromosomes for four of the five human genes were identified.

The validity of using comparative gene maps for locating homologous chromosomes in another species is supported by the findings of Youngblood and co-workers (1989). They are the only research group that has used the available cDNA probes for the steroidogenic enzymes and associated proteins to detect homologous mouse genes. Genomic DNA from a series of recombinant-inbred mouse strains were screened to map the $p450_{scc}$ gene to mouse chromosome 9 near the *P450-1* locus. These results agree with the determination made using the comparative gene mapping data presented by Lyon (1988): mouse chromosome 9 is homologous to the human $P450_{scc}$ gene chromosomal segment. The availability of cDNA probes, however, minimizes the importance of comparative gene mapping for the study of genes encoding steroidogenic enzymes.

IV. New Methods for Identifying Genes Controlling Spermatogenesis

Two recent reviews have emphasized the breadth of molecular biology applications for studying reproductive biology (Stewart et al., 1988; Butler and McDonough, 1989). This author proposes that the following methods, using new molecular technology, will provide

Table 4–2. Location of Mouse Chromosomal Segments Homologous to Human Genes Encoding Steroidogenic Enzymes and Associated Proteins Involved in Testosterone Biosynthesis

Steroidogenic enzyme and associated protein	Gene location	
	Human chromosome	Mouse chromosome
$P450_{scc}$	15	9[a]
$P450_{c17}$	10	?[b]
adrenodoxin reductase	17cen→q25	11[c]
adrenodoxin	11q13→qter	9[c]
17βHSD	17cen→q25	11[c]
3βHSD	1	?

[a]Data from Youngblood et al., 1989.

[b]Insufficient data.

[c]Mouse chromosome location based on comparative mapping information (Lyon, 1988; Lalley et al., 1989; Nadeau, 1989) and the human gene map (McAlpine et al., 1989).

important insights about the genetic regulation of spermatogenesis.

A. *Transgenic Mice and Insertional Mutagenesis*

Transgenic mice are defined as "animals that have integrated foreign DNA into their germ line as a consequence of experimental introduction of DNA" (Palmiter and Brinster, 1985). Transgenic mice are used to study the function and regulation of genes integrated into the host genome. In brief, the production of transgenic mice involves microinjecting cloned foreign DNA into the pronucleus of fertilized mouse eggs. Foreign DNA can also be integrated into the host genome by infection with retroviruses. In some cases, the foreign DNA is microdissected chromosomal segments located by cytogenetics (Richa and Lo, 1989). The eggs are then transferred to pseudopregnant female mice. The successful incorporation of the foreign DNA in the offspring genome is monitored by probing with the cDNA used for injection (for description see Camper, 1987).

In some cases, the integration site of the transgene mutates an endogenous gene; this process is referred to as insertional mutagenesis (Camper, 1987; Rinchik, 1987). The next step would be to investigate the reproductive system in mice carrying the transgene. This form of mutagenesis also provides a molecular tag for the mutated gene since the foreign DNA can be detected by cDNA probes.

The minimal knowledge concerning the molecular basis of the genetic regulation of spermatogenesis limits the use of transgenic mice to the random generation of insertional mutations and subsequent screening for spermatogenic defects in the progeny.

B. *Gene Targeting*

Recently, foreign DNA injected into cells was demonstrated to integrate into the host DNA via homologous recombination (Capecchi, 1989a, b). This major advance in molecular technology allows the investigator to insert an exogenous cloned gene into the *correct* position in the host DNA. The one requirement is that the gene of interest be cloned. Furthermore, the actions of the cloned gene can be defined by mutating the gene before insertion into the host DNA. Briefly, the mutated gene is microinjected or electroporated into a pluripotent, embryo-derived stem (ES) cell where it recombines with the ES genome at a site recognized due to homologous DNA sequences. The mutant ES cell is then microinjected into a mouse blastocyst which is then placed into a pseudopregnant female mouse. The offspring are chimeric mice consisting of cells carrying either the normal DNA or DNA incorporating the mutated gene. The heterozygous offspring are inbred to produce mice homozygous for the mutated gene. This powerful method may revolutionize the understanding of the genetic control of spermatogenesis since several genes or gene products have been described but without concurrent information about their function (see section II.E).

C. *Random Clone Analysis*

An almost universal characteristic of the gene mutations impacting spermatogenesis is that only the mutant phenotypes have been recognized—the protein products have not been isolated. Consequently, gene targeting is not applicable for these gene mutations because of the requirement for cloned genes. This requirement is circumvented in random clone analysis (Rinchik, 1987). In this method, randomly selected DNA segments are cloned and used to analyze the linkage between known genetic loci and restriction fragment length polymorphisms identified by the selected random clone. Most of the gene mutations have been mapped to specific chromosomal locations in the mouse. It is possible that the random clone will be located close to the gene involved in spermatogenesis. If so, the clone can be used to select and analyze DNA near the undefined gene locus.

An example of how genes from other research interests can assist in understanding the molecular basis of gene mutations is provided by Tan and co-workers (1990). Recent studies concluded that the *W* gene locus is allelic with the proto-oncogene *c-kit*. The *c-kit* gene encodes a transmembrane tyrosine protein kinase receptor. The activity of this protein product was decreased in mice either heterozygous or homozygous for W^{42} alleles—$W^{42}/+$ mice have small gonads (Green, 1989). A comparison of the nucleotide sequences of normal *c-kit* gene and the *c-kit* gene in the mutant mice revealed a missense mutation. Because mice heterozygous for the W^{42} mutant allele are affected, the authors concluded that the mutant allele was producing a protein that impaired the ability of the tyrosine protein kinase receptor to bind an unknown ligand. This information may provide direction for future experiments designed to understand the importance of the proto-oncogene *c-kit* for primordial germ cells and for defining the defects in the other *W*-sterile alleles.

V. Conclusions

The necessary experimental framework for elucidating the genes controlling specific steps in spermatogenesis is now available. Genes impacting spermatogenesis have been identified by mutation; genes that are expressed at specific times in spermatogenesis have been reported, though their function is unknown in many cases; germ cells within the testis can be isolated and purified; and the developmental stages of spermatogenesis have been defined. The new molecular technology provides the experimental tools needed to define the molecular basis of genes regulating spermatogenesis.

The recently described panel of genes that are expressed in specific germ cells can be used for targeted mutagenesis to understand their function in inbred mice. The actions of the mutated genes would have to be defined at biochemical, ultrastructural, and functional levels. Chimeric mice could be used to ascertain if the gene is affecting germ cells directly or indirectly via an action on nongerminal cells within the testis.

The goal is to define the genes controlling spermatogenesis. Several germ cell specific DNA probes are already available (Hecht, 1987). Armed with this basic information, contraceptives could be developed by targeting genes or proteins essential for defined steps of spermatogenesis. The contraceptive effect could be reversed by ceasing treatment; the next generation of germ cells would be unaffected. Conversely, diagnostic tests for male infertility could integrate probes for the specific genes regulating spermatogenesis. The results could indicate if therapy would be effective in reversing the infertility. The therapies for abnormal spermatogenesis could be developed and tested based on the homologies between mouse and human spermatogenesis and genomes. Finally, our knowledge of mammalian developmental biology would be increased by understanding how spermatogenesis can continually produce millions of male gametes on a daily basis.

ACKNOWLEDGMENTS

I thank Pierce Goetz for the excellent illustrations. Mr. Goetz was a graduate student in the Biomedical Illustration program at The University of Texas Southwestern Medical Center at Dallas. The author was supported by a Research Career Development Award from the National Institute of Child Health and Human Disease (KO4 HD00725).

REFERENCES

Baumann, N. A., Jacque, C. M., Pollet, S. A., and Harpin, M. L. (1968). Fatty acid and lipid composition of the brain of a myelin deficient mutant, the "quaking" mouse. Eur. J. Biochem. 4:340–344.

Beamer, W. G., Cunliffe-Beamer, T. L., Shultz, K. L., Langley, S. H., and Roderick, T. H. (1988). Juvenile spermatogonial depletion (*jsd*): A genetic defect of germ cell proliferation of male mice. Biol. Reprod. 38:899–908.

Bellvé, A. R. (1979). The molecular biology of mammalian spermatogenesis, in *Oxford Reviews of Reproductive Biology* (Finn, C. A., ed.) Vol. 1, pp. 159–261, Clarendon Press, Oxford.

Bellvé, A. R., and O'Brien, D. A. (1983). The mammalian spermatozoon: Structure and temporal assembly, in *Mechanism and Control of Animal Fertilization* (Hartmann, J. F., ed.) pp. 55–137, Academic Press, New York.

Bennett, W. I., Gall, A. M., Southard, J. L., and Sidman, R. L. (1971). Abnormal spermiogenesis in quaking, a myelin-deficient mutant mouse. Biol. Reprod. 5:30–58.

Bode, V. C. (1984). Ethylnitrosourea mutagenesis and the

isolation of mutant alleles for specific genes located in the *T* region of mouse chromosome 17. Genetics 108:457–470.

Braun, R. E., Behringer, R. R., Peschon, J. J., Brinster, R. L., and Palmiter, R. D. (1989). Genetically haploid spermatids are phenotypically diploid. Nature 337:373–376.

Bryan, J.H.D. (1977a). Spermatogenesis revisited. III. The course of spermatogenesis in a male-sterile pink-eyed mutant type in the mouse. Cell Tiss. Res. 180:173–186.

Bryan, J.H.D. (1977b). Spermatogenesis revisited. IV. Abnormal spermiogenesis in mice homozygous for another male-sterility-inducing mutation, *hpy* (hydrocephalic-polydactyl). Cell Tiss. Res. 180:187–201.

Bulfield, G. (1981). Inborn errors of metabolism in the mouse. Symp. Zool. Soc. (Lond.) 47:643–665.

Burgoyne, P. S. (1987). The role of the mammalian Y chromosome in spermatogenesis. Development 101 (Suppl):133–141.

Butler, W. J., and McDonough, P.G. (1989). The new genetics: Molecular technology and reproductive biology. Fertil. Steril. 51:375–386.

Camper, S. A. (1987). Research applications of transgenic mice. BioTechniques 5:638–650.

Capecchi, M. R. (1989a). The new mouse genetics: Altering the genome by gene targeting. Trends in Genetics 5:70–76.

Capecchi, M. R. (1989b). Altering the genome by homologous recombination. Science 244:1288–1292.

Chubb, C. (1992). Oligotriche and quaking gene mutations: phenotypic effects on mouse spermatogenesis and testicular steroidogenesis. J. Androl. (In press).

Chubb, C. (1989). Genetically defined mouse models of male infertility. J. Androl. 10:77–88.

Chubb, C., and Nolan, C. (1987). Mouse hybrid sterility and testicular function. Biol. Reprod. 36:1343–1348.

Chubb, C., and Nolan, C. (1985). Animal models of male infertility: Mice bearing single-gene mutations that induce infertility. Endocrinology 117:338–346.

Chubb, C., and Nolan, C. (1984). Genetic control of steroidogenesis and spermatogenesis in inbred mice. Ann. N.Y. Acad. Sci. 438:519–522.

Chung, B-C., Picado-Leonard, J., Haniu, M., Bienkowski, M., Hall, P. F., Shively, J. E., and Miller, W. L. (1987). Cytochrome P450c17 (steroid 17α-hydroxylase/17,20 lyase): Cloning of human adrenal and testis cDNAs indicates the same gene is expressed in both tissues. Proc. Natl. Acad. Sci. (USA) 84:407–411.

Chung, B-C., Matteson, K. J., Voutilainen, R., Mohandas, T. K., and Miller, W. L. (1986). Human cholesterol side-chain cleavage enzyme, P450scc: cDNA cloning, assignment of the gene to chromosome 15, and expression in the placenta. Proc. Natl. Acad. Sci. (USA) 83:8962–8966.

Cole, A., Meistrich, M. L., Cherry, L. M., and Trostle-Weige, P. K. (1988). Nuclear and manchette development in spermatids of normal and *azh/azh* mutant mice. Biol. Reprod. 38:385–401.

Coniglio, J. G., Grogan, W. M. Jr., Harris, D. G., and Fitzhugh, M. L. (1975). Lipid and fatty acid composition of testes of quaking mice. Lipids 10:109–112.

De Kretser, D. M., and Kerr, J. B. (1988). The cytology of the testis, in *The Physiology of Reproduction* (Knobil, E., Neill, J. D., Ewing, L. L., Greenwald, G. S., Markert, C. L., and Pfaff, D. W., eds.) pp. 837–932, Raven Press, New York.

Desjardins, C. (1985). Morphological, physiological, and biochemical aspects of male reproduction, in *Reproductive Toxicology* (Dixon, R. L., ed.) pp. 131–146, Raven Press, New York.

Distel, R. J., Kleene, K. C., and Hecht, N. B. (1984). Haploid expression of a mouse α-tubulin gene. Science 224:68–70.

Duchen, L. W., and Strich, S. J. (1968). An hereditary motor neuron disease with progressive denervation of muscle in the mouse: The mutant "wobbler." J. Neurol. Neurosurg. Psychiat. 31:535–542.

Eddy, E. M. (1988). The spermatozoon, in *The Physiology of Reproduction* (Knobil, E., Neill, J.D., Ewing, L. L., Greenwald, G.S., Markert, and C. L., Pfaff, D. W., eds.) pp. 27–68, Raven Press, New York.

Festing, M.F.W. (1989). Inbred strains of mice, in *Genetic Variants and Strains of the Laboratory Mouse* (Lyon, M. F., and Searle, A. G. eds.) 2nd ed., pp. 636–648, Oxford University Press, Oxford.

Fitch, W. M., and Atchley, W. R. (1985). Evolution in inbred strains of mice appears rapid. Science 228:1169–1175.

Forejt, J., and Ivânyi, P. (1975). Genetic studies on male sterility of hybrids between laboratory and wild mice (*Mus musculus* L.). Genet. Res. 24:189–206.

Green, M. C. (1989). Catalog of mutant genes and polymorphic loci, in *Genetic Variants and Strains of the Laboratory Mouse* (Lyon, M. F., and Searle, A. G., eds.) 2nd ed., pp. 12–403, Oxford University Press, Oxford.

Guraya, S. S. (1987). *Biology of Spermatogenesis and Spermatozoa in Mammals,* pp. 72–302, Springer-Verlag, New York.

Handel, M. A. (1987). Genetic control of spermatogenesis in mice, in *Results and Problems in Cell Differentiation* (Hennig, W., ed.) pp. 1–62, Springer-Verlag, Berlin.

Handel, M. A., Lane, P. W., Schroeder, A. C., and Davisson, M. T. (1988). New mutation causing sterility in the mouse. Gamete Res. 21:409–423.

Handel, M. A., Washburn, L. L., Rosenberg, M. P., and Eicher, E. M. (1987). Male sterility caused by p^{6H} and *qk* mutations is not corrected in chimeric mice. J. Exp. Zool. 243:81–92.

Handel, M. A., and Dawson, M. (1981). Effects on spermiogenesis in the mouse of a male sterile neurological mutation, Purkinje cell degeneration. Gamete Res. 4:185–192.

Hash, D. C., and Wolfe, H. G. (1979). Pink-eyed dilution alleles affect negative surface charges of mouse spermatozoa. Develop. Genet. 1:61–68.

Hecht, N. B. (1987). Detecting the effects of toxic agents on spermatogenesis using DNA probes. Env. Health Perspectives 74:31–40.

Hecht, N.B., Bower, P.A., Waters, S.H., Yelick, P.C., and Distel, R. J. (1986). Evidence for haploid expression of mouse testicular genes. Exp. Cell Res. 164:183–190.

Hollander, W. F., Bryan, J. H. D., and Gowen, J. W. (1960). A male sterile pink-eyed mutant type in the mouse. Fertil. Steril. 11:316–324.

Hunt, D. M., and Johnson, D. R. (1971). Abnormal spermiogenesis in two pink-eyed sterile mutants in the mouse. J. Embryol. Exp. Morph. 26:111–121.

Irons, M. J., and Clermont, Y. (1982a). Kinetics of fibrous sheath formation in the rat spermatid. Am. J. Anat. 165:121–130.

Irons, M. J., and Clermont, Y. (1982b). Formation of the outer dense fibers during spermiogenesis in the rat. Anat. Rec. 202:463–471.

Johnson, D. R., and Hunt, D. M. (1971). Hop-sterile, a mutant gene affecting sperm tail development in the mouse. J. Embryol. Exp. Morph. 25:223–236.

Justice, M. J., and Bode, V. C. (1986). Induction of new mutations in a mouse *t*-haplotype using ethylnitrosourea mutagenesis. Genet. Res. 47:187–192.

Kelly, N. J., and Woolley, D. M. (1980). The effects of the Wobbler gene on germ-cell differentiation in the mouse testis. J. Physiol. 303:11P.

Kierszenbaum, A. L., Ueda, H., Ping, L., Abdullah, M., and Tres, L. L. (1988). Antibodies to rat Sertoli cell secretory proteins recognize antigenic sites in acrosome and tail of developing spermatids and sperm. J. Cell Sci. 91:145–153.

Lalley, P. A., Davisson, M. T., Graves, J.A.M., O'Brien, S. J., Womack, J. E., Roderick, T. H., Creau-Goldberg, N., Hillyard, A. L., Doolittle, D. P., and Rogers, J. A. (1989). Report of the committee on comparative mapping. Cytogenet. Cell Genet. 51:503–532.

Leestma, J. E., and Sepsenwol, S. (1980). Sperm tail axoneme alterations in the Wobbler mouse. J. Reprod. Fertil. 58:267–270.

Lewis, S. E., Turchin, H. A., and Wojtowicz, T. E. (1978). Fertility studies of complementing genotypes at the albino locus of the mouse. J. Reprod. Fertil. 53:197–202.

Lorence, M. C., Corbin, C. J., Kamimura, N., Mahendroo, M. S., and Mason, J. I. (1990). Structural analysis of the gene encoding human 3β-hydroxysteroid dehydrogenase/Δ^{5-4}-isomerase. Mol. Endocrinol. 4:1850–1855.

Lyon, M. F. (1989). Rules for nomenclature of inbred strains, in *Genetic Variants and Strains of the Laboratory Mouse* (Lyon, M. F., and Searle, A. G., eds.) 2nd ed., pp. 632–635, Oxford University Press, Oxford.

Lyon, M. F. (1988). Mouse chromosome atlas. Mouse News Letter 81:20–41.

Lyon, M. F. (1986). Male sterility of the mouse *t*-complex is due to homozygosity of the distorter genes. Cell 44:357–363.

McAlpine, P. J., Shows, T. B., Boucheix, C., Stranc, L. C., Berent, T. G., Pakstis, A. J., and Douté, R. C. (1989). Report of the nomenclature committee and the 1989 catalog of mapped genes. Cytogenet. Cell Genet. 51:13–66.

Meistrich, M. L., Trostle-Weige, P. K., and Russell, L. D. (1990). Abnormal manchette development in spermatids of *azh/azh* mutant mice. Am. J. Anat. 188:74–86.

Melcangi, R. C., Celotti, F., Ballabio, M., Castano, P., Poletti, A., Milani, S., and Martini, L. (1988). Ontogenetic development of the 5α-reductase in the rat brain: Cerebral cortex, hypothalamus, purified myelin and isolated oligodendrocytes. Develop. Brain Res. 44:181–188.

Melvold, R. W. (1974). The effects of mutant *p*-alleles on the reproductive system in mice. Genet. Res. 23:319–325.

Miller, W. L. (1988). Molecular biology of steroid hormone synthesis. Endocrine Reviews 9:295–318.

Morel, Y., Picado-Leonard, J., Wu, D-A, Chang, C-Y., Mohandas, T. K., Chung, B-C., and Miller, W. L. (1988). Assignment of the functional gene for human adrenodoxin to chromosome 11q13→qter and of adrenodoxin pseudogenes to chromosome 20cen→q13.1. Am. J. Hum. Genet. 43:52–59.

Moutier, R. (1976). New mutations causing sterility restricted to the male in rats and mice, in *The Laboratory Animal in the Study of Reproduction* (Antikatzides, Th., Erichsen, S., and Spiegel, A., eds.) pp. 115–117, Gustav Fischer Verlag, Stuttgart.

Mullen, R. J., Eicher, E. M., and Sidman, R. L. (1976). Purkinje cell degeneration, a new neurological mutation in the mouse. Proc. Nat. Acad. Sci. (USA) 73:208–212.

Nadeau, J. H. (1989). Maps of linkage and synteny homologies between mouse and man. Trends in Genetics 5:82–86.

Nakayama, H., Kuroda, H., Onoue, H., Fujita, J., Nishimune, Y., Matsumoto, K., Nagano, T., Suzuki, F., and Kitamura, Y. (1988). Studies of *Sl/Sl*^{d}↔+/+ mouse aggregation chimaeras. II. Effect of the steel locus on spermatogenesis. Development 102:117–126.

O'Brien, D. A. (1987). Stage-specific protein synthesis by isolated spermatogenic cells throughout meiosis and early spermiogenesis in the mouse. Biol. Reprod. 37:147–157.

O'Brien, D. A., and Bellvé, A. R. (1980). Protein constituents of the mouse spermatozoon. II. Temporal synthesis during spermatogenesis. Develop. Biol. 75:405–418.

Olds-Clarke, P., and Peitz, B. (1985). Fertility of sperm from *t*/+ mice: Evidence that +- bearing sperm are dysfunctional. Genet. Res. 47:49–52.

Palmiter, R. D., and Brinster, R. L. (1985). Transgenic mice. Cell 41:343–345.

Peltoketo, H., Isomaa, V., Mäentausta, O., and Vihko, R. (1988). Complete amino acid sequence of human placental 17β-hydroxysteroid dehydrogenase deduced from cDNA. FEBS Lett. 239:73–77.

Popko, B., Puckett, C., Lai, E., Shine, H. D., Readhead, C., Takahashi, N., Hunt, S. W. III, Sidman, R. L., and Hood, L. (1987). Myelin deficient mice: Expression of myelin basic protein and generation of mice with varying levels of myelin. Cell 48:713–721.

Richa, J., and Lo, C. W. (1989). Introduction of human DNA into mouse eggs by injection of dissected chromosome fragments. Science 245:175–177.

Rinchik, E. M. (1987). Molecular analysis of heritable mouse mutations. Env. Health Perspectives 74:41–48.

Russell, E. S. (1985). A history of mouse genetics. Annu. Rev. Genet. 19:1–28.

Russell, W. L., Kelly, E. M., Hunsicker, P. R., Bangham, J. W., Maddux, S. C., and Phipps, E. L. (1979). Specific-locus test shows ethylnitrosourea to be the most potent mutagen in the mouse. Proc. Natl. Acad. Sci. (USA) 76:5818–5819.

Sawyer, J. R., and Hozier, J. C. (1986). High resolution of mouse chromosomes: Banding conservation between man and mouse. Science 232:1632–1635.

Searle, A. G. (1982). The genetics of sterility in the mouse, in *Genetic Control of Gamete Production and Function* (Crosignani, P. G., and Rubin, B. L., eds.) pp. 93–114, Academic Press, London.

Sidman, R. L., Green, M. C., and Appel, S. H. (1965). *Catalog of the Neurological Mutants of the Mouse.* Harvard University Press, Cambridge.

Sidman, R. L., Dickie, M. M., and Appel, S. H. (1964). Mutant mice (quaking and jimpy) with deficient myelination in the central nervous system. Science 144:309–311.

Silver, L. M. (1985). Mouse *t* haplotypes. Annu. Rev. Genet. 19:179–208.

Silvers, W. K. (1979). *The Coat Colors of Mice.* Springer-Verlag, New York.

Slaughter, G. R., Meistrich, M. L., and Means, A. R. (1989). Expression of RNAs for calmodulin, actins, and tubulins in rat testis cells. Biol. Reprod. 40:395–405.

Sotomayor, R. E., and Handel, M. A. (1986). Failure of acrosome assembly in a male sterile mouse mutant. Biol. Reprod. 34:171–182.

Stewart, H. J., Jones, D.S.C., Pascall, J. C., Popkin, R. M., and Flint, A.P.F. (1988). The contribution of recombinant DNA techniques to reproductive biology. J. Reprod. Fertil. 83:1–57.

Tan, J.C., Nocka, K., Ray, P., Traktman, P., and Besmer, P. (1990). The dominant W^{42} spotting phenotype results from a missense mutation in the *c-kit* receptor kinase. Science 247:209–212.

The, V. L., Lachance, Y., Labrie, C., Leblanc, G., Thomas, J. L., Strickler, R. C., and Labrie, F. (1989). Full length cDNA structure and deduced amino acid sequence of human 3,β-hydroxy-5-ene steroid dehydrogenase. Mol. Endocrinol. 3:1310–1312.

Thomas, K. H., Wilkie, T. M., Tomashefsky, P., Bellvé, A. R., and Simon, M. I. (1989). Differential gene expression during mouse spermatogenesis. Biol. Reprod. 41:729–739.

Tremblay, Y., Ringler, G. E., Morel, Y., Mohandas, T. K., Labrie, F., Strauss, J.F. III, and Miller, W. L. (1989). Regulation of the gene for estrogenic 17-ketosteroid reductase lying on chromosome 17cen→q25. J. Biol. Chem. 264:20458–20462.

Varnum, D. S. (1983). Blind-sterile: A new mutation on chromosome 2 of the house mouse. J. Heredity 74:206–207.

Willison, K., and Ashworth, A. (1987). Mammalian spermatogenic gene expression. Trends in Genetics 3:351–355.

Wolfe, H. G., Erickson, R. P., and Schmidt, L. C. (1977). Effects on sperm morphology by alleles at the pink-eyed dilution locus in mice. Genetics 85:303–308.

Youngblood, G. L., Nesbitt, M. N., and Payne, A. H. (1989). The structural genes encoding $P450_{scc}$ and $P450_{arom}$ are closely linked on mouse chromosome 9. Endocrinology 125:2784–2786.

CHAPTER FIVE

Mullerian Inhibiting Substance

Mary M. Lee, Tatsuo Kuroda, and Patricia K. Donahoe

I. Introduction

Mullerian inhibiting substance is a testicular hormone that causes involution of the mullerian ducts in male fetuses during normal genital tract differentiation (reviewed in Josso et al., 1977; Donahoe et al., 1982, 1987; Josso and Picard, 1986; Cate et al., 1990). In early fetal development, the gonads are bipotential and the mullerian and wolffian ducts are both present in the urogenital ridge: the former are the anlagen of the uterus, fallopian tubes, and the upper third of the vagina, and the latter are the embryologic precursors of the epididymis, vas deferens, and seminal vesicles. During normal female reproductive tract development, the mullerian ducts develop spontaneously, and the wolffian ducts atrophy in the absence of testosterone. In males, the wolffian ducts are stimulated to differentiate by testosterone, and the mullerian ducts regress under the influence of another hormone secreted by Sertoli cells. This hormone, termed mullerian inhibiting substance (MIS), has been identified as a 140 kDa glycoprotein composed of two identical 70 kDa subunits. The MIS gene, which has been cloned and sequenced (Cate et al., 1986a), is a member of the transforming growth factor-β (TGF-β) family of proteins that are all involved in growth and differentiation (Sporn et al., 1987; Massague, 1990). The purification and characterization of MIS, including its gene regulation, posttranslational modification, and mechanism of action are reviewed in this chapter. Current and potential areas of MIS research, including newly discovered functions of the hormone, are discussed.

Alfred Jost proposed that a testicular factor distinct from testosterone caused regression of the mullerian duct, the anlagen of the uterus, fallopian tube, and vagina, and was necessary for normal reproductive tract development in the male (Jost, 1947, 1953). He removed the gonads of young, sexually undifferentiated embryos from the hardy wild rabbit indigenous to the Bois de Boulogne, then provided hormonal replacement by implanting either testicular tissue or testosterone. Whereas the presence of testicular tissue resulted in both differentiation of the wolffian ducts and regression of the mullerian ducts, testosterone alone caused only differentiation of the wolffian ducts but not regression of the mullerian ducts, resulting in the growth of both male and female genital tract structures. These experiments demonstrated that another testicular factor besides testosterone was responsible for mullerian duct involution during normal male sexual differentiation. This factor was first termed "mullerian inhibitor" by Jost (1953), and later, mullerian inhibiting substance (Josso, 1972; Donahoe et al., 1977a; Cate et al., 1986a) or anti-mullerian hormone (Jossò et al., 1977).

Although MIS has a clearly defined role in sexual differentiation, the testicular production of MIS beyond the stage of genital tract differentiation, and the presence of MIS in graafian follicles of the pubertal and mature ovary, suggest that MIS has additional functions (discussed in Donahoe et al., 1982, 1987; Josso and

Picard, 1986). During fetal development, MIS may play a role in testicular descent (Hutson and Donahoe, 1986) and in inhibition of surfactant production and lung maturation (Catlin et al., 1988, 1990). MIS can inhibit oocyte meiosis in the ovary (Takahashi et al., 1986b; Ueno et al., 1988), and has been postulated to have a similar role in germ cell meiotic arrest and Sertoli cell mitotic arrest in the testis (Jost, 1972; Taketo et al., 1991). Purified MIS can cause germ cell depletion and formation of seminiferous cordlike structures in cultures of rat ovaries (Vigier et al., 1987), and can convert ovarian steroidogenesis from the production of estradiol to testosterone by inhibiting aromatase activity (Vigier et al., 1989). Female transgenic mice that chronically express MIS not only lacked mullerian structures, but also were agonadal (Behringer et al., 1990). Germ cells were present at birth, but rapidly degenerated, and follicles transiently formed tubular structures, then also degenerated. Unexpectedly, two of the male mice had feminized external genitalia and abnormal wolffian duct development. Thus, MIS may have a number of pleiotropic roles in addition to its well-characterized effects on involution of the mullerian ducts during male genital tract differentiation.

II. Bioassay for Mullerian Inhibiting Substance

Mullerian duct regression occurs early in embryonic development. In the rat, Sertoli cell differentiation and seminiferous tubule formation start on day 13 postcoitus when MIS mRNA and protein are first detectable in the testes (reviewed in Josso and Picard, 1986; Donahoe et al., 1984). By day 15 postcoitus, hyaluronidase activity is increased, and the extracellular matrix constituents start to disappear as dissolution of the duct occurs (Hayashi et al., 1982). Laminin, heparin sulfate proteoglycan, type IV collagen, and fibronectin disappear progressively (Ikawa et al., 1984), and by day 16, the basement membrane has disappeared and hyaluronidase activity is lost. On day 17, mesenchymal condensation is the only remnant of the mullerian duct (Trelstad et al., 1982).

In humans, by 51 days postovulation, male fetal genital tracts have signs of ductal regression that are not reversed by prolonged incubation in the absence of testicular tissue (Taguchi et al., 1984). Co-culture of the urogenital ridge of female fetuses less than 25 mm in crown–rump length with testicular tissue results in regression of the mullerian duct, while the ducts of fetuses greater than 30 mm in crown–rump length are not inhibited by MIS (Josso et al., 1977). Thus, the short period of responsiveness of the mullerian duct to MIS has rendered the study of MIS particularly difficult.

A bioassay to moniter MIS activity in vitro, first developed by Picon (1969), was a crucial step in advancing the study of this hormone. Fetal rat urogenital ridges were excised and placed in organ culture for 3 days, then examined histologically for regression of the mullerian duct. Picon found a critical period of sensitivity at 14.5 days postcoitus when the mullerian ducts of both male and female rat fetuses were identical histologically and responded similarly in culture. At 14.5 days the mullerian ducts developed in the absence of testes and regressed when co-cultured with testes. The urogenital ridges of older female fetuses exhibited incomplete regression of the mullerian duct when cultured with testes producing MIS, whereas the mullerian ducts of older male fetuses involuted even in the absence of co-cultured testicular tissue. Thus, the sensitivity of the mullerian ducts to the inhibitory effects of MIS depends primarily on the age of the urogenital ridge. Biochemical and cellular changes characteristic of ductal regression proceed irreversibly if exposure to MIS has occurred before 15 days, even if MIS is subsequently removed (Picon, 1969; Donahoe et al., 1977a). Conversely, ducts that have not been exposed to MIS by 15 days gestation no longer involute in response to the hormone (Picon, 1969; Donahoe et al., 1977a).

Donahoe and colleagues (1977a) modified Picon's bioassay to establish a graded organ culture assay for MIS bioactivity in either conditioned media or tissue (Fig. 5–1). In this assay, urogenital ridges of 14.5 day female rat embryos are placed on agar-coated stainless steel grids, and incubated with the tissue fragment to be tested or with conditioned media containing MIS. Because only female fetal calf serum is used in the media, testosterone is added to

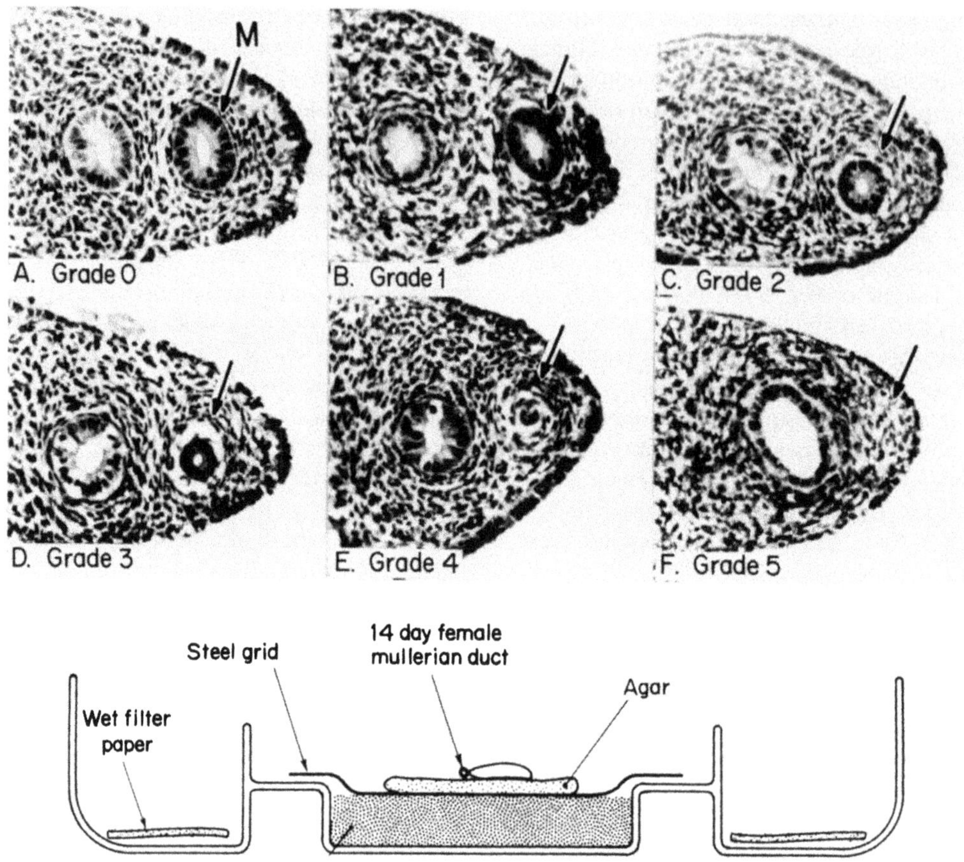

Figure 5–1. Organ culture assay for mullerian inhibiting substance. A schematic of the organ culture assay dish is depicted in the lower drawing. Photomicrographs of the graded stages of mullerian duct regression, from (0) no regression to (5+) complete regression are shown in the upper portion. The mullerian duct is labeled "M." ×200 magnification.

the media at 10^{-9} M, a concentration that by itself does not cause regression of the mullerian duct but both enhances the effect of MIS (Ikawa et al., 1982), and stimulates growth of the wolffian duct, thus facilitating morphologic comparison. After 72 hours incubation in humidified 5% CO_2/95% air at 37°C, the specimen is aligned with an additional layer of agar, then fixed, and embedded in paraffin. Eight-μm cross sections of the cephalic end of the duct are cut and stained with hematoxylin and eosin. Regression of the mullerian duct is graded from 0 (no regression) to 5 (complete regression) by at least two independent observers. The assay requires 1 to 2 μg/mL of MIS for ductal regression to occur. Because of the differing sensitivity of the cephalic and caudal ends of the urogenital ridge to MIS, Josso and colleagues (1977), using a similar assay, quantitate ductal regression by determining the extent (length) of the regression and grade the regression as "complete, near-complete, incomplete, and no regression." Baker and colleagues (1990) have set up a similar assay using the murine urogenital ridge. The organ culture assay of mullerian duct regression, although cumbersome, is still the most specific assay for MIS activity, and the one with which all other assays are compared.

More recently, another bioassay has been established based on the observation that MIS inhibits ovarian aromatase activity (Vigier et al., 1989). In this assay, fetal ovaries are incubated with MIS and tritiated androstenedione for 3 days. The aromatase activity of the gonads is quantitated by measuring the amount of tritiated water generated during the incubation. The inhibition of aromatase activity correlates linearly with the log concentration of MIS in the culture, and therefore, offers an alternative method of quantitating MIS bioactivity.

III. Immunological Detection of Mullerian Inhibiting Substance

The low sensitivity, technical difficulty, and length of time required for the bioassays encouraged the development of a number of more sensitive immunological assays for the detection of MIS. Because MIS was initially purified from neonatal calf testes, the first sandwich-type solid-phase assays developed recognized only bovine MIS (Vigier et al., 1982; Necklaws et al., 1986). Moreover, the impure MIS used in the calibration of these assays resulted in inconsistencies in the quantitation and standardization of MIS. Thus, a more sensitive liquid-phase competition type radioimmunoassay (RIA) was developed using anti-MIS polyclonal antibodies that cross-reacted with nonruminant MIS by immunohistochemical staining (Vigier et al., 1985). This RIA still was unable to detect human MIS. Cloning of the human MIS complementary DNA (cDNA) and gene (Cate et al., 1986a), however, enabled the production of human recombinant MIS (rMIS) on a large scale, thus facilitating the development of polyclonal and monoclonal antibodies to human rMIS. These antibodies were then used to establish enzyme linked immunosorbant assays (ELISA) for the detection of human MIS (Hudson et al., 1990; Josso et al., 1990; Baker et al., 1990).

In the ELISA, Mullerian inhibiting substance can be detected in human serum at the low nanogram/mL range, with a sensitivity as low as 0.4 fmol (Hudson et al., 1990). Using this ELISA, MIS can be detected in the serum of normal newborns, infants, children, and adults (Fig. 5-2). Serum levels of MIS in young males are 10 to 70 nanograms/mL at birth, increase

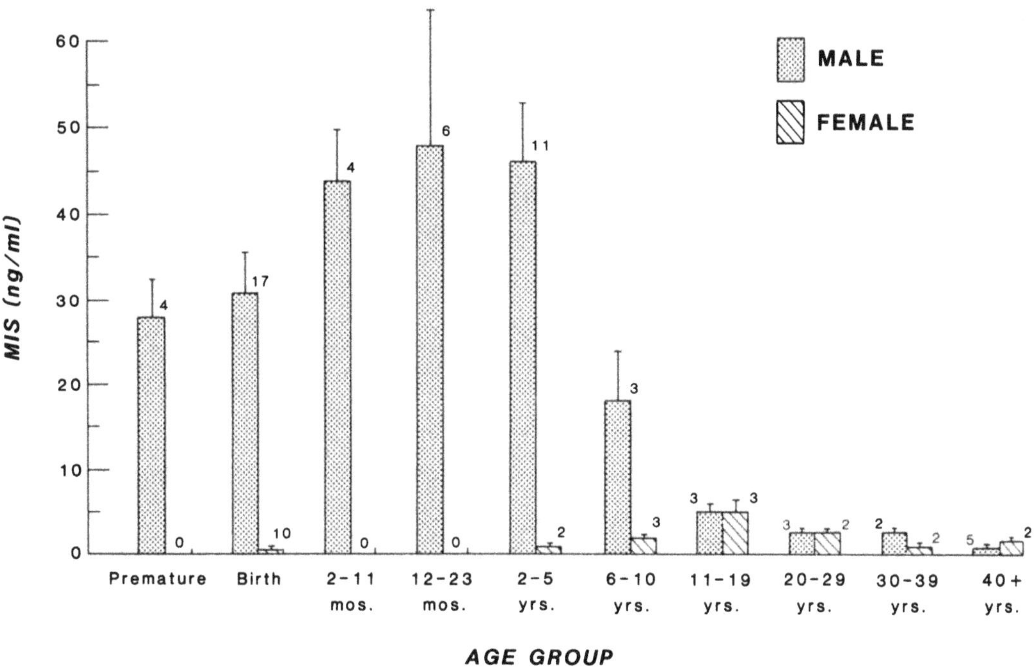

Figure 5-2. Serum levels of mullerian inhibiting substance (MIS) from infancy to adulthood, as measured by ELISA. The numbers of patients assayed for each group are shown over the bars. In men, MIS serum levels peak during the first year of life and are detectable through adulthood. In women, MIS levels are first detectable in early puberty and persist at low levels until menopause. (Hudson et al., 1990, reprinted with permission.)

slightly during the first year, then fall to a lower level until age 10, when they decrease further to the basal levels (2 to 5 nanograms/mL) found from adolescence through adulthood (Hudson et al., 1990). In contrast to the high levels in serum, MIS is barely detectable in the urine of newborn males. In females, MIS cannot be detected in serum at birth, or during early childhood. During the second decade, however, MIS levels gradually rise to those seen in the adult male (i.e., 2 to 5 nanograms/mL), which then persist through the reproductive years. Similar levels were found in adult ovarian follicular fluid harvested during procedures for in vitro fertilization (C. Richards and D. MacLaughlin, personal communication).

The ELISA has proven to be useful for the evaluation of patients with inappropriately elevated MIS levels. One female patient with an MIS-secreting sex cord tumor, which was composed of Sertoli cells, had serum values as high as 3.2 mg/mL (500 to 1000 times normal) (Gustafson et al., 1992). The elevated MIS values correlated with recurrence of the tumor, and hence served as a tumor marker. Two other female patients with granulosa cell tumors had elevated MIS serum values (Gustafson et al., 1992). These sensitive assays for MIS have been useful in the evaluation of congenital intersex abnormalities (M. M. Lee and P. K. Donahoe, unpublished data) and may also prove useful in diagnosing gonadal dysfunction in disorders of infertility.

IV. Biology of Mullerian Inhibiting Substance

A. Biochemical and Molecular Characterization

Blanchard and Josso (1974) identified fetal Sertoli cells as the source of MIS, and hence used fetal bovine testes for purification of the hormone. Donahoe and colleagues (1977c) subsequently demonstrated that postnatal bovine testes continued to produce bioactive MIS and purified MIS from postnatal bovine testes (reviewed in Donahoe et al., 1982; MacLaughlin et al., 1991). The freshly collected testes were minced and incubated in media with 5mM benzamidine. The crude MIS protein fraction was precipitated in 30 to 45 percent ammonium sulfate from concentrated incubation media, then desalted on a Sephadex G-25 column, and further purified by sequential ion-exchange chromatography. This was then followed by serial lectin and dye affinity chromatography to achieve several thousand-fold purification of MIS (Budzik et al., 1980, 1983). These highly purified preparations provided the antigens for the development of monoclonal antibodies to bovine MIS, which in turn served as absorbants for the immunoaffinity purification of bovine MIS (Picard and Josso, 1984; Shima et al., 1984) and as the basis for the development of the immunoassays described previously (Vigier et al., 1985; Necklaws et al., 1986). Bovine MIS was also purified to near homogeneity using immunoaffinity chromatography following ion exchange chromatography (Shima et al., 1984; Picard and Josso, 1984). Bovine MIS cDNAs were then isolated using oligonucleotide probes derived from peptide sequences of purified MIS (Cate et al., 1986a). Using bovine cDNA as a hybridization probe, the human MIS gene was isolated and expressed in Chinese hamster ovary (CHO) cells in an expression vector carrying the SV40 early promoter (Cate et al., 1986a). The recombinant protein is purified by either serial column chromatography or by immunoaffinity chromatography from CHO cell conditioned media and is active in the organ culture assay (Cate et al., 1986b).

Purified bovine MIS is a 140-kDa dimer of two identical 70-kDa subunits linked by disulfide bonds (Fig. 5–3) (Budzik et al., 1983; Picard and Josso, 1984). On reducing sodium dodecylsulfate (SDS)-polyacrylamide gels, MIS has an electrophoretic pattern consisting of a major doublet at approximately 70 kDa and a minor band at 55 kDa (Picard and Josso, 1984; Shima et al., 1984). Purified human rMIS also has a predominant 70-kDa band, but the minor band at 57 kDa is more prominent (5 to 20 percent of total protein) than in the bovine preparations, and a third minor species of 12.5 kDa is also observed (Pepinsky et al., 1988; Cate et al., 1989). On nonreduced gels, the 12.5-kDa fragment migrates as a dimer of 25 kDa and the 57-kDa fragment migrates as a complex in the region of the intact protein. Protein sequencing indicates that the 70- and 57-kDa species of MIS have identical N-terminal sequences of Leu-Arg-Ala-Glu-Glu, whereas the 12.5-kDa species has a sequence of Ser-Ala-Gly-Ala-Thr.

Thus, the cleavage site is near the carboxyl terminus of MIS between Arg 427 and Ser 428 (see Fig. 5–3), generating the 57-kDa and the 12.5-kDa species (Pepinsky et al., 1988; Cate et al., 1989). The cleavage site is located at a position similar to the cleavage sites found in TGF-β and related proteins (Derynck et al., 1985), but the pattern of amino acids around the site resembles the monobasic consensus cleavage site in prosomatostatin (Benoit et al., 1987), rather than the dibasic site found in the TGF-β family of proteins (Sporn et al., 1987; Massague, 1990).

TGF-β is produced as a dimeric precursor that requires cleavage and dissociation of the complex to release the bioactive disulfide-linked C-terminal fragment (Derynck et al., 1985; Sporn et al., 1987). The in vivo activation mechanism of TGF-β is not known, but in vitro, TGF-β can be activated by boiling, acid treatment, or plasmin cleavage (Sporn et al., 1987; Massague, 1990). Because full-length MIS is active in the organ culture assay, it was originally thought that MIS did not require cleavage for activation; however, the presence of a monobasic cleavage site at position 427 provided evidence that like TGF-β and inhibin, MIS may also undergo proteolytic processing to become biologically active. Plasmin or trypsin treatment of human rMIS generates 57- and 12.5-kDa fragments that when incubated together, have activity similar to untreated MIS in the organ culture assay (Pepinsky et al., 1988). When the plasmin-treated preparation of MIS is immunoaffinity purified with a monoclonal antibody specific for the N-terminus, the C-terminal dimer co-purifies with the N-terminal dimer, indicating that the fragments remain in noncovalent association (Cate et al., 1989). The complex remains active in the bioassay when it is dissociated with 1 percent sodium deoxycholate, but unlike TGF-β, it loses bioactivity when dissociated with acidification or boiling. When Arg 427 at the cleavage site is converted to threonine by site-directed mutagenesis, the mutated dimeric MIS secreted from CHO cells can not be cleaved by plasmin and is inactive in the organ culture assay, indicating that proteolytic processing of MIS is obligatory (Cate et al., 1989, 1990). More recent work has shown that cleaved MIS, dissociated under acid conditions followed by immediate neutralization, retains biological activity. Furthermore, studies using purified C- and N-terminal fragments indicate that the biological activity of MIS resides in the C-terminal moiety. Addition of the N-terminal fragment does not augment or antagonize the biological activity of the C-terminal fragments (D. T. MacLaughlin and P. K. Donahoe, unpublished data).

Mullerian inhibiting substance is composed of 560 amino acids with a 24-amino acid leader sequence consisting of a 16- or 17-amino acid signal peptide and a 7- or 8-amino acid putative pro sequence (see Fig. 5–3) (Cate et al., 1990). Bovine and human MIS share about 78 percent homology, with conservation of all 12 cyste-

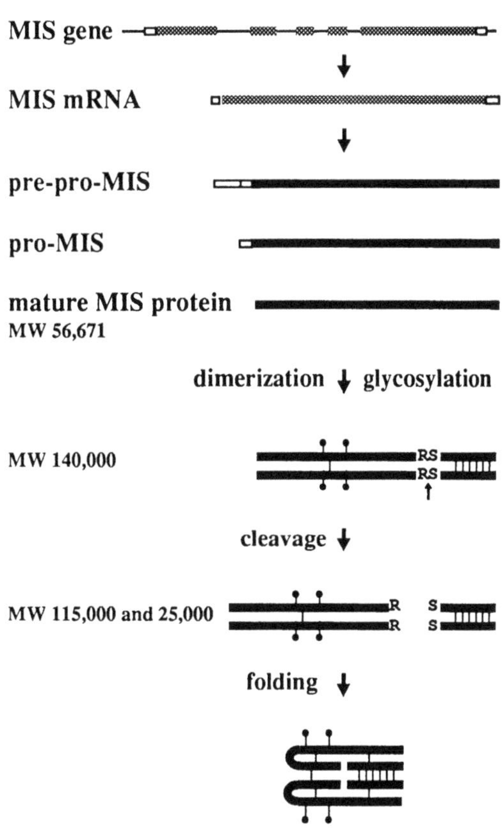

Figure 5–3. Schematic of the mullerian inhibiting substance (MIS) gene and protein. The human MIS gene is composed of five exons that encode a prehormone from which putative leader and pro sequences are removed. The mature protein then undergoes glycosylation and dimerization. Activation of the protein is postulated to occur by cleavage at position 427 and by conformational changes of the protein.

ines. The C-terminus shows the strongest homology with 108 of the last 112 amino acids conserved (Cate et al., 1990). Bovine MIS contains a high proportion of hydrophobic amino acids and both basic and acidic amino acids are equally represented (Picard et al., 1986b). The carbohydrate content of MIS has been estimated to be between 8.3 percent (Budzik et al., 1980) and 13.5 percent (Picard et al., 1986b) and all of the consensus N-linked and O-linked glycosylation sites are found in the N-terminal portion of the molecule. Although the removal of N-linked complex carbohydrates with Endoglycosidase F does not appear to affect the biological activity of MIS (Cate et al., 1990), the role of glycosylation in the action or processing of MIS remains to be determined.

Secretion of nascent rMIS is very slow in vitro, with 50% of MIS still in the CHO cell 5 hours after pulse-labeling with [^{35}S]-cysteine (Cate et al., 1990). The secretion of MIS from cultured bovine Sertoli cells is also slow, requiring approximately 48 hours of cycloheximide treatment before MIS levels in culture media are suppressed (Vigier et al., 1985). Although the majority of intracellular MIS is in the rough endoplasmic reticulum by immunohistochemistry (Tran and Josso, 1982; Hayashi et al., 1984), MIS is also detectable in the Golgi apparatus (Hayashi et al., 1984). Thus, MIS is probably synthesized and packaged in the rough endoplasmic reticulum, then transported to the Golgi apparatus for further glycosylaton and secretion.

Bovine MIS cDNAs have been isolated independently by two laboratories using different approaches. Cate and colleagues (1986a) synthesized degenerate oligonucleotide probes from amino acid sequences derived from typtic digests of purified protein to screen a λgt10 cDNA library made with poly (A$^+$) messenger RNA (mRNA) from newborn calf testes. Picard and colleagues (1986a) also isolated a partial MIS cDNA by screening a λgt11 cDNA expression library from newborn calf testes with an anti-idiotype polyclonal antibody raised against purified bovine MIS.

Cate and colleagues (1986a) isolated a cosmid clone containing the human MIS gene by using the full-length MIS bovine cDNA as a hybridization probe to screen a human genomic library. The human MIS gene (see Fig. 5–3), composed of five exons, exhibits 75 percent homology with the bovine gene, and the human promotor shares 67 percent homology with the bovine promoter. Both the bovine and human genes have a high CG content of 72 percent in the exons and 68 percent in the introns and flanking regions. The 5′ untranslated regions of the bovine and human MIS mRNAs are relatively short (10 nucleotides in length) and, therefore, are probably not involved in translational control. The 3′ untranslated region of the human mRNA contains a 30-nucleotide insert, two nucleotides downstream of the TGA codon, that is not found in the bovine mRNA. The mouse (Munsterberg and Lovell-Badge, 1991) and rat (Haqq et al., 1992) MIS genes have also been cloned and show 68–73% overall homology to the bovine and human genes. The human MIS gene has been mapped to the short arm of chromosome 19 using somatic cell hybrid technology and in situ hybridization (Cohen-Hagenauer et al., 1987).

The human gene, inserted into an expression vector carrying the SV40 early promoter has been expressed in CHO cells deficient in dihydrofolate reductase along with a plasmid containing mouse dihydrofolate reductase cDNA. Cell lines have been obtained by methotrexate amplification that produce up to 3 to 5 mg/L/day of MIS (Epstein et al., 1989). The protein from the transfected CHO cell conditioned media is purified by ion exchange and carbohydrate affinity chromatography or by immunoaffinity chromatography, and is bioactive in the organ culture assay (Cate et al., 1986a). An amplified CHO cell line (B9) expressing high levels of biologically active human rMIS has been adapted to grow on stainless steel coils in modified bioreactors (Epstein et al., 1989), further increasing the production of MIS. The availability of human rMIS has proven to be essential in studying the complex processing and activation of MIS by the embryonic urogenital ridge.

Mullerian inhibiting substance has been found to be a member of a large multigene family that now includes the TGF-βs (Derynck et al., 1985), inhibin/activins (Mason et al., 1985), *Xenopus* Vg-1 (Weeks and Melton, 1987), decapentaplegia complex (Gelbart et al., 1985), and bone morphogenesis factors (Wozney et al., 1988). Members of this gene family have 20 to 40 percent C-terminal homology, particularly around six or seven highly conserved cysteine residues (Gelbart et al., 1985). The N-terminal domains are not homologous. TGF-β, activin, and inhibin have large latent precursor forms that are cleaved at a dibasic site to produce bioactive cysteine-rich C-terminal dimers and N-terminal dimers with unknown function (Sporn et al., 1987; Massague, 1990). The protein products of the Vg-1 and decapentaplegia genes are also thought to be cleaved at a dibasic site; MIS, however, has a monobasic cleavage site. For all members of this family, only the N-terminal fragments are glycosylated.

Although all of these proteins are important in developmental processes, such as embryogenesis, and act as regulators of cell growth and differentiation, many differences have been observed among the proteins. For example, MIS is found in the rat ovary in the periluminal and cumulus granulosa cells in preantral, antral, and preovulatory follicles, with a fall in its levels just before ovulation (Ueno et al., 1989a). Although inhibin is found in the same granulosa cells and wanes before ovulation, it is also detected in basilar granulosa cells and in cells of the corpus luteum (Woodruff et al., 1988). Neither TGF-β1 nor TGF-β2 is recognized by an ELISA for MIS (Hudson et al., 1990). TGF-β1 over a wide dose range fails to act as either an MIS agonist or antagonist in the organ culture bioassay (W. Russell and P.K. Donahoe, unpublished data), indicating absence of functional cross reactivity. TGF-β causes germinal vesicle breakdown in preovulatory cumulus enclosed oocytes when given to rats stimulated by pregnant mare's serum gonadotropin (PMSG) (Feng et al., 1988). In contrast, MIS has no effect on oocytes of either PMSG or human chorionic gonadotropin (hCG)-stimulated rats (S. Ueno and P.K. Donahoe, unpublished data), although it inhibits both denuded and cumulus-enclosed unstimulated oocytes (Ueno et al., 1988). TGF-β, inhibin/activin, and MIS, therefore, are related but distinct proteins that appear to play discrete functional roles.

B. Mechanism of Action of Mullerian Inhibiting Substance

The mechanism of action of MIS is not yet fully understood. It is likely, however, that in vivo MIS must be proteolytically processed to exhibit full activity because a noncleavable mutant of MIS has no biological activity in the bioassay (Cate et al., 1989, 1990). In addition to the requisite processing steps for MIS activation, the presence of both N- and C-terminal fragments may be important for activity, in contrast to TGF-β (Sporn et al., 1987; Massague, 1990). Isolated N-terminal and C-terminal dimers, when incubated alone in the organ culture, are inactive, but when assayed together, cause regression of the mullerian duct (R. Cate and P.K. Donahoe, unpublished data). The N- and C-terminal fragments have also been tested in a bioassay that measures the ability of MIS to suppress aromatase activity in cultured rat ovaries, and thereby convert ovarian steroid production from estradiol to testosterone (Vigier et al., 1989). Preliminary evidence suggests that in this assay the C-terminal fragment is the active moiety, although co-culture with the N-terminal fragment potentiates its activity (R. Cate and N. Josso, personal communication). The different results obtained in the aromatase and the mullerian duct assays may be due to differential sensitivity of the two bioassays.

The activation of MIS probably occurs extracellularly because most of the protein secreted from testes is unprocessed. It is then presumably processed, perhaps near its target area, to become fully biologically active. Latent TGF-β, for example, in a co-culture system mimicking cellular interactions in vessel walls, was produced by endothelial cells and pericytes, both of which were required before activation of TGF-β and inhibition of endothelial cell growth could be observed (Antonelli-Orlidge et al., 1989). The microenvironment of the two cell types combined thus provided the requisite milieu for activation that neither cell type alone could provide. A similar cooperative effect may be occurring in the activation of MIS at the urogenital ridge, and at other sites of possible MIS action.

The specific binding of the TGF-β1 precursor to the mannose 6-phosphate receptor and the recent identification of mannose specific phosphorylation of its side chains (Purchio et al., 1988) suggest a possible targeting of the proteins in this gene family to lysosomes. The acidic and protease-rich environment of the lysosome may permit cleavage and dissociation to occur. Whether trafficking of the TGF-β1 precursor to the lysosome serves to scavenge, or to process and activate it, remains to be clarified.

The effects of MIS may occur through receptor-mediated transmembrane signal transduction, possibly involving a dephosphorylation event (Hutson et al., 1984). A potential interaction between MIS and epidermal growth factor (EGF) was investigated after similarities were noted between MIS and a membrane-bound phosphotyrosyl protein phosphatase that reverses EGF-stimulated phosphorylation (Brautigan et al., 1981). This phosphatase is activated by ethylenediamine tetraacetic acid (EDTA) and fluoride (Brautigan et al., 1981), and inhibited by zinc and vanadate, compounds that have been noted to have similar effects on MIS. EDTA and fluoride mimicked the effects of MIS on the mullerian duct, and zinc and vanadate inhibited MIS-induced regression (Hutson et al., 1984). EGF, with 1mM manganese as a co-factor, was therefore tested in the organ culture assay and was found to inhibit MIS-stimulated ductal regression (Hutson et al., 1984). Other growth factors with receptor tyrosine kinase activity, such as insulin and platelet-derived growth factor, did not inhibit MIS activity (Hutson et al., 1984; Budzik et al., 1985). Nerve growth factor and fibroblast growth factor also had no effect (Budzik et al., 1985). Thus, the interaction between MIS and EGF is specific. A basic tripeptide glycyl-L-histidyl-L-lysine, known as liver cell growth factor, however, inhibited MIS even in the absence of divalent cation, although addition of 1 mM manganese reduced the concentration of the peptide required for MIS inhibition (Budzik et al., 1985). None of these growth factors had any effects on the mullerian duct in the absence of MIS.

Subsequently, bovine MIS was tested on an EGF-receptor rich cell line (A-431, a human vulvar carcinoma line) that requires EGF for growth in soft agar. Purified bovine MIS was found to inhibit A-431 colony formation in a concentration-dependent manner (Coughlin et al., 1987). Bovine MIS also inhibited EGF-induced membrane tyrosine phosphorylation in A-431 cells without competing for EGF receptor binding sites and without increasing phosphotyrosine phosphatase activity (Coughlin et al., 1987). Recombinant MIS purified by dye or carbohydrate affinity chromatography also specifically inhibited EGF receptor autophosphorylation in both isolated membrane preparations and in intact cells (Cigarroa et al., 1989).

The antagonistic effects of EGF and MIS have been confirmed in several systems. Mullerian inhibiting substance induces and EGF inhibits regression of the mullerian duct. In the oocyte, MIS inhibits while EGF stimulates germinal vesicle breakdown. In A-431 cells, EGF stimulates cell proliferation and autophosphorylation of its receptor and MIS inhibits these EGF-stimulated effects (Coughlin et al., 1987; Cigarroa et al., 1989). Mullerian inhibiting substance and EGF also have antagonistic effects on EGF receptor autophosphorylation in fetal rat lung fibroblasts (E. Catlin and P. K. Donahoe, unpublished data). Thus, in a number of systems, EGF and MIS have opposing effects on cellular growth and differentiation.

A rat hepatocyte glycoprotein has been described as having antagonistic effects to insulin, similar to those of MIS and EGF (Auberger et al., 1989). In rat hepatoma cells, this 63-kDa secretory protein inhibits insulin stimulated thymidine incorporation, insulin receptor tyrosine kinase activity, and receptor autophosphorylation. Besides having these common functional properties, the hepatocyte protein also has sequence homology to MIS, and therefore may belong to the same protein family.

C. Receptor Characterization

Characterization of the MIS receptor has been a challenging task because of difficulties in labeling MIS, and in identifying an abundant source of receptor. An analysis of morphologic events observed during mullerian duct regression, namely breakdown of the basement membrane and migration of labeled epithelial cells into the mesenchyme in the rat and into the mesonephros of the chick reveals complex interactions of mesenchymal–epithelial events (reviewed in Donahoe et al., 1984). Mullerian inhibiting substance receptors, therefore, may be present in both epithelial and mesenchymal cells of the mullerian duct. Recent studies, using iodinated purified MIS, have demonstrated specific MIS binding to A-431 cells (N. Uitvlugt and P.K. Donahoe, unpublished data), and to fetal rat lung fibroblasts (Catlin et al., 1990). If the binding activity is confirmed to be specific for MIS, these two cell types may offer a more abundant source of MIS receptor than the fetal undifferentiated urogenital ridge. Anti-idiotypic antibodies raised against monoclonal antibodies to MIS (Taylor et al., 1989) that exhibit receptor binding should also be of value in the identification of the receptor.

Several molecular approaches are being used to isolate and clone the MIS receptor. Expression cloning using iodinated intact MIS to screen urogenital ridge libraries has been unsuccessful thus far (M. Gustafson and P.K. Donahoe, unpublished data). Recent data showing that the C-terminal fragment is the active moiety, indicate that intact MIS was probably a poor ligand. Therefore, ongoing efforts to identify the receptor will concentrate on screening with iodinated C-terminus. The MIS receptor is likely to be homologous to the recently cloned activin (Mathews and Vale, 1991) and TGF-β_2 receptors (Lin et al., 1992). These receptors have a single membrane spanning domain and are putative serine–threonine kinases. Degenerate oligonucleotide primers complementary to homologous regions of the activin and TGF-β_2 receptors were designed for polymerase chain reaction amplification of the MIS receptor from a urogenital ridge cDNA library. Several clones identified by this method are potential MIS receptors and are currently being further characterized (W.W. He and P.K. Donahoe, unpublished data).

V. Mullerian Inhibiting Substance in the Testis

A. Ontogeny and Cellular Localization

The undifferentiated gonad is induced to form aggregates of pre-Sertoli cells that encompass germ cells and form seminiferous cords by day 13 postcoitus in the rat. The immature Sertoli cells start to produce MIS at day 13.5 postcoitus, before the production of testosterone by Leydig cells (Josso et al., 1977). Mullerian inhibiting substance continues to be produced throughout gestation during which time the germ cells enter a state of meiotic and mitotic arrest. Jost (1972) suggested that a Sertoli cell factor like MIS might prevent germ cells from entering meiotic prophase.

Neonatal rat testes stain intensely for MIS until 2 days after birth, then staining intensity gradually decreases to a much lower level by the fifth postnatal day (Kuroda et al., 1990). The low level of MIS immunoreactivity persists through the 20th postnatal day, when it again declines to the barely detectable level seen in adult animals.

Recent studies using a rat genomic probe for MIS (Haqq et al., 1992) indicate that MIS has two differentially polyadenylated mRNA transcripts that are developmentally regulated (Lee et al., 1992). A larger transcript is prominent by 14 days postcoitus in the rat and gradually declines throughout gestation, then disappears after birth. A smaller transcript is barely detectable in the 14-day fetal testis, but increases progressively to maximal levels just before birth, then declines gradually but remains detectable through puberty. The two species differ by the extent of polyadenylation, and therefore may be differentially translated or differ in stability.

Studies in humans, using a sensitive immunoassay for human MIS (Hudson et al., 1990), indicate that MIS levels are high in the male until 10 years of age when they begin to decline to the low levels detectable in the serum of adults. The function that MIS subserves in adult testes is unknown but it seems likely that MIS may exert inhibitory control over spermatogenesis or Sertoli cell mitosis, similar to its postulated role in the ovary. Recent immunohistochemical studies in a sex-reversed mouse strain are consistent with this hypothesis. The B6.Y^{DOM} mice develop ovotestes with poorly organized testis chords and abnormal germ cell differentiation (Taketo et al., 1991). Germ cells that are arrested at the prespermatogonial prophase are surrounded by MIS-positive cells, in contrast to more mature germ cells in meiotic prophase that are surrounded by MIS negative cells. The pathogenesis of the sex reversal appears to be the delayed onset and progression of testicular differentiation, including MIS production.

Early studies attempting to localize MIS were performed on bovine testes, since monoclonal antibodies were first raised against bovine MIS. Tran and Josso (1982) and Hayashi and co-workers (1984), using monoclonal antibodies directed to bovine MIS, demonstrated that the majority of MIS is in the cisternae of rough endoplasmic reticulum in the cytoplasm of bovine Sertoli cells. Mullerian inhibiting substance also localizes to the Golgi bodies but not the nuclei of Sertoli cells, and is not found in germ cells (Hayashi et al., 1984).

B. Gene Expression and Regulation

Until recently, little was known about the regulation of MIS production in either testes or ovaries because of the lack of reactivity across species of the available antibodies, and also because of the difficulty in maintaining the production of MIS by Sertoli cells in culture. However, the newly developed polyclonal antibodies to human rMIS cross react with rat, dog, and mouse, as well as human MIS. Moreover, the rat MIS gene has recently been cloned and sequenced (Haqq et al., 1992). Using these new polyclonal antibodies and a rat genomic MIS probe, the ontogeny and regulation of MIS in rat testes and ovaries have been studied semiquantitatively and delineated in more detail.

An early study demonstrated that neonatal testes from rats treated with luteinizing hormone–releasing hormone (LHRH) antiserum in utero had greater MIS bioactivity in the organ culture bioassay than testes from vehicle-treated rats (Bercu et al., 1979). This effect was reversed by further treatment of the newborn pups with follicle-stimulating hormone (FSH) for 5 days after birth. Treatment with hCG (primarily LH activity) or water had no effect.

Subsequent attempts to modulate MIS production in vitro were unsuccessful. LaQuaglia and colleagues (1986) demonstrated an increase in cyclic adenosine monophosphate (cAMP) concentrations in the culture media of Sertoli cells and testicular tissue fragments after FSH treatment, but MIS concentrations in the media after treatment with FSH, LH, testosterone, progesterone, estrogen, or gonadotropin-releasing hormone (GnRH) were no different from control media. Under culture conditions, the production of MIS by the Sertoli cells and testicular fragments fell dramatically with time, complicating the in vitro study of MIS production and regulation. Vigier and colleagues (1985) also found a rapid decline in MIS production by Sertoli cells in culture and a lack of effect of testosterone or FSH on MIS secretion. In cultured human fetal Sertoli cells, dibutyryl cAMP increased MIS mRNA levels fivefold, but hCG and FSH had no effect (Voutilainen and Miller, 1987). In contrast, when adult granulosa cells were cultured, MIS mRNA levels were stimulated ninefold by dibutyryl cAMP and fivefold by hCG, but FSH, insulin, and progesterone had no effect.

To examine the possibility that FSH and LH modulate MIS production without the prenatal injections of LHRH antiserum (Bercu et al., 1979), Sprague-Dawley rats delivered from timed pregnant dams were injected twice a day with either porcine FSH, human LH, or sterile vehicle (Kuroda et al., 1990). The testes of animals treated with FSH showed less staining on the second to fourth postnatal day than did testes from the same age animals treated with vehicle alone. By the fifth postnatal day, when MIS immunoreactivity had dropped constitutively to very low levels, no significant difference could be detected between FSH and vehicle treated testes. Testes treated with LH showed no significant difference in immunoreactive MIS compared with that found in the testes of control animals. By Northern blot analysis, MIS mRNA was also less abundant in testes treated with FSH than with LH, or vehicle (Kuroda et al., 1990). Similar results were obtained in embryos injected at day 19 with gonadotropins, and delivered 2 days later (Kuroda et al., 1991). These observations support the hypothesis first suggested by Bercu and colleagues (1979) that FSH has an inhibitory effect on MIS production. Furthermore, testes of testosterone and hCG-treated fetuses had decreased immunostaining of intact MIS, but increased staining with antibodies specific for the N- and C-terminal fragments (Kuroda et al., 1991). In contrast to FSH-treated animals, no differences were detected in the abundance of MIS mRNA. These data suggest that testosterone and hCG/LH may have a novel role in regulating posttranslational processing of MIS, but additional studies are necessary to confirm this hypothesis.

It will be of interest to determine whether the DNA binding proteins such as Sry (Gubbay et al., 1990; Sinclair et al., 1990) involved in testicular determination are transcriptional regulators of the MIS gene. To address this possibility, regulatory consensus nucleotide sequences were identified in the upstream region of the MIS gene. A sequence with one mismatch for the cAMP responsive element was found in the human but not the bovine promoter. An SPl binding site was found in the bovine gene and was also found further upstream in the human gene; however, the high GC content of the MIS gene makes it difficult to determine whether the CCGCCC sequence is important in transcriptional control of MIS (Cate et al., 1990). These nucleotide sequences and other consensus-repeated sequences with characteristics of dyad symmetry that are conserved across species are being studied to identify factors that regulate the rat MIS gene. FSH and its second messengers deserve particular study as do the sex steroid hormones that either stimulate (progestins, androgens) or inhibit (estrogens) MIS action (Ikawa et al., 1982).

VI. Potential Roles of Mullerian Inhibiting Substance

A. Defects in Mullerian Duct Regression

Defects in mullerian duct regression are multifactorial. There is evidence for both end-organ unresponsiveness to MIS despite normal production of the hormone, and insufficient or deficient production of MIS. The testes of affected dogs with an autosomal recessive defect of retained mullerian ducts and nonaffected heterozygous animals produce equivalent amounts of bioactive MIS (Meyers-Wallen et al., 1989). Thus, the defect in this strain of dogs appears to result from a receptor defect rather than a lack of production of bioactive hormone. A recent analysis by Guerrier and colleagues (1989) of human patients with retained mullerian duct syndrome revealed that bioactive MIS was detectable in testicular biopsies in two of six patients tested, although MIS mRNA was detectable in all of the testicular biopsies. Of these patients, three siblings who lacked bioactive or immunoreactive MIS were further studied (Knebelmann et al., 1991). Molecular studies subsequently identified a point mutation in the fifth exon of the MIS gene that resulted in premature termination of translation. Thus, the syndrome of persistent mullerian ducts is a heterogeneous disorder resulting either from peripheral resistance to MIS or from a deficiency of the bioactive hormone.

B. Descent of the Testes

A triphasic model for descent of the testes has been proposed. The first stage is the nephric displacement of the gonad early in embryogenesis, which occurs in both males and females. The second stage is the transabdominal movement of the testes as a result of rapid gubernacular growth, and the third stage is movement through the inguinal canal (reviewed in Hutson and Donahoe, 1986). The second stage occurs after completion of mullerian duct regression, when MIS secretion is still elevated, and is proposed to be under MIS control, while the third stage is thought to be mediated by androgens. These were based on clinical observations that patients with androgen resistance who had normal production of bioactive MIS, often had inguinal testes, whereas patients with the persistent mullerian duct syndrome, in which testosterone production is normal but either MIS or its receptor is defective, often had abdominal testes. In support of this, lower MIS bioactivity has been demonstrated in the testes of infants with undescended testes compared with age-matched infants with normally descended testes (Donahoe et al., 1977b).

Josso and co-workers (1983) argue that the cryptorchidism seen in patients with the persistent mullerian duct syndrome is secondary to mechanical restraint from the retained ductal structures, which limits movement of the testes, rather than from a direct role for MIS. Male offspring of rabbits immunized with MIS during early gestation, before differentiation of the genital tracts, had persistence of the mullerian ducts, but normal testicular descent (Tran et al., 1986). These experiments, however, did not conclusively exclude a role for MIS in testicular descent because the timing of the immunization was selected to affect mullerian development, and may not have been optimal for inhibiting testicular descent.

C. Inhibition of Oocyte Meiosis

In addition to its well-defined role in male sexual differentiation, the ontogeny (Takahashi et al., 1986a; Ueno et al., 1989b) and function

(Takahashi et al., 1986b; Ueno et al., 1988) of MIS in the ovary has also been recently examined.

MIS has been detected in bovine follicular fluid by RIA (Necklaws et al., 1986), and has also been localized to the granulosa cells of bovine ovaries throughout reproductive life by immunohistochemical staining (Takahashi et al., 1986a). A comparatively low level of MIS mRNA was detected in the adult bovine ovary by Northern analysis (Cate et al., 1986a), although MIS mRNA was present in primary cultures of human adult granulosa cells (Voutilainen and Miller, 1987). No immunoreactive MIS has been detected in prenatal or newborn rat ovaries (Ueno et al., 1989b). By 7 days of age, however, MIS could be seen in preantral and antral follicles and localized clearly to the cumulus cells surrounding the oocyte in the preantral follicle as well as the periluminal granulosa cells, but not to cells lining the basement membranes of antral follicles (Ueno et al., 1989b). Neither atretic follicles, nor primary follicles, stained for MIS. When the studies were repeated in the cycling adult ovary (Ueno et al., 1989a), the same cells stained in the preantral, antral, and preovulatory follicles throughout estrous, midestrous, diestrous, and proestrous (Fig. 5–4). Late in proestrous, however, the staining in the preovulatory follicles

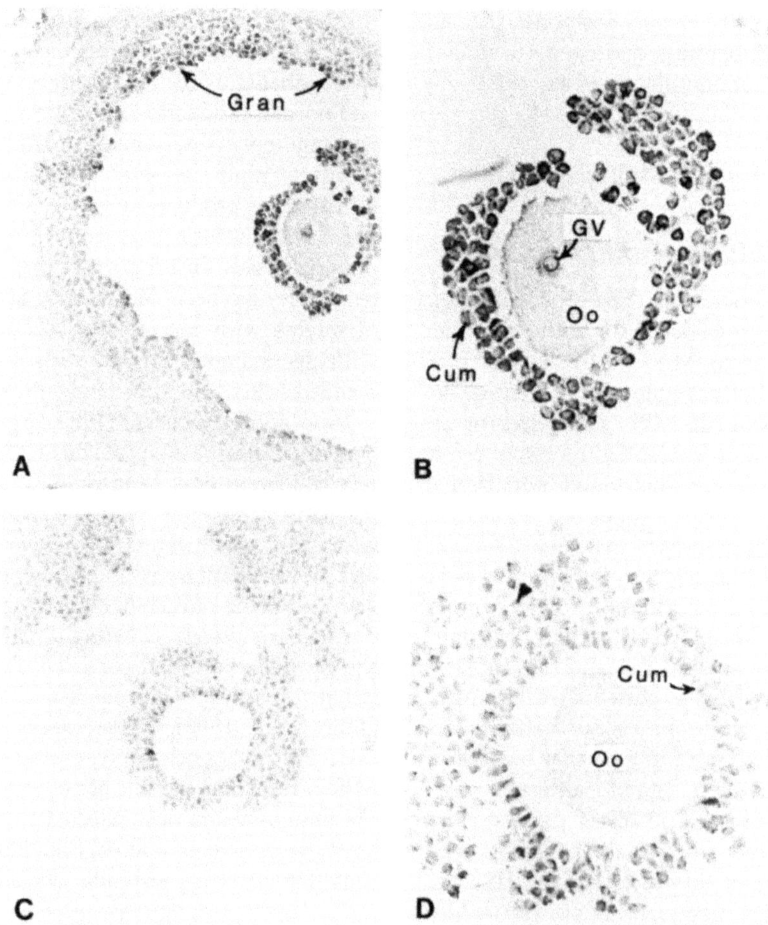

Figure 5–4. Immunohistochemical staining for mullerian inhibiting substance (MIS) in preovulatory follicles in time cycled rat ovaries. During proestrous (10 AM), when the germinal vesicle is present, MIS stains intensely in luminal granulosa cells (**A**) and in cumulus oophorus (**A** and **B**). Soon after the LH surge (11 PM), the germinal vesicle is no longer seen, and MIS staining is decreased. **A** and **C**, ×200 magnification. **B** and **D**, ×400 magnification. (Ueno et al., 1989a, reprinted with permission.)

waned, as progesterone levels rose just before ovulation when the oocyte resumes meiotic division. Corpus luteum staining was never seen. Cumulus cells surrounding oocytes removed from ovaries of animals pretreated with hCG for 72 hours stained less intensely for MIS than did the cumulus cells from untreated animals. The difficulty in studying MIS production in ovaries, however, lies in the different stages of developing follicles in the ovary.

By bioassay, rat ovaries have bioactive MIS from 3 days after birth, when the rat oocyte becomes arrested in late prophase in the dictyate or diffuse diplotene stage, until just before ovulation when meiosis resumes with breakdown of the germinal vesicles and expulsion of the first polar body (Ueno et al., 1989b). Purified bovine MIS, in a concentration-dependent, reversible, and cAMP-independent manner, inhibited resumption of meiosis in both denuded and cumulus-enclosed oocytes harvested from 25 to 27-day-old prepubertal unstimulated Sprague-Dawley rat ovaries (Takahashi et al., 1986b). Germinal vesicle breakdown was used as a measure of resumption of meiotic division. Similar results were observed using partially purified human rMIS. When rMIS was purified further with the initial immunoaffinity methods, this effect was lost, but it could be restored with the addition of low concentrations of a stabilizing detergent, NP40, which alone had no effect on either regression of the mullerian duct or inhibition of oocyte meiosis (Ueno et al., 1988). This oocyte meiosis inhibitory effect of human rMIS could be reversed by anti-MIS antibody and by EGF, thus attesting to its specificity (Ueno et al., 1988). The requirement for high concentrations of purified rMIS and NP40 indicate that activation of latent MIS may be required for this effect or that NP40 may make MIS available by reducing the aggregation that occurs when highly purified rMIS is subjected to affinity chromatography.

D. Antiproliferative Effects

The possibility that a fetal regressor, such as MIS, might be used as a therapeutic tumor regressor has been proposed (Donahoe et al., 1979, Cate et al., 1986b). The mechanisms responsible for the process of regression might provide a model for the development of targeted biological modifiers with reduced toxicity and increased specificity. Because the mullerian duct is the embryologic precursor of the uterus, fallopian tube, and vagina, it seemed reasonable to test tumors from these sites as potential MIS targets. Ovarian epithelial carcinomas were also studied as they resemble histologically tissues derived from the embryonic mullerian ducts (Scully, 1970), and they originate from the coelomic epithelium covering the ovary, the same epithelium that invaginates to form the mullerian ducts. These tumors, which represent 95 percent of ovarian carcinomas in the human, occur in the postmenopausal ovary in which there is no longer any MIS production. Growth of human tumors of mullerian duct origin were shown to be inhibited by partially purified bovine MIS, using established cell lines of gynecologic origin in colony inhibition or in microcytotoxicity assays, and in vivo in nude mice (reviewed in Cate et al., 1986b). These preparations of bovine MIS, purified by ion exchange, carbohydrate and dye affinity chromatography, also inhibited colony growth of a large number of primary ovarian, endometrial, and tubal carcinomas derived from patients (Fuller et al., 1985).

The method of purification of MIS, however, which may affect activation of the molecule, has significant consequences on its growth inhibitory properties. Thus, a variety of established gynecologic human tumor cell lines were not affected by human rMIS immunoaffinity purified by the initial methods (Wallen et al., 1989). When primary human tumors were studied in human tumor colony forming assays in soft agar, only 3 of 13 ovarian cancer explants and one small cell carcinoma of the lung were significantly inhibited by high concentrations of immuno-affinity purified rMIS. This human rMIS was purified by sequential ion exchange, carbohydrate, and immunoaffinity chromatography, then eluted by chaotropic salts, and was shown to be 85 percent pure by polyacrylamide gel electrophoresis, N-terminal amino acid sequencing, and amino acid composition analysis. When the human endometrial cell line HTB-111 was implanted under the subrenal capsules of irradiated CD-1 mice, rMIS that was purified by these previous affinity chromatography methods showed insignificant inhibition of tumor growth, whereas rMIS, purified by ion exchange and carbohydrate affinity chromatography, caused significant inhibition of tumor growth (A. K. Goodman and P. K. Donahoe, unpublished data).

The recent addition of a salt elution step before the immunoaffinity column has resulted in increased purification of the rMIS to 95% (R. Ragin, personal communication). The salt wash step removes low molecular weight proteins derived from CHO cells that have growth-stimulatory properties. This highly purified preparation of MIS is active in all the *in vivo* and *in vitro* anti-proliferative assays described and has also inhibited growth of nonreproductive tract tumors (Chin et al., 1991).

E. Mullerian Inhibiting Substance in the Developing Lung

Newborn respiratory distress syndrome, which is thought to result from a deficiency of pulmonary surfactant, has a higher incidence in male neonates than in females (Torday et al., 1981). In fetal rat lung cells, androgens have been shown to inhibit the synthesis of phosphatidylcholine, a major component of surfactant (Torday, 1985). Because testosterone is synergistic with MIS in the organ culture assay (Ikawa et al., 1982), and MIS is present at high serum concentrations in the fetus during the period of lung maturation, the effect of MIS on pulmonary surfactant production was studied (Catlin et al., 1988). Lung fragments from 17.5 day gestation female fetuses were incubated with fetal testes or ovaries, or with nanomolar concentrations of bovine MIS or picomolar concentrations of human rMIS. The testicular fragments and the crude bovine and human rMIS preparations all suppressed the accumulation of disaturated phosphatidylcholine compared with lungs cultured with ovaries or buffer (Catlin et al., 1988). Similar experiments were repeated in vivo, by intrauterine injection of human rMIS in nanomolar concentrations in the dorsal subcutaneous tissue of 19-day rat fetuses (Catlin et al., 1990). The delivery of MIS was verified by ELISA of serum from buffer-treated and MIS-treated fetuses. Lungs were harvested from the fetuses 48 and 72 hours after the injection, then extracted and assayed for disaturated phosphatylcholine. The lungs of female fetuses treated with human rMIS showed a marked diminution of phosphotidylcholine in comparison to that of buffer-injected controls (Catlin et al., 1990). This potent fetal regressor is being further investigated as an inhibitor of surfactant production in the male fetus and as a potential culprit in the increased incidence of neonatal respiratory distress syndrome in males. The finding of MIS activity in the fetal lung provides an additional, and possibly more abundant source for the purification and cloning of the MIS receptor.

VII. Summary

Mullerian inhibiting substance, a Sertoli cell hormone originally thought to be essential only for regression of the mullerian ducts in normal male sexual differentiation, has now been shown to have a number of extra-mullerian functions, particularly inhibition of growth. Although much is known about the biochemistry and molecular biology of MIS, its mechanism of action remains to be elucidated and its receptor has not been cloned. An understanding of the regulation of this potent fetal growth inhibitor will provide further insights concerning the control of growth and differentiation, and may provide a therapeutic model for the control of reproductive tract cancers.

REFERENCES

Antonelli-Orlidge, A., Saunders, K. B., Smith, S. R., D'Amore, P. A. (1989). An activated form of transforming growth factor-β is produced by cocultures of endothelial cells and pericytes. Proc. Natl. Acad. Sci. 86:4544–4548.

Auberger, P., Falquerho, L., Contreres, J. O., Pages, G., Le Cam, G., Rossi, B., and Le Cam, A. (1989). Characterization of a natural inhibitor of the insulin receptor tyrosine kinase: cDNA cloning, purification and anti-mitogenic activity. Cell 58:631–640.

Baker, M. L., Metcalfe, S. A., and Hutson, J. M. (1990). Serum levels of Mullerian inhibiting substance in boys from birth to 18 years, as determined by enzyme immunoassay. J. Clin. Endocrinol. Metab. 70:11–15.

Behringer, R. R., Cate, R. L., Froelick, G. J., Palmiter, R. D., and Brinster, R. L. (1990). Abnormal sexual development in transgenic mice chronically expressing Mullerian inhibiting substance. Nature 345:167–170.

Benoit, R., Ling, N., and Esch, F. (1987). A new Prosomatostatin-derived peptide reveals a pattern for prohormone cleavage at monobasic sites. Science 238:1126–1129.

Bercu, B. B., Morikawa, Y., Jackson, I.M.D., and Donahoe, P. K. (1979). Inhibition of Mullerian Inhibiting Substance secretion by FSH. Pediat. Res. 13:246–249.

Blanchard, M. G., and Josso, N. (1974). Source of anti-Mullerian hormone synthesized by the fetal testis: Mullerian inhibiting activity of fetal bovine Sertoli cells in tissue culture. Pediatr. Res. 8:968–971.

Brautigan, D. L., Bornstein, P., and Gallis, B. (1981). Phosphotyrosyl-protein phosphatase: Specific inhibition by Zn^{2+}. J. Biol. Chem. 256:6519–6522.

Budzik, G. P., Donahoe, P. K., and Hutson, H. M. (1985). A possible purification of Mullerian inhibiting substance and a model for its mechanism of action, in *Developomental Mechanisms: Normal and Abnormal* (Lash, J. W., ed.) pp. 207–223, Alan R. Liss Inc., New York.

Budzik, G. P., Powell, S. M., Kamagata, S., and Donahoe, P. K. (1983). Mullerian inhibiting substance fractionation by dye affinity chromatography. Cell 34:307–314.

Budzik, G. P., Swann, D. A., Hayashi, A., and Donahoe, P. K. (1980). Enhanced purification of Mullerian inhibiting substance by lectin affinity chromatography. Cell 21:909–915.

Cate, R. L., Donahoe, P. K., and MacLaughlin, D. T. (1990). Mullerian inhibiting substance, in *Handbook of Experimental Pharmacology,* Vol. 95/II *Peptide Growth Factors and Their Receptors* (Sporn, M. B., and Robert, A. B., eds.) pp. 179–210, Springer-Verlag, Heidelberg.

Cate, R. L., Rezaie, A. R., Ehrenfels, C., and Pepinsky, R. B. (1989). Proteolytic processing of human Mullerian inhibiting substance, in *Development and Function of the Reproductive Organs* (Josso, N., ed.) Vol. III. Ares-Serona Symposia, Rome, Italy.

Cate, R. L., Mattaliano, R. J., Hession, C., Tizard, R., Farber, N. M., Cheung, A., Ninfa, E. G., Frey, A. Z., Gash, D. J., Chow, E. P., Fisher, R. A., Bertonis, J. M., Torres, G., Wallner, B. P., Ramachandran, K. L., Ragin, R. C., Manganaro, T. F., MacLaughlin, D. T., and Donahoe, P. K. (1986a). Isolation of the bovine and human genes for Mullerian Inhibiting Substance and expression of the human gene in animal cells. Cell 45:685–698.

Cate, R. L., Ninfa, E. G., Pratt, D. J., MacLaughlin, D. T., and Donahoe, P. K. (1986b). Development of Mullerian inhibiting substance as an anti-cancer drug. Cold Spring Harbor Symposium 51:641–647.

Catlin, E. A., Powell, S. M., Manganaro, T. F., Hudson, P. L., Ragin, R. C., Epstein, J., and Donahoe, P. K. (1990). Sex-specific fetal lung development and mullerian inhibiting substance. Am. Rev. Respir. Dis. 141:466–470.

Catlin, E. A., Manganaro, T. F., and Donahoe, P. K. (1988). Mullerian Inhibiting Substance depresses accumulation in vitro of disaturated phosphatidylcholine in fetal rat lung. Am. J. Obstet. Gynecol. 159:1299–1303.

Chin, T., Parry, R. L., and Donahoe, P. K. (1991). Human Mullerian inhibiting substance inhibits tumor growth in vitro and in vivo. Cancer Res. 51:2101–2106.

Cigarroa, F. G., Coughlin, J. P., Donahoe, P. K., White, M. F., Uitvlugt, N., and MacLaughlin, D. T. (1989). Human recombinant Mullerian inhibiting substance inhibits epidermal growth factor receptor tyrosine kinase. Growth Factors 1:179–191.

Cohen-Hagenauer, O., Picard, J. Y., Mattei, M. G., Serero, S., Nguyen, V. C., de Tand, M. F., Guerrier, D., Hors-Cayla, M. C., Josso, N., and Frezal, J. (1987). Mapping of the gene for Anti-Mullerian Hormone to the short arm of human chromosome 19. Cytogenet. Cell Genet. 44:2–6.

Coughlin, J. P., Donahoe, P. K., Budzik, G. P., and MacLaughlin, D. T. (1987). Mullerian Inhibiting Substance blocks autophosphorylation of the EGF receptor by inhibiting tyrosine kinase. Mol. Cell Endocrinol. 49:75–86.

Derynck, R., Jarrett, J. A., Chen, E. Y., Eaton, D. H., Bell, J. R., Assoian, R. K., Roberts, A. B., Sporn, M. B., and Goeddel, D. V. (1985). Human transforming growth factor-β complementary DNA sequence and expression in normal and transforming cells. Nature 316:701–705.

Donahoe, P. K., Cate, R., MacLaughlin, D. T., Epstein, J., Fuller, A. F., Takahashi, M., Coughlin, J. P., Ninfa, E. G., and Taylor, L. A. (1987). Mullerian Inhibiting Substance: Gene structure and mechanism of action of a fetal regressor. Rec. Prog. Horm. Res. 43:431–467.

Donahoe, P. K., Budzik, G. P., Trelstad, R. L., Schwartz, B. R., Fallat, M. E., and Hutson, J. M. (1984). Molecular dissection of Mullerian duct regression, in *The Role of Extracellular Matrix in Development* (Trelstad, R. L., ed.) pp. 573–595, Alan R. Liss, New York.

Donahoe, P. K., Budzik, G. P., Trelstad, R., Mudgett-Hunter, M., Fuller, A., Hutson, J. M., Ikawa, H., Hayashi, A., and MacLaughlin, D. (1982). Mullerian-inhibiting substance: An update. Rec. Prog. Horm. Res. 38:279–330.

Donahoe, P. K., Swann, D. A., Hayashi, A., and Sullivan, M. D. (1979). Mullerian duct regression in the embryo correlated with cytotoxic activity against human ovarian cancer. Science 205:913–915.

Donahoe, P. K., Ito, Y., and Hendren W. H. (1977a). A graded organ culture assay for the detection of Mullerian Inhibiting Substance. J. Surg. Res. 23:141–148.

Donahoe, P. K., Ito, Y., Morikawa, Y., and Hendren, W. H. (1977b). Mullerian inhibiting substance in human testes after birth. J. Pediatr. Surg. 12:323–330.

Donahoe, P. K., Ito, Y., Price, J. M., and Hendren, W. H. (1977c). Mullerian inhibiting substance activity in bovine fetal, newborn and prepubertal testes. Biol. Reprod. 16:238–243.

Epstein, J., DesJardins, E. J., Hudson, P. L., and Donahoe, P. K. (1989). Stainless steel mesh supports high density cell growth and production of recombinant mullerian inhibiting substance. In vitro Cellular and Developmental Biology 25:213–216.

Feng, P., Catt, K. I., and Knecht, M. (1988). Transforming growth factor-B stimulates meiotic maturation of the rat oocyte. Endocrinology 122:181–186.

Fuller, A. F. Jr., Guy, D. R. Budzik, G. P., and Donahoe,

P. K. (1982). Mullerian inhibiting substance inhibits colony growth of a human ovarina carcinoma cell line. J. Clin. Endocrinol. Metab. 54:1051–1055.

Gelbart, W. M., Irish, V. F., St. Johnston, R. D., Hoffmann, F. M., Blackman, R. K., Segal, D., Posakony, L. M., and Grimaila, R. (1985). The decapentaplegic gene complex in *Drosophila melanogaster*. Cold Spring Harbor Symposium 50:119–125.

Gubbay, J., Collignon, J., Koopman, P., Capel, B., Economou, A., Munsterberg, A., Vivian, N., Goodfellow, P., and Lovell-Badge, R. (1990). A gene mapping to the sex-determining region of the mouse Y chromosome is a member of a novel family of embryonically expressed genes. Nature 346:245–250.

Guerrier, D., Tran, D., Vanderwinden, J. M., Hideux, S., van Outryve, L., Legeai, L., Bouchard, M., van Vliet, G., De Laet, M. H., Picard, J. Y., Kahn, A., and Josso, N. (1989). The persistent mullerian duct syndrome: A molecular approach. J. Clin. Endocrinol. Metab. 68(1)46–52.

Gustafson, M. L., Lee, M. M., Scully, R. E., Moncure, A. C., Hirakawa, T., Goodman, A., Muntz, H. G., Donahoe, P. K., MacLaughlin, D. T., and Fuller, A. F. (1992). Mullerian inhibiting substance as a marker for ovarian sex-chord tumor. N. Engl. J. Med. 326:466–471.

Haqq, C., Lee, M. M., Tizard, R., Wyst, M., DeMarinas, J., Donahoe, P. K., and Cate, R. L. (1992). Isolation of the rat gene for Mullerian inhibiting substance. Genomics (in press).

Hayashi, M., Shima, H., Hayashi, K., Trelstad, R. L., and Donahoe, P. K. (1984). Immunocytochemical localization of Mullerian inhibiting substance in the rough endoplasmic reticulum and Golgi apparatus in Sertoli cells of the neonatal calf testis using a monoclonal antibody. J. Histochem. Cytochem. 32:649–654.

Hayashi, A., Donahoe, P. K., Budzik, G. P., and Trelstad, R. L. (1982). Periductal and matrix glycosaminoglycans in rat Mullerian duct development and regression. Dev. Biol. 92:16–26.

Hudson, P. L., Dougas, I., Donahoe, P. K., Cata, R. L., Epstein, J., Pepinsky, R. B., and MacLaughlin, D. T. (1990). An immunoassay to detect human Mullerian inhibiting substance in males and females during normal development. J. Clin. Endocrinol. Metab. 70:16–22.

Hutson, J. M., and Donahoe, P. K. (1986). The hormonal control of testicular descent. Endocrine Reviews 7:270–283.

Hutson, J. M., Fallat, M. E., Kamagata, S., Donahoe, P. K., and Budzik, G. P. (1984). Phosphorylation events during Mullerian duct regression. Science 223:586–589.

Ikawa, H., Trelstad, R. L., Hutson, J. M., Manganaro, T. F., and Donahoe, P. K. (1984). Changing patterns of fibronectin, laminin, type IV collagen, and a basement membrane proteoglycan during rat Mullerian duct regression. Dev. Biol. 102:260–263.

Ikawa, H., Hutson, J. M., Budzik, G. P., MacLaughlin, D. T., and Donahoe, P. K. (1982). Steroid enhancement of Mullerian duct regression. J. Pediatr. Surg. 17:453–458.

Josso, N., Legai, L., Forest, M. G., Haussain, J.-L., and Brauner, R. (1990). An enzyme linked immunoassay for anti-mullerian hormone: A new tool for the evaluation of testicular function in infants and children. J. Clin. Endocrinol. Metab. 70:23–27.

Josso, N., and Picard, J. Y. (1986). Anti-Mullerian hormone. Physiol. Rev. 66:1038–1090.

Josso, N., Fekete, C., Cachin, O., Nezelof, C., and Rappaport, R. (1983). Persistence of Mullerian ducts in male pseudohermaphroditism, and its relationship to cryptorchidism. Clin. Endocrinol. 19:247–258.

Josso, N., Picard, J. Y., and Tran, D. (1977). The anti-Mullerian hormone. Rec. Prog. Horm. Res. 33:117–167.

Josso, N. (1972). Permeability of membranes to the Mullerian-inhibiting substance synthesized by the human fetal testis in vitro: A clue to its biochemical nature. J. Clin. Endocrinol. 34:265–270.

Jost, A. (1972). A new look at the mechanisms controlling sex differentiation in mammals. J. Hopkins Med. J. 130:38–53.

Jost, A. (1953). Problems of fetal endocrinology: The gonadal and hypophyseal hormones. Rec. Prog. Horm. Res. 8:379–418.

Jost, A. (1947). Recherches sur la differenciation sexuelle de l'embryon de lapin. Arch. Anat. Microsc. Morphol. Exp. 36:271–315.

Knebelmann, B., Boussin, L., Guerrier, D., Legeai, L., Kahn, A., Josso, N., and Picard, J. Y. (1991). Anti-Mullerian hormone Bruxelles: A nonsense mutation associated with the persistent Mullerian duct syndrome. Proc. Natl. Acad. Sci. USA. 88:3767–3771.

Kuroda, T., Lee, M. M., Ragin, R. C., Hirobe, S., and Donahoe, P. K. (1991). Mullerian inhibiting substance production and cleavage is modulated by gonadotropins and steroids. Endocrinology 129:2985–2992.

Kuroda, T., Lee, M. M., Haqq, C. M., Powell, D. M., Manganaro, T. F., and Donahoe, P. K. (1990). Mullerian inhibiting substance ontogeny and its modulation by follicle-stimulating hormone in the rat testes. Endocrinology 127:1825–1832.

LaQuaglia, M., Shima, H., Hudson, P., Takahashi, M., and Donahoe, P. K. (1986). Sertoli cell production of mullerian inhibiting substance in vitro. J. Urol 136:219–224.

Lee, M. M., Cate, R. L., Donahoe, P. K., and Waneck, G. L. (1992). Developmentally regulated polyadenylation of two discrete messenger ribonucleic acids for Mullerian inhibiting substance. Endocrinology 130:847–853.

Lin, M. Y., Wang, X. F., Ng-Eaton, E., Weinberg, R. A. and Lodish, M. F. (1992). Expression cloning of the TGF-β type II receptor, a functional transmembrane serine/threonine kinase. Cell 68:1–20.

MacLaughlin, D. T., Epstein, J., and Donahoe, P. K. (1991). Bioassay, purification, cloning and expression of Mullerian inhibiting substance. Methods Enzymol. 198:358–369.

Mason, A. J., Hayflick, J. S., Ling, N., Esch, F., Ueno, N., Ying, S., Guillemin, R., Niall, H., and Seeberg, P. (1985). Complementary DNA sequences of ovarian follicular fluid inhibin show precursor structure and homology with transforming growth factor-β. Nature 318:659–663.

Massague, J. (1990). The transforming growth factor-β family. Annu. Rev. Cell. Biol. 6:597–641.

Mathews, L. S., and Vale, W. W. (1991). Expression cloning of an activin receptor, a predicted transmembrane serine kinase. Cell 65:973–982.

Meyers-Wallen, V. N., Donahoe, P. K., Ueno, S., Manganaro, T. F., and Patterson, D. J. (1989). MIS is present in testes of dogs with PMDS. Biol. Reprod. 41(5):879–886.

Munsterberg, A., and Lovell-Badge, R. (1991). Expression of the mouse anti-Mullerian hormone gene suggests a role in both male and female sexual differentiation. Development 113:613–624.

Necklaws, E. C., LaQuaglia, M. P., MacLaughlin, D. T., Hudson, P., Mudgett-Hunter, M., and Donahoe, P. K. (1986). Detection of Mullerian inhibiting substance in biological samples by a solid phase sandwich radioimmunoassay. Endocrinology 118:791–796.

Pepinsky, R. B., Sinclair, L. K., Chow, E. P., Mattaliano, R. J., Manganaro, T. F., Donahoe, P. K., and Cate, R. L. (1988). Proteolytic processing of Mullerian inhibiting substance produces a transforming growth factor β-like fragment. J. Biol. Chem. 263:18961–18964.

Picard, J. Y., Benarous, R., Guerrier, D., Josso, N., and Kahn, A. (1986a). Cloning and expression of cDNA for anti-mullerian hormone. Proc. Natl. Acad. Sci. (USA) 83:5464–5468.

Picard, J. Y., Goulut, C., Bourillon, R., and Josso, N. (1986b). Biochemical analysis of bovine testicular anti-Mullerian hormone. FEBS Lett. 195:73–76.

Picard, J. Y., and Josso, N. (1984). Purification of testicular anti-Mullerian hormone allowing direct visualization of the pure glycoprotein and determination of yield and purification factor. Mol. Cell. Endocrinol. 34:23–29.

Picon, R. (1969). Action du testicule foetal sur le developpement in vitro des canaux de Muller chez le rat. Arch. Anat. Micros. Morphol. Exp. 58:1–19.

Purchio, A. F., Cooper, J. A., Brunner, A. M., Lioubin, M. N., Gentry, L. E., Kovacina, K. S., Roth, R. A., and Marquardt, H. (1988). Identification of mannose 6-phosphate in two asparagine-linked sugar chains of recombinant transforming growth factor-β 1 precursor. J. Biol. Chem. 263:14211–14215.

Scully, R. E. (1970). Recent progress in ovarian cancer. Hum. Pathol. 1:73–98.

Shima, H., Donahoe, P. K., Budzik, G. P., Kamagata, S., Hudson, P. L., and Mudgett-Hunter, M. (1984). Production of monoclonal antibodies for affinity purification of bovine Mullerian inhibiting substance activity. Hybridoma 3:201–214.

Sinclair, A. H., Berta, P., Palmer, M. S., Hawkins, J. R., Griffiths, B. L., Smith, M. J., Foster, J. W., Frischauf, A. M., Lovell-Badge, R., and Goodfellow, P. N. (1990). A gene from the human sex-determining region encodes a protein with homology to a conserved DNA-binding motif. Nature 346:240–244.

Sporn, M. B., Roberts, A. B., Wakefield, L. M., and de Crombrugghe, B. (1987). Some recent advances in the chemistry and biology of transforming growth factor-β. J. Cell Biol. 105:1039–1045.

Taguchi, O., Cunha, G. R., Lawrence, W. D., and Robboy, S. J. (1984). Timing and irreversibility of Mullerian duct inhibition in the embryonic tract of the human male. Dev. Biol. 106:394–398.

Takahashi, M., Hayashi, M., Manganaro, T. F., and Donahoe, P. K. (1986a). The ontogeny of Mullerian inhibiting substance in granulosa cells of the bovine ovarian follicle. Biol. Reprod. 35:447–453.

Takahashi, M., Koide, S. S., and Donahoe, P. K. (1986b). Mullerian inhibiting substance as oocyte meiosis inhibitor. Mol. Cell. Endocrinol. 47:225–234.

Taketo, T., Saeed, J., Nishioka, Y., and Donahoe, P. K. (1991). Delay of testicular differentiation in the B6.Y^{DOM} ovotestis demonstrated by immunocytochemical staining for Mullerian inhibiting substance. Dev. Biol. 146:386–395.

Taylor, L. A., Levin, R. K., Donahoe, P. K., and MacLaughlin, D. T. (1989). Development of an anti-idiotypic antibody as a probe for the Mullerian inhibiting substance (MIS) receptor. Growth Inhibitory and Cytotoxic Polypeptides 155–168.

Torday, J. S., Nielsen, H. C., Fencl, M. D., and Avery, M. E. (1981). Sex differences in fetal lung maturation. Am. Rev. Respir. Dis. 123:205–208.

Torday, J. S. (1985). Dihydrotestosterone inhibits fibroblast-pneumonocyte factor-mediated synthesis of saturated phosphatidylcholine by fetal rat lung cells. Biochem. Biophys. Acta 835:23–28.

Tran, D., Picard, J. Y., Vigier, B., Berger, R., and Josso, N. (1986). Persistence of Mullerian ducts in male rabbits passively immunized against bovine anti-Mullerian hormone during fetal life. Dev. Biol. 116:160–167.

Tran, D., and Josso, N. (1982). Localization of anti-Mullerian hormone in the rough endoplasmic reticulum of the developing bovine Sertoli cell using immunocytochemistry with a monoclonal antibody. Endocrinology 111:1562–1567.

Trelstad, R. L., Hayashi, A., Hayashi, K., and Donahoe, P. K. (1982). The epithelial-mesenchymal interface of the male rat Mullerian duct: Loss of basement membrane integrity and ductal regression. Dev. Biol. 92:27–40.

Ueno, S., Kuroda, T., MacLaughlin, D. T., Ragin, R. C., Manganaro, T. F., and Donahoe, P. K. (1989a). Mullerian inhibiting substance in the adult rat ovary during various stages of the estrous cycle. Endocrinology 125:1060–1066.

Ueno, S., Takahashi, M., Manganaro, T. F., Ragin, R. C., and Donahoe, P. K. (1989b). Cellular localization of Mullerian inhibiting substance in the developing rat ovary. Endocrinology 124:1000–1006.

Ueno, S., Manganaro, T. F., and Donahoe, P. K. (1988). Human recombinant Mullerian inhibiting substance inhibition of rat oocyte meiosis is reversed by epidermal growth factor in vitro. Endocrinology 123:1652–1659.

Vigier, B., Forest, M. G., Eychenne, B., Bezard, J., Garrigou, O., Robel, P., and Josso, N. (1989). Anti-Mullerian hormone produces endocrine sex reversal of fetal ovaries. Proc. Natl. Acad. Sci. (USA) 86:3684–3688.

Vigier, B., Watrin, F., Magre, S., Tran, D., and Josso, N. (1987). Purified bovine AMH induces a characteristic freemartin effect in fetal rat prospective ovaries exposed to it in vitro. Development 100:43–55.

Vigier, B., Picard, J. Y., Campargue, J., Forest, M. G., Heyman, Y., and Josso, N. (1985). Secretion of anti-Mullerian hormone by immature bovine Sertoli cells in primary culture, studied by a competition-type radioimmunoassay: Lack of modulation by either FSH or testosterone. Mol. Cell. Endocrinol. 43:141–150.

Vigier, B., Legeai, L., Picard, J., and Josso, N. (1982). A sensitive radioimmunoassay for bovine anti-Mullerian hormone, allowing its detection in male and freemartin fetal serum. Endocrinology 111:1409–1411.

Voutilainen, R., and Miller, W. L. (1987). Human Mullerian inhibitory factor messenger RNA is hormonally regulated in the fetal testis and in adult granulosa cells. Mol. Endocrinol. 1:604–608.

Wallen, J. W., Cate, R. L., Kiefer, D. M., Rieman, M. W., Martinez, D., Hoffman, R. M., Donahoe, P. K., Von Hoff, D. D., Pepinsky, B., and Oliff, A. (1989). Minimal antiproliferative effects of recombinant Mullerian inhibiting substance on gynecological tumor cell lines and tumor explants. Cancer Res. 49:2005–2011.

Weeks, D. L., and Melton, D. A. (1987). A maternal mRNA localized to the vegetal hemisphere in *Xenopus* eggs codes for a growth factor related to TGF-β. Cell 51:861–867.

Woodruff, T. K., D'Agostino, J. B., Schwartz, N. B., and Mayo, K. E. (1988). Dynamic changes in inhibin messenger RNAs in rat ovarian follicles during the reproductive cycle. Science 239:1296–1299.

Wozney, J. M., Rosen, V., Celeste, A. J., Mitsock, L. M., Whittiers, M. J., Kriz, R. W., Hewick, R. M., and Wang, E. A. (1988). Novel regulators of bone formation: Molecular clones and activities. Science 242:1528–1534.

CHAPTER SIX

Design and Function of the Microcirculation

Claude Desjardins

I. Introduction

The circulatory system delivers blood at pressures that are adequate to perfuse organs and thereby regulates the microenvironment of individual cells within each tissue. If all vessels in every tissue were completely filled with blood, and had constant flow, the total volume of blood would be enormous: significantly more energy would be required to propel blood through the vasculature (Wiedeman et al., 1981). Instead, the efficiency of the cardiovascular system is enhanced in two ways. First, the total volume of blood in the vascular compartment is limited and controlled (Guyton, 1991). Second, capillary perfusion is regulated to coincide with the metabolic needs of cells at rest, and perfusion can be increased in direct proportion to the work performed by individual cells (Duling, 1990). Tissue perfusion is governed by controlling the number of microvessels that are perfused at any instant and by regulating the number of red blood cells that flow through perfused capillary segments (Desjardins and Duling, 1987). Therefore, cell performance and tissue perfusion are intimately coupled by the microcirculation and linked to the fraction of the cardiac output that is available for tissue perfusion.

The summed interaction of three forms of vasomotor control—neural, endocrine, and local—determines the fraction of the cardiac output that is delivered to a tissue at rest and during work. Neurogenic and endocrine mechanisms exert "remote" control over the blood supply delivered to a tissue by directing adjustments in the diameter of arteries that supply organs with temporally distinct functional needs (Folkow and Neil, 1971). Local flow control mechanisms, in contrast, are intrinsic to the organ and operate within the terminal vascular bed of each tissue. The local control of blood flow is achieved through regulatory interactions among vascular smooth muscle cells, endothelial cells, and the metabolic needs of the tissue's parenchymal cells. Cell-to-cell communication among the vascular and parenchymal elements of each tissue provides a general principle explaining how local changes in capillary perfusion are regulated to influence the performance of cells in a complex gland such as the testis.

The purpose of this chapter is to analyze how the form and the function of the vasculature are integrated to limit or control the performance of cells in the interstitial and tubular compartments of the testis. Attention is focused first on the structural organization of the vasculature and its relationship to the vascular control of capillary perfusion. Consideration then shifts to local vascular mechanisms that may limit or control the performance of Leydig cells and seminiferous tubules by regulating the delivery of blood-borne solutes. Communication among the vascular and parenchymal cells within the testis illustrates how multiple cell types are integrated to support the complementary processes of steroidogenesis and spermatogenesis.

II. Design of the Testicular Vasculature

The primary function of the testicular vasculature is to ensure the exchange of heat and solutes. These exchanges are subserved by several vascular specializations that make the design of the macrocirculation and microcirculation a critical aspect of testicular performance. The structural features of the macrocirculation and microcirculation are presented separately to emphasize the contributions of each vascular element to testicular function.

A. Macrocirculation

The macrocirculation consists of a series of distributing and collecting vessels that conduct blood to and from the testes (Berne and Levy, 1981). The distribution of blood to the testis and its adnexa is achieved by the systematic division of feed arteries that arise from the spermatic artery to supply the testis, epididymis, and cremaster muscle (Harrison, 1949). Blood is returned from the testes by an extensive network of collecting venules. Venules converge near the inguinal canal to form the spermatic vein (Harrison, 1949). Venous drainage differs within and between species. The significance of this variation is not apparent, but the potential for the preferential delivery of testicular products should be considered when samples of venous effluent are removed to estimate the concentration of materials secreted by the testes (Chubb and Desjardins, 1982).

1. ARTERIAL INFLOW

Testes are supplied with blood by a primary feed artery (Chubb and Desjardins, 1982), but collateral arteries have been identified in men (Jarow, 1990) and experimental animals (Free, 1977). The spermatic artery is remarkably long relative to the feed arteries supplying other organs. The unusual length of this artery accommodates the descent of the testis into the scrotal sac and allows for the physical displacement of the testis outside of the body. The remarkable length of the feed artery, coupled with its prominent supply of smooth muscle, suggests that this vascular segment behaves as a resistor to modulate pressure or the flow of blood, or both, as defined by the Poiseuille equation (Folkow and Neil, 1971). Indeed, the contractile response of smooth muscle cells in the wall of the spermatic artery makes this vascular segment well adapted to modify resistance and autoregulate arterial inflow to the testis (Davis, 1990).

2. ARTERIOVENOUS INTERACTIONS

The long and extensively coiled feed artery is intimately associated with a network of anastomotic veins that give rise to the prominent pampiniform plexus (Harrison, 1949). This macrovascular arrangement facilitates the exchange of heat between counterflowing streams of blood that enter and exit the testes of mammals with scrotal testes (Setchell and Brooks, 1988). Within the pampiniform plexus, the arteries and veins are separated only by the thickness of the vascular wall (Harrison, 1949). Blood entering the testis is cooled several degrees below core body temperature as a result of the intimate association of macrovessels within the pampiniform plexus (Setchell and Brooks, 1988).

The adventitial layers of arterial and venous vessels are in direct contact in the testes of eutherian mammals (Waites, 1970). This structural arrangement may have functional implications for diffusional shunting of lipophilic substrates, but preliminary tests indicate that shunting may be limited (Setchell and Brooks, 1988). Nevertheless, locally produced vasoactive products could be transferred to receptors within the arterial wall to regulate smooth muscle cell contractility (Lew et al., 1989; Lew and Duling, 1990).

Direct anatomical shunts may exist between the arterial inflow and venous outflow in both experimental (Weerasooriya and Yamamoto, 1985) and farm animals (Noordhuizen-Stassen et al., 1985). The functional importance of these shunts has not been established, but their role must be considered because as much as 40 percent of the total blood supplied by the testicular artery may not pass through the tissue (Setchell and Brooks, 1988).

3. VENOUS OUTFLOW

The organization of venules within the pampiniform plexus is not homogenous. Variation exists in the diameter of venous vessels and the amount of intramural smooth muscle within each vessel. The absence of valves in collecting venules and interindividual variation in venular smooth muscle may contribute to the presence of palpable varicoceles in about 15 percent of the human male population (Pryor and Howards, 1988). The contractile properties of smooth muscle in the veins of the pampiniform plexus are undefined. The presence of smooth muscle in the pampiniform plexus suggests that these cells may have a role in regulating venular diameter and flow within the spermatic cord. Direct assessments of resistance in the vessels of the pampiniform plexus will be required to test this possibility.

B. Microcirculation

The primary feed artery that supplies the testis gives rise to a series of repeating centrifugal arteries near the surface of the testis (Chubb and Desjardins, 1982). These first-order arterioles penetrate the testicular parenchyma before branching into higher order arterioles (Chubb and Desjardins, 1982). The penultimate arteriolar branches give rise to a dual set of terminal arterioles that direct blood to separate intertubular and peritubular capillaries (Muller, 1957). This dual set of terminal arterioles has important functional implications since arterioles can selectively regulate the perfusion of capillaries supplying cells in the interstitial and tubular compartments of the testis.

1. STRUCTURAL PLAN

The organization of the testicular microcirculation shares many of the design elements for a capillary bed as proposed by Chambers and Zwiefach (Fig. 6–1). This generalization should be tempered by the realization that the ulti-

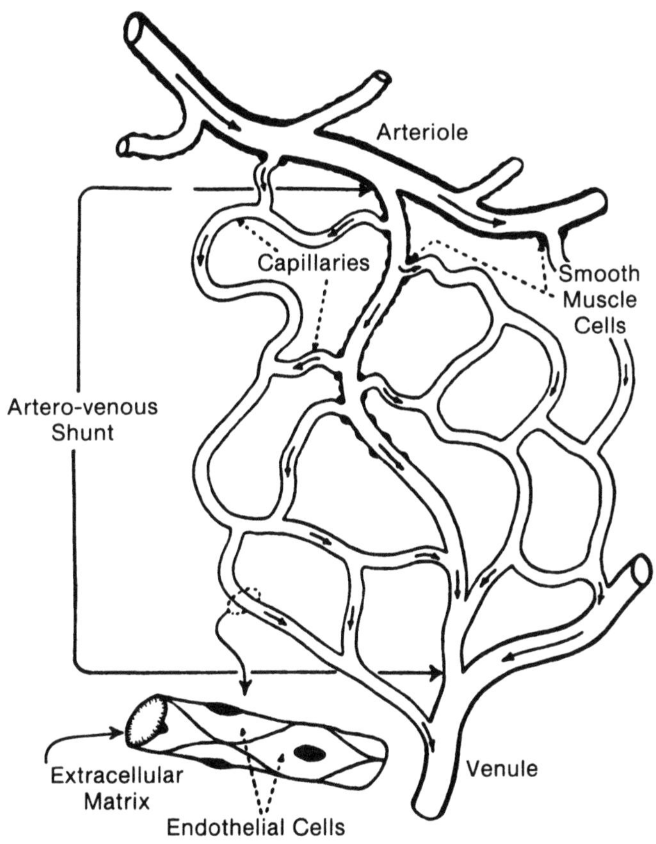

Figure 6–1. Schematic illustration of the structural organization of the microcirculation, modified from Chambers and Zwiefach. (From Am. J. Anat. 75:173, 1944; reproduced with permission.)

mate organization of testicular microvessels depends on the structure and function of the parenchymal cells that they support (Wiedeman et al., 1981).

2. ARTERIOLES

Arterioles are sometimes referred to as "resistance" vessels as moment-to-moment changes in arteriolar diameter control the perfusion of capillary beds (Granger et al., 1983). At any instant, external or local stimuli can change the diameter of the arteriolar wall and thereby control flow to the network of capillary segments supplied by individual arterioles. The responsiveness of smooth muscle cells to neural or locally produced stimuli makes the terminal segment of the arteriolar system the cellular site of flow control within the tissue (Granger et al., 1983).

3. CAPILLARIES

The smallest vessels of the microcirculation consist of a single layer of simple squamous epithelial cells joined by intercellular junctions (Fig. 6–1). The endothelial tube serves as the cellular site of solute exchange in all tissues (Granger et al., 1983). The luminal face of the endothelial tube is in direct contact with blood, whereas the opposite side, or the adluminal surface, is exposed to interstitial fluid while attached to a basal lumina. The endothelium, therefore, controls the exchange of blood-borne solutes with the underlying parenchymal cells.

Capillary endothelial cells are not simply selective barriers to diffusion. Instead, solute exchange is accomplished by multiple mechanisms including endocytosis, exocytosis, transcytosis, metabolism, and biosynthesis (Simionescu and Simionescu, 1984). The structure and function of capillary endothelia differ between organs, within parts of the same organ, and even along sequential segments of a single capillary bed. Diversity in the form and function of capillaries makes it difficult to generalize about the specific permeability characteristics of intertubular and peritubular capillaries in the testes as they have not been established in any model.

4. VENULES

The region between capillaries and venules is not structurally discrete, as noted in Figure 6–1. Continuous increments in the diameter of the venular segments are evident with each successive convergence in the venous circuit. The diameter of initial postcapillary venular segments (10 to 20 μm, inner diameter) is significantly larger than that of "true" capillaries (Fig. 6–1).

Venules regulate the volume of blood within a tissue, and the endothelium of specific venous segments contains receptors for the entire repertoire of white blood cells (deBono, 1977). The endothelial cells of postcapillary venules consist of highly specialized intercellular junctions that can be modified to allow attached lymphocytes to migrate from the lumen to the interstitial space. The specific molecular basis of the attachment and migration of lymphocytes in the testes is unknown, but some basic aspects of this process have been identified in vitro (Ryan, 1985). The medial portion of the wall of the larger venules contains scattered smooth muscle cells that influence venous tone.

5. ARTERIOVENOUS SHUNTS

In certain tissues, arterioles can be connected directly to venules by microvascular segments known as arteriovenous anastomoses, as shown in Figure 6–1. The inner diameter of these vascular connections is on the order of 10 to 15 μm. Smooth muscle cells are associated with the initial segment of the arteriolar end of the shunt. Then, the vessels acquire the morphological features of venules at their terminus. Arteriovenous anastomoses do not appear to be present in the human testis (Kormano, 1970). They might be present, however, in several species of experimental animals (Weerasooriya and Yamamoto, 1985), based on the redistribution of indicators used to assess total blood flow through the testis (Free, 1977). The presence of arteriovenous channels in the testis would lead to bias estimates of regional blood flow through the tissue when measurements are based on indicator–dilution techniques.

III. Microvascular Control of Testicular Function

A. Arteriolar Regulation of Capillary Perfusion

Terminal arterioles control the perfusion of individual capillary units and the performance of parenchymal cells by regulating the supply of

Po_2, hormones, and other substrates required to support the steroidogenic and spermatogenic functions of the testis (Desjardins, 1989). Regulation of arteriolar vasomotor tone by mechanisms that are intrinsic to the testes stabilizes pressure within capillary segments in the face of alterations in perfusion pressure (Sweeney et al., 1991). Significant reductions or elevations in perfusion can be expected to have a marked influence on the flux of solutes across the capillary membrane and affect fluid exchange. Perfusion pressure, estimated by impaling arterioles at each branch order in the testis, is carefully controlled as a major reduction in total pressure occurs in the arteriolar system supplying blood to the testes (Sweeney et al., 1991).

The external and local factors that control arteriolar vasomotor tone within the testis are not well known. The potential for the local control of vasomotor tone appears to be great, however, as the cells that direct the steroidogenic and spermatogenic function of the tissue release a host of vasoactive metabolites. Important metabolites include prostanoids and peptides with the capacity to provoke either long-term or short-term changes in arteriolar tone (Kasson and Tuchel, 1989; Pickering et al., 1989). In addition, adenosine, a powerful vasodilatory metabolite, is produced within the testis (Stiles et al., 1986; Conti et al., 1989), and testicular cells synthesize an organ-specific form of angiotensin converting enzyme (Jackson et al., 1988). This enzyme directs the formation of biologically active angiotensin II within testicular microvessels. The capacity of such locally produced, vasoactive substances to regulate local changes in arteriolar tone, and thereby influence the selective perfusion of capillary beds, is an important undefined aspect of microvessel function in the testis.

Evidence suggesting that capillary perfusion is controlled within the testis is based on the observation that the filling of peritubular capillary beds was appreciably less than the filling of intertubular capillary beds (Kormano, 1970). Local variation in capillary perfusion within the testis can be attributed to regional differences in the metabolic requirements of parenchymal cells along the length of a seminiferous tubule (Free, 1977). Such differences could stem from temporally restricted critical periods in cell metabolic demand occasioned by mitosis or meiosis within well-defined segments of a seminiferous tubule. Furthermore, an obligatory integration between vascular demand and cell performance can be inferred from the episodic activation of Leydig cell steroidogenic activity by luteinizing hormone (Fig. 6–2) (Sisk and Desjardins, 1986).

The proposition that local vascular mechanisms may be important in regulating arteriolar tone and capillary perfusion within the testis is consistent with the finding that capillary perfusion is controlled long after organs are removed from the body and vasomotor control mechanisms are divorced from external control owing to nerve or endocrine influences (Johnson, 1986). Provocative stimuli that elicit changes in the vasomotor tone of terminal arterioles include alterations in tissue Po_2 levels, selective increases in the metabolic demands of parenchymal cells for glucose, and modifications of venous pressure (Gorczynski et al., 1978). The local or intrinsic control of tissue perfusion demonstrates that arteriolar vaso-

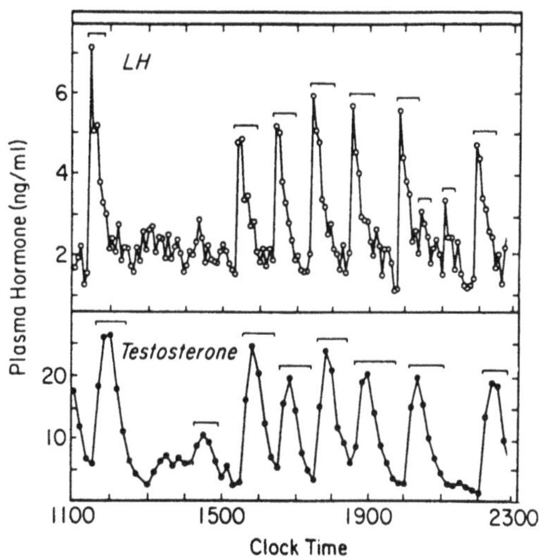

Figure 6–2. Plasma concentrations of luteinizing hormone (LH; *upper panel*) and testosterone (*lower panel*) in consecutive samples of systemic blood removed at 5-minute intervals from an adult male ferret. The concentration of hormone in individual samples has been plotted to illustrate the temporal coordination between the discharge of pituitary luteinizing hormone and the secretion of testosterone by the testis. (Modified from Sisk and Desjardins, Endocrinology 119:1195, 1986; reproduced with permission.)

motor tone can be autoregulated in accordance with the specific metabolic needs of the tissue's parenchymal cells. In the testis, ischemia is caused by torsion or elevated venous pressure arising from varicocele. Either circumstance might impair testicular function due to alterations in the local vasomotor mechanisms that control arteriolar tone.

Heterogeneity in capillary perfusion due to local vascular control could interfere with estimates of total blood flow that are based on the instantaneous distribution of microspheres or other indicator–dilution techniques. Variation in indices of capillary perfusion has not been acknowledged in estimates of hormone production by the testis, although regional inequalities in capillary perfusion have been recognized in the brain (Lassen, 1978) and skeletal muscle (Laughlin and Armstrong, 1982).

B. Regulation of Solute Exchange

The endothelium of testicular capillaries is on the order of 0.2 to 0.3 μm thick, but the number of plasmalemmal vesicles and transendothelial channels appears to be quite variable (Fawcett et al., 1970). Fenestrae are absent from exchange vessels in the testis, making this gland unique among visceral tissues that typically contain fenestrated capillaries (Simionescu and Simionescu, 1984). Whole-tissue estimates of capillary permeability (Setchell and Brooks, 1988), based on vascular markers with differing molecular weights and charges, confirm ultrastructural studies that failed to disclose fenestrae within microvessels of the testis (Fawcett et al., 1970). Despite well-established evidence showing that solute flux is appreciably greater in fenestrated capillaries, fenestrae do not have any appreciable influence on the permeability of water-soluble macromolecules (Simionescu and Simionescu, 1984).

Direct measurements of hydrostatic pressure in the hamster testis (Sweeney et al., 1991) established that perfusion pressure is closely regulated: changes in systemic blood pressure failed to evoke any significant change in hydrostatic pressure measured at several points in the

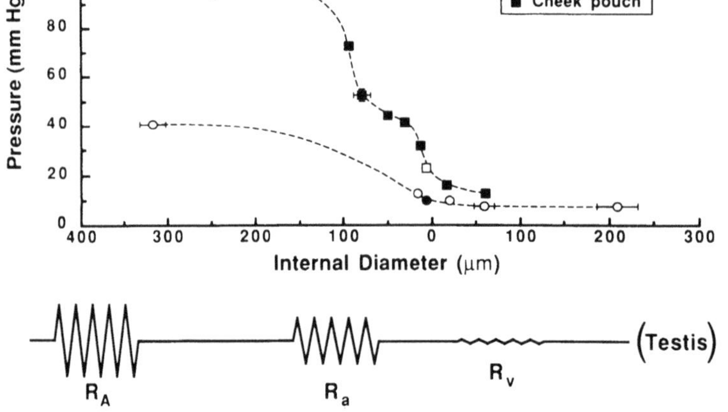

Figure 6–3. Hydrostatic pressure in the testis and cheek pouch of anesthetized hamsters prepared for microvascular estimates of hemodynamic resistance. The relative magnitude of the hemodynamic resistance in the testis and cheek pouch is shown above and below the graph. Resistance symbol dimensions are proportional to the total vascular resistance for indicated segments of each vascular bed. R_A, arterial resistance; R_a, arteriolar resistance; R_v, venular resistance. (Modified from Sweeney et al., Am. J. Physiol. 260:H1581, 1991; reproduced with permission.)

arteriolar network (Fig. 6–3). Hydrostatic pressures are relatively low in the testis, as compared with other capillary beds. Low testicular capillary pressure is a consequence of a remarkably large precapillary to postcapillary resistance ratio in the microvascular network of the testis, an assertion based on direct estimates of hydrostatic pressure in the testis and cheek pouch of hamsters.

Analysis of the distribution of hemodynamic resistance across the microvascular network of the hamster testis and cheek pouch indicates that postcapillary resistance accounts for somewhat less than 5 percent of the total resistance to blood flow through the testis. In the cheek pouch, however, about 17 percent of the vascular resistance is of postcapillary origin (Sweeney et al., 1991). In the testis, the low postcapillary resistance is associated with a relatively high resistance in the feed artery and appreciably less arteriolar resistance, compared with estimates made in the cheek pouch (Fig. 6–3).

Relatively low capillary pressure may subject the testes to certain pathophysiological consequences if venous pressure is elevated (Sweeney et al., 1991). For example, varicocele could alter convective transport through testicular capillaries, as shown in Figure 6–4. According to this diagram, an increase in venous pressure may be expected to enhance the exchange of interstitial fluid and the production of testicular lymph. Such a circumstance would be expected to produce significant alterations in the local concentration of substrates within the interstitial compartment of the tissue.

Endothelial cells have been isolated and maintained in culture (Ryan, 1985). This advance permits direct assessments of the biosynthetic and metabolic characteristics of cells isolated from specific segments of the vascular tree within the testis. The opportunity to dissect specific microvascular segments from the testis of experimental animals and men makes it possible to exploit the testis to investigate the unique functions of endothelial cells present in functionally distinct vascular segments: arterioles, capillaries, and venules.

The endothelium mediates the bidirectional exchange of solutes between plasma and fluids present in insterstitial spaces (Simionescu and Simionescu, 1984). Endothelial cells are highly differentiated, with organelles capable of directing endocytosis and exocytosis. Solute transport is accomplished by transcytosis or vesicular transport across endothelial cells. The magnitude of the process is unknown for tes-

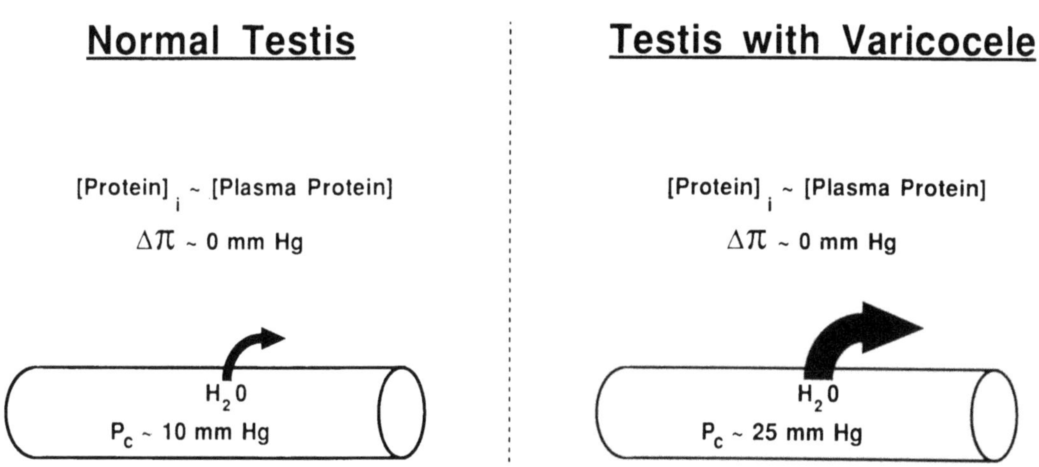

Figure 6–4. Schematic diagram illustrating the regulation of capillary fluid filtration in a normal testis (**A**), and after the experimental induction of varicocele (**B**). In the normal testis, capillary hydrostatic pressure is low but adequate to support fluid filtration and the exchange of solutes directing steroidogenesis and spermatogenesis. After the induction of an experimental varicocele, capillary hydrostatic pressure would increase owing to increments in venous pressure. Capillary fluid filtration would be increased, and this would alter the concentration of solutes in the interstitial compartment that are required for the production of steroid hormones or sperm. (Modified from Sweeney et al., Am. J. Physiol. 260:H1581, 1991; reproduced with permission.)

ticular capillaries, but it is rapid and extensive in the endothelium of other tissues such as the brain, myocardium, and kidneys (Brown, 1989).

Evidence for the existence of rapid and effective bidirectional exchange of solutes in the testis is based on the temporal coordination of luteinizing hormone (molecular weight = 28,000) and testosterone in blood-borne pulses, shown in Figure 6–2. The interval between the episodic release of luteinizing hormone from the anterior pituitary gland and the appearance of testosterone in the systemic blood is remarkably brief. The integration between peptide and steroid hormone pulses, at the level of the Leydig cell, implies that the flux of physiologically relevant hormones may be faster than current estimates of transvascular exchange in any tissue with a continuous type of epithelium.

In the testis and other endocrine glands, the bulk of the secreted products reach the systemic blood by the venous effluent, not by lymphatic vessels (Setchell and Brooks, 1988). The microvascular epithelium of the testis may be differentiated to account for the unusual permeability to macromolecules. Future studies of this endothelium should uncover the special structural and functional properties that facilitate the exchange of permeant molecules with diverse physical properties (size, shape, charge, chemical nature), as well as the magnitude and mechanisms for the exchange of critical solutes directing testicular function.

C. Function of Postcapillary Venules

Leukocytes possess an array of cell surface receptors that mediate their adhesion to the luminal surface of venules. The expression of cell surface adhesion molecules facilitates cell attachment to the vascular wall, migration across the wall, and subsequent interaction with macromolecules found within the extracellular matrix of the interstitial space (Vane et al., 1990). An inflammatory-like response, characterized by a separation of luminal endothelial cells in postcapillary venules, can be provoked by human chorionic gonadotropin (hCG) (Bergh et al., 1988). Changes in microvessel permeability are profound but relatively brief (Setchell, 1986). The most prominent alterations in permeability are localized in venular endothelial cells of the testes (Bergh et al., 1988). Similar postcapillary venular responses can be evoked by inflammatory agents in other tissues (Svensjo and Grega, 1986). These findings suggest that venular endothelial cells play an active role in regulating the porosity of the microvascular membrane to macromolecules in the testes and other tissues.

V. Regulation of the Interstitial Microenvironment

An ever-growing list of locally secreted chemical effectors are thought to play a role in either steroidogenesis, spermatogenesis, or both (Sharpe, 1986; Bardin, 1988). Rapid advances have been made in identifying possible interactions between different types of testicular cells because in vitro cell culture systems facilitate the identity of the sites of synthesis, the sites of action, and the biochemical characteristics of the molecules involved (Skinner, 1987). Eventually, certain aspects of these cell-to-cell interactions must be defined in situ to establish the physiological significance of the interactions. An improved understanding of the physical and chemical properties of the testicular interstitial space is required to predict the movement of effector molecules from their site of release to their site of action within the testis.

An inescapable feature of the testicular microenvironment is the determination of the rate of formation and removal of interstitial fluid by microvessel permeability (Sweeney et al., 1991). Fluid movement in the interstitial space is regulated by hydrostatic pressure and structural elements within the interstitium. Diffusion represents the primary, if not the only, mechanism for the exchange of regulatory molecules between cells in the interstitial space and cells in adjacent seminiferous tubules (Berne and Levy, 1981). Diffusion is a random process constrained by time, distance, and temperature. Although only a small amount of a chemical effector may be required to effect a change in cell function, the information content of effector molecules is usually large, and signal transduction is facilitated by the presence of specific receptors. Chemical effectors involved in cell-to-cell interactions in the testes can be diluted by the remarkable volume of fluid present in interstitial spaces or by the relatively rapid turnover of fluids in this compartment (Setchell, 1986).

Important questions remain unanswered about the physiochemical determinants of fluid homeostasis in the interstitial compartment of the testis. The mechanisms that limit or regulate fluid formation and movement will influence the distribution of paracrine effectors from their cellular sites of release to their cellular sites of action. The present consideration of fluid mass–balance in the testis emphasizes that cell-to-cell signaling in a fluid microenvironment, as in the testes, is a complex process. The physiochemical components of the process must be highly organized to overcome the limitations imposed by temporal and spatial variations in the formation and convection of fluid within the interstitial space.

VI. Conclusions

Tissue perfusion cannot be described adequately by estimates of regional or total blood flow through the testis. Inequalities in the distribution of blood within capillary beds of the testis can be caused by multiple factors. These include the structural geometry of the vascular bed, the neural and myogenic mechanisms that affect vasomotor tone in arterial or arteriolar segments of the vasculature, and the temporal differences in the metabolic demands of parenchymal cells located within the interstitial and tubular compartments of the tissue. Moment-to-moment differences in capillary perfusion can evoke compensating variations in the formation and convection of fluid within interstitial spaces. Temporal and spatial variation in the perfusion of intertubular or peritubular capillary beds will also have a significant impact on the processing of chemical effectors within the testicular microenvironment. The distribution of hormones, growth factors, and other paracrine molecules within the interstitial space must be subject to precise regulation so that their physiological effects can be manifested. The integrated performance of each component of the testicular vasculature ensures that solute exchange and fluid formation are carefully regulated to support the performance of steroidogenic and spermatogenic cells.

REFERENCES

Bardin, C. W. (1988). Pro-opiomelanocortin peptides in reproductive biology. Hosp. Pract. pp. 109–127.

Bergh, A., Damber, J. E., and Widmark, A. (1988). Hormonal control of testicular blood flow, microcirculation and vascular permeability, in *Molecular and Cellular Endocrinology of the Testis* (B. A. Cook and R. M. Sharpe, eds.), pp. 123–133. Raven Press, New York.

Berne, R. M., and Levy, M. N. (1981). *Cardiovascular Physiology,* pp. 94–122, C. V. Mosby Co., St. Louis.

Brown, D. (1989). Membrane recycling and epithelial cell function. Am. J. Physiol. 256:F1–F12.

Chambers, R., and Zwiefach, B. W. (1944). The topography and function of the mesenteric capillary circulation. Am. J. Anat. 75:173–205.

Chubb, C., and Desjardins, C. (1982). Vasculature of the mouse, rat, and rabbit testis-epididymis. Am. J. Anat. 165:357–372.

Conti, M., Boitani, C., Demanno, D., Migliaccio, S., Monoaco, L., and Szymeczek, C. (1989). Characterization and function of adenosine receptors in the testis. Ann. NY Acad. Sci. 564:39–47.

Davis, J. R. (1990). Myogenic tone of the rat testicular subcapsular artery has a role in autoregulation of testicular blood supply. Biol. Reprod. 42:727–735.

deBono, D. (1977). Lymphocytes and the microcirculation. Adv. Microcirc. 7:68–95.

Desjardins, C. (1989). The microcirculation of the testis. Ann. NY Acad. Sci. 564:243–249.

Desjardins, C., and Duling, B. R. (1987). Microvessel hematocrit: Measurement and implications for capillary oxygen transport. Am. J. Physiol. 252:H494–503.

Duling, B. R. (1990). Metabolic control of striated muscle blood flow, in *The Lung: Scientific Foundations* (R. Crystal and J. West, eds.), vol. 2, pp. 1497–1506. Raven Press, New York.

Fawcett, D. W., Leak, L. V., and Heidger, P. M. (1970). Electron microscopic observations on the structural components of the blood–testis barrier. J. Reprod Fert. (Suppl.) 10:105–122.

Folkow, B., and Neil, E. (1971). *Circulation,* pp. 265–306, Oxford Univ. Press, London.

Free, M. J. (1977). Blood supply to the testis and its role in local exchange and transport of hormones, in *The Testis* (A. D. Johnson and W. R. Gomes, eds.), vol. 4, pp. 39–90. Academic Press, New York.

Gorczynski, R. J., Kltizman, B., and Duling, B. R. (1978). Interrelations between contracting striated muscle and precapillary microvessels. Am. J. Physiol. 235:H494–H504.

Granger, H. J., Borders, J. L., Meininger, G. A., Goodman, A. H., and Barnes, G. E. (1983). Microcirculatory control systems, in *The Physiology and Pharmacology of the Microcirculation* (N. A. Mortillaro, ed.), vol. 1, pp. 209–236. Academic Press, New York.

Guyton, A. C. (1991). Blood pressure control—Special role of the kidneys and body fluids. Science 252:1813–1817.

Harrison, R. G. (1949). The comparative anatomy of the blood supply of the mammalian testis. Proc. Zool. Soc. (Lond.) 119:325–344.

Jackson, B., Cubela, R. B., Sakaguchi, K., and Johnston, C. I. (1988). Characterization of angiotensin converting enzyme (ACE) in the testis and assessment of the in vivo effects of the ACE inhibitor perindopril. Endocrinology 123:50–55.

Jarow, J. P. (1990). Intratesticular arterial anatomy. J. Androl. 11:255–259.

Johnson, P. C. (1986). Autoregulation of blood flow. Circ. Res. 59:483–495.

Kasson, B. G., and Tuchel, T. L. (1989). Hormonal regulation of rat testicular arginine vasopressin receptors. Endocrinology 124:2777–2784.

Kormano, M. (1970). An experimental technique of in vivo high resolution microangiography. Br. J. Radiol. 43:180–183.

Lassen, N. (1978). Brain, in *Peripheral Circulation* (Johnson, P. C., ed.), pp. 337–358. J. Wiley and Sons, New York.

Laughlin, M. H., and Armstrong, R. B. (1982). Muscular blood flow distribution patterns as a function of running speed in rats. Am. J. Physiol. 243:H296–H306.

Lew, M. J., Rivers, R. J., and Duling, B. R. (1989). Arteriolar smooth muscle responses are modulated by an intramural diffusion barrier. Am. J. Physiol. 257:H10–H16.

Lew, M. J., and Duling, B. R. (1990). Arteriolar reactivity in vivo is influenced by an intramural diffusion barrier. Am. J. Physiol. 259:H574–H581.

Muller, I. (1957). Kanalchen-und Capillararchitekonik des Rattenhodens. Z. Zellforsch 45:522–537.

Noordhuizen-Stassen, E. N., Charbon, G. A., deJong, F. H., and Wensing, C.J.G. (1985). Functional arterio-venous anastomoses between the testicular artery and the pampiniform plexus in the spermatic cord of rams. J. Reprod. Fert. 75:193–201.

Pickering, B. T., Birkett, S. D., Guldenaar, S.E.F., Nicholson, H. D., Worley, R.T.S., and Yavachev, L. (1989). Oxytocin in the testis: What, where and why? Ann. NY Acad. Sci. 564:198–209.

Pryor, J. L., and Howards, S. S. (1988). Varicocele, in *Contemporary Management of Impotence and Infertility* (E. Tanagho, T. Lue, and R. McClure, eds.), pp. 247–264. Williams & Wilkins, Baltimore.

Ryan, U.S. (1985). The endothelial surface and responses to injury. Federation Proc. 45:101–108.

Setchell, B. P., and Brooks, D. E. (1988). Anatomy, vasculature, innervation and fluids of the male reproductive tract, in *The Physiology of Reproduction* (E. Knobil, and J. D. Neill, eds.), pp. 753–836. Raven Press, New York.

Setchell, B. P. (1986). The movement of fluids and substances in the testis. Aust. J. Biol. Sci. 39:193–207.

Sharpe, R. M. (1986). Paracrine control of the testis. Clin. Endocrinol. Metab. 15:158–178.

Simionescu, M., and Simionescu, N. (1984). Ultrastructure of the microvascular wall: Functional correlations, in *Handbook of Physiology* (E. M. Renkin, and C. C. Michel, eds.), vol. 4, pt. 1, pp. 41–164. American Physiological Society, Bethesda, Maryland.

Sisk, C. L., and Desjardins, C. (1986). Pulsatile release of luteinizing hormone and tetosterone in male ferrets. Endocrinology 119:1195–1203.

Skinner, M. K. (1987). Cell–cell interactions in the testis. Ann NY Acad. Sci. 513:158–178.

Stiles, G. L., Pierson, G., Suany, S., and Parsons, W. J. (1986). The rat testicular A_1 adenosine receptor-adenylate cyclase system. Endocrinology 119:1845–1851.

Svensjo, E., and Grega, G. J. (1986). Evidence for endothelial cell-mediated regulation of macromolecular permeability by postcapillary venules. Federation Proc. 45:89–95.

Sweeney, T. E., Rozum, J. S., Desjardins, C., and Gore, R. W. (1991). Microvascular pressure distribution in the hamster testis. Am. J. Physiol. 260:H1581–1589.

Vane, J. R., Anggard, E. E., and Botting, R. M. (1990). Regulatory functions of the vascular endothelium. New Engl. J. Med. 323:27–34.

Waites, G.M.H. (1970). Temperature regulation and the testis, in *The Testis* (A. D. Johnson, W. R. Gomes, and N. L. Van Demark, eds.), vol. 1, pp. 241–279. Academic Press, New York.

Weerasooriya, T. R., and Yamamoto, T. (1985). Three-dimensional organization of the vasculature of the rat spermatic cord and the testis. Cell. Tissue Res. 241:317–323.

Wiedeman, M. P., Tuma, A. F., and Mayrovitz, H. N. (1981). *An Introduction to the Microcirculation,* pp. 99–139, Academic Press, New York.

CHAPTER SEVEN

Leydig Cells: Structure and Function

Larry L. Ewing* and Diane S. Keeney

I. Introduction

This is a selective review of two major issues regarding the mammalian Leydig cell. First, the newly emerging body of data shows that Leydig cell function involves the production of polypeptides and is not restricted to testosterone production. Second, there is a striking correlation between Leydig cell structure and steroidogenic function. The latter includes a discussion of the genes encoding the enzymes involved in testosterone production, the spatial organization of these steroid biosynthesizing enzymes, the movement of Leydig cell sterol/steroid precursors and products into, about, and out of the Leydig cell, and the trophic effect of luteinizing hormone (LH) on gene expression, enzyme biosynthesis, cytoplasmic organelle biogenesis/turnover, and Leydig cell volume, number, and proliferation. We have confined ourselves largely to rodent species and, whenever possible, to discussing the earliest and latest papers on a topic. Consequently, whenever possible, we have referred the reader to important contemporary and comprehensive reviews for details.

II. Function of Leydig Cells

A. Peptides

It has become increasingly apparent during the past decade that, in addition to steroids, mammalian Leydig cells synthesize and secrete a variety of peptide products. Great interest has been generated in these peptidic Leydig cell products because of evidence (Aoki and Fawcett, 1978; Sharpe, 1988) suggesting regiospecific intratesticular communication between Leydig cells and seminiferous tubules capable of modulating the well-known pituitary control of testicular function. At present, the status of our knowledge concerning the mode of synthesis, rate of secretion, intratesticular function and regulation of peptidic synthesis, and secretion is fragmentary. Therefore, in this review we present evidence only for the presence of peptide products of Leydig cells. Included are those peptides that meet three criteria: (1) those that exist in the testes in concentration higher than in peripheral blood or that are secreted by a purified Leydig cell preparation in vitro, or both; (2) those that have been shown to reside in Leydig cells by immunocytochemical techniques; and (3) those for which messenger RNA (mRNA) exists in the testis and/or specifically in Leydig cells.

1. PROOPIOMELANOCORTIN-DERIVED PEPTIDES

Proopiomelanocortin (POMC) is a polyprotein that gives rise to adrenocorticotropin (ACTH), melanocyte-stimulating hormone (MSH), endorphin, corticotropin-like intermediate lobe peptide (CLIP), and lipotropins (Douglass et al., 1984). Please see Bardin and co-workers (1987)

*Dr. Ewing participated in the preparation of this chapter but died during the production of this book.

for a detailed review of the evidence that peptides of this family are produced by Leydig cells. Briefly, β-endorphin immunoreactivity was first detected in rat testes in 1980 (Sharp et al., 1980). Subsequently, immunostainable β-endorphin was localized to Leydig cells in the rat (Tsong et al., 1982b), mouse, hamster, guinea pig and rabbit (Tsong et al., 1982a); and immunostainable ACTH was localized to rat Leydig cells (Tsong et al., 1982a); and immunostainable γ-MSH was localized to mouse Leydig cells (Shaha et al., 1984). Evidence for local synthesis of these POMC-derived peptides came forth in 1984 when mRNA encoding POMC was identified in rat testes and localized by in situ hybridization to Leydig cells (Pintar et al., 1984). Recently, the concept that rat Leydig cells produce β-endorphin, ACTH, and γ-MSH has been confirmed (DeBold et al., 1988). Interestingly, testicular POMC-like mRNA is 150 to 200 nucleotides shorter than that found in the hypothalamus and pituitary (Bardin et al., 1987). The concentration of POMC mRNA in Leydig cells is, however, less than that in testicular macrophages and seminiferous tubules (Li et al., 1991). The presence of POMC mRNA and immunoreactive ACTH and β-endorphin in highly purified testicular macrophages and Leydig cells suggests that these POMC-derived peptides are synthesized de novo in both cell types (Li et al., 1991).

At this juncture, very little is known about the function of POMC-derived peptides in the testis. In the mouse, POMC mRNA is present in only 2 percent of interstitial cells in the testis; these are associated spatially with stage IX–XII seminiferous tubules containing meiotic prophase spermatocytes and early spermatids that also contain POMC mRNA (Gizang-Ginsberg and Wolgemuth, 1985). Such spatial organization suggests that these peptides might be acting in a paracrine fashion to modulate seminiferous tubule function. One example of such a paracrine effect was the demonstration that in the neonatal rat, β-endorphin inhibits follicle-stimulating hormone (FSH)-stimulated Sertoli cell division by a mechanism involving one or more membrane-associated guanosine triphosphate (GTP) binding proteins (Orth and Boehm, 1990). Also, it has been suggested that β-endorphin might exert an autocrine effect on Leydig cells because it interacts synergistically with human chorionic gonadotropin (hCG) to stimulate in vitro testosterone production by Leydig cells obtained from 20-day-old rats (Bardin et al., 1987). This was not confirmed when adult rat Leydig cells were cultured (Clements et al., 1988), nor when adult rat testes were perfused (Juniewicz et al., 1988) or perifused (Margioris et al., 1989) in vitro with β-endorphin. Juniewicz and colleagues (1988) showed that rabbit and guinea pig, but not rat, hamster, and dog testes secrete testosterone in response to ACTH.

2. ENKEPHALINS AND DYNORPHINS

Opiate peptides, in addition to β-endorphin, are derived from two polyprotein families including preproenkephalin (preproenkephalin A) and preprodynorphin (preproenkephalin B). Historically, these opiate peptides have been localized in pituitary, adrenal medulla, and peptidergic neurons, and associated with receptors mediating analgesia, euphoria, and sedation (Bolander, 1989). Although fragmentary at this juncture, evidence is accumulating that opioid peptides derived from preproenkephalin and preprodynorphin are produced in mammalian testes. Immunoreactive dynorphin A and dynorphin B have been quantified in rat, guinea pig, and rabbit testes (Cox et al., 1987; Douglass et al., 1987) and [leu^5] enkephalin in rat and guinea pig testes (Cox et al., 1987). Immunostainable met-enkephalin and leu-enkephalin, but not α-neoendorphin or dynorphin (1–17), were localized to Leydig cells and seminiferous tubules by Engelhardt and colleagues (1986). Immunostainable dynorphin B, α-neoendorphin, dynorphin A, and dynorphin A (1–8) were localized to interstitial cells (probably Leydig cells), but not seminiferous tubules, by Douglass and colleagues (1987). Evidence for local synthesis of these opioid peptides was forthcoming when mRNA encoding proenkephalin (Kilpatrick et al., 1985; Douglass et al., 1987) and prodynorphin (Douglass et al., 1987; McMurray et al., 1989) were identified in rat testes. Recently, Collard and colleagues (1990) used cell fractionation, Northern blot analysis, and in situ hybridization to show that the primary site of synthesis of prodynorphin mRNA in the rat testis was the Sertoli cell. In view of these results, reinterpretation of earlier immunocytochemical studies is necessary. To our knowledge, mRNA encoding proenkephalin has not yet been localized directly to Leydig cells by in situ hybridization techniques. Moreover, the function of these opioid peptides in the testes remains unknown.

3. CORTICOTROPIN-RELEASING FACTOR

Immunoreactive corticotropin-releasing factor (CRF) has been quantified in testes of mouse, sheep, humans, and guinea pig (Hollander et al., 1983) and in rat (Hollander et al., 1983; Yoon et al., 1988). Yoon and colleagues (1988) reported immunoreactive CRF in both Leydig cells and seminiferous tubules. Subsequently, Audhya and co-workers (1989) reported that Leydig cells, but not seminiferous tubules, contain immunoreactive CRF. Clearly, this controversy requires resolution. Aside from microheterogeneity at position 39, the chemical structure of ovine testicular CRF is identical to ovine hypothalamic CRF (Audhya et al., 1989). Evidence for local synthesis of CRF was provided by the discovery of CRF mRNA in rat testis (Thompson et al., 1987). To our knowledge, however, this mRNA has not yet been localized directly to Leydig cells by in situ hybridization techniques. Although the function of CRF in the testis remains unknown, it has been suggested that it may have local antireproductive effects by inhibiting LH-stimulated testosterone secretion (Dufau et al., 1989), and may play a role in the regulation of testicular opioid peptides (Eskeland et al., 1989).

4. OXYTOCIN

Oxytocin and arginine-vasopressin are nonapeptides synthesized from preprooxytocin and preprovasopressin, respectively, which are encoded by two closely related but independent genes (Douglass et al., 1984). In this section, we focus only on oxytocin because controversy exists regarding the localization of arginine-vasopressin in Leydig cells (Guldenaar and Pickering, 1985; Mezey et al., 1986; Clements et al., 1988). See Pickering and colleagues (1989) for a detailed review of the evidence that oxytocin is produced by Leydig cells and exerts a paracrine function in the testis.

Briefly, oxytocin immunoreactivity was first detected in rat and human testes in 1984 (Nicholson et al., 1984). Immunostainable oxytocin was localized to Leydig cells in situ (Guldenaar and Pickering, 1985) and after their purification on Percoll gradients (Yeung et al., 1988). Oxytocin mRNA is expressed in the bovine testis at relatively high levels; these transcripts were localized to seminiferous tubules by in situ hybridization (Ang et al., 1991). Oxytocin mRNA is expressed in the rat testis, but its abundance is extremely low and can be detected only after amplification by polymerase chain reaction (Foo et al., 1991). Because of the low abundance of transcripts, it is unclear which cell type expresses oxytocin in the rat testis. Nicholson and Hardy (1992) have provided the most direct evidence to date that rat Leydig cells can biosynthesize oxytocin de novo. These investigators demonstrated the presence of immunoreactive oxytocin in the cellular extracts and culture medium of highly purified Leydig cells; no oxytocin was detected in the cellular extracts and culture medium of Sertoli cells. They also showed that LH can regulate the secretion of this peptide in vitro because the amount of oxytocin accumulated in the culture medium was twice as great in Leydig cell cultures treated with LH as in untreated cultures.

Apparently, oxytocin has only a small effect on testicular steroidogenesis (Adashi and Hsueh, 1981). Oxytocin probably acts via myoid cells to stimulate seminiferous tubule contraction (Pickering et al., 1989). At this juncture, it is unclear how or whether seminiferous tubule contractions influence male reproduction.

5. RENIN-ANGIOTENSIN

The renin-angiotensin system classically includes: (1) renin that is an aspartyl proteinase produced in the kidney and secreted into the blood; (2) angiotensinogen that is a tetradecapeptide substrate produced in liver and secreted into blood where renin cleaves it thus forming; (3) the decapeptide angiotensin I (AI); and (4) angiotensin-converting enzyme (ACE) that is a dipeptidyl peptidase that cleaves the carboxyterminal dipeptide of AI releasing angiotensin II (AII) that is a potent vasoconstrictor and powerful aldosterone secretogogue (Dzau and Pratt, 1986).

Although incomplete at this time, evidence is accumulating that this renin-angiotensin system is present in the mammalian testis. At present the data are strongest for renin and ACE, less compelling for AI and AII, and indirect for angiotensinogen. Renin immunoreactivity is present in rat testis (Naruse et al., 1984). Immunostainable renin was localized to Leydig cells first in 1983 (Parmentier et al., 1983) and most recently in 1986 (Deschepper et al., 1986). Evidence for local synthesis of renin was presented in 1984 when renin mRNA was identified in mouse testis (Field et al., 1984). This was confirmed and extended to the rat testis (Dzau et al., 1987) where renin was localized by in situ hybridization to rat Leydig cells (Deschepper et al., 1986).

Angiotensin I is produced by cultured mouse Leydig tumor cells (MA-10); most is retained in the cell lysate (Pandey and Inagami, 1986).

Angiotensin-converting enzyme activity was first discovered in rat testis in 1971 (Cushman and Cheung, 1971), and rabbit testis in 1982 (El-Dorry et al., 1982). These latter investigators showed that, although catalytically similar, the testicular enzyme was smaller and differed in its NH_2 and COOH termini from ACE purified from lung. This was confirmed when Ehlers and colleagues (1989) cloned the gene for human testicular ACE. Evidence for local synthesis of ACE was presented first in 1982 when mRNA from mature rabbit testis was shown to direct the in vitro synthesis of an ACE that was 35 percent smaller than the pulmonary enzyme (El-Dorry et al., 1982). The intratesticular location of ACE has been controversial. Pandey and co-workers claim that ACE is localized in Leydig cells (Pandey et al., 1984; Pandey and Inagami, 1986). Others claim that ACE is associated primarily with germ cells (Strittmatter and Snyder, 1984; Langford et al., 1991). Recently, the specific localization of ACE in germ cells has been confirmed in mice that were made transgenic for a DNA construct encoding the reporter gene β-galactosidase and the testis-specific promoter for murine ACE (Langford et al., 1991). The transgene was expressed exclusively in the testis, and the highest expression occurred in elongating spermatids.

Angiotensin II is produced by cultured mouse Leydig tumor cells (MA-10) and secreted into the medium (Pandey and Inagami, 1986). The only evidence for a testicular angiotensinogen is that AI and AII are produced de növo by MA-10 cells cultured in vitro (Pandey and Inagami, 1986).

Although incomplete, these data prompted the following hypothesis (Pandey and Inagami, 1986): renin in Leydig cells converts angiotensinogen to AI. Two alternatives exist to explain subsequent steps in the cascade. One is that AI is converted in the Leydig cell to AII, which is then secreted. Alternatively, AI is secreted and immediately converted to AII by ACE located in the extracellular environment.

The function of AII in the testis is unknown. Recent evidence, however, has localized AII receptors to rat, monkey, and human Leydig cells (Millan and Aguilera, 1988) rather than germ cells, suggesting that AII exerts some effect on Leydig cell function. This idea is supported by Dufau and colleagues (1989) who showed that AII, acting through high affinity AII receptors, inhibited adenylate cyclase activity in Leydig cell membranes and hCG-stimulated testosterone secretion by rat Leydig cells in vitro.

In summary, the evidence presented above strongly supports the idea that Leydig cells secrete biologically important peptides including POMC-derived peptides, enkephalins and dynorphins, CRF, oxytocin, and AII. The challenge is to elucidate their intratesticular function and the regulation of their synthesis and secretion. Undoubtedly, other equally important Leydig cell peptidic secretogogues await discovery. The remainder of this chapter is devoted to a discussion about the important and better understood role of Leydig cells in producing testosterone.

B. Testosterone

1. BIOLOGICAL CONSEQUENCES OF TEMPORAL CHANGES IN TESTOSTERONE DURING THE LIFE CYCLE

The most well-understood function of Leydig cells in male mammals is the biosynthesis and secretion of testosterone in well-defined epochs during the life cycle (Ewing et al., 1980); Figure 7–1 shows such changes for the human male. In those species that have been studied thoroughly (Ewing et al., 1980), the surge of testosterone in fetal males is responsible for the

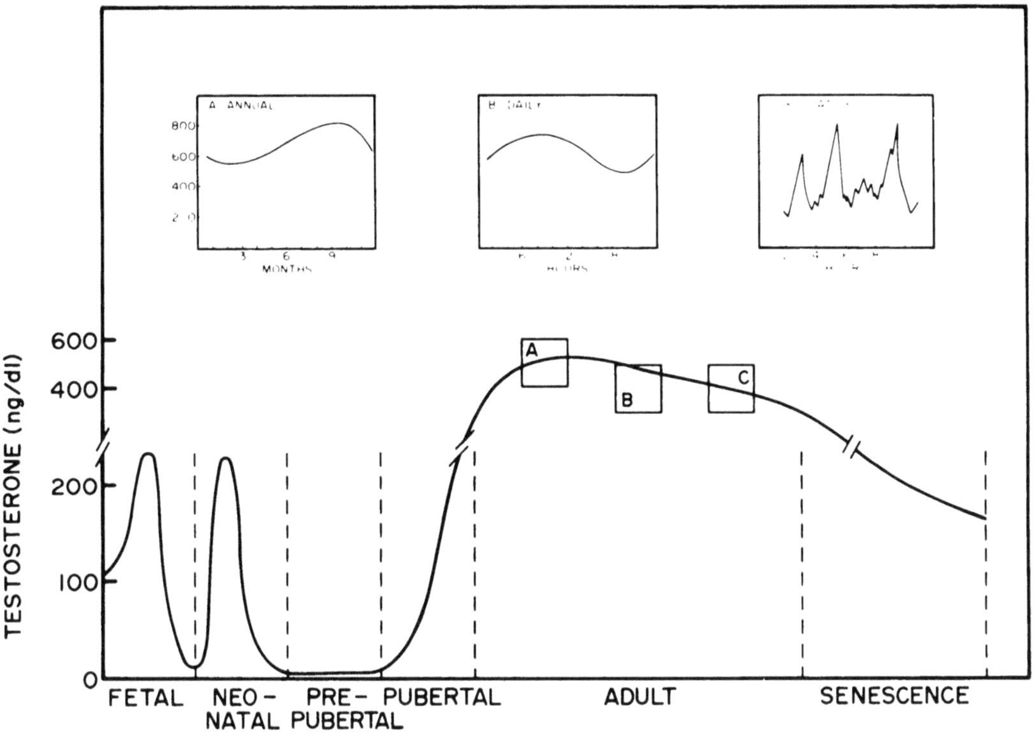

Figure 7–1. Concentration of testosterone in peripheral blood of the human male at different times in the life cycle. The peak of testosterone in the peripheral blood of the fetus occurs between 12 and 18 weeks of gestation. The peak of testosterone in the peripheral blood of the neonate occurs at approximately 2 months. The pubertal increase in testosterone concentration is variable occurring between 12 and 17 years depending on the individual. Testosterone concentration in the adult reaches its maximum during the second or third decade and then declines slowly through the fifth decade. The rate of decline occurring during senescence varies depending on the individual. Insets **A**, **B**, and **C** show the annual, daily, and pulsatile changes in testosterone concentration in peripheral blood of the human male. (From Ewing et al., 1980, by courtesy of the editor, Vol 22, *International Review of Physiology.*)

differentiation and development of the reproductive tract. The neonatal surge is responsible for "imprinting" androgen-dependent tissues so that they respond appropriately to testosterone during adulthood; the pubertal surge is responsible for masculinization and the onset of male sexual behavior and spermatogenic function. Finally, the decline of testosterone production with advancing age represents the loss and/or dysfunction of Leydig cells during senescence. Albeit attenuated in human males, testosterone concentration in peripheral blood exhibits circannual, diurnal, and rapid pulsatile changes. Except for the obvious importance of the circannual cycles in seasonally breeding species, the consequences of the latter three episodic changes in testosterone blood concentration are largely unknown.

III. Steroidogenic Enzymes: Intracellular Location and Genes

Figure 7–2 depicts the metabolic pathways for the conversion of cholesterol to testosterone. There follows a discussion of the steroidogenic enzymes, their intracellular location and their genes. The enzymes and genes involved in converting cholesterol to steroidal hormones in general (Waterman and Simpson, 1989; Miller,

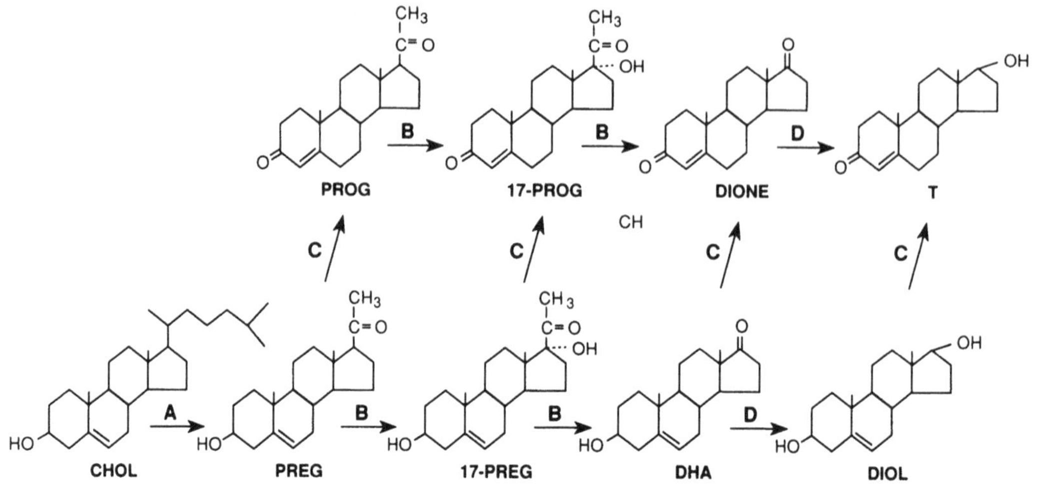

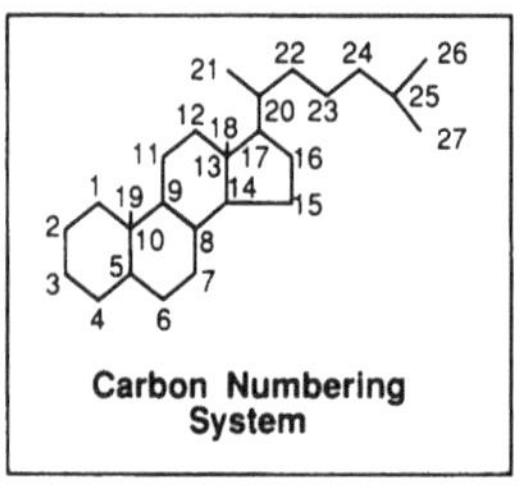

	Steroidogenic Enzymes	Abbreviation	Cofactors	Location
A	Cholesterol side-chain cleavage enzyme	$P450_{scc}$	NADPH	Mitochondrial inner membrane
B	17α-hydroxylase / C17-20 lyase	$P450_{17\alpha}$	NADPH	Smooth endoplasmic reticulum
C	3β-hydroxysteroid dehydrogenase / $\Delta^{5\text{-}4}$ isomerase	3βHSD	NAD^+	Smooth endoplasmic reticulum
D	17β-hydroxysteroid dehydrogenase	17βHSD	$NADP^+$	Smooth endoplasmic reticulum

Chubb 41127a MDB 11-88

Figure 7–2. The metabolic pathways for the conversion of cholesterol to pregnenolone. CHOL, cholesterol; PREG, pregnenolone; 17-PREG, 17α-hydroxypregnenolone; DHA, dehydroepiandrosterone; DIOL, androstenediol; PROG, progesterone; 17-PROG, 17α-hydroxyprogesterone; DIONE, androstenedione; T, testosterone. (From Chubb and Ewing, 1991, by courtesy of the editor, *Infertility in the Male.*)

1988) and to testosterone in particular (Hall, 1988) have been the subject of recent reviews. A synopsis of this information follows.

A. Cholesterol Side Chain Cleavage Cytochrome P450

Cholesterol is transported into the inner mitochondrial membrane where cholesterol side chain cleavage cytochrome P450 ($P450_{scc}$) resides. Purification of a single protein and identification of a single gene encoding $P450_{scc}$ (Miller, 1988; Hall, 1988) suggest that a single polypeptide carries out three separate reactions including hydroxylation of the C-22 carbon, hydroxylation of the C-20 carbon and scission of the C-20–C-22 bond at a single catalytic site to produce pregnenolone. Each of these reactions requires a pair of electrons. Although not established completely for Leydig cells, evidence (Waterman and Simpson, 1989; Miller, 1988) suggests that generic electron transport proteins exist for all mitochondrial cytochrome P450s. Thus, the matrix side of the Leydig cell inner mitochondrial membrane contains an electron transfer chain in which a ferredoxin reductase transfers electrons from NADPH to a ferredoxin, which in turn transfers them to $P450_{scc}$. Complementary DNAs (cDNAs) encoding $P450_{scc}$ have been isolated and characterized from bovine adrenocortical (Morohashi et al., 1984), human adrenocortical, testicular (Chung et al., 1986), and rat ovarian (McMasters et al., 1987) cDNA libraries. The human gene encoding $P450_{scc}$ maps to chromosome 15 (Chung et al., 1986; Miller, 1988). Complementary DNAs encoding ferredoxin-$NADP^+$ reductase (adrenodoxin reductase) have been isolated from bovine adrenocortical (Hanukoglu et al., 1987) and human adrenocortical and testicular (Solish et al., 1988) cDNA libraries. Evidence indicates that these transcripts are encoded by a single gene that is located on chromosome 17 in humans (Solish et al., 1988; Miller, 1988). Ferredoxin (adrenodoxin) has also been cloned from bovine adrenocortical (Okamura et al., 1985) and human placental (Mittal et al., 1988) cDNA libraries, and maps to chromosome 11 in humans (Miller, 1988).

B. 3β-Hydroxysteroid Dehydrogenase/$\Delta^{5\rightarrow4}$ Isomerase

Pregnenolone produced by the $P450_{scc}$ moves into the smooth endoplasmic reticulum where it is converted to testosterone via either a 4-ene or a 5-ene pathway, depending on the species (Chubb and Ewing, 1991). If the 4-ene pathway is used, pregnenolone is bound to the 3β-hydroxysteroid dehydrogenase/$\Delta^{5\rightarrow4}$ isomerase that utilizes NAD^+ as a co-factor to catalyze the dehydrogenation of the C-3 hydroxyl (3βHSD activity) and rearrangement of the double bond between C-5 and C-6 to the double bond between C-4 and C-5 (isomerase activity) (Hall, 1988) to produce progesterone.

A single 46,500 molecular weight protein was purified from rat adrenal (Ishii-Ohba et al., 1986b) and rat testicular (Ishii-Ohba et al., 1986a) microsomes that exhibited both 3βHSD and isomerase activities, suggesting both enzyme activities reside in a single protein. NAD^+ was required as a co-factor for 3βHSD. Either NAD^+ or NADH was required for isomerase activity, probably as an allosteric effector rather than a hydrogen donor (Ishii-Ohba et al., 1987). Recently, it was shown for the human placental enzyme that the dehydrogenase and isomerase activities co-purify, but that the pregnene and androstene dehydrogenase activities are expressed at one site and the C-21 and C-19 isomerase activities are expressed at a second site on a single enzyme protein (Thomas et al., 1990). Full length cDNAs encoding the human placental (Luu-The et al., 1989a) and rat testicular (Lorence et al., 1991) 3β-hydroxy-5-ene steroid dehydrogenases were obtained. The cDNA encoding the placental enzyme was used to localize the gene encoding this enzyme to human chromosome 1 (Bérubé et al., 1990). Transfection and expression of the cDNAs encoding the human placental and rat testicular enzymes in COS-1 cells (transformed monkey kidney) proved that a single enzyme protein catalyzes both dehydrogenase and isomerase activities (Lorence et al., 1990, 1991). It is now apparent that multiple isoforms of 3βHSD exist in various tissues and species, and new members of this gene family continue to be identified. Although the nomenclature is not yet standard, at this time the predominant isoforms of 3βHSD found in the gonads are type I in the rat (Lorence et al., 1991; Zhao et al., 1991) and mouse (Bain et al., 1991) and type II in humans (Rhéaume et al., 1991). At present, it is unclear how many genes may encode for the multitude of unique 3βHSD transcripts that exhibit tissue-specific and species-specific expression.

C. 17α-Hydroxylase/17,20-Lyase Cytochrome P450

Slaunwhite and Samuels (1956) were the first to demonstrate that rat testicular microsomes contained enzyme activity(ies) capable of cleaving the side chain from the C-21 *progesterone* to produce the C-19 androgen, androstenedione. Nakajin and Hall (1981) provided the first evidence that a single cytochrome P450 protein ($P450_{17\alpha}$) catalyzes two mixed function oxidase reactions, namely the 17α hydroxylation of progesterone to 17α-hydroxyprogesterone and the cleavage of the bond between C-17 and C-20 to convert 17α-hydroxyprogesterone to androstenedione. Apparently, this testicular $P450_{17\alpha}$ enzyme is encoded by a single gene that has been localized to human chromosome 10 (Chung et al., 1987; Miller, 1988). Complementary DNAs have been isolated and characterized first for bovine adrenocortical $P450_{17\alpha}$ (Zuber et al., 1986a), and then for human adrenocortical and testicular (Chung et al., 1987) and rat testicular (Nishihara et al., 1988; Fevold et al., 1989) $P450_{17\alpha}$. Transfection and expression of bovine adrenocortical (Zuber et al., 1986b) and rat testicular (Namiki et al., 1988; Fevold et al., 1989) cDNAs in nonsteroidogenic COS-1 cells proved that both 17α-hydroxylase and 17,20-lyase activities reside in a single protein. Finally, it was shown that a single active site in this enzyme catalyzes both 17α-hydroxylase and 17,20-lyase activities (Onoda et al., 1987).

The Leydig cell smooth endoplasmic reticulum contains, in addition to $P450_{17\alpha}$, a P450 reductase (flavoprotein) that accepts two electrons from NADPH and transfers them one at a time to the $P450_{17\alpha}$ enzyme (Hall, 1988). The NADPH–cytochrome P450 reductase has been purified from rat testes (Inano and Tamaoki, 1986b). Apparently, the $P450_{17\alpha}$ shows no absolute specificity for a specific cytochrome P450 reductase because reductase purified from rat or pig liver transfers electrons effectively from NADPH to the purified testicular $P450_{17\alpha}$ (Nakajin and Hall, 1981). A cDNA encoding NADPH cytochrome P450 reductase has been cloned from rat liver (Gonzalez and Kasper, 1982).

D. 17β-Hydroxysteroid Dehydrogenase

Androstenedione is converted to testosterone by a 17-oxidoreductase enzyme (also denoted 17-ketosteroid reductase, 17β-hydroxysteroid dehydrogenase, 17βHSD) that resides in the smooth endoplasmic reticulum of Leydig cells (Inano and Tamaoki, 1986a; Kurosumi et al., 1986). Inano and co-workers (1980; Inano and Tamaoki, 1986a) purified the enzyme to homogeneity from porcine testes and reported that it is a single $NAD(P)^+/NAD(P)H$ dependent, non-P450, monomeric molecule that catalyzes both the reduction of androstenedione and the oxidation of testosterone. Apparently, the direction of the reaction is dependent on substrate and product concentration (Hall, 1988). There is evidence, however, for isoforms of 17βHSD in the testis (Hall, 1988; Inano and Tamaoki, 1986a). Bogovich and Payne (1980) reported that in rat testes the 17-ketosteroid reductase and 17βHSD activities represent separate and distinct enzymes.

Although testicular 17βHSD has not yet been cloned, cDNAs encoding 17βHSD have been isolated from human placental cDNA libraries (Peltoketo et al., 1988; Luu-The et al., 1989b). The human genome contains two genes encoding 17βHSD that occur in tandem and map to chromosome 17 (Luu-The et al., 1989b, 1990). Both genes are potentially transcribed and translated, but they encode proteins of different sizes, 214 and 327 amino acids. The predicted molecular weight of the larger protein (Luu-The et al., 1989b) is consistent with that of 17βHSD purified from porcine testis (Inano et al., 1980). The expression of these placental 17βHSD cDNAs in cells that lack endogenous 17βHSD activity has not yet been reported; however, there is evidence to suggest that placental 17βHSD that is responsible for the formation of estrogens (C18 steroids) is distinct from the testicular enzyme(s) responsible for the formation of androgens (C19 steroids). The presence of different isoforms of 17βHSD in different tissues could readily explain the following observations. First, individuals having a deficiency of testicular 17βHSD can exhibit normal 17βHSD activity in extragonadal tissues (Wilson et al., 1987). Second, an antiserum raised against human placental 17βHSD failed to recognize 17βHSD in various steroidogenic tissues including human testis, mouse and rat Leydig and adrenocortical tumor cells, and bovine adrenocortical cells, although these tissues are known to express relatively high levels of 17βHSD activity (Milewich et al., 1990). Molecular cloning and expression of 17βHSD from different tissues would prove whether 17βHSD is similar to 3βHSD in that the predominant placental isoform of 3βHSD (human type I) differs from that in testis (human type II) (Rhéaume et al., 1991). Interestingly, the amino acid sequence of human placental 17βHSD is similar to the *Rhizobium meliloti* nodG protein, *Klebsiella aerogenes* ribitol dehydrogenase, *Escherichia coli* glucitol-6-phosphate dehydrogenase, *Streptomyces coelicolor* actIII protein, *Drosophila melanogaster* alcohol dehydrogenase, *Pseudomonas pseudoalcaligenes* dihydrodiol dehydrogenase, and rat corticosteroid 11β-dehydrogenase (Baker, 1990). One implication of this analysis is that 17βHSD is one member of a large family of related dehydrogenases that may be derived from a common ancestor gene (Baker, 1990).

IV. Spatial Organization of Steroid Biosynthesis

Herein we briefly recapitulate earlier reviews (Ewing et al., 1980; Ewing and Zirkin, 1983;

Hall, 1988) and then update our knowledge on the relationship between Leydig cell structure and function. Ewing and Zirkin (1983) presented the following rationale for the highly ordered spatial arrangement of steroid synthesizing enzymes in membranes of the Leydig cell. They stated, "first, sequestering steroidogenic enzymes in membranes in all likelihood facilitates the interaction of hydrophobic sterol or steroid molecules with reactants such as oxygen or hydrogen dissolved in the aqueous milieu of the Leydig cell cytoplasm. Second, a membrane-bound multienzyme sequence such as that required to convert pregnenolone to testosterone may exist in a highly ordered state to reduce substrate diffusion time, thus preventing leakage of biosynthetic intermediates from the enzyme-membrane complex. Third, separation of competing steroidogenic reactions in space limits the potential products produced by the testis. Fourth, control points are established when a substrate is separated in space from the enzyme that is required to catalyze its conversion into product. Fifth, alteration in the amount or chemical characteristics of Leydig cell cytoplasmic organelle membranes might influence profoundly the activity of the steroidogenic enzymes sequestered therein." These observations prompted Ewing and Zirkin (1983) to hypothesize that Leydig cell ultrastructure must be tightly coupled to testosterone-producing capacity. A discussion of this idea follows.

A. The Cellular Level

The bulk of testosterone-producing capacity in the mammal resides in a tiny cellular compartment, the Leydig cell. Numerous studies from independent laboratories confirm that the number of rat Leydig cells per testis ranges from 18 to 34 $\times 10^6$ with an average of 26 $\times 10^6$ (Mori and Christensen, 1980; Ewing and Zirkin, 1983; Tapanainen et al., 1984; Zirkin and Ewing, 1987; Kerr et al., 1987; Mendis-Handagama et al., 1988). Apparently, these 26 $\times 10^6$ Leydig cells represent a slowly turning over, static population of cells (Keeney et al., 1990) that comprise about 3 percent of testicular volume in the rat; this translates into a total volume of 40 μL per testis and the volume of an average rat Leydig cell of approximately 1600 μm^3 (Mori and Christensen, 1980; Keeney et al., 1988; Mendis-Handagama et al., 1987, 1988). Mendis-Handagama and colleagues (1988) reported a significant, linear and positive correlation between Leydig cell volume per gram testis or the volume of an average Leydig cell and testosterone-secreting capacity by testes of three species perfused in vitro. By contrast, testosterone-secreting capacity was not correlated with Leydig cell number per gram testis, or the volume of blood vessels, elongated spindle-shaped cells, and macrophages. These results show that differences in testosterone-secreting capacity among species are due largely to differences in the volume, rather than the number, of Leydig cells per testis. Clearly, regulation of Leydig cell volume dictates total testosterone-producing capacity.

B. Cytoplasmic Organelles

Similar stereologic studies at the electron microscopic level (Kerr et al., 1979; Zirkin et al., 1980; Mori and Christensen, 1980; Mendis-Handagama et al., 1988) have reported the proportional volume of a rat Leydig cell that is occupied by subcellular organelles. For example, Mendis-Handagama and colleagues (1988) showed that smooth endoplasmic reticulum and mitochondria represented 40 percent and 13 percent, respectively, of Leydig cell volume in the rat. By comparison, rough endoplasmic reticulum and peroxisomes occupied 1 and <1 percent, respectively, of Leydig cell volume.

A significant, linear and positive correlation ($r = 0.99$) between the percentage of Leydig cell volume occupied by smooth endoplasmic reticulum and testosterone secretion by testes of five species perfused in vitro was discovered by Zirkin and colleagues (1980), proving that Leydig cell structure and steroidogenic function are tightly coupled. This prompted a similar but more detailed study comparing 118 factors associated with structure and steroidogenic function of hamster, rat, and guinea pig testes by Mendis-Handagama and colleagues (1988). In this latter study, the differences in Leydig cell volume among species were due largely to differences in the volume of smooth endoplasmic reticulum and peroxisomes; these were the only ultrastructural measures correlated with testosterone production. It is noteworthy that we never have observed a positive correlation between mitochondrial membranes and testosterone-secreting capacity; this despite the fact that the $P450_{scc}$ enzyme resides in the inner mitochondrial membrane and is purported to be the rate-limiting step in testosterone biosynthesis (Hall, 1988). We can only conclude that $P450_{scc}$ is a minor component of the mitochondrion and that its presence or absence does not affect mitochondrial membrane volume/surface area.

The fact that the quantity of smooth endoplasmic reticulum is correlated with testosterone-secreting capacity suggests that the enzymes required to convert pregnenolone to testosterone constitute a significant portion of smooth endoplasmic reticulum. Surprisingly, the enzymes required to convert pregnenolone to testosterone, namely, 3βHSD/$\Delta^{5\rightarrow4}$ isomerase, $P450_{17\alpha}$ and 17βHSD, are not distributed evenly throughout Leydig cell smooth endoplasmic reticulum. Wing and colleagues (1984) concluded that 3βHSD/$\Delta^{5\rightarrow4}$ isomerase is associated primarily with perimitochondrial smooth endoplasmic reticulum that is intimately associated with mitochondrial membranes. See Section VI.3 for a detailed discussion of the data supporting this conclusion. This observation, if true, would suggest that pregnenolone produced in the inner mitochondrial membrane by the action of $P450_{scc}$ would have immediate access to 3βHSD/$\Delta^{5\rightarrow4}$ isomerase. This exciting possibility awaits direct proof by ultrastructural immunocytochemistry.

Why is peroxisome volume of Leydig cells correlated with testosterone secretion per Leydig cell in the three species described previously (Mendis-Handagama et al., 1988)? This correlation cannot be accounted for by a generalized hypertrophy of cytoplasmic organelles in species producing more testosterone because no such positive, linear correlation was observed between testosterone secretion and the volume of outer and inner mitochondrial membranes, lysosomes, ribosomal aggregates, multivesicular bodies, and so forth (Mendis-Handagama et al., 1988).

In an effort to answer why the volume of peroxisomes in Leydig cells is correlated with testosterone production, Mendis-Handagama and colleagues (1990b) completed a comprehensive morphological study of peroxisomes in rat Leydig cells. They confirmed the observations of Reddy and Svoboda (1972) that peroxisomes in Leydig tumor cells are pleomorphic and can be divided into small, intermediate and large structures that assume circular, oval, and dumbbell shapes in two dimensions. This is consistent with the idea that peroxisomes exist as a network rather than discrete organelles (Gorgas, 1985). Mendis-Handagama and colleagues (1990b) further showed that Leydig cell peroxisomes contain sterol carrier protein 2 (SCP_2), as has been shown previously for liver (Keller et al., 1989). It is noteworthy that liver peroxisomes have been shown to contain 3-hydroxy-3-methylglutaryl coenzyme A reductase (Keller et al., 1985), a key regulatory enzyme in cholesterol biosynthesis, as well as several enzymes involved in the synthesis of cholesterol from lanosterol including dihydrolanosterol oxidase, steroid-14-reductase, steroid-8-isomerase, and steroid-3-ketoreductase (Appelkvist et al., 1990). Thus, it seems likely that peroxisomes play an important role in either producing cholesterol perhaps destined for testosterone production or facilitating cholesterol transport to mitochondria where the $P450_{scc}$ enzyme converts it into pregnenolone. This hypothesis is supported by the observation that a single injection of LH in an adult rat results in a transient and specific increase in the volume of peroxisomes per Leydig cell and the peroxisomal content of immunoreactive SCP_2, coincident with the increase in testosterone secretion (Mendis-Handagama et al., 1990c).

C. *Enzyme Topology and Membrane Microenvironment*

The compartmentalization of Leydig cell steroidogenic enzymes into ever smaller microenvironments extends beyond the cellular level and cytoplasmic organelles into a precise membrane topology within these organelles, which undoubtedly has profound consequences on testosterone production. Unfortunately, only fragmentary information is available for the Leydig cell so any discussion of membrane topology of steroidogenic enzymes depends heavily on analogy to liver and adrenal (Nelson and

Strobel, 1988). However briefly, this topic has been recently reviewed for the Leydig cell (Hall, 1986, 1988; Chubb and Ewing, 1991).

Chubb and Ewing (1991) stated that "the active sites of $P450_{scc}$ and $P450_{17\alpha}$ are on the mitochondrial matrix side of the mitochondrial inner membrane and the cytoplasmic side of the smooth endoplasmic reticulum membrane, respectively. Recent research has provided further evidence that $P450_{scc}$ is bound to the inside half of the mitochondrial inner membrane in adrenals (presumed true for the testis). Moreover, the active sites of the cytochrome P450 enzymes face the lipid membrane and the sterol/steroid molecules enter the active sites from the membrane. The electron donor proteins are either within the mitochondrial matrix (adrenodoxin reductase and adrenodoxin) or membrane bound (P450 reductase). The active sites of 3βHSD and 17βHSD are also believed to be exposed on the cytoplasmic side of the smooth endoplasmic reticulum membrane."

Chubb and Ewing (1991) continue by stating that, "an understanding of the biophysics of steroids explains how the predicted intramembrane organization of steroidogenic enzymes may be the most appropriate. The C-3 hydroxyl of steroids intercalates between the polar heads of membrane phospholipids and the remainder of the steroid molecule penetrates the nonpolar chains. Consequently, steroids with hydroxyl groups would be near the membrane surface. In addition, the steroids would be segregated in the cytoplasmic surfaces (instead of the luminal surfaces) of the smooth endoplasmic reticulum because pregnenolone enters the membrane from the cytoplasm. In other words, the steroids and enzyme active sites would be juxtaposed. The hydrophobicity of steroids is directly correlated with the length of the hydrophobic side chain attached to the C-17 carbon. On the basis of the above guidelines and reported steroid partition coefficients, pregnenolone would be designated as the most lipophilic steroid and steroid hydrophobicity would decrease with the addition of hydroxyl groups at C-17, the removal of C-3 hydroxyl groups, and the reduction of the steroid nucleus to 19 carbons. This evidence suggests that steroids become more hydrophilic during the conversion of pregnenolone to testosterone, and thus their passage through the cytoplasm during the secretory process is facilitated."

Examples of important consequences of this enzyme membrane topology follow. One example is the sequestration of the $P450_{scc}$ enzyme in the inner mitochondrial membrane behind the aqueous barrier formed between the inner and outer mitochondrial membranes (Stevens et al., 1985). Clearly, this represents a widely discussed hormonally regulated control point of testosterone production (Hall, 1988; Dufau, 1988). In yet another example, the topology of the multienzyme sequence of microsomal enzymes responsible for converting pregnenolone to testosterone in the Leydig cell assures that 17α-hydroxyprogesterone produced endogenously from progesterone is preferentially accumulated in the microsomal membrane (Kühn-Velten et al., 1988). Only this membrane-associated pool of 17α-hydroxyprogesterone is directly accessible to the $P450_{17\alpha}$ enzyme; this results in the functional coupling of the 17α-hydroxylase and 17,20-lyase activities of $P450_{17\alpha}$ and efficient conversion of progesterone to androstenedione (Kühn-Velten et al., 1988). Further, the efficiency with which these two enzymic activities are coupled is inversely proportional to the concentration of the substrate progesterone (Kühn-Velten, 1990).

Finally, it follows that subtle changes in the lipid constituents of membranes containing steroidogenic enzymes may have profound effects on testosterone production. This was first shown for the testis by Samuels et al. (1975). Cooke (1989) showed that phospholipids containing choline were inhibitors and those containing serine were activators of 3βHSD/$\Delta^{5\rightarrow4}$ isomerase activity. An exquisite specificity was shown by the discovery (Cooke, 1989) that phosphatidyl choline containing saturated acyl chains of 8, 10, 12, and 14 carbons decreased while those containing 16, 18, and 20 carbons essentially were without effect on 3βHSD/$\Delta^{5\rightarrow4}$ isomerase activity. Clearly, membrane lipid content profoundly alters the activity of specific steroidogenic enzymes.

Taken together, these results show clearly that Leydig cell structure is closely related to steroidogenic function and raise the issue of the movement of sterols/steroids into, about, and out of Leydig cells. This topic is discussed in the next section.

V. Movement of Leydig Cell Sterol Precursors and Steroid Products

A. Sources of Cholesterol for Testosterone Biosynthesis

Cholesterol, which is an obligatory intermediate in testosterone biosynthesis, is derived from de novo biosynthesis in Leydig cells, from intracellular stores, and from lipoprotein-mediated endocytosis (Hall, 1988). The contribution of cholesterol from these three sources to steroid biosynthesis has been elucidated elegantly in the MA-10 clonal strain of cultured mouse Leydig cells by Ascoli and Freeman (1985). They showed that in the unstimulated basal state, progesterone production (MA-10 cells do not produce testosterone) is low, free cholesterol is two to four times higher than esterified cholesterol stores, and most intracellular cholesterol is derived from de novo synthesis. This state changes dramatically with hCG stimulation, however. During the first 4 hours after hCG stimulation, progesterone is derived primarily from esterified intracellular cholesterol stores; between 4 and 8 hours after hCG stimulation, intracellular stores are turned over and now 34 and 65 percent of the progesterone produced is derived from de novo synthesis and low density lipoprotein-mediated endocytosis, respectively.

Freeman (1987), again with MA-10 cells, showed that in the unstimulated basal state the pool of nonesterified cholesterol used for progesterone synthesis resides within the plasma membrane. Apparently, stimulation with hCG simultaneously causes plasma membrane cholesterol stores to be internalized directly and diverts both newly synthesized and lipoprotein-derived cholesterol from the storage into the progesterone biosynthetic pathway, thus depleting plasma membrane cholesterol.

Finally, Freeman (1989) showed that cholesterol can enter and leave MA-10 cell membranes without causing mass changes in protein or phospholipid content. He also presented evidence that although there is a high concentration of cholesterol in mitochondrial membranes, this cholesterol is not used as substrate for steroidogenesis. Apparently, plasma membrane cholesterol is incorporated directly into steroid hormone without equilibrating with cholesterol in mitochondrial membranes.

Two cautions are worthy of note. First, MA-10 cells may not be representative of fully differentiated mouse Leydig cells with an intact steroidogenic pathway. Second, there are likely to be significant species differences in the transport and source of cholesterol used in steroidogenesis (Quinn et al., 1985). With these caveats in mind, these results are compatible with the idea that mitochondrial cholesterol is inert as a substrate for steroidogenesis in the Leydig cell. This emphasizes the important role of the intracellular transport of cholesterol from the plasma membrane to the mitochondria and the bridging of the aqueous barrier between the outer and inner mitochondrial membranes that prevents the free access of cholesterol to the $P450_{scc}$ enzyme.

B. Intracellular Cholesterol Transport

Three topics are discussed in this section: (1) proteins that may be involved in intracellular cholesterol transport; (2) the role of cytoskeletal elements in intracellular cholesterol transport; and (3) the mechanism by which cholesterol bridges the aqueous barrier between the outer and inner mitochondrial membranes. Once again, the information on Leydig cells is fragmentary and we must rely on data derived in part from the adrenal. The reader is referred to several excellent reviews for a complete discussion of these topics (Lambeth and Stevens, 1984; Pedersen, 1984; Scallen and Vahouny, 1985; Hall, 1985, 1988).

1. CHOLESTEROL TRANSPORT PROTEINS

The role of *sterol carrier protein* 2 (SCP_2; nonspecific lipid transfer protein; nsLTP) in intracellular cholesterol transport in liver and adrenal has been reviewed by Scallen and Vahouny (1985). In the testis, SCP_2 is localized exclusively in Leydig cells (van Noort et al., 1988). As in the adrenal, SCP_2 stimulates pregnenolone production by mitochondria isolated from Leydig tumor cells (van Noort et al., 1988). Apparently, SCP_2 does not transport cholesterol by binding it in a 1:1 molar complex but instead may form a dimer complex that enables lipids to redistribute between donor and acceptor membranes (van Amerongen et al., 1989).

Sterol carrier protein 2 does not affect directly the association of cholesterol with $P450_{scc}$ (Scallen and Vahouny, 1985). Interestingly, SCP_2 is localized primarily in peroxisomes in rat liver (Keller et al., 1989) and Leydig cells (Mendis-Handagama et al., 1990b), suggesting a role for peroxisomes in intracellular cholesterol transport. This may explain the correlation between testosterone-secreting capacity and Leydig cell peroxisome content discovered by Mendis-Handagama and co-workers (1988).

The role of *steroidogenic activator peptide* (SAP) in the intracellular cholesterol transport in steroidogenic tissues has been reviewed by Pedersen (1984). Discovery of SAP resulted from the search for the cycloheximide-sensitive protein responsive to cyclic adenosine 3′,5′-monophosphate (cAMP), which is responsible for the increased cholesterol metabolism to pregnenolone in steroidogenic tissues (Pedersen and Brownie, 1983). Steroidogenic activator peptide was discovered in the testis (Pedersen, 1987) and purified from rat H-540 Leydig tumor cells (Pedersen and Brownie, 1987). Subsequently, it was proposed that SAP is derived by cleavage from the carboxyl-terminal end of a minor heat shock protein (glucose related protein 78; GRP78) by a Ca^{++}-protease-dependent proteolysis (Mertz and Pedersen, 1989). Steroidogenic activator peptide may bridge the aqueous barrier between the outer and inner mitochondrial membranes directly because it causes spectral changes characteristic of increased cholesterol binding to $P450_{scc}$ (Pedersen and Brownie, 1983).

Yanagibashi and colleagues (1988) purified an *8.2 kd protein* from bovine adrenal cortex that stimulated both the entry of cholesterol into the outer and the transport of cholesterol from the outer to the inner mitochondrial membranes. Subsequently, it was shown that this protein, which binds to peripheral-type benzodiazepine receptors in mitochondria, was *des-(Gly-Ile)-endozepine* (Besman et al., 1989). Endozepines or diazepam-binding inhibitors (DBI) are proteins that inhibit binding of diazepam to diazepam receptors. This is of great interest because peripheral-type benzodiazepine receptors occur in high concentration in rat testicular interstitium (reviewed by Snyder et al., 1987), because immunoreactive endozepine is present in the rat testis (Gray et al., 1986), and because endozepine mRNA is expressed in rat testis (Mocchetti et al., 1986). Rhéaume and co-workers (1990) used immunocytochemistry and *in situ* hybridization to demonstrate specific localization of octadecaneuropeptide (a tryptic fragment of endozepine) to interstitial cells of the rat testis.

Recently, Papadopoulos and colleagues (1990a) have shown that specific ligand binding to peripheral-type benzodiazepine receptors stimulates steroid production in both MA-10 mouse Leydig tumor and purified rat Leydig cells. These results corroborate earlier studies of the adrenal (Yanagibashi et al., 1988; Besman et al., 1989) and are consistent with localization of the benzodiazepine receptor in the outer mitochondrial membrane of the Leydig cell. The stimulation of Leydig cell steroidogenesis was proposed to result from increased mitochondrial uptake of cholesterol and/or increased transfer of cholesterol from the outer to the inner mitochondrial membrane by a cycloheximide-insensitive mechanism; direct stimulation of $P450_{scc}$ activity was ruled out (Papadopoulos et al., 1990a). Papadopoulos and colleagues (1991a) demonstrated that the benzodiazepine flunitrazepam, a specific agonist of the peripheral-type benzodiazepine receptor, inhibited hormonally stimulated intramitochondrial cholesterol transport and steroid biosynthesis in mouse adrenal and Leydig tumor cells, thus providing evidence that steroidogenesis stimulated by ACTH or hCG is mediated, at least in part, by the peripheral-type benzodiazepine receptor. These investigators propose that DBI or its processing products may mediate this effect of the peripheral-type benzodiazepine receptor on hormone-stimulated steroid biosynthesis (Papadopoulos et al., 1991a,b). Snyder and co-workers (1987), however, caution that the affinity of endozepine for the peripheral-type benzodiazepine receptor is 1,000 times less than that of some porphyrins. Instead, they (Snyder et al., 1987) suggest that porphyrins are the endogenous ligand for the peripheral-type benzodiazepine receptor and that the endozepine-like activity in extracts of six different tissues studied could be accounted for by their porphyrin content. Moreover, Snyder and co-workers suggest that the peripheral-type benzodiazepine receptor that is located on the outer mitochondrial membrane corresponds to the voltage-dependent anion channel (VDAC; Porin), which admits molecules less than 1 kd. Clearly, this area requires additional research to clarify the function of these interesting peripheral-type benzodiazepine receptors and their endogenous ligands.

Leydig cells contain a *family* of cAMP inducible, cycloheximide-sensitive 28 kd proteins similar to those found in adrenal cortex and corpus luteum (Pon et al., 1986). This was confirmed recently by Stocco and Kilgore (1988) in the MA-10 Leydig tumor cell and in adult rat Leydig cells. More recently, members of a family of 30-kd phosphoproteins have been described that accumulate in mitochondria of hormonally stimulated mouse Leydig and adrenocortical (Epstein and Orme-Johnson, 1991) and MA-10 mouse Leydig tumor (Stocco and Sodeman, 1991) cells coincident with the acute stimulation of steroidogenesis. It was suggested that these proteins may be instrumental in transporting cholesterol across the aqueous barrier separating the inner and outer mitochondrial membranes to the active site of the $P450_{scc}$ enzyme in the inner mitochondrial membrane. This interesting family of proteins must be purified to homogeneity before their function can be proven.

2. CYTOSKELETAL ELEMENTS

Almost all of the studies of the role of the cytoskeleton in steroidogenesis have been carried out on the adrenal and have focused on the mechanism by which ACTH stimulates corticosteroid production (Hall, 1984a, b). It is clear, however, that rat Leydig cells contain actin and are endowed richly with intermediate (10 nm) filaments of the vimentin type (Russell et al., 1987).

In 1979 Hall and colleagues showed that antiactin entrapped in rat Leydig cells by liposomes inhibited LH-stimulated testosterone production and $P450_{scc}$ activity, and increased inner mitochondrial membrane cholesterol content. They concluded that microfilaments in Leydig cells somehow promote transport of cholesterol into mitochondria. This has been borne out by extensive studies in adrenal cortical (Hall, 1984a, b, and 1988) and MA-10 Leydig tumor cells (Nagy and Freeman, 1990). Apparently, microfilaments influence the rate of cholesterol transport into the inner mitochondrial membrane by altering cell shape (rounding), resulting in a closer spatial association between the ultrastructural organelles involved in steroidogenesis (Soto et al., 1986; Bilinska, 1989). Interestingly, these cytoskeletal effects on cholesterol transport do not require the synthesis of new protein (Privalle et al., 1983) but apparently do require a Ca^{++}-calmodulin-dependent protein kinase (Papadopoulos et al., 1990b). This enzyme complex phosphorylates cytoskeletal elements including microfilaments. Inhibitors of calmodulin inhibit not only these phosphorylation events, but cholesterol transport to mitochondria and adrenocortical steroidogenesis as well. The finding that one-third of lipid droplets were tightly bound to intermediate filaments in adrenocortical cells (Almahbobi and Hall, 1990) provides further evidence for a role of cytoskeletal elements in cholesterol transport.

3. MECHANISM OF INTRACELLULAR CHOLESTEROL TRANSPORT

In summary, the intracellular transport of cholesterol from the metabolizable pool in the Leydig cell plasma membrane to the inner mitochondrial membrane where it is converted to pregnenolone by the $P450_{scc}$ enzyme probably consists of two steps. The first is the intermitochondrial transport of cholesterol to the vicinity of the outer mitochondrial membrane. This step is dependent on the cytoskeleton but independent of nascent protein synthesis. The second step is the intramitochondrial transport of cholesterol from the vicinity of the outer mitochondrial membrane into the active site of $P450_{scc}$ that is embedded in the inside half of the inner mitochondrial membrane. There are several characteristics of this second step in the process. First, the cholesterol derived from the metabolizable pool in the plasma membrane must negotiate the aqueous barrier between the outer and inner mitochondrial membranes. Second, this cholesterol, which is destined for steroidogenesis, apparently does not mix with cholesterol in the mitochondrial membrane. Third, this step of cholesterol transport is independent of the cytoskeleton but does require de novo protein synthesis.

C. Intracellular Transport of Steroids

Pregnenolone produced by the $P450_{scc}$ enzyme is the most lipophilic of the steroid intermediates in the testosterone biosynthetic pathway. Moreover, pregnenolone must move out of the inner mitochondrial membrane and into the smooth endoplasmic reticulum where 3βHSD/$\Delta^{5\to4}$ isomerase converts it to progesterone. Because the enzyme 3βHSD/$\Delta^{5\to4}$ isomerase interfaces with the cytoplasm of the Leydig cell, pregnenolone probably enters the smooth endoplasmic reticulum from the cytoplasm.

At this juncture, it is unknown how this is accomplished. In an earlier section of this chapter, we presented evidence suggesting that 3βHSD/$\Delta^{5\to4}$ isomerase is concentrated in a perimitochondrial compartment of smooth endoplasmic reticulum, which is intimately associated with mitochondria. Clearly, such a spatial organization would greatly reduce diffusion distances and diminish the aqueous barrier to pregnenolone movement.

Also, a pregnenolone binding protein has been purified from the adrenal cortex (Strott, 1977) that may serve as a pregnenolone transporter. Although the function of this protein remains to be established, it has been purified from guinea pig adrenal glands and shown to be regulated by a phosphorylation/dephosphorylation mechanism in adrenal fasciculata and reticularis (Driscoll et al., 1990). The fact that this pregnenolone-binding protein was reported to be absent in adrenal glomerulosa cells and testicular Leydig cells (Driscoll et al., 1990) suggests that its function is not vital to steroid biosynthesis.

The enzymes in the smooth endoplasmic reticulum that metabolize pregnenolone to testosterone are arranged in a highly ordered multienzyme sequence, as discussed earlier in this chapter, which reduces substrate diffusion and assures a rapid and efficient conversion of pregnenolone to testosterone. The final issue is how testosterone is secreted from the Leydig cell.

D. Secretion of Testosterone

Very little is known about the mechanism of testosterone secretion (Ewing et al., 1980; Chubb and Ewing, 1991). The C_{19} testosterone is less lipophilic than cholesterol and pregnenolone; consequently, free testosterone diffuses readily between membranes and cytoplasm (Chubb and Ewing, 1991). Turner and associates (1984) showed that the concentration of testosterone is highest in the interstitium of the rat testis, and that testosterone diffuses down a concentration gradient either into the seminiferous tubules and rete testis fluid or into testicular venous effluent and peripheral blood. Interestingly, the interstitium of the rat testis contains a high concentration of albumin (Christensen et al., 1985). Albumin binds testosterone with low affinity but high capacity (Ewing et al., 1976). Moreover, when Leydig cell testosterone secretion is inhibited by treating rats with subdermal, steroid-filled implants, exogenous testosterone emanating from the implants is concentrated in interstitial fluid in a dose-dependent manner (Zirkin et al., 1989).

Taken together, these results suggest that testosterone diffuses from its highest concentration at the site of synthesis into the interstitium, that albumin acts as a macromolecular sink trapping free testosterone in the interstitium and thus removing it from the interstitial reservoir, and that testosterone moves out of the interstitium by diffusing down a concentration gradient of free testosterone.

Experimental proof for the idea that testosterone-binding macromolecules in the testicular interstitium facilitate testosterone secretion was obtained in 1976 (Ewing et al., 1976). These investigators showed that substitution of 70,000 mol wt Dextran or ovalbumin, neither of which binds testosterone, for albumin in media used to perfuse rabbit testes in vitro caused a drastic reduction in testosterone production despite maximal stimulation with LH. This discovery was confirmed recently (Melsert et al., 1989) when albumin purified from rat testicular interstitial fluid resulted in a dose-dependent increase in LH-stimulated steroidogenesis by isolated rat Leydig cells cultured in vitro. Apparently, binding of testosterone to albumin in the interstitium is required to lower the free testosterone concentration and prevent the well-known short-loop direct inhibition of testosterone on its own biosynthesis (Darney and Ewing, 1981; Payne, 1990).

E. Summary Hypothesis

Our knowledge of the movement of Leydig cell sterol/steroid precursors and products into, about, and out of Leydig cells is fragmentary.

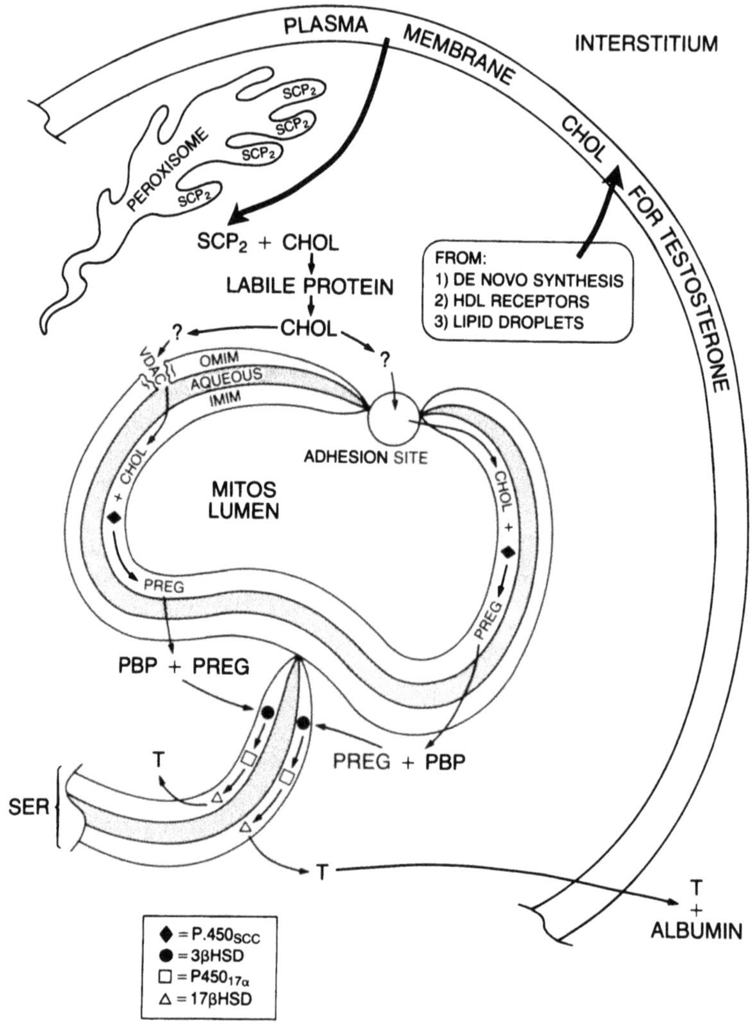

Figure 7–3. Summary of the movement of Leydig cell sterol/steroid precursors about and testosterone out of the Leydig cell. CHOL, cholesterol; SCP_2, sterol carrier protein 2; VDAC, voltage-dependent anion channel; OMIM, outer mitochondrial membrane; IMIM, inner mitochondrial membrane; Aqueous, hydrophilic aqueous layer between OMIM and IMIM; MITOS LUMEN, lumen of mitochondrion; PREG, pregnenolone; PBP, pregnenolone binding protein; T, testosterone; SER, smooth endoplasmic reticulum.

Figure 7–3 represents our hypothesis of this complicated process constructed from bits of knowledge gleaned from studies with liver, adrenal cortical, granulosa, and Leydig cells. We recognize that this hypothesis will require revision in the future as gaps in our knowledge are filled.

The first step in this process is the intermitochondrial transport of cholesterol from its primary source(s) to the vicinity of the outer mitochondrial membrane. It seems clear that the primary metabolizable pool of cholesterol destined for testosterone production is stored in the Leydig cell plasma membrane (Freeman, 1987). Also, it is clear that SCP_2 is stored in peroxisomes in Leydig cells (Mendis-Handagama et al., 1990b). We hypothesize that SCP_2 is released from peroxisomes when cholesterol is required for increased testosterone production (e.g., after LH stimulation). This SCP_2 release may be associated with changes in microfilament configuration. Sterol carrier protein acts permissively to extract cholesterol from plasma membrane and/or to intercept it from other sources before membrane deposition. Sterol carrier protein 2 probably acts as a transient collisional complex between donor and acceptor membranes (van Amerongen et al., 1989).

The second step, namely the intramitochondrial cholesterol transport process, can be achieved without mixing the donated cholesterol with mitochondrial membrane cholesterol (Freeman, 1989), implying that specialized regions of mitochondrial membrane must be involved. Also, this step requires a cycloheximide-sensitive labile protein (Hall, 1988). As shown in Figure 7–3, two specialized regions of the mitochondrial membrane are the adhesion sites and the voltage-dependent anion channel; either, neither, or both may be involved in the intramitochondrial transport of cholesterol. There are three likely candidates for the labile protein factor: SAP (Pedersen and Brownie, 1983), des-(Gly-Ile) endozepine (Besman et al., 1989), and the family of cAMP inducible, cycloheximide-sensitive 28 kd proteins (Pon et al., 1986); again all, one, or none of these proteins may be involved in the intramitochondrial transport of cholesterol.

The end result is that providing more cholesterol to the $P450_{scc}$ enzyme causes increased pregnenolone production. Figure 7–3 shows that pregnenolone then diffuses the short distance to the 3βHSD, which we hypothesize is sequestered in a perimitochondrial compartment of the smooth endoplasmic reticulum intimately associated with the outer mitochondrial membrane (Wing et al., 1984). It remains to be seen whether this diffusion process is sufficient to account for all the pregnenolone subsequently converted to testosterone or whether a pregnenolone-binding protein is required to transport pregnenolone.

Pregnenolone is rapidly and efficiently converted to testosterone by the smooth endoplasmic reticulum-bound multienzyme complex that prevents the loss of intermediates into the cytoplasm (Fig. 7–3). The end product of this process is testosterone, which probably diffuses out of the Leydig cell down a concentration gradient of free testosterone. Testosterone then binds to albumin in the interstitial space, reducing the interstitial fluid-free testosterone concentration and preventing the short-loop direct inhibition of testosterone by testosterone itself (Darney and Ewing, 1981).

VI. Trophic Effects of Luteinizing Hormone on Leydig Cells

It is too early to discuss at length the regulation of Leydig cell peptide products because their discovery and function still are emerging. It is beyond the scope of this review to discuss the effects of growth factors, effector molecules such as arginine vasopressin, and hormones other than LH, on Leydig cell function. Therefore, we confine our remarks to the effects of LH on Leydig cell structure and function with an emphasis on the trophic rather than acute effects of LH.

A. Acute Effects of Luteinizing Hormone

Leydig cell testosterone secretion is dependent on LH. The pituitary gland secretes pulses of LH into the peripheral circulation; LH binds to specific receptors on the Leydig cell plasma membrane and initiates a cascade of events resulting in the acute (i.e., minutes) stimulation of testosterone secretion. This process has been reviewed recently (Hall, 1988; Dufau, 1988) and is described briefly. The reader is referred to the above reviews for details. The LH receptor complex activates two enzymes in the Leydig cell plasma membrane, the stimulatory GTP binding protein and adenylate cyclase, to increase cAMP formation. Cyclic AMP binds to the regulatory subunit of protein kinase A; this binding is presumed to result in the phosphorylation of specific proteins. These events occur within seconds to minutes and result in the following: (1) depletion of intracellular lipid droplets that store cholesterol esters, (2) stimulation of cholesterol hydrolase to produce free cholesterol, (3) changes in cytoskeletal elements, (4) transport of cholesterol to the inner mitochondrial membrane, site of the $P450_{scc}$ enzyme, and (5) stimulation of RNA and protein synthesis.

Other acute actions of LH include regulating the number of its own receptors. In certain experimental conditions, LH can produce so-called lesions in the steroidogenic pathway, independently of its effects on LH receptors, that interfere with LH-stimulated testosterone production. The precise nature and biological significance of these lesions is controversial (Payne et al., 1987; Dufau, 1988).

B. Trophic Effects of Luteinizing Hormone

1. GENES

Recent evidence suggests that trophic LH stimulation maintains Leydig cell steroidogenic enzymes by regulating gene transcription (Payne, 1990; Miller, 1988). Miller and colleagues (see review Miller, 1988) investigated the developmental expression of mRNA encoding $P450_{scc}$ and $P450_{17\alpha}$ in human fetal testes. Message encoding these two enzymes was most abundant in testicular tissue from 14 to 16 weeks gestation and diminished significantly thereafter. They proposed that fetal $P450_{scc}$ and $P450_{17\alpha}$ mRNA were regulated by hCG because this developmental pattern of gene expression coincided temporally with maximal testicular content and secretion of testosterone and circulating levels of fetal hCG, and because fetal testosterone secretion is known to be responsive to hCG. This idea was supported by studies of cultured human fetal testicular cells where hCG and cAMP, the intracellular mediator of hCG, stimulated mRNA levels of $P450_{scc}$ and adrenodoxin (Miller, 1988). Dufau and colleagues (see review Dufau et al., 1989) reported similar results for fetal and adult rat Leydig cells. Human CG induced changes in the level of mRNA encoding $P450_{17\alpha}$, which paralleled changes in the activity and content of this enzyme. Payne and co-workers (see review Payne, 1990) showed that cAMP analogs stimulated the accumulation of $P450_{scc}$mRNA in MA-10 mouse Leydig tumor cells. In primary cultures of mouse Leydig cells, cAMP was required to maintain optimal levels of mRNA encoding $P450_{17\alpha}$ and $P450_{scc}$, although the basal level of expression of $P450_{17\alpha}$ was lower and more responsive to cAMP (Payne and Sha, 1991). The role of cAMP in transcriptional regulation of steroid hydroxylases is well established because the genes encoding these enzymes have been characterized; current investigations are targeted at identifying cAMP responsive sequences in the 5′-flanking regions of these genes and transacting nuclear proteins required for cAMP-mediated transcriptional activation (Waterman et al., 1990). In contrast, regulatory sequences important for transcriptional regulation of the genes encoding 3βHSD and 17βHSD have not been identified. Recent evidence, however, supports a role for cAMP in the transcriptional regulation of 3βHSD/$\Delta^{5\to4}$ isomerase in cultured mouse (Payne and Sha, 1991) and rat (Keeney and Mason, 1992) Leydig cells. Thus, current evidence indicates that LH/hCG and cAMP can regulate positively the expression of several enzymes involved in testosterone biosynthesis; these include $P450_{17\alpha}$, $P450_{scc}$, adrenodoxin, and 3βHSD/$\Delta^{5\to4}$ isomerase.

2. ENZYMES

Leydig cells require trophic stimulation to maintain the activity and amount of their steroidogenic enzymes (Miller, 1988; Hall, 1988). When rats are deprived of LH by hypophysectomy (Payne et al., 1987) or by treatment with steroid-filled Silastic implants to inhibit endogenous LH secretion (Wing et al., 1984; Klinefelter et al., 1987), there is a gradual loss of the mitochondrial and microsomal enzymes $P450_{scc}$ and $P450_{17\alpha}$. Exogenous LH or hCG can increase the activity and the de novo synthesis of $P450_{scc}$ and $P450_{17\alpha}$ (Payne et al., 1987).

3. CYTOPLASMIC ORGANELLES

The topic of cytoplasmic organelles was reviewed in 1983 (Ewing and Zirkin, 1983) and again in 1989 (Ewing, 1989). Leydig cell ultrastructure is altered after prolonged deprivation of LH; one of the earliest and most profound changes is a striking diminution in the amount of smooth endoplasmic reticulum in the cytoplasm (Christensen and Gillim, 1969; Dym and Raj, 1977; Ewing and Zirkin, 1983). Zirkin and colleagues (1982) used stereological techniques and in vitro testicular perfusion to show a striking correlation between the loss and restoration of smooth endoplasmic reticulum surface area and the testosterone-secreting capacity of the testis coincident with treatments that inhibited and then restored endogenous LH (Fig. 7–4). Importantly, exogenous LH from an Alzet miniosmotic pump, but not FSH, prolactin, thyroid-stimulating hormone (TSH), or growth hormone, maintained both the surface area of smooth endoplasmic reticulum and testosterone-secreting capacity in rats treated with steroid-filled Silastic subdermal implants

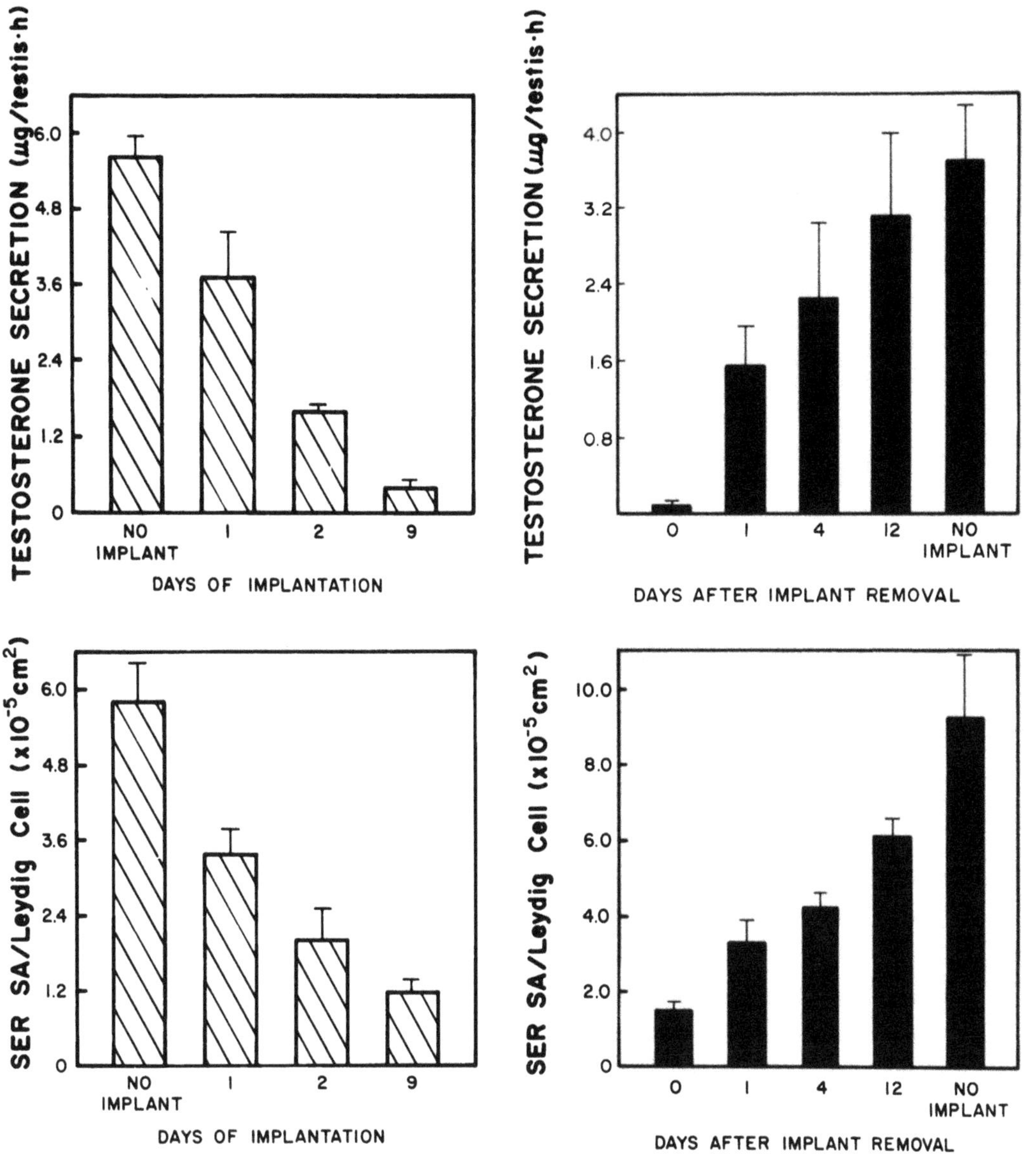

Figure 7–4. Testosterone secretion by in vitro perfused testes *(upper panels)* and surface area (SA) of smooth endoplasmic reticulum (SER) *(lower panels)* of rats implanted with testosterone–estradiol Silastic capsules for 0, 1, 2, and 9 days *(left panels; cross-hatched bars)* or implanted for 2 weeks and then allowed to recover from treatment for 0, 1, 4, and 12 days after implant removal *(right panels; solid bars)*. NO IMPLANT, intact controls. The T above each bar represents the standard error of the mean. (From Ewing and Zirkin, 1983, by courtesy of the editor of *Recent Progress in Hormone Research.*)

designed to inhibit endogenous LH production (Ewing and Zirkin, 1983).

In a subsequent experiment, Wing and colleagues (Wing et al., 1984) tested the effect of withdrawing endogenous LH production for 0, 3, 5, 7, and 10 days on both smooth endoplasmic reticulum surface area and the activities of 3βHSD/$\Delta^{5\rightarrow 4}$ isomerase, $P450_{17\alpha}$, and 17βHSD. Predictably, the smooth endoplasmic reticulum surface area diminished over 10 days coincident with a diminution in $P450_{17\alpha}$ and 17βHSD enzyme activity; however, there was no change in 3βHSD activity within the 10 days (Fig. 7–5).

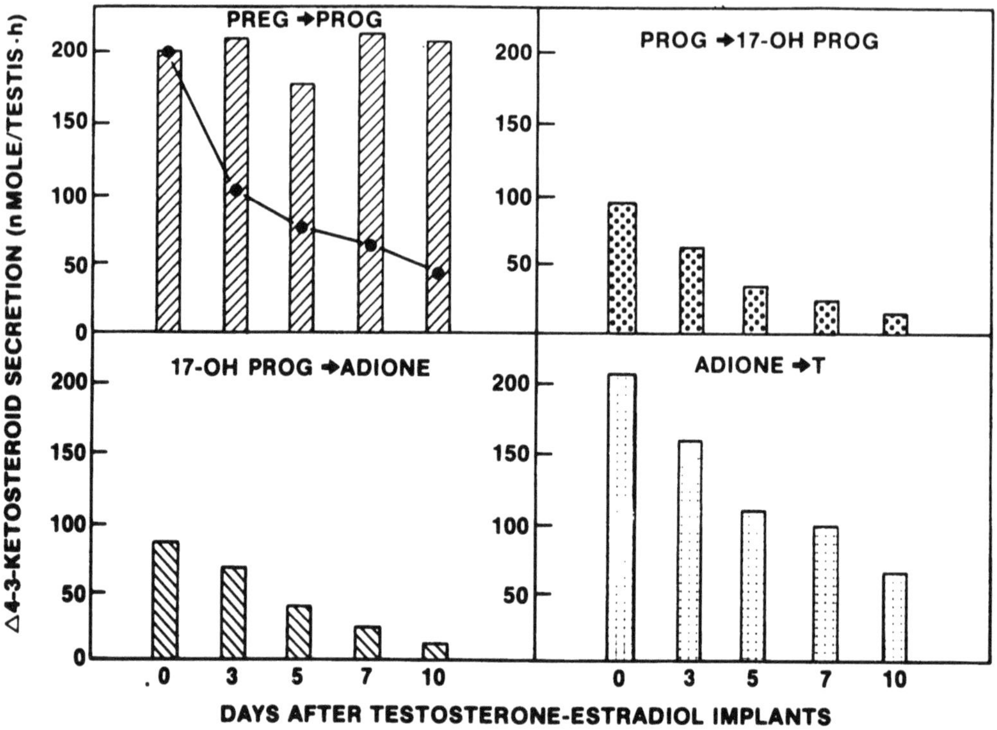

Figure 7–5. The effect of implanting testosterone–estradiol-filled Silastic capsules for 0, 3, 5, 7, and 10 days on the capacity of in vitro perfused rat testes to convert PREG to PROG *(upper left),* PROG to 17-OH PROG *(upper right),* 17-OH PROG to ADIONE *(lower left),* and ADIONE to T *(lower right).* The *solid line* in the upper left panel equals the surface area of smooth endoplasmic reticulum per testis. PREG, pregnenolone; PROG, progesterone; 17-OH PROG, 17α-hydroxyprogesterone; ADIONE, androstenedione; T, testosterone. (Wing et al., 1984, by courtesy of the editor of *Endocrinology.*)

Taken together, these results show that the surface area of smooth endoplasmic reticulum is correlated highly with Leydig cell testosterone-secreting capacity. Clearly, one important trophic effect of LH is to regulate the biogenesis and/or turnover of smooth endoplasmic reticulum. Surprisingly, there must be a subcompartment of smooth endoplasmic reticulum where 3βHSD is concentrated that is less dependent on LH. We have hypothesized that this is a perimitochondrial compartment that is intimately associated spatially with the outer mitochondrial membrane to reduce pregnenolone diffusion distances between the $P450_{scc}$ enzyme in the inner mitochondrial membrane (see Fig. 7–3) and 3βHSD in the smooth endoplasmic reticulum.

4. CELL VOLUME

Leydig cell size can be expressed either as the volume or the mass of a typical or average Leydig cell; volume and mass units are interchangeable as the specific gravity of the testis is 1.0 (Mori and Christensen, 1980). The volume of an average Leydig cell reflects both differentiated structure and steroidogenic function because changes in the volume of an average Leydig cell are associated with changes in smooth endoplasmic reticulum and the capacity for testosterone secretion (Christensen and Gillim, 1969; Dym and Raj, 1977; Ewing and Zirkin, 1983).

Dym and Raj (1977) treated rats with anti-LH serum to neutralize LH bioactivity in the peripheral circulation. After 5 days of treatment, the cross-sectional area of Leydig cell profiles decreased by 50 percent. Most recently, Keeney and co-workers (1988) showed that the volume of an average Leydig cell decreased significantly (50 percent) within 1 week and was only 20 percent of that in control rats after 16 weeks of treatment with subdermal steroid-filled Silastic capsules designed to inhibit endogenous LH production. Continuous infusion of exogenous LH for 5 days, but not other pituitary hormones, simultaneously with either implantation of steroid-filled Silastic capsules or hypophysectomy, maintained total Leydig cell volume per testis similar to that in untreated controls (Ewing and Zirkin, 1983).

Sharpe (1988) suggested that paracrine testicular factors modulate the responsiveness of Leydig cells to LH. Recently, Keeney and Ewing (1990) compared the volume of Leydig cells in testes of pituitary-intact and hypophysectomized rats with and without exogenous testosterone treatment; the latter treatment resulted in testosterone levels sufficient to maintain spermatogenesis. Results of this study showed that both Leydig cell volume per testis and the volume of an average Leydig cell decreased 70 to 85 percent in all treated rats compared with untreated intact controls whether the treatment resulted in deprivation of LH alone or of all pituitary hormones and whether spermatogenesis was arrested or maintained by exogenous testosterone. These results were interpreted to mean that LH is the only factor required to maintain Leydig cell volume in rats; they do not rule out the possibility that other factors modulate the effects of LH on Leydig cell volume.

5. CELL NUMBER

The literature is replete with data derived from inaccurate methods that have been used to estimate the number of Leydig cells per testis. In 1984, Sterio introduced the first of a series of new stereological methods (Sterio, 1984; Gundersen, 1986) devised to count cells or their nuclei directly in three-dimensional space through the use of serial sections. However, even these counting methods are subject to errors introduced by differential shrinkage of tissue from the fresh to the processed state when comparing control and atrophied testes (Mendis-Handagama and Ewing, 1990a).

Recently, we used these stereological techniques with shrinkage corrections to test the effect of inhibiting endogenous LH secretion for 16 weeks on Leydig cell number per testis in rats in three separate studies (Keeney et al., 1988, 1990; Mendis-Handagama and Ewing, 1990a). The percentage decline in Leydig cell numbers due to LH deprivation varied from a low of 15 to a high of 40 percent in these three studies. In one of these studies (Keeney et al., 1990), endogenous LH production was restored after 16 weeks of LH deprivation by removing steroid-filled Silastic subdermal implants that had been used to inhibit LH production. In that study, total Leydig cell number per testis was marginally but not significantly ($P = 0.06$) decreased in rats treated with steroid implants for 16 weeks compared to controls (18.4 ± 2.2 vs $25.4 \pm 1.2 \times 10^6$); within 3 weeks after implant removal and the restoration of endogenous LH production, the numbers of Leydig cells/testis were identical ($26.8 \pm 2.8 \times 10^6$) to intact control rats.

It is noteworthy that species differences in the regulation of Leydig cell number per testis may exist, particularly in photoperiodic seasonal breeders. For example, both Hardy and associates (1987) and Mendis-Handagama and Ewing (1990a) showed dramatic and highly significant reductions in Leydig cell numbers per testis in hamsters maintained on short days for 10 weeks compared with long-day controls.

We conclude that depriving rats of LH causes a slow and variable decline in the number of Leydig cells per testis that may or may not reach statistical significance even after 16 weeks. This loss is nearly imperceptible compared with that which occurs in the atrophied testes of the photoperiodic responsive hamster on short days or to the dramatic changes in rat Leydig cell volume and steroidogenic function after LH deprivation.

6. CELL PROLIFERATION

It is known from cell kinetic studies that the difference between the number of cells entering the population (cell production) and the number leaving the population (cell loss) determines whether the size of the population will change (Cameron, 1971). In growing populations, cell production exceeds cell loss, resulting in a net increase in total cell number. By this definition, Leydig cells in the pubertal rat comprise a growing population; they increase in number by 100-fold during the first 8 weeks of postnatal life (Hardy et al., 1989). By contrast, Leydig cells in the adult rat comprise a slowly renewing population because their numbers remain relatively constant: that is, cell production equals cell loss.

In adult mice, the rate of Leydig cell division

is so slow that not all Leydig cells renew within a lifetime (Cameron, 1971); similar results were obtained for adult rats (Teerds et al., 1989). Mitotic Leydig cells are observed only rarely in adult rats (Fouquet and Kann, 1987). Recently, Keeney and co-workers (1990) used the more sensitive ^{3}H-thymidine labeling index to show that the percentage of Leydig cells labeled with ^{3}H-thymidine was less than 0.1 percent in intact control adult rats. These investigators showed that even this low labeling index became undetectable when rats were treated with steroid-filled subdermal Silastic implants designed to inhibit LH secretion for 16 weeks. However, removal of these steroid-filled implants after 16 weeks, which caused a return of LH to control levels coincident with a transient stimulation of serum concentrations of FSH, resulted in a 5.4-fold increase in the ^{3}H-thymidine labeling index within 1 week of implant removal. This increase was sustained from 1 to 4 weeks and returned to the low control values at 6 weeks after implant removal. Moreover, these investigators showed that exogenous LH but not exogenous FSH stimulated a similar increase in Leydig cell labeling index in rats in which endogenous LH was inhibited with steroid-filled Silastic implants for 16 weeks. Taken together, these results suggest that LH is mitogenic for rat Leydig cells.

VII. Summary

This review introduces evidence strongly supporting the concept that Leydig cells secrete biologically important peptides including POMC-derived peptides, enkephalins, dynorphins, CRF, oxytocin, and AII. Undoubtedly other important Leydig cell peptidic secretogogues await discovery. The challenge in the coming decade will be to expand our limited knowledge of the intratesticular function and the regulation of synthesis and secretion of these interesting molecules.

Next, we discussed the evidence for the purification to homogeneity and cloning of the genes for the enzymes required to convert cholesterol to testosterone. One interesting aspect of this research is that a single protein may catalyze more than one reaction. For example, the $P450_{17\alpha}$ enzyme catalyzes two mixed function oxidase reactions, namely the 17α-hydroxylation of progesterone to 17α-hydroxyprogesterone and cleavage of the bond between C-17 and C-20 to convert 17α-hydroxyprogesterone to androstenedione. The $3\beta HSD/\Delta^{5\to4}$ isomerase catalyzes two reactions as well: dehydrogenation of 3β-hydroxysteroids to 3-ketosteroids and isomerization of the double bond of 5-ene-steroids to make 4-ene-steroids.

Then we discussed the amazing correlation between Leydig cell structure and steroidogenic function probably resulting from the lipophilic nature of the sterol/steroid substrates of the steroidogenic enzymes, the latter as a consequence being sequestered in the hydrophobic environment of the cytoplasmic organelle membranes.

We hypothesize that the metabolizable pool of cholesterol, which is destined for testosterone production, is stored in the plasma membrane of the rat Leydig cell. Sterol carrier protein 2, which is stored in the peroxisomes, is released into the cytoplasm by a cytoskeletal-mediated event when the demand for cholesterol is increased (e.g., after LH stimulation). The SCP2 acts permissively to extract cholesterol from the plasma membrane or to intercept it before deposition in the Leydig cell membrane; SCP2 then donates this cholesterol, probably by a labile protein factor, to the $P450_{scc}$ enzyme, thus bridging the aqueous barrier between the outer and inner mitochondrial membranes. Pregnenolone, the product of this reaction, diffuses the short distance to the $3\beta HSD/\Delta^{5\to4}$ isomerase enzyme sequestered in a perimitochondrial compartment of smooth endoplasmic reticulum. Progesterone is rapidly metabolized to testosterone by a multienzyme complex preventing the leakage of substantial amounts of intermediates into the cytoplasm of the Leydig cell. Testosterone probably then diffuses out of the Leydig cell down a concentration gradient of free testosterone maintained in part by the huge store of albumin in the rat interstitium, which binds testosterone avidly and prevents the short-loop inhibitory feedback of testosterone on its own synthesis.

In keeping with the concept of the importance of structure to steroidogenic function, we finally focused on the trophic effect of LH on the Leydig cell in the rat. We showed that in addition to regulating steroidogenic enzyme gene expression and enzyme content, LH also regulates the biogenesis/turnover of smooth endoplasmic reticulum and the volume of the Leydig cell. We see that the effects of LH on Leydig cell mitogenesis and number of Leydig cells per testis are less clear at this juncture.

ACKNOWLEDGMENTS

Supported in part by NIH Grant HD-07204, NIH Training Grants HD-07276 and HD-07190, and the Hopkins Population Center Grant HD-06268. We thank Mrs. Jeanne Whitaker for her help in the preparation of the manuscript.

REFERENCES

Adashi, E. Y., and Hsueh, A.J.W. (1981). Direct inhibition of testicular adrogen biosynthesis by arginine vasopressin: Mediation through pressor-selective testicular recognition sites. Endocrinology 109:1793–1795.

Almahbobi, G., and Hall, P. F. (1990). The role of intermediate filaments in adrenal steroidogenesis. J. Cell Sci. 97:679–687.

Ang, H.-L., Ungefroren, H., De Bree, F., Foo, N.-C., Carter, D., Burbach, J. P., Ivell, R., and Murphy, D. (1991). Testicular oxytocin gene expression in seminiferous tubules of cattle and transgenic mice. Endocrinology 128:2110–2117.

Aoki, A., and Fawcett, D. W. (1978). Is there a local feedback from the seminiferous tubules affecting activity of the Leydig cells? Biol. Reprod. 19:144–158.

Appelkvist, E.-L., Reinhart, M., Fischer, R., Billheimer, J., and Dallner, G. (1990). Presence of individual enzymes of cholesterol biosynthesis in rat liver peroxisomes. Arch. Biochem. Biophys. 282:318–325.

Ascoli, M., and Freeman, D. A. (1985). Sources of cholesterol used for steroid biosynthesis in cultured Leydig tumor cells, in *Lipoprotein and Cholesterol Metabolism in Steroidogenic Tissues* (Strauss, J. F., III and Menon, K.M.J., eds.) pp. 21–32, George F. Stickley Co., Philadelphia.

Audhya, T., Hollander, C. S., Schlesinger, D. H., and Hutchinson, B. (1989). Structural characterization and localization of corticotropin releasing factor in testis. Biochim. Biophys. Acta 995:10–16.

Bain, P. A., Yoo, M., Clarke, T., Hammond, S. H., and Payne, A. H. (1991). Multiple forms of mouse 3β-hydroxysteroid dehydrogenase/Δ^5-Δ^4 isomerase and differential expression in gonads, adrenal glands, liver, and kidneys of both sexes. Proc. Natl. Acad. Sci. (USA) 88:8870–8874.

Baker, M. E. (1990). A common ancestor for human placental 17β-hydroxysteroid dehydrogenase, *Streptomyces coelicolor* act III protein, and *Drosaphila melanogaster* alcohol dehydrogenase. FASEB J. 4:222–226.

Bardin, C. W., Chen, C-L. C., Morris, P. L., Gerendai, I., Boitani, C., Liotta, A. S., Margioris, A., and Krieger, D. T. (1987). Proopiomelanocortin-derived peptides in testis, ovary and tissues of reproduction. Rec. Prog. Horm. Res. 43:1–28.

Bérubé, D., The, V. L., Lachance, Y., Gagné, R., and Labrie, F. (1990). Assignment of the human 3β-hydroxysteroid dehydrogenase gene (HSDB3) to the p13 band of chromosome 1. Cytogenet. Cell Genet. 52:199–200.

Besman, M. J., Yanagibashi, K., Lee, T. D., Kawamura, M., Hall, P. F., and Shively, J. E. (1989). Identification of des-(Gly-Ile)-endozepine as an effector of corticotropin-dependent adrenal steroidogenesis: Stimulation of cholesterol delivery is mediated by the peripheral benzodiazepine receptor. Proc. Natl. Acad. Sci. (USA) 86:4897–4901.

Bilińska, B. (1989). Visualization of the cytoskeleton in Leydig cells in vitro effect of luteinizing hormone and cytoskeletal disrupting drugs. Histochem. 93:105–110.

Bogovich, K., and Payne, A. H. (1980). Purification of rat testicular microsomal 17-ketosteroid reductase evidence that 17-ketosteroid reductase and 17β-hydroxysteroid dehydrogenase are distinct enzymes. J. Biol. Chem. 255:5552–5559.

Bolander, F. F. (1989). *Molecular Endocrinology*, Academic Press, New York.

Cameron, I. L. (1971). Cell proliferation and renewal in the mammalian body, in *Cellular and Molecular Renewal in the Mammalian Body* (Cameron, I. L., and Thrasher, J. D., eds.) pp. 45–85, Academic Press, New York.

Chang, C.-Y., Wu, D.-A., Lai, C.-C., Miller, W. L., and Chung, B.-C. (1988). Cloning and structure of the human adrenodoxin gene. DNA 7:609–615.

Christensen, A. K., Komorowski, T. E., Wilson, B., Ma, S.-F., and Stevens, R. W., III. (1985). The distribution of serum albumin in the rat testis, studied by electron microscope immunocytochemistry on ultrathin frozen sections. Endocrinology 116:1983–1996.

Christensen, A. K., and Gillim, S. W. (1969). The correlation of fine structure and function in steroid-secreting cells, with emphasis on those of the gonads, in *The Gonads* (McKerns, K. W., ed.) pp. 415–488, Appleton-Century-Crofts, New York.

Chubb, C., and Ewing, L. L. (1991). Leydig Cell, in *Infertility in the Male* (Lipshultz, L. I., and Howards, S. S., eds.) 2nd edition. Mosby Yearbook Medical Publishers, Inc., Chicago.

Chung, B.-C., Picado-Leonard, J., Haniu, M., Bienkowski, M., Hall, P. F., Shively, J. E., and Miller, W. L. (1987). Cytochrome P450c17 (steroid 17α-hydroxylase/$_{17,20}$ lyase): Cloning of human adrenal and testis cDNAs indicates the same gene is expressed in both tissues. Proc. Natl. Acad. Sci. (USA) 84:407–411.

Chung, B.-C., Matteson, K. J., Voutilainen, R., Mohandas, T. K., and Miller, W. L. (1986). Human cholesterol side-chain cleavage enzyme, $P450_{scc}$: cDNA cloning, assignment of the gene to chromosome 15, and expression in placenta. Proc. Natl. Acad. Sci. (USA) 83:8962–8966.

Clements, J. A., He, L., Risbridger, G. P., Smith, A. I., and Funder, J. W. (1988). Pituitary peptides in the ovary and testis, in *Progress in Endocrinology* (Imura, H., et al., eds.) pp. 469–474, Excerpta Medica, Amsterdam.

Collard, M. W., Day, R., Akil, H., Uhler, M. D., and Douglass, J. O. (1990). Sertoli cells are the primary site of prodynorphin gene expression in rat testis: regulation of mRNA and secreted peptide levels by cyclic adenosine 3′,5′-monophosphate analogs in cultured cells. Mol. Endocrinol. 4:1488–1496.

Cooke, G. M. (1989). Identification and mechanism of action of phospholipids capable of modulating rat testicular microsomal 3β-hydroxysteroid dehydrogenase-isomerase activity in vitro. Biol. Reprod. 41:438–445.

Cox, B. M., Rosenberger, J. G., and Douglass, J. (1987). Chromatographic characterization of dynorphin and [Leu^5] enkephalin immunoreactivity in guinea pig and rat testis. Regulatory Peptides 19:1–12.

Cushman, D. W., and Cheung, H. S. (1971). Concentrations of angiotensin-converting enzyme in tissues of the rat. Biochim. Biophys. Acta 250:261–265.

Darney, K. J., and Ewing, L. L. (1981). Autoregulation of testosterone secretion in perfused testes. Endocrinol. 109:993–995.

DeBold, C. R., Nicholson, W. E., and Orth, D. N. (1988). Immunoreactive proopiomelanocortin (POMC) peptides and POMC-like messenger ribonucleic acid are present in many rat non-pituitary tissues. Endocrinol. 122:2648–2657.

Deschepper, C. F., Mellon, S. H., Cumin, F., Baxter, J. D., and Ganong, W. F. (1986). Analysis by immunocytochemistry and in situ hybridization of renin and its mRNA in kidney, testis, adrenal and pituitary of the rat. Proc. Natl. Acad. Sci. (USA) 83:7552–7556.

Douglass, J., Civelli, O., and Herbert, E. (1984). Polyprotein gene expression: Generation of diversity of neuroendocrine peptides. Ann. Rev. Biochem. 53:665–715.

Douglass, J., Cox, B., Quinn, B., Civelli, O., and Herbert, E. (1987). Expression of the prodynorphin gene in male and female mammalian reproductive tissues. Endocrinol. 120:707–713.

Driscoll, W. J., Lee, Y. C., and Strott, C. A. (1990). Regulation of adrenocortical pregnenolone-binding protein activity by phosphorylation/dephosphorylation. Phosphatase-mediated inactivation is reversed by cytosolic kinase. J. Biol. Chem. 265:12306–12311.

Dufau, M. L. (1988). Endocrine regulation and communicating functions of the Leydig cell. Ann. Rev. Physiol. 50:483–508.

Dufau, M. L., Ulisse, S., Khanum, A., Buczko, E., Kitamura, M., Fabbri, A., and Namiki, M. (1989). LH action in the Leydig cell: modulation by angiotensin II and corticotropin releasing hormone, and regulation of $P450_{17\alpha}$ mRNA. J. Steroid Biochem. 34:205–217.

Dym, M., and Raj, H.G.M. (1977). Response of adult rat Sertoli cells and Leydig cells to depletion of luteinizing hormone and testosterone. Biol. Reprod. 17:676–696.

Dzau, V. J., Ellison, K. E., Brody, T., Ingelfinger, J., and Pratt, R. E. (1987). A comparative study of the distributions of renin and angiotensinogen messenger ribonucleic acids in rat and mouse tissues. Endocrinol. 120:2334–2338.

Dzau, V. J., and Pratt, R. E. (1986). Renin-angiotensin system: biology, physiology, and pharmacology, in *The Heart and Cardiovascular System* (Fozzard, H. A., et al., eds.) pp. 1631–1662, Raven Press, New York.

Ehlers, M.R.W., Fox, E. A., Strydom, D. J., and Riordan, J. F. (1989). Molecular cloning of human testicular angiotensin-converting enzyme: The testis isozyme is identical to the c-terminal half of endothelial angiotensin-converting enzyme. Proc. Natl. Acad. Sci. (USA) 86:7741–7745.

El-Dorry, H. A., Bull, H. G., Iwata, K., Thornberry, N. A., Cordes, E. H., and Soffer, R. L. (1982). Molecular and catalytic properties of rabbit testicular dipeptidyl carboxypeptidase. J. Biol. Chem. 257:14128–14133.

Engelhardt, R. P., Saint-Pol, P., Tramu, G., and Leonardelli, J. (1986). Immunohistochemical localization of enkephalin-like peptides during testicular development in rats. Arch. Andrology 17:49–56.

Epstein, L. F., and Orme-Johnson, N. R. (1991). Acute action of luteinizing hormone on mouse Leydig cells: Accumulation of mitochondrial phosphoproteins and stimulation of testosterone biosynthesis. Mol. Cell. Endocrinol. 81:113–126.

Eskeland, N. L., Lugo, D. I., Pintar, J. E., and Schachter, B. S. (1989). Stimulation of β-endorphin secretion by corticotropin-releasing factor in primary rat Leydig cell cultures. Endocrinol. 124:2914–2919.

Ewing, L. L. (1989). The trophic effect of luteinizing hormone on the rat Leydig cell. J. Am. College Toxicol. 8:473–485.

Ewing, L. L., and Zirkin, B. (1983). Leydig cell structure and steroidogenic function. Recent Prog. Hormone Res. 39:599–635.

Ewing, L. L., Davis, J. C., and Zirkin, B. R. (1980). Regulation of testicular function: a spatial and temporal view, in *Reproductive Physiology III International Review of Physiology* (Greep, R. O., ed.) Vol. 22, pp. 41–115, University Park Press, Baltimore.

Ewing, L. L., Chubb, C. E., and Robaire, B. (1976). Macromolecules, steroid binding and testosterone secretion by rabbit testes. Nature 264:84–86.

Fevold, H. R., Lorence, M. C., McCarthy, J. L., Trant, J. M., Kagimoto, M., Waterman, M. R., and Mason, J. I. (1989). Rat P450 17α from testis: Characterization of a full-length cDNA encoding a unique steroid hydroxylase capable of catalyzing both Δ^4- and Δ^5-steroid-17,20-lyase reactions. Mol. Endocrinol. 3:968–975.

Field, L. J., McGowan, R. A., Dickinson, D. P., and Gross, K. W. (1984). Tissue and gene specificity of mouse renin expression. Hypertension 6:597–603.

Foo, N.-C., Carter, D., Murphy, D., and Ivell, R. (1991). Vasopressin and oxytocin gene expression in rat testis. Endocrinology 128:2118–2128.

Fouquet, J.-P., and Kann, M.-L. (1987). Mitotic activity in monkey and rat Leydig cells. Reprod. Nutrition Develop. 27:1053–1062.

Freeman, D. A. (1989). Plasma membrane cholesterol: Removal and insertion into the membrane and utilization as substrate for steroidogenesis. Endocrinology 124:2527–2534.

Freeman, D. A. (1987). Cyclic AMP mediated modification of cholesterol traffic in Leydig tumor cells. J. Biol. Chem. 262:13061–13068.

Gonzalez, F. J., and Kasper, C. B. (1982). Cloning of DNA complementary to rat liver NADPH-cytochrome c (P-450) oxidoreductase and cytochrome P-450b mRNAs. J. Biol. Chem. 257:5962–5968.

Gorgas, K. (1985). Serial section analysis of mouse hepatic peroxisomes. Anat. Embryol. 172:21–32.

Gizang-Ginsberg, E., and Wolgemuth, D. J. (1985). Localization of mRNAs in mouse testes by in situ hybridization: Distribution of α-tubulin and developmental stage specificity of pro-opiomelanocortin transcripts. Dev. Biol. 111:293–305.

Gray, P. W., Glaister, D., Seeburg, P. H., Guidotti, A., and Costa, E. (1986). Cloning and expression of cDNA for human diazepam binding inhibitor, a natural ligand of an allosteric regulatory site of the γ-aminobutyric acid type A receptor. Proc. Natl. Acad. Sci. (USA) 83:7547–7551.

Guldenaar, S.E.F., and Pickering, B. T. (1985). Immunocytochemical evidence for the presence of oxytocin in rat testis. Cell Tissue Res. 240:485–487.

Gundersen, H.J.G. (1986). Stereology of arbitrary particles. A review of unbiased number and size estimators and the presentation of some new ones, in memory of William R. Thompson. J. Microsc. 143:3–45.

Hall, P. F. (1988). Testicular steroid synthesis: organization and regulation, in *The Physiology of Reproduction* (Knobil, E., and Neill, J. et al., eds.) Vol. 1, pp. 975–998, Raven Press, New York.

Hall, P. F. (1986). Cytochromes P-450 and the regulation of steroid synthesis. Steroids 48:131–196.

Hall, P. F. (1985). The role of the cytoskeleton in the supply of cholesterol for steroidogenesis, in *Lipoprotein and Cholesterol Metabolism in Steroidogenic Tissues* (Strauss, J. F., III, and Menon, K.M.J., eds.) pp. 207–217, George F. Stickley Co., Philadelphia.

Hall, P. F. (1984a). The role of the cytoskeleton in hormone action. Can. J. Biochem. Cell Biol. 62:653–665.

Hall, P. F. (1984b). On the mechanism of action of ACTH: The role of actin. Endocrine Res. 10:431–461.

Hall, P. F., Charpponnier, C., Nakamura, M., and Gabbiani, G. (1979). The role of microfilaments in the response of Leydig cells to luteinizing hormone. J. Steroid. Biochem. 11:1361–1366.

Hanukoglu, I., Gutfinger, T., Haniu, M., and Shively, J. E. (1987). Isolation of a cDNA for adrenodoxin reductase (ferredoxin-NADP$^+$ reductase). Implications for mitochondrial cytochrome P-450 systems. Eur. J. Biochem. 169:449–455.

Hardy, M. P., Zirkin, B. R., and Ewing, L. L. (1989). Kinetic studies on the development of the adult population of Leydig cells in testes of the pubertal rat. Endocrinology 124:762–770.

Hardy, M. P., Mendis-Handagama, S.M.L.C., Zirkin, B. R., and Ewing, L. L. (1987). Photoperiodic variation of Leydig cell numbers in the testis of the golden hamster: A possible mechanism for their renewal during recrudescence. J. Exp. Zool. 244:269–276.

Hollander, C. S., Audhya, T., Nakane, T., Kanie, N., Kageyama, N., Kuwayama, A., Golbe, L., and Schlesinger, D. (1983). Distribution, biosynthesis and physiological role of corticotropin-releasing factor in the human: An overview. Trans. Assoc. Am. Phys. 96:122–130.

Inano, H., and Tamaoki, B.-I. (1986a). Testicular 17β-hydroxysteroid dehydrogenase: Molecular properties and reaction mechanism. Steroids 48:3–26.

Inano, H., and Tamaoki, B.-I. (1986b). Purification of NADPH-cytochrome P-450 reductase from microsomal fraction of rat testes, and its chemical modification by tetranitromethane. J. Steroid. Biochem. 25:21–28.

Inano, H., Tamaoki, B.-I., Hamana, K., and Nakagawa, H. (1980). Amino acid composition and immunochemical properties of porcine testicular 17β-hydroxysteroid dehydrogenase. J. Steroid Biochem. 13:287–295.

Ishii-Ohba, H., Inano, H., and Tamaoki, B.-I. (1987). Testicular and adrenal 3β-hydroxy-5-ene-steroid dehydrogenase and 5-ene-4-ene isomerase. J. Steroid Biochem. 27:775–779.

Ishii-Ohba, H., Inano, H., and Tamaoki, B.-I. (1986a). Purification and properties of testicular 3β-hydroxy-5-ene-steroid dehydrogenase and 5-ene-4-ene isomerase. J. Steroid Biochem. 25:555–560.

Ishii-Ohba, H., Saiki, N., Inano, H., and Tamaoki, B.-I. (1986b). Purification and characterization of rat adrenal 3β-hydroxysteroid dehydrogenase with steroid 5-ene-4-ene-isomerase. J. Steroid Biochem. 24:753–760.

Juniewicz, P. E., Keeney, D. S., and Ewing, L. L. (1988). Effect of adrenocorticotropin and other pro-opiomelanocortin-derived peptides on testosterone secretion by the in vitro perfused testis. Endocrinol. 122:891–898.

Keeney, D. S., and Mason, J. I. (1992). Expression of testicular 3β-hydroxysteroid dehydrogenase/$\Delta^{5\to4}$-isomerase: Regulation by luteinizing hormone and forskolin in Leydig cells of adult rats. Endocrinology 130 (in press).

Keeney, D. S., and Ewing, L. L. (1990). Effects of hypophysectomy and alterations in spermatogenic function on Leydig cell volume, number, and proliferation in adult rats. J. Andrology 11:367–378.

Keeney, D. S., Mendis-Handagama, S.M.L.C., Zirkin, B. R., and Ewing, L. L. (1988). Effect of long term deprivation of luteinizing hormone on Leydig cell volume, number and steroidogenic capacity of the rat testis. Endocrinol. 123:2906–2915.

Keeney, D. S., Sprando, R. L., Robaire, B., Zirkin, B. R., and Ewing, L. L. (1990). Reversal of long-term LH deprivation on testosterone secretion and Leydig cell volume, number and proliferation in adult rats. J. Endocrinol. 127:47–58.

Keller, G. A., Scallen, T. J., Clarke, D., Maher, P. A., Krisans, S. K., and Singer, S. J. (1989). Subcellular localization of sterol carrier protein-2 in rat hepatocytes: Its primary localization to peroxisomes. J. Cell Biol. 108:1353–1361.

Keller, G. A., Barton, M. C., Shapiro, D. J., and Singer, S. J. (1985). 3-hydroxy-3-methylglutaryl-coenzyme A reductase is present in peroxisomes in normal rat liver cells. Proc. Natl. Acad. Sci. (USA) 82:770–774.

Kerr, J. B., Bartlett, J.M.S., and Sharpe, R. M. (1987). Origin of regenerating Leydig cells in the testis of the adult rat: An ultrastructural, morphometric and hormonal assay study. Cell Tissue Res. 249:367–377.

Kerr, J. B., Rich, K. A., and deKretser, D. M. (1979). Alterations of the fine structure and androgen secretion of the interstitial cells in the experimentally cryptorchid rat testis. Biol. Reprod. 20:409–422.

Kilpatrick, D. L., Howells, R. D., Noe, M., Bailey, L. C., and Udenfriend, S. (1985). Expression of preproenkephalin-like mRNA and its peptide products in mammalian testis and ovary. Proc. Natl. Acad. Sci. (USA) 82:7467–7469.

Klinefelter, G. R., Hall, P. F., and Ewing, L. L. (1987). Effect of luteinizing hormone deprivation in situ on steroidogenesis of rat Leydig cells purified by a multistep procedure. Biol. Reprod. 36:769–783.

Kühn-Velten, W. N. (1990). A microcompartmentation analysis of intermediate leakage response to substrate excess in a membrane-bound bifunctional enzyme: local control of hydroxyprogesterone channeling efficiency during cytochrome P450XVII-catalyzed androgen biosynthesis. J. Cellular Biochem. 43:149–159.

Kühn-Velten, N., Lessmann, M., Förster, M.E.C., and Staib, W. (1988). Specific accumulation of 17α-hydroxyprogesterone in microsomal membranes during the process of cytochrome P-450 (C-17)-catalysed androgen biosynthesis. Biochem. J. 256:53–59.

Kurosumi, M., Ishimura, K., Yoshinaga, T., Fujita, H., and Tamaoki, B.-I. (1986). Immunocytochemical localization of 17β-hydroxysteroid dehydrogenase in porcine testis. Histochemistry 85:287–289.

Lambeth, J. D., and Stevens, V. L. (1984). Cytochrome P-450 scc: Enzymology, and the regulation of intramitochondrial cholesterol delivery to the enzyme. Endocrine Research 10:283–309.

Langford, K. G., Shai, S.-Y., Howard, T. E., Kovac, M. J., Overbeek, P. A., and Bernstein, K. E. (1991). Transgenic mice demonstrate a testis-specific promoter for angiotensin-converting enzyme. J. Biol. Chem. 266:15559–15562.

Li, H., Hedger, M. P., Clements, J. A., and Risbridger, G. P. (1991). Localization of immunoreactive β-endorphin and adrenocorticotropic hormone and pro-opiomelanocortin mRNA to rat testicular interstitial tissue macrophages. Biol. Reprod. 45:282–289.

Lorence, M. C., Naville, D., Graham-Lorence, S. E., Mack, S. O., Murry, B. A., Trant, J. M., and Mason, J. I. (1991). 3β-Hydroxysteroid dehydrogenase/$\Delta^{5\rightarrow4}$-isomerase expression in rat and characterization of the testis isoform. Mol. Cell. Endocrinol. 80:21–31.

Lorence, M. C., Murry, B. A., Trant, J. M., and Mason, J. I. (1990). Human 3β-hydroxysteroid dehydrogenase/$\Delta^{5\rightarrow4}$ isomerase from placenta: Expression in nonsteroidogenic cells of a protein that catalyzes the dehydrogenation/isomerization of C21 and C19 steroids. Endocrinology 126:2493–2498.

Luu-The, V., Labrie, C., Simard, J., Lachance, Y., Zhao, H.-F., Couet, J., Leblanc, G., and Labrie, F. (1990). Structure of two tandem human 17β-hydroxysteroid dehydrogenase genes. Mol. Endocrinol. 4:268–275.

Luu-The, V., Lachance, Y., Labrie, C., Leblanc, G., Thomas, J. L., Strickler, R. C., and Labrie, F. (1989a). Full length cDNA structure and deduced amino acid sequence of human 3β-hydroxy-5-ene steroid dehydrogenase. Mol. Endocrinol. 3:1310–1312.

Luu-The, V., Labrie, C., Zhao, H.-F., Couet, J., Lachance, Y., Simard, J., Leblanc, G., Coté, J., Bérubé, D., Gagné, R., and Labrie, F. (1989b). Characterization of cDNAs for human estradiol 17β-dehydrogenase and assignment of the gene to chromosome 17: Evidence for two mRNA species with distinct 5′-termini in human placenta. Mol. Endocrinol. 3:1301–1309.

Margioris, A. N., Koukoulis, G., Grino, M., and Chrousos, G. P. (1989). In vitro-perifused rat testes secrete β-endorphin and dynorphin: Their effect on testosterone secretion. Biol. Reprod. 40:776–784.

McMasters, K. M., Dickson, L. A., Shamy, R. V., Robischon, K., Macdonald, G. J., and Moyle, W. R. (1987). Rat cholesterol side-chain cleavage enzyme (P-450_{scc}): Use of a cDNA probe to study the hormonal regulation of P-450_{scc} mRNA levels in ovarian granulosa cells. Gene 57:1–9.

McMurray, C. T., Devi, L., Calavetta, L., and Douglass, J. O. (1989). Regulated expression of the prodynorphin gene in the R2C Leydig tumor cell line. Endocrinol. 124:49–59.

Melsert, R., Hoogerbrugge, J. W., and Rommerts, F.F.G. (1989). Albumin, a biologically active protein acting on Leydig cells. Mol. Cell. Endocrinol. 64:35–44.

Mendis-Handagama, S.M.L.C., and Ewing, L. L. (1990a). Sources of error in the estimation of Leydig cell numbers in control and atrophied mammalian testes. J. Microsc., 159:73–82.

Mendis-Handagama, S.M.L.C., Zirkin, B. R., Scallen, T. J., and Ewing, L. L. (1990b). Studies on peroxisomes of the adult rat Leydig cell. J. Andrology. 11:270–278.

Mendis-Handagama, S.M.L.C., Watkins, P. A., Gelber, S. J., Scallen, T. J., Zirkin, B. R., and Ewing, L. L. (1990c). Luteinizing hormone causes rapid and transient changes in rat Leydig cell peroxisome volume and intraperoxisomal sterol carrier protein-2 content. Endocrinology 127:2947–2954.

Mendis-Handagama, S.M.L.C., Zirkin, B. R., and Ewing, L. L. (1988). Comparison of components of the testis interstitium with testosterone secretion in hamster, rat and guinea pig testes perfused in vitro. Am. J. Anat. 181:12–22.

Mendis-Handagama, S.M.L.C., Risbridger, G. P., and deKretser, D. M. (1987). Morphometric analysis of the components of the neonatal and the adult rat testis interstitium. Int. J. Androl. 10:525–534.

Mertz, L. M., and Pedersen, R. C. (1989). Steroidogenesis activator polypeptide may be a product of glucose regulated protein 78 (GRP78). Endocrine Research 15:101–115.

Mezey, E., Seidah, N. G., Chretien, M., and Brownstein, M. J. (1986). Demonstration of the vasopressin associated glycopeptide in the brain and peripheral tissues of the Brattleboro rat. Neuropeptides 7:79–85.

Milewich, L., Carr, B. R., Frenkel, R. A., Marrero, M., MacDonald, P. C., and Mason, J. I. (1990). 17β-Hydroxysteroid oxidoreductase of human fetal and adult tissues: Immunological cross-reactivity with an anti-human placental cytosolic 17β-hydroxysteroid oxidoreductase antibody. Placenta 11:95–108.

Millan, M. A., and Aguilera, G. (1988). Angiotensin II receptors in testes. Endocrinology 122:1984–1990.

Miller, W. L. (1988). Molecular biology of steroid hormone synthesis. Endocrine Rev. 9:295–318.

Mittal, S., Zhu, Y.-Z., and Vickery, L. E. (1988). Molecular

cloning and sequence analysis of human placental ferredoxin. Arch. Biochem. Biophys. 264:383–391.
Mocchetti, I., Einstein, R., and Brosius, J. (1986). Putative diazepam binding inhibition peptide: cDNA clones from rat. Proc. Natl. Acad. Sci. (USA) 83:7221–7225.
Mori, H., and Christensen, A. K. (1980). Morphometric analysis of Leydig cells in the normal rat testis. J. Cell Biol. 84:340–354.
Morohashi, K., Fujii-Kuriyama, Y., Sogawa, K., Hirose, T., Inayama, S., and Omura, T. (1984). Molecular cloning and nucleotide sequence of cDNA for mRNA of mitochondrial cytochrome P-450(SCC) of bovine adrenal cortex. Proc. Natl. Acad. Sci. (USA) 81:4647–4651.
Nagy, L., and Freeman, D. A. (1990). Effect of cholesterol transport inhibitors on steroidogenesis and plasma membrane cholesterol transport in cultured MA-10 Leydig tumor cells. Endocrinology 126:2267–2276.
Nakajin, S., and Hall, P. F. (1981). Microsomal cytochrome P-450 from neonatal pig testis purification and properties of a C_{21} steroid side-chain cleavage system (17α-hydroxylase-$C_{17,20}$ lyase). J. Biol. Chem. 256:3871–3876.
Namiki, M., Kitamura, M., Buczko, E., and Dufau, M. L. (1988). Rat testis P-450 17α cDNA: The deduced amino acid sequence, expression and secondary structural configuration. Biochem. Biophys. Res. Comm. 157:705–712.
Naruse, M., Naruse, K., Shizume, K., and Inagami, T. (1984). Gonadotropin-dependent renin in the rat testes. Proc. Soc. Expt. Biol. Med. 177:337–342.
Nelson, D. R., and Strobel, H. W. (1988). On the membrane topology of vertebrate cytochrome P-450 proteins. J. Biol. Chem. 263:6038–6050.
Nicholson, H. D., and Hardy M. P. (1992). Luteinizing hormone differentially regulates the secretion of testicular oxytocin and testosterone by purified adult rat Leydig cells in vitro. Endocrinology 130:671–677.
Nicholson, H. D., Swann, R. W., Burford, G. D., Wathes, D. C., Porter, D. G., and Pickering, B. T. (1984). Identification of oxytocin and vasopressin in the testis and in adrenal tissue. Regulatory Peptides 8:141–146.
Nishihara, M., Winters, C. A., Buzko, E., Waterman, M. R., and Dufau, M. L. (1988). Hormonal regulation of rat Leydig cell cytochrome P-450 17α mRNA levels and chracterization of a partial length rat P-450 17α cDNA. Biochem. Biophys. Res. Comm. 154:151–158.
Okamura, T., John, M. E., Zuber, M. X., Simpson, E. R., and Waterman, M. R. (1985). Molecular cloning and amino acid sequence of the precursor form of bovine adrenodoxin: Evidence for a previously unidentified COOH-terminal peptide. Proc. Natl. Acad. Sci. (USA) 82:5705–5709.
Onoda, M., Haniu, M., Yanagibashi, K., Sweet, F., Shively, J. E., and Hall, P. F. (1987). Affinity alkylation of the active site of C_{21} steroid side-chain cleavage cytochrome P-450 from neonatal porcine testis: A unique cysteine residue alkylated by 17-(bromoacetoxy) progesterone. Biochemistry 26:657–662.
Orth, J., and Boehm, R. (1990). Endorphin suppresses FSH-stimulated proliferation of isolated neonatal Sertoli cells by a pertussis toxin-sensitive mechanism. Anat. Rec. 226:320–327.
Pandey, K. N., and Inagami, T. (1986). Regulation of renin angiotensins by gonadotropic hormones in cultured murine Leydig tumor cells, release of angiotensin but not renin. J. Biol. Chem. 261:3934–3938.
Pandey, K. N., Misono, K. S., and Inagami, T. (1984). Evidence for intracellular formation of angiotensins: Coexistence of renin and angiotensin-converting enzyme in Leydig cells of rat testis. Biochem. Biophys. Res. Comm. 122:1337–1343.
Papadopoulos, V., Nowzari, F. B., and Krueger, K. E. (1991a). Hormone-stimulated steroidogenesis is coupled to mitochondrial benzodiazepine receptors. Tropic hormone action on steroid biosynthesis is inhibited by flunitrazepam. J. Biol. Chem. 266:3682–3687.
Papadopoulos, V., Berkovich, A., Krueger, K. E., Costa, E., and Guidotti, A. (1991b). Diazepam binding inhibitor and its processing products stimulate mitochondrial steroid biosynthesis via an interaction with mitochondrial benzodiazepine receptors. Endocrinology 129:1481–1488.
Papadopoulos, V., Mukhin, A. G., Costa, E., and Krueger, K. E. (1990a). The peripheral-type benzodiazepine receptor is functionally linked to Leydig cell steroidogenesis. J. Biol. Chem. 265:3772–3779.
Papadopoulos, V., Brown, A. S., and Hall, P. F. (1990b). Calcium-calmodulin-dependent phosphorylation of cytoskeletal proteins from adrenal cells. Mol. Cell. Endocrinol. 74:109–123.
Parmentier, M., Inagami, T., Pochet, R., and Desclin, J. C. (1983). Pituitary-dependent renin-like immunoreactivity in the rat testis. Endocrinology 112:1318–1323.
Payne, A. H., and Sha, L. (1991). Multiple mechanisms for regulation of 3β-hydroxysteroid dehydrogenase/$\Delta^{5\rightarrow\Delta4}$-isomerase, 17$\alpha$-hydroxylase/$C_{17-20}$ lyase cytochrome P450, and cholesterol side-chain cleavage cytochrome P450 messenger ribonucleic acid levels in primary cultures of mouse Leydig cells. Endocrinology 129:1429–1435.
Payne, A. H. (1990). Hormonal regulation of cytochrome P450 enzymes, cholesterol side-chain cleavage and 17α-hydroxylase/$C_{17\text{-}20}$ lyase in Leydig cells. Biol. Reprod. 42:399–404.
Payne, A. H., Anakwe, O. O., Hales, D. B., Georgiou, M., Perkins, L. M., and Quinn, P. G. (1987). Regulation of mitochondrial and microsomal cytochrome P-450 enzymes in Leydig cells, in *Regulation of Ovarian and Testicular Function, Advances in Experimental Medicine and Biology* (Mahesh, V. B., Dhindsa, D. S., Anderson, E., and Kalra, S. P., eds.) pp. 467–487, Plenum Press, New York.
Pedersen, R. C. (1987). Steroidogenesis activator polypeptide (SAP) in the rat ovary and testis. J. Steroid Biochem. 27:731–735.
Pedersen, R. C. (1984). Polypeptide activators of cholesterol side-chain cleavage. Endocrine Res. 10:533–561.
Pedersen, R. C., and Brownie, A. C. (1987). Steroidogenesis-activator polypeptide isolated from a rat Leydig cell tumor. Science 236:188–190.
Pedersen, R. C., and Brownie, A. C. (1983). Cholesterol side-chain cleavage in the rat adrenal cortex: Isolation of a cycloheximide-sensitive activator peptide. Proc. Natl. Acad. Sci. (USA) 80:1882–1886.
Peltoketo, H., Isomaa, V., Mäentausta, O., and Vihko, R. (1988). Complete amino acid sequence of human pla-

cental 17β-hydroxysteroid dehydrogenase deduced from cDNA. FEBS Lett. 239:73–77.

Pickering, B. T., Birkett, S. D., Guldenaar, S.E.F., Nicholson, H. D., Worley, R.T.S., and Yavachev, L. (1989). Oxytocin in the testis: what, where, and why? in *Regulation of Testicular Function Signaling Molecules and Cell-Cell Communication* (Ewing, L. L., and Robaire, B., eds.) Ann. N. Y. Acad. Sci., Vol. 564, pp. 198–209, The New York Academy of Sciences, New York.

Pintar, J. E., Schachter, B. S., Herman, A. B., Durgerian, S., and Krieger, D. T. (1984). Characterization and localization of proopiomelanocortin messenger RNA in the adult rat testis. Science 225:632–634.

Pon, L. A., Epstein, L. F., and Orme-Johnson, N. R. (1986). Acute cAMP stimulation in Leydig cells: Rapid accumulation of a protein similar to that detected in adrenal cortex and corpus luteum. Endocrine Res. 12:429–446.

Privalle, C. T., Crivello, J. F., and Jefcoate, C. R. (1983). Regulation of intramitochondrial cholesterol transfer to side-chain cleavage cytochrome P-450 in rat adrenal gland. Proc. Natl. Acad. Sci. (USA) 80:702–706.

Quinn, P. G., Georgiou, M., and Payne, A. H. (1985). Differences in hydroxysterol metabolism between rat and mouse Leydig cells, in *Lipoprotein and Cholesterol Metabolism in Steroidogenic Tissues* (Strauss, J. F., III, and Menon, K.M.J., eds.) pp. 243–246, George F. Stickley Co., Philadelphia.

Reddy, J. K., and Svoboda, D. (1972). Microbodies in Leydig cell tumors of rat testis. J. Histochem. Cytochem. 20:793–803.

Rhéaume, E., Lachance, Y., Zhao, H.-F., Breton, N., Dumont, M., de Launoit, Y., Trudel, D., Luu-The, V., Simard, J., and Labrie, F. (1991). Structure and expression of a new complementary DNA encoding the almost exclusive 3β-hydroxysteroid dehydrogenase/ Δ^5-Δ^4 isomerase in human adrenals and gonads. Mol. Endocrinol. 5:1147–1157.

Rhéaume, E., Tonon, M.-C., Smih, F., Simard, J., Désy, L., Vaudry, H., and Pelletier, G. (1990). Localization of the endogenous benzodiazepine ligand octadecaneuropeptide in the rat testis. Endocrinology 127:1986–1994.

Russell, L. D., Amlani, S. R., Vogl, A. W., and Weber, J. E. (1987). Characterization of filaments within Leydig cells of the rat testis. Am. J. Anat. 178:231–240.

Samuels, L. T., Bussmann, L., Matsumoto, K., and Huseby, R. A. (1975). Organization of androgen biosynthesis in the testis. J. Steroid Biochem. 6:291–296.

Scallen, T. J., and Vahouny, G. V. (1985). The participation of sterol carrier proteins in cholesterol biosynthesis, utilization and intracellular transfer, in *Lipoprotein and Cholesterol Metabolism in Steroidogenic Tissues* (Strauss, J. F., III, and Menon, K.M.J., eds.) pp. 219–236, George F. Stickley Co., Philadelphia.

Shaha, C., Liotta, A. S., Krieger, D. T., and Bardin, C. W. (1984). The ontogeny of immunoreactive β-endorphin in fetal, neonatal and pubertal testes from mouse and hamster. Endocrinology 114:1584–1591.

Sharp, B., Pekary, A. E., Meyer, N. V., and Hershman, J. M. (1980). β-endorphin in male rat reproductive organs. Biochem. Biophys. Res. Comm. 95:618–623.

Sharpe, R. M. (1988). Endocrinology and Paracrinology of the Testis, in *Physiology and Toxicology of Male Reproduction* (Lamb J. C., and Foster, P.M.D., eds.) pp. 71–102, Academic Press, New York.

Slaunwhite, W. R., and Samuels, L. T. (1956). Progesterone as a precursor of testicular androgens. J. Biol. Chem. 220:341-352.

Snyder, S. H., Verma, A., and Trifiletti, R. R. (1987). The peripheral-type benzodiazepine receptor: A protein of mitochondrial outer membranes utilizing porphyrins as endogenous ligands. FASEB J. 1:282–288.

Solish, S. B., Picado-Leonard, J., Morel, Y., Kuhn, R. W., Mohandas, T. K., Hanukoglu, I., and Miller, W. L. (1988). Human adrenodoxin reductase: Two mRNAs encoded by a single gene on chromosome 17cen→q25 are expressed in steroidogenic tissues. Proc. Natl. Acad. Sci. (USA) 85:7104–7108.

Soto, E. A., Kliman, H. J., Strauss, J. F., III, and Paavola, L. G. (1986). Gonadotropins and cyclic adenosine-3′5′-monophosphate (cAMP) alter the morphology of cultured human granulosa cells. Biol. Reprod. 34:559–569.

Sterio, D. C. (1984). The unbiased estimation of number and sizes of arbitrary particles using the disector. J. Microsc. 134:127–136.

Stevens, V. L., Tribble, D. L., and Lambeth, J. D. (1985). Regulation of mitochondrial volumes in rat adrenal cortex by ether stress. Arch. Biochem. Biophys. 242:324–327.

Stocco, D. M., and Sodeman, T. C. (1991). The 30-kDa mitochondrial proteins induced by hormone stimulation in MA-10 Leydig tumor cells are processed from larger precursors. J. Biol. Chem. 266:19731–19738.

Stocco, D. M., and Kilgore, M. W. (1988). Induction of mitochondrial proteins in MA-10 Leydig tumour cells with human choriogonadotropin. Biochem. J. 249:95–103.

Strittmatter, S. M., and Snyder, S. H. (1984). Angiotensin-converting enzyme in the male rat reproductive system: Autoradiographic visualization with [^{3}H] captopril. Endocrinology 115:2332–2341.

Strott, C. A. (1977). A pregnenolone-binding protein in soluble fraction of guinea pig adrenal cortex. J. Biol. Chem. 252:464–470.

Tapanainen, J., Kuopio, T., Pelliniemi, L. J., and Huhtaniemi, I. (1984). Rat testicular endogenous steroids and number of Leydig cells between the fetal period and sexual maturity. Biol. Reprod. 31:1027–1035.

Teerds, K. J., De Rooij, D. G., Rommerts, F.F.G., Van Der Tweel, I., and Wensing, C.J.G. (1989). Turnover time of Leydig cells and other interstitial cells in testes of adult rats. Arch. Androl. 23:105–111.

Thomas, J. L., Myers, R. P., Rosik, L. O., and Strickler, R. C. (1990). Affinity alkylation of human placental 3β-hydroxy-5-ene-steroid dehydrogenase and steroid 5→4-ene isomerase by 2α-bromoacetoxyprogesterone: evidence for separate dehydrogenase and isomerase sites on one protein. J. Steroid Biochem. 36:117–123.

Thompson, R. C., Seacholtz, A. F., and Herbert, E. (1987). Rat corticotropin-releasing hormone gene: Sequence and tissue specific expression. Mol. Endocrinol. 1:363–370.

Tsong, S.-D., Phillips, D. M., Halmi, N., Krieger, D., and Bardin, C. W. (1982a). β-endorphin is present in the male reproductive tract of five species. Biol. Reprod. 27:755–764.

Tsong, S.-D., Phillips, D., Halmi, N., Liotta, A. S., Margioris, A., Bardin, C. W., and Krieger, D. T. (1982b). ACTH and β-endorphin-related peptides are present in multiple sites in the reproductive tract of the male rat. Endocrinology 110:2204–2206.

Turner, T. T., Jones, C. E., Howards, S. S., Ewing, L. L., Zegeye, B., and Gunsalus, G. L. (1984). On the androgen microenvironment of maturing spermatozoa. Endocrinology 115:1925–1932.

Van Amerongen, A., Demel, R. A., Westerman, J., and Wirtz, K.W.A. (1989). Transfer of cholesterol and oxysterol derivatives by the non-specific lipid transfer protein (sterol carrier protein 2): A study on its mode of action. Biochim. Biophys. Acta 1004:36–43.

van Noort, M., Rommerts, F.F.G., Van Amerongen, A., and Wirtz, K.W.A. (1988). Regulation of sterol carrier protein 2 (SCP_2) levels in the soluble fraction of rat Leydig cells. Kinetics and the possible role of calcium influx. Mol. Cell. Endocrinol. 56:133–140.

Velletri, P.A. Aquilano, D. R., Bruckwick, E., Tsai-Morris, C. H., Dufau, M. L., and Lovenberg, W. (1985). Endocrinological control and cellular localization of rat testicular angiotensin-converting enzyme (EC 3.4.15.1). Endocrinology 116:2516–2522.

Waterman, M. R., Lund, J., Ahlgren, R., Wu, D., and Simpson, E. R. (1990). Regulation of steroid hydroxylase gene expression and steroid hormone biosynthesis. Biochem. Soc. Trans. 18:26–28.

Waterman, M. R., and Simpson, E. R. (1989). Regulation of steroid hydroxylase gene expression is multifactorial in nature. Rec. Prog. Horm. Res. 45:533–566.

Wilson, S. C., Hodgins, M. B., and Scott, J. S. (1987). Incomplete masculinization due to a deficiency of 17β-hydroxysteroid dehydrogenase: Comparison of prepubertal and peripubertal siblings. Clin. Endocrinol. 26:459–469.

Wing, T.-Y., Ewing, L. L., and Zirkin, B. R. (1984). Effects of luteinizing hormone withdrawal on Leydig cell smooth endoplasmic reticulum and steroidogenic reactions which convert pregnenolone to testosterone. Endocrinology 115:2290–2296.

Yanagibashi, K., Ohno, Y., Kawamura, M., and Hall, P. F. (1988). The regulation of intracellular transport of cholesterol in bovine adrenal cells: Purification of a novel protein. Endocrinology 123:2075–2082.

Yeung, W.S.B., Guldenaar, S.E.F., Worley, R.T.S., Humphrys, J., and Pickering, B. T. (1988). Oxytocin in Leydig cells: An immunocytochemical study of Percoll-purified cells from rat testes. Cell Tiss. Res. 253:463–468.

Yoon, D. J., Sklar, C., and David, R. (1988). Presence of immunoreactive corticotropin-releasing factor in rat testis. Endocrinology 122:759–761.

Zhao, H.-F., Labrie, C., Simard, J., de Launoit, Y., Trudel, C., Martel, C., Rhéaume, E., Dupont, E., Luu-The, V., Pelletier, G., and Labrie, F. (1991). Characterization of rat 3β-hydroxysteroid dehydrogenase/Δ^5-Δ^4 isomerase cDNAs and differential tissue-specific expression of the corresponding mRNAs in steroidogenic and peripheral tissues. J. Biol. Chem. 266:583–593.

Zirkin, B. R., Santulli, R., Awoniyi, C. A., and Ewing, L. L. (1989). Maintenance of advanced spermatogenic cells in the adult rat testis: Quantitative relationship to testosterone concentration within the testis. Endocrinology 124:3043–3049.

Zirkin, B. R, and Ewing, L. L. (1987). Leydig cell differentiation during maturation of the rat testis: A stereological study of cell number and ultrastructure. Anat. Rec. 219:157–163.

Zirkin, B. R., Dykman, D. D., Kromann, N., Cochran, R. C., and Ewing, L. L. (1982). Inhibition and recovery of testosterone secretion in rats are tightly coupled to quantitative changes in Leydig cell smooth endoplasmic reticulum. Ann. N.Y. Acad. Sci. 383:17–28.

Zirkin, B. R., Ewing, L. L., Kromann, N., and Cochran, R. C. (1980). Testosterone secretion by rat, rabbit, guinea pig, dog and hamster testes perfused in vitro: Correlation with Leydig cell ultrastructure. Endocrinology 107:1867–1874.

Zuber, M. X., John, M. E., Okamura, T., Simpson, E. R., and Waterman, M. R. (1986a). Bovine adrenocortical cytochrome $P\text{-}450_{17\alpha}$. Regulation of gene expression by ACTH and elucidation of primary sequence. J. Biol. Chem. 261:2475–2482.

Zuber, M. X., Simpson, E. R., and Waterman, M. R. (1986b). Expression of bovine 17α-hydroxylase cytochrome P-450 cDNA in non-steroidogenic (COS 1) cells. Science 234:1258–1261.

CHAPTER EIGHT

Regulation of Spermatogenesis in the Adult Mammal: Gonadotropins and Androgens

Barry R. Zirkin

I. Introduction

A. Why Another Review of Gonadotropins and Androgens?

Sixty years ago, Smith (1930) first demonstrated that the removal of the rat pituitary gland resulted in the cessation of sperm formation. In the years since, numerous publications have dealt with the roles of gonadotropins and of androgens in regulating mammalian spermatogenesis, among which have been a number of outstanding reviews (Steinberger, 1971; Lostroh, 1976; Fritz, 1978; Parvinen, 1982; Sharpe, 1988; Matsumoto, 1989). The sheer number of these publications, and the length of time over which they have appeared, attest to the considerable interest in, and importance of, the issue of the hormonal regulation of mammalian spermatogenesis.

At this time, recent reviews have begun to focus more on the paracrine and autocrine factors involved in gonadal regulation and less on gonadotropins and androgens (Sharpe, 1983; Saez et al., 1987; Sharpe, 1988; Bellvé and Zheng, 1989; Risbridger and de Kretser, 1989). Yet, six decades after the seminal observations by Smith (1930) on the central role of the pituitary, and despite the considerable attention that the hormonal regulation of spermatogenesis has received over the years, careful reading of the literature indicates that much of what we "know" is soft; that, in fact, there is much to be confirmed and/or learned about gonadotropin and androgen effects on spermatogenesis.

And so, this review of the gonadotropin and androgen regulation of spermatogenesis is offered, in which the bases for long-held dogma relating to the roles of luteinizing hormone (LH), follicle-stimulating hormone (FSH), and testosterone in regulating spermatogenesis in the adult mammal are critically examined and some of the remaining questions that are receiving, or require, scrutiny are elucidated. As has been the case in many past reviews, I emphasize studies of the rat because there has been so much work done with this species. Comparisons with other species, however, particularly nonhuman primates and humans, are made because the literature is replete with indications of significant species differences in the regulation of spermatogenesis.

B. Overview of Spermatogenesis in the Adult Mammal

The reader is referred to Ewing and colleagues (1980) and deKretser and Kerr (1988) for reviews of mammalian spermatogenesis.

The mammalian testis is composed of two major compartments: the seminiferous tubules, within which Sertoli cells reside and spermatogenesis occurs; and the interstitium, within which reside the Leydig cells, the testicular cells primarily responsible for testosterone production. Mammalian spermatogenesis encompasses three phases: (1) the proliferation and renewal of spermatogonia by mitosis; (2) meiosis; and (3) the metamorphosis of the products of meiosis, the spermatids, to maturity during spermiogenesis. Spermatogonial proliferation, meiosis, and spermiogenesis occur throughout the mature testis, and thus spermatogonia, spermatocytes, and spermatids coexist within the seminiferous epithelium. Because new waves of spermatogonial divisions are initiated at periodic, fixed intervals before the completion of a previous round of spermatogenesis, any given area of seminiferous tubule will contain one or two cohorts of each of spermatogonia, spermatocytes, and spermatids. These cells associate with each other as they divide and/or differentiate, forming, in the rat, a series of 14 distinct, constant, and predictable cellular associations over time (Fig. 8–1). Each of the 14 distinguishable cellular associations in the rat testis defines a "stage" of the cycle of the seminiferous epithelium. The proliferation and/or differentiation of the spermatogonia, spermatocytes, and spermatids of a given stage cause a succession of the 14 stages to appear, over time, in any given area of the tubule. In the rat, the appearance of the 14 stages in succession defines one cycle of the seminiferous epithelium, and requires approximately 12 to 13 days; four such cycles are required to produce mature spermatids from proliferating spermatogonia. Thus, in the adult rat, a round of spermatogenesis is completed in approximately 48 to 52 days, the specific time dependent on the species.

C. The "Initiation," "Maintenance," and "Restoration" of Spermatogenesis

In general, publications dealing with the hormonal regulation of spermatogenesis have addressed the requirements for *initiating* spermatogenesis at puberty, for *maintaining* spermatogenesis in the adult testis, and/or for *restoring* spermatogenesis after experimental manipulation of the adult testis. Unfortunately, because these terms are not carefully defined in some publications, it is not always obvious which of these aspects of spermatogenesis is being examined. Moreover, in many publications, little attempt is made to define and distinguish carefully between the *qualitative* and *quantitative* effects of gonadotropins and androgens on spermatogenesis.

This review focuses on the gonadotropin and androgen requirements for maintaining or restoring spermatogenesis in the adult testis, with some comparisons made to the requirements of initiating spermatogenesis. The *initiation* of spermatogenesis will refer to the completion of the first round of spermatogenesis; that is, the first production of mature spermatids in the

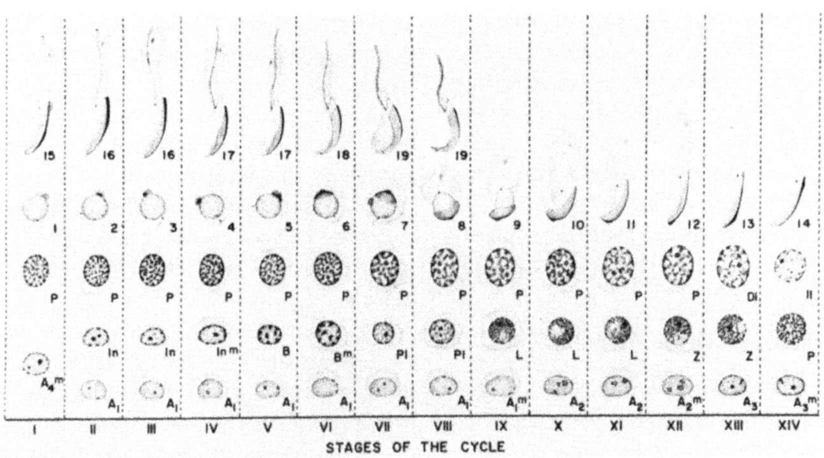

Figure 8–1. Representation of the stages of the cycle of the seminiferous epithelium in the rat. Each of the 14 columns, numbered I–XIV, represents a distinct, constant, and predictable association of spermatogonia, spermatocytes, and spermatids, which defines one stage of the cycle. (From Perey, B., Clermont, Y., and Leblond, C.P., 1961, Am. J. Anat. 108:47.)

testes of pubertal animals. In the rat, the most mature spermatid steps are first seen in histological sections at about day 45 of life (Clermont and Perey, 1957). The appearance of these cells indicates that spermatogenesis is *qualitatively* complete; that is, all of the cell types that are characteristic of spermatogenesis, from spermatogonia to mature spermatids, are represented in the testes. Spermatogenesis is not *quantitatively* complete at this time, however, because increasing numbers of mature spermatids populate the testes subsequent to day 45, until a steady state of daily sperm production and spermiation is reached. This occurs at about day 75 (Robb et al., 1978), at which time spermatogenesis in the rat can be said to be *quantitatively* complete. The *maintenance* of spermatogenesis will refer to the requirements for sustaining the qualitative or quantitative production of all germ cells in the mature adult testis.

The testicular content of advanced spermatids normally remains constant until old age, at which time there may be a decline. Experimental manipulation of the adult animal before senescence can also cause the loss of advanced spermatids, as well as of other germ cell types. For example, withdrawal of LH by any of a number of means (e.g., hypophysectomy, LH-suppressing steroid implants, immunologic methods, gonadotropin-releasing hormone (GnRH) angonists, GnRH antagonists) can cause the seminiferous epithelium to lose or diminish its ability to produce and/or sustain particular germ cell types. The *qualitative restoration* of spermatogenesis refers to the restoration to the adult testis of all of the cell types that characterize spermatogenesis; the *quantitative restoration* of spermatogenesis refers to the restoration of germ cells to numbers that are equivalent to those in the testes of control animals.

D. Quantification of Germ Cells

Assessment of the quantitative maintenance or restoration of spermatogenesis depends on methods to quantify germ cells. Most quantitative analyses of spermatogenesis have been based on counts of particular germ cells in cross sections of seminiferous tubules (Clermont and Morgentaler, 1955; Wing and Christensen, 1982). For such analyses, raw counts of the number of nuclei of particular germ cells types are obtained, and these are corrected for section thickness and nuclear diameter. In some species (the rat, for example), possible shrinkage of the seminiferous tubules can be taken into account by expressing the cell counts per Sertoli cell (or per Sertoli cell nucleolus) because the Sertoli cell number remains constant (Clermont and Morgentaler, 1955). In species in which the Sertoli cell number is not constant (the horse, for example: Johnson and Thompson, 1983; Johnson and Tatum, 1989), a Sertoli cell-based shrinkage correction cannot be used.

Quantification of germ cells by counts in tubule cross sections has provided important data on the relative numbers of particular germ cells at specific stages of the cycle. It should be noted, however, that because only cells with spherical nuclei can be analyzed reliably by these methods, the end products of spermatogenesis, the advanced spermatids, cannot be enumerated because these cells have elongated nuclei. Moreover, only *relative,* not absolute cell numbers can be obtained, and thus such methods provide only a quantitative index of changes in germ cell numbers.

Other methods are available by which advanced spermatogenic cells can be quantified throughout the testis. Advanced spermatids are resistant to the shearing forces of homogenization, and thus their number per testis can be assessed by hemacytometric counting of testicular homogenates (Robb et al., 1978). Quantification of advanced spermatids serves as a surrogate for assessing total daily sperm production per testis, and thus this method can be used to assess the extent to which spermatogenesis in the adult testis is maintained or restored. A disadvantage of this technique is that it cannot be used to assess stage-specific events, or to examine cells other than advanced spermatids.

Improvements in stereologic methods over the last decade (Gundersen, 1986) now make it feasible to determine reliably the absolute number of germ cells of given types per testis and/or at particular stages of the cycle. By these newer methods, cells or their nuclei are counted directly in serial sections, and thus cells of any size and shape can be counted. Among the unbiased stereologic estimators that have been introduced, one, the disector (Sterio, 1984), was recently used to quantify Leydig cells throughout the rat testis (Keeney et al., 1988; Mendis-Handagama et al., 1989). As yet, there have been no reports in the literature in which an unbiased stereologic method has been used to quantify cells within seminiferous tubules.

E. Complicating Issues to be Considered in Assessing the Hormonal Regulation of Spermatogenesis

It is apparent from these descriptions that mammalian spermatogenesis and its analysis are complex. This complexity has seriously compromised our ability to interpret the results of many of the studies that have examined the hormonal regulation of spermatogenesis. Several difficulties are encountered. First, the factors that control stem cell behavior, spermatogonial proliferation, meiotic divisions, and spermiogenesis may differ. Second, the requirements for maintaining spermatogenesis may differ from the requirements for restoring spermatogenesis in the adult and/or for initiating spermatogenesis in the pubertal animal. Third, hormones may be required only at specific and perhaps brief phases of spermatogenesis. For example, Steinberger and Duckett (1967) reported that daily injections of estradiol given to rats from birth through day 5 had profound qualitative effects on spermatogenesis (assessed by observation of testis sections) that persisted through adult life. Likewise, Almiron and coworkers (1984) showed that in rats treated with testosterone from birth or from day 5 through day 35, many tubules became arrested at meiotic prophase, whereas in rats treated similarly from day 10 or 15 through day 35, spermatogenesis was qualitatively similar to that in controls. Fourth, yet another layer of complexity derives from the well-established fact that the primary intratubular targets of gonadotropins and androgens are not the germ cells themselves, but rather Leydig, myoid, and Sertoli cells (Hansson et al., 1975; Means et al., 1976; Fritz, 1978; Sanborn et al., 1983; Bardin et al., 1988; Ritzen et al., 1989). Thus, particular germ cells at some stages of the cycle may be more or less susceptible to FSH and/or androgens because of the effects of these hormones on the Sertoli cells with which the germ cells are associated. Indeed, there are reported differences in androgen receptor concentrations, FSH binding, and biochemical responses to FSH by seminiferous tubules at different stages of the cycle (Parvinen, 1982; Isomaa et al., 1985), presumably reflecting the different stage-specific properties of Sertoli cells. Finally, results obtained and conclusions reached by simple observation of testis sections cannot readily be compared from study to study; and data derived by estimating *relative* changes in germ cell numbers may be useful for assessing experimental treatments *within* a given study, but not necessarily *between* studies.

Some of these issues were raised years ago by Steinberger, who in his highly quoted review of 1971, stated: "First, we should consider the question whether the hormonal mechanisms responsible for maintenance of spermatogenesis in an adult testis are the same as those responsible for initiation of this process in immature animals or animals with posthypophysectomy regression of the testis; second, we should ask whether the same factors are essential for qualitative and for quantitative maintenance of the spermatogenic process; and third, we should keep in mind the possibility that a hormone may be essential only during a very specific stage of the spermatogenic development, which may be quite brief. For example, a hormonal factor may be essential for division of the spermatogonial stem cell, but all subsequent divisions of the spermatogonia and their differentiation may proceed autonomously."

We would argue that there is considerable confusion in the literature that has resulted from the failure of some papers to identify and address these issues carefully.

II. Androgen Microenvironment of Testicular Compartments

It is well established that spermatogenesis is an androgen-dependent process. The following discussion focuses on the microenvironment within the testis because it is this androgen-containing milieu to which Sertoli and germ cells are exposed.

Testosterone is the major secretory product of the Leydig cell (Ewing and Zirkin, 1983).

Turner and colleagues (1984) reported that dihydrotestosterone (DHT) in the interstitial and seminiferous tubule fluids of the rat testis represents less than 5 percent of the concentration of testosterone. Moreover, testosterone is the predominant androgen bound by rat seminiferous tubule cell nuclei (Wright and Frankel, 1979). We focus on intratesticular testosterone in this chapter because results such as these suggest (though they do not prove) that testosterone is the active androgen in the seminiferous tubules.

Comhaire and Vermeulen (1976) and Turner and associates (1984), using micropuncture methods, reported that the testosterone concentration in interstitial fluid of the rat (70 to 150 ng/mL) is far greater than in cardiac blood serum (1 to 2 ng/mL), and that the testosterone concentration in seminiferous tubule fluid (50 to 90 ng/mL) is significantly lower than in interstitial fluid, but still several orders of magnitude higher than in blood serum. More recently, Maddocks and Setchell (1989a), using a push-pull cannula method, also reported that the testosterone level in interstitial fluid of untreated rats is significantly greater than in peripheral blood. As would be predicted from these observations, the testosterone concentration in rete testis fluid of the rat also is high relative to that in blood serum (Harris and Bartke, 1975).

Taken together, these results, illustrated in Figure 8–2 (from Turner et al., 1984) indicate that there is a far higher local concentration of testosterone within the interstitium and the seminiferous tubules of the testis than in the periphery. This suggests either that androgen-dependent organs, such as the prostate, require far less androgen for their maintenance than does the testis, or that far less testosterone is needed within the testis to maintain spermatogenesis than is normally present. Figure 8–2 also shows that seminiferous tubule fluid testosterone concentration exceeds that of andro-

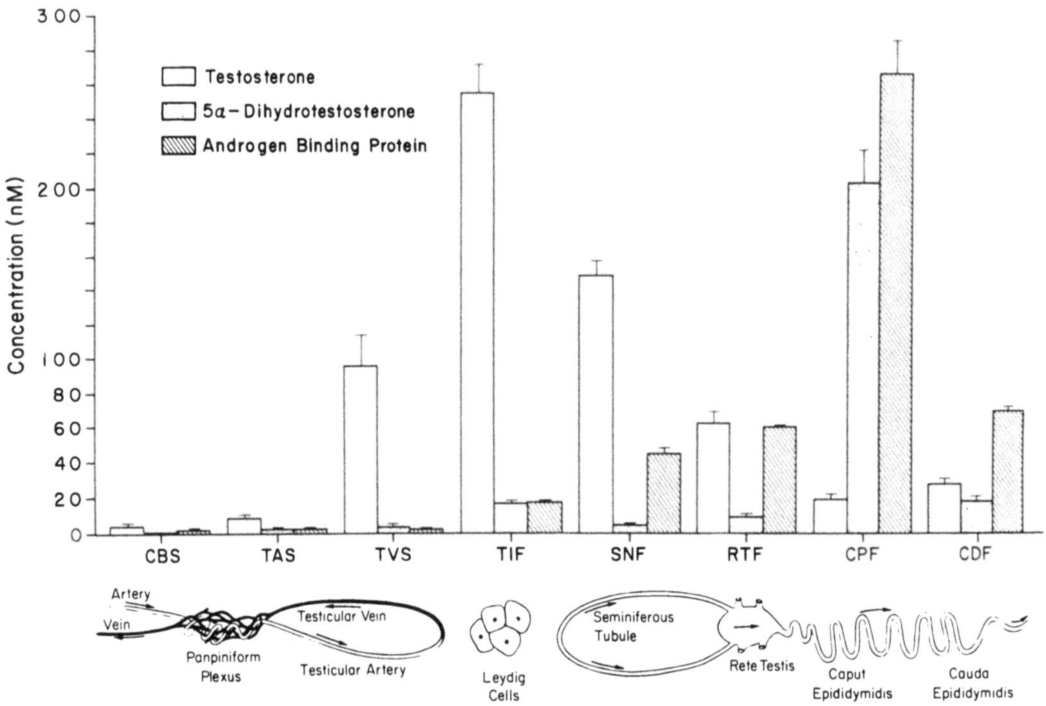

Figure 8–2. Concentrations of testosterone, dihydrotestosterone, and androgen-binding protein in adult rat cardiac blood serum (CBS), testicular arterial serum (TAS), testicular venous serum (TVS), testicular interstitial fluid (TIF), seminiferous tubule fluid (SNF), rete testis fluid (RTF), caput epididymal fluid (CPF), and cauda epididymal fluid (CDF). (From Turner et al., 1984.)

gen binding protein (ABP). This suggests that testicular ABP concentration alone cannot account for the retention of androgens in the seminiferous tubules or rete testis.

The observations of a high concentration of testosterone in interstitial fluid and a downhill gradient from the interstitial fluid to the seminiferous tubule fluid might simply result from local production of testosterone by Leydig cells in the interstitial compartment and then diffusion of testosterone into the seminiferous tubules. According to this hypothesis, the high intratesticular testosterone concentrations would not necessarily be due to the ability of testicular compartments to concentrate testosterone, but rather to the fact that testosterone is produced locally. If this were true, the administration of testosterone to animals in which local production of testosterone by Leydig cells was suppressed would be expected to result in the equilibration of testosterone concentration in the blood serum, interstitial fluid, and seminiferous tubule fluid, and not in high intratesticular concentrations relative to the periphery.

Zirkin and colleagues (1989) tested this by administering testosterone to intact male rats by sustained release polydimethylsiloxane implants (PDS) of increasing sizes. Testosterone administration in this manner inhibited LH secretion and, therefore, Leydig cell testosterone production, and thus there was little or no local production of testosterone in the implanted rats. The results of measuring testosterone concentrations in blood serum, interstitial fluid, and seminiferous tubule fluid 2 months after administering implants are shown in Figure 8–3. In Figure 8–3, serum, interstitial fluid, and seminiferous tubule fluid testosterone concentrations are plotted against the lengths of the testosterone-filled capsules that were implanted. In each case, linear relationships between implant length and testosterone concentration were seen, but the slopes of the dose-response lines (0.3, 0.5, and 1.3 for serum, seminiferous tubule fluid, and interstitital fluid, respectively) were significantly different from each other. Thus, with each capsule size, the testosterone concentration in the interstitial fluid was consistently higher than in the seminiferous tubule fluid, and the testosterone concentrations in the interstitial and seminiferous tubule fluids were consistently higher than the

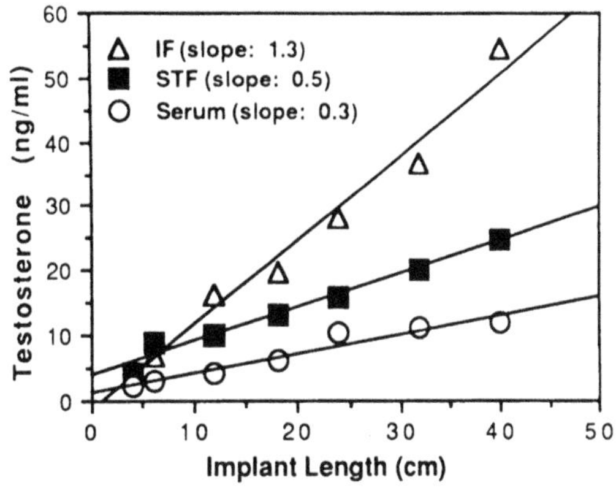

Figure 8–3. Linear regression analysis between the lengths of testosterone-filled polydimethylsiloxane capsules implanted into adult male rats for 2 months and testosterone concentrations in serum, interstitial fluid (IF), and seminiferous tubule fluid (STF). (From Zirkin et al., 1989.)

concentration in serum. Turner and colleagues (1985a,b) similarly had shown that testosterone became concentrated in the interstitial and seminiferous tubule fluids of hypophysectomized rats that received testosterone.

Taken together, these results indicate that there is a mechanism, as yet unknown, by which testosterone is able to become concentrated within testicular compartments of the testis even under circumstances in which there is no local production of testosterone by Leydig cells. It should be noted, however, that this conclusion is not consistent with the results of Maddocks and Setchell (1989b), who used push-pull cannulae to obtain interstitial fluid from rats that had received testosterone via PDS implants. In the Maddocks and Setchell (1989b) study, no differences were seen in testosterone concentration in the interstitial fluid regardless of the size of the implants that were used; but plasma testosterone concentrations increased relative to implant size. Maddocks and Setchell (1989a) pointed out that fluids obtained by push-pull cannulae may be diluted during collection, and therefore, that testosterone concentrations might be underestimated by this method. Nonetheless, their finding argues against an intratesticular concentrating mechanism for testosterone. Also, Turner (1988) reported restricted uptake of exogenously administered androgens from blood plasma into intratesticular compartments, which also is inconsistent with our results. Thus, the issue of whether testosterone can be concentrated from the periphery in compartments of the testis is controversial. Moreover, if indeed testosterone can be concentrated, the mechanism by which this occurs is unknown.

III. Maintenance of Spermatogenesis in the Adult Mammal

In this part of the review, we examine the involvement of androgens and gonadotropins in the qualitative and quantitative maintenance of spermatogenesis in the adult mammalian testis. The focus is on the rat. Later in the chapter, comparisons between the rat and the primate are made.

A. *To What Extent Is Testosterone Able to Maintain Spermatogenesis in the Absence of Endogenous Testosterone Production?*

Luteinizing hormone binds specifically to Leydig cells in the testicular interstitium, thereby stimulating the biosynthesis of testosterone (for reviews, please see Ewing and Zirkin, 1983; and Chapter 7). It has been known for several decades that androgens are necessary prerequisites for the maintenance of established spermatogenesis in adult mammals. Numerous studies have shown that the administration of increasing amounts of testosterone to intact rats (Robaire et al., 1979) and other mammalian species (Ewing et al., 1973, 1976) causes a biphasic response in the number of advanced spermatids that are maintained in the testes; high testosterone doses result in the quantitative maintenance of advanced spermatids despite the absence of detectable LH, whereas low testosterone doses result in reduced numbers of advanced spermatids.

Nonetheless, there has been a long-standing debate about how much testosterone is needed to maintain spermatogenesis. Turner and colleagues (1984) reported that the average testosterone concentration in rat seminiferous tubule fluid is approximately 50 ng/mL (or 1.7×10^{-7}M). As indicated, this concentration is considerably higher than in blood serum (approximately 1 to 2 ng/mL or 3 to 7 $\times 10^{-9}$M), and also considerably higher than the testosterone concentration required to saturate testicular androgen receptors (3×10^{-9}M; Wright et al., 1982). This suggests that the testosterone concentration in the seminiferous tubules may be considerably in excess of the concentration that is required to maintain spermatogenesis. Indeed, there is now substantial evidence that established spermatogenesis can be qualitatively, and indeed quantitatively, maintained in the intact adult rat testis at only a fraction of the intratesticular testosterone concentration found in the intact adult male. The evidence for this is reviewed next.

In an important publication, Cunningham and Huckins (1979) administered testosterone propionate to intact rats, a procedure that produced a 30-fold reduction in the intratesticular concentration of testosterone with little effect on serum FSH, and examined testicular histology 39 days later. They reported that qualitatively complete spermatogenesis persisted through at least the 39 days. Several studies since Cunningham and Huckins (1979) similarly reported that qualitatively complete spermatogenesis is able to persist in the presence of no more than 15 to 20 percent of the testicular testosterone concentration of controls (Huang and Nieschlag, 1986; Huang and Boccabella, 1988). Moreover, Rea and co-workers (1986) reported that when rats treated with a GnRH antagonist simultaneously received exogenous testosterone by subcutaneous implants for 30 days, the relative numbers of spermatogenic cells at stage VII of the cycle of the seminiferous epithelium (i.e., type A spermatogonia, preleptotene spermatocytes, pachytene spermatocytes, and step 7 spermatids), counted in seminiferous tubule cross sections, remained unchanged from their control values despite a reduction in intratesticular testosterone concentration of approximately 85 percent.

Before discussing more recent data addressing the extent to which spermatogenesis can be maintained by exogenously administered testosterone, it should be noted that the conclusion that spermatogenesis can be maintained at relatively low intratesticular testosterone concentrations rests on the ability of investigators to reliably measure intratesticular testosterone concentrations both in untreated and experimentally manipulated animals. This issue is controversial. For example, Sharpe (1987) and Rommerts (1988) pointed to discrepancies in reports of intratesticular testosterone concentrations as measured in testicular fluids obtained from anesthetized rats by push-pull cannulae versus testicular fluids collected from harvested testes by "drip" or centrifugation methods. The latter "postmortem" methods are often used because of the relatively large amounts of testicular fluids that can be collected. According to Sharpe and colleagues (Sharpe et al, 1988a; Maddocks and Sharpe, 1989), however, a major problem is that Leydig cells may continue to secrete testosterone even after the blood supply to the testis has been severed, thereby causing testosterone concentrations to rise artifactually in testes in which there are testosterone-producing Leydig cells. If true, this would predict that testosterone concentration in testicular fluids of control (untreated) rats would increase in relationship to the time required for the collection of fluids from testes harvested postmortem. Indeed, Maddocks and Sharpe (1989) reported that testosterone concentrations were considerably lower in interstitial fluid collected with push-pull cannulae than in fluid collected by a lengthy postmortem drip method; and that testosterone concentrations obtained from postmortem testicular fluid were lower when fluids were obtained rapidly (minutes) than when they were obtained slowly (hours).

Do these results invalidate the conclusion that spermatogenesis can be maintained at testosterone concentrations that are considerably below those of control rats? What is the relationship between intratesticular testosterone concentration and the extent to which spermatogenesis can be maintained? These issues are discussed next.

Two possible relationships between intratesticular testosterone concentration and the extent to which spermatogenesis can be maintained have been postulated. First, there could be a threshold concentration of testosterone below which the ability of the seminiferous epithelium to sustain spermatogenesis would decline precipitously. Alternatively, germ cell production might respond in dose-response fashion to graded reductions in testosterone concentration. We (Zirkin et al., 1989) examined these alternative possibilities by implanting intact rats with testosterone-containing PDS implants of increasing sizes, and relating testosterone concentration within compartments of the testes to the number of advanced spermatids per testis that were present 2 months after the initiation of the steroid treatment. In our study, the quantification of advanced spermatid number per testis was used as a surrogate for assessing daily sperm production throughout the testes, without regard to specific stages of the cycle. Intratesticular fluids were collected from isolated testes by a *rapid* postmortem centrifugation method that had been developed by Turner and colleagues (1984); collection of interstitial fluid required about 15 minutes from the time the rats were killed, and seminiferous tubule fluid collection required approximately 1 hour. Our rationale for using this method was that Turner and colleagues (1984) reported that the androgen concentrations that they obtained by this technique were identical to those obtained by in vivo micropuncture of anesthetized rat testes.

The administration of testosterone to rats by sustained-release PDS capsules caused pituitary LH production, and therefore, endogenous testosterone production by Leydig cells, to be rendered nondetectable, with little effect on serum FSH. In addition, testosterone was clamped within the seminiferous tubule fluid at concentrations that were directly related to the sizes of the implanted testosterone-filled capsules. The results that we obtained, shown in Figure 8–4, indicated: (1) that significant numbers of advanced spermatids were maintained at a seminiferous tubule fluid testosterone concentration of approximately 4 ng/mL (1.4×10^{-8}M), a value reduced by 90 percent from the average testosterone concentration in control rats, but approximately that in blood serum; and (2) that quantitatively complete spermatogenesis was maintained at testosterone concentrations within the seminiferous tubule fluid of 15 to 20 ng/mL, representing a reduction of 70 to 80 percent from the concentration in controls, but a molar concentration of testosterone (5.2 to 6.8×10^{-8}M) still an order of magnitude higher than the testosterone concentration required to saturate testicular androgen receptors.

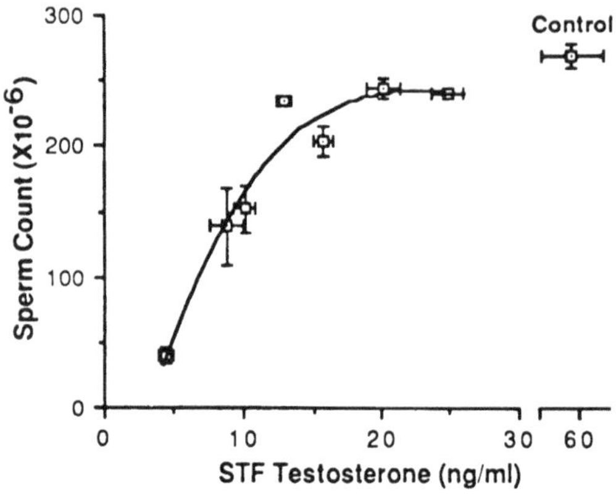

Figure 8–4. Relationship between the mean number of advanced spermatids per testis (Sperm Count) and the mean testosterone concentration in seminiferous tubule fluid (STF) 2 months after implantation of testosterone-filled polydimethylsiloxane capsules of increasing sizes into intact adult male rats. (From Zirkin et al., 1989.)

Figure 8–4 shows that the ability of the testis to maintain its quantitatively normal complement of advanced spermatids was extremely sensitive to reductions in seminiferous tubule fluid testosterone concentration from approximately 15 ng/mL; graded decreases in testicular advanced spermatid content were seen, suggestive of a dose-response relationship between seminiferous fluid testosterone concentration and the maintenance of spermatogenesis. A recent publication of Sun and associates (1989), which related daily sperm production to interstitial fluid testosterone concentration, similarly demonstrated that daily sperm production levels were extremely sensitive to a narrow testosterone dose range. At this juncture, it is not known whether specific cell types throughout the testis, specific cells at particular stages of the cycle, or all cells at particular stages show graded responses to testosterone concentration changes.

As discussed previously, testosterone effects on germ cells undoubtedly are tranduced by Sertoli (and perhaps peritubular) cells (Skinner and Fritz, 1985; Bardin et al., 1988). Unfortunately, we know relatively little about the effects of testosterone withdrawal on adult Sertoli cell function, or indeed how changes in Sertoli cell function might affect germ cells. These important issues clearly require careful analysis, and fortunately are now amenable to study both in vitro and in vivo.

The observation that spermatogenesis can be maintained quantitatively in intact rats at a seminiferous tubule fluid testosterone concentration of approximately 20 ng/mL (6.8×10^{-8}M) suggests that a testosterone concentration significantly in excess of that required to saturate testicular androgen receptors (3×10^{-9}M) may be required within the seminiferous tubules to maintain spermatogenesis quantitatively. If true, this calls into question the mechanism by which testosterone functions within the testis, and perhaps suggests that factors in addition to testosterone participate in defining the mechanism of action of testosterone in the seminiferous epithelium. One wonders, for example, whether testosterone function within the testes may be unrelated to its binding to androgen receptor. Indeed, previous studies have shown that the biological response of a target cell to an androgen may depend on events proximal or distal to steroid binding to the androgen receptor (Brown and Migeon, 1987).

B. To What Extent Is Testosterone Able to Maintain Spermatogenesis in the Absence of All Pituitary Factors?

The studies outlined earlier showed clearly that spermatogenesis can be quantitatively maintained with exogenously administered testosterone alone under circumstances in which LH, and therefore, endogenous testosterone production, are nondetectable but in which FSH is essentially unaffected. This leads to the question: Are FSH and/or other pituitary factors required in addition to testosterone for the maintenance of spermatogenesis?

Dym and colleagues (1979) reported that the administration of FSH antiserum to adult rats for 14 days had relatively little effect on testis weight or germ cell number, indicating a relative insensitivity of the adult rat testis to withdrawal of immunoassayable FSH. The significance of this result is underscored by its striking contrast to the effects of passive immunization against FSH in the immature rat. For example, Chemes and colleagues (1979) treated groups of newborn rats with antiserum to FSH from the day of birth to day 32 and examined these rats at day 33. They reported reductions from control values of 25 to 30 percent in the relative numbers of preleptotene and pachytene spermatocytes, and of 90 percent in the relative numbers of step 7–8 spermatids in rats that had received FSH antisera. These results indicate a requirement for FSH for the initiation of spermatogenesis in the rat, which contrasts with the situation in adults in which antisera to FSH had relatively little effect (Dym et al., 1979).

The relative lack of effect of FSH antisera on spermatogenesis in the adult rat is consistent with studies that have shown a decrease in the stimulatory effects of FSH on Sertoli cell functions, such as cyclic AMP accumulation, protein kinase activation, and RNA and protein synthesis, as rats mature (Van Sickle, 1981; Hugly et al., 1988; Roberts and Griswold, 1989). Taken together, these findings support the oft-stated contention in the literature that FSH is not required for the maintenance of established spermatogenesis in the adult rat, but is required for the initiation of spermatogenesis at puberty. It should be noted, however, that Dym and colleagues (1979) reported that the relative numbers of preleptotene spermatocytes at stage VII of the cycle of adult rats were diminished somewhat in rats that were immunized against FSH, an observation that is rarely mentioned in the literature. If true, however, this would suggest that spermatogonia, the precursors of preleptotene spermatocytes, may be affected somewhat by short-term FSH withdrawal, and perhaps that decreased numbers of advanced spermatids would have been seen by Dym and associates (1979) at later times if their study had been carried out for longer than 14 days.

If indeed FSH and other pituitary factors play little or no role in maintaining adult spermatogenesis, it should be possible to maintain spermatogenesis quantitatively in hypophysectomized rats by administering a high enough dose of testosterone at the time of hypophysectomy. Despite the substantial number of studies that have addressed this issue, the literature is replete with contrasting conclusions. For example, numerous studies of hypophysectomized rats (only some of which used quantitative methods) concluded that spermatogenesis could be qualitatively but not quantitatively maintained by exogenously administered testosterone (Clermont and Harvey, 1967; Steinberger, 1971; Elkington and Blackshaw, 1974; Chowdhury and Tcholakian, 1979; Bartlett et al., 1989a). In other studies, exogenously administered testosterone alone failed to maintain even qualitatively normal spermatogenesis in hypophysectomized rats (see review by Steinberger, 1971). Finally, Robaire and Zirkin (1981) reported that the administration of testosterone by PDS implants to hypophysectomized rats resulted in the quantitative maintenance of advanced spermatogenic cell number in the testis (although testis weights were reduced by about 20 percent from the control value in the latter study, an observation that is not consistent with the conclusion that spermatogenesis was quantitatively maintained).

The discrepant findings in the older literature make it impossible to come to definitive conclusions about the requirements for pituitary factors (except LH) in maintaining spermatogenesis. What accounts for the vastly different results in different publications? Among the possibilities that occurred to us are the following: (1) the success of hypophysectomy can be questioned in some studies; (2) in some studies, testosterone was administered at the time of hypophysectomy and in others at some time thereafter; thus, some studies of spermatogenesis "maintenance" were, in fact, studies of spermatogenesis restoration; (3) the amount of administered androgen that reached the milieu of the testis was not determined in most studies, and thus may have differed; and (4) the methods used for examining treatment effects (observation in many studies vs. quantitation in others) have varied from study to study.

Santulli and colleagues (1990) recently completed a study in which comparisons were made between the extent to which spermatogenesis could be maintained by testosterone in rats that lacked all pituitary factors (hypophysectomized) and rats that lacked LH but not FSH (testosterone implanted). For this study, testosterone was administered to hypophysectomized rats at the time of hypophysectomy, as well as to intact rats, by PDS implants designed to clamp testosterone at comparable and increasing concentrations within the seminiferous tubule fluid. Figure 8–5 shows that the

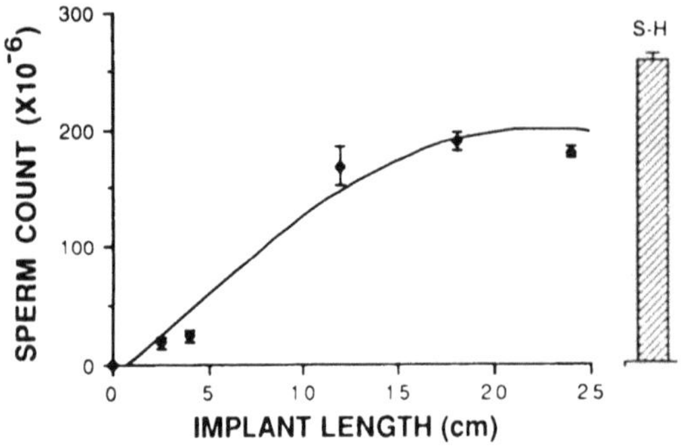

Figure 8–5. Relationship between the lengths of testosterone-filled polydimethylsiloxane capsules implanted into adult hypophysectomized rats at the time of hypophysectomy and the number of advanced spermatids per testis (Sperm Count) 2 months later. The *bar graph* shows the mean sperm count for sham-hypophysectomized (S-H) rats. (From Santulli et al., 1990.)

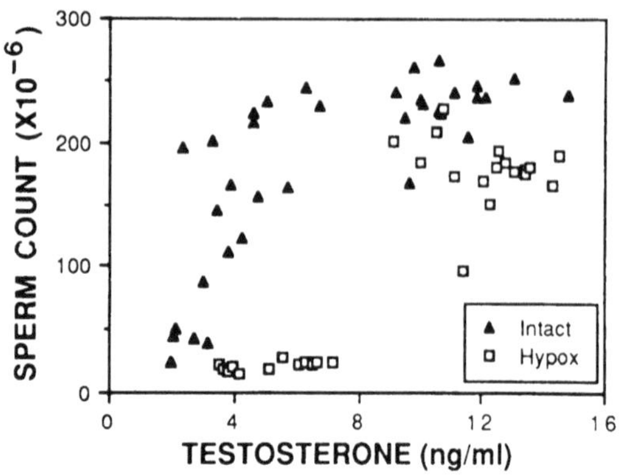

Figure 8–6. Comparison of the ability of exogenously administered testosterone to maintain spermatogenesis in *individual* hypophysectomized (Hypox) and intact rats. The number of advanced spermatids per testis (Sperm Count) is plotted against the serum testosterone concentration. (From Santulli et al., 1990.)

numbers of advanced spermatids per testis that were maintained in the hypophysectomized rats were related to the sizes of the implants that were administered, with steep changes in advanced spermatid number per testis over a narrow testosterone dose range, as in intact rats (see Fig. 8–4). Notably, the maximum number of advanced spermatids that could be maintained in the hypophysectomized rats was only 60 percent of, and significantly less than, the control value; whereas the number maintained in intact rats was not significantly different from the control value. This difference is illustrated in Figure 8–6, which compares the ability of exogenously administered testosterone to maintain spermatogenesis in individual hypophysectomized and intact rats; at any given serum testosterone concentration, the number of advanced spermatids per testis maintained in the intact rats was considerably greater than the number in the hypophysectomized rats. These results were not due to differences in intratesticular testosterone because the maximum seminiferous tubule fluid testosterone concentration in the hypophysectomized rats (about 25 ng/mL) was greater than the testosterone concentration found to be sufficient to maintain spermatogenesis quantitatively in the intact rats (about 15 ng/mL).

Thus, intratesticular testosterone concentrations that were as high as, or higher than, those that maintained spermatogenesis in intact rats lacking LH failed to maintain spermatogenesis in rats lacking all pituitary factors. Possible explanations for these results follow. (1) It is possible that the lack of FSH in hypophysectomized rats affects the ability of testosterone to maintain spermatogenesis. For example, as indicated earlier, Dym and co-workers (1979) reported some effect of FSH withdrawal on germ cell number in adult rats. Moreover, testosterone and FSH have been reported to act synergistically in the adult rat to stimulate germ cell development, to inhibit germ cell degeneration, or both (Russell and Clermont, 1977; Bartlett et al., 1989a; Sharpe, 1989). (2) Pituitary factors in addition to LH and FSH may be required for the maintenance of adult spermatogenesis, perhaps in part because of their indirect effect(s) on testicular paracrine factors (Bardin et al., 1988). These important issues merit further study.

C. *Are Nonsteroidal Leydig Cell Factors Involved in the Maintenance of Spermatogenesis?*

When either intact or hypophysectomized rats receive exogenous testosterone, Leydig cells, although considerably diminished in volume and steroidogenically suppressed compared with controls, are not lost from the interstitial compartment of the testis (Keeney et al., 1988). Are such Leydig cells capable of producing nonsteroidal products that are involved in maintaining spermatogenesis?

Two recent publications have examined this issue (Sharpe et al., 1988b; Sprando et al., 1990). In both, rats received a single injection of ethane 1,2-dimethanesulphonate (EDS), an agent reported to destroy Leydig cells specifically. At the time of EDS injection, the rats also were provided with testosterone by subdermal injection of testosterone esters (Sharpe et al., 1988b) or by testosterone-filled PDS implants (Sprando et al., 1990). Besides providing testosterone to the seminiferous epthelium to replace the endogenous testosterone lost when Leydig cells were eliminated, the administration of testosterone in the two studies prevented the reappearance of Leydig cells by inhibiting LH.

Sharpe and colleagues (1988b) reported that the relative numbers of germ cells at stage VII of the cycle of the seminiferous epithelium were similar in the treated and control rats for up to 10 weeks despite the absence of Leydig cells, and thus concluded that Leydig cell products other than testosterone are not essential for normal spermatogenesis. (It should be noted, however, that these investigators reported a three to fourfold increase in the relative number of degenerating germ cells at stages XIV-I, but did not emphasize this result.) Sprando and colleagues (1990) also reported that spermatogenesis appeared normal in rats injected with EDS and implanted with PDS testosterone-filled capsules, but that the advanced spermatid number per testis was significantly lower in the EDS implanted rats than in controls. One possible explanation for these results is that Leydig cell products other than testosterone may indeed be required for the maintenance of spermatogenesis, a conclusion that is not consistent with that of Sharpe and coworkers (1988b). An alternative explanation, however, is that although its major effect may well be on Leydig cells, EDS may also have a cytotoxic effect on spermatogenesis that is unrelated to the diminished intratesticular testosterone that follows the EDS-induced elimination of Leydig cells. This possibility merits further study because EDS is widely purported to be highly specific for Leydig cells.

D. Are There Different Requirements for Regulating Germ (or Sertoli) Cells at Different Stages of the Cycle of the Seminiferous Epithelium?

The possibility that the regulation of spermatogenic cells may differ at different stages of the cycle has been suggested. For example, studies of microdissected and cultured rat seminiferous tubule segments (Parvinen, 1982) have indicated that there are small but significant differences in testosterone concentration at different stages of the cycle; maximal concentrations were reported for stages VII-XI, with particularly high concentrations at stage VIII. Whether or not the germ cells (or Sertoli cells) at different stages require these greater or lesser testosterone concentrations is not known. Nuclear androgen receptor concentrations also have been reported to vary with the cycle (Isomaa et al., 1985); androgen receptor concentrations were found to be highest at stages IX–I. The relationship, if any, of androgen receptor concentration to endogenous testosterone concentration or to the testosterone concentrations required for germ cells to be maintained at particular stages, however, is also not known. Finally, stage-specific differences in FSH binding and in the biochemical responses of the tubules to added FSH also are reported to differ (Parvinen, 1982). Taken together, these data suggest that there may indeed be stage-specific requirements for the quantitative maintenance of spermatogenesis. This important possibility is far from established, however, in part because the dissection and subsequent culture of tubule segments and/or of germ cells may or may not reflect the in vivo condition.

Fortunately, an animal model was described recently that might be used to validate and extend these important in vitro results. Morales and Griswold (1987) reported success in producing rats with high degrees of synchrony of the stages of the cycle, using a regimen of vitamin A deprivation of young rats followed by retinol feeding some time thereafter. These rats hold promise of allowing in vivo evaluation of stage-specific aspects of the regulation of spermatogenesis, including, for example, the stage-specific transcriptional activity of Sertoli and germ cells (Morales et al., 1987, 1989), and the stage-specific intratesticular testosterone and androgen receptor concentrations (Bartlett et al., 1989b).

E. *Are There Species Differences in the Hormonal Regulation of the Maintenance of Spermatogenesis?*

Most of our present concepts of the hormonal regulation of spermatogenesis derive from studies of the rat. Perusal of the literature, however, suggests that there may be significant species differences (Sharpe, 1989). Because particularly striking differences have been reported between rodents and primates, we will focus on this comparison.

Reported differences between rodents and primates include the following. (1) The seminiferous epithelium of the primate responds differently to hypophysectomy than does the rat; for example, hypophysectomy in the monkey results in an epithelium composed only of Sertoli cells and early spermatogonia (Marshall et al., 1986), whereas the seminiferous epithelium of the hypophysectomized rat contains some germ cells as advanced as step 7 spermatids (Clermont and Morgentaler, 1955). (2) Studies have suggested that active or passive immunization against FSH leads to a significant suppression of sperm production in primates (Murty et al., 1979; Wickings et al., 1980; Wickings and Nieschlag, 1980; Sharpe, 1989), whereas passive immunization against FSH in adult rats (Dym et al., 1979) is reported to have relatively little effect on spermatogenesis. (3) The responsiveness of primate spermatogenesis to FSH is also suggested by a report that significant increases in the number of spermatogonia occur in the seminiferous epithelium of monkeys treated with homologous FSH over a 28-day period, resulting subsequently in increased numbers of spermatocytes and spermatids (van Alphen et al., 1988). Taken together, these results suggest that the hormonal regulation in the primate and rat may indeed be fundamentally different.

Sharpe (1989), however, recently suggested that "the differences between rats and primates may not be so great as they first appear." What are the similarities? First, the administration of testosterone to hypophysectomized (Marshall et al., 1986) or stalk-sectioned (Marshall et al., 1983) monkeys has been shown to maintain spermatogenesis qualitatively but not quantitatively, which also is the case with hypophysectomized rats (Sun et al., 1989; Santulli et al., 1990). Second, the observation that there is synergistic interaction of testosterone and FSH in maintaining spermatogenesis in hypophysectomized adult rats (Russell and Clermont, 1977; Bartlett et al., 1989a; Sharpe, 1989) is consistent with reports indicating that the maintenance of primate spermatogenesis requires FSH (above). Third, there is evidence that FSH has an effect on the numbers of spermatogonia and thus on the numbers of cells that enter meiosis in both primates (van Alphen et al., 1988) and rats (Steinberger, 1971).

Based on this evidence, Sharpe (1989) postulated that in both rats and primates, FSH might determine the number of cells that are available to enter meiosis, and that testosterone might then be responsible for meiosis and the subsequent development of mature spermatids. Sharpe (1989) stated: "Therefore, FSH on its own would be ineffective in maintaining spermatogenesis as continued development of the later germ cells would fail due to the absence of androgen support . . . (and) testosterone on its own would be effective in maintaining some degree of spermatogenesis . . . as the supply of differentiated spermatogonia is only reduced, not abolished, when FSH levels are suppressed." According to this thesis, the primary difference between rats and primates would be the *extent* to which FSH is involved in the regulation of spermatogenesis, with the dependence on FSH greater in primates than in the rat. Furthermore, FSH would not be required for the qualitative maintenance of spermatogenesis in either case, but might be required for its quantitative maintenance in both cases. This hypothesis is worthy of critical examination.

IV. Restoration of Spermatogenesis in the Adult Mammal

Spermatogenesis can be suppressed in the mammalian testis by the withdrawal of endogenous LH. As with spermatogenic maintenance, a great deal of attention has been paid to elucidating the gonadotropin and/or androgen requirements to restore spermatogenesis, both qualitatively and quantitatively. Unfortunately, considerable uncertainty also remains about these requirements.

In this section, we review the current status of our knowledge of the involvement of gonadotropins and androgens in the qualitative and quantitative restoration of spermatogenesis. We focus again on the rat, but later make comparisons between the rat and other species, particularly the human.

A. Restoration of Spermatogenesis After Posthypophysectomy Testicular Regression

Luteinizing hormone withdrawal, which causes spermatogenesis suppression and testicular regression, can be accomplished by a variety of methods, including: hypophysectomy, steroid administration, GnRH agonists, GnRH antagonists, or immunization against LH or GnRH. For example, a well-established effect of hypophysectomy in the rat is the rapid loss of advanced spermatids as well as the loss of substantial numbers of earlier germ cell types from the seminiferous epithelium (Clermont and Harvey, 1967; Chowdhury and Tcholakian, 1979), resulting from the inhibition of gonadotropin-dependent Leydig cell testosterone production. Figure 8–7, based on data in Clermont and Morgentaler (1955), illustrates the

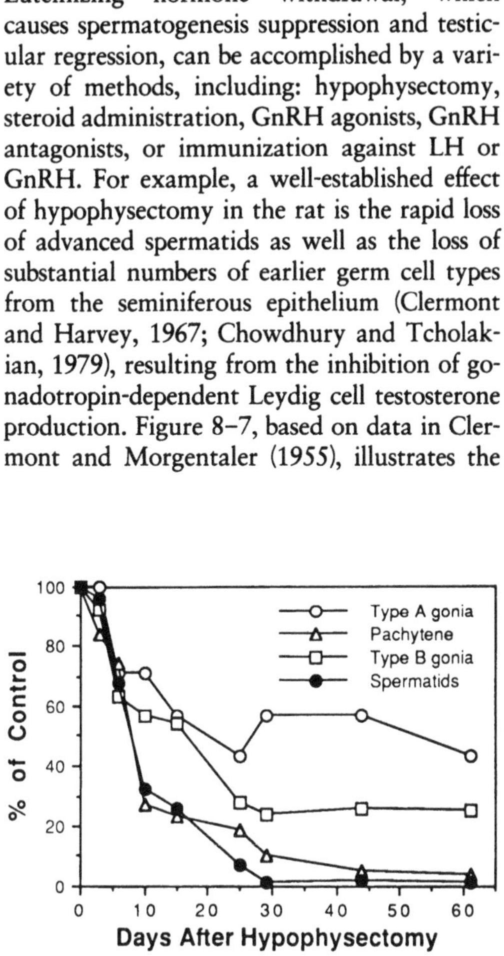

Figure 8–7. Illustration of the timing and extent of the loss of particular spermatogenic cell types (type A spermatogonia, pachytene spermatocytes, type B spermatogonia, and steps 1–7 spermatids) from the adult rat testis after hypophysectomy. (Drawn from data presented in Clermont and Morgentaler, 1955.)

timing and extent of the loss of particular spermatogenic cell types from the adult rat testes after hypophysectomy. Figure 8–7 shows rapid, extensive losses of steps 1–7 spermatids and pachytene spermatocytes, a somewhat slower and less extensive loss of type B spermatogonia, and a relatively slow loss of type A spermatogonia.

Numerous studies have examined the effects of androgen and of FSH replacement on the restoration of cells in the rat testes after periods of posthypophysectomy regression. In some studies, the administration of LH, testosterone, or FSH individually to such rats failed to restore any advanced spermatogenic cells to the testes (Lostroh, 1976). A number of studies, however, have shown clearly that advanced spermatogenic cells can be restored to the testes of chronically hypophysectomized rats with exogenously administered testosterone alone (Buhle et al., 1982; Huang et al., 1987; Awoniyi et al., 1990), although not necessarily to all rats or to all seminiferous tubules of particular rats. Finally, combinations of LH, testosterone, and FSH consistently have been shown to restore at least some advanced cells to posthypophysectomy regressed rat testes, suggesting that synergistic interaction or sequential effects of LH (or testosterone) and FSH may be required for the reinitiation and/or completion of spermatogenesis (Lostroh, 1976; Elkington and Blackshaw, 1974; Huang et al., 1987). As yet, there have been no studies in which spermatogenesis was restored *quantitatively* in chronically hypophysectomized rats, although some studies presented evidence that greater numbers of spermatogenic cells were restored to the posthypophysectomy regressed rat testis by the administration of both FSH and testosterone than by the administration of testosterone alone (Elkington and Blackshaw, 1974).

It is obvious from this discussion that the results of studies of the hormonal requirements to restore quantitatively normal spermatogenesis in hypophysectomized adult rats have been highly variable. What accounts for this variability? Among the possible explanations are the following: (1) The time between hypophysectomy and hormone replacement therapy has varied widely from study to study, and this timing probably has important consequences for the success of restoring spermatogenesis. (2)

Relatively few studies used quantitative methods to assess the roles of specific hormones in the restoration of spermatogenesis. (3) There have been few studies in which spermatogenesis restoration was evaluated with respect to the dose of the administered hormones; and in most studies, no indication was given as to how much of the administered hormone (testosterone, for example) was delivered to the testis. (4) The FSH preparations used in many studies undoubtedly contained significant LH contamination, thus confounding the effects of FSH from those of LH and testosterone. (5) It is still not clear how best to administer FSH (by injection? by continuous administration? by pulsatile administration?); and most studies did not consider whether exogenously administered FSH retained its biologic activity, particularly when the administered FSH was from another species.

At this juncture, therefore, what can be concluded about the hormonal requirements to restore spermatogenesis after hypophysectomy-induced spermatogenesis suppression? It is clear that the restoration of spermatogenesis depends at minimum on testosterone because the administration of testosterone to azoospermic, hypophysectomized rats in some experiments was able to restore spermatogenesis at least qualitatively. However, for all the reasons presented earlier, and particularly because of possible difficulties with the administration of FSH and other pituitary factors, determination of the hormonal requirements for restoring spermatogenesis quantitatively in the posthypophysectomy rat testis has not been accomplished. Indeed, recent studies have approached the issue of the hormonal requirements to restore spermatogenesis in testes caused to regress not by hypophysectomy but by more selective elimination of pituitary factors. These studies are described next. We then return to further consider the requirements of restoring spermatogenesis after hypophysectomy.

B. Restoration of Spermatogenesis in Testes Caused to Regress by Inhibition of Endogenous Luteinizing Hormone but not Follicle-Stimulating Hormone

In a recent study, Awoniyi and colleagues (1989b) examined the extent to which testosterone was able to restore spermatogenesis in rat testes caused to regress by withdrawal of endogenous LH but not FSH. To this end, rats were implanted with PDS capsules containing testosterone and estradiol. The rationale for using this combination was that testosterone and estradiol implants had been shown to act synergistically to reversibly inhibit spermatogenesis by suppressing LH production without affecting serum testosterone, estradiol, or FSH concentrations (Ewing et al., 1977; Robaire et al., 1979). Because endogenous FSH was unaffected after testosterone–estradiol administration, the use of these implants provided an experimental model in which attention could be focused on the testosterone requirements for restoration of spermatogenesis.

As illustrated in Figure 8–8, Awoniyi and colleagues (1989b) showed that azoospermia occurred in a period of 8 weeks in adult rats that received a subdermal, 2.5-cm testosterone-filled PDS capsule and a 0.1-cm estradiol-filled capsule. Removal of the implants at week 8 resulted in the quantitative restoration of spermatogenesis by 8 weeks later (Fig. 8–8). Next, rats induced to become azoospermic with testosterone–estradiol PDS implants for 8 weeks were subsequently administered additional tes-

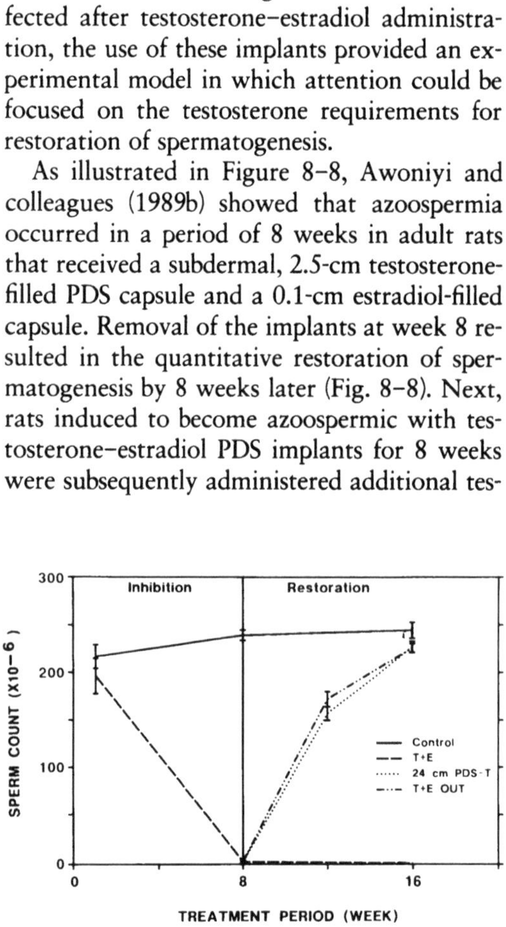

Figure 8–8. Effects on the number of advanced spermatids per testis (Sperm Count) of implanting adult rats with polydimethylsiloxane (PDS) capsules containing testosterone and estradiol (T+E) for 8 weeks (denoted "Inhibition"), and then removing the implants (T+E OUT) or administering additional testosterone-containing capsules (+24 cm PDS-T) to the implanted rats for another period of 8 weeks (denoted "Restoration"). (From Awoniyi et al., 1989b.)

tosterone-filled PDS implants (24 cm in total length), while the original implants were left in place. Figure 8–8 shows that spermatogenesis was restored quantitatively in these rats 8 weeks after the additional implants were administered. Thus, Awoniyi and associates (1989b) succeeded in quantitatively restoring advanced spermatogenic cells to azoospermic testes in rats in which endogenous steroidogenic function had been suppressed by suppression of LH but in which endogenous FSH had been unaffected. Measurements of testosterone concentration in testicular fluids (Fig. 8–9) showed that the testosterone concentration in the seminiferous tubule fluid of the rats that had received testosterone–estradiol implants for 8 weeks was reduced by more than 80 percent compared with that in controls (from approximately 60 to 10 ng/mL), was restored to control values 8 weeks after the removal of the implants, but rose only to 40 percent of the control value (about 24 ng/mL) in implanted rats that received additional testosterone.

Taken together, these data showed that advanced spermatogenic cell numbers could be quantitatively restored to the implanted rats with only 40 percent of the seminiferous tubule fluid testosterone concentration of controls, indicating for the first time that the quantitative restoration of spermatogenesis in the rat does not require the high testosterone concentration found in controls. These results were reminiscent of data discussed earlier showing that the quantitative maintenance of spermatogenesis similarly requires only a fraction of the seminiferous tubule fluid testosterone concentration of intact rats. Again, it must be emphasized that the testosterone concentration found able to restore spermatogenesis (24 ng/mL) is about 10 times greater than the concentration in blood serum, and represents a molar concentration (8.3×10^{-8} M) that is far in excess of the concentration required to saturate testicular androgen receptors. As yet, we have no explanation for this observation.

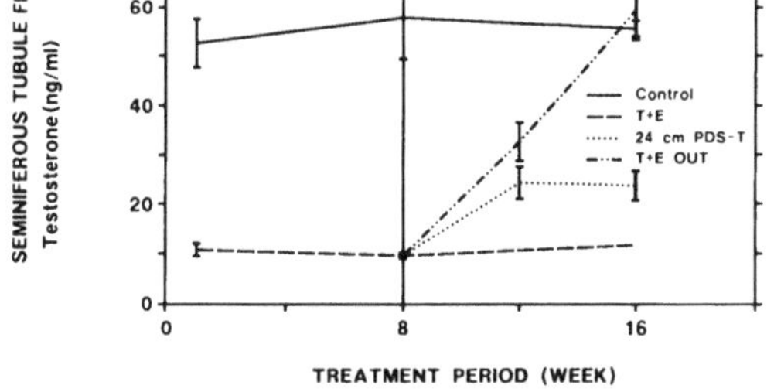

Figure 8–9. Measurements of testosterone concentration in the seminiferous tubule fluid of rats: (a) that received testosterone–estradiol (T+E) implants for 8 weeks; (b) that received T+E implants for 8 weeks and then whose implants were removed for 8 weeks (T+E out); and (c) that received T+E implants for 8 weeks and then received additional testosterone-containing implants for 8 weeks (+24 cm PDS-T). (From Awoniyi et al., 1989b.)

C. Comparison of Testosterone Effects on Spermatogenesis Restoration in Testes Caused to Regress by Hypophysectomy Versus by Withdrawal of Luteinizing Hormone Alone

As discussed earlier, for a number of reasons, previous studies may have failed to quantitatively restore spermatogenesis in rat testes made azoospermic by hypophysectomy. Among these, an obvious, testable possibility is that insufficient testosterone was delivered to the testes in those studies. To test this, we designed an experiment to compare the effects of comparable, increasing doses of testosterone on the restoration of spermatogenesis in rats made azoospermic by hypophysectomy or by testosterone–estradiol implantation of intact rats (Awoniyi et al., 1990). In particular, we critically compared the extent to which comparable, increasing testosterone concentrations within the seminiferous tubule fluid were able to restore advanced spermatogenic cells in rats made azoospermic by withdrawal of all pituitary hormones (hypophysectomy) or of LH alone (testosterone–estradiol implantation).

Rats made azoospermic by hypophysectomy for 4 weeks or by testosterone–estradiol implantation for 8 weeks received testosterone-filled PDS implants of 4 to 24-cm length for an additional 8 weeks. In both the hypophysectomized and testosterone–estradiol implanted rats, *qualitative* restoration of spermatogenesis was seen at testosterone concentrations of 5 to 10 ng/mL; that is, 8 weeks after testosterone was first administered, advanced spermatogenic cells were seen in the testes of both groups. In both groups, the numbers of advanced spermatids increased with increasing doses of testosterone, but, as illustrated in Figure 8–10, the extent to which advanced spermatids were restored differed significantly in the two groups. In the testosterone–estradiol implanted rats, spermatogenesis was *quantitatively* restored to control values with the 18- and 24-cm additional testosterone-filled capsules; this occurred at seminiferous tubule fluid testosterone concentrations of 20 to 25 ng/mL. Although capsules of these sizes produced comparable testosterone concentrations in the seminiferous tubule fluid of the hypophysectomized rats, the numbers of advanced spermatids restored to the testes of these rats were only 20 percent of the control value.

The explanation for this difference in the testosterone responsiveness of the seminiferous epithelium in the two groups of rats is not known, but several possibilities occurred to us: *First,* the absence of FSH in the hypophysectomized rats and its presence in rats implanted with testosterone–estradiol capsules suggest that FSH might be required for the restoration of spermatogenesis in the posthypophysectomy-regressed rat testis (Huang et al., 1987). *Second,* if the absence of pituitary factors (? FSH) in the hypophysectomized rats progressively eliminates spermatogonial stem cells (Sharpe, 1989), the restoration of spermatogenesis with testosterone would be more or less successful depending on the duration of hypophysectomy. If true, this might also explain some of the discrepant results in the literature regarding the extent to which spermatogenesis was restored in testosterone-replaced hypophysectomized rats. *Third,* pituitary factors in addition to LH and FSH, also absent in hypophysectomized rats, might be required for the restoration of spermatogenesis. For example, prolactin may potentiate the action of testosterone on spermatogenesis in the testosterone–estradiol implanted rats and thus may play a role in the quantitative restoration of spermatogenesis in these rats but not in hypophysectomized rats in which prolactin is absent. *Fourth,* there is now a great deal of evidence for the existence of local paracrine factors in the mammalian testis that modulate the function of various germ cells. It is possible that the production of these factors depends on pituitary factors that are removed by hypophysectomy but not by testosterone–estradiol implants.

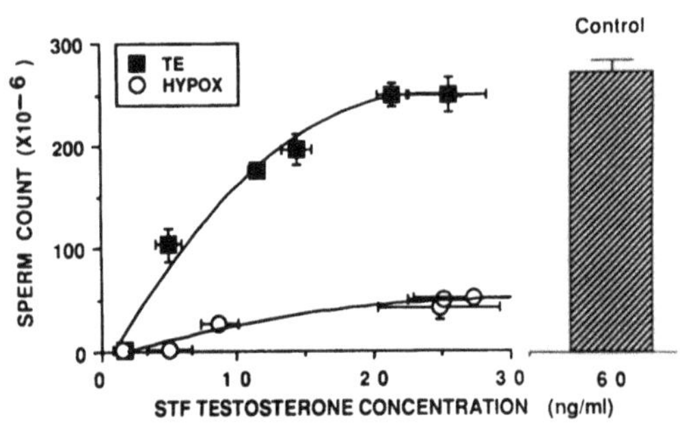

Figure 8–10. Relationship between the mean number of advanced spermatids per testis (Sperm Count) and the mean seminiferous tubule fluid (STF) testosterone concentration in rats that first were made azoospermic by testosterone–estradiol implants (TE) or hypophysectomy (HYPOX) and subsequently that received testosterone-filled polydimethylsiloxane implants of increasing sizes for 8 weeks. (From Awoniyi et al., 1990.)

It seemed most likely to us that FSH, in addition to testosterone, was required for the quantitative restoration of spermatogenesis and therefore, that the explanation for the differential effect of testosterone spermatogenesis restoration in hypophysectomized rats and rats that had received testosterone–estradiol implants was that the former rats lacked FSH, but that FSH levels in the testosterone–estradiol rats were not significantly different from the level in controls. Awoniyi and associates (1989a) tested this possibility by examining the ability of testosterone to restore spermatogenesis to azoospermic rats in which LH alone or both LH and FSH had been suppressed immunologically. To this end, rats were actively immunized against LH or GnRH to cause azoospermia, and then were administered testosterone by PDS implants. The objective was to determine whether exogenously administered testosterone could elicit the quantitative restoration of advanced spermatogenic cells in azoospermic rats in which LH alone or both LH and FSH had been suppressed. With this design, any potential problems associated with the long-term administration of FSH were avoided.

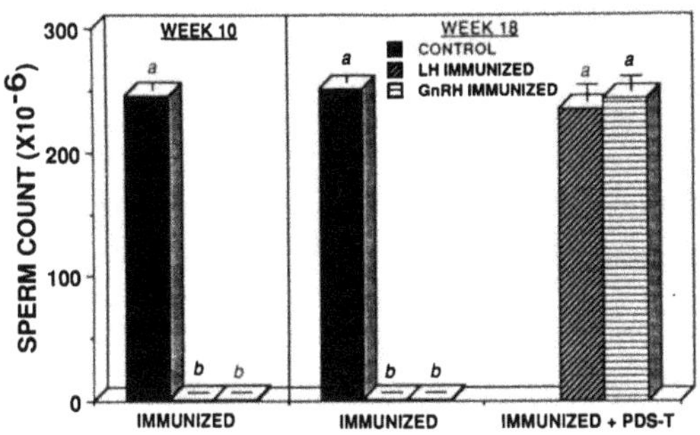

Figure 8–11. Testicular advanced spermatid content (Sperm Count) of control rats and LH- and GnRH-immunized rats at weeks 10 and 18; and of LH- and GnRH-immunized rats that received testosterone-filled polydimethylsiloxane capsules from weeks 10–18 of immunization. Different letters above bars indicate significant differences ($P < 0.01$). (From Awoniyi et al., 1989a.)

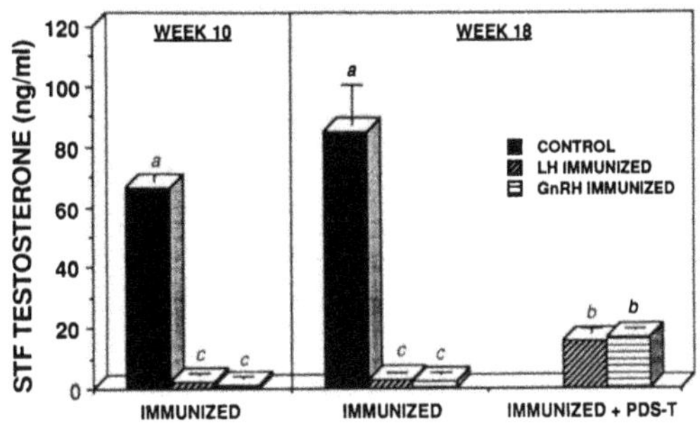

Figure 8–12. Testicular seminiferous tubule fluid (STF) testosterone concentration of control rats and LH- and GnRH-immunized rats at weeks 10 and 18; and of LH- and GnRH-immunized rats that received testosterone-filled polydimethylsiloxane capsules from weeks 10–18 of immunization. Different letters above bars indicate signficant differences ($P < 0.01$). (From Awoniyi et al., 1989a.)

As illustrated in Figure 8–11, active immunization of rats against LH or GnRH for 10 weeks resulted in azoospermia in both cases. Awoniyi and co-workers (1989b) then administered 24-cm testosterone capsules to both groups of immunized rats. Not surprisingly, spermatogenesis was quantitatively restored by 8 weeks when the LH immunized rats received testosterone supplementation (Fig. 8–11). Most surprisingly, however, the quantitative restoration of spermatogenesis also was achieved in the GnRH-immunized rats that received testosterone supplementation (Fig. 8–11). These latter results suggested that FSH in fact may not be required for the quantitative restoration of spermatogenesis in the rat, provided that sufficient concentrations of testosterone are present within the testes. Examination of the intratesticular testosterone concentrations in these rats (Fig. 8–12) revealed that the initial LH or GnRH immunizations reduced testosterone in the seminiferous tubule fluid to extremely low values compared with controls, as expected, and that implantation of the 24-cm testosterone-filled capsules caused testosterone concentrations of both the LH and the GnRH-immunized rats to rise significantly to about 20 ng/mL, approximately the same concentration as found to be required to maintain spermatogenesis quantitatively in intact rats (Zirkin et al., 1989).

Taken together, these unexpected results demonstrated for the first time that testosterone alone is capable of restoring quantitatively normal spermatogenesis in the regressed testes of rats in which there was no radioimmunoassayable serum FSH. A caveat that should be noted, however, is the possibility that small amounts of *bioactive* FSH may be all that is required for spermatogenesis restoration (Storring et al., 1981); we did not assay for *bioactive* FSH. This caveat aside, the results of Awoniyi and colleagues (1989a) were in contrast with numerous reports in the literature that have suggested that the restoration of spermatogenesis in the hypophysectomy-induced regressed rat testes requires FSH as well as testosterone. The obvious question raised by this finding is: Why has the administration of testosterone in previous studies of hypophysectomized rats failed to restore spermatogenesis quantitatively? At this juncture, we simply do not know.

D. Are There Species Differences in the Hormonal Regulation of the Restoration of Spermatogenesis?

The relative contributions of LH, FSH, and testosterone to the restoration of spermatogenesis are unclear for any species, including the rat. However, there are reports of species differences in the requirements to restore spermatogenesis, and therefore, conclusions derived from studies of the rat may not be applicable to mammals in general. In this part of the review, we discuss the hormonal requirements for the restoration of spermatogenesis in other species, primarily in humans because of the substantial amounts of recent data on human spermatogenesis.

An experimental approach used for restoration studies in humans has been to examine the effects of selective gonadotropin replacement in men with experimentally induced hypogonadotropic hypogonadism. Exogenously administered testosterone of particular doses in humans can suppress both FSH and LH, resulting in severe oligospermia or azoospermia. Matsumoto (1989) reported that in men made azoospermic in this way, the administration of human chorionic gonadotropin (hCG) or human luteinizing hormone (hLH) caused marked increases in sperm production, although serum FSH levels remained undetectable. These studies demonstrated that FSH was not required for the *qualitative* restoration of spermatogenesis in men who had been deprived of gonadotropins for periods of 4 to 9 months. This result is consistent with the report that FSH is also not required for the qualitative restoration of spermatogenesis in hypophysectomized rats (Awoniyi et al., 1990) or in rats actively immunized against GnRH (Awoniyi et al., 1989a). When human FSH was administered to men made azoospermic by exogenously administered testosterone (Matsumoto, 1989), however, sperm production was stimulated, again resulting in the qualitative but not quantitative restoration of spermatogenesis. These results indicated a synergistic interaction of FSH and testosterone *at the doses that were used.* Some studies of rats also have indicated a synergistic interaction of FSH and testosterone on spermatogenesis, as outlined previously.

Matsumoto (1989) addressed more directly the possibility of synergistic interactions of testosterone and FSH on spermatogenesis restoration in humans by administering hCG to men to induce azoospermia and then administering hFSH to these men. This combined treatment restored sperm numbers *to control values,* demonstrating that the restoration of *quantitatively* normal spermatogenesis was possible in azoospermic humans if appropriate doses of both FSH and LH (and, therefore, testosterone) were administered. These results do not necessarily mean that the quantitative restoration of spermatogenesis in the human *requires* both FSH and testosterone, however. For example, dose-response studies with testosterone comparable with those conducted with rats (Awoniyi et al., 1986b) have not been conducted in humans, and therefore, it is not known whether spermatogenesis might be restored in humans in the absence of FSH if high enough testosterone doses were used. Thus, at this juncture, whether or not there are significant differences in the mechanism by which spermatogenesis is restored in rats and humans remains uncertain.

V. Summary

The major issues that received attention in this review are summarized below.

1. Although six decades have passed since the discovery by Smith (1930) that the pituitary plays a central role in the regulation of spermatogenesis, careful reading of the literature indicates that there is much left to be confirmed or learned about the roles of gonadotropins and androgens in this process.
2. There is considerable confusion in the literature resulting from the failure of numerous papers to distinguish carefully between the *maintenance* and the *restoration* of spermatogenesis, and between the *qualitative* and *quantitative* effects of hormones on spermatogenesis.
3. There is compelling evidence for far higher concentrations of testosterone within the interstitium and the seminiferous tubules of rat testes than in the periphery, suggesting either that androgen-dependent organs, such as the prostate, require far less androgen for their maintenance than do the testes, or that far less testosterone is needed within the testes to maintain spermatogenesis. Indeed, evidence is discussed that strongly supports the contention that established spermatogenesis can be quantitatively maintained

in the intact adult rat at only a fraction of the intratesticular testosterone concentration found in intact controls. The minimum intratesticular testosterone concentration that apparently is required to maintain spermatogenesis, however, is much greater than the testosterone concentration required to saturate testicular androgen receptors, thus calling into question the mechanism by which testosterone functions to maintain spermatogenesis.

4. Evidence for the possibility that testosterone may become concentrated within compartments of the testes even under circumstances in which there is little or no local production of testosterone by Leydig cells is discussed.
5. Whereas the administration of testosterone to intact rats lacking LH, but not FSH, can maintain spermatogenesis quantitatively, the administration of comparable amounts of testosterone to hypophysectomized rats fails to maintain spermatogenesis quantitatively. This, in combination with evidence that testosterone and FSH may act synergistically in the adult rat to stimulate germ cell development and/or to inhibit germ cell degeneration, suggests that FSH and perhaps other pituitary factors, in addition to testosterone, may be involved in the maintenance of spermatogenesis in rats. Evidence for the involvement of FSH in regulating the maintenance of spermatogenesis in primates is also discussed.
6. Spermatogenesis can be quantitatively *restored* in rats made azoospermic by suppressing LH but not FSH. As with the maintenance of spermatogenesis, spermatogenesis restoration does not require the high testosterone concentrations found in the testes of control rats.
7. In contrast, spermatogenesis can be only qualitatively, not quantitatively, restored to rats made azoospermic by hypophysectomy. This indicates a significant difference in the responsiveness of the seminiferous epithelium of rats lacking LH versus rats lacking all pituitary factors.
8. The administration of testosterone alone is capable of restoring quantitatively normal spermatogenesis in the azoospermic testes of rats in which both LH and FSH are suppressed. This calls into question the numerous reports in the literature that have suggested that the restoration of spermatogenesis in the regressed rat testis requires FSH as well as testosterone.

REFERENCES

Almiron, I., Domene, H., and Chemes, H. E. (1984). The hormonal regulation of premeiotic steps of spermatogenesis in the newborn rat. J. Androl. 5:235–242.

Awoniyi, C. A., Sprando R. L., Santulli, R., Chandrasheker, V., Ewing, L. L., and Zirkin, B. R. (1990). Restoration of spermatogenesis by exogenously administered testosterone in rats made azoospermic by hypophysectomy or by withdrawal of luteinizing hormone alone. Endocrinology 127:177–184.

Awoniyi, C. A., Santulli, R., Chandrashekar, V., Schanbacher, B. D., and Zirkin, B. R. (1989a). Quantitative restoration of advanced spermatogenic cells in adult male rats made azoospermic by active immunization against luteinizing hormone or gonadotropin-releasing hormone. Endocrinology 125:1303–1309.

Awoniyi, C. A., Santulli, R., Sprando, R. L., Ewing, L. L., and Zirkin, B. R. (1989b). Restoration of advanced spermatogenic cells in the experimentally regressed rat testis: Quantitative relationship to testosterone concentration within the testis. Endocrinology 124:1217–1223.

Bardin, C. W., Cheng, C. Y., Musto, N. A., and Gunsalus, G. L. (1988). The Sertoli cell, in *The Physiology of Reproduction* (Knobil, E., and Neill, J., eds.) pp. 933–974, Raven Press, New York.

Bartlett, J.M.S., Weinbauer, G. F., and Nieschlag, E. (1989a). Differential effects of FSH and testosterone on the maintenance of spermatogenesis in the adult hypophysectomized rat. J. Endocrinol. 121:49–58.

Barlett, J.M.S., Weinbauer, G. F., and Nieschlag, E. (1989b). Quantitative analysis of germ cell numbers and relation to intratesticular testosterone following vitamin A-induced synchronization of spermatogenesis in the rat. J. Endocrinol. 123:403–412.

Bellvé, A. R., and Zheng, W. (1989). Growth factors as autocrine and paracrine modulators of male gonadal functions. J. Reprod. Fertil. 85:771–793.

Brown, T. R., and Migeon, C. J. (1987). Androgen insensitivity syndromes: Paradox of phenotypic feminization with the male genotype and normal testicular androgen secretion, in *Hormone Resistance And Other Endocrine Parodoxes* (Cohen, M. P., and Foa, P. O.,

eds.) Vol. 1, pp. 157–203, Springer-Verlag, New York.
Buhl, A. E., Cornette, J. C., Kirton, K. T., and Yuan, Y.-D. (1982). Hypophysectomized male rats treated with polydimethylsiloxane capsules containing testosterone: Effects on spermatogenesis, fertility and reproductive tract concentrations of androgens. Biol. Reprod. 27:183–188.
Chemes, H. E., Dym, M., and Raj, H.G.M. (1979). The role of gonadotropins and testosterone on initiation of spermatogenesis in the immature rat. Biol. Reprod. 21:241–249.
Chowdhury, A. K., and Tcholakian, R. K. (1979). Effects of various doses of testosterone propionate on intratesticular and plasma testosterone levels and maintenance of spermatogenesis in adult hypophysectomized rats. Steroids 34:151–162.
Clermont, Y., and Harvey, S. C. (1967). Effects of hormones on spermatogenesis in the rat. Ciba Found. Colloq. Endocrinol. 16:173–196.
Clermont, Y., and Perey, B. (1957). Quantitative study of the cell population of the seminiferous tubules in immature rats. Am. J. Anat. 100:241–267.
Clermont, Y., and Morgentaler, H. (1955). Quantitative study of spermatogenesis in the hypophysectomized rat. Endocrinology 57:369–382.
Comhaire, F. H., and Vermeulen, A. (1976). Testosterone concentration in the fluids of seminiferous tubules, the interstitium and the rete testis of the rat. J. Endocrinol. 70:229–235.
Cunningham, G. R., and Huckins, C. (1979). Persistence of complete spermatogenesis in the presence of low intratesticular concentrations of testosterone. Endocrinology 105:177–186.
de Kretser, D. M., and Kerr, J. B. (1988). The cytology of the testis, in *The Physiology of Reproduction* (Knobil, E., and Neill, J., eds.) pp. 837–932, Raven Press, New York.
Dym, M., Raj, H.G.M., Lin, Y. C., Chemes, H. E., Kotite, N. J., Nayfeh, S. N., and French, F. S. (1979). Is FSH required for maintenance of spermatogenesis in adult rats? J. Reprod. Fertil. (Suppl) 26:175–181.
Elkington, J.S.H., and Blackshaw, A. W. (1947). Studies in testicular function. I. Quantitative effects of FSH, LH, testosterone and dihydrotestosterone on restoration and maintenance of spermatogenesis in the hypophysectomized rat. Aust. J. Biol. Sci. 27:47–57.
Ewing, L. L., and Zirkin, B. R. (1983). Leydig cell structure and steroidogenic function. Recent Progr. Horm. Res. 39:599–635.
Ewing, L. L., Davis, J. C., and Zirkin, B. R. (1980). Regulation of testicular function: A spatial and temporal view. Int. Rev. Physiol. (Greep, R. O., ed.) Vol. 22, pp. 41–115, University Park Press, Baltimore, Md.
Ewing, L. L., Desjardins, C., Irby, D. C., and Robaire, B. (1977). Synergistic interaction of testosterone and oestradiol inhibits spermatogenesis in rats. Nature 269:409–411.
Ewing, L. L., Schanbacher, B., Desjardins, C., and Chaffee, V. (1976). The effect of subdermal testosterone filled polydimethylsiloxane implants on spermatogenesis in rhesus monkeys *(Macaca mulata)*. Contraception 13:583–596.
Ewing, L. L., Desjardins, C., and Stratton, L. G. (1973). Testosterone polydimethylsiloxane implants and contraception in male rabbits, in *Temporal Aspects of Therapeutics* (Urquhart, J., and Yates, F. E., eds.) pp. 165–180, Plennum Press, New York.
Fritz, I. B. (1978). Sites of action of androgens and follicle stimulating hormone on cells of the seminiferous tubule. Biochem. Actions Hormones V:249–281.
Gundersen, H.J.G. (1986). Stereology of arbitrary particles. A review of unbiased number and size estimators and the presentation of some new ones. J. Microsc. 143:3–45.
Hansson, V., Weddington, J.S.C., McLean, W. S., Smith, A. A., Nayfeh, S. N., French, F. S., and Ritzen, E. M. (1975). Regulation of seminiferous tubular function by FSH and androgen. J. Reprod. Fertil. 44:363–375.
Harris, M. E., and Bartke, A. (1975). Maintenance of rete testis fluid testosterone and dihydrotestosterone levels by pregnenolone and other C_{21} steroids in hypohysectomized rats. Endocrinology 96:1396–1402.
Huang, H.F.S., and Boccabella, A. V. (1988). Dissociation of qualitative and quantitative effects of the suppression of testicular testosterone upon spermatogenesis. Acta Endocrinol. 118:209–217.
Huang, H.F.S., Marshall, G. R., Rosenberg, R., and Nieschlag, E. (1987). Restoration of spermatogenesis by high levels of testosterone in hypophysectomized rats after long-term regression. Acta Endocrinol. 116:433–444.
Huang, H.F.S., and Nieschlag, E. (1986). Suppression of the intratesticular testosterone is associated with quantitative changes in spermatogonial populations in intact adult rats. Endocrinology 118:619–627.
Hugly, S., Roberts, K., and Griswold, M. (1988). Transferrin and sulfated glycoprotein-2 messenger ribonucleic acid levels in the testis and isolated Sertoli cells of hypophysectomized rats. Endocrinology 122:1390–1396.
Isomaa, V., Parvinen, M., Janne, O. A., and Bardin, C. W. (1985). Nuclear androgen receptors in different stages of the seminiferous epithelial cycle and the interstitial tissue of rat testis. Endocrinology 116:132–137.
Johnson, L., and Tatum, M. E. (1989). Temporal appearance of seasonal changes in numbers of Sertoli cells, Leydig cells, and germ cells in stallions. Biol. Reprod. 40:994–999.
Johnson, L., and Thompson, D. L. (1983). Age-related and seasonal variation in the Sertoli cell population, daily sperm production and serum concentrations of follicle-stimulating hormone, luteinizing hormone and testosterone in stallions. Biol. Reprod. 29:777–789.
Keeney, D. S., Mendis-Handagama, S.M.L.C., Zirkin, B. R., and Ewing, L. L. (1988). Effect of long-term deprivation of luteinizing hormone on Leydig cell volume, Leydig cell number, and steroidogenic capacity of the rat testis. Endocrinology 123:2906–2915.
Lostroh, A. J. (1976). Hormonal control of spermatogenesis, in *Regulatory Mechanisms of Male Reproductive Physiology* (Spilman, C. H., Lobl, T. J., and Kirton, K. T., eds.) pp. 13–27, Excerpta Medical, Amsterdam.
Maddocks, S., and Setchell, B. P. (1989a). Testosterone concentrations in testicular interstitial fluid collected with a push-pull cannula or by drip-collection from adult rats given testosterone or aminoglutethimide. J. Endocrinol. 121:303–309.

Maddocks, S., and Setchell, B. P. (1989b). Effect of a single injection of human chorionic gonadotrophin on testosterone levels in testicular interstitial fluid, and in testicular and peripheral blood in adult rats. J. Endocrinol. 121:311–316.

Maddocks, S., and Sharpe, R. M. (1989). Dynamics of testosterone secretion by the rat testis: Implications for measurement of the intratesticular levels of testosterone. J. Endocrinol. 122:323–329.

Marshall, G. R., Jockenhovel, F., Ludecke, D., and Nieschlag, E. (1986). Maintenance of complete but quantitatively reduced spermatogenesis in hypophysectomized monkeys by testosterone alone. Acta Endocrinol. 133:424–431.

Marshall, G. R., Wickings, E. J., Ludecke, D. K., and Nieschlag, E. (1983). Stimulation of spermatogenesis in stalk-sectioned Rhesus monkeys by testosterone alone. J. Clin. Endocrinol. 57:152–159.

Matsumoto, A. M. (1989). Hormonal control of human spermatogenesis, in *The Testis,* 2nd ed. (Burger, H., and deKretser, D., eds.) pp. 181–196, Raven Press, New York.

Means, A. R., Fakunding, J. L., Huckins, C., Tindall, D. J., and Vitale, R. (1976). Follicle-stimulating hormone, the Sertoli cell, and spermatogenesis. Recent Prog. Horm. Res. 32:477–527.

Mendis-Handagama, S.M.L.C., Keeney, D. S., Hardy, M. P., and Ewing, L. L. (1989). Application of the disector method to enumerate cells in the testis. Ann. N.Y. Acad. Sci. 564:86–98.

Morales, C. R., Alcivar, A. A., Hecht, N. B., and Griswold, M. D. (1989). Specific mRNAs in Sertoli and germinal cells of testes from stage synchronized rats. Mol. Endocr. 3:725–733.

Morales, C., Hugly, S., and Griswold, M. D. (1987). Stage-dependent levels of specific mRNA transcripts in Sertoli cells. Biol. Reprod. 36:1035–1046.

Morales, C., and Griswold, M. D. (1987). Retinol-induced stage synchronization in seminiferous tubules of the rat. Endocrinology 121:432–434.

Murty, G.S.R.C., Sheela Rani, C. S., Moudgal, N. R., and Prasad, M.R.N. (1979). Effect of passive immunization with specific antiserum to FSH on the spermatogenic process and fertility of adult male bonnet monkeys *(Macaca radiata).* J. Reprod. Fertil. (Suppl) 26:147–163.

Parvinen, M. (1982). Regulation of the seminiferous epithelium. Endocrinol. Rev. 3:404–417.

Perey, B., Clermont, Y., and Leblond, C. P. (1961). The wave of the seminiferous epithelium in the rat. Am. J. Anat. 108:47–48.

Rea, M. A., Marshall, G. R., Weinbauer, G. F., and Nieschlag, E. (1986). Testosterone maintains pituitary and serum FSH and spermatogenesis in gonadotrophin-releasing hormone antagonist-suppressed rats. J. Endocrinol. 108:101–107.

Risbridger, G. P., and de Kretser, D. M. (1989). Paracrine regulation of the testis, in *The Testis,* 2nd ed. (Burger, H., and de Kretser, D. M., eds.) pp. 255–268, Raven Press, New York.

Ritzen, E. M., Hansson, V., and French, F. S. (1989). The Sertoli cell, in *The Testis,* 2nd ed. (Burger, H., and de Kretser, D. M., eds.) pp. 269–302, Raven Press, New York.

Robaire, B., and Zirkin, B. R. (1981). Hypophysectomy and simultaneous testosterone replacement effects on male rat reproductive tract and epididymal Δ^4-5α-reductase and 3α-hydroxysteroid dehydrogenase. Endocrinology 109:1225–1233.

Robaire, B., Ewing, L. L., Irby, D. C., and Desjardins, C. (1979). Interactions of testosterone and estradiol-17β on the reproductive tract of the male rat. Biol. Reprod. 21:455–463.

Robb, G. W., Amann, R., and Killian, K. J. (1978). Daily sperm production and epididymal sperm reserves of pubertal and adult rats. J. Reprod. Fertil. 54:103–107.

Roberts, K., and Griswold, M. D. (1989). Testosterone induction of cellular proteins in cultured Sertoli cells from hypophysectomized rats and rats of different ages. Endocrinology 125:1174–1179.

Rommerts, F.F.G. (1988). How much androgen is required for maintenance of spermatogenesis? J. Endocrinol. 116:7–9.

Russell, L. D., and Clermont, Y. (1977). Degeneration of germ cells in normal, hypophysectomized and hormone treated hypophysectomized rats. Anat. Rec. 187:347–355.

Saez, J. M., Perrard-Sapori, M. H., Chatelain, P. G., Tabone, E., and Rivarola, M. A. (1987). Paracrine regulation of testicular function. J. Steroid Biochem. 27:317–329.

Sanborn, B. M., Wagle, J. R., Steinberger, A., and Lamb, D. J. (1983). Sertoli cell as an androgen target, in *Recent Advances in Male Reproduction: Molecular Basis and Clinical Implications* (D'Agata, R., Lipsett, M. B., Polosa, P., and van der Molen, H. J., eds.) pp. 69–78, Raven Press, New York.

Santulli, R., Sprando, R. L., Awoniyi, C. A., Ewing, L. L., and Zirkin, B. R. (1990). To what extent can spermatogenesis be maintained in the hypophysectomized adult rat testis with exogenously administered testosterone? Endocrinol. 126:95–101.

Sharpe, R. M. (1989). Follicle-stimulating hormone and spermatogenesis in the adult male. J. Endocrinol. 121:405–407.

Sharpe, R. M. (1988). Endocrinology and paracrinology of the testis, in *Physiology and Toxicology of Male Reproduction* (Lamb, J. C., and Foster, M. D., eds.) pp. 71–102, Academic Press, New York.

Sharpe, R. M. (1987). Testosterone and spermatogenesis. J. Endocrinol. 113:1–2.

Sharpe, R. M. (1983). Local control of testicular function. Quart. J. Exper. Physiol. 68:265–287.

Sharpe, R. M., Donachie, K., and Cooper, I. (1988a). Re-evaluation of the intratesticular level of testosterone required for quantitative maintenance of spermatogenesis in the rat. J. Endocrinol. 117:19–26.

Sharpe, R. M., Fraser, H. M., and Ratnasooriya, W. D. (1988b). Assessment of the role of Leydig cell products other than testosterone in spermatogenesis and fertility in adult rats. Int. J. Androl. 11:507–523.

Skinner, M. K., and Fritz, I. B. (1985). Testicular peritubular cells secrete a protein under androgen control that modulates Sertoli cell functions. Proc. Natl. Acad. Sci. (USA) 82:114–118.

Smith, P. E. (1930). Hypophysectomy and a replacement therapy in the rat. Am. J. Anat. 45:205–256.

Sprando, R. L., Santulli, R., Awoniyi, C. A., Ewing, L. L.,

and Zirkin, B. R. (1990). Does ethane 1,2-dimethanesulphonate (EDS) have a direct cytotoxic effect on the seminiferous epithelium of the rat testis? J. Androl. 11:344–352.

Steinberger, E. (1971). Hormonal control of mammalian spermatogenesis. Physiol. Rev. 51:1–22.

Steinberger, E., and Duckett, G. E. (1967). Hormonal control of spermatogenesis. J. Reprod. Fertil. (Suppl) 2:75–87.

Sterio, D. C. (1984). The unbiased estimation of number and sizes of arbitrary particles using the disector. J. Microsc. 134:127–136.

Storring, P. L., Zaidi, A. A. Mistry, Y. G., Froysa, B., Stenning, B. E., and Diczfalusy, E. (1981). A comparison of preparations of highly purified human pituitary follicle-stimulating hormone: Differences in the follicle-stimulating hormone potencies as determined by in vivo bioassay, in vitro bioassay and immunoassay. J. Endocrinol. 91:353–362.

Sun, T.-T., Irby, D. C., Robertson, D. M., and deKretser, D. M. (1989). The effects of exogenously administered testosterone on spermatogenesis in intact and hypophysectomized rats. Endocrinology 125:1000–1010.

Turner, T. T. (1988). Transepithelial movement of ^{3}H-androgen in seminiferous and epididymal tubules: A study using in vivo micropuncture and in vivo microperifusion. Biol. Reprod. 39:399–408.

Turner, T. T., Ewing, L. L., Jones, C. E., Howards, S. S., and Zegeye, B. (1985a). Androgens in male rat reproductive tract fluids: Hypophysectomy and steroid replacement. Am. J. Physiol. 248:E274–E280.

Turner, T. T., Ewing, L. L., Jones, C. E., Howards, S. S., and Zegeye, B. (1985b). Androgens in various fluid compartments of the rat testis and epididymis after hypophysectomy and gonadotropin supplementation. J. Androl. 6:353–358.

Turner, T. T., Jones, C. E., Howards, S. S., Ewing, L. L., Zegeye, B., and Gunsalus, G. L. (1984). On the androgen microenvironment of maturing spermatozoa. Endocrinology 115:1925–1932.

van Alphen, M.M.A., van de Kant, H.J.G., and de Rooij, D. G. (1988). Follicle-stimulating hormone stimulates spermatogenesis in the adult monkey. Endocrinology 123:1449–1455.

Van Sickle, M., Oberwetter, J. M., Birnbaumer, L., and Means, A. R. (1981). Developmental changes in the hormonal regulation of rat testis Sertoli cell adenylylcyclase. Endocrinol. 109:1270–1280.

Wickings, E. J., Usadel, K. H., Dathe, G., and Nieschlag, E. (1980). The role of follicle stimulating hormone in testicular function of the mature rhesus monkey. Acta Endocrinol. (Copenh) 95:117–128.

Wickings, E. J., and Nieschalg, E. (1980). Suppression of spermatogenesis over two years in Rhesus monkeys actively immunized with follicle stimulating hormone. Fertil. Steril. 34:269–274.

Wing, T.-Y., and Christensen, K. (1982). Morphometric studies on rat seminiferous tubules. Am. J. Anat. 165:13–25.

Wright, W., Chan, K., Sundaram, K., and Bardin, C. W. (1982). New observation on androgen action: Androgen receptor stabilization and antisteroid effects of LHRH agonists, in *Hormones and Cancer* (Leavitt, W. W., ed.) pp. 325–336, Plenum Press, New York.

Wright, W. W., and Frankel, A. I. (1979). Endogenous androgen concentrations in nuclei isolated from seminiferous tubules of mature rat testes. J. Steroid Biochem. 10:633–640.

Zirkin, B. R., Santulli, R., Awoniyi, C. A., and Ewing, L. L. (1989). Maintenance of advanced spermatogenic cells in the adult rat testis: Quantitative relationship to testosterone concentration within the testis. Endocrinology 124:3043–3049.

CHAPTER NINE

The Cell Biology of the Sertoli Cell

C. Wayne Bardin, Glen L. Gunsalus, and C. Yan Cheng

The testis can be divided functionally into interstitium and seminiferous tubules, which are responsible for the production of androgens and spermatozoa, respectively. The linkage of the androgen-producing cells with the germ cell-producing elements of the testis in vertebrates is, on one hand, a functional necessity, and, on the other, a regulatory paradox. The necessity for the Leydig cells to be adjacent to the seminiferous tubular epithelium derives from the notion that high concentrations of androgens are required for differentiation of spermatozoa. Such an idea is supported by numerous studies, including those showing that a large dose of androgens maintained spermatogenesis in hypophysectomized animals. The paradox of this association lies in the different pathways through which Leydig cells and seminiferous epithelium are regulated.

Leydig cells respond to luteinizing hormone (LH) by a rapid rise in androgen secretion in a manner similar to the response of the adrenal cortex to adrenocorticotropic hormone (ACTH). By contrast, the hormonal control of the seminiferous epithelium is not characterized by rapid responses because its primary product, spermatozoa, needs weeks to develop. Furthermore, germ cells develop in large cohorts that are arranged in groups that are in the same stage of development along the length of the seminiferous tubule. The cells of each cohort are connected by cytoplasmic bridges until the final steps of spermatogenesis and thus develop in unison. The fact that multiple cohorts with the same morphological stage of differentiation are grouped together in the seminiferous epithelium forms the basis for the spermatogenic cycle (discussed in more detail in Chapter 14). This cyclic maturation of multiple cohorts of germ cells at the same time assures the continued release of millions of mature sperm on a daily basis. In view of the complex and prolonged process of germ cell differentiation, it is easy to understand how minor disruptions in this process would result in decreased sperm production. As a consequence, the regulatory hormones (follicle-stimulating hormone [FSH], LH, prolactin, and testosterone), other molecules from the blood, along with autocrine, paracrine, and cytocrine factors from multiple testicular cells, function to maintain homeostasis in the seminiferous tubule. Rapid changes in any of the regulators must be compensated so that sperm production is not interrupted. This mode of regulation stands in sharp contrast to the regulation of Leydig cell function where wide fluctuations in testosterone secretion reflect changes in LH secretion. The purpose of this chapter is to review some of the new insights into how Sertoli cells participate in the maintenance of homeostasis within the testis. Sertoli cells secrete many regulators that comprise parts of the ultrashort loop system that operates between the cells of the seminiferous epithelium, and of the short loop system between the seminiferous epithelium and the interstitial cells.

The identification of potential testicular autocrine, paracrine, and cytocrine regulators has come from isolation of products directly from

Table 9–1. Criteria Used to Establish that a Molecule Found in an Organ is an Important Autocrine or Paracrine Regulator of Cellular Function

1. The molecule should be synthesized and secreted in biologically active form.
2. It should have action on local cells.
3. It should be released from cells in response to appropriate physiological signals.
4. Its removal (with an antibody or an antagonist) and replacement both in vivo and in vitro should influence an important physiological event.

testicular fluids or by use of antibodies and complement DNA (cDNA) probes directed against molecules present in other tissues that could be regulatory molecules in the testes. These approaches have been complementary in that many possible molecules have been demonstrated either to be present in the testes or to have the capacity to act on testicular cells. Using the rigorous criteria outlined in Table 9–1, however, relatively few of these molecules have been shown to be essential or important regulatory molecules. As a consequence some investigators doubt whether any are important (Sharpe, 1985). The present review examines recent evidence for autocrine, paracrine, and cytocrine regulators of testicular function that are produced by and act on Sertoli cells. Some of these molecules are established regulators, some are possible regulators, and others are likely candidates for isolation and identification. Not all important intratesticular regulatory systems can be covered in this review. We have, therefore, selected examples that illustrate some of the problems and progress in this field.

I. The Structure of Sertoli Cells

There are a number of reviews that describe the structural features of Sertoli cells and attempt to relate them to Sertoli cell function (Dym and Fawcett, 1970; Fawcett, 1975; Setchell, 1978). The structure of the nucleus and the Golgi apparatus is consistent with synthesis and rapid secretion of numerous proteins without intermediate storage in secretory granules. One of the most interesting features of Sertoli cells is the unique intracellular junctional specializations that maintain the structural integrity of the seminiferous tubular epithelium, provide a low resistance pathway for electrical coupling of adjacent Sertoli cells, and constitute a component of the blood–testis barrier. This barrier segregates the epithelium of the tubule into an exterior or basal compartment containing spermatogonia as well as preleptotene spermatocytes, and an interior or adluminal compartment that contains the remaining spermatocytes and spermatids. There are other features of Sertoli cells that also relate to their function in the seminiferous tubules, including: (1) numerous primary lysosomes and autophagic vacuoles that participate in the phagocytotic function of these cells (Chemes, 1986); (2) smooth endoplasmic reticulum in some species suggestive of steroid synthesis and metabolism; (3) cyclic alternations of Sertoli cell morphology along the length of the seminiferous tubule indicating that these cells participate along with the germ cells in the spermatogenic cycle; and (4) an intimate relation with germ cells consistent with the postulate that Sertoli cells support germ cell maturation (Bardin et al., 1988).

II. Secretory Products of Sertoli Cells

Sertoli cells secrete a complex fluid into the lumina of the seminiferous tubule behind the blood–testis barrier (Setchell, 1978; Bardin et al., 1988). In view of the unusual ion and protein composition of the tubular fluid, it is clear that even small molecules in lymph and serum do not readily enter the seminiferous tubules and vice versa. A variety of data show that there is differential transport of various factors from the serum into tubular and rete testis fluid. Thus, one functional consequence of the blood–testis barrier is to maintain a gradient of ions, other small molecules, and proteins between blood and tubular fluid.

Although many studies indicate that the ion gradient is maintained by ion pumps and that gradients of small molecules, such as steroids, are a result of slow diffusion, it is not clear whether proteins can move from the blood into the tubular fluid by selective transport mechanisms analogous to those used to transport gamma globulin across the epithelia of the gut. It appears more likely that most of the proteins in tubular fluid are secreted by Sertoli cells (Mather et al., 1983). Thus, the Sertoli cells secrete products that create the environment for developing germ cells and that provide the signaling systems between the various testicular cell types. These secretory products maintain the homeostasis within seminiferous tubules that is essential for germ cell maturation (Skinner, 1987).

Of the many proteins produced by Sertoli cells, only a few have been isolated and the functions of many of these are not known. Fractionation of the proteins in Sertoli cell culture media by both size and charge separation revealed that this cell type can secrete over 300 radiolabeled proteins (Roberts and Griswold, 1989), only 30 to 50 of which can be detected with protein staining (Cheng et al., 1986). Some of these proteins appear to be tissue specific, whereas others are the same as those normally found in serum. Skinner and Griswold (1980) first demonstrated that Sertoli cells secrete transferrin. Subsequently, Wright and colleagues showed that in addition to transferrin, cultured Sertoli cells synthesize and secrete multiple serum proteins (Wright et al., 1981). Although it had been known for a long time that tubular and rete testis fluid contain serum proteins, it had been assumed that these were transported from blood (Setchell, 1978). The demonstration that Sertoli cells can synthesize such proteins provided an alternative explanation for the presence of such products in the adluminal compartment. Thus, Sertoli cells appear to be able to secrete many of the essential proteins necessary for maintenance and maturation of germ cells in the adluminal compartment. Table 9–2 lists many of the proteins secreted by Sertoli cells. For convenience they are divided into signaling proteins, which communicate between cells, and environmental proteins, which maintain the environment in the luminal compartment of the seminiferous tubule in the same way that proteins in blood maintain the environment in the basal compartment of the testis and in other organs that do not have a barrier to their entry.

Table 9–2. Examples of Sertoli Cell Proteins and Peptides

Environmental	Signaling
Androgen binding protein	Inhibins
Cortisol binding globulin	Activins
Transferrin	TGF-β_1
Ceruloplasmin	MIH
Retinol-binding protein	TGF-α/EGF
Clusterin/SGP-2/SP-40,40	IGF-I
Testibumin	Leydig cell stimulator
Testin 1 and 2	FGF
α_2-macroglobulin	Met-enkephalin
Heat-shock proteins	
Bio-matrix	
SPARC	
Cycle protein-2	

EGF, epidermal growth factor; FGH, fibroblast growth factor; IGF-1, insulinlike growth factor-1; MIH, Mullerian inhibitory hormone; SPARC, secreted protein, acidic and rich in cysteine; SPG-2, sulfated glycoprotein-2; TGF, transforming growth factor.

A. *Sertoli Cell Products That Maintain the Environment of the Tubule*

1. ANDROGEN-BINDING PROTEIN

Androgen-binding protein (ABP) was one of the first Sertoli cell-specific proteins to be identified and extensively studied (Hansson et al., 1975). This Sertoli cell product is important because it provided a prototype for the study of other proteins produced by this cell. A significant feature of this protein is that it is secreted both into the blood and tubular lumen; this fact allowed the demonstration that Sertoli cells are polarized with respect to secretion. Approximately 80 percent of ABP in the rat (rABP) is secreted into the lumen of the seminiferous tubule and the other 20 percent is released into the blood, probably from the base of Sertoli cells (Gunsalus et al., 1980). When germ cells are lost, apical secretion is markedly reduced and basal secretion is not altered (Morris et al., 1987a; Gunsalus et al., 1985). Other factors that facilitate apical and basal secretion are summarized in Figure 9–1A (Serio et al., 1988). The potential for studying the factors that regulate bidirectional secretion by Sertoli cells was enhanced by the development of technology to

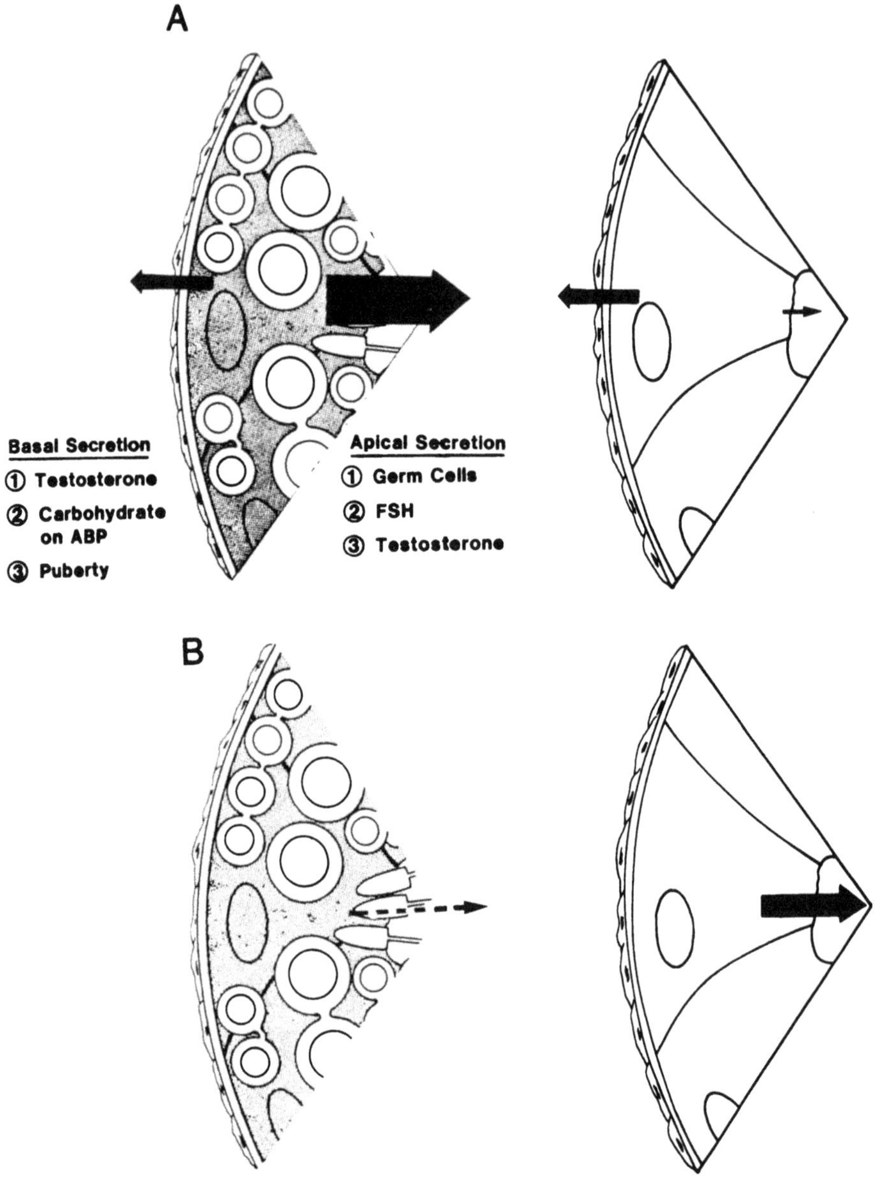

Figure 9–1. Protein secretion by Sertoli cells. (**A**) The bidirectional secretion of androgen-binding protein (ABP) by Sertoli cells in a pie-shaped section of seminiferous tubule with *(left)* and without *(right)* germ cells. The secretion of ABP into the lumen of the tubule (apical secretion) is markedly reduced by loss of germ cells in the tubule on the right; secretion of ABP from the base of Sertoli cells into blood and lymph is not affected by loss of germ cells. The width of the *arrows* indicates the portion of the ABP secreted in either direction. Apical secretion is regulated by germ cells, FSH, and testosterone; basal secretion is regulated by testosterone, nature of the carbohydrates on ABP, and changes in Sertoli cells that occur at puberty. (**B**) The unidirectional secretion of testins. In the presence of germ cells, little or no testins are secreted; loss of germ cells increases the apical secretion of this protein. Basal secretion of testins has not been detected.

grow Sertoli cells in bicameral chambers. Two groups of investigators have examined the secretion of rABP and transferrin in such chambers. Using substantially different culture techniques, these groups have obtained divergent results (Hadley et al., 1987; Janecki and Steinberger, 1987). Nonetheless, the early observations on rABP gave rise to the concept that some proteins are secreted bidirectionally and that the physiology of the seminiferous tubule can be studied by measuring such proteins in the blood and reproductive tract (Gunsalus et al., 1981).

In the adults of many species, but not the rat, the liver synthesizes and secretes into the blood an extracellular steroid-binding protein that has remarkable similarity to the ABP produced by the testis (Bardin et al., 1981). The proteins synthesized by the liver are known by several designations, including testosterone estradiol binding globulin (TeBG), sex hormone binding globulin (SHBG), and steroid binding protein (SBG). Studies in the rabbit, human, and ram show that within a single species the ABPs and TeBGs have identical Stokes radii and immunologic properties. In most species, however, TeBG and ABP can be distinguished on sodium dodecylsulfate (SDS) gels where they exhibit minor variations in the sizes or distribution patterns of subunits; or on concanavilin A (Con A) columns where different distributions of binding and nonbinding forms are noted because of their differential carbohydrate composition. Recent studies have also shown that in the rat an ABP is made by the fetal (but not the adult) liver and this protein is similar to rABP in the adult testis. The description of these ABPs in both the blood and the testis of several species has been covered in several reviews (Bardin et al., 1988; Forest and Pugeat, 1986).

Insights into the subtle differences between TeBGs and ABPs are evident from studies of cDNAs and genes for these proteins. In the human, TeBG and ABP cDNAs are identical, as determined by hybridization (Reventos et al., 1988) and sequence analysis (Hammond et al., 1989; Gershagen et al., 1989). Several studies indicate that the messenger RNAs (mRNAs) coding for both proteins are from a single gene (Hammond et al., 1989). These considerations suggest that TeBG and ABP in the human are the same protein. Their subtle structural differences, therefore, probably result from differential posttranslational processing of the same protein in different organs. In the rat, analysis of testicular ABP cDNAs (Joseph et al., 1987; Reventos et al., 1988) and the presence of a single gene for ABP also suggest that the protein in the testis and fetal liver are identical (Joseph et al., 1988).

In the rat, the fetal brains of both sexes have high concentrations of rABP by day 19 (Carreau et al., 1990). Evidence for rABP synthesis in the brain was not obtained using immunoprecipitation, suggesting that either the rate of synthesis was below the detection limit of the method used, or there is receptor-mediated uptake of this steroid-binding protein in the brain. There are a variety of studies that support the uptake of steroid-binding proteins by brain (Toran-Allerand, 1980), as well as other tissues (Bordin and Petra, 1980). Furthermore, the finding of TeBG uptake by the testis raises the possibility that both synthesis of rABP and uptake of the same protein from the blood can occur in a single organ (Sakiyama et al., 1988). In this regard, it is of note that small amounts of rABP mRNA are present in the brain postnatally, suggesting that this protein can be produced by this organ (Wang et al., 1989). These observations, along with those demonstrating rABP and TeBG uptake by a variety of other issues, suggest that internalization of steroid-binding proteins may be important for their mechanisms of action as well as degradation.

2. METAL-BINDING PROTEINS

Transferrin

Transferrin was one of the first proteins to be identified as a secretory product of Sertoli cells (Skinner and Griswold, 1980). It was identified in cell culture medium by its ability to bind to iron, cross react with transferrin antibodies, and migrate with serum transferrin on two-dimensional gels (Wright et al., 1981; Skinner et al., 1984). Results from more recent studies support the postulate that the function of transferrin is to transport iron across the blood–testis barrier so as to maintain this ion in the tubular compartment in a soluble and nonreactive form (Griswold et al., 1989). Iron is an essential element for cellular growth and development of all procaryotic and eucaryotic cells. As a consequence, a mechanism is required to transport iron to germ cells that are behind the blood–testis barrier. The transport of iron into the seminiferous tubules is believed to proceed as follows: Each transferrin molecule in serum binds two molecules of ferric iron; the diferric transferrin binds to transferrin receptors that are distributed over the basal portions of Sertoli cells (Djakiew et al., 1986); and the receptor–ligand complex is internalized and transferred to membrane vesicles. Subsequent acidification of these vesicles allows release of the iron but retains the bound apotransferrin–receptor complex that is recycled to the plasma membrane on the base of Sertoli cells. The newly released iron is then transported across the Sertoli cell and is released into the tubular lumen where it is then bound to transferrin of Sertoli cell origin (Serio et al., 1988). The iron–transferrin complex then binds to transferrin receptors on germ cells (Wauben-Penris et al., 1986; Serio et al., 1988), thus completing iron transport across the seminiferous tubular epithelium to germ cells that are distant from the systemic circulation (Morales et al., 1987; Toebosch et al., 1987).

Transferrin and its mRNA in Sertoli cells are regulated by vitamin A, testosterone, and FSH (Perez-Infante et al., 1986; Huggenvik et al., 1987). In intact animals the transferrin receptor mRNA in Sertoli cells is regulated acutely by iron; FSH and testosterone regulate this receptor in hypophysectomized animals (Roberts and Griswold, 1990).

Holmes and associates (1982) demonstrated that 60 percent of the transferrin in human semen originates from Sertoli cells. These and subsequent investigations have indicated that transferrin appears to be a good marker for Sertoli cell function in humans (Orlando et al., 1985; Foresta et al., 1986).

Ceruloplasmin

A metal-binding protein for copper is also believed to be synthesized by Sertoli cells. Wright and colleagues (1981) identified an immunoreactive serum protein in Sertoli cell media that migrated with ceruloplasmin on two-dimensional gels. Subsequently, Skinner and Griswold (1983) isolated a ceruloplasminlike protein from Sertoli cell media. The identity, function, and regulation of this protein have not been as extensively studied as transferrin.

3. SULFATED GLYCOPROTEIN-1

Sulfated glycoprotein-1 (SGP-1) is secreted by Sertoli cells and by epididymal cells into the tubules of the reproductive tract (Griswold et al., 1989). Analysis of the predicted amino acid sequence from the SGP-1 cDNA indicated that this is the rat homologue of the human protein, saposin percursor. This protein codes for four polypeptides of approximately 8 kd each that are, in turn, designated saposins A, B, C, and D. Saposins metabolize sphingolipids in lysosomes. Each of the saposins is believed to bind individual groups of glycosphingolipids and retain them in solution so that degradative enzymes can carry out their specific functions. In Sertoli cells, the saposin precursor is localized in lysosomes as it is in other tissues. This protein, however, is also found in high concentrations in seminiferous tubular fluid, although to date its product peptides, saposins A, B, C, and D, have not been identified in this fluid. The function of this peptide precursor in the tubules of the male reproductive tract has not been determined (Griswold et al., 1989).

4. CLUSTERIN/SULFATED GLYCOPROTEIN-2/ SP-40,40

Sometimes the study of a protein in multiple laboratories is reminiscent of the story of the blind men and the elephant. Each blind man touched a different part of the elephant and thus insisted that it should be characterized by a feature different from that of his colleagues.

Sulfated glycoprotein-2 (SGP-2, first designated Dag) has been extensively studied by Griswold and colleagues (Sylvester et al., 1984; Collard and Griswold, 1987). These investigators demonstrated that this protein is made both by Sertoli and epididymal cells, and is secreted into the tubular lumen where it is free and bound to spermatozoa. On sperm, it is located primarily on tails below the midpiece and on the acrosome. Analysis of the protein structure and the predicted amino acid sequence from its cDNA indicates that SGP-2 is a dimer of molecular weight 43,000 and 47,000 disulfide-linked subunits, respectively. Both subunits are derived from a single precursor protein. Because sequence analysis indicated a similarity to human apolypoprotein A-1, a role in lipid transport was postulated.

In a parallel set of studies, Fritz and his colleagues isolated an abundant dimeric protein from ram rete testis fluid that was designated clusterin because it elicits aggregation of erythrocytes and spermatozoa of several species in vitro (Fritz et al., 1983; Blaschuk and Fritz, 1984). Immunohistochemical studies revealed that clusterin is localized in the seminiferous tubules and the epithelial cells of the rete testis and epididymis (Tung and Fritz, 1985).

N-terminal amino acid analyses by Cheng and colleagues (1988c) indicated that ram clusterin is a heterodimer composed of two 40,000 mol wt subunits. Additional studies showed that an antibody to a 43,000 mol wt protein (designated CMB-21) in Sertoli cell culture medium cross reacted with ram clusterin. A rat protein was isolated from culture medium based on its ability to aggregate red cells. Its N-terminal amino acid sequence and the amino acid sequence predicted from its cDNAs were similar to the N-terminal sequence of ram clusterin and identical to the sequence of SGP-2 (Cheng et al., 1988b).

Other independent studies identified an mRNA that increased many fold during regression of the prostate after androgen withdrawal and in tissues undergoing reorganization. This mRNA, termed androgen-repressed prostate message-2 (TRPM-2) (Montpetit et al., 1986; Leger et al., 1987), codes for a protein (TRPP-2) that was believed to be involved with programmed cell death (Buttyan et al., 1989). Bettuzzi and co-workers (1989) identified the same mRNA in prostate and called it androgen-repressed message (ARM). Androgen-repressed message was identified in many tissues from both male and female rats; in some tissues there was an increase and in some a decrease after gonadectomy. Both TRPM-2 and ARM are identical to the mRNA for SGP-2 (Buttyan et al., 1989; Bettuzzi et al., 1989). In some instances there is a dramatic increase in this protein or mRNA level in cells and tissues undergoing reorganization, cell death, or tissue regression (Duguid et al., 1989; Buttyan et al., 1989; Grima et al., 1990).

In still another laboratory, SP-40,40 was isolated from the renal glomeruli of patients with glomerulonephritis. SP-40,40 is a serum protein that is a component of the C5b-9 complex of complement (Murphy et al., 1988). Another laboratory isolated the same protein, which they termed complement cytolysis inhibitor (CLI) from affinity purified soluble C5b-9 complex (Jenne and Tischopp, 1989). The fact that the predicted amino acid sequence of SP-40,40 determined from its cDNA has a 77 percent amino acid sequence identity to SGP-2 suggests that SP-40,40 is the human homologue of the protein isolated from the reproductive tract fluids of rat and ram (Kirszbaum et al., 1989; Jenne and Tischopp, 1989). In a variety of bioassays, SP-40,40 has been shown to inhibit complement-mediated cell lysis (Choi et al., 1989).

Cell aggregation, lipid transport, programmed cell death, and inhibition of complement-mediated cell lysis are among the proposed functions of clusterin/SGP-2. It has been suggested, based on studies in humans, that this protein protects sperm from complement-mediated cell lysis in the male and female reproductive tracts (Jenne and Tischopp, 1989). Such a function might also be served by this protein in any tissue that is undergoing reorganization. Because the terminal components of the complement have not been detected in seminal fluid, however, it is possible that the function of this protein in the reproductive tract is unrelated to the complement system. A possibility recently suggested by Kirszbaum and colleagues (1989) is that clusterin/SGP-2 binds to and inhibits a factor that stimulates the acrosome reaction in spermatozoa.

In summary, clusterin/SGP-2 is an example of a protein that has been studied in several species by numerous investigators, many of whom have assigned it a different name and/or function.

5. TESTIBUMIN

Testibumin is a protein identified in Sertoli cell media that increases dramatically in response to testosterone and FSH. Structural and immunological analysis indicates that this protein is related to albumin and alpha-fetoprotein (Cheng and Bardin, 1986; Cheng et al., 1987). Immunostainable testibumin is identifiable in Sertoli cells and in the epithelium of the epididymis, as well as the uterus and corpora lutea (Shaha et al., 1988). In view of its localization in the male reproductive tract and its structural relation to albumin, we postulated that testibumin may be the albumin homologue in the reproductive tract. Recent studies, however, showed that partially purified testibumin does not have the Leydig cell stimulating activity that is attributed to albumin (Melsert et al., 1989). These observations, coupled with the fact that this protein is also found in the blood of adult female rats, indicates that unlike rABP, this protein is not synthesized exclusively in testes. Nevertheless, it is a useful marker of Sertoli cell function (Rossi et al., 1989).

6. TESTINS

Among the abundant proteins secreted into Sertoli cell culture media are two structurally and immunologically related proteins with a molecular weight of 37,000 and 40,000 (Cheng and Bardin, 1987). These monomeric proteins have virtually identical NH_2-terminal amino acid sequences, except that the larger protein has three extra NH_2-terminus amino acids (Cheng et al., 1989). As these proteins appear not to have been isolated before, they were given the names testin-1 and testin-2 for the smaller and larger molecules, respectively. These molecules are highly concentrated in the testis and are synthesized by Sertoli cells as demonstrated by immunoprecipitation experiments. The distribution of testins in reproductive tract tissues and fluids was compared with those of testibumin and androgen-binding protein; unlike these latter Sertoli cell products, testins are not transported to the epididymis. Although the amount of testin secreted by Sertoli cell-enriched cultures is similar to that of testibumin and ABP, the concentrations of testins in rete testis fluid are seven orders of magnitude less than those of testibumin and ABP (Cheng and Bardin, 1987; Cheng et al., 1989). These observations suggested that the secretion of testins in vivo may be suppressed by germ cells. This postulate was supported by the following observations: (1) there is ten times more testin secreted by tubules from immature rats than from those of adult rats; (2) the secretion of testins by isolated Sertoli cells is orders of magnitude higher than isolated tubular segments in vitro; and (3) there is an increase in the testicular content of testins after depletion of germ cells by busulfan treatment. Thus, unlike ABP whose secretion of testins into the seminiferous tubular lumen is dramatically increased by destruction of germ cells (Fig. 9–1B), these results suggest that testins are unique testicular proteins that can be used to study Sertoli cell/germ cell interactions in the seminiferous tubular epithelium (Cheng et al., 1988a).

7. α_2-MACROGLOBULIN

Of all the proteins in the tubular lumen, no protease inhibitors have been extensively studied except for the low molecular weight (6,500) inhibitor associated with spermatozoa (Zaneveld et al., 1975). This is of physiologic significance as the process of spermatogenesis in the seminiferous tubule must be associated with the release of proteases, including the serine protease acrosin, that could severely damage the seminiferous tubules and the rest of the genital tract if not inactivated. It was, therefore, reasoned that protease inhibitors should be relatively abundant in the tubular fluid, and that the same molecules would likely be secreted by Sertoli cells.

Using an assay based on the hydrolysis of protein by trypsin, an abundant inhibitor was identified in media from Sertoli cell-enriched cultures (Cheng et al., 1990). This protein was purified to apparent homogeneity by high-performance liquid chromatography (HPLC). Analysis of two separate batches of purified protein yielded a single NH_2-terminal sequence that was identical to the NH_2-terminal sequence of rat serum α_2-macroglobulin. The purified protease inhibitor from the testis inhibited a wide spectrum of proteases, such as trypsin, papain, and thermolysin.

Analysis of the subunit structure of the purified protease inhibitor revealed that it migrated on SDS gels as a band of molecular weight 170,000 or 340,000 under reducing and nonreducing conditions, respectively. Thus, the two monomers appeared to be linked by disulfide bonds. The molecular weight of the native protein as estimated by gel permeation HPLC was 680,000. Thus, the subunit structure of the purified protease inhibitor from the testis is identical to that of α_2-macroglobulin in the serum, which is also a tetrimer composed of two pairs of disulfide-bonded monomers (Barrett et al., 1979; Gehring et al., 1987).

The synthesis and secretion of α_2-macroglobulin by cultured Sertoli cells was confirmed by immunoprecipitation and radioimmunoassay. To obtain evidence for the presence of testicular α_2-macroglobulin in the adluminal compartment of the testis, seminiferous tubular fluid and rete testis fluid were collected by micropuncture. Immunoblots indicated that both fluids contain immunoreactive α_2-macroglobulin. In rete testis fluid, immunoassayable testicular α_2-macroglobulin constitutes about 10 percent of the total protein. Analysis of immunoblots indicated that the concentration of testicular α_2-macroglobulin in the seminiferous tubular fluid is similar to that in rete testis fluid. The demonstration of immunoreactive α_2-macroglobulin in rete testis and seminiferous tubular fluid, coupled with immunoprecipitation studies, provides compelling evidence that this protease inhibitor is secreted by Sertoli cells behind the blood–testis barrier into the tubular lumen. Other inhibitors were sought; α_1-antitrypsin, the prominent protease inhibitor in the systemic circulation, does not appear to be secreted by Sertoli cells (Cheng et al., 1990).

α_2-Macroglobulin inhibits almost all known proteases examined to date, irrespective of their catalytic mechanisms. Its proposed mechanism of action involves cleavage of a susceptible peptide bond by the attacking protease, resulting in a conformational change in α_2-macroglobulin that traps the protease (Barrett and Starkey, 1973; Harpel, 1973). Once the protease is bound to α_2-macroglobulin, the complex is cleared from the circulation by a receptor-mediated mechanism that is ten times faster than for the unliganded inhibitor. Receptors for α_2-macroglobulin have been demonstrated on fibroblast, microphages, and tumor cells. In the testis, the most likely site of such a receptor would be the Sertoli cells because this is a major site for phagocytosis in the seminiferous tubular epithelium.

In addition to its protease inhibitor activity, α_2-macroglobulin is also known to stimulate the growth of a variety of cells in vitro (Salomon et al., 1982; Feldman and Pizzo, 1984). This activity may be explained in part by the finding that α_2-macroglobulin also binds a variety of growth factors including transforming growth factor (TGF-β) (O'Connor-McCourt and Wakefield, 1987; Huang et al., 1988), which also is a product of Sertoli cells (discussed later). The interaction of TGF-β with α_2-macroglobulin appears to have features in common with that of the protease-inhibitor interactions described previously. That is, TGF-β is entrapped by and covalently coupled to α_2-macroglobulin. It is, therefore, proposed that α_2-macroglobulin scavenges growth factors as well as proteases that are released at sites of injury.

Based on the assumption that the actions of α_2-macroglobulin in the testis are the same as those in other organs, we propose a bifunctional role for this protein. It is secreted into the adluminal compartment where it limits damage of proteases released by degenerating spermatozoa and the seminiferous tubular epithelium. In addition, this protein also may maintain TGF-β in an inactive or latent form so that the specific actions of this factor can be exerted in only a limited portion of the epithelium.

8. CYCLIC PROTEIN-2

Some Sertoli cell proteins are secreted primarily at specific stages of the cycle of the seminiferous epithelium (Parvinen, 1982; Wright et al., 1983; Shabanowitz et al., 1986). Of such proteins, cyclic protein-2 (CP-2) is of interest because of the magnitude of its fluctuation from stage to stage and because of its mechanism of regulation (Wright et al., 1989). Cyclic protein-2 was identified as a major secretory product of stage IV–VII tubules in vitro (Wright et al., 1983). The protein was isolated from seminiferous tubular fluid by four chromatographic steps (Wright and Luzarraga, 1986). Using an antiserum against the protein, it was demonstrated that maximal synthesis of CP-2 was at stages VI and VIIa,b and minimal at stage XII.

To understand the regulation of CP-2, its secretion was compared with that of transferrin. A study of cultured tubules from 17- and 75-day-old rats indicated that both proteins increase with age; the increase in CMB-2 synthesis as a function of age was much greater than that of transferrin. Transferrin synthesis and secretion of Sertoli cells were stimulated by hormones and vitamins, whereas CP-2 was not. Finally, the age-dependent increase in CP-2 secretion by tubules was not observed in Sertoli cell culture; in such cultures more transferrin is secreted. These data indicate that the regulation of CP-2 appears to be different from many other Sertoli cell products. This problem appears to be regulated by paracrine factors secreted by germ cells or by direct contact with germ cells (Wright et al., 1989). The mechanism by which germ cells regulate this aspect of Sertoli cell function is not understood.

9. SECRETED PROTEIN, ACIDIC AND RICH IN CYSTEINE

Secreted protein, acidic and rich in cysteine (SPARC) is synthesized in vitro by endothelial cells, vascular smooth muscle cells, fibroblasts, and several transformed cell lines. Secretion of SPARC by cells is increased in vitro in response to injury or to conditions that stimulate cell proliferation, reorganization, or migration (Sage et al., 1984, 1986; Mason et al., 1986). In a survey of mouse tissues by in situ hybridization, SPARC mRNA was demonstrated in the testicular capsule and in Leydig cells (Holland et al., 1987). Immunostainable SPARC was then demonstrated in Leydig cells and in Sertoli cells. The latter cells stained only when they were associated with late-stage elongated spermatids, suggesting that SPARC is also highly related to the spermatogenic cycle. Western blot analysis confirmed that immunoreactive SPARC (43,000 mol wt) was present in extracts of purified Sertoli cells and Leydig cells as well as in culture media from both of these cell types. The finding that purified SPARC bound Ca^{+2}, Fe^{2+}, and Cu^{2+} (Vernon and Sage, 1989) suggests that in the testis, SPARC may sequester or transport certain metallic ions. Recent studies also indicate that SPARC induces changes in shape of certain nongonadal cell types.

B. Sertoli Cell Products That Signal Other Cells

Some of the products of Sertoli cells that stimulate or inhibit other cells are listed in Table 9–2. Few of these peptides have actually been isolated directly from biological fluids derived from Sertoli cells. Furthermore, in most instances the biological activities of these signaling peptides have been demonstrated only in vitro. Thus, important or essential roles for many of these molecules in the testes have yet to be established in vivo.

1. Insulinlike Growth Factors

The role of insulin and insulinlike growth factor-1 (IGF-1) in the regulation of testicular function was first suggested by a variety of physiological studies. Humans and animals with naturally occurring or experimentally induced diabetes have abnormal testicular function, which in some instances can be corrected by insulin replacement therapy. These studies suggested that insulin could affect directly both gonadotroph and Leydig cell functions (Charreau et al., 1978; Rodriguez-Rigau, 1980; Murray et al., 1981). Furthermore, the early studies of testicular cells maintained in serum-free culture emphasized the essential role of insulinlike peptides. These studies used insulin at microgram per milliliter concentrations (Mather and Phillips, 1984), thus obscuring the ability to distinguish whether insulin or IGF-1 was preferred by a given cell type.

The role of a growth hormone or a growth hormone-stimulated product, such as IGF-1, in testicular function was suggested by the observation that children and animals with growth hormone (GH) deficiency have delayed puberty (Tanner and Whithouse, 1975; Kulin et al., 1981; Laron, 1984; Hochereau-de Reviers et al., 1987). The suggestion that this disruption of sexual maturation was due to a deficiency of IGF-1 rather than GH was based on studies showing that a reduction of both GH and IGF-1 secretion was associated with delayed puberty (Hochereau-de Reviers et al., 1987; Geffner et al., 1987). Thus, studies in humans and animals suggested that insulin and IGF-1 in blood are important for testicular function, but they did not determine if these peptides are also locally produced and whether their actions are on single or multiple cells.

Secretion of IGF-1

Studies from several laboratories indicate that immunoreactive IGF-1 is released into media from rat and pig Sertoli cells (Rizen, 1983; Tres et al., 1986; Smith et al., 1987; Chatelain et al., 1987; Saez et al., 1988). Immunostainable IGF-1 was also localized in human Sertoli cells (Vannelli et al., 1989). By contrast, in immature rats immunostainable IGF-1 was present in Sertoli cells, Leydig cells, and germ cells; during the third and fourth postnatal week, immunostainable material decreased in Leydig and Sertoli cells so that in adult rats staining was only in germ cells (Hansson et al., 1989). Partial purification of peptides from immature Sertoli cell media yielded products with the same physicochemical, receptor binding, and biological activity as pure human IGF-1 and recombinant IGF-1 (Saez et al., 1988).

The major stimulators of immunoreactive IGF-1 secretion by pig Sertoli cells were fibroblast growth factors (Chatelain et al., 1987). Interestingly, neither FSH nor GH had any stimulatory effects when the cells were cultured on plastic. When these cells were cultured on extracellular matrix, however, they secreted IGF-1 in response to epidermal growth factor (EGF) and fibroblast growth factor (FGF), but not to FSH and GH. Follicle-stimulating hormone was able to stimulate IGF-1 when the pig Sertoli cells were pretreated for 24 hours with either FGF or EGF. Similarly, FSH and GH also had only a minor effect on IGF-1 secretion by rat Sertoli cells (Tres et al., 1986). These results suggest that Sertoli cells can secrete IGF-1 and that GH has little, if any, effect on the regulation of IGF-1 secretion by these cells as is the case for most tissues other than the liver (Saez et al., 1988). Thus, in the testes IGF-1 secretion appears to be regulated by peptides that are known to act on the Sertoli cell, many of which are also produced locally.

Receptors and Actions of IGF-1 and Insulin

Although the sites of production in the testes appear to decline in adult animals (Hansson et al., 1989), the distribution of IGF-1 receptors and sites of action remains diverse. Using a monoclonal antibody, immunostainable IGF-1 receptor was localized to Sertoli cells, Leydig cells, secondary spermatocytes, and early spermatids (Vannelli et al., 1989). IGF-1 receptors were also demonstrated on pig and rat Leydig and Sertoli cells using radiolabeled ligands (Borland et al., 1984; Handelsman et al., 1985; Lin et al., 1986; Kasson and Hsueh, 1987; Saez et al., 1988; Oonk and Grootegoed, 1988); the properties of these receptors were similar to IGF-1 receptors in other tissues (Rechler and Nissley, 1985). In contrast to IGF-1 receptors, insulin receptors were found on Leydig but not Sertoli cells from immature pigs. In the rat, however, insulin receptors were demonstrated on both Leydig (Lin et al., 1986; Rigaudiere et al., 1988) and Sertoli cells (Oonk and Grootegoed, 1987). Several investigators have not been able to demonstrate effects at nanomolar concentrations of insulin on Sertoli cells and have used such observations to argue against the existence of insulin receptors on these cells. For example, Saez and colleagues (1988) have argued that the actions of IGF-1 at nanomolar concentrations and insulin at micromolar concentrations on porcine Leydig and Sertoli cells suggest that the actions of insulin on testicular cells in this species are by the IGF-1 receptor. Such an argument is supported only if one can exclude the possibility of insulin proteolysis when this hormone is added to such cultures in serum-free media.

In conclusion, IGF-1 is produced by Sertoli and possibly other testicular cells, and may exert both autocrine and paracrine effects in the testes. Cultures of both rat (Borland et al., 1984; Mita et al., 1985; Oonk, 1987; Kasson and Hseuh, 1987) and pig (Jaillard et al., 1987; Saez et al., 1988, 1989) testicular cells respond to IGF-1 (insulin and IGF-2) by enhanced differentiated function. The data collected to date do not resolve the extent to which insulin and IGF-1 in the blood may also be important regulators of testicular function. The apparent need of testicular cells for insulinlike peptides that are not under the control of serum glucose or GH appears to be served by Sertoli and possibly other cells (Daughaday and Rotwein, 1989).

2. INHIBIN AND ACTIVIN

In the testes, inhibin subunits are produced by Sertoli cells throughout the life of the animal. Leydig cells appear to produce lesser amounts of these subunits but only during selected periods of differentiation (Lee et al., 1989; Shaha et al., 1989). The structure and function of inhibin as a hormone is reviewed elsewhere in this volume and in recent reviews (Vale et al., 1988; Burger et al., 1989; Bardin et al., 1989). In this chapter, only testicular sites of action are considered.

Actions on Germs Cells

The first source of highly purified gonadal inhibin from a male animal was rete testis fluid, which in turn originated from tubular fluid (Bardin et al., 1987b). In view of this, it was possible that inhibin could have paracrine effects on germ cells (Franchimont et al., 1981). This possibility was tested using partially and highly purified inhibin in intact adult mice and hamsters. The purified material decreased the numbers of intermediate and B_1 spermatogonia to 87 and 91 percent of control values, respectively. It is possible that these effects would have been larger in hypophysectomized animals. These results suggest that inhibin may have an effect on spermatogonial development (van Dissel-Emilliani et al., 1989).

Table 9–3. Factors in Testicular Fluids that Influence Leydig Cells

End point of assay	Assayed on	Source of factor	Stimulated (inhibited) by	Mol. wt. kD	Reference
Inhibitors					
Testosterone, LH stim.[a]	LC[b]	STM	FSH	40–50	Syed et al., 1985 Syed et al., 1987 Syed et al., 1988 Syed et al., 1989
Testosterone hCG binding	LC	Porcine SCM[d]	(FSH)	24–30	Benahmed et al., 1986a
LH binding	LC	Mouse SCM[d]		30–100	Takase et al., 1988
Testosterone hCG stim.	LC (Percoll)[c]	STM, IF		30–100	Vihko and Huhtaniemi, 1989
Aromatase	LC (Percoll)	STM, SCM	FSH	>10	Papadopoulos et al., 1987a
Stable Stimulators					
Testosterone, B/ LH stim.	LC	STM		5–10	Syed et al., 1986
Testosterone		IF			Sharpe and Cooper, 1984
Sensitive Stimulators					
Testosterone, B/ LH stim.	LC (Percoll)	IF		>30	Sharpe et al., 1984 Sharpe and Bartlett, 1985 Sharpe et al., 1988
Testostrone hCG stim.	Co-culture porcine	ST/LC	FSH	>10	Benahmed et al., 1985
Testosterone	LC porcine	SCM	FSH		Benahmed et al., 1986b
Testosterone	Co-culture	ST/LC			Parvinen et al., 1984
Estradiol (Aromatase)					Papadopoulas et al., 1987a
Testosterone, B/ LH stim.	LC (Percoll)	STM, SCM	Germ cells	10–50	Papadopoulas et al., 1987b
Testosterone, B/ LH stim.	LC (Percoll)	SCM rat	FSH	10–30	Verhoeven and Cailleau, 1985
		STM human	FSH		Verhoeven and Cailleau, 1985
Testosterone, B/ LH stim.	LC (Percoll)	SCM	FSH	>1	Janecki et al., 1985
Testosterone, B/ LH stim.	LC	IF		50	Jansz and Ponerantz, 1987
Testosterone, B/ LH stim.	Co-culture LC	LC/SC SCM	FSH		Liu and Dahl, 1988

[a]In the example shown on the first line of this table the inhibitor in seminiferous tubular media (STM) reduced testosterone secretion of LH stimulated (stim.) Leydig cells (LC); the inhibitor concentration was increased by FSH.

[b]LC, Leydig cell; STM, seminiferous tubular media; SCM, Sertoli culture media; IF, interstitial fluid; ST, seminiferous tubule; SC, Sertoli cell; B, basal secretion; LH stim., stimulated by LH.

[c]Percoll was used to prepare highly purified LC.

[d]The species source of the factor is indicated when it is other than the rat.

Actions on Leydig Cells

Because a variety of Sertoli cell factors have been shown to regulate Leydig cell function (Tables 9–3 and 9–4), the effects of inhibin and activin on androgen production in vitro were examined. Suspensions of testicular cells were prepared from testes removed from neonatal or adult hypophysectomized rats. Inhibin isolated from ram rete testis fluid and activin from porcine follicular fluid had no effect on basal testosterone secretion. When administered with LH, inhibin shifted the dose–response curve to the left and activin shifted the curve to the right. Thus, inhibin increased and activin decreased the sensitivity of Leydig cells from fetal or hormone-regressed animals to LH (Hsueh et al., 1987).

In another study, the effects of activin and inhibin on androstenedione secretion were studied using primary Leydig cell-enriched cultures established from the testes of 20-day-old rats. Androstenedione was measured as it is the primary androgen secreted by Leydig cells from rats of this age. Three days of inhibin treatment produced a modest but significant stimulatory effect on basal androstenedione secretion. Inhibin also enhanced the androstenedione response (1.6 to 1.9-fold) to a submaximal dose of human chorionic gonadotropin (hCG) (P. Morris, unpublished). By contrast, the same dose of inhibin suppressed the androgen response to a maximal dose of hCG. These results suggest that the period of exposure of Leydig cells, as well as the amount of gonadotropin present, affects the response of the cell to inhibin. Activin, regardless of dose or time in culture, had no effect on basal androstenedione secretion but did suppress by one-third the androstenedione response to all doses of hCG at 24 hours of culture. These studies led us to the postulate that inhibin has a biphasic effect on androgen production by immature Leydig cells that is dependent on the dose of hCG, and that the opposing actions of inhibin and activin are evident in different time frames.

Table 9–4. Testicular Products Which Are Possibly of Sertoli Cell Origin that Can Modulate Leydig Cell Functions

Growth factor	Action	Reference
IGF-1, IGF-2	S	Kasson and Hsueh, 1987
FGF	S	Sordoillet et al., 1988
Inhibin	S	Hsueh et al., 1987
		Morera et al., 1987
		Lin et al., 1987
TGF-β	I	Saez et al., 1989
		Hsueh et al., 1987
Activin	I	Lin et al., 1989
Interleukin	I	Calkins et al., 1988
EGF/TGF-α	S or I	Ascoli and Segaloff, 1989
LHRH	S or I	Saez et al., 1983

S, stimulator; I, inhibitor; S or I can stimulate or inhibit depending on the dose and duration of action.

A study of the action of these peptides on Leydig cells isolated from adult rats indicated that activin inhibits hCG-induced testosterone secretion. The mechanism by which activin exerted this effect is by inhibition of cyclic adenosine monophosphate (cAMP) production and by suppression of both progesterone and pregnenolone conversion to testosterone. Similar effects were produced by TGF-β. By contrast, inhibin had no effect on hCG-induced testosterone secretion by adult Leydig cells but did antagonize the effects of activin (Lin et al., 1989).

The growth of Leydig cells derived from a testicular tumor (R2C) was also inhibited by large doses of activin (10 ng/mL), whereas only 0.1 ng/mL TGF-β was a potent inhibitor of those cells. By contrast, 1.8 ng/mL activin inhibited steroid secretion; a similar effect was produced by 0.04 ng/mL TGF-β. Inhibin had no effect on this cell line (Gonzalez-Manchon and Vale, 1989).

The studies of Hsueh (1987), Lin (1989), and Morris (unpublished) and associates, which all used nontransformed Leydig cells, demonstrate the actions of inhibin and activin on Leydig cells. In all these experiments, activin is an inhibitor of LH/hCG action regardless of the stage of Leydig cell differentiation. Inhibin appears to facilitate the action of LH/hCG only in undifferentiated Leydig cells although it is able to inhibit the effect of activin on both immature and adult Leydig cells. These observations suggest that the action of inhibin may be dependent on the state of Leydig cell differentiation.

Actions on Sertoli Cells

The effects of activin on inhibin secretion by Sertoli cells were examined in vitro. If activin is present during the first 24 hours of culture, then this peptide elevates basal inhibin secretion and shifts the FSH dose–response to the left. If, however, activin is not added until the second day of culture, then this peptide suppresses basal inhibin secretion and shifts the dose–response to FSH to the right. In both instances, activin does not affect the inhibitory concentration of 50 percent (IC_{50}). These differing responses of Sertoli cells to activin, which are dependent on the time that the peptide is added, are reminiscent of some of the differentiating effects of TGF-β (Massague, 1987; Roberts et al., 1988). Interestingly, TGF-β had no effect on Sertoli cells when tested under the same conditions as activin (Morris et al., 1990b). These observations suggest that activin can regulate inhibin secretion. This is one of several possible mechanisms that assures inhibin secretion over that of activin.

3. TRANSFORMING GROWTH FACTOR-β

Transforming growth factor-β (TGF-β) is secreted by Sertoli cells as evidenced by partial purification of peptides that have the same physical properties, receptor binding, and biological activity as TGF-β_1 (Skinner and Moses, 1989; Saez et al., 1989). That this peptide can be synthesized and secreted by Sertoli cells was suggested by the finding of TGF-β mRNA by Northern blot analysis in extracts of these cells isolated from pubertal rats. This mRNA was also found in myoid but not germ cells (Skinner and Moses, 1989).

No effect of TGF-β_1 on Sertoli cells was observed as judged by inhibin, transferrin, or radiolabeled protein secretion (Skinner and Moses, 1989; Morris et al., 1990b). By contrast, myoid cells respond to TGF-β_1 by increasing radiolabeled protein synthesis and colony formation. Co-cultures of myoid and Sertoli cells form large cell clusters in response to TGF-β_1, a phenomenon also associated with the actions of FSH and extracellular matrix in such cultures. Recent studies from a number of laboratories have suggested that the differentiating actions of some growth factors including TGF-β are mediated by regulation of the metabolism of extracellular matrices (Rizzino, 1988). These observations suggest a mechanism by which TGF-β_1 could have a major effect on differentiation of the seminiferous tubule.

As noted above, TGF-β has inhibitory effects on proliferation of and steroid synthesis by a Leydig cell line (R2C) derived from a testicular tumor. Inhibitory effects of this peptide have also been observed by several laboratories that have studied the actions of this peptide on neonatal pig Leydig cells (Avallet et al., 1987; Saez et al., 1989), neonatal rat testicular cells (Fauser and Hsueh, 1988), and adult rat Leydig cell cultures (Lin et al., 1987). The findings of inhibitory actions of TGF-β are amazingly consistent considering the diverse sources of these observations. There is general agreement that this peptide inhibits hCG-stimulated testosterone secretion. Part of this effect is probably due to reduced hCG/LH receptors and cAMP production, and part due to actions of TGF-β at specific sites distal to cAMP generation in the steroid synthetic pathway. The necessity or significance of this potential Leydig cell modulator and how it is regulated in the intact testis remain to be established.

4. TRANSFORMING GROWTH FACTOR-α

One of the major advances in the study of Sertoli and other testicular cells was the development of procedures for their culture in serum-free medium (Mather and Sato, 1979). One of the components of the serum-free media used in many of these studies was EGF that acts by the same receptor as TGF-α. The availability of murine EGF permitted extensive studies on the action of this peptide on both Sertoli and Leydig primary cells and cell lines as a component of the originally proposed culture media (Mather et al., 1982).

As early as 1984, Elson and colleagues (1984) demonstrated EGF bioactivity in male reproductive tissues and fluids. These observations were confirmed by other laboratories (Verhoeven and Cailleau, 1985; Holmes et al., 1986). In a recent study, media from rat Sertoli and myoid cell cultures were also found to contain a peptide that bound to EGF receptors and supported growth of an EGF-dependent cell line. The biochemical properties of this peptide, however, were distinct from those of murine EGF. Analysis of RNA extracts of myoid and Sertoli cells by Northern blotting demonstrated TGF-α mRNA but not EGF mRNA. The secretion of TGF-α by both cell types was confirmed by Western blotting (Skinner et al., 1989c).

TGF-α stimulated myoid cell proliferation, migration, and colony formation. Co-cultures of myoid and Sertoli cells responded to TGF-α by the formation of large balls of cells and by increased transferrin secretion. Sertoli cells cultured alone showed no proliferative response and no increase in transferrin secretion in response to TGF-α (Skinner et al., 1989c). This lack of response was explained by the fact that no EGF receptors could be demonstrated on Sertoli cells. Such an interpretation argues that the effects of TGF-α and EGF on Sertoli cells are mediated through myoid cells. An alternative explanation is that not all Sertoli cell products are responsive to these peptides and that EGF receptors were below the detection limits of the assay used.

The most extensive studies of this peptide on gonadal cells have been those by Ascoli and Segaloff using the MA-10 cell line, which was derived from a mouse Leydig cell tumor (Ascoli and Segaloff, 1989). Epidermal growth factor is not a mitogen for MA-10 cells but it does modulate, by high affinity EGF receptors, their ability to secrete steroids. It is of interest that EGF has different effects on MA-10 cells depending on its duration of action. For example, between 0 and 1 hour of exposure, EGF attenuates hCG-activated steroid synthesis by reducing hCG-activated adenylate cyclase activity; between 1 and 8 hours of exposure, EGF activates steroid biosynthesis without increasing cAMP, which in turn potentiates the increase in steroidogenesis by LH and cAMP analogues; between 8 and 48 hours, EGF decreases the levels of LH receptors (Ascoli and Segaloff, 1989). Many of the effects of EGF on MA-10 cells have been duplicated with TGF-α. The mechanisms by which the peptides bring about these effects are not known. Nonetheless, these observations illustrate that a growth factor can have many effects other than as a mitogen and that the actions of a single peptide on a single cell type can be very complex.

5. LEYDIG CELL REGULATORS

Aoki and Fawcett (1978) demonstrated that focal damage to the seminiferous tubules from locally injected antiandrogen resulted in morphological changes in the Leydig cells adjacent to the damaged tubules. Subsequent studies showed that tubular cell damage induced by a variety of agents affected Leydig cell morphology. In most instances, the damage appeared to result in hypertrophy of Leydig cells but in others there was atrophy (Bergh, 1983). In addition, in many of the studies the Leydig cells in animals with damaged testes appeared to be hypersensitive to stimulation by a variety of gonadotropic hormones such as LH and hCG (Ishida et al., 1987). These studies supported the postulate that a factor produced by Sertoli cells can modulate both the function and morphology of Leydig cells. A particularly impressive study showed morphological effects in vitro (Tabone et al., 1984). When cultured alone in chemically defined media, porcine Leydig cells regress by loss of smooth endoplasmic reticulum (SER) and swelling of mitochondria; these changes were not prevented by FSH or LH treatment. Co-culture of Leydig with Sertoli cells maintained the SER in Leydig cells at the same levels as in those of intact animals. Addition of FSH to the co-culture stimulates Leydig cell development and the steroidogenic response to hCG. Such observations have led numerous investigators to search in various unfractionated biological fluids for factors, which affect Leydig cell function. Such studies have led to the identification of both inhibitory and stimulatory activities of Leydig cells (Table 9–3). In parallel studies, investigators have also studied the effects of various growth factors that were originally identified in other tissues but are known to be, or at least are suspected to be, produced in Sertoli cells. Some of these factors stimulate Leydig cell function and some are inhibitory (Table 9–4).

Leydig Cell Inhibitor Activities

Interstitial fluid and media from Sertoli cells as well as seminiferous tubules contain factors that inhibit various aspects of Leydig cell function including testosterone secretion, cAMP production, and LH/hCG binding (Table 9–3). Such studies suggested that the inhibitor is a protein between 20 and 100 kd. Syed and coworkers (1987) demonstrated inhibitor production by tubules in stages VIII–XI. By contrast, Vihko and Huhtaniemi (1989) demonstrated that inhibitor, which reduced testosterone production, was secreted primarily by tubules at stage IX–I of the spermatogenic cycle and inhibitor for cAMP production at all stages. In addition to changes with the stage of the tubule, hormones appear to modulate inhibitor activity. One experiment showed that inhibitor secretion decreased with testicular damage, presumably as FSH increased (Syed et al., 1986). This possibility was supported by a study showing that FSH suppresses a Leydig cell inhibitor and in so doing, sensitizes Leydig cell to LH/hCG stimulation (Benahmed et al., 1986a). By contrast, treatment of immature rats with FSH markedly increased inhibitor activity (Syed et al., 1989). None of the studies on Leydig cell inhibitors have provided insight as to whether there is more than one molecular entity with this activity. As noted earlier, the study of such inhibitor activities is complicated by observations showing that some of the growth factors produced by Sertoli cells can inhibit Leydig cell function (Table 9–4). Whether one or several growth factors contribute part or all of the inhibitory activity that has been assayed in testicular fluids remains to be established.

Leydig Cell Stimulator Activities

Many different investigators have demonstrated stimulatory activities in media from cultured seminiferous tubules and Sertoli cells as well as in interstitial fluid. Two groups of investigators have demonstrated biological activities that appear to be stable to heat, acids, or organic solvents and that are capable of increasing both basal and LH-stimulated testosterone secretion. The molecular weights of these stable stimulators are between 5,000 and 10,000 (Table 9–3).

Other investigators have identified a group of activities in these same fluids that are heat and enzyme sensitive, and that appear to be proteins with molecular weights between 10,000 and 60,000. These factors also can stimulate both basal and LH-stimulated testosterone secretion (Table 9–3). In some experiments these factors appear to be produced by tubules that are in stages VII–VIII of the cycle. In many studies the concentration of this factor is increased by FSH (Table 9–3), whereas in others this factor is regulated by germ cells rather than FSH (Papadopoulos et al., 1987b). Other studies suggest that there are multiple Leydig cell stimulators produced by the testes; some exert their effects by cAMP, some independent of cAMP, and others directly on particular functions such as aromatase activity (Carreau et al., 1988) or growth stimulation (Brown et al., 1982). One of the major obstacles to studies of stimulatory factors in various testicular fluids is that one cannot be sure whether one or multiple biological activities are being assayed and whether these are already identified (Table 9–4) or unknown proteins or peptides. As a consequence, Leydig cell stimulatory activities are prime candidates for chemical characterization. Finally, it is not certain how effects of testicular damage are mediated on Leydig cells by suppression of an inhibitor, increase of a stimulator, or both.

III. Leydig Cell Products that Signal Sertoli Cells

Testosterone was the first Leydig cell product that was shown to regulate Sertoli and other testicular cells. Other steroids that are either precursors or metabolites of testosterone have also been proposed as intratesticular regulators. Recently, an increasing number of proteins and peptides have been identified as Leydig cell products. In addition to the previously identified products (Table 9–5), Morris and colleagues (1987b, 1990a) have shown that Leydig cell culture media from both primary cells and Leydig cell lines contain many proteins that remain to be identified.

A. Proopiomelanocortin

PRESENCE AND SECRETION

Research on proopiomelanocortin (POMC)-derived peptides in the testis (Bardin et al., 1984;

Table 9–5. Leydig Cell Proteins/Peptides

Proopiomelanocortin	Bardin et al., 1987a
Prodynorphin	McMurray et al., 1989
Proenkephalin	Garrett and Douglass, 1989
Oxytocin/vasopressin	Pickering et al., 1989
Renin-angiotensin	Parmentier et al., 1983
Activin/inhibin	Lee et al., 1989
Gastrin	Gnessi et al., 1989
CRF	Yoon et al., 1988
SPARC	Vernon and Sage, 1989

CRF, corticotropin-releasing factor; SPARC, secreted protein, acidic and rich in cysteine.

Kreiger et al., 1984; Bardin et al., 1987a,b) began with the chance demonstration of immunostainable POMC-derived peptides in Leydig cells (Spera et al., 1988). A detailed study of the POMC-derived peptides isolated from rat testes confirmed the existence of these materials, and suggested that the posttranslational processing of POMC in this organ was similar to that in rat brain (Margioris et al., 1983) where acetylation of these peptides is either substantially reduced (O'Donohue et al., 1981; Zakarian and Smyth, 1982) or absent (Weber et al., 1981; Evans et al., 1982). Such results, however, differ from those in the rat intermediate pituitary in which both α-MSH and β-endorphin are α-*N*-acetylated. The relative concentrations of POMC-derived peptides in the testes were very low compared with those in the pituitary (Tsong et al., 1982b; Bardin et al., 1987a). These observations suggest that testes do not contribute to the concentrations of these peptides in blood; a conclusion in keeping with their putative paracrine and autocrine functions.

Immunoassayable ACTH and β-endorphin were also demonstrated in the interstitial fluid that surrounds Leydig cells in vivo (Valenca and Negro-Vilar, 1986), and β-endorphin was demonstrated in media from Leydig cell cultures (Fabbri et al., 1988; Eskeland et al., 1989). These latter findings suggested that POMC-derived peptides can be secreted by Leydig cells both in vivo and in vitro. Other studies indicated that POMC peptides are in the seminiferous tubular epithelium. Immunostainable POMC-derived peptides were first demonstrated in germ cells, but unexpectedly they were localized in nuclei (Cheng et al., 1984) rather than in cytoplasm as is the case for Leydig cells, which secrete POMC peptides (Tsong et al., 1982a,b).

THE RELATION OF POMC mRNA AND SECRETED PEPTIDES IN TESTIS

Although these studies were consistent with the synthesis of POMC in the testes, studies were undertaken to identify POMC mRNA in rat testes. The predominant POMC mRNA in testes was approximately 150 to 200 bases shorter than in the pituitary and hypothalamus (Chen et al., 1986; Chen and Madigan, 1987; Gizang-Ginsberg and Wolgemuth, 1987). Short POMC mRNA was also detected in hamster, mouse, ovine, monkey, and human testes (Chen et al., 1984, 1986; Bardin et al., 1987a). The localization of POMC mRNA in Leydig cells was supported by several lines of evidence. First, POMC-like mRNA was demonstrated in extracts from mouse Leydig cells (Chen et al., 1984) and from a transplantable murine Leydig cell tumor (Melner and Puett, 1984). Second, POMC and mRNA was also localized in rat Leydig cells by in situ cDNA hybridization (Krieger et al., 1984; Pintar et al., 1984; Gizang-Ginsberg and Wolgemuth, 1985). In addition to Leydig cells, isolated germ cells contain POMC mRNA (Gizang-Ginsberg and Wolgemuth, 1987). Since there are many more germ cells than Leydig cells in a testis, the dominant site of POMC mRNA localization is in the former cell type.

As noted earlier, the POMC mRNA in the testis is approximately 150 to 200 bases shorter than that in the pituitary and hypothalamus (Chen et al., 1984; Chen and Madigan, 1987). Both large and small heterogeneous POMC-like mRNAs have now been demonstrated in several extrapituitary tissues. Because there is only one POMC gene in the rat and most other species, heterogeneous-sized POMC mRNA could arise for a variety of reasons including transcription occurring at different initiation sites, altered splicing of precursor RNA, multiple polyadenylation sites, and poly(A) tails of different lengths. In the testes, both different transcription start sites and poly(A) tails of different lengths are responsible for the shorter mRNA (Lacaze-Masmonteil et al., 1987; Jeannotte et al., 1987; Kilpatrick et al., 1987). In spermatids, both different transcription start sites and poly(A) tails of different lengths are responsible for the shorter POMC mRNA relative to that in pituitary (Kilpatrick et al., 1987). In other germ cells and Leydig cells only different transcription start sites are believed to contribute to the shorter mRNA (Chen and Boitani, unpublished data).

The initial measurements of POMC mRNA indicated that its concentration in the testes was the same as in the hypothalamus (Chen et al., 1984), although the concentration of POMC peptides was 500- to 1000-fold higher in the hypothalamus (Bardin et al., 1987a). Insight into a possible reason for this marked discrepancy between the concentrations of mRNA and peptide in these tissues was suggested by the demonstration that the shorter mRNA in testis results primarily from the use of an alternate start site in exon-3 of the POMC gene. Because this start site is located in the sequence that corresponds to γ-MSH peptide, there is no known signal sequence at the N-terminus of the peptide that would be produced from the short mRNA, and recent studies show that this peptide cannot be secreted (Clark et al., 1989). It is of interest that immunoreactive POMC peptides in germ cell nuclei correlate with the absence of a signal sequence (Chen et al., 1984). These findings question whether there is a function for intracellular peptides that would normally be expected to act on cell surface receptors. The recent observation of Dunbar and colleagues (1989) on the cytocrine effects of "intracellular" interleukin-3, however, suggests a regulatory role for peptides that become targeted to an intracellular compartment.

The contradiction posed by the shortened POMC mRNA in testes must be rationalized with the unequivocal evidence for cytoplasmic localization (Tsong et al., 1982a,b) and secretion of POMC peptides by Leydig cells both in vivo and in vitro (Valenca and Negro-Vilar, 1986; Fabbri et al., 1988). The most likely explanation is that the testis produces a small amount of authentic normal-sized POMC mRNA that is not usually detected on Northern blots because it represents as little as 1/1000 of that of the shorter form (Lacaze-Masmonteil et al., 1987). Studies by Schachter and co-workers (personal communication) also favor this explanation. These results indicate that the concentration of the major class of POMC mRNA in the testes and other reproductive organs may not relate directly to the concentrations of POMC peptides that are secreted.

REGULATION OF POMC

In the fetal mouse and hamster, immunoreactive β-endorphin was detected in primitive Leydig cells on the day that these cells first appear in the testis. The intensity and number of staining cells progressively increased throughout fetal life, declined after birth, and increased again as the animals progressed through puberty (Shaha et al., 1984). These results suggest that the number and intensity of immunostainable Leydig cells are developmentally regulated with peaks at birth and again after puberty. This time-course correlates closely with the acquisition and loss of Leydig cell enzymes required for energy metabolism and testosterone synthesis. These results correlating immunostainable material with testicular maturation suggest that the synthesis of POMC peptides is regulated by gonadotropins.

The direct regulation of POMC by gonadotropins was demonstrated in a series of studies in which hCG increased the immunostainable β-endorphin in prepubertal testes (Shaha et al., 1984). Immunoreactive β-endorphin and ACTH present in the interstitial fluid surrounding Leydig cells also declined after hypophysectomy and increased after hCG treatment. Short-term and long-term treatments with superactive gonadotropin-releasing hormone (GnRH) agonist stimulated and suppressed, respectively, the β-endorphin concentrations in rat interstitial fluid parallel with changes in testosterone levels (Valenca and Negro-Vilar, 1986). Corticotropin-releasing factor (CRF), which is also produced by Leydig cells, stimulated release of β-endorphin from primary Leydig cell cultures; the effect of CRF was blocked by specific CRF antagonists (Eskeland et al., 1989). The fact that steroid synthesis inhibitors increase β-endorphin secretion suggests that testosterone may inhibit release of this peptide (Fabbri et al., 1988). Additional results indicate that the content of POMC-like mRNA in the testes did not increase as it does in the anterior pituitary after adrenalectomy, suggesting that glucocorticoids are not primary regulators of POMC in the testes as they are in the anterior pituitary (Chen and Madigan, 1987).

THE PARACRINE EFFECTS OF POMC-DERIVED PEPTIDES IN THE TESTES

Even before it was recognized that Leydig cells secrete POMC-derived peptides, it was known that ACTH increased the rate of growth of the TM_4 Sertoli cell line (Bardin et al., 1987a). It was subsequently demonstrated that synthetic ACTH, β-MSH, α-MSH, and des-acetyl α-MSH stimulated cAMP synthesis in Sertoli cells; this effect is rapid, occurring within minutes to hours after addition of the peptides (Bardin et al., 1984; Mather et al., 1984; Boitani et al., 1985, 1986). The action of these peptides, like that of isoproterenol and FSH, was potentiated by forskolin and the phosphodiesterase inhibitor, methylisobutalxanthine (MIX) (Boitani et al., 1986). These peptides shifted the FSH dose–response curve to the left increasing the sensitivity of these cells to this gonadotropin.

The apparent potencies of α-MSH and its des-acetyl derivative could not be distinguished from one another when measured on Sertoli cells (Boitani et al., 1986). Similar results were obtained when aromatase, a Sertoli cell enzyme, was used as an end point rather than cAMP. These observations suggest that the ACTH/MSH family of peptides modulate Sertoli cell function by the cAMP system. The finding that Sertoli cells are equally responsive to α-MSH and des-acetyl-α-MSH was unexpected as the cells of the central nervous system, adrenal cortex, and melanocytes show differential responses to these peptides. Studies on a series of MSH derivatives with a wide range of effects on frog melanocytes indicated that the potencies of these analogues are similar to α-MSH when assayed on Sertoli cells. These observations suggest that the receptors for these peptides in the testes are strikingly different from those on amphibian melanocytes.

In contrast to α-MSH/ACTH, β-endorphin appears to suppress Sertoli cell function as evidenced by reduced FSH-stimulated cAMP accumulation and protein secretion. A series of in vivo experiments on neonatal rats using opiate antagonists provided the first evidence suggesting that β-endorphin and possibly other endogenous testicular opiates may have important suppressive effects on Sertoli cells (Fig. 9–2). These opiate antagonists facilitated compensatory hypertrophy and Sertoli cell protein secretion after unilateral castration of neonatal rats (Gerendai et al., 1983, 1986). In addition, naloxone increased Sertoli cell division in explant cultures of fetal testis (Orth, 1986). As the hypertrophy after unilateral orchidectomy of neonatal rats is believed to occur because of a rise in serum FSH, the opiate antagonists appear to block or negate an endogenous inhibitor of FSH. The increase in testicular size and rABP secretion after antagonist administration is consistent with the hypothesis that β-endorphin may inhibit Sertoli cell growth and secretion during early testicular development. It follows, therefore, that endogenous opiates antagonize FSH, which is the major stimulus for Sertoli cell division during fetal and neonatal life.

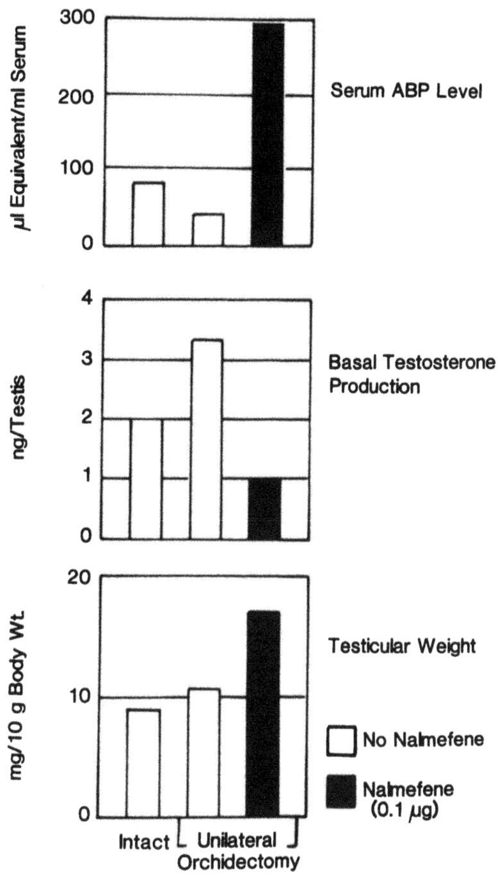

Figure 9–2. An in vivo study showing that intratesticular opiates regulate Sertoli and Leydig cell function. Five-day-old rats were injected intratesticularly (on the right) with or without the opiate antagonist, nalmefene, and the left testis was removed. Animals were killed at 11 days of age. Serum androgen binding protein (ABP), in vitro testosterone secretion from the right testis, and testicular weight were measured. The action of this opiate antagonist was to facilitate testicular hypertrophy (by Sertoli division) and Sertoli cell ABP secretion as well as to inhibit Leydig cell testosterone secretion. Naloxone had the same effects. These data support the postulate that the actions of testicular opiates are to inhibit Sertoli cells and stimulate Leydig cells under the conditions tested (Gerendai et al., 1983, 1986).

If one considers that an opiate antagonist administered either in vivo or in vitro is active for only a brief period of time and that Sertoli cell division results from such treatment, then one must assume that Sertoli cell division is being chronically held in check by endogenous opiates, and the blockade of its action for even a short time allows for the onset of mitosis in Sertoli cells. Because Sertoli cell division can occur only in fetal and neonatal life before closure of the blood–testis barrier, the number of these cells and testicular size are fixed at puberty. In view of this, testicular size is regulated by the opposing actions of FSH and β-endorphin. These concepts are summarized in Table 9–6. The observation of opiate receptors on pubertal Sertoli cells and the inhibition of FSH-induced rABP secretion in vitro by an opiate agonist suggest that the action of opiates on Sertoli cells can continue once Sertoli cell division is complete (Fabbri et al., 1985).

Taken together, the results from in vitro and in vivo experiments suggest that peptides derived from different portions of POMC have differential effects on Sertoli cells. As noted earlier, ACTH and MSH-like peptides stimulate production of cAMP in Sertoli cells over a short time frame. By contrast, the inhibitory effect of β-endorphin on Sertoli cell proliferation and protein are slow or "tonic" relative to those of the peptides with MSH activity. Thus, the biological consequences of POMC-derived peptides in the testes depend on the amount of the precursor that is made, which of its component peptides are present in biologically active form (acetylated or nonacteylated), the duration of time over which they act, and the presence of hormone(s) with which they interact.

Table 9–6. Sertoli Cells as a Determinant of Testicular Size

1. Before puberty, Sertoli cell division stops; testicular size is fixed as each Sertoli cell supports a fixed number of germ cells.
2. FSH stimulates division; if its action is reduced before puberty, the testes will be small.
3. Testicular opiates inhibit Sertoli cell division; if their actions are blocked, the testes will be large.
4. Therefore: Sertoli cell number and thus testicular size are determined by opposing actions of FSH and opiates.

THE AUTOCRINE EFFECTS OF POMC-DERIVED PEPTIDES

An effect of opiate antagonists and Leydig cell function was also observed in unilaterally castrated neonatal and adult rats (Gerendai et al., 1984, 1986). In neonatal animals, these antagonists inhibited basal testosterone secretion from the remaining testes (see Fig. 9–1); this effect was also observed in adult male animals as judged by reduced testosterone secretion both in vivo and in vitro. These effects of opiate antagonists on Leydig cell function implied that opiates either directly or indirectly facilitate testosterone secretion. Although there appears to be an explanation for the actions of opiate antagonists on Sertoli cells, their action on Leydig cells is not understood in view of the short time that the antagonist remains in the treated testis as contrasted to the prolonged time over which the antagonist appears to exert its biological effect. The results suggest, nonetheless, that opiates help regulate the increase in androgen secretion that occurs after unilateral orchidectomy.

To investigate the direct effects on β-endorphin in vitro, adenylyl cyclase activity in a TM_3 Leydig cell line was evaluated; treatment with β-endorphin resulted in an increase in adenylyl cyclase activity. By contrast, in primary immature Leydig cells from 20-day-old rats this peptide depresses adenylyl cyclase activity (Morris and Bardin, 1987). In spite of the inconsistency of the results between experiments with cell lines and primary cells, the findings are consistent with a direct action of an opiate on Leydig cells (Bardin et al., 1987a). In addition, a variety of experiments indicate that under some conditions β-endorphin can stimulate both basal and hCG-stimulated androgen secretion by primary rat Leydig cells, an observation that is consistent with in vivo experiments using the antagonists noted previously. These studies suggest that β-endorphin may function as an autocrine modulator that enhances the sensitivity of Leydig cells to LH (Bardin et al., 1987a).

There is now considerable evidence that POMC is produced in a variety of tissues. It has been, however, most extensively studied in the testis. In fact, it is one of the few such factors produced by the testis that fulfills the criterion of a local regulator as outlined in Table 9–1.

B. Oxytocin and Vasopressin

Oxytocin and vasopressin have been identified by radioimmunoassay in the testes of the human, rat, mouse, and dog. Oxytocin has also been identified in cat and rabbit. Oxytocin and vasopressin, like peptides, have been isolated from testicular extracts (Pickering et al., 1989). Testes of Brattleboro rats contain an arginine vasopressinlike peptide, even though these animals lack this peptide in the hypothalamus, pituitary, and blood. Blood levels of oxytocin are higher in sheep testicular venous blood than in testicular artery, and explants of testis (a dog and human) release immunoreactive oxytocin into the media for a number of days (Pickering et al., 1986). Northern blots detected vasopressin and oxytocin mRNAs in extracts of rat and pig testes (Rehbein et al., 1986).

A variety of experiments indicate that oxytocin is synthesized and secreted by Leydig cells. These include immunostaining, disappearance of immunoassayable oxytocin from ethane-1,2,-dimethanesulphonate (EDS)-treated animals, and secretion of immunoactive oxytocin by Leydig cell-enriched cultures (Nicholson et al., 1987; Pickering et al., 1989).

Adashi and Hsueh (1981) demonstrated that neurohypophysial hormones inhibited androgen biosynthesis in Leydig cells, presumably through an inhibitory receptor that couples to adenylate cyclase through the inhibitory regulatory membrane protein N_1 (Adashi et al., 1984). Subsequently, other investigators have shown a stimulatory action of vasopressin (Tahri-Joutei et al., 1987) or have concluded that it was inactive (Sharpe and Cooper, 1987).

In another series of studies, Pickering and co-workers (1989) demonstrated that oxytocin increases both low and high frequency contractions. A study using synthetic analogues was conducted to ascertain the type of receptor in the seminiferous tubules. An agonist (Gly^7-oxytocin) that is specific for oxytocin receptor in rat uterus was inactive on the seminiferous tubules. A V_2 agonist was inhibitory, whereas a V_1 agonist both increased tubule contractility and augmented the action of oxytocin. Based on these studies, oxytocin is believed to act on the seminiferous tubule by way of the V_1 receptor (Pickering et al., 1989). The actual demonstration of this receptor, however, has not been performed. In a separate set of studies, it has been clearly demonstrated that androgen action on the seminiferous tubule is also essential for the action of oxytocin (Nicholson et al., 1987).

Whether oxytocin acts on myoid cells or myoid plus Sertoli cells is not entirely clear. It is known, however, that a Sertoli cell product can facilitate the release of oxytocin from Leydig cells (Pickering et al., 1989).

IV. Myoid Cell Products that Signal Sertoli Cells

Experiments from our own and other laboratories show that myoid cells secrete many products in vitro (Mather et al., 1983). The demonstration that at least one of these products, which is increased by androgens, modulates Sertoli cell function, posed the question as to whether the effects of testosterone on Sertoli cells are direct or indirect (Holmes et al., 1984; Skinner and Fritz, 1985a,b; Verhoeven and Cailleau, 1988). Skinner and associates identified Sertoli cell stimulatory activity in fractionated myoid cell culture media that they termed P-Mod-S. This protein stimulates Sertoli cell transferrin, androgen-binding protein, lactalbumin, and radiolabeled protein, but not plasminogen activator secretion (Skinner and Fritz, 1986). P-Mod-S was purified by heparin affinity and reverse phase chromatography. Two forms of this protein were identified with similar biological activities. P-Mod-S(A) and P-Mod-S(B) have molecular weights of 56,000 and 59,000 and distinct amino acid compositions, respectively. The staining pattern of these isoforms indicated that P-Mod-S(A) is more heterogeneous. The purified forms of P-Mod-S have a greater effect on Sertoli cell function than any other previously identified cellular regulator (Skinner and Fritz, 1986; Skinner et al., 1988).

The mechanism of action of P-Mod-S was investigated. Unlike FSH, which stimulated cAMP levels after 60-minute and 72-hour treatments, P-Mod-S had no effect on cAMP levels at 60 minutes and only a small effect at 72 hours. P-Mod-S stimulated cAMP levels and had no effect on inositol phosphate hydrolysis. Both forms of P-Mod-S had similar mechanisms of action (Norton and Skinner, 1989).

The interaction of myoid and Sertoli cells in the testes has been compared with the mesenchymal–epithelial interactions in other tissues, such as prostate and mammary gland, where androgens stimulate and inhibit epithelial function, respectively. Recent studies have demonstrated that a stromal cell product of the prostate is similar to the androgen-induced product of myoid cells with regard to transferrin production by Sertoli cells. The prostatic mediator appears to have the same size and biological actions as P-Mod-S. These observations suggest that mesenchymal–epithelial interactions play a role in androgen-supported regulation of testicular function and that mesenchymal cells from different androgen-dependent organs produce identical or similar mediators of androgen action (Skinner and Fritz, 1986; Skinner et al., 1989a). It will be of interest if the same mesenchymal hormone mediates androgen-induced regression of mammary epithelium.

Recently, LH and hCG were shown to increase inhibin blood levels in vivo (Sharpe et al., 1988). Because neither LH nor testosterone has any direct action on Sertoli cell inhibin secretion, a better understanding of the action of this gonadotropin was sought. Both P-Mod-S(A) and P-Mod-S(B) stimulated inhibin secretion by Sertoli cells after 7 days in culture. The effects of P-Mod-S were additive to those of FSH. P-Mod-S provides one mechanism whereby LH and testosterone could influence inhibin secretion (Skinner et al., 1989b), even though the time for the rise in inhibin after LH in vivo is much faster than P-Mod-S treatment in vitro.

In addition to P-Mod-S, myoid cells also produce several other autocrine/paracrine regulators. These include TGF-β, TGF-α, and matrix proteins (Table 9–7). It is anticipated that many other such factors will be identified as secretory products of myoid cells in view of the large number of proteins that this cell produces (Mather et al., 1983; Shabanowitz et al., 1986).

Table 9–7. Myoid Cell Proteins/Peptides

P-Mod-S	Skinner et al., 1988 Skinner et al., 1989b
TGF-β	Saez et al., 1989
TGF-α	Skinner et al., 1989a
Bio-Matrix	Mather et al., 1982

V. Germ Cell Products that Signal Sertoli Cells

The postulate that germ cells produce factors that modulate Sertoli cell function is based on the interaction of these cells that is required to maintain the spermatogenic cycle (Parvinen, 1982) and polarized secretion of fluid and proteins by Sertoli cells (Bardin et al., 1988). The recent demonstration that short-term cultures of primary spermatocytes and spermatids can be maintained in culture for over 24 hours provided the opportunity to study the secretory products of these cells (Djakiew and Dym, 1988; Le Magueresse et al., 1988). The stimulatory effect of media from germ cells on Sertoli cell transferrin secretion is shown in Fig. 9–3. In addition to transferrin, germ cell products also increase ABP secretion and suppress aromatase activity (Le Magueresse et al., 1986; Le Magueresse and Jegou, 1988). The presence of mRNA for β-nerve growth factor (NGF) receptors in Sertoli cells and its down regulation by

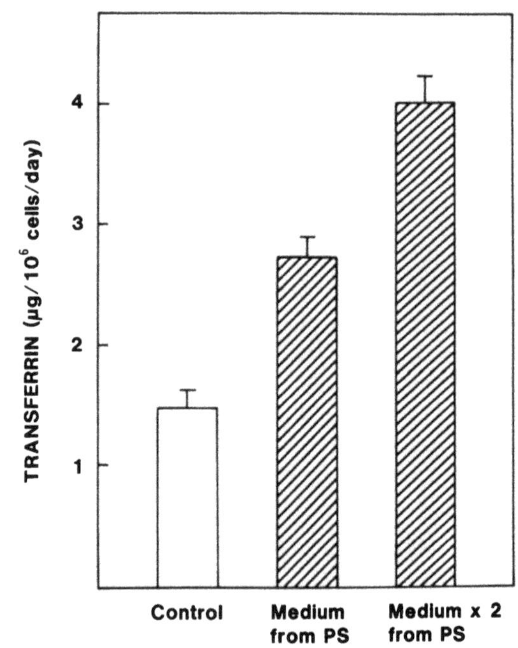

Figure 9–3. The effects of germ cell products on Sertoli cells in vitro. Media from cultures of primary spermatocytes (PS) were incubated with primary Sertoli cell cultures and transferrin accumulation was measured. X2 = concentrated medium from primary spermatocytes. (From Djakiew and Dym, 1988.)

testosterone suggest the germ cell product NGF is involved in the interaction of germ and Sertoli cells, with the testicular function being strongly modulated by testosterone from Leydig cells (Persson et al., 1990). These bioassays suggest that there are multiple germ cell products that affect Sertoli cells. The purification of these germ cell products has just begun; at present it is not known whether these effects on Sertoli cells are by known regulators or by unique molecular entities.

VI. Conclusion

Millions of years ago, biologically active peptides were adapted by primitive organisms as humoral messengers during the early phases of chemical evolution. Although some changes in their structures have occurred with time, there has been considerable constraint in the scope of peptide diversification. As organisms became more complex and additional endocrine, paracrine, and autocrine mediators were needed, new functions were assumed by existing molecules. This diversity has been achieved by a number of mechanisms. For example, diverse peptides may be derived from the same precursor molecule, and such peptides may have either similar or opposing actions as is the case with POMC-derived peptides. Another mechanism is gene duplication followed by limited evolution so that a series of related molecules are generated that are active either alone or through the formation of homodimers or heterodimers. Such peptides may also have similar or opposing functions and their functions may change with the state of differentiation as is the case for inhibin, activin, TGF-β, and other members of this peptide superfamily. Still another mechanism of diversification exists when a peptide is secreted from different cell types in response to different stimulators as is the case of IGF-1 from hepatic and Sertoli cells. These and other mechanisms greatly expand the informational potential of the genome in that a single gene may represent a unit that is eventually expressed as multiple signals that control a pattern of activity or behavior with a given organ. Thus, the regulation of differentiated function in the testis is not only dependent on hormones in blood (the long-loop systems), but also on ultrashort-loop and short-loop feedback systems that use autocrine and paracrine modulators made up of signaling systems that are used in many other organs.

ACKNOWLEDGMENTS

This work was supported by NIH Grant HD13541.

REFERENCES

Forest, M. G., and Pugeat, M., eds. (1986). *Binding Proteins of Steroid Hormones,* Vol. 149, London–Paris, John Libbey Eurotext Ltd.

Adashi, E. Y., Resnick, C. E., Cronin, M. J., and Hewlett, E. L. (1984). Antigonadal activity of the neurohypophysial hormones in cultured rat testicular cells: Abolition by pertussis toxin. Endocrinology 115:839–841.

Adashi, E. Y., and Hsueh, A.J.W. (1981). Direct inhibition of testicular androgen biosynthesis revealing antigonadal activity of neurohypophysial hormones. Nature 293:650–652.

Aoki, A., and Fawcett, D. W. (1978). Is there a local feedback from the seminiferous tubules affecting activity of the Leydig cells. Biol. Reprod. 19:144–158.

Ascoli, M., and Segaloff, D. L. (1989). Regulation of the differentiated functions of Leydig tumor cells by epidermal growth factor. Ann. NY Acad. Sci. 564:99–115.

Avallet, O., Vigler, M., Perrard-Sapori, M. H., and Saez, J. M. (1987). Transforming growth factor β inhibits Leydig cell functions. Biochem. Biophys. Res. Commun. 146:575.

Bardin, C. W., Morris, P. L., Shaha, C., Feng, Z. M., Rossi, V., Vaughan, J., Vale, W. W., Voglmayr, J., and Chen, C.-L.C. (1989). Inhibin structure and function in the testis. Ann. NY Acad. Sci. 564:10–23.

Bardin, C. W., Cheng, C. Y., Musto, N. A., and Gunsalus, G. L. (1988). The Sertoli cell, in *The Physiology of Reproduction* (Knobil, E., Neill, J. D., Ewing, L. L., Greenwald, G. S., Markert, C. L., and Pfaff, D. W., eds.) Vol. 1, pp. 933–974, Raven Press, New York.

Bardin, C. W., Chen, C.-L.C., Morris, P. L., Gerendai, I., Boitani, C., Liotta, A. S., Margioris, A., and Krieger, D. T. (1987a). Pro-opiomelanocortin-derived peptides in testis, ovary and tissues of reproduction. Recent Prog. Horm. Res. 43:1–28.

Bardin, C. W., Morris, P. L., Chen, C.-L.C., Shaha, C., Voglmayr, J., Rivier, J., Spiess, J., and Vale, W. W. (1987b). Testicular inhibin: Structure and regulation by FSH, androgens and EGF, in *Inhibin: Non-Steroidal Regulation of Follicle Stimulating Hormone Secretion, Serono Symposia,* Vol. 42 (Burger, H. G., deKretser, D. M., Findlay, J. K., and Igarashi, M., eds.) pp. 179–190, Raven Press, New York.

Bardin, C. W., Shaha, C., Mather, J., Salomon, Y., Margioris, A. N., Liotta, A. S., Gerendai, I., Chen, C.-L., and Krieger, D. T. (1984). Identification and possible function of pro-opiomelanocortin-derived peptides in the testis. Ann. NY Acad. Sci. 438:346–364.

Bardin, C. W., Musto, N., Gunsalus, G., Kotite, N., Cheng, S.-L., Larrea, F., and Becker, R. (1981). Extracellular androgen binding proteins. Ann. Rev. Physiol. 43:189–198.

Barrett, A. J., Brown, M. A., and Sayers, C. A. (1979). The electrophoretically slow and fast forms of the α_2-macroglobulin molecule. Biochem. J. 181:401–418.

Barrett, A. J., and Starkey, P. M. (1973). The interaction of α_2-macroglobulin with proteinases: Characteristics and specificity of the reaction and a hypothesis concerning its molecular mechanisms. Biochem. J. 133:709–724.

Benahmed, M., Morera, A. M., and Chauvin, M. A. (1986a). Evidence for a Sertoli cell, FSH-suppressible inhibiting factor(s) of testicular steroidogenic activity. Biochem. Biophys. Res. Commun. 139:169–178.

Benahmed, M., Tabone, E., Grendt, C., Sanchez, P., Chauvin, M. A., and Morera, A. M. (1986b). Paracrine control of Leydig cell activity by FSH dependent proteins from Sertoli cells: An in vitro study. J. Steroid Biochem. 24:311–315.

Benahmed, M., Reventos, J., Tabone, E., and Saez, J. M. (1985). Cultured Sertoli cell-mediated FSH stimulatory effect on Leydig cell steroidogenesis. Am. J. Physiol. 248:E176–E181.

Bergh, A. (1983). Paracrine regulation of Leydig cells by the seminiferous tubules. Int. J. Androl. 6:57–65.

Bettuzzi, S., Hiipakka, R. A., Gilna, P., and Liao, S. (1989). Identification of an androgen-repressed mRNA in rat ventral prostate as coding for sulfated glycoprotein-2 by cDNA cloning and sequence analysis. Biochem. J. 257:293–296.

Blaschuk, O., and Fritz, I. B. (1984). Isoelectric forms of clusterin isolated from rat rete testis fluid and from secretions of primary cultures of ram and rat Sertoli cell-enriched preparations. Can. J. Biochem. Cell Biol. 62:456–461.

Boitani, C., Mather, J. P., and Bardin, C. W. (1986). Stimulation of cAMP production in rat Sertoli cells by α-MSH and des-acetyl α-MSH. Endocrinology 118:1513–1518.

Boitani, C., Chen, C.-L.C., Margioris, A. N., Morris, P. L., and Bardin, C. W. (1985). Pro-opiomelanocortin-derived peptides in the testis: Evidence for a possible role in Leydig and Sertoli cell function. Med. Biol. 63:251–258.

Bordin, S., and Petra, P. H. (1980). Immunocytochemical localization of the sex steroid-binding protein (SBP) of plasma in tissues of the adult monkey, *Macaca nemestrina.* Proc. Natl. Acad. Sci. (USA) 77:5678–5682.

Borland, K., Mita, M., Oppenheimer, C. L., Blinderman, L. A., Massague, J., Hall, P. F., and Czech, M. P. (1984). The actions of insulin-like growth factors I and II on cultured rat Sertoli cells. Endocrinology 114:240–246.

Brown, K. D., Blakeley, D. M., Henville, A., and Setchell, B. P. (1982). Rete testis fluid contains a growth factor for cultured fibroblasts. Biochem. Biophys. Res. Commun. 105:391–397.

Burger, H. G., McLachlan, R. I., Robertson, D. M., Bremner, W. J., and De Kretser, D. M. (1989). Inhibin and the regulation of testicular function. Ann. NY Acad. Sci. 564:1–9.

Buttyan, R., Olsson, C. A., Pintar, J., Chang, C., Bandyk, M., and Sawczuk, I. S. (1989). Induction of the TRPM-2 gene in cells undergoing programmed death. Mol. Cell. Biol. 9:3473–3481.

Calkins, J. H., Siegel, M. M., Nankin, H. R., and Lin, T. (1988). Interleukin-1 inhibits Leydig cell steroidogenesis in primary culture. Endocrinology 123:1605–1609.

Carreau, S., Papadopoulos, V., and Drosdowsky, M. A. (1988). Stimulation of adult rat Leydig cell aromatase activity by a Sertoli cell factor. Endocrinology 122:1103–1109.

Charreau, E. H., Calvo, J. C., Tesone, M., De Souza Valle, L. B., and Baranao, J. L. (1978). Insulin regulation of Leydig cells luteinizing hormone receptors. J. Biol. Bhem. 253:2504–2506.

Chatelain, P. G., Naville, D., and Saez, J. M. (1987). Somatomedin-C/insulin-like growth factor-I-like material secreted by porcine Sertoli cells in vitro: Characterization and regulation. Biochem. Biophys. Res. Commun. 146:1009–1017.

Chemes, H. (1986). The phagocytic function of Sertoli cells: A morphological, biochemical, and endocrinological study of lysosomes and acid phosphatase localization in the rat testis. Endocrinology 119:1673–1681.

Chen, C.-L.C., Mather, J. P., Morris, P. L., and Bardin, C. W. (1984). Expression of pro-opiomelanocortin-like gene in the testis and epididymis. Proc. Natl. Acad. Sci. (USA) 81:5672–5675.

Chen, C.-L.C., Chang, C.-C., Krieger, D. T., and Bardin, C. W. (1986). Expression and regulation of pro-opiomelanocortin-like gene in the ovary and placenta: Comparison with the testis. Endocrinology 118:2382–2389.

Chen, C.-L.C., and Madigan, M. (1987). Regulation of testicular proopiomelanocortin-like gene expression. Endocrinology 121:590–596.

Cheng, C. Y., Grima, J., Stahler, M. S., Guglielmotti, A., Silvestrini, B., and Bardin, C. W. (1990). Sertoli cell synthesizes and secretes a protease inhibitor, α_2-macroglobulin. Biochemistry 29:1063–1068.

Cheng, C. Y., Grima, J., Stahler, M. S., Lockshin, R. A., and Bardin, C. W. (1989). Testins are two structurally related Sertoli cell secretory proteins whose secretion is tightly coupled to the presence of germ cells. J. Biol. Chem. 264:21386–21393.

Cheng, C. Y., Bardin, C. W., Boitani, C., Grima, J., and Morris, I. (1988a). Testins are Sertoli cell proteins that may be used to study cell–cell interactions within the seminiferous epithelium, in *Progress in Endocrinology* (Imura, H., Shizume, K., and Yoshida, S., eds.) Vol. 2,

pp. 1157–1162, Elsevier Science Publishers B.V., Amsterdam.
Cheng, C. Y., Chen, C.-L. C., Feng, Z. M., Marshall, A., and Bardin, C. W. (1988b). Rat clusterin isolated from primary Sertoli cell-enriched culture medium is sulfated glycoprotein-2 (SGP-2). Biochem. Biophys. Res. Commun. 155:398–404.
Cheng, C. Y., Mathur, P. P. and Grima, J. (1988c). Structural analysis of clusterin and its subunits in ram rete testis fluid. Biochemistry 27:4079–4088.
Cheng, C. Y., and Bardin, C. W. (1987). Identification of two testosterone-responsive testicular proteins in Sertoli cell-enriched culture medium whose secretion is suppressed by cells of the intact seminiferous tubule. J. Biol. Chem. 262:12768–12779.
Cheng, C. Y., Grima, J., Lee, W. M., and Bardin, C. W. (1987). The distribution of rat testibumin in the male reproductive tract. Biol. Reprod. 37:875–885.
Cheng, C. Y., and Bardin, C. W. (1986). Rat testicular testibumin is a protein responsive to follicle stimulating hormone and testosterone that shares immunodeterminants with albumin. Biochemistry 25:5276–5288.
Cheng, C. Y., Mather, J. P., Byer, A. L., and Bardin, C. W. (1986). Identification of hormonally responsive proteins in primary Sertoli cell culture medium by high performance liquid chromatography. Endocrinology 118:480–488.
Cheng, M. C., Clements, J. A., Smith, A. I., Lolait, S. J., and Funder, J. W. (1984). *N*-Acetyl endorphin in rat spermatogonia and primary spermatocytes. J. Clin. Invest. 75:832–835.
Choi, N.-H., Mazda, T., and Tomita, M. (1989). A serum protein SP40,40 modulates the formation of membrane attack complex of complement on erythrocytes. Mol. Immunol. 26:835–840.
Clark, A.J.L., Lavender, P. M., Coates, P., Johnson, M. R., and Rees, L. H. (1989). The short pro-opiomelanocortin mRNA fails to produce secreted peptide in permanently transfected GH3 cell lines. Endocrine Society Meeting Abstract No. 954.
Collard, M. W., and Griswold, M. D. (1987). Biosynthesis and molecular cloning of sulfated glycoprotein-2 secreted by rat Sertoli cells. Biochemistry 26:3297–3303.
Daughaday, W. H., and Rotwein, P. (1989). Insulin-like growth factors I and II. Peptide, messenger ribonucleic acid and gene structures, serum, and tissue concentrations. Endocrine Rev. 10:68–90.
Djakiew, D., and Dym, M. (1988). Pachytene spermatocyte proteins influence Sertoli cell function. Biol. Reprod. 39:1193–1205.
Djakiew, D., Hadley, M. A., Byers, S. W., and Dym, M. (1986). Transferrin-mediated transcellular transport of ^{59}Fe across confluent epithelial sheets of Sertoli cells grown in bicameral cell culture chambers. J. Androl. 7:355–366.
Duguid, J. R., Bormont, C. W., Liu, N., and Tourtellotte, W. W. (1989). Changes in brain gene expression shared by scrapie and Alzheimer disease. Proc. Natl. Acad. Sci. (USA) 86:7260–7264.
Dunbar, C. E., Browder, T. M., Abrams, J. S., and Nienhuis, A. W. (1989). COOH-terminal-modified interleukin-3 is retained intracellularly and stimulates autocrine growth. Science 245:1493–1496.
Dym, M., and Fawcett, D. W. (1970). The blood–testis barrier in the rat and the physiological compartmentation of the seminiferous epithelium. Biol. Reprod. 3:308–326.
Elson, S. D., Browne, C. A., and Thorburn, G. D. (1984). Identification of epidermal growth factor-like activity in human male reproductive tissue and fluids. J. Clin. Endocrinol. Metab. 58:589–594.
Eskeland, N. L., Lugo, D. I., Pintar, J. E., and Schachter, B. S. (1989). Stimulation of β-endorphin secretion by corticotropin-releasing factor in primary rat Leydig cell cultures. Endocrinology 124:2914–2929.
Evans, C. J., Lorenz, R., Weber, E., and Barchas, J. D. (1982). Variants of alpha-melanocyte stimulating hormone in rat brain and pituitary: Evidence that acetylated α-MSH exists only in the intermediate lobe of pituitary. Biochem. Biophys. Res. Commun. 106:910–919.
Fabbri, A., Knox, G., Buczko, E., and Dufau, M. L. (1988). β-endorphin production by the fetal Leydig cell: Regulation and implications for paracrine control of Sertoli cell function. Endocrinology 122:749–756.
Fabbri, A., Tsai-Morris, C. H., Luna, S., Fraioli, F., and Dufau, M. L. (1985). Opiate receptors are present in the rat testis. Identification and localization in Sertoli cells. Endocrinology 117:2544–2546.
Fauser, B.C.J.M., and Hsueh, A.J.W. (1988). Effect of transforming growth factor-β on human chorionic gonadotropin induced testosterone production by cultured rat testicular cells. Life Sci 43:1363–1370.
Fawcett, D. W. (1975). Ultrastructure and function of the Sertoli cell, in *Handbook of Physiology, Section 7: Endocrinology, Volume V: Male Reproductive System* (Hamilton, D. W., and Greep, R. O., eds.) pp. 21–55, American Physiological Society, Washington, D.C.
Feldman, S. R., and Pizzo, S. V. (1984). Comparison of the binding of chicken α-macroglobulin and ovomacroglobulin to the mammalian α_2-macroglobulin receptor. Arch. Biochem. Biophys. 235:267–275.
Foresta, C., Manoni, F., Businaro, V., Donadel, C., Indino, M., and Scandellari, C. (1986). Possible significance of transferrin levels in seminal plasma of fertile and infertile men. J. Androl. 7:77–82.
Franchimont, P., Croze, F., Demoulin, A., Bologne, R., and Hustin, J. (1981). Effect of inhibin on rat deoxyribonucleic acid (DNA) synthesis in vivo and in vitro. Acta Endocrinol. (Copenh) 98:312.
Fritz, I. B., Burdzy, K., Setchell, B., and Blaschuk, O. (1983). Ram rete testis fluid contains a protein clusterin which influences cell–cell interactions in vitro. Biol. Reprod. 28:1173–1188.
Garrett, J. E., and Douglass, J. O. (1989). Human chorionic gonadotropin regulates expression of the proenkephalin gene in adult rat Leydig cells. Mol. Endocrinol. 3:2093–2100.
Geffner, M. E., Golde, D. W., Lippe, B. M., Kaplan, S. A., Bersch, N., and Li, C. H. (1987). Tissues of the Laron dwarf are sensitive to insulin-like growth factor I but not to growth hormone. J. Clin. Endocrinol. Metab. 64:1042–1046.
Gehring, M. R., Shiels, B. R., Northemann, W., de Bruijn, M.H.L., Kan, C. C., Chain, A. C., Noonan, D. J., and

Fey, G. H. (1987). Sequence of rat liver α_2-macroglobulin and acute phase control of its messenger RNA. J. Biol. Chem. 262:446–454.

Gerendai, I., Shaha, C., Gunsalus, G. L., and Bardin, C. W. (1986). The effects of opioid receptor antagonists suggest that testicular opiates regulate Sertoli and Leydig cell function in the neonatal rat. Endocrinology 118:2039–2044.

Gerendai, I., Shaha, C., Thau, R., and Bardin, C. W. (1984). Do testicular opiates regulate Leydig cell function? Endocrinology 115:1645–1647.

Gerendai, I., Nemeskeri, A., and Csernus, A. (1983). Naloxone has a local effect on the testis of immature rats. Andrologia 15:398–403.

Gershagen, S., Lundwall, A., and Fernlund, P. (1989). Characterization of the human sex hormone binding globulin (SHBG) gene and demonstration of two transcripts in both liver and testis. Nucleic Acids Res 17:9245–9258.

Gizang-Ginsberg, E., and Wolgemuth, D. J. (1987). Expression of the proopiomelanocortin gene is developmentally regulated and affected by germ cells in the male mouse reproductive system. Proc. Natl. Acad. Sci. (USA) 84:1600–1604.

Gizang-Ginsberg, E., and Wolgemuth, D. J. (1985). Localization of mRNA in mouse testes by in situ hybridization: Distribution of alpha-tubulin and development stage specificity of pro-opiomelanocortin transcripts. Devel. Biol. 111:293–305.

Gnessi, L., Isidori, A., Bolotti, M., Altamura, S., Ulisse, S., Jannini, E. A., Fabbri, A., and Spera, G. (1989). Identification of immunoreactive gastrin-releasing peptide related substances in adult rat Leydig cells. Endocrinology 124:558–560.

Gonzalez-Machon, C., and Vale, W. (1989). Activin-A, inhibin and transforming growth factor-β modulate growth of two gonadal cell lines. Endocrinology 125:1666–1672.

Grima, J., Zwain, I., Lockshin, R. A., Bardin, C. W., and Cheng, C. Y. (1990). Diverse secretory patterns of clusterin by epididymis and prostate undergoing cell regression following orchiectomy. Endocrinology 126:2989–2997.

Griswold, M. D., Heckert, L., Karl, A. F., Law, L., Roberts, K., Siiteri, J. E., Stallard, B., Tsuruta, J., and Sylvester, S. R. (1989). Sertoli cell secretory proteins in normal and synchronized testes, in *Development and Function of the Reproductive Organs* (Josso, N., ed.) Vol. III, pp. 207–217, Serono Symposium Review Aves-Serono Symposia, Rome.

Gunsalus, G. L., Carreau, S., Porto, C., Bardin, C. W. (1991). ABP/TeBG: Steroid hormone transporter or signal transducer? *Advances in Reproductive Biology,* Chinese Academy of Sciences, pp. 96–105.

Gunsalus, G. L., Musto, N. A., Bardin, C. W., Kunz, H. W., and Gill III, T. J. (1985). Rats homozygous for the *grc* complex have defective transport of androgen binding protein to the epididymis, but normal secretion into the blood. Biol. Reprod. 33:1057–1063.

Gunsalus, G. L., Larrea, F., Musto, N. A., Becker, R. R., Mather, J. P., and Bardin, C. W. (1981). Androgen binding protein as a marker for Sertoli cell function. J. Steroid Biochem. 15:99–106.

Gunsalus, G. L., Musto, N. A., and Bardin, C. W. (1980). Bidirectional release of a Sertoli cell product, androgen binding protein, into the blood and seminiferous tubule, in *Testicular Development, Structure and Function* (Steinberger, A., and Steinberger, E., eds.) pp. 291–297, Raven Press, New York.

Hadley, M. A., Djakiew, D., Byers, S. W., and M. Dym (1987). Polarized secretion of androgen-binding protein and transferrin by Sertoli cells grown in a bicameral culture system. Endocrinology 120:1097–1103.

Hammond, G. L., Underhill, D. A., Rykse, H. M., and Smith, C. L. (1989). The human sex hormone-binding globulin contains exons for androgen-binding protein and two other testicular messenger RNAs. Mol. Endocrinol. 3:1869–1876.

Handelsman, D. J., Spaliviero, J. A., Scott, C. A., and Baxter, R. C. (1985). Identification of insulin-like growth factor-I and its receptor in the rat testis. Acta Endocrinol. 120:543–549.

Hansson, H.-A., Billig, H., and Isgaard, J. (1989). Insulin-like growth factor I in the developing and mature rat testis: Immunohistochemical aspects. Biol. Reprod. 40:1321–1328.

Hansson, V., Ritzen, E. M., French, F. S., and Nayfeh, S. N. (1975). Androgen transport and receptor mechanisms in testis and epididymis, in *Handbook of Physiology, Section 7: Endocrinology* (Hamilton, D. W., and Greep, R. O., eds.) Vol. V, Male Reproductive System, pp. 173–201, American Physiological Society, Washington, D.C.

Harpel, P. C. (1973). Studies on human plasma α_2-macroglobulin-enzyme interactions. Evidence for proteolytic modification of the subunit chain structure. J. Exp. Med. 138:508–521.

Hochereau-de Reviers, M. T., de Reviers, M. M., Monet-Kuntz, C., Perreau, C., Fontaine, I., and Viguier-Martinez, M. C. (1987). Testicular growth and hormonal parameters in the male Snell dwarf mouse. Acta Endocrinol. 115:399–405.

Holland, P.W.H., Harper, S. J., McVey, J. H., and Hogan, B.L.M. (1987). In vivo expression of mRNA for the Ca^{++}-binding protein SPARC (osteonectin) revealed by in situ hybridization. J. Cell. Biol. 105:473–482.

Holmes, S. D., Spotts, G., and Smith, R. G. (1986). Rat Sertoli cells secrete a growth factor that blocks epidermal growth factor (EGF) binding to its receptor. J. Biol. Chem. 261:4076–4080.

Holmes, S. D., Lipshultz, L. I., and Smith, R. G. (1984). Regulation of transferrin secretion by human Sertoli cells cultured in the presence or absence of human peritubular cells. J. Clin. Endocrinol. Metab. 59:1058–1062.

Holmes, S. D., Lipshultz, L. I., and Smith, R. G. (1982). Transferrin and gonadal dysfunction in man. Fertil. Steril. 38:600–604.

Hsueh, A.J.W., Dahl, K. D., Vaughan, J., Tucker, E., Rivier, J., Bardin, C. W., and Vale, W. (1987). Hetero- and homo-dimers of inhibin subunits have different paracrine action in the modulation of LH-stimulated androgen biosynthesis. Proc. Natl. Acad. Sci. (USA) 84:5082–5086.

Huang, S. S., O'Grady, P., and Huang, J. S. (1988). Human transforming growth factor $\beta.\alpha_2$-macroglobulin complex is a latent form of transforming growth factor β. J. Biol. Chem. 263:1535–1541.

Huggenvik, J. I., Idzerda, R. L., Haywood, L., Lee, D. C., McKnight, G. S., and Griswold, M. D. (1987). Transferrin messenger ribonucleic acid: Molecular cloning and hormonal regulation in rat Sertoli cells. Endocrinology 120:332–340.

Ishida, H., Risbridger, G. P., and deKretser, D. M. (1987). Variation in the effect of a nongonadotropic Leydig cell stimulating factor in testicular interstitial fluid after exposure of the testis to a single episode of heat treatment. J. Androl. 8:247–252.

Jaillard, C., Chatelain, P. G., and Saez, J. M. (1987). In vitro regulation of pig Sertoli cell growth and function. Effects of fibroblast growth factor and somatomedin-C. Biol. Reprod. 37:665–674.

Janecki, A., and Steinberger, A. (1987). Bipolar secretion of androgen-binding protein and transferrin by Sertoli cells cultured in a two-compartment culture chamber. Endocrinology 120:291–298.

Janecki, A., Jakubowiak, A., and Lukaszyk, A. (1985). Stimulatory effect of Sertoli cell secretory products on testosterone secretion by purified Leydig cells in primary culture. Mol. Cell. Endocrinol. 42:235–243.

Jansz, G. F., and Ponerantz, D. K. (1987). The effect of spermatogenic disruption on the ability of testicular fluid to stimulate androgen production by normal Leydig cells. Biol. Reprod. 36:807–815.

Jeannotte, L., Burbach, J.P.H., and Drouin, J. (1987). Unusual proopiomelanocortin ribonucleic acids in extrapituitary tissues: Intronless transcripts in testes and long poly(A) tails in hypothalamus. Mol. Endocrinol. 1:749–757.

Jenne, D. E., and Tischopp, J. (1989). Molecular structure and functional characterization of a human complement cytolysis inhibitor found in blood and seminal plasma: Identity to sulfated glycoprotein 2, a constituent of rat testis fluid. Proc. Natl. Acad. Sci. (USA) 86:7123–7127.

Joseph, D. R., Hall, S. H., Conti, M., and French, F. S. (1988). The gene structure of rat androgen-binding protein: Identification of potential regulatory deoxyribonucleic acid elements of a follicle-stimulating hormone-regulated protein. Mol. Endocrinol. 2:3–13.

Joseph, D. R., Hall, S. H., and French, F. S. (1987). Rat androgen-binding protein: Evidence for identical subunits and amino acid sequence homology with human sex hormone-binding globulin. Proc. Natl. Acad. Sci. (USA) 84:339–343.

Kasson, B. G., and Hsueh, A.J.W. (1987). Insulin-like growth factor-1 augments gonadotropin-stimulated androgen biosynthesis by cultured rat testicular cells. Mol. Cell. Endocrinol. 52:27–34.

Kilpatrick, D. L., Borland, K., and Jin, D. F. (1987). Differential expression of opioid peptide genes by testicular germ cells and somatic cells. Proc. Natl. Acad. Sci. (USA) 84:5695–5699.

Kirszbaum, L., Sharpe, J. A., Murphy, B., d'Apice, A.J.F., Classon, B., Hudson, P., and Walker, I. D. (1989). Molecular cloning and characterization of the novel, human complement-associated protein, SP-40,40: A link between the complement and reproductive systems. EMBO J. 8:711–718.

Krieger, D. T., Margioris, A. N., Liotta, A. S., Shaha, C., Gerendai, I., Pintar, J., and Bardin, C. W. (1984). Proopiomelanocortin (POMC)-derived peptides in the rodent male reproductive tract, in *Opioid Modulation of Endocrine Function* (Delitala, G., Motta, M., and Serio, M., eds.) pp. 223–235, Raven Press, New York.

Kulin, H. E., Samoslik, E., Santen, R., and Santner, S. (1981). The effect of growth hormone on the Leydig cell response to chorionic gonadotropin in boys with hypopituitarism. Clin. Endocrinol. 45:463–472.

Lacaze-Masmonteil, T., deKeyzer, Y., Luton, J.-P., Kahn, A., and Bertagna, X. (1987). Characterization of proopiomelanocortin transcripts in human nonpituitary tissues. Proc. Natl. Acad. Sci. (USA) 84:7261–7265.

Laron, Z. (1984). Laron-type dwarfism (hereditary somatomedin deficiency). A review. Adv. Int. Med. Pediatr. 51:117–140.

Lee, W., Mason, A. J., Schwall, R., Szonyi, E., and Mather, J. P. (1989). Secretion of activin by interstitial cells in the testis. Science 243:396.

Leger, J. G., Montpetit, M. L., and Tenniswood, M. R. (1987). Characterization and cloning of androgen-repressed mRNAs from rat ventral prostate. Biochem. Biophys. Res. Commun. 147:196–203.

La Magueresse, B., and Jegou, B. (1988). Paracrine control of immature Sertoli cells by adult germ cells, in the rat (an in vitro study): Cell–cell interactions within the testis. Mol. Cell. Endocrinol. 58:65–72.

Le Magueresse, B., Pineau, C., Guillou, F., and Jegou, B. (1988). Influence of germ cells upon transferrin secretion by rat Sertoli cells in vitro. J. Endocrinol. 118:13–16.

La Magueresse, B., Le Gac, F., Loir, M., and Jegou, B. (1986). Stimulation of rat Sertoli secretory activity in vitro by germ cells and residual bodies. J. Reprod. Fertil. 77:489–498.

Lin, T., Calkins, J. H., Morris, P. L., Vale, W., and Bardin, C. W. (1989). Regulation of Leydig cell function in primary culture by inhibin and activin. Endocrinology 125:2134–2140.

Lin, T., Blaisdell, J., and Haskell, J. F. (1987). Transforming growth factor-β inhibits Leydig cell steroidogenesis in primary culture. Biochem. Biophys. Res. Commun. 146:387–394.

Lin, T., Haskell, J., Vinson, N., and Terracio, L. (1986). Characterization of insulin and insulin-like growth factor 1 receptors of purified Leydig cells and their role in steroidogenesis in primary culture: A comparative study. Endocrinology 119:1641–1647.

Lin, Y., and Dahl, K. D. (1988). A factor(s) produced by Sertoli cell stimulates androgen biosynthesis by Leydig cells in neonatal rats. Scientia Sinica 31:817–827.

Margioris, A., Liotta, A. S., Vaudry, H., Bardin, C. W., and Krieger, D. T. (1983). Characterization of immunoreactive pro-opiomelanocortin-related peptides in rat testes. Endocrinology 113:663–671.

Mason, I. J., Taylor, A., Williams, J. G., Sage, H., and Hogan, B.L.M. (1986). Evidence from molecular cloning that SPARC, a major product of mouse embryo parietal endoderm, is related to an endothelial cell culture shock glycoprotein of Mr 43,000 EMBO J. 5:1465–1472.

Massague, J. (1987). The TGF-β family of growth and differentiation factors. Cell 49:437.

Mather, J. P., and Phillips, D. M. (1984). Primary culture of testicular somatic cells, in *Methods for Serum-Free Culture of Cells of the Endocrine System* (Barnes, D. W., Sirbasku, D. A., and Sato, G. H., eds.) pp. 29–46, Alan R. Liss, Inc., New York.

Mather, J. P., Morris, P. L., Perez-Infante, V., Salomon, Y. A., Nakhla, A., Gunsalus, G. L., and Bardin, C. W. (1984). Hormonal control of Sertoli cell function, in *Endocrinology: Proceedings of the 7th International Congress of Endocrinology* (Labire, F., and Proulx, L., eds.) pp. 615–618, Elsevier, Amsterdam.

Mather, J. P., Gunsalus, G. L., Musto, N. A., Cheng, C. Y., Parvinen, M., Wright, W., Perez-Infante, V., Margioris, A., Liotta, A., Becker, R., Krieger, D. T., and Bardin, C. W. (1983). The hormonal and cellular control of Sertoli cell secretion. J. Steroid Biochem. 19:41–51.

Mather, J. P., Zhuang, L.-Z., Perez-Infante, V., and Phillips, D. M. (1982). Culture of testicular cells in hormone-supplemented serum-free medium, in *The Cell Biology of the Testis* (Bardin, C. W., and Sherins, R. J., eds.) Vol. 383, pp. 44–68, New York Academy of Sciences, New York.

Mather, J. P., and Sato, G. (1979). The growth of mouse melanoma cells in serum-free hormone supplemented medium. Exp. Cell Res. 120:191–200.

McMurray, C. T., Devi, L., Calavetta, L., and Douglass, J. O. (1989). Regulated expression of the prodynorphin gene in R2C Leydig tumor cell line. Endocrinology 124:49–59.

Melner, M. H., and Puett, D. (1984). Evidence for the synthesis of multiple proopiomelanocortin-like precursors in murine Leydig tumor cells. Arch. Biochem. Biophys. 232:197–199.

Melsert, R., Hoogerbrugge, J. W., and Rommerts, F.F.G. (1989). Albumin, a biologically active protein acting on Leydig cells. Mol. Cell. Endocrinol. 64:35–44.

Mita, M., Borland, K., Price, J. M., and Hall, P. F. (1985). The influence of insulin and insulin-like growth factor-1 on hexose transport by Sertoli cells. Endocrinology 116:987–992.

Montpetit, M. L., Lawless, K. R., and Tenniswood, M. R. (1986). Androgen repressed messages in the rat ventral prostate. Prostate 8:25–36.

Morales, C., Sylvester, S. R., and Griswold, M. D. (1987). Transport of iron and transferrin synthesis by the seminiferous epithelium of the rat in vivo. Biol. Reprod. 37:995–1005.

Morera, A. M., Benahmed, M., Cochet, C., Chauvin, M. A., Chambaz, E., and Revol, A. (1987). A TGFβ-like peptide is a possible intratesticular modulator of steroidogenesis. Ann. NY Acad. Sci 513:494–496.

Morris, I. D., Bardin, C. W., Musto, N. A., Thau, R. B., and Gunsalus, G. L. (1987). Evidence suggesting that germ cells influence the bidirectional secretion of androgen binding protein by the seminiferous epithelium demonstrated by selective impairment of spermatogenesis with busulfan. Int. J. Androl. 10:691–700.

Morris, P. L., Cheng, C. Y., and Bardin, C. W. (1990a). Chorionic gonadotropin (hCG), prolactin (PRL) and epidermal growth factor (EGF) stimulate the secretion of proteins by Leydig cell-enriched cultures. Endocrinology, in press.

Morris, P. L., Vale, W. W., and Bardin, C. W. (1990b). Activin modulation of inhibin secretion by Sertoli cells in vitro. Endocrinology, in press.

Morris, P. L., and Bardin, C. W. (1987). Intratesticular opiates as regulators of Leydig cell function. Ann. NY Acad. Sci 513:380–382.

Morris, P. L., Cheng, C. Y., and Bardin, C. W. (1987). Identification of prolactin-responsive proteins in Leydig cell-enriched culture media. Ann. N.Y. Acad. Sci. 513:390–392.

Murphy, B. F., Kirszbaum, L., Walker, I. D., and d'Apice, A.J.F. (1988). SP-40,40, a newly identified normal human serum protein found in the SC56-9 complex of complement and in the immune deposits in glomerulonephritis. J. Clin. Invest. 81:1858–1864.

Murray, F. T., Orth, J., Gunsalus, G. L., Weisz, J., Li, J. B., Jefferson, L. S., Musto, N. A., and Bardin, C. W. (1981). The pituitary–testicular axis in the streptozotocin diabetic male rat: Evidence for gonadotroph, Sertoli cell and Leydig cell dysfunction. Int. J. Androl. 4:265–280.

Nicholson, H. D., Worley, R.T.S., Guldenaar, S.E.F., and Pickering, B. T., (1987). Ethan-1,2-dimethanesulphonate reduces testicular oxytocin content and seminiferous tubule movements in the rat. J. Endocrinol. 112:311–316.

Norton, J. N., and Skinner, M. K. (1989). Regulation of Sertoli cell function and differentiation through the actions of a testicular paracrine factor P-Mod-S. Endocrinology 124:2711–2719.

O'Connor-McCourt, M. D., and Wakefield, L. M. (1987). Latent transforming growth factor-β in serum. A specific complex with α_2-macroglobulin. J. Biol. Chem. 262:14090–14099.

O'Conohue, T. L., Charlton, C. G., Thoa, M. B., Helke, C., Moody, T. W., Pera, A., Williams, A., Miller, R. L., and Jacobowtiz, D. M. (1981). Release of α-melanocyte stimulating hormone into rat and human cerebrospinal fluid in vivo and from rat hypothalamus slices in vitro. Peptides 2:93–100.

Oonk, R. B., and Grootegoed, J. A. (1988). Insulin-like growth factor-1 (EGF-1) receptors on Sertoli cells from immature rats and age-dependent testicular binding of IGF-1 and insulin. Mol. Cell. Endocrinol. 55:33–44.

Oonk, R. B. (1987). Effects of FSH, insulin and IGF-1 on rat Sertoli cells. Thesis, University of Rotterdam, Rotterdam.

Oonk, R. B., and Grootegoed, J. A. (1987). Identification of insulin receptors on rat Sertoli cells. Mol. Cell. Endocrinol. 49:51–62.

Orlando, C., Caldini, A. L., Barni, T., Wood, W. G., and Strasburger, C. J. (1985). Ceruloplasmin and transferrin in human seminal plasma: Are they an index of seminiferous tubular function. Fertil. Steril. 43:280–284.

Orth, J. M. (1986). FSH-induced Sertoli cell proliferation in the developing rat is modified by β-endorphin. Endocrinology 119:1876–1878.

Papadopoulos, V., Kamtchousing, P., Drosdowsky, M. A., and Carreau, S. (1987a). Spent media from immature seminiferous tubules and Sertoli cells inhibit adult rat Leydig cell aromatase activity. Horm. Metab. Res. 19:62–64.

Papadopoulos, V., Kamtchouing, P., Drosdowsky, M. A., Hochereau deReviers, M. T., and Carreau, S. (1987b). Adult rat Sertoli cells secrete a factor or factors which modulate Leydig cell function. J. Endocrinol. 114:459–467.

Parmentier, M., Inagami, T., Pochet, R., and Desclin, J. C. (1983). Pituitary-dependent renin-like immunoreactivity in the rat testis. Endocrinology 112:1318–1323.

Parvinen, M. (1982). Regulation of the seminiferous epithelium. Endocr. Rev. 3:404–417.

Parvinen, M., Nikula, H., and Huhtaniemi, I. (1984). Influence of rat seminiferous tubules on Leydig cell testosterone production in vitro. Mol. Cell. Endocrinol. 37:331–336.

Perez-Infante, V., Bardin, C. W., Gunsalus, G. L., Musto, N. A., Rich, K. A., and Mather, J. P. (1986). Differential regulation of testicular transferrin and androgen-binding protein in primary cultures of rat Sertoli cells. Endocrinology 118:383–392.

Persson, H., Lievre, C.A.-L., Söder, O., Villar, M. J., Metsis, M., Olson, L., Ritzen, M., and Hökfelt, T. (1990). Expression of β-nerve growth factor receptor mRNA in Sertoli cells downregulated by testosterone. Science 247:704–707.

Pickering, B. T., Birkett, S. D., Guldenaar, S.E.F., Nicholson, H. D., Worley, R.T.S., and Yavachev, L. (1989). Oxytocin in the testis: What, where, and why. Ann. NY Acad. Sci. 564:198–209.

Pickering, B. T., Birkett, S. D., Charlton, H. M., Guldenaar, S.E.F., Nicholson, H. D., O'Shaughnessy, P. J., Swann, R. W., Wathes, D. C., and Worley, R.T.S. (1986). Neurohypophysial peptides in the gonads, in *Neuroendocrine Molecular Biology* (Fink, G., Harmar, A. J., and McKerns, K. W., eds.) pp. 231–237, Plenum Press, New York.

Pintar, J. E., Schachter, B. S., Herman, A. B., Durgerian, S., and Krieger, D. T. (1984). Characterization and localization of proopiomelanocortin in mRNA in the adult rat. Science 225:632–634.

Rechler, M. M., and Nissley, S. P. (1985). The nature and regulation of the receptors for insulin-like growth factors. Ann. Rev. Physiol. 47:425–442.

Rehbein, M., Hillers, M., Mohr, E., Ivell, R., Morley, S., Schmale, H., and Richter, D. (1986). The neurohypophysial hormones vasopressin and oxytocin: Precursor structure, synthesis and regulation. Biol. Chem. Hoppe-Seyler 367:695–704.

Reventos, J., Hammond, G. L., Crozat, A., Brooks, D. E., Bunsalus, G. L., Bardin, C. W., and Musto, N. A. (1988). Hormonal regulation of rat androgen-binding protein (ABP) messenger ribonucleic acid and homology of human testosterone-estradiol-binding globulin and ABP complementary deoxyribonucleic acids. Mol. Endocrinol. 2:125–132.

Rigaudiere, N., Loubassou, S., Grizard, G., and Boucher, D. (1988). Characterization of insulin binding and comparative action of insulin and insulin-like growth factor I on purified Leydig cells from the adult rat. Int. J. Androl. 11:165–178.

Ritzen, E. M., (1983). Chemical messengers between Sertoli cells and neighbouring cells. J. Steroid Biochem. 19:499–504.

Rizzino, A. (1988). Transforming growth factor-β: Multiple effects on cell differentiation and extracellular matrices. Devel. Biol. 130:411–422.

Roberts, A. B., Flanders, K. C., Kondaiah, P., Thompson, N. L., Van Obberghen-Schilling, E., Wakefield, L., Rossi, P., De Crombrugghe, B., Heine, U., and Sporn, M. B. (1988). Transforming growth factor β: Biochemistry and roles in embryogenesis, tissue repair and remodeling, and carcinogenesis. Recent Prog. Horm. Res. 44:157.

Roberts, K. P., and Griswold, M. D. (1990). Characterization of rat transferrin receptor cDNA: The regulation of transferrin receptor mRNA in testes and in Sertoli cells in culture. Mol. Endocrinol., in press.

Roberts, K. P., and Griswold, M. D. (1989). Testosterone induction of cellular proteins in cultured Sertoli cells from hypophysectomized rats and rats of different ages. Endocrinology 125:1174–1179.

Rodriguez-Rigau, J. L. (1980). Diabetes and male reproductive function. J. Androl. 1:105–110.

Rossi, V., Cheng, C. Y., Gunsalus, G. L., Bardin, C. W., and Spitz, I. M. (1989). The differential effects of the indazole-carboxylic acid derivative, tolnidamine, on Sertoli cell protein secretion. J. Androl. 10:466–471.

Saez, J. M., Avallet, O., Naville, D., Perrard-Sapori, M. H., and Chatelain, P. G. (1989). Sertoli-Leydig cell communications. Ann. NY Acad. Sci. 564:210–231.

Saez, J. M., Chatelain, P. G., Perrard-Sapori, M.-H., Jaillard, C., and Naville, D. (1988). Differentiating effects of somatomedin-C/insulin-like growth factor I and insulin on Leydig and Sertoli cell functions. Reprod. Nutr. Develop. 28:989–1008.

Saez, J. M., Benahmed, M., Reventos, J., Bommelaer, M. C., Monbrial, C., and Haour, F. (1983). Hormonal regulation of pig Leydig cells in culture. J. Steroid Biochem. 19:375–384.

Sage, H., Johnson, C., and Bornstein, P. (1984). Characterization of a novel serum albumin-binding glycoprotein secreted by endothelial cells in culture. J. Biol. Chem. 259:3993–4007.

Sakiyama, R., Pardridge, W. M., and Musto, N. A. (1988). Influx of testosterone-binding globulin (TeBG) and TeBG-bound sex steroid hormones into rat testis and prostate. J. Clin. Endocrinol. Metab. 67:98–103.

Salomon, A., Bano, M., Smith, K. B., and Kidwell, W. R. (1982). Isolation and characterization of a growth factor (embryonin) from bovine fetuin which resembles α_2-macroglobulin. J. Biol. Chem. 257:14093–14101.

Serio, M., Vannelli, G. B., Barni, T., Orlando, C., Forti, G., and Balboni, G. C. (1988). Sertoli cell proteins in the human seminiferous tubule, in *Cell-to-Cell Communication of Endocrinology* (Piva, F., Bardin, C. W., Forti, G., and Motta, M., eds.) pp. 155–165, Raven Press, New York.

Setchell, B. P. (1978). *The Mammalian Testis.* Ithaca, New York, Cornell University Press.

Shabanowitz, R. B., DePhilip, R. M., Crowell, J. A., Tres, L. L., and Kierszenbaum, A. L. (1986). Temporal appearance and cyclic behavior of Sertoli cell-specific secretory proteins during the development of the rat seminiferous tubule. Biol. Reprod. 35:745–760.

Shaha, C., Morris, P. L., Chen, C.-L.C., Vale, W., and Bardin, C. W. (1989). Immunostainable inhibin subunits are in multiple types of testicular cells. Endocrinology 125:1941–1950.

Shaha, C., Cheng, C. Y., Phillips, D. M., Talwar, G. P., and Bardin, C. W. (1988). Localization of immunoreactive testibumin in the testis and epididymis of adult rats. Int. J. Androl. 11:547–559.

Shaha, C., Liotta, A. S., Krieger, D. T., and Bardin, C. W. (1984). The ontogeny of immunoreactive β-endorphin in the fetal, neonatal, and pubertal testes from mouse and hamster. Endocrinology 114:1584–1591.

Sharpe, R. M., Kerr, J. B., and Maddocks, S. (1988). Evidence for a role of the Leydig cells in control of the intratesticular secretion of inhibin. Mol. Cell. Endocrinol. 60:243–247.

Sharpe, R. M., and Cooper, I. (1987). Comparison of the effects on purified Leydig cells of four hormones (oxytocin, vasopressin, opiates, and LHRH) with suggested paracrine roles in the testis. J. Endocrinol. 113:89–96.

Sharpe, R. M., Kerr, J. B., Cooper, I., and Bartlett, J.M.S. (1986). Intratesticular factors and testosterone secretion: The effect of treatment with ethane dimethanesulphonate (EDS) and the induction of seminiferous tubule damage. Int. J. Androl. 9:285–298.

Sharpe, R. M., and Barlett, J.M.S. (1985). Stimulation of Leydig cell function by a polypeptide present in testicular interstitial fluid. Med. Biol. 63:245–250.

Sharpe, R. M., and Cooper, I. (1984). Intratesticular secretion of a factor(s) with major stimulatory effects on Leydig cell testosterone secretion in vitro. Mol. Cell. Endocrinol. 37:159–168.

Sharpe, R. M., Cooper, I., and Bartlett, J.M.S. (1984). Intratesticular production of a high MW factor(s) which stimulates Leydig cell testosterone secretion in vitro. INSERM Colloq 123:413–418.

Skinner, M. K., and Moses, H. L. (1989). Transforming growth factor β gene expression and action in the seminiferous tubule: Peritubular cell–Sertoli cell interactions. Mol. Endocrinol. 3:625–635.

Skinner, M. K., Cailleau, J., Heyns, W., and Verhoeven, G. (1989a). Prostatic stromal cells and testicular peritubular cells produce similar paracrine mediators of androgen action. Endocrinology, in press.

Skinner, M. K., McLachlan, R. I., and Bremner, W. J. (1989b). Stimulation of Sertoli cell inhibin secretion by the testicular paracrine factor PModS. Mol. Cell. Endocrinol. 66:239–249.

Skinner, M. K., Takacs, K., and Coffey, R. J. (1989c). Transforming growth factor-α-gene expression and action in the seminiferous tubule: Peritubular cell–Sertoli cell interactions. Endocrinology 124:845–854.

Skinner, M. K., Fetterolf, P. M., and Anthony, C. T. (1988). Purification of a paracrine factor, P-Mod-S, produced by testicular peritubular cells that modulates Sertoli cell function. J. Biol. Chem. 263:2884.

Skinner, M. K. (1987). Cell–cell interactions in the testis. Ann. NY Acad. Sci. 513:158–171.

Skinner, M. K., and Fritz, I. B. (1986). Identification of a non-mitogenic paracrine factor involved in mesenchymal–epithelial cell interactions between testicular peritubular cells and Sertoli cells. Mol. Cell. Endocrinol. 44:85.

Skinner, M. K., and Fritz, I. B. (1985a). Testicular peritubular cells secrete a protein under androgen control that modulates Sertoli cell functions. Proc. Natl. Acad. Sci. (USA) 82:114.

Skinner, M. K., and Fritz, I. B. (1985b). Androgen stimulation of Sertoli cell function is enhanced by peritubular cells. Mol. Cell. Endocrinol. 40:115.

Skinner, M. K., Cosand, W. L., and Griswold, M. D. (1984). Purification and characterization of testicular transferrin secreted by rat Sertoli cells. Biochem. J. 218:313–320.

Skinner, M. K., and Griswold, M. D. (1983). Sertoli cells synthesize and secrete a ceruloplasmin-like protein, Biol. Reprod. 28:1225–1230.

Skinner, M. K., and Griswold, M. D. (1980). Sertoli cells synthesize and secrete transferrin-like protein. J. Biol. Chem. 255:9523–9525.

Smith, E. P., Svoboda, M. E., Van Wyk, J. J., Kierszenbaum, A. L., and Tres, L. L. (1987). Partial characterization of a somatomedin-like peptide from the medium of cultured rat Sertoli cells. Endocrinology 120:186–193.

Sordoillet, C., Chauvin, M. A., Revol, A., Morera, A. M., and Benahmed, M. (1988). Fibroblast growth factor is a regulator of testosterone secretion in cultured immature Leydig cells. Mol. Cell. Endocrinol. 58:283–286.

Spera, G., E. Di Benedetto, Falaschi, P., Fabbri, A., Lauro, G., and Bolotti, M. (1988). Immunohistochemical localization of β-endorphin in hyperplastic interstitial tissue of the human testis. J. Androl. 9:78–81.

Syed, V., Lindh, M., Khan, S. A., and Ritzen, E. M. (1989). Hormonal regulation of a rat seminiferous tubule factor which inhibits LH action on interstitial cells. Int. J. Androl. 12:464–472.

Syed, V., Khan, S. A., Lindh, M., and Ritzen, E. M. (1988). Mechanism of action of the factor(s) secreted by rat seminiferous tubules and inhibiting interstitial cell testosterone production in vitro. Acta Endocrinologia (Copenh) 119:427–434.

Syed, V., Khan, S. A., Lindh, M., and Ritzen, E. M. (1987). Ontogeny and cellular origin of a rat seminiferous tubule factor(s) that inhibits LH-dependent testosterone production by interstitial cells in vitro. Int. J. Androl. 10:711–720.

Syed, V., Karpe, B., Ploen, L., and Ritzen, E. M. (1986). Regulation of interstitial cell function by seminiferous tubules in intact and cryptorchid rats. Int. J. Androl. 9:271–284.

Syed, V., Khan, S. A., and Ritzen, E. M. (1985). Stage-specific inhibition of interstitial cell testosterone secretion by rat seminiferous tubules in vitro. Mol. Cell. Endocrinol. 40:257–264.

Sylvester, S. R., Skinner, M. K., and Griswold, M. D. (1984). A sulfated glycoprotein synthesized by Sertoli cells and by epididymal cells is a component of the sperm membrane. Biol. Reprod. 31:1087–1102.

Tabone, E., Benahmed, M., Reventos, J., and Saez, J. M. (1984). Interactions between immature porcine Leydig and Sertoli cells in vitro. Cell Tissue Res. 237:357–362.

Tahri-Joutei, A., Latreille, M. T., and Pointis, G. (1987). The dynamics of the testosterone response of perfused mouse Leydig cells to hCG and arginine vasopressin. Biochem. Biophys. Res. Commun. 146:216–223.
Takase, M., Tsutsui, K., and Kawashima, S. (1988). Inhibitory effect of Sertoli cell-cultured media on LH binding to mouse Leydig cells in culture. Endocrinol. Japon. 35:285–293.
Tanner, J. M., and Whitehouse, R. H. (1975). A note on the bone age at which patients with true isolated growth hormone deficiency enter puberty. J. Clin. Endocrinol. Metab. 41:788–790.
Toebosch, A.M.W., Kroos, M. J., and Grootegoed, J. A. (1987). Transport of transferrin-bound iron into rat Sertoli cells and spermatids. Int. J. Androl. 10:753–764.
Toran-Allerand, C. D. (1980). Coexistence of α-fetoprotein, albumin and transferrin immunoreactivity in neurons of the developing mouse brain. Nature 286:733–735.
Tres, L. L., Smith, E., Van Wyk, J. J., and Kierszenbaum, A. L. (1986). Immunoreactive sites and accumulation of somatomedin-C in rat Sertoli-spermatogenic cell cultures. Exp. Cell Res. 162:33–50.
Tsong, S. D., Phillips, D. M., Halmi, N., Krieger, D., and Bardin, C. W. (1982a). β-Endorphin is present in the male reproductive tract of five species. Biol. Reprod. 27:755–764.
Tsong, S.-D., Phillips, D. M., Halmi, N., Liotta, A. J., Margioris, A., Bardin, C. W., and Krieger, D. T. (1982b). ACTH and β-endorphin related peptides are present in multiple sites in the reproductive tract of the male rat. Endocrinology 110:2204–2206.
Tung, P. S., and Fritz, I. B. (1985). Immunolocalization of clusterin in the ram testis, rete testis, and excurrent ducts. Biol. Reprod. 33:177–186.
Vale, W., Rivier, C., Hsueh, A., Campen, C., Meunier, H., Bicsak, T., Vaughan, J., Corrigan, A., Bardin, W., Sawchenko, P., Petraglia, F., Yu, J., Plotsky, P., Spiess, J., and Rivier, J. (1988). Chemical and biological characterization of the inhibin family of protein hormones. Recent Prog. Horm. Res. 44:1–34.
Valenca, M. M., and Negro-Vilar, A. (1986). Pro-opiomelanocortin-derived peptides in testicular interstitial fluid: Characterization and changes in secretion after human chorionic gonadotropin or leuteinizing hormone-releasing hormone analog treatment. Endocrinology 118:32–37.
Vannelli, R. G., Barni, T., Orlando, C., Natale, A., Serio, M., and Balboni, G. C. (1989). IGF-1 and IGF-1 receptor in human testis: An immunohistochemical study. Fertil. Steril., in press.
van Dissel-Emilliani, F.M.F., Grootenhuis, A. J., de Jong, F. H., and de Rooij, D. G. (1989). Inhibin reduces spermatogonial numbers in testes of adult mice and Chinese hamsters. Endocrinology 125:1898–1903.
Verhoeven, G., and Calilleau, J. (1988). Testicular peritubular cells secrete a protein under androgen control that inhibits induction to aromatase activity in Sertoli cells. Endocrinology 123:2100.
Verhoeven, G., and Cailleau, J. (1987). A Leydig cell stimulatory factor produced by human testicular tubules. Mol. Cell. Endocrinol. 49:137–147.
Verhoeven, G., and Cailleau, J. (1985). A factor in spent media from Sertoli cell-enriched cultures that stimulates steroidogenesis in Leydig cells. Mol. Cell. Endocrinol. 40:57–68.
Vernon, R. B., and Sage, H. (1989). The calcium-binding protein SPARC is secreted by Leydig and Sertoli cells of the adult mouse testis. Biol. Reprod. 40:1329–1340.
Vihko, K. K., and Huhtaniemi, I. (1989). A rat seminiferous epithelial factor that inhibits Leydig cell cAMP and testosterone production: Mechanism of action, stage-specific secretion, and partial characterization. Mol. Cell. Endocrinol. 65:119–127.
Wang, Y.-M., Bayliss, D. A., Milhorn, D. E., and Joseph, D. R. (1989). The androgen-binding protein is expressed throughout the rat brain of both sexes. Program of the 19th Annual Meeting of The Society for Neuroscience, Phoenix, AZ 843 (Abstract 339.11).
Wauben-Penris, P.J.J., Strous, G. J., and Van Der Donk, H. A. (1986). Transferrin receptors of isolated rat seminiferous tubules bind both rat and human transferrin. Biol. Reprod. 35:1227–1234.
Weber, E., Evans, C. J., and Barchas, J. D. (1981). Acetylated and nonacetylated forms of β-endorphin in rat brain and pituitary. Biochem. Biophys. Res. Commun. 103:982–989.
Wright, W. W., Zabludoff, S. D., Erickson-Lawrence, M., and Karzai, A. W. (1989). Germ cell-Sertoli cell interactions. Studies of cyclic protein-2 in the seminiferous tubule. Ann. NY Acad. Sci. 564:173–185.
Wright, W. W., and Luzarranga, M. L. (1986). Isolation of cyclic protein-2 from rat seminiferous tubule fluid and Sertoli cell culture medium. Biol. Reprod. 35:761–772.
Wright, W. W., Parvinen, M., Musto, N. A., Gunsalus, G. L., Phillips, D. M., Mather, J. P., and Bardin, C. W. (1983). Identification of stage-specific proteins synthesized by rat seminiferous tubules. Biol. Reprod. 29:257–270.
Wright, W. W., Musto, N. A., Mather, J. P., and Bardin, C. W. (1981). Sertoli cells secrete both testis-specific and serum proteins. Proc. Natl. Acad. Sci. (USA) 78:7565–7569.
Yoon, D. J., Sklar, C., and David, R. (1988). Presence of immunoreactive corticotropin-releasing factor in rat testis. Endocrinology 122:757–761.
Zakarian, S., and Smyth, D. G. (1982). β-endorphin is processed differently in specific regions of rat pituitary and brain. Nature 296:250–252.
Zaneveld, L.J.D., Polakoski, K. L., Schumacher, G.F.B., Rifkin, D. B., and Shaw, E. (1975). The proteolytic enzyme systems of mammalian genital tract secretions and spermatozoa, in *Proteases and Biological Control: Conferences on Cell Proliferation* (Reich, R. E., ed.) Vol. 2, pp. 683–705, Cold Spring Harbor Laboratory, Cold Spring Harbor, NY.

CHAPTER TEN

Inhibin and Inhibin-Related Proteins

David M. Robertson, Gail P. Risbridger, and David M. de Kretser

I. Introduction

It has been recognized for many years that the gonads produce a nonsteroidal factor termed inhibin that preferentially suppresses pituitary follicle-stimulating hormone (FSH) secretion. Progress in its isolation, characterization, and clarification of its physiology, however, has only occurred since 1985 after the development of sensitive and reliable in vitro bioassay methods that provided a means to monitor the purification of inhibin and to provide baseline information on its physiology and mechanism of action. During this period inhibin has been purified from the gonadal extracts of several species, its amino acid sequence determined by cloning techniques, and, more recently, expressed in human cell lines. Its role in the physiological control of FSH secretion has been clarified with the development of sensitive radioimmunoassay methods. Much of the early work concentrated in females, particularly with the isolation and characterization of inhibin from ovarian follicular fluid sources; however, inhibin with similar characteristics to that isolated in females has also been isolated from male gonadal extracts.

In addition to inhibin, other proteins that affect FSH have been identified in gonadal extracts. These are activin, a protein with high homology, but opposing action, to inhibin, that stimulates FSH secretion in vitro, and a protein structurally different from inhibin termed FSH suppressing protein (FSP) or follistatin that behaves like a weak inhibin agonist.

In this review, the isolation, characterization, and cloning of inhibin, activin, and FSP are detailed before a discussion of the endocrine and paracrine roles of these proteins in the mammalian testes. Only key references have been cited; for recent reviews, including references to females, see dePaolo et al. (1991) and de Kretser and Robertson (1989).

II. Isolation and Characterization

A. Inhibin

The initial isolation of inhibin from follicular fluid was monitored by in vitro bioassay procedures based on the suppression of FSH cell content or FSH release by pituitary cells in culture (discussed in further detail later). The major form of inhibin, with a molecular weight of 30 to 32 kd, has been isolated from bovine, porcine, and ovine follicular fluids, as well as ovine rete testis fluid and rat Sertoli cell extracts (review de Kretser and Robertson, 1989). A 30 to 32 kd inhibin consists of two disulfide-linked subunits termed α and β of molecular weight 20 to 22 kd and 13 to 16 kd respectively, as assessed on sodium dodecylsulfate (SDS)-polyacrylamide gel electrophoresis. Porcine 32 kd inhibin has been isolated as two structurally related forms, A and B, of similar in vitro bioactivity, with identical α subunits but 85 percent homologous β subunits (β_A and β_B) (Ling et al., 1985). It is suspected that there may be species differences in the expression of inhibin A and B as inhibin B has not been identified in either ovine or bovine follicular fluids, whereas inhibin B has been found in porcine follicular fluid. Human recombinant 34 kd inhibin A with comparable properties to 31 kd bovine inhibin A has been recently produced (Tierney et al., 1991) in sufficiently high amounts to enable in vivo studies.

The 30 to 32 kd inhibin is microheterogeneous. Two to three molecular weight forms in the 30 to 34 kd range and six to eight forms with pl values ranging from 6.0 to 7.2 have been identified in bovine and ovine inhibin, and these differences are attributed to differences in glycosylation of the α subunit. Their biological activities are similar but not identical (Robertson et al., 1987a).

Inhibin isolated from male sources (Bardin et al., 1987) is similar to that identified in follicular fluid, although in addition to 30 kd inhibin, which is identical to the female form, inhibin with a NH_2-terminally truncated α subunit has been isolated from ovine rete testis fluid, probably as a result of differential processing. It is unclear whether this processing is an artifact of the purification process.

A larger (58 kd) form of bovine inhibin also has been isolated from bovine follicular fluid (Robertson et al., 1985) consisting of a dimer of an extended α subunit (43 kd) and the β subunit. It is believed that 58 kd inhibin or a larger molecular weight form is the major form of inhibin found in bovine follicular fluid and that at least in this species, 30 to 32 kd inhibin isolated from follicular fluid is in fact a product of the purification process. Incubation of 58 kd inhibin with serum results in the formation of a 30-kd protein believed to be 30 to 32 kd inhibin, suggesting that 58 kd inhibin is secreted by the ovary and processed outside the follicle to form both 30 to 32 kd inhibin and the NH_2-terminal α subunit sequence (termed α_N subunit). The released α_N subunit may have a biological role distinct from inhibin as recently shown in studies that active immunization of ewes with the α_N subunit resulted in reduced fertility, although follicular development was apparently normal (Findlay et al., 1989). The 58-kd inhibin is similarly active in vitro to 30 to 32 kd inhibin, although it has yet to be established if 58 kd inhibin is not initially processed by the pituitary to the biologically active 30 to 32 kd form.

Using monoclonal antibodies directed to bovine α and β inhibin subunits, Miyamoto and colleagues (1986) identified a range of high molecular forms of inhibin (120 kd, 108 kd, 88 kd, and 65 kd) in bovine follicular fluid in addition to the 55 kd and 32 kd forms, which may represent larger precursors as anticipated from its complementary DNA (cDNA) structure. Knight and associates (1989), using a similar approach, identified immunoactive proteins of 116 kd, 44 kd, and 25 kd in bovine follicular fluid. In addition to the 55 to 58 kd and 30 to 32 kd bioactive forms of inhibin, a number of immunoactive, bioinactive (at least in the pituitary cell bioassay) inhibin subunit proteins have been identified. Knight and associates (1989) isolated a 27 kd (nonreduced 20 kd reduced) protein with one NH_2 terminal sequence, identical to the NH_2-sequence of the 20 kd α subunit. High levels of the free α subunit in addition to a 45-kd protein were also identified by immunoblotting in plasma, follicular fluid, and granulosa culture medium from rat and cow. In addition to the monomeric α subunit, two fragments of the α subunit have been identified in bovine follicular fluid, αN (25 kd nonreduced, 26 kd reduced) and a disulfide-linked dimer of the Pro- region of the α subunit precursor and the αC subunit (27 kd nonreduced, 20 and 6 kd reduced) (Sugino et al., 1989; Robertson et al., 1989) (Fig. 10–1). This latter protein (termed Pro-α_C subunit) was not anticipated as inhibin forms with this α subunit extension have not been identified previously. It is presumed that the α_N and Pro-α_C subunit fragments originate from a single protein sequence, Pro-α_N-α_C subunit (molecular weight ~45 kd), from which the α_N subunit is proteolytically excised during the purification procedure or during natural biosynthesis. It is hypothesized that this α subunit precursor binds to the corresponding (hypothetical) β subunit precursor, before disulfide exchange and subsequent processing to form the dimeric 58 kd and 30 to 32 kd inhibin. The presence of immunoactive inhibin precursor proteins in significant amounts in biological fluids has drawn attention to the limited specificity of current immunoassay methods used to measure inhibin. After the isolation and NH_2-terminal sequencing of bovine and porcine inhibin, the precursor sequences of α and β subunits of in-

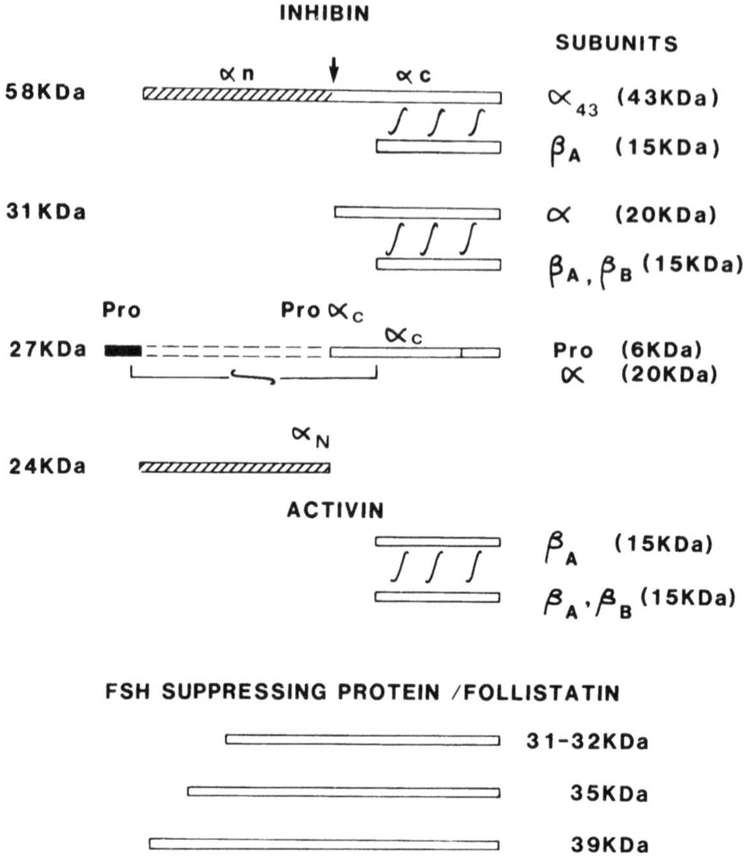

Figure 10–1. A schematic diagram of the structure of inhibin and inhibin-related proteins.

hibin from these and other (human, ovine, rat) species were deduced from its nucleotide sequence after cDNA and genomic cloning (for review, see Forage et al., 1987) (Fig. 10–2). As anticipated from these structural studies, the α_N and α_C sequences of the α subunit were identified as well as the pre- and pro- regions of the α subunit. Two potential glycosylation sites were also identified supporting the conclusion that the α subunit is glycosylated. The β subunit precursor sequence (including porcine β_A and β_B) is considerably larger (398 amino acids) than the mature β subunit (116 amino acids).

B. Activin (Inhibin β Subunit Dimers)

After the isolation of inhibin from follicular fluid, proteins were isolated from side fractions that, in contrast to inhibin, stimulated FSH release from pituitary cells in vitro (Vale et al., 1986; Ling et al., 1986) and in vivo (Schwall et al., 1989). These proteins, termed activins, were identified as inhibin β subunit dimers ($\beta_A\beta_A$, $\beta_A\beta_B$) and termed activin A and activin AB, respectively. Recently, the inhibin β subunit monomer has been isolated from follicular fluid (Robertson et al., 1992) and shown to exhibit reduced (17%) FSH-stimulating activity by pituitary cells in vitro in comparison with activin. It is unclear if this activity is a consequence of dimerization during culture or if it is inherent to the activity of the monomer. The activin gene has been cloned and expressed, exhibiting similar properties to the native molecule (Schwall et al., 1988b, 1989).

Activin exhibits a number of additional biological activities, namely the induction of erythroid differentiation in Friend leukemia cells (Eto et al., 1987), the modulation of the induction of hemoglobin accumulation in a human erythroleukemic cell line, and the proliferation

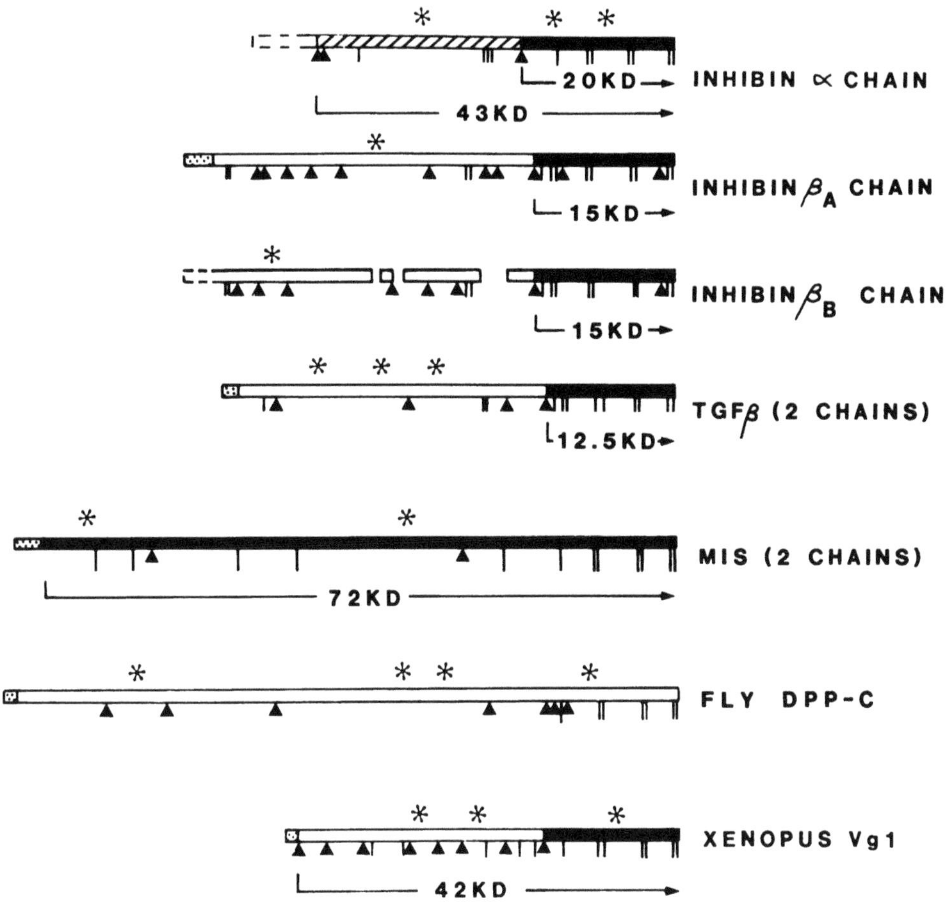

Figure 10–2. cDNA amino acid sequences of the precursor molecules of inhibin and structurally related proteins. Pair of dibasic amino acids *(triangle)*; potential glycosylation site *(asterisks)*.

of erythroid progenitor cells in human bone marrow cultures (Yu et al., 1987). Of particular interest was the observation that activin can induce the formation of axial structures during early embryogenesis in *Xenopus* (Thomsen et al., 1990) and the chick (Mitrani et al., 1990). Activin appears to play a major paracrine role within the ovary with stimulatory effects on aromatase activity (Hutchinson et al., 1987) and inhibition of progesterone secretion by granulosa cells (Sugino et al., 1988a). Activin stimulates granulosa FSH receptor numbers (Hasegawa et al., 1988). Activin receptors have been identified in rat granulosa cell cultures with kd of 3×10^{-10}M and their numbers are stimulated by FSH (Sugino et al., 1988b). The identification of activin receptors in the pituitary and the testes has not yet been reported. Recently, the activin receptor has been identified by expression cloning (Mathews and Vale, 1991). The cDNA codes for a protein consisting of a ligand-binding extracellular domain, a single membrane-spanning domain and an intracellular domain. The sequence of the intracellular domain is closely related to putative transmembrane serine/threonine specific protein kinase. In transfected cells, activin B and inhibin A but not TGFβ competed with activin A.

C. Follicle-Stimulating Hormone Suppressing Protein/Follistatin

Another class of proteins suppressing FSH secretion in vitro has been identified in follicular fluid, which is structurally distinct from inhibin. Several forms have been isolated from the pig (32 and 35 kd termed follistatin; Ueno et al., 1987) and the cow (31, 35, and 39kd, termed FSH-suppressing protein; Robertson et al., 1987b). Recombinant forms of human FSP have been produced (Inouye et al., 1991). All forms have identical NH_2-terminal sequences with in vitro bioactivities of 5 to 30 percent that of inhibin.

The FSP precursor sequence and genomic organization have been determined in the pig, human, and rat (Shimasaki et al., 1989). The FSP is a product of a single gene sequence with the different molecular weight forms produced by alternative splicing of the precursor messenger RNA (mRNA). The precursor protein sequence consists of three of the four homologous domains encoded by separate exons highly conserved between species, which shows sequence homology to human epidermal growth factor and human pancreatic trypsin inhibitor. The FSP gene is expressed in granulosa cells in the rat ovary as well as the kidney and the brain, and because the human cDNA library used to clone FSP in these cloning studies was derived from mRNA isolated from the testis of a human adult male, it is expressed in the human male testis too. These results suggest that FSP is not exclusively a gonadal product.

The FSP is less potent (5 to 30 percent) than inhibin in vitro, with suppressive effects like inhibin on FSH basal release and cell content, luteinizing hormone–releasing hormone (LHRH)-stimulated release of FSH and LH, and limited effects on basal release and cell content of LH, suggesting that FSP is a weak inhibin agonist (Robertson et al., 1990). Recent evidence suggests that, as for inhibin and activin, FSP may play a paracrine role in the ovary by modulating the effects of FSH on steroid synthesis (Xiao et al., 1990). FSP has also been shown to bind activin with high affinity resulting in a neutralization of activin activity (Nakamura et al., 1990), thus providing a basis for the FSH-suppressing activity of FSP (Kogawa et al., 1991). These data were supported by studies by Corrigan et al. (1991) who showed that FSH secretion by rat pituitary cells in vitro was suppressed in the presence of an antibody to activin B.

To date, FSP has not been identified in plasma or testes, although DePaolo and colleagues (1991) established that FSP is active in vivo. At this stage, the biological importance of FSP in the regulation of FSH secretion remains unclear.

STRUCTURAL HOMOLOGIES

A comparison of sequence homologies between the α and β subunits within species and with those of other proteins have revealed some surprising findings (Fig. 10–2). First, the α, βA, and βB subunits are structurally similar as evident from the similarity in relative positions of the cysteine residues of each subunit. The mature βA subunit sequence is totally conserved between porcine, bovine, human, and rat species with one amino acid difference in the sheep (Forage et al., 1987). The α subunit as expected was highly (85 percent), but not totally, conserved. Second, a high similarity in sequence homology was observed between the relative positions of the cysteine residues in the βA subunit and the subunits of other known growth factors, namely transforming growth factor-β (TGFβ) (Derynck et al., 1985), mullerian inhibitory substance (MIS) (Cate et al., 1986), and two potential transcripts of the decapentaplegic gene complex involved in segment formation in the developing *Drosophilia* (Padgett et al., 1987), and the VG_1 gene in *Xenopus* (Weekes and Melton, 1987). These latter observations have generated considerable interest, as inhibin, TGFβ, and MIS exhibit inhibitory activities (a novel observation among growth factors) suggesting that these structurally similar proteins may also share similar biological activities with inhibin and activin.

III. Inhibin Assays

The development of robust and sensitive in vitro bioassays for inhibin has been of fundamental importance to the isolation of inhibin. Using either FSH cell content or basal FSH release by rat pituitary cells in culture as assay end points (Scott et al., 1980), assays of requisite sensitivity and precision were developed that were able to discriminate between specific and nonspecific (toxic) effects by the co-monitoring of LH levels, which were only marginally affected under the assay conditions used.

These assay methods have been applied extensively in the quantitation of inhibin levels in Sertoli cell cultures and testes extracts; however, they have a number of drawbacks, namely their lack of sensitivity in application to serum from animals under normal physiological conditions and their lack of specificity as these procedures detect activin and FSP in addition to inhibin. The sensitivity of in vitro bioassays has been dramatically improved with the use of sheep pituitary cells in culture (Tsonis et al., 1986), although even with this method, it is difficult to detect inhibin in serum from men. As discussed previously, rat pituitary cells are sensitive to inhibin, activin, and FSP, and thus unless the levels of these proteins in a test sample are known, the specificity of the assay is a serious limitation. In some fluids, such as bovine follicular fluid, inhibin is in excess in comparison with activin and FSP, and thus the bioassay values are likely to be accurate; however, this information is not known in other tissues or fluids.

To improve assay sensitivity and specificity, radioimmunoassays have been developed using either purified inhibin (McLachlan et al., 1986; Hasegawa et al., 1987) or a 26 to 33 amino acid sequence of the NH_2-terminal of the inhibin α subunit as tracer and antigen (Rivier and Vale, 1987), which are applicable to serum, cell culture medium, and testes extracts. Initial studies in this laboratory, using an antiserum raised to purified inhibin, established that these assays did not significantly cross react with activin, TGFβ, MIS, FSP, or inhibin subunits isolated after reductive alkylation. In studies in which in vitro bioactivities and immunoactivities in the fractionation of follicular fluid were compared, however, an additional immunoactive protein Pro-α_C subunit (discussed earlier) was identified. Two proteins of molecular weight 25 kd and 45 kd have been identified in bovine serum by immunoblotting with α subunit antisera, which are likely to be the α subunit and its precursor, Pro-α_N-α_C subunit (45 kd protein) (Knight et al., 1989). Their presence in serum may seriously compromise the specificity of these radioimmunoassays.

To establish if both in vitro bioassays and radioimmunoassays are seriously compromised by the presence of these cross reactants (e.g., activin and FSP in the in vitro bioassay and Pro-α_C/Pro-α_N-α_C subunit in the radioimmunoassay), serum was collected from women undergoing gonadotropin stimulation at various stages of the menstrual cycle and after menopause, and in normal and castrated men (Robertson et al., 1988a). Inhibin levels were measured by in vitro bioassay based on ovine pituitary cell cultures and by radioimmunoassay. Levels could not be detected in postmenopausal women and castrated men, as expected, as the gonads are the presumed major source of inhibin. In normal men and gonadotropin-stimulated women, a similar pattern of both activities was observed during the follicular phase, although in the luteal phase the ratio of biological to immunological (B:I) activities fell by 50 percent. In male serum the B:I ratio was lower still and this fall may in part reflect the interference of cross reactant(s) in one of the assays or that inhibin in vitro bioactivity levels in serum are too low to be reliably measured. These results suggest that the levels of cross reactants in female serum are probably not high; however, the study highlights the need for more specific assays.

IV. Mechanism of Action

A. *In Vivo Actions of Inhibin*

The limited availability of sufficient stocks of purified inhibin has prevented detail studies on the in vivo action of inhibin. It is anticipated that large stocks of inhibin produced by recombinant methods will soon be available for more detailed studies.

Purified bovine and porcine inhibin administered to sheep and rats suppresses basal serum FSH levels with minimal effects on serum LH levels (Findlay et al., 1987; Ying, 1988). Administration of human recombinant inhibin to male and female rats (Robertson et al., 1991; Rivier et al., 1991) resulted in a similar suppressive effect on FSH but not LH levels as previously seen with native inhibin or bovine follicular fluid. These results confirm earlier findings observed with crude extracts and are supported by the in vitro findings discussed earlier. As an alternative approach, active and passive inhibin immunization of rabbits, sheep, and rats (in some physiological situations) have resulted in a corresponding increase in serum

FSH, supporting the importance of inhibin in regulating serum FSH levels. Recently, recombinant activin A administered to immature rats resulted in a doubling in serum FSH levels as well as an increase in circulating concentrations of red blood cells and hemoglobin (Schwall et al., 1989). Porcine FSP administered to rats suppressed FSH but not LH secretion with a similar potency compared to inhibin to that seen in vitro (De Paolo et al., 1991).

B. Action on Pituitary Cells In Vitro

Studies exploring the mode of action of purified inhibin on pituitary cells in culture, primarily from the rat, have identified several potentially different sites of action (Farnworth et al., 1988a, b). The addition of inhibin to pituitary cells in culture isolated from adult rats results in a prompt dose-related suppression in FSH secretion within 6 hours with total suppression at high doses. The corresponding cell content of FSH, which in the absence of inhibin slowly decreases with time in culture, decreases at more rapid rates in the presence of inhibin. The consequence of these combined effects of inhibin on basal release and cell content is that by 10 days of culture, the pituitary cell is neither producing nor storing FSH. A similar but less marked effect of inhibin on LH release and cell content is also observed in this culture system; however, because the release of LH under basal conditions is much more limited compared with that of FSH, the overall effect of inhibin on LH basal release is small. Inhibin concentration giving a 50 percent suppression in response (IC_{50}) has been calculated for the dose-response effects of inhibin in suppressing FSH and LH basal release and cell content with the lowest IC_{50} for the effect of inhibin on basal FSH release. Based on the IC_{50}, inhibin primarily affects basal FSH release and as basal FSH release is closely coupled to FSH synthesis, inhibin also suppresses FSH synthesis. Indeed, the action of inhibin in vitro is similar to the effects of the protein synthesis inhibitor, cycloheximide in this system (Fukuda et al., 1987). At higher concentrations, and presumably by a separate and unknown mechanism, inhibin promotes the intracellular degradation of both FSH and LH.

Additional studies have explored the effects of inhibin on the gonadotropin releasing factor (GnRH)-stimulated release of FSH and LH (Farnworth et al., 1988b). Using the same culture system described previously, the cell culture was initially exposed to inhibin for 3 days followed by a 4-hour incubation period in the presence of graded doses of GnRH. The interaction between inhibin and GnRH on the amount of FSH and LH released from the culture was determined. Inhibin was able to antagonize GnRH-stimulated FSH and LH release with an IC_{50} higher than that obtained for basal FSH release and similar to effects of inhibin on FSH and LH cell content. These findings suggest that, based on differences in IC_{50}, inhibin also has a separate action on suppressing GnRH-stimulated release. Using the same experimental design with rat pituitary cells in culture, inhibin suppresses both the number of GnRH receptors per cell and GnRH-stimulated increase in GnRH receptor numbers (Wang et al., 1988). This effect of inhibin is not attributed to competition with GnRH for GnRH receptors. The time course of this effect on GnRH receptors does not parallel the rapid effects of inhibin on FSH release, suggesting that inhibin's action of suppressing GnRH-stimulated gonadotropin release is not by reducing GnRH receptor numbers. Recent studies (Miller et al., 1988; Attardi et al., 1989) have shown that purified or partially purified inhibin suppresses FSHβ subunit expression with variable effects on FSHα subunit expression.

Inhibin's action on rat pituitary gonadotropin synthesis and release in vitro is also influenced by steroids (Campen and Vale, 1988). Testosterone and other androgens (5α dihydrotestosterone and androstenedione) have a small stimulatory effect on basal FSH release and this effect is suppressed by inhibin. Testosterone also potentiates the action of inhibin to suppress GnRH-stimulated FSH and LH release. Estradiol was without effect (Campen and Vale, 1988). These observations are currently based on data generated in rats and it is expected that there will be species differences. For example, in sheep pituitary cell cultures, estradiol has a pronounced inhibitory effect on FSH release, which is not observed in the rat.

The action of activin is less clearly defined. Activin stimulates FSH, but not LH release and synthesis, resulting in a three to fourfold increase in FSH over a 3-day culture period. The effect of activin and GnRH on FSH release is additive, probably operating through separate mechanisms, as activin continues to stimulate FSH release in cultures that are GnRH insensitive as a result of prior treatment with a maximally desensitizing dose of a GnRH agonist (Schwall et al., 1988a).

V. Production of Inhibin and Related Proteins

The testes are the major source of circulating inhibin in the male, as castration results in a rapid decline in serum inhibin levels with an initial half-time of 15 minutes (Robertson et al., 1988b). Using bioassays or radioimmunoassays to measure inhibin, the major site of production has been shown to be the Sertoli cell (Steinberger and Steinberger, 1976; Bicsak et al., 1987) and mRNA for the α, β_A, and β_B subunits of inhibin has been detected in the testes (Meunier et al., 1988).

Relatively little is known about the stored and secreted forms of inhibin produced by Sertoli cells. A 30 kd bioactive material has been identified in Sertoli cell culture medium (Bicsak et al., 1987; Grootenhuis et al., 1989). Furthermore, an alteration has been observed in the B:I activity of inhibin after FSH stimulation (Toebosch et al., 1988; Risbridger et al., 1989b). Thus, with the addition of increasing doses of FSH, there is a progressive decline in the B:I activity of the inhibin secreted into the media, irrespective of whether Sertoli cells or tubule segments were cultured. This phenomenon was investigated further by fractionating the media on reverse phase high performance liquid chromatography (HPLC) (Risbridger et al., 1989b), which identified two peaks of bioactivity and immunoactivity of inhibin (I and II) in media from seminiferous tubule cultures. The addition of FSH led to a proportionally greater increase in the immunoactivity in peak II, resulting in a dramatic decrease in the B:I ratio. Preparative polyacrylamide gel electrophoresis and electroelution of these peaks I and II indicated molecular weights of 31 kd and 27 kd, respectively. It seems likely that the 31 kd band represents inhibin, but the nature of the 27 kd material is not known. The reasons for the decline in B:I ratio of peak II may be due to (1) the production of activin under FSH stimulation, which then co-elutes in the same region as the 27 kd material on reverse phase HPLC and thus reduces the bioactivity of peak II, (2) a change in the degree of glycosylation of inhibin thus altering bioactivity, or (3) the material in peak II may represent a protein that cross reacts with the antiserum in the inhibin RIA, but has diminished bioactivity.

Few studies have attempted to measure the production of inhibin by bioassay and immunoassay simultaneously. Because both these assays are subject to interference by other inhibin-related peptides, it seems appropriate to monitor closely the B:I ratio after hormonal stimulation of cultured Sertoli cells or experimental manipulations in vivo.

The secretion of activin by Sertoli cells is yet to be definitively established, but Grootenhuis and colleagues (1990) noted a fraction of Sertoli cell culture media that stimulated FSH. Lee and colleagues (1989), however, suggested that interstitial cells contain mRNA for the β subunit in vitro. More recently, Risbridger and associates (1989a) have reported the presence of inhibin immunoactivity and bioactivity and the expression of α subunit mRNA in purified adult rat Leydig cell cultures. Under basal conditions, inhibin bioactivity and immunoactivity is secreted into the media in a 20-hour culture. After LH stimulation, however, there was a threefold increase in immunoactivity but no detectable rise in bioactivity of inhibin. The FSH did not significantly affect inhibin or testosterone production, attesting to the purity of the Leydig cell preparation without significant contamination from Sertoli cells. The fall in B:I ratio after LH stimulation of Leydig cell inhibin production may be due to: (1) the stimulation of α subunit secretion, or (2) the production of activin that would reduce the biopotency of the co-secreted inhibin. Clearly, activin alone is not produced after LH stimulation, because one would expect a significant decrease in bioactive inhibin production in the presence, compared with the absence, of LH. The reason for the decrease in B:I ratio after LH stimulation of Leydig cell inhibin production requires further investigation. Although these data challenge the classic concept that the Sertoli cells are the sole source of inhibin production in the testes, they may explain previous observations that LH/hCG administration to adult male rats or humans results in a significant increase in serum inhibin (McLachlan et al., 1988a; Sharpe et al., 1988; Drummond et al., 1989). The more pertinent question is: Why are inhibin and related peptides produced by more than one type of cell in the testes?

VI. Control of Inhibin Secretion

A. *In Vivo Observations*

The recent evidence that both Sertoli cells and Leydig cells have the capacity to produce immunoactive and bioactive inhibin requires a reconsideration of the data concerning the changes in inhibin levels in the circulation under differing gonadotropic stimuli. The studies by Au and associates (1985) demonstrated that after hypophysectomy there was a profound decline in the testicular content of inhibin measured by bioassay. This data has recently been supported by the 60 percent decrease in circulating inhibin levels after hypophysectomy in the rat (Sun et al., 1989). The persistent low inhibin production in the absence of pituitary stimulation, however, indicates a constitutive ability of the testes to produce inhibin and is consistent with the ability of Sertoli cells in culture to produce inhibin for long periods without any hormonal stimulation (Le Gac and de Kretser, 1982; Ultee-van Gessel et al., 1986). Clear evidence that the pituitary stimulation was due to gonadotropic hormones emerged from the observations by Sheckter and colleagues (1988) who showed that the levels of inhibin in the circulation in men with hypogonadotropic hypogonadism were low and subsequently raised during GnRH stimulation. Several studies have clearly documented that FSH is implicated in the control of inhibin secretion. McLachlan and colleagues (1988b) demonstrated that replacement of FSH in men in whom endogenous gonadotropic secretion had been suppressed by testosterone treatment, significantly stimulated inhibin levels in the circulation. These data supported the observations of Au and associates (1985) that a 3-day treatment with FSH resulted in a stimulation of the testicular inhibin content in hypophysectomized rats. Further evidence that FSH and LH stimulated inhibin secretion was the increase in levels after the treatment of men with clomiphene citrate (Tenover et al., 1988). The involvement of FSH in the control of inhibin secretion agrees with a number of studies (Le Gac and de Kretser, 1982; Risbridger et al., 1989a) demonstrating that Sertoli cells responded to FSH by an increase in both bioactive and immunoactive inhibin. Although Sertoli cells are known to have androgen receptors (Verhoeven and Cailleau, 1988), the majority of studies have concluded that testosterone plays little role in the control of inhibin secretion by the Sertoli cells (de Kretser and Robertson, 1989). In addition, FSH-induced Sertoli cell secretion can be inhibited by very high pharmacological doses of testosterone (10^{-5}M) (Gonzales et al., 1988).

The failure of testosterone to influence Sertoli cell inhibin secretion raised some difficulties in our interpretation of several studies that have demonstrated that LH can significantly influence inhibin levels in the circulation. McLachlan and colleagues (1988b) demonstrated that inhibin levels in men in whom pituitary gonadotropic secretion has been suppressed by testosterone treatment, rose significantly when LH or human chorionic gonadotropin (hCG) was added to this regime. Furthermore, studies in the rat demonstrated that hCG stimulation resulted in a peak of inhibin secretion of 24 hours after the initial injection of hCG (Sharpe et al., 1988; Drummond et al., 1989). Conclusive evidence that this stimulation involved the Leydig cells emerged from the studies of Drummond and co-workers (1989), which showed that destruction of the Leydig cells by the cytotoxin, ethane dimethane sulfonate (EDS), resulted in the loss of the ability of hCG to stimulate the peak of inhibin secretion in the rat. The recent demonstrations that the Leydig cells, in fact, have the capacity to produce inhibin, now explain these observations.

B. *In Vitro Control of Inhibin Secretion by Sertoli Cells*

Follicle-stimulating hormone (FSH) is the principal regulator of Sertoli cell inhibin production. Initially this was demonstrated using immature cultures of Sertoli cells or tubules, but adult rat tubule segments also respond to FSH stimulation by increased synthesis of bioactive or immunoactive inhibin (Le Gac and de Kretser, 1982; Gonzales et al., 1988). This stimulation of inhibin production is through a cyclic adenosine monophosphate (cAMP)-dependent mechanism (Bicsak et al., 1987). More recently, FSH has been shown to stimulate α subunit mRNA without influencing β subunit mRNA that is thought to be expressed constitutively by Sertoli cells with no effect on β subunit mRNA expression (Toebosch et al., 1988). The differential regulation of the levels of α subunit mRNA has also been demonstrated in vivo by determining the effects of hypophysectomy and selective FSH replacement in both immature and adult rats (Krummen et al., 1989). The mechanisms of achieving differential regulation of the α subunit mRNA by FSH may involve an alteration in: (1) the rate of transcription of the mRNA, (2) the stability of the mRNA, or (3) the efficiency of translation of the message. These possibilities are currently under investigation (Krummen et al., 1989), and the differential regulation of the α subunit mRNA provides a system to produce the diverse inhibin products from the Sertoli cell.

Although the inhibin α subunit appears to be FSH controlled, the regulation of the expression of the β subunit is unknown. This is interesting because the β subunit is less abundant in the rat testes. Meunier and colleagues (1988) have reported that the level of inhibin α subunit mRNA is 10 times greater than the β_A or β_B subunit mRNA, and it would be reasonable to predict that regulation of the inhibin dimer secretion is closely coupled to the rate-limiting β subunit rather than the abundant α subunit production.

C. In Vitro Control of Leydig Cell Inhibin and Activin Production

As a result of reports that LH/hCG administration to intact adult rats or men caused a significant stimulation of serum or testicular inhibin levels (Drummond et al., 1989; McLachlan et al., 1988b; Sharpe et al., 1988), it was suggested that LH stimulated Sertoli cell inhibin production by an indirect action of the Leydig cells. Leydig cells themselves, however, can produce inhibin with LH, but not FSH, stimulation (Risbridger et al., 1989a). The action of LH in the regulation of inhibin production is also consistent with the observation that patients with Klinefelter's syndrome have high levels of serum inhibin associated with increased LH levels, yet the seminiferous epithelium is virtually devoid of Sertoli cells (de Kretser et al., 1989b). With the identification of the inhibin-related products that result from LH stimulation of the Leydig cells, it will be possible to investigate thoroughly the differential production and regulation of the inhibin subunits. Preliminary data from Roberts and colleagues (1989), however, suggest that the production of α subunit mRNA in Leydig cells from immature rat testis is regulated by LH-like hormones, whereas the β_A and β_B subunit mRNA is not.

D. Regulation of Inhibin by Local Factors

The postulate that germ cell associations may play an important role in the paracrine regulation of Sertoli cell inhibin production was initially supported by the immunohistochemical localization of inhibin to some tubules but not others (Merchenthaler et al., 1987). Subsequently, the expression of the α and β_B subunits of inhibin and the production of immunoactive inhibin have been shown to be stage-specific (Bhasin et al., 1989; Gonzales et al., 1989), thus supporting this hypothesis. Further support for this hypothesis is provided by the observations of Au et al. (1987) and Allenby et al. (1991), who have suggested that late spermatids stimulate inhibin production by Sertoli cells. In vitro data provided by Pineau et al. (1991) suggested that pachytene spermatocytes and round spermatids may stimulate inhibin secretion indirectly.

The possibility that testosterone synthesized by Leydig cells may locally influence inhibin production has been extensively investigated, but the evidence to date is inconclusive. Using immature Sertoli cell cultures, Le Gac and de Kretser (1982) and Bicsak and associates (1987) showed no effect of testosterone on bioactive or immunoactive inhibin secretion, respectively. Yet Verhoeven and Franchimont (1983) reported a stimulation of testosterone on bioactive inhibin production, whereas Morris and co-workers (1988) reported minor inhibitory effects on immunoactive inhibin secretion by Sertoli cells.

There are a variety of other hormones, produced locally in the testes, that have the potential to act as paracrine or autocrine regulators of Sertoli cell inhibin production. These include the epidermal growth factors (EGF) and the insulin-like growth factor 1 (IGF-1), and receptors for these hormones are located on Sertoli cells. Although EGF stimulates immunoactive inhibin secretion by immature Sertoli cells in vitro, IGF-1 has no effect on either bioactive or immunoactive inhibin (Morris et al., 1988; Toebosch et al., 1988). β-Endorphin (which is produced by Leydig cells and/or germ cells) had a small inhibitory effect on FSH-stimulated inhibin production by Sertoli cells. Altogether, these limited data suggest that in addition to the regulation of inhibin production by FSH, paracrine or autocrine factors may also modulate inhibin production; the mechanisms by which this may occur remain unknown.

E. Secretion of Inhibin

The observations demonstrating that the Sertoli cell secretes its product in both basal and luminal directions led to the studies by Handelsman and associates (1989). These studies have shown that inhibin is predominantly secreted in a luminal direction. These in vitro data supported early evidence that the Sertoli cell secreted inhibin into the lumen of the seminiferous tubules, because the levels of inhibin within the testes increased progressively over a 24-hour period after efferent duct ligation (Au et al., 1984). Several investigators, however, have postulated that the levels of inhibin in testicular interstitial fluid, represent a secretion of inhibin across the basal aspect of the Sertoli cell and secretion by Leydig cells. These data suggest that the in vitro model using bicameral chambers, which establish a predominantly luminal secretion of inhibin, may not represent the in vivo situation. The recent demonstration, however, that the Leydig cells have the capacity to produce inhibin may provide an explanation for the higher levels of inhibin in the interstitial fluid. This explanation should also be viewed with caution as the levels of inhibin in testicular interstitial fluid do not decrease after destruction of the Leydig cells by the cytotoxin EDS (de Kretser et al., 1989a).

The route by which inhibin leaves the testes is of interest. The levels of inhibin in the testicular vein are threefold higher than those found in peripheral venous blood in the rat (de Kretser et al., 1987; Gonzales et al., 1989). The ratio is somewhat higher in humans where testicular vein levels are 10-fold higher than in peripheral venous blood (Ishida et al., 1990). The difference in these ratios may reflect the organization of the intertubular tissue of the rat, which contains large sinusoidal networks that contribute to testicular interstitial fluid and thus contain very high levels of inhibin, whereas in humans, a small central lymphatic is present in the intertubular areas. A recent study by Maddocks and Sharpe (1989) has raised the possibility that the route of entry of inhibin into the bloodstream involves that fraction of inhibin secreted to the lumen of the tubule. They propose that the luminal secretion is reabsorbed through the vascular plexus surrounding the rete testis.

F. Local Actions Within the Testes

The significant secretion of inhibin into the lumen of the seminiferous tubules has raised the possibility that inhibin exerts important local actions within the seminiferous epithelium. Impure preparations of inhibin have been demonstrated to inhibit the incorporation of tritiated thymidine into spermatogonia (van Dissel-Emiliani et al., 1989). Because the preparations used in these experiments consisted of partially purified follicular fluid inhibin, a potential exists that other substances may have been present that could have influenced the division of spermatogonia. This view is supported by the results of Mather et al. (1990) who showed that activin stimulated spermatogonial mitoses in co-cultures of this cell type with immature Sertoli cells. They were unable to confirm an action of inhibin in this system. Further studies are required using pure preparations to demonstrate an action on mitosis. The demonstration that pure preparations of inhibin have the capacity to stimulate mitotic division by 3T3 cells, however, raises the possibility of an action on mitosis within the seminiferous epithelium (Hedger et al, 1988). It is possible that inhibin may influence the meiotic process in men in view of our recent demonstration that inhibin can inhibit germinal vesicle breakdown in mouse oocytes (O et al., 1989). This action, which prevents the resumption of meiosis, may have a parallel role in men.

Hsueh and colleagues (1987) used purified preparations of inhibin and activin to examine their effect on interstitial cells maintained for 3 days in culture and then exposed to peptides for another 3 days. Inhibin stimulated, but activin suppressed LH-stimulated steroid production by these interstitial cell preparations from immature (8-to-10-day) or adult hypophysectomized rats. Although this led to the postulate that these two members of the inhibin family of peptides have opposing regulatory actions on Leydig cell steroidogenesis, Payne and colleages (1988) suggested that Leydig cells from hypophysectomized rats were of limited utility as a model to examine adult testicular physiology. We have, therefore, tested the action of bovine inhibin A, bovine activin A, bovine FSP, and the inhibin α subunit products α_C and α_N on testosterone production over a period of 20 hours by Percoll-purified Leydig cells from adult rats. Our results fail to demonstrate any significant effect of these hormones on normal adult rat Leydig cells maintained in short-term culture. Lin and associates (1989) have also observed no effect of inhibin alone on Leydig cells from adult rats, although activin can inhibit testosterone production after 2 days, but not 1 day, in culture and this effect is antagonized by inhibin. The longer term effects of these hormones, which may involve their actions as growth factors, require further investigation.

Using a different culture system, Morris and co-workers (1989) report that inhibin has a biphasic effect on primary immature (25-day-old) rat Leydig cell cultures being initially stimulatory and then inhibitory. Activin, however, was inhibitory during the first day of culture, but then had no significant effect with 3 more days of culture. These authors, furthermore, report that the amount of gonadotropin present in these cultures also modified these responses, so that inhibin enhanced the androstenedione response to a submaximal dose of hCG, but suppressed the response to maximal hCG stimulation. Clearly the effects of inhibin and activin are dependent on (1) the length of culture, (2) the absence or presence of gonadotropin and the dose used, and (3) the type of culture system used, that is, whether regressed or nonregressed, or fetal, immature, or adult populations of Leydig cells were prepared.

VII. Patterns of Inhibin Levels in Blood

Assessments of multiple samples of inhibin in the circulation of normal men and men with testicular disorders have shown stable levels with little evidence of episotic fluctuations (Morrow, personal communication). The castration of rats leads to a rapid decline of inhibin levels in the circulation with an initial half-time of approximately 14 minutes (Robertson et al., 1988b). Similar data have been obtained in men undergoing orchiectomy for prostatic cancer in whom the inhibin levels rapidly declined with an initial half-time of approximately 4 minutes (Ishida et al., 1990). Both of these studies demonstrate that the disappearance of measurable levels of inhibin in the bloodstream after castration is consistent with the concept that the testis is the major source of inhibin in the circulation. This observation is of importance as mRNA for the inhibin subunits has been found at multiple sites (for reviews see Ying, 1988; de Kretser and Robertson, 1989).

VIII. The Role of Inhibin in the Control of Follicle-Stimulating Hormone

There is considerable evidence that in addition to inhibin, testosterone is involved in the control of FSH secretion. Some of the data concerning the suppressive action of testosterone on FSH was used as a cogent argument that there was no need to postulate the existence of inhibin (Decker et al., 1981). These data demonstrated that testosterone had the capacity to suppress both FSH and LH in a dose-dependent manner (de Kretser et al., 1987). The pattern of suppression of the gonadotropins, however, differed. In the rat, increasing doses of testosterone caused a progressive and complete suppression of LH to undetectable levels, whereas suppression of FSH progressed to levels similar to those in normal rats, but despite continued increase in testosterone levels, no further suppression of FSH occurred (de Kretser et al., 1987). The suppression of FSH levels that occurred at testosterone levels ranging from 20 to 30 ng/mL was found in the normal range and these were considerably above levels found in hypophysectomized rats.

Recent studies using the cytotoxin, EDS, lend further support to the concept that both inhibin and testosterone are involved in the control of FSH secretion. When the Leydig cells are destroyed by EDS in normal rats, the levels of testosterone decline rapidly and this is accompanied by a rise of FSH to levels approximately 50 percent of those seen in castrated animals (Jackson and Morris, 1977). We contend that inhibin maintains a tonic inhibitory action on FSH secretion in these EDS-treated rats, a view that is supported by our observations that inhibin levels in the circulation remain unchanged for 5 to 7 days after EDS and then rise significantly (de Kretser et al., 1989a). If, on the other hand, EDS is given to cryptorchid rats in which FSH levels are already approximately 50 percent of castrated rats, due presumably to the lowering of testicular and serum inhibin levels (Gonzales et al., 1989), the removal of testosterone enables FSH levels to rapidly reach the castrated range (O'Leary et al., 1986). These levels subsequently return to the elevated pretreatment range of the cryptorchid rats after 2 to 3 weeks when testosterone concentrations rise after regeneration of the Leydig cells. Further support for the concept that FSH is controlled by both testosterone and inhibin comes from the studies of Dubey and colleagues (1987), who noted that if GnRH secretion is controlled by the pulsatile administration of that factor to monkeys with arcuate nucleus lesions, testosterone alone could not control the postcastration rise of FSH. In recent studies using this model in monkeys, they have demonstrated that a crude preparation of porcine follicular fluid inhibin had the capacity to prevent the postcastration rise of FSH (Abeyawardene and Plant, 1989).

Such a dual control of FSH in men is also supported by recent studies using antisera to the α subunit of inhibin to neutralize passively circulating inhibin levels. These studies demonstrated that in immature rats, the passive neutralization of inhibin levels caused FSH to rise significantly (Culler and Negro-Vilar, 1988; Rivier et al., 1988). If, however, the same antiserum was given to adult male rats, no change in FSH levels occurred. These data suggested that inhibin was more important in the control of FSH secretion in immature animals than in the adult. In a recent study, however, Culler (1989) reported that the same antiserum, when given to adult rats in which testosterone had been lowered by the use of the Leydig cell cytotoxin EDS, elicited a significant rise in FSH. These data again point to a complex interrelationship between testosterone and inhibin in the control of FSH secretion. The nature of the mechanisms involved await the availability of recombinant inhibin to enable appropriate in vivo physiological studies to be designed.

REFERENCES

Abeyawardene, S. A., and Plant, T. M. (1989). Institution of combined treatment with testosterone and charcoal-extracted porcine follicular fluid immediately after orchidectomy prevents the postcastration hypersecretion of follicle-stimulating hormone in the hypothalamus-lesioned Rhesus monkey *(Macaca mulatta)* receiving an invariant intravenous gonadotropin-releasing hormone infusion. Endocrinology 124:1310–1318.

Allenby, G., Foster, P.M.D., and Sharpe, R.M. (1991). Evidence that secretion of immunoactive inhibin by seminiferous tubules from adult rat testis is regulated by specific germ cell types: Correlation between in vivo and in vitro studies. Endocrinology 128:467–476.

Attardi, B., Keeping, H. S., Winters, S. J., Kotsuji, F., Maurer, R. A., and Troen, P. (1989). Rapid and profound suppression of messenger ribonucleic acid encoding follicle-stimulating hormone β by inhibin from primate Sertoli cells. Mol. Endocrinol. 3:280–287.

Au, C. L., Robertson, D. M., and de Kretser, D. M. (1984). Relationship between testicular inhibin content and serum FSH concentrations in rats after bilateral efferent duct ligation. J. Reprod. Fert. 72:351–356.

Au, C. L., Robertson, D. M., and de Kretser, D. M. (1985). Effects of hypophysectomy and subsequent FSH and testosterone treatment on inhibin production by adult rat testes. J Endocrinol 105:1–6.

Au, C. L., Robertson, D. M., and de Kretser, D. M. (1987). Changes in testicular inhibin after a single episode of heating of rat testes. Endocrinology 120:973–977.

Bardin, C. W., Morris, P. L., Chen, C. L., Shaha, C., Voglmayr, J., Rivier, J., Spiess, J., and Vale, W. W. (1987). Testicular inhibin: Structure and regulation by FSH, androgens and EGF, in *Inhibin Non-Steroidal Regulation of Follicle Stimulating Hormone Secretion* (Burger H. G., de Kretser, D. M., Findlay, J. K., and Igarashi, M., eds.) pp. 179–190, Serono Symposia, Raven Press, New York.

Bhasin, S., Krummen, L. A., Swerdloff, R. S., Morelos, B. S., Kim, W. H., Di Zerega, G. S., Ling, N., Esch, F., Shimasaki, S., and Toppari, J. (1989). Stage dependent expression of inhibin α and β-B subunits during the cycle of the rat seminiferous epithelium. Endocrinology 124:987–991.

Bicsak, T., Vale, W., Vaughan, J., Tucker, E., Cappel, S., and Hsueh, A.J.W. (1987). Hormonal regulation of inhibin production by cultured Sertoli cells. Mol. Cell Endocrinol. 49:211–217.

Campen, C. A., and Vale, W. (1988). Interaction between purified ovine inhibin and steroids on the release of gonadotropins from cultured rat pituitary cells. Endocrinology 123:1320–1328.

Cate, R. L., Mattaliano, R., Hession, C., Tizard, R., Farber, N. M., Cheung, A., Ninfa, E. G., Frey, A. Z., Gash, D. J., Chew, E. P., Fisher, R. A., Bertonis, J. M., Torres, G., Wallner B. P., Ramachandran, K. L., Ragin, R. C., Manganaro, T. F., MacLaughlin, D. T., and Donahoe, P. K. (1986). Isolation of the bovine and human genes for mullerian inhibiting substance and expression of the human gene in animal cells. Cell 45:686–694.

Corrigan, A. Z., Bilezikjian, L. M., Carroll, R. S., Bald, L. N., Schmeizer, C. H., Fendly, B. M., Mason, A. J., Chin, L. W., Schwall, R. H., and Vale, W. (1991). Evidence for an autocrine role of activin B within rat anterior pituitary cultures. Encrinology 28:1682–1684.

Culler, M. D. (1989). Abolition of endogenous testosterone secretion is necessary to reveal a suppressive effect on endogenous inhibin on FSH secretion in the adult male rat. Proc. 71st Meeting of the Endocrine Soc. Abs 595, p. 171.

Culler, M. D., and Negro-Vilar, A. (1988). Passive immunoneutralization of endogenous inhibin: Sex-related differences in the role of inhibin during development. Mol. Cell. Endocrinol. 58:263–273.

Decker, M. H., Loriaux, D. L., and Cutler, G. B. (1981). A seminiferous tubular factor is not obligatory for regulation of plasma follicle-stimulating hormone in the rat. Endocrinology 108:1035–1039.

de Kretser, D. M., and Robertson, D. M. (1989). The isolation and physiology of inhibin and related proteins. Biol. Reprod. 40:33–47.

de Kretser, D. M., O'Leary, P. O., Irby, D. C., and Risbridger, G. P. (1989a). Inhibin secretion is influenced by Leydig cells: Evidence from studies using the cytotoxin ethane dimethane sulphonate (EDS). Int. J. Androl. 12:273–280.

de Kretser, D. M., McLachlan, R. I., Robertson, D. M., and Burger, H. G. (1989b). Serum inhibin levels in normal men and men with testicular disorders. J. Endocrinol. 120:517–523.

de Kretser, D. M., Au, C. L., and Robertson, D. M. (1987). The physiology of inhibin in the male, in *Inhibin—Non Steroidal Regulation of Follicle Stimulating Hormone Secretion* (Burger, H. G., de Kretser, D. M., Findlay, J. K., and Igarashi, M., eds.) pp. 149–161, Serono Symposia, Raven Press, New York.

De Paolo, L. V., Bicsak, T. A., Erickson, G. F., Shimasaki, S., and Ling, N. (1991). Follistatin and activin: A potential intrinsic regulatory system within diverse tissues. Proc. Soc. Exp. Biol. Med. 198:500–512.

De Paolo, L. V., Shimonaka, M., Schwall, R. H., and Ling, N. (1991). In vivo comparison of the follicle-stimulating hormone suppressing activity of follistatin and inhibin in ovariectomised rats. Endocrinology 128:668–674.

Derynck, R., Jarret, J. A., Chen, E. Y., Eaton, D. H., Bell, J. R., Assoian, R. K., Roberts, A. B., Sporn, M. B., and Goeddel, D. V. (1985). Human transforming growth factor β complementary DNA sequence and expression in normal and transformed cells. Nature 316:701–705.

Drummond, A. E., Risbridger, G. P., and de Kretser, D. M. (1989). The involvement of Leydig cells in the regulation of inhibin secretion by the testis. Endocrinology 125:510–515.

Dubey, A. K., Zeleznik, A. J., and Plant, T. M. (1987). In the Rhesus monkey *(Macca mulatta),* the negative feedback regulation of follicle stimulating hormone secretion by an action of testicular hormone directly at the level of the anterior pituitary gland cannot be accounted for by either testosterone or estradiol. Endocrinology 121:2229–2237.

Eto, Y., Tsuji, T., Takezawa, M., Takano, S., Yokagawa, Y., and Shibai, H. (1987). Purification and characterization of erythroid differentiation factor (EDF) isolated from human leukaemia cell line THP-1. Biochem. Biophys. Res. Commun. 142:1095–1103.

Farnworth, P. G., Robertson, D. M., de Kretser, D. M., and Burger, H. G. (1988a). Effects of 31 kDa bovine inhibin on FSH and LH in rat pituitary cells in vitro: Actions under basal conditions. Endocrinology 122:207–213.

Farnworth, P. G., Robertson, D. M., de Kretser, D. M., and Burger, H. G. (1988b). Effects of 31 kDa bovine inhibin on FSH and LH in rat pituitary cells in vitro: antagonism of GnRH agonists. J. Endocrinol. 119:233–241.

Findlay, J. K., Tsonis, C. G., Doughton, B., and Forage, R. G. (1989). Immunization against His_1-Arg_{166} (N) of the alpha subunit of inhibin impairs fertility in sheep. Endocrinology 124:3122–3124.

Findlay, J. K., Robertson, D. M., and Clarke, I. J. (1987). Influence of dose and route of administration of bovine follicular fluid and the suppressive effect of purified bovine inhibin (M_r 31 000) on plasma FSH concentrations in ovariectomised ewes. J. Reprod. Fert. 80:455–461.

Forage, R. G., Brown, R. W., Ring, J. M., Stewart, A. G., Milborrow, H. M., Oliver, K. J., Atrache, B. T., Devine, P. L., Hudson, G. C., Goss, N. H., Tolstoshev, P., Robertson, D. M., Doughton, B., de Kretser, D. M., Burger, H. G., and Findlay, J. K. (1987). The cloning and expression of inhibin genes: Subunit use as a fecundity vaccine, in *Inhibin Non-Steroidal Regulation of Follicle Stimulating Hormone Secretion* (Burger, H. G., de Kretser, D. M., Findlay, J. K., and Igarashi, M., eds.) pp. 89–103, Serono Symposia, Raven Press, New York.

Fukuda, M., Miyamoto, K., Hasegawa, Y., Ibuki, Y., and Igarashi, M. (1987). Action mechanisms of inhibin in vitro—Cycloheximide mimics inhibin actions on pituitary cells. Mol. Cell. Endocrinol. 51:41–50.

Gonzales, G., Risbridger, G. P., and de Kretser, D. M. (1989). In vivo and in vitro production of inhibin by cryptorchid testis from adult rats. Endocrinology 124:1661–1668.

Gonzales, G. F., Risbridger, G. P., and de Kretser, D. M. (1988). In vitro synthesis and release of inhibin in response to FSH stimulation by isolated segments of seminiferous tubules from normal adult male rats. Mol. Cell Endocrinol. 59:179–185.

Grootenhuis, A. J., Steenbergen, J., Timmerman, M. A., Dorsman, A.N.R.D., Schaaper, W.M.M., Meloen, R. H., and de Jong, F. H. (1989). Inhibin and activin-like activity in fluids from male and female gonads: Different molecular weight forms and bioactivity/immunoactivity ratios. J. Endocrinol. 122:293–301.

Grootenhuis, A. J., Timmerman, M. A., Hordjik, P. L., and de Jong, F. H. (1990). Inhibin in immature rat Sertoli cell conditioned medium: a 32 KDa alpha beta-B dimer. Mol. Cell Endocrinol. 1:109–116.

Handelsman, D. J., Spalivero, J. A., Kidston, E., and Robertson, D. M. (1989). Highly polarized secretion of inhibin by Sertoli cells in vitro. Endocrinology 125:721–729.

Hasegawa, Y., Miyamoto, K., Abe, Y., Nakamura, T., Sugino, H., Eto, Y., Shibai, H., and Igarashi, M. (1988). Induction of follicle stimulating hormone receptors by erythroid differentiating factor on rat granulosa cell. Biochem. Biophys. Res. Commun. 156:668–674.

Hasegawa, Y., Miyamoto, K., Igarashi, M., Yanaka, T., Sasaki, K., and Iwamura, S. (1987). Changes in serum concentrations of inhibin during the estrous cycle of the rat, pig and cow, in *Inhibin Non-Steroidal Regulation of Follicle Stimulating Hormone Secretion* (Burger, H. G., de Kretser, D. M., Findlay, J. K., Igarashi, M., eds.) pp. 119–133, Serono Symposia, Raven Press, New York.

Hedger, M. P., Drummond, A. E., Robertson, D. M., Risbridger, G. P., and de Kretser, D. M. (1989). Inhibin and activin regulate [^{3}H] thymidine uptake by rat thymocytes and 3T3 cells in vitro. Mol. Cell Endocrinol. 61:133–138.

Hsueh, A. J., Dahl, K. D., Vaughan, J., Tucker, E., Rivier, J., Bardin, C. W., and Vale, W. (1987). Heterodimers and homodimers of inhibin subunits have different paracrine action in the modulation of luteinizing hormone-stimulated androgen biosynthesis. Proc. Natl. Acad. Sci. (USA) 84:5082–5086.

Hutchinson, L. A., Findlay, J. K., de Vos, F. L., and Robertson, D. M. (1987). Effects of bovine inhibin, transforming growth factor β and bovine activin A on granulosa cell differentiation. Biochem. Biophys. Res. Commun. 146:1405–1412.

Inouye, S., Guo, Y., De Paolo, L., Shimonaka, M., Ling, N., and Shimasaki, S. (1991). Recombinant expression of human follistatin with 315 and 288 amino acids: Chemical and biological comparison with native porcine follistatin. Endocrinology (in press).

Ishida, H., Tashiro, H., Watanabe, M., Fujii, T., Yoshida, H., Imamura, K., Minowada, S., Shinohara, M., Fukutani, K., Aso, Y., and de Kretser, D. M. (1990). Measurements of inhibin concentrations in men: Study of changes after castration and comparison with androgen levels in testicular tissue, spermatic vein blood and peripheral venous blood. J. Clin. Endocr. Metabolism 70:1019–1022.

Jackson, C. M., and Morris, I. D. (1977). Gonadotrophin levels in male rats following impairment of Leydig cell function by ethylene dimethane sulphonate. Andrologia 9:29–32.

Knight, P. G., Beard, A. J., Wrathall, J.H.M., and Castillo, R. J. (1989). Evidence that the bovine ovary secretes large amounts of monomeric inhibin α subunit and its isolation from bovine follicular fluid. J. Mol. Endocrinol 2:189–200.

Kogawa, K., Nakamura, T., Sugino, K., Takio, K., Titani, K., and Sugino, H. (1991). Activin-binding protein is present in the pituitary. Endocrinology 128:1434–1440.

Krummen, L. A., Toppari, J., Kim, W. H., Morelos, B. S., Ahmad, N., Swerdloff, R. S., Ling, N., Shimasaki, S., Esch, F., and Bhasin, S. (1989). Regulation of testicular inhibin subunit messenger ribonucleic acid levels in vivo: Effects of hypophysectomy and selective follicle stimulating hormone replacement. Endocrinology 125:1630–1637.

Lee, W., Mason, A. J., Schwall, R., Szonyi, E., and Mather, J. P. (1989). Secretion of activin by interstitial cells in the testis. Science 243:396–398.

Le Gac, F., and de Kretser, D. M. (1982). Inhibin production by Sertoli cell cultures. Mol. Cell. Endocrinol. 28:487–498.

Lin, T., Calkins, J., Morris, P., Vale, W. W., and Bardin, C. W. (1989). Regulation of Leydig cell function in primary culture by inhibin and activin. Endocrinology 125:2134–2140.

Ling, N., Ying, S. Y., Ueno, N., Shimasaki, S., Esch, F., Hotta, M., and Guillemin, R. (1986). Pituitary FSH is released by a heterodimer of the β subunits of the two forms of inhibin. Nature 321:779–782.

Ling, N., Ying, S. Y., Ueno, N., Esch, F., Denorary, L., and Guillemin, R. (1985). Isolation and partial characterization of a M_r 32,000 protein with inhibin activity from porcine follicular fluid. Proc. Natl. Acad. Sci. (USA) 82:7217–7221.

Maddocks, S., and Sharpe, R. M. (1989). The route of secretion of inhibin from the rat testis. J. Endocrinol. 120:R5-R8.

Mather, J. P., Attie, K. M., Woodruff, T. K., Rice, G. C., and Phillips, D. M. (1990). Activin stimulates spermatogonial proliferation in germ cell—Sertoli cell cocultures from immature rat testis. Endocrinology 127:3206–3214.

Mathews, L. S., and Vale, W. W. (1991). Expression cloning of an activin receptor, a predicted transmembrane serine kinase. Cell 65:973–982.

McLachlan, R. I., Matsumoto, A. M., Burger, H. G., de Kretser, D. M., and Bremner, W. J. (1988b). Relative roles of follicle-stimulating hormone and luteinizing hormone in the control of inhibin secretion in normal men. J. Clin. Invest. 82:1–5.

McLachlan, R. I., Robertson, D. M., Burger, H. G., and de Kretser, D. M. (1986). The radioimmunoassay of bovine follicular fluid inhibin. Mol. Cell Endocrinol. 46:175–185.

Meunier, H., Rivier, C., Evans, R. M., and Vale, W. (1988). Gonadal and extragonadal expression of inhibin α, β_A and β_B subunits in various tissues predicts diverse functions. Proc. Natl. Acad. Sci. 85:247–251.

Merchenthaler, I., Culler, M. D., Petruz, P., and Negro-Vilar, A. (1987). Immunocytochemical localisation of inhibin in rat and human reproductive tissues. Mol. Cell. Endocrinol. 54:239–243.

Miller, W. L., Lin, L. W., Phillips, C. L., Wu, J. C., Laws, S. C., Beggs, M. J., Guzman, K., and Milsted, A. (1988). Regulation of the subunit mRNAs of follicle stimulating hormone by inhibin, estradiol and progesterone, in *CONRAD International Workshop: Nonsteroidal Gonadal Factors* (Hodgen, G., Rosenwaks, Z., Spieler, J., eds.) pp. 110–124, Jones Institute Press, Virginia.

Mitrani, E., Ziv, T., Thomsen, G., Shimoni, Y., Melton, D. A., and Bril, A. (1990). Activin can induce the formation of axial structures and is expressed in the hypoblast of the chick. Cell 63:495–501.

Miyamoto, K., Hasegawa, Y., Fukuda, M., and Igarashi, M. (1986). Demonstration of high molecular weight forms of inhibin in bovine follicular fluid (bFF) by using monoclonal antibodies to bFF 32K inhibin. Biochem. Biophys. Res. Comm. 136:1103–1109.

Morris, P. L., Vale, W. W., Cappel, S., and Bardin, C. W. (1988). Inhibin production by primary Sertoli cell-enriched cultures: Regulation by FSH, androgens and EGF. Endocrinology 122:717–725.

Nakamura, T., Takio, K., Eto, Y., Shibai, H., Titani, and Sugino, H. (1990). Activin-binding protein from rat ovary is follistatin. Science 247:836–838.

O, W-S., Robertson, D. M., and de Kretser, D. M. (1989). Inhibin as an oocyte meiotic inhibitor. Mol. Cell Endocrinol. 62:307–311.

O'Leary, P. O., Jackson, A. E., Averill, S., and de Kretser, D. M. (1986). The effects of ethane dimethane sulphonate (EDS) on bilaterally cryptorchid rat testes. Mol. Cell Endocrinol. 45:183–190.

Padgett, R. W., St. Johnston, R. D., and Gelbart, W. M. (1987). A transcript from a *Drosophila* pattern gene predicts a protein homologous to the transforming growth factor β family. Nature 325:81–84.

Payne, D. W., Holtzclau, D., and Adashi, E. (1988). The steroidogenic characteristics of primary testicular cell cultures from adult hypophysectomised rats: Enhanced formation of C_{21} steroids. Biol. Reprod. 39:581–591.

Pineau, C., Sharpe, R. M., Saunders, P.T.K., Gérard, N., and Jégou, B. (1990). Regulation of Sertoli cell inhibin production and of inhibin α subunit mRNA levels by specific germ cell types. Mol. Cell. Endocrinol. 72:13–22.

Risbridger, G. P., Clements, J. A., Robertson, D. M., Drummond, A. E., Muir, J., Burger, H. G., and de Kretser, D. M. (1989a). Immuno- and bioactive inhibin and inhibin α subunit expression in rat Leydig cell cultures. Mol. Cell Endocrinol. 66:119–122.

Risbridger, G. P., Hancock, A., Robertson, D. M., Hodgson, Y., and de Kretser D. M. (1989b). Follitropin (FSH) stimulation of inhibin biological and immunological activities by seminiferous tubules and Sertoli cell cultures from immature rats. Mol. Cell Endocrinol. 67:1–9.

Rivier, C., Schwall, R., Mason, A., Burton, L., and Vale, W. (1991). Effect of recombinant inhibin on gonadotropin secretion during proestrus in the rat. Endocrinology 128:2223–2228.

Rivier, C., Cajander, S., Vaughan, J., Hsueh, A.J.W., and Vale, W. (1988). Age-dependent changes in physiological action, content, and immunostaining of inhibin in male rats. Endocrinology 123:120–126.

Rivier, C., and Vale, W. (1987). Inhibin: Measurement and role in the immature female rat. Endocrinology 120:1688–1690.

Roberts, V., Meunier, H., Vaughan, J., Rivier, J., Rivier, C., Vale, W., and Sawchenko, P. (1989). Production and regulation of inhibin subunits in pituitary gonadotropes. Endocrinology 124:552–554.

Robertson, D. M., Foulds, L. M., Prisk, M., and Hedger, M. P. (1992). Inhibin/activin β subunit monomer: Isolation and characterisation. Endocrinology (in press).

Robertson, D. M., Prisk, M., McMaster, J. W., Irby, D. C., Findlay, J. K., and de Kretser, D. M. (1991). Serum FSH-suppressing activity of human recombinant inhibin A in male and female rats. J. Reprod. Fert. 91:321–328.

Robertson, D. M., Farnworth, P. G., Clarke, L., Jacobsen, J., Cahir, N. F., Burger, H. G., and de Kretser, D. M. (1990). Effect of bovine 35kDa FSH-suppressing protein on FSH and LH in rat pituitary cells in vitro: Comparison with bovine 31kDa inhibin. J. Endocrinol. 124:417–423.

Robertson, D. M., Giacometti, M., Foulds, L. M., Lahnstein, J., Goss, N. H., Hearn, M.T.W., and de Kretser, D. M. (1989). Isolation of inhibin α-subunit precursor proteins from bovine follicular fluid. Endocrinology 125:2141–2149.

Robertson, D. M., Tsonis, C. G., McLachlan, R. I., Handelsman, D. J., Leask, R., Baird, D. T., McNeilly, A. S., Hayward, S., Healy, D. L., Findlay, J. K.., Burger, H. G., and de Kretser, D. M. (1988a). Comparison of inhibin immunological and in vitro biological activities in human serum. J. Clin. Endocrinol. Metabl. 67:438–448.

Robertson, D. M., de Vos, F., McLachlan, R. I., Salamonsen, L. A., Hearn, M.T.W., Morgan F. J., Findlay, J. K., Burger, H. G., and de Kretser, D. M. (1987a). Purification of inhibin and its isoforms, in *Inhibin Non-Steroidal Regulation of Follicle Stimulating Hormone Secretion* (Burger, H. G., de Kretser, D. M., Findlay, J. K., Igarashi, M., eds.) pp. 17–33, Serono Symposia, Raven Press, New York.

Robertson, D. M., Klein, R., de Vos, F. L., McLachlan, R.

I., Wettenhall, R.E.H., Hearn, M.T.W., Burger, H. G., and de Kretser, D. M. (1987b). The isolation of polypeptides with FSH suppressing activity from bovine follicular fluid which are structurally different to inhibin. Biochem. Biophys. Res. Commun. 149:744–749.

Robertson, D. M., Foulds, L. M., Leversha, L., Morgan, F. J., Hearn, M.T.W., Burger, H. G., Wettenhall, R.E.H., and de Kretser, D. M. (1985). Isolation of inhibin from bovine follicular fluid. Biochem. Biophys. Res. Comm. 126:220–226.

Robertson, D. M., Hayward, S., Irby, D. C., Jacobsen, J. V., Clarke, L. J., McLachlan, R. I., and de Kretser, D. M. (1988b). Radioimmunoassay of rat serum inhibin: Changes after PMSG stimulation and gonadectomy. Mol. Cell Endocrinol. 58:1–8.

Schwall, R. H., Schmelzer, C. H., Matsuyama, E., and Mason, A. J. (1989). Multiple actions of recombinant activin-A in vivo. Endocrinology 125:1420–1424.

Schwall, R. H., Szonyi, E., Mason, A. J., and Nikolics, K. (1988a). Activin stimulates secretion of follicle-stimulating hormone from pituitary cells desensitized to gonadotropin-releasing hormone. Biochem. Biophys. Res. Commun. 151:1099–1104.

Schwall, R. H., Nikolics, K., Szonyi, E., Gorman, C., and Mason, A. J. (1988b). Recombinant expression and characterisation of human activin A. Mol. Endocrinol. 2:1237–1243.

Scott, R. S., Burger, H. G., and Quigg, H. (1980). A simple and rapid in vitro bioassay for inhibin. Endocrinology 107:1536–1542.

Sharpe, R. M., Kerr, J. B., and Maddocks, S. (1988). Evidence for a role of the Leydig cells in control of the intratesticular secretion on inhibin. Mol. Cell. Endocrinol. 60:243–247.

Sheckter, C. B., McLachlan, R. I., Tenover, J. S., Matsumoto, A. M., Burger, H. G., de Kretser, D. M., and Bremner, W. J. (1988). Serum inhibin concentrations rise during GnRH treatment of men with idiopathic hypogonadotrophic hypogonadism. J. Clin. Endocrinol. Metab. 67:1221–1224.

Shimasaki, S., Koga, M., Buscaglia, M. L., Simmons, D. M., Bicsak, T. A., and Ling, N. (1989). Follistatin gene expression in the ovary and extragonadal tissue. Mol. Endocrinol. 3:651–659.

Steinberger, A., and Steinberger, E. (1976). Secretion of an FSH-inhibiting factor by cultured Sertoli cells. Endocrinology 99:918–921.

Sugino, K., Nakamura, T., Takio, J., Titiani, K., Miyamoto, K., Hasegawa, Y., Igarashi, M., and Sugino, H. (1989). Inhibin alpha-subunit monomer is present in bovine follicular fluid. Biochem. Biophys. Res. Commun. 159:1323–1329.

Sugino, H., Nakamura, T., Hasegawa, Y., Miyamoto, K., Abe, Y., Igarashi, M., Eto, Y., Shibai, H., and Titiani, K. (1988a). Erythroid differentiation factor can modulate follicular granulosa cell functions. Biochem. Biophys. Res. Commun. 153:281–288.

Sugino, H., Nakamura, T., Hasegawa, Y., Miyamoto, K., Igarashi, M., Eto, Y., Shibai, H., and Titiani, K. (1988b). Identification of a specific receptor for erythroid differentiating factor on follicular granulosa cell. J. Biol. Chem. 263:15249–15252.

Sun, Y-T., Robertson, D. M., Gonzales, G., Risbridger, G. P., and de Kretser D. M. (1989). The effect of testosterone on serum immunoactive inhibin concentrations in intact and hypophysectomised male rats. J. Reprod. Fert. 87:795–801.

Tenover, J. S., McLachlan, R. I., Burger, H. G., de Kretser, D. M., and Bremner, W. J. (1988). Decreased serum inhibin levels in normal elderly men: Evidence for a decline in Sertoli cell function with aging. J. Clin. Endocrinol. Metab. 67:455–459.

Thomsen, G., Woolf, T., Whitman, M., Sokol, S., Vaughan J., Vale, W., and Melton, D. A. (1990). Activins are expressed early in *Xenopus* embryogenesis and can induce axial mesoderm and anterior structures. Cell 63:485–493.

Tierney, M. L., Goss, N. H., Tomkins, S. M., Kerr, D. B., Pitt, D. E., Forage, R. G., Robertson, D. M., Hearn, M. T. W., and de Kretser, D. M. (1990). Physicochemical and biological characterisation of recombinant human inhibin A. Endocrinology 126:3268–3270.

Toebosch, A.M.W., Robertson, D. M., Trapman, J., Klaassen, P., de Paus, R. A., de Jong, F. H., and Grootegoed, J. A. (1988). Effects of FSH and IGF_1 on immature Sertoli cells: Inhibin α and β subunit mRNA levels and inhibin secretion. Mol. Cell Endocrinol. 55:101–105.

Tsonis, C. G., McNeilly, A. S., and Baird, D. T. (1986). Measurement of exogenous and endogenous inhibin in sheep serum using a new and extremely sensitive bioassay for inhibin based on inhibition of ovine pituitary PSH in vitro. J Endocrinol. 110:341–352.

Ueno, N., Ling, N., Ying, S. Y., Esch, F., Shimasaki, S., and Guillemin, R. (1987). Isolation and partial characterization of follistatin: A single-chain Mr 35000 monomeric protein that inhibits the release of follicle stimulatory hormone. Proc. Natl. Acad. Sci. (USA) 84:8282–8286.

Ultee-van Gessel, A. M., Leemborg, F. G., de Jong, F. H., and van der Molen, H. J. (1986). In vitro secretion of inhibin-like activity by Sertoli cells from normal and prenatally irradiated immature rats. J. Endocrinol. 109:411–418.

van Dissel-Emiliani, F.M.F., Grootenhuis, A. J., de Jong, F. H., and de Rooij, D. G. (1989). Inhibin reduces spermatogonial numbers in testis of adult mice and Chinese hamsters. Endocrinology 125:1899–1903.

Vale, W., Rivier, J., Vaughan, J., McClintock, R., Corrigan, A., Woo, W., Karr, D., and Spiess, J. (1986). Purification and characterization of an FSH releasing protein from porcine ovarian follicular fluid. Nature 321:776–779.

Verhoeven, G., and Callieau, J. (1988). Follicle-stimulating hormone and androgens increase the concentrations of androgen receptor in Sertoli cells. Endocrinology 122:1541–1550.

Verhoeven, G., and Franchimont, P. (1983). Regulation of inhibin secretion by Sertoli cell-enriched cultures. Acta Endocrinol. 102:136–143.

Wang, Qi-Fa., Farnworth, P. G., Findlay, J. K., and Burger, H. G. (1988). The effect of pure 31 kDa bovine inhibin on the specific binding of GnRH to rat anterior pituitary cells in culture. Endocrinology 123:2161–2166.

Weeks, D. L., and Melton, D. A. (1987). A maternal mRNA localised to the vegetal hemisphere in *Xeno-*

pus eggs codes for a growth factor related to TGFβ. Cell 51:861–867.
Xiao, S., Findlay, J. K., and Robertson, D. M. (1990). The effect of bovine activin and FSH suppressing protein/follistatin on FSH induced differentiation of rat granulosa cell in vitro. Mol. Cell Endocrinol. 69:1–8.
Yu, J., Shao Li-en, Lemas, V., Yu, A. L., Vaughan, J., Rivier, J., and Vale, W. (1987). Importance of FSH-releasing protein and inhibin in erythrodifferentiation. Nature 330:765–767.

CHAPTER ELEVEN

The Seminiferous Tubular Microenvironment

Barry T. Hinton and Terry T. Turner

I. Introduction

Although we often speak of the seminiferous tubular microenvironment as if it were a singular thing, in reality the tubules are surrounded by and contain a series of microenvironments, each playing a necessary role in the development of spermatozoa. The developing germ cell is faced with several different microenvironments within the seminiferous epithelium and another that bathes the luminal spermatozoa and apical membranes of Sertoli cells. Even this latter microenvironment may change from one area of the tubular lumen to another and may exert different influences on germ cells or Sertoli cells along the way. Knowledge of seminiferous tubule physiology is now much more extensive than it was 20 years ago when specific studies of the intratubular microenvironment began (Tuck et al., 1970; Levine and Marsh, 1971); nevertheless, much of what is presently known about the seminiferous microenvironment is still limited to the microenvironment in the seminiferous tubule lumen. For example, we know much less about the microenvironment surrounding the developing germ cells within the seminiferous epithelium. The knowledge that is developing within this latter area comes primarily from cell biological studies, and this discussion attempts to integrate these developments with what is known about the physiology of the testicular parenchyma.

Other chapters in this volume deal specifically with testicular microvasculature (Chapter 6), the biology of Leydig and Sertoli cells (Chapters 7 and 9, respectively), and the cellular structure of seminiferous tubules (Chapter 14). Because the intratubular microenvironment is a product of the functions of all these components, it will be necessary to our argument to present some information on these subjects within this chapter as well.

II. Fluid Compartments of the Testes: Anatomical and Physiological Features

All soft tissues in the body are in reality a series of solutions bound by membranes. It is the relationship between solutions divided by cell membranes that forms the basis of physiology, and it is the separation of these solutions by specialized membranes within a cell that forms the basis of cell biology. Thus, compartmentalization of fluid in any organ is important to its function, and understanding how one fluid compartment can influence adjacent compartments is fundamental to the understanding of how any organ works in vivo. The compartments containing freely flowing fluid in the testes are the vascular compartment, the interstitial compartment (for this discussion, to include any testicular lymphatic space), and the intraluminal compartment. In the present discussion, we deal briefly with each of these fluid spaces as the characteristics of the seminiferous tubular microenvironment depend greatly on the relationships between them.

A. The Vascular Compartment

The vasculature is the first source of fluid, nutrients, electrolytes, and humoral signaling molecules for any organ, and as "the source," it should be among the first features examined for regulatory input into an organ's function. This is as true for the testes as it is for any organ, yet, aside from the constitution of the blood within it, our knowledge about the testicular vascular compartment remains relatively rudimentary. In this section, we discuss some of what is known and not known about the testicular vasculature, and how it relates to the other testicular compartments. Much of this dicussion focuses on the microvasculature because it is at this level that the vasculature has an impact on the biology of specific cell populations.

The architecture of the testicular vasculature has been recently reviewed (Setchell and Brooks, 1988), and only salient features of the microvascular anatomy are presented here. The blood vessels serving the seminiferous tubules never penetrate those tubules; rather, they serve the interstitial space directly within the testis (Fig. 11–1). In the rat, both peritubular (circumferential) and intertubular (longitudinal) capillaries enshroud each tubule. Precapillary arterioles and postcapillary venules generally, but not always, run in parallel with the tubules (see Setchell and Brooks, 1988). The human testicular microvasculature also surrounds the seminiferous tubule, but in a less rigid pattern than that of the laboratory species studied. In humans, a distinction between intertubular and peritubular capillaries is not readily made, but casts of human testicular microvasculature still indicate the close encirclement of seminiferous tubules by testicular capillaries (Suzuki and Nagano, 1986).

Blood flow in testicular microvessels has not been widely studied. Total organ blood flow to the rat testis, when measured by the radiolabeled microsphere distribution technique, is approximately 30 mL/min/100 g of tissue (Hurt et al., 1987), which, for example, is much less than total blood flow to the kidney (500 to 600 mL/min/100 g tissue), similar to blood flow to the liver (20 to 30 mL/min/100 g tissue), but more than the blood flow to resting smooth muscle (5 to 6 mL/min/100 g tissue) in the same animals. This total organ blood flow is relatively stable after testicular stimulation with trophic hormones or even after treatment with some common vasodilatory drugs (see Setchell and Brooks, 1988). These findings have led some to believe that blood flow to the testes is relatively constant. By extension, that capillary blood flow is also constant.

Capillary blood flow at the level of the individual capillaries in the rat testis is, in fact, highly variable. During in vivo microscopy, blood flow in specific vessels of the microvasculature can easily be observed to slow, halt, increase, or redirect to other microvessels (as at branches in the microvasculature where flow down one branch diminishes and the flow in the alternate branch increases). These alterations in capillary blood flow can only be observed with micromethods and are not reflected in older testicular capillary blood flow studies that were based on whole organ blood flow. Likewise, traditional calculations of capillary blood flow based on measurements of total organ blood flow have little meaning when addressing the actual rate of blood movement down individual capillaries or even whole capillary beds that may surround a length of seminiferous tubule.

Capillary blood flow (by which we mean flow within a given testicular capillary or capillary bed) is important to the underlying seminiferous tubule(s) because that blood flow is responsible for providing the tubule with its bathing interstitial fluid, its metabolic nutrients, physiological buffers, hormonal signals, and O_2/CO_2 exchange capacity. Because variability in the provision of these vital factors can affect all cells served by the capillary (e.g., Leydig cells, peritubular myoid cells, Sertoli cells, germ cells), blood flow within the capillary is a key factor in the manufacture and maintenance of the seminiferous tubular microenvironment.

Not only have oscillations and interruptions of capillary blood flow been observed by investigators studying the testicular microenvironments in vivo (Turner and Hinton, unpublished observations), but variations in local blood flow have been recorded by laser Doppler flowmetry

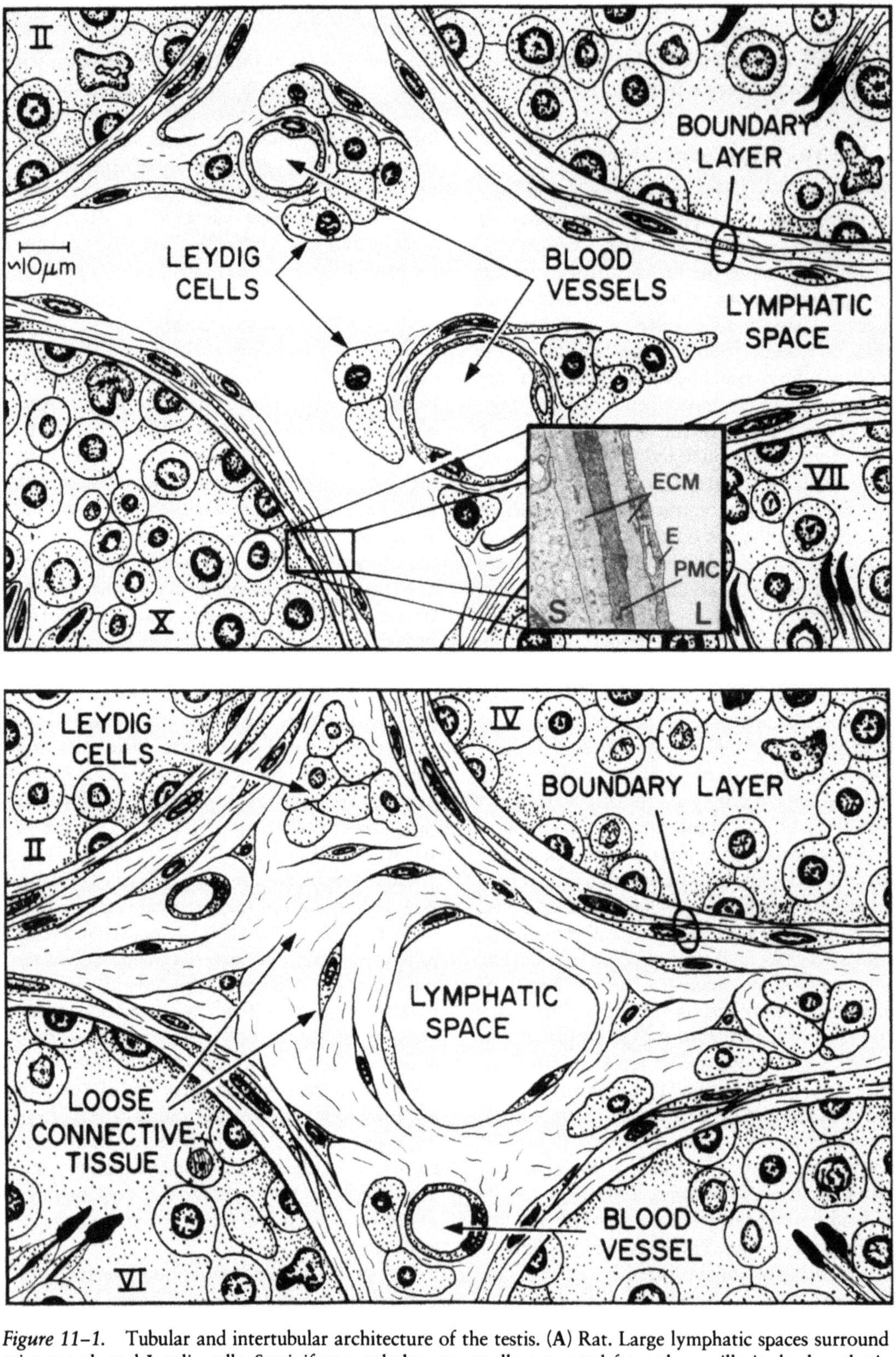

Figure 11–1. Tubular and intertubular architecture of the testis. (**A**) Rat. Large lymphatic spaces surround microvessels and Leydig cells. Seminiferous tubules are usually separated from the capillaries by lymphatic space and the boundary layer *(see insert)*, which consists of lymphatic endothelium (E), peritubular myoid cells (PMC), and extracellular matrix (ECM). S, Sertoli cell; L, lymphatic space. (**B**) Human. Loose connective

and associated with acute changes (within seconds) in relative hematocrit and capillary blood-cell velocity within single capillaries (Damber et al., 1986). Qualitative changes in testicular capillary blood flow have been reported after treating rats with human chorionic gonadotropin (hCG; Widmark et al., 1986) and luteinizing hormone–releasing hormone (LHRH) agonist (Damber et al., 1987a), as have hCG-induced changes in testicular vascular permeability and interstitial fluid accumulation (Setchell and Sharpe, 1981; Widmark et al., 1986). This increase in testicular vascular permeability is associated with an increase in polymorphonuclear leukocytes in the testes, and the changes in vascular permeability are due to lymphocyte-mediated increases in venule permeability (Bergh et al., 1987). It is possible that neither the vascular permeability changes nor the capillary blood flow changes occur in the absence of Leydig cells (Setchell and Rommerts, 1985; Damber et al., 1987b), which implies an important, still undefined, relationship between Leydig cells and the vascular endothelium or intravascular leukocytes.

Any interpretation of the above experiments has to be done with caution, however, as the common mechanism for eliminating Leydig cells in the testes has been to use ethane dimethyl sulfonate (EDS). The effects of EDS on Leydig cells are clear (Molenaar et al., 1985), but its effects on other cells in the testis, or in the body at large for that matter, have not been rigorously explored. Present knowledge of EDS toxicity is too limited to assume that the effect of EDS treatment of the testis is due to the absence of Leydig cells alone (Sprando et al., 1990).

It should also be appreciated that changes in capillary blood flow and vascular permeability induced by the treatments mentioned previously have never occurred within a time frame consistent with the native fluctuations seen in microvascular flow in vivo. Nevertheless, these studies do demonstrate induced changes in local flow and vascular permeability around the seminiferous tubules. Greater permeability of bioactive compounds from the vasculature; more or less dilution of these compounds by increases in interstitial fluid; variable removal of paracrine, endocrine, or metabolic products; and local changes in oxygen tension can all have effects on localized cell populations and potentially contribute to the peritubular and intratubular seminiferous microenvironment.

Another aspect of understanding how the microvasculature relates to the seminiferous tubular microenvironment is to appreciate the three-dimensional aspect of testicular architecture and to realize that diffusion and flow around tubules is also in three dimensions. It is tempting to visualize the relationships of cell types and fluid compartments in the testes as being in two dimensions only, as illustrated in Figure 11–1 and as commonly seen in testicular histology cross sections. This view is correct within its limits, but is not complete with regard to how the vasculature serves the seminiferous epithelium. It must be appreciated that there is flow within the capillary as it runs along the seminiferous tubule and that that tubule and its microenvironment are everchanging from one stage of the seminiferous cycle to the next (Fig. 11–2). Our understanding of the control of the seminiferous microenvironments along the tubule will not be complete until we understand the control of capillary blood flow along the tubule.

In addition to blood flow within the vasculature, there is fluid flow outside the capillary, in the interstitial and lymphatic space (Fig. 11–2). This means that molecules exiting the vasculature at any given point do not necessarily address only adjacent Leydig cells, peritubular myoid cells, or Sertoli cells as seen in cross section (Fig. 11–1), but may also be available to longitudinally remote cells after further transport in lymph or interstitial fluid (Fig. 11–2).

tissue surrounds microvessels and Leydig cells. Leydig cells secrete products into the interstitial fluid space of the loose connective tissue. The boundary layer around each tubule consists of peritubular myoid cells and extracellular matrix. In both panels, different stages of the cycle of the seminiferous epithelium are indicated by Roman numerals. Note that different stages of the cycle may be equally adjacent to common microvessels or Leydig cells. (Illustration adapted from Fawcett, 1973, Adv. Biosci. 10:81. Electron micrograph of boundary layer courtesty of M. Dym.)

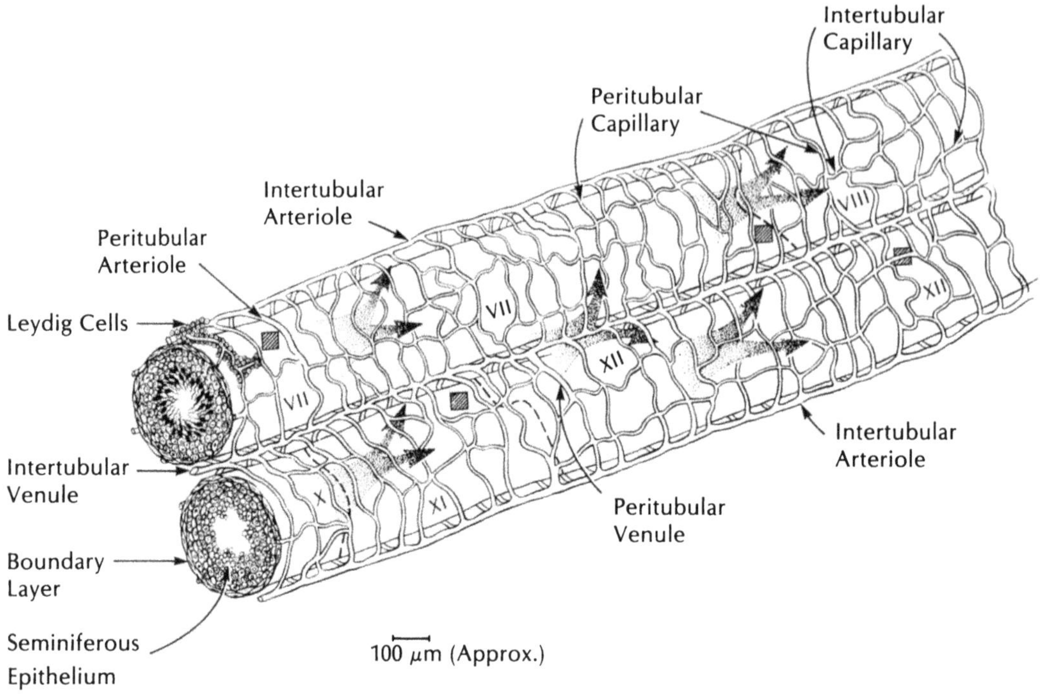

Figure 11–2. Schematic illustration of two rat seminiferous tubules and their supporting vasculature. In each tubule, boundaries of the stages of the cycle of the seminiferous epithelium are indicated with *broken lines* and are drawn to reflect the average proportionate length of those numbered stages illustrated. The tubules are transected at a point half-way along the average length of stage VII (the upper tubule) and stage X (the lower tubule). The microvasculature is drawn exploded from the tubules for purposes of illustration. Connecting vessels to other lateral tubules are not drawn in for sake of simplicity. *Broad arrows* indicate possible direction of interstitial flow to illustrate how molecules appearing in the interstitium over one stage of the cycle of the seminiferous epithelium might rapidly be smeared over an adjacent stage(s) in that same or different tubules. Interstitial flow in vivo is from distal to proximal pole of the testis, and actual direction of flow would depend on the orientation of the tubules. *Hatched squares* illustrate surface areas equivalent to the surface areas of the proposed *unit segments* of the seminiferous epithelium as proposed by Perey (1961). Illustration of Leydig cells is limited for the sake of clarity.

Thus, products from the vasculature surrounding stage X of the rat seminiferous epithelium may be spread over adjacent stages of the cycle within the same tubule, or even to adjacent tubules, which may also be in a different stage of the cycle (Figs. 11–1 and 11–2).

If local changes in capillary blood flow can be imagined to have stage-specific effects on the microenvironment and ultimately on stages of spermatogenesis, then control of local blood flow must be available within the anatomical dimensions of the different stages of the spermatogenic cycle. Figure 11–2 illustrates schematically, and roughly to scale for the rat, that the stages of the cycle of the seminiferous epithelium vary on a scale of millimeters; that is, stage VII averages 3.2 mm in length along the tubule and stage XI averages 0.6 mm (Perey et al., 1961). The vessels of the seminiferous microvasculature are on the order of 30 to 100 μm apart. Thus, these capillaries can easily serve discrete stages of the cycle of the seminiferous epithelium (Fig. 11–2).

Perey and colleagues (1961) hypothesized that the smallest divisible unit of a stage of the cycle of the seminiferous epithelium was the *unit-segment,* which they calculated to have a surface area of approximately 2000 μm^2. Several square areas of this dimension have been plotted on the seminiferous tubules illustrated in Figure 11–2: (1) to show their dimension relative to the other features of the seminiferous tubule and its attendant microvasculature and (2) to illustrate that even these discrete unit-segments may be served by a capillary or relatively few capillaries.

B. Interstitial Compartment

There are species differences in the interstitial architecture of the testis, primarily relating to the distribution of the lymphatic space versus loose connective tissue, of which opposing examples are the rat and the human (Fig. 11–1). In some species there are also large differences in the proportion of interstitial tissue that consists of Leydig cells and connective tissue (see DeKretser and Kerr, 1988). The most studied species, the rat, has a peritubular capillary network invested with Leydig cells, but lymphatic space between seminiferous tubules comprises approximately 40 percent of the testicular volume (see Mendis-Handagana et al., 1988, and Fig. 11–1A). In the case of the human, peritubular capillaries and Leydig cells are surrounded by loose connective tissue, and the lymphatic drainage consists of more discrete channels (Fig. 11–1B).

An interesting feature of the testicular interstitium, which has only recently begun to receive attention, is the boundary layer surrounding each tubule. In the rat, this layer consists of extracellular matrix (ECM) on which the Sertoli cells and spermatogonia rest, the peritubular myoid cells, and the lymphatic endothelium (Fig. 11–1A). In the human, the boundary layer does not include the lymphatic endothelium (Fig. 11–1B).

The total interstitial compartment of the rat testis can be illustrated by fluorescent light microscopy of the interstitial tissue after exposure to a fluorescein-coupled antibody to rat serum albumin (Fig. 11–3; Christensen et al., 1985). Albumin detected in this manner fills the testicular interstitial space, but is excluded by the Leydig cells and from the seminiferous tubules above the Sertoli cell-Sertoli cell tight junctions.

The boundary layer does not exclude albumin from the base of the seminiferous tubule (Christensen et al., 1985; Fig. 11–3) or from other molecules of similar size. It is believed that the basolateral membrane of the Sertoli cells and the entire cell membrane of spermatogonia and preleptotene primary spermatocytes are exposed to relatively unrestricted products from the vasculature. The boundary layer does potentially provide each tubule with a relatively unstirred layer where proteins or peptides might be retained longer around a portion of tubule than would otherwise be possible in the more freely flowing interstitial or lymphatic fluid. When Christensen and associates (1985) attempted to flush albumin from the interstitial space of the rat testis, they noted that the boundary layer retained albumin because it was an area relatively inaccessible to their interstitial flushing medium.

In recent years, the ECM and peritubular myoid cells of the boundary layer have been shown to be important to Sertoli cells in vitro and may well play a role in the regulation of the seminiferous tubule microenvironment. The ECM has been shown to be profoundly important to the development and expression of differential cell structure and function in many tissues and this appears to be the case in the testes as well. Laminin, type IV collagen, and heparan sulfate proteoglycan have been localized in the basement membrane (see Fig. 11–1) on which the Sertoli cells and spermatogonial cells rest (Hadley and Dym, 1987), and these three ECM molecules and fibronectin have also been found in the matrix surrounding the peritubular myoid cells (Hadley and Dym, 1987). The ECM has been shown to be important to polarized Sertoli cell secretion, transepithelial transport, morphological appearance, testicular cord formation, and germ cell development in vitro (see Hadley et al., 1985; Hadley and Dym, 1987; Hadley et al., 1987). Recent questions have been raised about the importance of ECM to several aspects of cell function (Anthony and Skinner, 1989), and it is possible that ECM plays a permissive, not regulatory, role in Sertoli cell biology. ECM is known to be secreted in culture by both Sertoli cells and peritubular myoid cells (see Hadley and Dym, 1987), and the cooperation of these two cell types in vitro has also been demonstrated.

It has been shown that peritubular myoid cells enhanced Sertoli cell aggregation and androgen binding protein (ABP) synthesis in coculture, and it has subsequently been found that, under androgen regulation, peritubular

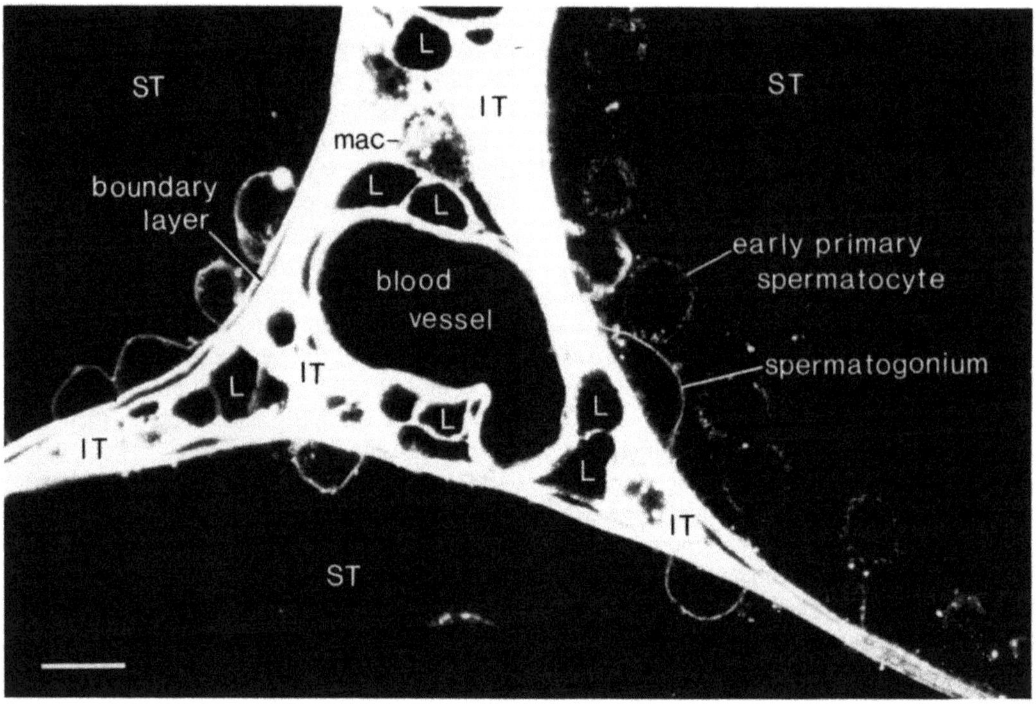

Figure 11–3. Rat testis interstitial space illuminated by fluorescent microscopy. A frozen section of rat testis was treated with fluorescein-labeled antirat serum albumin; thus the fluorescence seen in the micrograph indicates the total albumin space in the interstitial tissue (IT) between seminiferous tubules (ST) except for the vascular lumen that was cleared by perfusion with fixative. Albumin is excluded by the Leydig cells (L), but is endocytosed by macrophages (MAC). Fluorescent staining of albumin also appears in the basal compartment of the seminiferous epithelium around spermatogonia and early spermatocytes, but does not enter the adluminal compartment. Bar = 15 μm. (From Christensen et al., 1985; micrograph courtesy of K. Christensen.)

myoid cells secrete a factor termed P-Mod-S that, in turn, modulates Sertoli cell functions in vitro, such as secretion of ABP and transferrin. P-Mod-S has an apparent molecular weight of 50,000 under physiological conditions and may be an inducer substance similar to those found in the mesenchymal–epithelial interactions of embryological tissues (see Skinner et al., 1988). More recently, Anthony et al. (1991) have argued that P-Mod-S may be an important initiator of Sertoli cell differentiation at the onset of puberty.

Whether P-Mod-S is a specific inducer of Sertoli cell activity is unknown, and to our knowledge the molecule has not been localized in the testis itself; but its presence and testosterone-stimulated activity in myoid cell-Sertoli cell co-cultures argue that androgens regulate peritubular myoid cell secretion of a powerful inducer of Sertoli cell secretion. Thus, both seminiferous ECM as well as paracrine factors produced by peritubular myoid cells can potentially regulate Sertoli cell function and the microenvironment that results from that function.

C. Intraluminal Compartment

The constitution of the fluid in the seminiferous tubule intraluminal compartment is the net product of the vascular delivery of blood and its metabolic and hormonal constituents to the testis, the interstitial diffusion and transport of those constituents, the barrier or sifting property of the Sertoli cell–Sertoli cell tight junctions, and the secretory and reabsorptive properties of the Sertoli cells themselves. Estimates of fluid secretion rate into the intraluminal compartment of the seminiferous tubule vary widely. Values derived from observations on single rat seminiferous tubules in vivo have been between 0.5 and 1.0 μL/tubule/hour (Hinton and Setchell, 1978) (this calculates to approximately 7 to 15 μL/g/hour on the basis of 30 tubules/2 g testis). There are periods of fast and slow flow within a given tubule, as well as passage, then return–surge of fluids through a given point of observation in a tubule (see Setchell et al., 1978). Other estimates of seminiferous tubule fluid secretions derived from collection of rat rete testis fluid in vivo or seminiferous tubule fluid secretion in vitro in the rat have ranged from 20 to 60 μL/g/hour (Tuck et al., 1970; Cheung et al., 1977). The primate *(Macaque)* testis has been estimated to produce seminiferous tubule fluid at a rate of 10μL/g/hour (Waites and Einer-Jensen, 1974). The constitution of intraluminal fluid is discussed later in this chapter (section III.A.).

The Sertoli cells secrete the fluid of the intraluminal compartment in a manner dependent on a basolateral Na^+–K^+ translocating ATPase, with intraluminal Na^+ entering the lumen primarily through a paracellular route and K^+ entering the lumen by diffusion from the Sertoli cell cytoplasm (see Waites and Gladwell, 1982). The regulation of intraluminal fluid secretion remains uncertain, however, and electrolyte transport is only one activity of Sertoli cells in providing the intraluminal microenvironment. Sertoli cells synthesize and secrete proteins, sorting them for apical or basolateral secretion (Shabanowitz and Kierszenbaum, 1986; Ueda et al., 1988). They also produce cell membrane receptors (Sylvester and Griswold, 1984), and intracellular binding proteins (Kato et al., 1985) for factors provided by the vascular or interstitial compartments. The regulation of the Sertoli cell contribution to the seminiferous microenvironment is discussed in sections V. C. and V. D.

To understand the luminal compartment of the seminiferous microenvironment, it is also important to consider the adluminal, paracellular compartment within the seminiferous epithelium. The microenvironment surrounding the developing spermatocytes and spermatids above the Sertoli cell barrier is in a fluid space through which important Sertoli cell–germ cell communication occurs. It is this adluminal space, for example, to which transferrin-Fe^{++} is delivered by Sertoli cells in vivo (Morales et al., 1987), and to which Sertoli cell synthesized proteins are transported for incorporation into the developing sperm cell membrane (Sylvester et al., 1984). Also, there appears to be bilateral communication between germ cells and Sertoli cells, and it is the adluminal compartment into which germ cell-derived factors must be secreted to affect Sertoli cell function (Wright, 1988; Djakiew and Dym, 1988). The adluminal microenvironment is too small to be sampled and analyzed by conventional biochemical means, but information is being gained with immunocytochemical and radiocytochemical techniques that promise to reveal more about the potentially dynamic microenvironment within the adluminal compartment.

In conclusion, the vascular, interstitial, and intraluminal compartments are specialized physiological compartments within the testis. They form the microenvironments in which the events of Leydig, Sertoli, and germ cell biology must occur and through which these events must be communicated. These compartments are dynamic with regard to flow, a factor that can impinge on events requiring diffusion of substances from one compartment to another. This fact alone raises some challenge to simplified versions of cell–cell communication in the testis where it is often assumed that the fluid medium separating nearby cell populations (e.g., Leydig cells and Sertoli cells as seen in cross section) is static. The adluminal microenvironment is an important aspect of Sertoli and germ cell interaction, and it is entirely possible that these interactions cause the development of a series of different microenvironments around the different types of advanced germ cells still within the seminiferous epithelium.

III. Intraluminal Microenvironment

A. Fluid Milieu

The intraluminal microenvironment is perhaps more complicated than at first imagined as it is a highly dynamic, everchanging milieu. The advancement in our knowledge of the microenvironment has been very slow when we consider that seminiferous tubule fluid was recognized as early as the 17th century. In 1680, Borelli noted, "... The famous Auberio when he taught anatomy in Pisa, on the ides of May 1657, showed to myself, the famous Malpighi and others, that all the mass of a boiled testicle of a sheep consists of countless white columnar filaments, like worms, extending from both nervous envelopes to the median axis of the testicle. These fibers were full of white semen. Slightly squeezed they secreted semen drop by drop ..." (translated by Maquet, 1989). With the advent of a number of techniques it has been possible to measure the concentration of a number of elements, ions, organic solutes, and macromolecules within the seminiferous tubule. It should be realized, however, that these studies have provided only a clue as to what the microenvironment is at any one moment in time and in any seminiferous tubule as a whole. Such studies do not tell us what the microenvironment is around each developing germ cell and their associated Sertoli cells; presumably Sertoli cells are continually altering the fluid milieu that is in immediate contact with itself and germ cells as spermatogenesis proceeds. To complicate matters further, other cell types also influence the Sertoli cell in its formation of the fluid microenvironment (see section V. and Chapter 15). The point is, that although we have some understanding of the composition of the seminiferous tubule fluid, it only represents the net microenvironment; it is the product of a very complex series of cellular interactions.

The seminiferous tubule essentially secretes a potassium bicarbonate-rich environment that is subsequently modified to a sodium chloride environment as the fluid passes into the rete testis. Hence, one of the first important messages here is that rete testis fluid is not truly representative of the environment that surrounds the developing germ cells. This point is important as many investigators use rete testis fluid as a marker for Sertoli cell/seminiferous tubule/germinal epithelium secretion. The limitation of using rete testis fluid in this way should be realized. In addition, rat rete testis fluid (RTF) is often collected after 18- to 24-hour efferent duct ligation (EDL) and it is difficult to avoid contamination of lymph/interstitial fluid after puncture of the testicular capsule near the ligation point. Routine protein measurements on this fluid should be performed as a check for contamination; protein concentration is approximately 1 mg/mL in rat RTF and approximately 40 mg/mL in lymph/interstitial fluid (Hinton and Keefer, 1983). To avoid contamination of lymph/interstitial fluid and the possible deleterious effects of EDL, RTF can be collected by micropuncturing the efferent ducts just as they emerge from the normal nonligated testis (Olson and Hinton, 1985).

It is also difficult to compare products secreted by Sertoli cells in vitro with seminiferous tubule fluid (STF) collected in vivo. There are several factors that can account for these differences. The first has been discussed previously, that Sertoli cell secretion is influenced by a number of different cell types, and Sertoli cell cultures are generally performed in isolation. Second, it is difficult to replicate the in vivo situation in vitro, although investigators have made remarkable strides in their efforts to replicate in vivo conditions using bicameral chambers, co-culture systems, and appropriate ECM support for cells grown in culture. Third, many, if not all, Sertoli cell cultures are performed on cells isolated from immature animals ($\leq$20 days of age); therefore, data from such experiments should be interpreted carefully before extrapolations are made to the adult situation. It is recognized that technical difficulties in isolating pure, mature Sertoli cells may preclude the use of tissue from adult animals, but technical necessity may leave unanswered many questions about the realities of mature Sertoli cell function.

This review will not provide a list of all the substances found within the seminiferous tubule as much of this data has already been published and discussed extensively in other reviews (see Setchell and Waites, 1975; Waites and Gladwell, 1982; Setchell and Brooks, 1988). We will, however, attempt to relate the function of individual substances with the processes of the germinal epithelium.

B. Formation of Seminiferous Tubular Fluid

One of the major issues that has been partially resolved is the manner in which STF is formed. Tuck and colleagues (1970) presented the hypothesis that the seminiferous tubule secretes a primary fluid that mixes with fluid derived from the rete testis, which in turn forms a free-flow fluid; this hypothesis has been referred to as the "ebb and flow" hypothesis of STF formation. As studies from other investigators began to appear in the literature, however, it became apparent that the ebb and flow hypothesis was possibly not correct. To take a critical look at the Tuck data, it is important to consider the manner in which each fluid was collected. Primary fluid was collected by injecting mineral oil into a seminiferous tubule by micropuncture, allowing the seminiferous epithelium to secrete primary fluid that broke the oil column. This fluid secreted between the oil droplets was then collected by micropuncture. Free-flow fluid was collected from the seminiferous tubule by micropuncture and RTF was collected from a testis that previously had its efferent duct ligated for 18 hours. The data obtained are shown in Table 11–1.

From this data it seemed plausible that free-flow fluid was derived from mixing of primary fluid and RTF. Supporting evidence for this hypothesis came from measurements of the spermatocrit in STF and RTF (see Setchell and Waites, 1975; Waites and Gladwell, 1982), which suggested that the concentration of spermatozoa declined from STF to RTF; that is, there was dilution of STF as it entered the rete testis. There are some problems, however, with the ebb and flow hypothesis. It is not known if mineral oil injected into the seminiferous tubule caused cellular damage resulting in the high potassium measured in primary fluid. In addition, this study has to our knowledge never been repeated successfully. It is possible that the breaking of the mineral oil column was an artifact or the result of contractions of the seminiferous tubule caused by the surrounding layer of myoid cells. Furthermore, there is very little evidence that fluid within the rete testis, under normal physiological conditions, is able to move in a retrograde manner. There is probably a small amount of fluid that can do this but it appears doubtful that this occurs on a scale required to form free-flow fluid.

Table 11–1. Concentrations of Ions within Rat Seminiferous Tubular and Rete Testis Fluid

	K^+ (mM)	Na^+ (mM)	Cl^- (mM)
Primary fluid	112 ± 2	38 ± 3	62 ± 3
Free flow fluid	50 ± 5	108 ± 7	120 ± 5
Rete testis fluid	14 ± 1	143 ± 7	140 ± 2

Data from Tuck et al. (1971).

A subsequent study by Henning and Young (1971) demonstrated that when solutions of varying composition were introduced into the seminiferous tubule between oil columns, that solution never reached the ionic values measured for primary fluid, even after 120 minutes; instead the composition reached levels between that of primary and free-flow fluid. The authors concluded that although seminiferous tubules do secrete a potassium-rich fluid, whether this fluid mixes with rete testis fluid needed to be studied further.

In 1978, Setchell and colleagues (1978) put forward an alternative hypothesis. Briefly, these investigators postulated that seminiferous tubules secreted a potassium bicarbonate-rich environment that was subsequently modified to a sodium chloride-rich environment as the fluid passed into the rete testis. This hypothesis would imply that the epithelium of the rete testis plays a considerable role in modifying the fluid, an epithelium not previously considered to be very active. Setchell and co-workers (1978) also demonstrated that the concentrations of spermatozoa within rat STF and RTF were essentially equal and that organic solutes like inositol did not change in concentration as they moved from seminiferous tubule to rete testis. In contrast, these investigators and others (Turner et al., 1984) found the concentration of testosterone and protein in RTF to be lower than in STF, a finding that supports the previous observations that dilution of STF occurred as it entered the rete testis. A further complication is that the epithelium of the rete testis can actively absorb protein (Hinton and Keefer, 1983) and this may account for the protein concentration differences found by Setchell (1978) and Turner (1984) and associates. Whether testosterone is metabolized or transported into and/or across the rete epithelium remains to be studied.

In addition to ions, STF and RTF contain a number of organic solutes and proteins and many are synthesized by the Sertoli cells (see section V.D., Chapter 9; Bardin et al., 1988). With respect to the organic solutes, amino acids, inositol, and steroids predominate. Glucose enters the seminiferous tubule by a facilitated diffusion mechanism (see Setchell and Waites, 1975). Presumably a carrier system is needed as the testis is heavily dependent on glucose for its metabolic needs and the seminiferous tubular concentration of glucose is very low. Much of the glucose is oxidized but some is used to synthesize amino acids and inositol. Using specific transport carriers the testis can also transport amino acids, such as leucine, phenylalanine (see Setchell, 1986), and a neutral amino acid, α-aminoisobutyric acid (Hinton and Howards, 1982). Both transport and synthesis of amino acids contribute to the unusual amino acid composition of STF and RTF (Hinton, 1990), which is distinctly different from that of blood plasma. The precise role of each amino acid in sperm differentiation is still unknown, but the testis is affected in cases of amino acid deficiencies (see Setchell and Brooks, 1988).

Although much of the inositol measured within STF is synthesized from glucose, some is probably transported directly from blood (Hinton and Howards, 1982). Again, as with other organic solutes, the precise role of inositol in the testis is unknown, although it may be important as an osmoregulator as it is in the kidney (Nakanishi et al., 1989). Osmolytes, such as inositol and sorbitol, are considered to be nonperturbing solutes; that is, they do not affect macromolecular function (key enzymes, etc) when they are present in high concentrations. In contrast, ions like sodium, potassium, and chloride are classified as perturbing solutes because they interact with many co-factors, such as enzymes, and cause loss of function (Yancey et al., 1982). By using such nonperturbing solutes, important proteins can function over a wide range of osmotic pressures. It is possible that inositol acts as an osmoregulator in Sertoli cells and in the developing germ cells; however, further experimentation is needed to support this hypothesis.

There are several factors that can affect secretion of fluid into the seminiferous tubule. It appears that fluid secretion is not impressively regulated by gonadotropins or testosterone (see Setchell and Brooks, 1988) but is affected by several agents that affect kidney function, such as ouabain and acetazolamide (see Setchell, 1974; Cheung et al., 1977). The fact that fluid secretion can be inhibited by carbonic anhydrase inhibitors (e.g., acetazolamide) is somewhat paradoxical in view of the data of Cohen and colleagues (1976). These investigators had previously shown that there is little, if any, carbonic anhydrase present within Sertoli cells of the rat testis as revealed by histochemical staining. Waites and Gladwell (1982) postulated that the effect of the carbonic anhydrase inhibitors may be due to effects on pH secretion, which in turn will affect fluid secretion. Interestingly, fluid secretion can still occur under conditions of germ cell depletion and local heating or cryptorchidism (see reviews by Setchell and Waites, 1975; Waites and Gladwell, 1982; Setchell and Brooks, 1988), although in Sertoli cell-only rats the composition of STF is distinctly different from that of normal control animals and approximates that of blood plasma (see Setchell et al., 1978). Later studies by Muffly and co-workers (1985) did demonstrate, however, that the potassium concentration in STF of Sertoli cell-only and normal rats was not significantly different. It is possible that the differences observed between the two studies may be due to the age of the animals that were used by each group. Fluid secretion was measured in the above cases by the rate of fluid gain in an 18- to 24-hour ligated testis, by micropuncture, or by using an in vitro perfusion technique.

C. Fluid Movement Within the Seminiferous Tubule

Fluid movement within the seminiferous tubule is also more complicated than first imagined. If small oil droplets are microinjected into the seminiferous tubule and their behavior observed, the droplets are seen to move in both directions and at extremely variable rates, often pausing in one place for over 1 hour (see Setchell et al., 1978). There are several hypotheses that may account for this behavior. It is possible that fluid within the seminiferous tubule accumulates at a point in which the pressure therein is able to overcome the pressure within the rete testis. When this happens, the tubuli recti open and STF rapidly enters the rete; once the pressure within the seminiferous tubule declines, the pressure within the rete predominates and closes the tubuli recti to prevent further STF from entering the rete testis. Hence, the rapid movement of oil droplets may be due to the pressure changes that occur within the seminiferous tubule as STF enters the rete. As the rapid movement can occur in both directions, presumably this is the result of the influx of STF into the rete at both ends of the seminiferous tubule. Findings from Johnson and Howards (1976), showing that the intraluminal pressure within the proximal part of the epididymis is higher compared with the seminiferous tubule, would tend to support this hypothesis. An alternate hypothesis is that movement of fluid is controlled by the contraction of the myoid layer surrounding the seminiferous tubule and by the tunica albuginea. Because it has been shown that an oxytocinlike peptide is present in Leydig cells and that oxytocin causes contraction of myoid cells, oxytocin may be a paracrine regulator of seminiferous tubular contractions (see Pickering et al., 1989). If this second hypothesis is correct, then further studies should be performed to understand the manner in which fluid secretion and oxytocin release, and hence tubular contraction, are coordinated. Both hypotheses may be correct as it would also seem plausible that movement of fluid is influenced by contractions of the seminiferous tubule and tunica albuginea and as STF enters the rete.

IV. The Sertoli Cell (Blood–Testis) Barrier

A. Historical Perspectives

It was recognized during the early 1900s that the substance of the testis was not permeable to certain dyes after their systemic injection (see Setchell and Waites, 1975; Setchell and Brooks, 1988, for further references). This demonstration of different fluid compartments within the testis led to the concept of a barrier existing between blood and seminiferous tubule. Although these studies were confirmed by other investigators (see Setchell and Waites, 1975), it was not until the late 1960s that Kormano performed a series of elegant experiments showing that a blood–testis barrier was not at the level of the capillaries as previously shown for the brain, but was between the interstitial and intraluminal spaces (Kormano, 1967). In addition, as judged by penetration of dye, Kormano demonstrated that the blood–testis barrier was not present before puberty in rats (Kormano, 1967).

Setchell and colleagues (1969) were able to define more precisely the ability of the blood–testis barrier to exclude certain blood-borne molecules. These investigators systemically injected markers of several molecular weights, including ions, and measured their appearance in both rete testis fluid and testicular lymph. The composition of rete testis fluid was also compared with blood plasma. This study was later confirmed and extended by Tuck and associates (1970), Levine and Marsh (1971), and Howards and associates (1976) who showed that the fluid composition within the seminiferous tubule is distinctly different from blood plasma and that the movement of certain molecules across the seminiferous tubular epithelium was restricted.

Although Brökelmann (1963), Flickinger and Fawcett (1967), and Nicander (1967) recognized that adjacent Sertoli cell membranes were intimately associated (Fig. 11–4), the elaborate Sertoli cell tight junctional meshwork was not appreciated until improved fixation and staining procedures were used (Dym and Fawcett, 1970). At this time the Sertoli cell tight junctions were postulated to be the site of the blood–testis barrier. These latter workers concluded that the seminiferous epithelium could be considered to have two compartments: basal and adluminal. The basal compartment contains spermatogonia undergoing mitotic divisions that are exposed to the interstitial fluid environment. The adluminal compartment contains germ cells through their meiotic stages of development and their subsequent modification to spermatozoa. All germ cells within this compartment are exposed to the luminal fluid environment.

The original studies of Setchell and associates (1969) and Dym and Fawcett (1970) demonstrated: (1) that the composition of RTF is distinctly different from that of blood plasma, (2) that substances readily pass into testicular lymph but not rete testis (and presumably seminiferous tubule) from blood plasma, and (3) that the site of restriction of the passage of

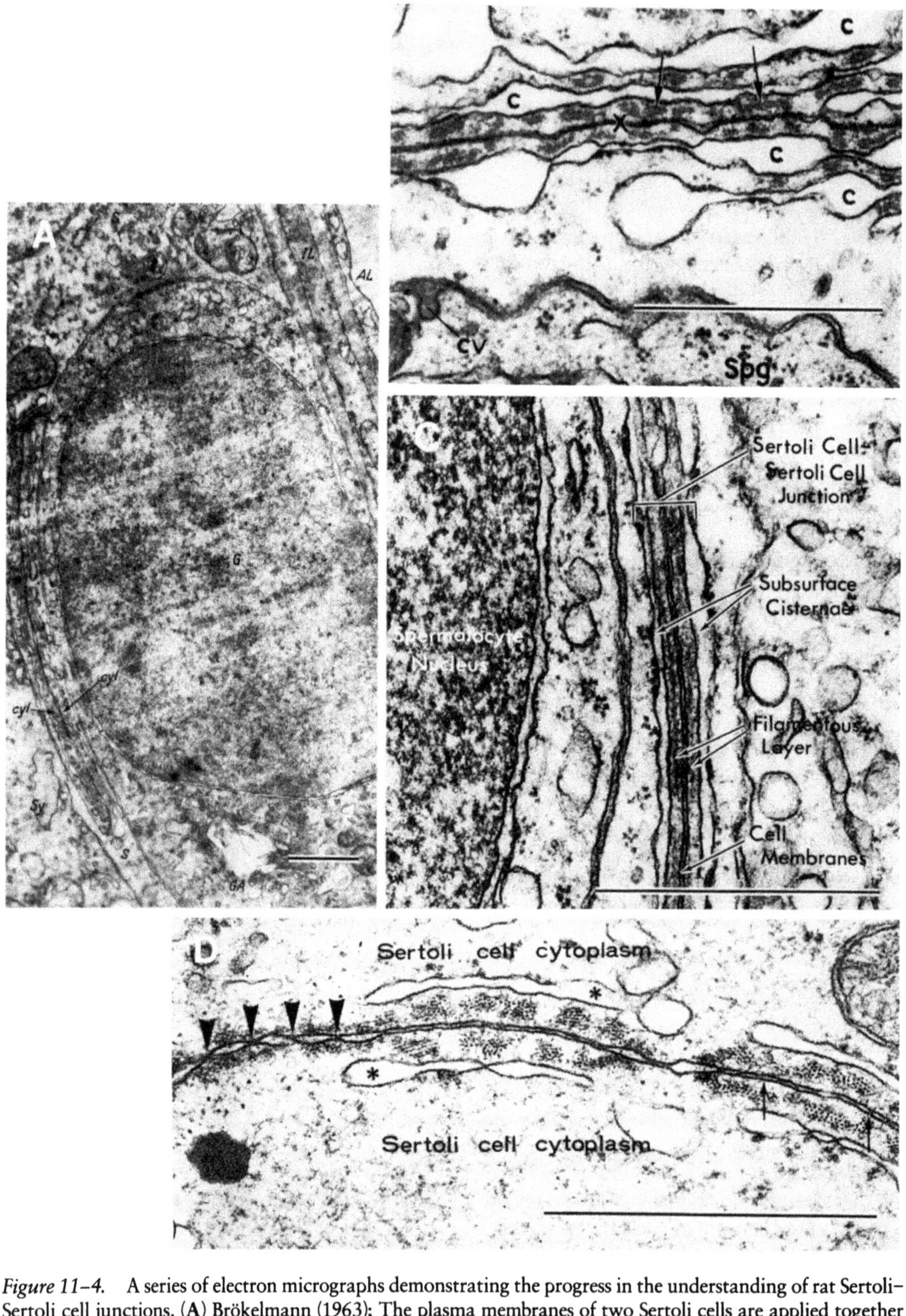

Figure 11–4. A series of electron micrographs demonstrating the progress in the understanding of rat Sertoli–Sertoli cell junctions. (**A**) Brökelmann (1963); The plasma membranes of two Sertoli cells are applied together (cyl) leaving an extracellular cleft of approximately 70Å. Sy, spermatocyte; G, spermatogonium; S, Sertoli cell; AL, adventitial layer; IL, interlamellar layer; GA, Golgi apparatus. (**B**) Nicander (1967); Narrow inter-Sertoli junction (X) with subsurface cisterns (c) and dense material *(arrows)*. Spg, spermatogonium; cv, coated vesicle.

blood-borne substances is the Sertoli cell junctions. Although Martinez-Palomo and Erlij (1975) suggested that the number of rows of junctional strands is not related to the degree of tightness of the barrier, the number of strands that comprise the Sertoli cell junction is impressive (Russell and Peterson, 1985; Fig. 11–5).

B. Conceptual Perspectives

Although the blood–testis barrier has been under considerable investigation, the ramifications of such a barrier are unknown. It probably has a role in maintaining an optimal environment for germ cell development, although it may not play a significant role in sequestering some testicular autoantigens (Yule et al., 1988). The blood–testis barrier does afford some protection to outside toxic agents; however, the precise mechanisms by which the testis is able to perform such tasks are unknown.

A restriction to our understanding of the blood–testis barrier is the terminology itself and the fact that many may think of the barrier as simply consisting of Sertoli cell–Sertoli cell tight junctions. The reality is that the tight junctions are only one aspect of the Sertoli cell's contribution to the maintenance of an intraluminal environment separate from blood and/or interstitial fluid. The blood–testis barrier is in reality a Sertoli cell barrier, a term originally proposed by Russell and Peterson (1985), and it is this terminology that we wish to advance in this presentation.

The Sertoli cell barrier may be viewed from different aspects as the Sertoli cell can act in many ways to construct the adluminal and luminal microenvironments. This view may be from a morphological, physiological, or immunological interest. The term Sertoli cell barrier is more precise than the term blood–testis

Figure 11–5. Electron micrograph of a freeze-fracture replica of Sertoli–Sertoli cell occluding junctions. Note that some strands are parallel and circumferential, and others are perpendicular. The E-face of one Sertoli cell overlies the P-face of another Sertoli cell. Over 100 rows of parallel strands can be observed at this junction. Bar = 1 μm. (Micrograph courtesy of L. Russell and reproduced with the permission of Alan R. Liss Inc.)

(C) Flickinger and Fawcett (1967); Comparison of Sertoli –spermatocyte and Sertoli cell–Sertoli cell junctions. In the former, junctions are approximately 200Å compared with 75 to 95Å seen in the latter. (D) Dym and Fawcett (1970); Tight junctions *(arrowheads)* and *narrow junctions (arrows)* are shown. *Asterisk* denotes the subsurface cistern of endoplasmic reticulum. Note in D how the Sertoli–Sertoli cell junctions form contact points *(arrows);* these were not observed in earlier micrographs (A, B, and C). It was appreciated from the earlier studies, however, that Sertoli cells formed intimate contacts with neighboring Sertoli cells. Each bar represents 100 μm. (Reproduced with the permission of Springer-Verlag Inc: A and B; the Wistar Press: C; Society for the Study of Reproduction: D.)

barrier and may be used accurately whether the investigator is studying: (1) morphological tight junctions, (2) the physiological activities of the Sertoli cells that contribute to the intraluminal microenvironment, and/or (3) the mechanisms by which the Sertoli cells are able to protect advanced germinal cells from immunological damage.

If the Sertoli cell barrier is viewed from a physiological perspective, it can be seen that the entire basolateral membrane of the Sertoli cell must be taken into account in addition to the tight junctions; that is, substances can enter the seminiferous tubule by a transcellular or paracellular pathway. Since the resistance across the seminiferous tubule is quite high (R_{in} = 198.0 ± 7.8 KΩ at testis temperature), this would suggest that many solutes move through the transcellular pathway rather than the paracellular pathway (Gladwell, 1977). The Sertoli cell basolateral membrane has a number of transporting proteins and ion channels inserted into its bilayer that selectively allow the entry of a number of substances. Thus, the ion channels and transport carriers are acting as a barrier themselves in addition to the restrictive nature of the lipid bilayer.

C. Developmental Perspectives

Investigations spanning more than a decade have led to the dogma that the Sertoli cell barrier forms at approximately 16 to 18 days of age in rats. Data supporting this observation comes from the morphological studies of Vitale and colleagues (1973; Fig. 11–6A) and Russell and associates (1989; Fig. 11–6B). Recent observations, however (see Setchell, 1986), suggest that a barrier to water-soluble markers develops fully at approximately day 44 (Fig. 11–6C). At first glance it would appear that these studies contradict each other, but they are probably both correct. If we view the Sertoli cell barrier from a morphological aspect, and if the end-point is the ability of the Sertoli cell tight junctions to prevent the transfer of electron opaque substances from entering the lumen, or if the integrity of the barrier is judged by hypertonic fixative induced-shrinkage then it is at day 18 (Vitale et al., 1973; Fig. 11–6A) or day 16 (Russell et al., 1989; Fig. 11–6B), respectively, that the Sertoli cell barrier forms. If the end-point is the ability of much smaller molecules, such as Cr-EDTA, to enter the seminiferous tubular lumen, however, then it is approximately day 40 or so before the Sertoli cell barrier is fully developed (Fig. 11–6C; Setchell, 1986; Setchell et al., 1988).

Because one of the major pieces of evidence for the existence of a Sertoli cell barrier is the ability of Sertoli cells to secrete a luminal fluid microenvironment that differs from blood plasma, then additional studies measuring the concentrations of a number of ions, organic solutes, and macromolecules within the rat seminiferous tubule during development would be informative. Such experiments would be difficult to perform because the luminal diameter of the seminiferous tubule is quite small during the initial stages of development. It may be possible to collect STF from days 25 to adult, however, since at day 25 almost all of the seminiferous tubules have open lumina and fluid is secreted into the tubule (see Russell et al., 1989). In this instance the end-point of the formation of the Sertoli cell barrier can be considered in terms of the day when the composition of STF in the developing testis is not significantly different from that of the adult testis. In determining the time of the establishment of the Sertoli cell barrier, it is important to consider our previous discussion: that the "barrier" separating the basal from the adluminal compartment should not be considered as a single entity; rather it is a summation of the morphological and physiological features of the Sertoli cell.

D. Integrity of the Sertoli Cell Barrier

1. INTEGRITY DURING SPERMATOGENESIS

One aspect that has received limited attention is the mechanism(s) by which germ cells are able to move from the basal compartment (below the tight junctions) to the adluminal compartment (above the tight junctions), yet still maintain a functional Sertoli cell barrier (both morphologically and physiologically). This problem was studied by Russell (1977, 1978) using electron microscropic techniques. Russell demonstrated that during stages VI, VII, and VIII of the rat seminiferous tubule cycle, preleptotene spermatocytes were located within the basal compartment but Sertoli cell

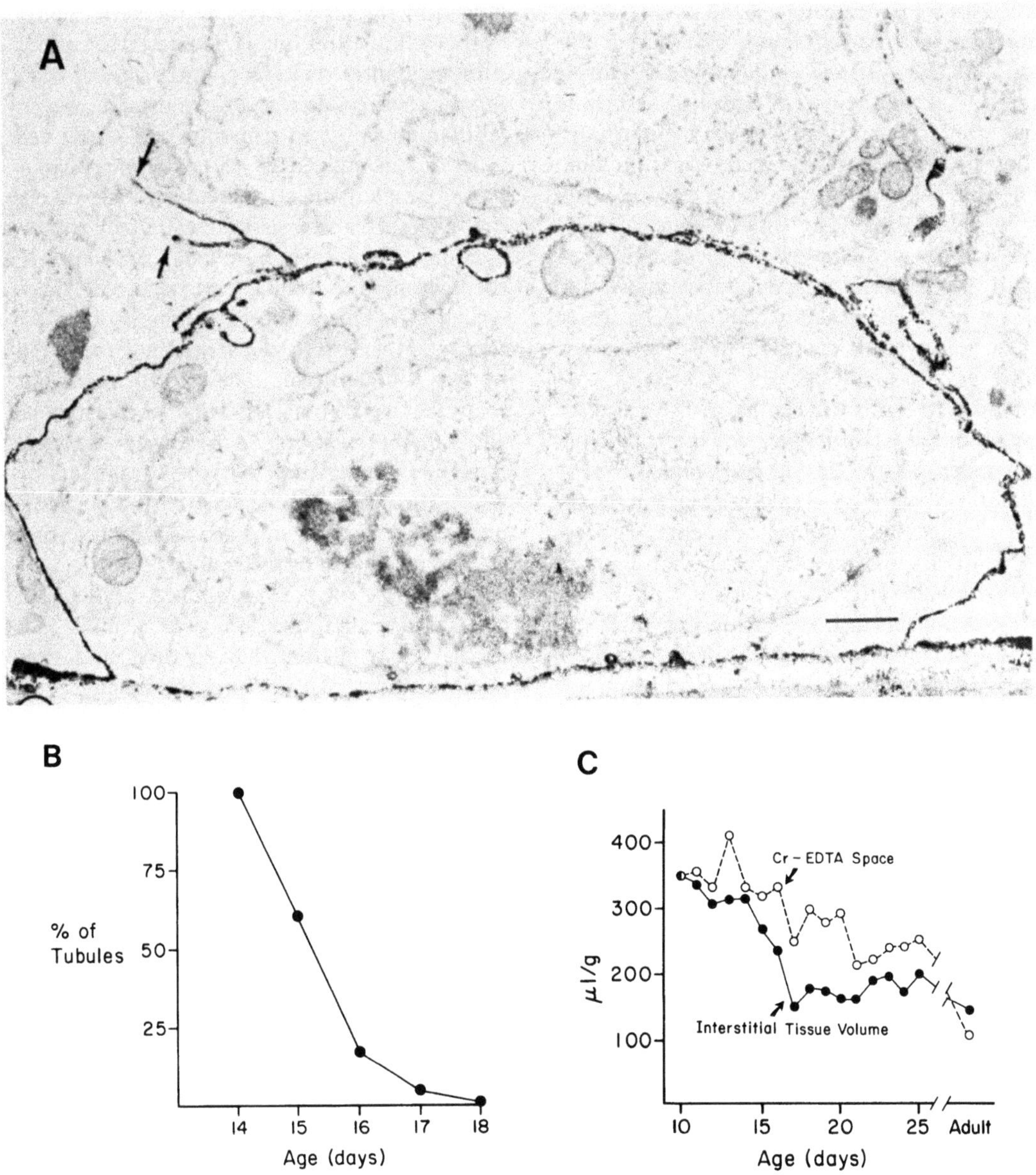

Figure 11–6. (**A**) Electron micrograph of basal aspect of seminiferous epithelium of an 18-day postnatal rat perfused with lanthanum nitrate. Note how the lanthanum has failed to penetrate the Sertoli–Sertoli cell junctions above the spermatogonia of this day-aged animal (line represents 100 μm; reproduced with the permission of The Wistar Press; Vitale et al., 1973). (**B**) Integrity of the rat Sertoli cell barrier at various ages as judged by hypertonic fixative-induced shrinkage. The X-axis shows the percentage of tubules showing shrinkage around adluminal germ cells and the Y-axis shows the age of the animal. Note that shrinkage was not observed in the day 18 rat, indicating that the Sertoli cell barrier was intact as measured by this technique (data has been taken from Russell et al., 1989). (**C**) The interstitial volume and Cr-EDTA space of the testes of rats of various ages. Note that even after 25 days the Cr-EDTA space is still not equal to that of the interstitial volume and is not until approximately day 45. This data would suggest that the Sertoli cell barrier is not fully intact until at least day 45 as measured by this technique (redrawn from Setchell, 1986).

processes were seen between the newly formed leptotene spermatocytes and basal lamina during late stage VII and early stage IX. During stages IX and X adjacent Sertoli cells met to form contacts and, finally, to form tight junctions. In addition, during these stages the tight junctions above the preleptotene spermatocytes were dissociating. The intermediate stage of this process was when the preleptotene spermatocytes showed tight junctions above and below them. The exact mechanism(s) by which this whole event is coordinated is not known; however, one of the factors that may be contributing to the restructuring of the germinal epithelium is plasminogen activator. Lacroix and colleagues (1981) and Fritz and Karmally (1983) had shown that plasminogen activator was most active in tubule segments at stages VII and VIII, and they suggested that plasminogen activator may cause dissociation of matrix components near the tubular wall that will assist in the movement of Sertoli cell cytoplasmic extensions; the result is that adjacent Sertoli cells will unite to form tight junctions. Further studies by Vihko and associates (1987) have shown that the secretion of plasminogen activator by Sertoli cells in stages VII and VIII is regulated by both hormones (e.g., follicle-stimulating hormone (FSH), retinoic acid, cyclic adenosine monophosphate (cAMP)) and adjacent germ cells.

The restructuring of the seminiferous tubule needs to be understood in morphological, cellular, and molecular biological terms. The signal that initiates this whole series of events to occur remains unknown as does the nature of the interaction between the premeiotic germ cells and the Sertoli cells that causes such initiation. Likewise, it is not known how the Sertoli cells maintain the junctional barriers during the reconstructuring process. These are just some of the issues that need to be addressed to obtain a better understanding of the dynamics of Sertoli cell junction formation.

2. INTEGRITY DURING INSULT

It is surprising how the Sertoli cell barrier can withstand a number of insults such as irradiation; heat, including cryptorchidism; varicocele; and vasectomy (see Turner and Howards, 1985; Setchell and Brooks, 1988) without loss of integrity. The integrity of the barrier was judged in the above cases by the permeability of the seminiferous tubule/rete testis to various substances after treatment. In the case of heating the testis or after cryptorchidism, however, lanthanum failed to penetrate the Sertoli cell junctions as judged by electron microscopic techniques, but the seminiferous tubule potential difference and resistance were reduced, indicating that a change in the permeability of the seminiferous tubule had occurred (Gladwell, 1977). This is another example of the integrity of the Sertoli cell barrier being measured by two different techniques after a particular treatment and obtaining what appears to be very different answers. As previously discussed, however, it is crucial that the parameters by which the integrity of the Sertoli cell barrier is measured are taken into consideration during the interpretation of the results.

Disruption of the Sertoli cell barrier can occur after immunization with seminiferous tubules, after efferent duct ligation, and after administration of cadmium, busulphan, cytochalasin D, or gossypol; the integrity of the barrier in each of these cases was assessed using a number of techniques (see Russell et al., 1989, for references). It has been shown, however, that there was neither disruption of the Sertoli cell nor blood–epididymis barrier as measured by lanthanum penetration of the Sertoli junctions and the permeability of the seminiferous and epididymal epithelium to radiolabeled inulin after administration of gossypol (Hoffer and Hinton, 1985).

It is interesting to note that a study by Fritz and colleagues (1983) has demonstrated that there is a defective Sertoli cell barrier in Tfm mice when judged by either hypertonic fixative-induced shrinkage or by measurement of the sucrose space. How this genetic defect is related to the impairment of the barrier is not known, although it would be of interest to know if this genetic condition is related to the synthesis and formation of the actin cytoskeleton that is associated with the zonulae occludens in normal animals (Weber et al., 1988). An earlier study by Nagano and Suzuki (1976) did not find disruption of Sertoli cell junctional strands in a testicular feminized human male but this does not mean that the permeability of the seminiferous tubules in this case was unaffected.

V. Regulation of the Seminiferous Tubule Microenvironment

There are several microenvironments within and immediately around the seminiferous tubule. They are those: (1) within the boundary tissue of each seminiferous tubule, (2) within the adluminal compartment surrounding late spermatocytes and spermatids, and (3) within the lumen of the seminiferous tubule. Only this later microenvironment has been partially characterized (see section III.), but it is known that vascular and interstitial factors, both hormonal and paracrine, have regulatory effects on all these microenvironments (Skinner et al., 1991).

A. Influences of the Vascular Compartment

In the beginning, there is blood flow. The vascular compartment presents blood to the testicular interstitium in pulses or oscillations of flow within the capillaries (Damber et al., 1986). The control of these pulses is not well understood, but the metabolic hypothesis of local control of blood flow states that the important factors in autoregulation of blood flow are the local tissue concentrations of O_2 and CO_2.

In the testis, Leydig cells may produce some other factor(s) important to the control of capillary blood flow as normal oscillations in flow as detected by laser Doppler flowmetry were abolished in animals whose Leydig cells had been eliminated with EDS (Damber et al., 1987b). These fluctuations in delivery of vascular fluid and solute are thought to be important in fluid exchange between the capillary and the interstitial space (see Damber et al., 1986), and, as already mentioned, it has been shown that the damping of these flow oscillations with hCG or with Leydig cell removal increases the formation rate of testicular interstitial fluid (Widmark et al., 1986; Damber et al., 1987b). The effect of EDS on testicular interstitial fluid formation had been previously reported by Setchell and Sharpe (1981). It has been cautioned previously that, aside from the physical removal of Leydig cells from the testis, the effects of EDS are not well established.

Another problem in interpreting these findings is that the time domain for acute, local changes in capillary blood flow is seconds, whereas the time domain for changes induced by experimental inhibitors is hours (Widmark et al., 1985; Damber et al., 1987a). Thus, although Leydig cells may indeed be involved in the in vivo regulation of capillary blood flow, the demonstration of this within a physiologically relevant paradigm is still lacking.

The possibility of local control of capillary flow is enticing not only because it would help explain the flow phenomena observed at the level of the capillaries, but because it may also be important to the regulation of the cycle of the seminiferous epithelium. The distribution and spacing of seminiferous capillaries would anatomically allow for vascular service to discrete stages of the cycle in the seminiferous epithelia (see Fig. 11–2) and could potentially even serve the Sertoli or Leydig cells within or bounding the putative unit-segments proposed by Perey and associates (1961; see Fig. 11–2).

Leydig cells seem to respond to Sertoli cell paracrine regulation in vitro (see section V.B.), and show morphological changes broadly dependent on the stage of the cycle of the seminiferous epithelium in the tubules adjacent to them. Thus, it is convenient to believe that Leydig cells can act in a stage-dependent manner as transducers for Sertoli cell signals back to the vasculature in a manner that would regulate local blood flow. The advantage of such a system is difficult to perceive when one considers that Sertoli cells might produce more efficiently their own paracrine signals directly to the microvasculature.

It is possible that any regulation of the microvasculature by Leydig cells is for the Leydig cells themselves, not for the seminiferous epithelium. In this regard, it is interesting to note that treatments that stimulate Leydig cells to produce testosterone (exposure to luteinizing hormone [LH], luteinizing hormone-releasing hormone agonist, or hCG) also cause a Leydig cell-dependent increase in rat vascular permeability (Setchell and Rommerts, 1985; Bergh et al., 1987; Damber et al., 1987a, 1987b). Although the detected changes in vascular permeability occur over hours and the stimulation of testosterone secretion occurs within minutes, it is possible that much earlier, more subtle changes than those detected by the above investigators occur locally in the vasculature. Such early changes might underly the later, gross changes in vascular permeability detected experimentally and might have more to do with Leydig cell function than with Sertoli cell function.

Gaps in our present knowledge are evident from this discussion. As the delivery of fluid and serum solutes to the interstitium and tubules is of primary importance to the resulting seminiferous microenvironment, it remains important to develop a better understanding of the regulation of the microvascular compartment. Understanding these vascular phenomena is basic to our understanding of how molecular signaling occurs in the testis.

For example, it is well known that pituitary gonadotropins are delivered to the testis by the vasculature. Luteinizing hormone passes into the interstitial space and stimulates Leydig cells to produce testosterone. Follicle-stimulating hormone likewise enters the interstitial compartment, transits the boundary layer, and binds to the Sertoli cell's basolateral membrane where it stimulates cAMP-regulated processes in those cells. The pattern in which these gonadotropins are delivered to the testis is important to testicular function, and implicitly, to the proper maintenance of the seminiferous microenvironment. Because LH is the most studied compound in this regard, the emphasis of the following discussion focuses on that gonadotropin.

First, if the gonadotropin were delivered from the pituitary in a nonpulsatile fashion, the fluctuating capillary blood flow itself would deliver the gonadotropin to the seminiferous tubule in a roughly pulsatile manner. Aside from that vascular compartment effect on hormone delivery, the release of LH from the pituitary in rats, humans, monkeys, ferrets, sheep, and other animals is known to actually occur in a pulsatile manner (see Desjardins, 1989; Turek and Van Cauter, 1988). The LH pulsatility has usually been shown to be closely linked with the pulsatile release of testosterone in the species studied, but the net effect of pulsatile LH secretion and pulsatile capillary blood flow on the true capillary delivery of gonadotropin to Leydig cells is unknown.

In a closely monitored study in the ferret, testosterone pulses in peripheral plasma followed LH pulses by approximately 10 minutes, and although time estimates are more difficult to make from the data in the rat, pulses of testosterone secretion appear to follow single LH pulses or the last of a train of LH pulses by a similar time period (see Sisk and Desjardins, 1986). With exogenously administered LH, it is possible to detect a response in testosterone secretion within 5 minutes of the administration (Fig. 11–7). Thus, the gonadotropin must arrive at the capillary, diffuse across the vas-

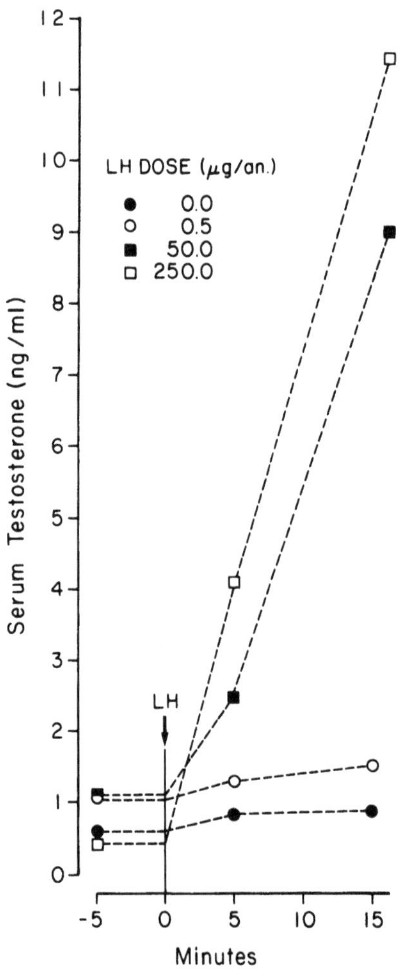

Figure 11–7. The response of the rat testis to graded doses of LH. Serum samples were obtained at 5 minutes before LH administration *(arrow)* and at 5 and 15 minutes afterward. The testis has the capacity to respond to LH in less than 5 minutes after administration of an exogenous pulse of gonadotropin. (T. T. Turner, work in progress.)

cular endothelium into interstitial fluid, bind to the Leydig cell membrane LH receptor, and stimulate the events of adenylate cyclase-dependent phosphorylation and the subsequent synthesis of testosterone. Then time must be allowed for diffusion of testosterone back out of the Leydig cell, movement back across the interstitial space, through the vascular endothelium, and into blood in sufficient quantities to enable detection of increases in testosterone concentrations within 5 minutes of the origin of the LH pulse.

The method by which LH first traverses the vascular endothelium and stimulates Leydig cells within perhaps 1 to 2 minutes is unknown, but it has not been questioned that LH movement is due to simple diffusion. When one considers the response times involved and the fact that LH is a 28,000 dalton peptide, it appears reasonable to invoke the possibility of some mechanism in addition to simple diffusion. Such endothelial events as transcytosis or receptor-mediated endocytosis may be involved in movements of gonadotropin from the vascular to the interstitial compartment in the testis, just as albumin is known to be transported across the vascular endothelium in other vessels.

B. Influence of the Interstitial Compartment

The interstitial compartment is important to the regulation of the intraluminal microenvironment because it is the immediate source of all pro-vita fluid, nutrients, and electrolytes that make Sertoli and germ cell function possible, as well as the endocrine and paracrine signals that potentially regulate that function. Although study of other cell types in the testicular interstitium (e.g., peritubular myoid cells, macrophages) is now providing new and interesting information on how the cell biology of the seminiferous tubule might be regulated, it is still believed that Leydig cells are the primary actors in the interstitial compartment. The function of Leydig cells has been recently reviewed (DeKretser and Kerr, 1988). It is known that these cells secrete testosterone upon stimulation by pituitary LH and that the testosterone is required for the maintenance of spermatogenesis. Testosterone is secreted into the peritubular lymph or interstitial fluid where it can exist in concentrations 100 to 200 times that in peripheral serum in the rat (Turner et al., 1984). Similar conditions are presumed to exist in other species including humans because testosterone concentrations in tissue biopsies are conventionally 50 to 100 times higher than peripheral serum (Steinberger, 1976).

Testosterone secreted from Leydig cells reaches the peritubular myoid cells, Sertoli cells, and germ cells by diffusion from a moving fluid. Thus, it appears unlikely that the steroids secreted by any given set of Leydig cells affect only laterally adjacent Sertoli cells or peritubular myoid cells (see section II. A.). If Leydig cells are important for stage-specific regulation of the seminiferous epithelium it is possible that nonsteroidal, paracrine products of the Leydig cells would have a more focused effect on adjacent Sertoli cells than would steroids. Paracrine factors presumably require cell surface receptors, thus specificity of action can be increased not only by local secretion from specific Leydig cells, but the reception of such secretions only by cells stimulated to express the appropriate receptors. Molecules that have the immunological characteristics of β-endorphins, adrenocorticotrophic hormone (ACTH), the enkephalins, and insulinlike growth factor have been localized in Leydig cells of rats and mice and have been shown to have direct effects on Sertoli cells and Leydig cells in purified cell culture systems (see Bardin, 1988; Bellve and Zheng, 1989). β-endorphins, for example, reduce cultured Sertoli cells' capacity to produce cAMP and androgen binding protein (ABP), and stimulate cultured Leydig cells' androgen secretion in response to hCG. Also, as described previously, Setchell and Rommerts (1985) and Damber and colleagues (1987b) have presented evidence that Leydig cells are important, presumably through the production of some paracrine factor, in the LH-induced perturbation of the testicular vascular endothelium. Thus, evidence is accumulating that Leydig cells are important for regulation of spermatogenesis in ways beyond the conventionally accepted secretion of testosterone. This has become part of the popular conception of cell–cell communication in the testis.

In contravention to this view, Sharpe and co-workers (1988) have found that spermatogenesis and fertility in rats rendered Leydig cell-free by EDS treatment can be maintained with androgens alone. Sprando and co-workers (1990) have shown that spermatogenesis can be maintained qualitatively but not quantitatively by androgens alone in rats treated similarly. This question remains to be resolved, but if it is true that spermatogenesis and fertility are maintained in the absence of Leydig cells and with nonpulsatile delivery of testosterone, this would strongly imply that local regulation of Leydig cells is not a necessary feature of the regulation of spermatogenesis nor the intratubular microenvironment.

The implication of these latter works is that although a plethora of immunoreactive substances may be found in the testis, and although these substances may elicit a cellular response under in vitro cell culture conditions, this does not automatically mean that these molecules have a physiological role in vivo or even in vitro when cells are left in their normal relationships.

As already noted, interstitial fluid flow poses a problem in understanding cell–cell communication in the testis. This flow will spread secreted molecules along the seminiferous tubules, perhaps across several different stages at the same time (see Fig. 11–2), and make it difficult for secreted products to be focused only to the laterally adjacent seminiferous tubule. With all the above problems considered, it is appropriate to question the degree to which Leydig cells have a meaningful influence on the local regulation of the seminiferous tubule and its microenvironment in vivo. It was discussed in section II.B. that the peritubular myoid cells have also been shown to secrete a factor in vitro that affects Sertoli cell function and that is regulated by androgens. As above, whether or not, or at least how, such factors actually operate in vivo remains to be determined.

A noncellular aspect of the interstitium, the ECM, was described earlier and may also be involved in regulating Sertoli cells and, thereby, the seminiferous microenvironment. The ECM of cells in general is a field of research in rapid development. It is generally proposed that ECM molecules confer tissue specificity and perhaps cell-function information to epithelial cells. This may also be the case with Sertoli cells. Type IV collagen and fibronectin have been localized in seminiferous ECM (see section II. B.), and these two ECM molecules, among others, are distinguished by possessing the tripeptide binding sequence, RGD (the single-letter codes for arginine, glycine, and aspartic acid). The RGD sequence is recognized by a class of cell surface receptors called integrins, which constitute a versatile recognition system for ECM in many cell types and have been shown in vitro to provide cells with anchorage to their ECM and signals for polarity, shape, position, or proliferation.

The integrin molecule has an extracellular region, a transmembrane region, and an intracellular region. The extracellular region binds to RGD, and this is hypothesized to cause conformational changes in the transmembrane region and the intracellular region. The intracellular region is proposed to be attached to cytoskeletal elements, and alterations in configuration are hypothesized to transduce biological information to the interior of cells. Thus, in Sertoli cells it is speculated that ECM information is conveyed through cell-surface receptor molecules that span the Sertoli cell basolateral membrane and connect to the Sertoli cell cytoskeleton. Sertoli cell cytoskeleton is important not only for maintaining cell shape, but for the maintenance and operation of the Sertoli cell barrier, the movement of cells in the adluminal compartment, and spermiation at the apical Sertoli cell surface.

Exactly how ECM causes intracellular changes through integrin or integrinlike receptors is not well understood, but it is possible that subtle, yet dynamic changes, in ECM properties (i.e., the characteristics of other molecules bound to ECM, the proportion of different ECM molecules in the matrix, etc.) can alter ECM–integrin interactions, thus causing changes in the cytoskeletal elements having an impact on the adluminal and intraluminal compartments. Alterations in the tight junctions of the Sertoli cell barrier or alteration in the movement of cells through the adluminal compartment can potentially change the characteristics of fluids that constitute the microenvironment of the seminiferous epithelium.

As we stated at the beginning of this section, the two sure things about the influence of the interstitial compartment of the seminiferous microenvironment are that Leydig cells produce testosterone under the influence of pituitary LH and that testosterone is required for spermatogenesis. Paracrine issues aside, testosterone exerts its influence primarily by the Sertoli cell that has specific androgen receptors and that has been shown to respond to testosterone in a number of ways (Bardin et al., 1988). The Sertoli cell and its influence on the seminiferous microenvironment is a major focus in the following section.

D. Influence of the Seminiferous Epithelium

The seminiferous epithelium is made up of Sertoli cells and developing germ cells, both of which may play a role in regulating the seminiferous microenvironment. We discuss the influences first of Sertoli cells in general, then the impact of germ cells on Sertoli cells, and, finally the influences of the stages of the cycle of the seminiferous epithelium on the seminiferous tubular microenvironment.

1. SERTOLI CELL INFLUENCES

Just as Leydig cell secretion of testosterone is known to be regulated by LH, Sertoli cell function is conventionally viewed as being under the primary control of FSH (Bardin et al., 1988; see Chapter 9) along with testosterone. FSH stimulates Sertoli cell glucose transport, energy metabolism, intracellular cAMP, the production or activation of intracellular enzymes, and the synthesis and secretion of a large panel of proteins including ABP, transferrin, plasminogen activator, and a rapidly expanding list of growth factors (Table 11–2). The secreted proteins and growth factors obviously affect the constitution of the adluminal and intraluminal microenvironments where the Sertoli cell products can bind to spermatozoa (Sylvester et al., 1984; O'Brian, 1991) and potentially be involved in sperm cell development. Other products are secreted from the Sertoli cell's basolateral membrane and potentially interact with ECM or cells in the interstitial compartment, whether they be Leydig cells, peritubular myoid cells, or even macrophages.

Table 11–2. Proteins and Growth Factors Secreted by Sertoli Cells

Protein/factor	Reference
Unnamed testis-specific proteins	Wilson and Griswold (1979) Exp. Cell Res. 123.127
	Kissinger et al. (1982) Biol. Reprod. 27:233
	Cheng et al. (1986) Endocrinol. 118:480
	Wright et al. (1981) Proc. Natl. Acad. Sci. USA 78:7565
Cyclic protein-2	Wright et al. (1983) Biol. Reprod. 29:257
SGP-2	Sylvester et al. (1984) Biol. Reprod. 31:1087
Transferrin	Holmes et al. (1983) Endocrinol. 117:1616
	Morales et al. (1987) Biol. Reprod. 37:995.
Androgen binding protein	Fritz et al. (1976) J. Reprod. Fert. 46:17
	Galdieri et al. (1984) J. Androl. 5:409
Ceruloplasmin-like protein	Skinner and Griswold (1983) Biol. Reprod. 28:1225
DNA–RNA stimulating protein	Tres et al. (1986) Endocrinol. 117:1896
Insulinlike growth factor	Benahmed et al. (1984) Molec. Cell Endocrinol. 50:69
Seminiferous growth factor	Feig et al. (1983) J. Cell Biol. 97:1435
Sertoli cell secreted growth factor	Buch et al. (1988) Fertil. Steril. 49:658
Transforming growth factor-α	Skinner et al. (1989) Endocrinol. 124:845
Transforming growth factor-β	Avellet et al. (1987) Biochem. Biophys. Res. Commun. 146:575
	Skinner and Moses (1989) Mol. Endocrinol. 3:703
Interleukin-1-like factor	Khan et al. (1987) Int. J. Androl. 10:495
LHRH-like factor	Sharpe (1985) Mol. Cell Endocrinol. 41:247

Some Sertoli cell products (e.g., inhibin) are secreted as hormonal factors important for the remote feedback regulation of pituitary FSH (see Bardin et al., 1988), but in vitro studies have shown that inhibin also can facilitate LH-induced testosterone secretion by Leydig cells (see Bardin, 1988). This finding has been interpreted to mean that inhibin may play a paracrine role in the interstitial compartment of the testis and Bhasin and colleagues (1989) have reported a stage-dependent expression of inhibin α- and β-B subunits in the rat seminiferous epithelium.

Other Sertoli cell products also enhance Leydig cell testosterone secretion in vitro. Unidentified factors (Parvinen et al., 1984; Bartlett et al., 1987), CMB-21 (Cheng et al., 1988), and insulinlike growth factor (see Bellve and Zheng, 1989) are examples of substances localized to Sertoli cells or that have been found in conditioned Sertoli cell culture medium and that enhance LH-induced testosterone production by Leydig cells in vitro. As was the case in the previous discussion of Leydig cell influences on Sertoli cells (section V. B.), there are problems here that confound our understanding of Sertoli cell influences on Leydig cells.

First, there appears to be a veritable pantheon of Sertoli cell factors that are claimed to provide local regulation of testosterone secretion. If this is true, nature has designed a highly redundant and complex system for local regulation. Although this may be the case, it is perhaps likely that only some of the factors listed in Table 11–2 have a physiological role in testicular regulation. It is hoped that further research will help separate the real actors from the stand-ins.

Sertoli cell secretions are potentially important to the microenvironment on either side of the seminiferous epithelium. Some of the Sertoli cell products listed in Table 11–2 are presumed to be important to the short-loop regulation of Leydig cell testosterone production; thus the secretions are primarily into the interstitial compartment and are important to that microenvironment. Alternatively, the secretions of the Sertoli cell into the adluminal and intraluminal compartment, along with the sieving characteristics of the Sertoli cell barrier, are important to the constitution of the intratubular microenvironment. Transferrin and other proteins (Sylvester et al., 1984; Morales et al., 1987), for example, are transported to the germ cells in the adluminal areas. ABP (see Bardin, 1988), a sulfated glycoprotein termed clusterin or SGP-2 (see Griswold et al., 1986), cyclic protein-2 (Wright et al., 1983) and other proteins (Olson and Hinton, 1985; Shabanowitz and Kierszenbaum, 1986; Ueda et al., 1988) are secreted into the seminiferous tubule lumen where they may have direct effects on spermatozoa or act in a functional way within the epididymis. Germ cells within the seminiferous epithelium are not only influenced by Sertoli cell products, but they, in turn, influence Sertoli cell function and indirectly affect their own microenvironment.

2. GERM CELL INFLUENCES

There are theoretical as well as evidentiary reasons for expecting cell–cell interactions between germ cells and Sertoli cells (Wright, 1988; Jégou, 1991), and it is believed that these interactions are as necessary as extratubular signals for the successful development of the male gamete. As already stated, Sertoli cells secrete products that bind to germ cells (e.g., transferrin, SGP-2, insulinlike growth factor), and Sertoli cell products can influence germ cell RNA and DNA synthesis (Rivarola et al., 1985) and glutathione synthesis (Li et al., 1989) in vitro. In turn, germ cells synthesize a variety of intracellular and membrane-associated proteins (see Djakiew and Dym, 1988) and stimulate Sertoli cell secretion in vitro. The ABP secretion, which is known to occur primarily in a proluminal direction (see Bardin, 1988), is stimulated by the presence of spermatocytes in co-culture with Sertoli cells (Galdieri et al., 1984). In vitro studies with immature Sertoli cells in bicameral diffusion chambers have shown that pachytene spermatocytes can alter apical secretion by Sertoli cells. Some evidence indicates the need for direct contact between Sertoli cells and germ cells for this effect (Janecki et al., 1988), whereas other evidence indicates a paracrine relationship between the two cell types (Djakiew and Dym, 1988). Whatever the case, it is clear that the intratubular microenvironment produced by the Sertoli cells, whether in the adluminal interstices or in the luminal compartment itself, is influenced by the presence of germ cells. It is also clear that Sertoli cell secretions and biochemical constitution are influenced by the different complex germ cell associations in the various stages in the cycle of the seminiferous epithelium.

3. INFLUENCE OF THE STAGE OF THE CYCLE OF THE SEMINIFEROUS EPITHELIUM

Many compounds have been localized in or found to be secreted by seminiferous tubules in a stage-dependent manner. Even a sample list of some of these compounds (Table 11–3) can be quite long and the implication is that the intratubular localization or the secretion of these compounds is regulated by the stages of the cycle in the seminiferous epithelium. If this were not the case, the compounds would appear randomly along all stages.

The biochemical or secretory characteristics of the stages of the rat cycle of the seminiferous epithelium have been discussed elsewhere (Bardin, 1988; Wright, 1988; Bhasin et al., 1989). From a synthesis of these reports and others listed in Table 11–3 it appears that the most dynamic stages of the cycle of the seminiferous epithelium, both morphologically and metabolically, are stages VII–VIII. These are the stages in the cycle just before and during the loss of mature germ cells from the apical Sertoli cell surface, the time when excess cytoplasm from the late spermatids is being reingested by the Sertoli cells, and when spermatocytes are moving through the Sertoli cell tight junctions to reach the adluminal space within the seminiferous epithelium. Plasminogen activator secretion, ABP secretion, somatomedin (insulinlike growth factor) secretion, and cyclic protein-2 secretion are highest in either stage VII or VIII. Also, stage VII–VIII tubules maximize testosterone secretion by hCG-stimulated Leydig cell in culture, and, in turn, stages VII–VIII appear particularly sensitive to androgen action. On the other hand, FSH receptors, FSH-stimulated adenylate cyclase, and inhibin secretion are

Table 11–3. Substances Whose Localization Within the Seminiferous Epithelium or Secretion by the Epithelium is Influenced by the Stage of the Seminiferous Epithelium

Compound/factor	Reference
Proteins (pattern of secretion)	Lacroix et al. (1981) Biol. Reprod. 25:143
	Ritzen et al. (1982) Mol. Cell Endocrinol. 25:25
	Wright et al. (1983) Biol. Reprod. 25:257
	Shabanowitz and Kierszenbaum (1986) Biol. Reprod. 35:745
Plasminogen activator	Lacroix et al. (1981) Biol. Reprod. 25:143
Androgen binding protein	Parvinen (1982) Endocr. Rev. 3:404
Meiosis-inducing substance	Parvinen (1982) Endocr. Rev. 3:404
Cyclic protein-2	Wright et al. (1983) Biol. Reprod. 29:257
Sulfated glycoproteins	Shabanowitz and Kierszenbaum (1986) Biol. Reprod. 35:745
mRNAs for sulfated glycoprotein	Morales et al. (1987) Biol. Reprod. 36:1035
Retinol-binding protein	Kato et al. (1985) Biol. Reprod. 32:173
Transferrin-like protein	Wright et al. (1983) Biol. Reprod. 29:257
mRNAs for transferrin	Morales et al. (1987) Biol. Reprod. 36:1035
Inhibin α and β subunits	Bhasin et al. (1989) Endocrinol. 124:987
Leydig cell influencing factor(s)	Parvinen et al. (1984) Mol. Cell Endocrinol. 37:331
Adenylate cyclase	Gordeladze et al. (1982) Arch. Androl. 6:43
Testosterone	Parvinen and Ruokenen (1982) J. Androl. 3:211
FSH receptor	Parvinen (1982) Endocr. Rev. 3:404
Somatomedin	Ritzen (1983) J. Steroid. Biochem. 19:499

lowest in stages VII–VIII. Thus, it can be interpreted that (1) stages VII–VIII influence Sertoli cells in such a way as to stimulate maximum secretion of several compounds into the adluminal and luminal microenvironments, (2) that this effect is mediated through an interaction of testosterone availability to the Sertoli cell, and (3) that testosterone availability is potentially affected by paracrine signals from the Sertoli cell to the Leydig cell (see section V.B. for a *caveat* regarding the necessity of Leydig cell regulation in maintaining normal spermatogenesis).

Finally, inhibin secretion, transferrin secretion, FSH binding, and FSH-stimulated adenylate cyclase activity are high in stages I–II, not VII–VIII, of the rat cycle of the seminiferous epithelium. Thus, different stages of the cycle appear to be more or less sensitive to gonadotropins or to testosterone in a cyclic manner and different proteins are secreted maximally by Sertoli cells in different stages (Wright et al., 1983), depending on the regulating signals those cells receive from the germ cell associations within the different stages. The exact nature of these signals, whether paracrine or direct cell–cell contact remains unresolved.

VI. Summary

Throughout this review the complex nature of the seminiferous tubular microenvironment and its regulation have been emphasized. Gaps in our knowledge of basic testicular function have been intentionally pointed out in an effort to define what is and what is not known. In fact, the formation and regulation of the seminiferous tubular microenvironment are probably even more complicated than current studies show. Interested investigators are faced with a complex biological puzzle that has had a little piece here and a little piece there pressed into place. The continuing challenge is to discover more pieces of the puzzle and to fit them all together in a way that presents a true picture of testicular cell biology.

When we look over the past 20 years from the original studies of Tuck and associates (1970) and Levine and Marsh (1971), great strides in understanding the seminiferous tubular microenvironment have been made. However, we are still far from understanding how the seminiferous tubule is able to produce those specialized cells that guarantee the continuation of the species.

REFERENCES

Anthony, C. T., Rosselli, M., and Skinner, M. K. (1991). Actions of the testicular paracrine factor (P-Mod-S) on Sertoli cell transferrin secretion throughout pubertal development. Endocrinology 129:353–360.

Anthony, C. T., and Skinner, M. K. (1989). Actions of extracellular matrix on Sertoli cell morphology and function. Biol. Reprod. 40:691–702.

Bardin, W. (1988). Pro-opiomelanocortin peptides in reproductive physiology. Hosp. Prac. 23:108–127.

Bardin, C. W., Cheng, C. Y., Musto, N. A., and Gunsalus, G. L. (1988). The Sertoli Cell, in *Physiology of Reproduction* (Knobil, E., and Neil, J., eds.) Vol. 1, pp. 933–974, Raven Press, New York.

Bartlett, J. M., Wu, F.C.W., and Sharpe, R. M. (1987). Enhancement of Leydig-cell testosterone secretion by isolated seminiferous tubules during co-perfusion in vitro—Comparison with static co-culture systems. Int. J. Androl. 10:603–617.

Bellve, A. R., and Zheng, W. (1989). Growth factors as autocrine and paracrine modulators of male gonadal functions. J. Reprod. Fert. 85:771–793.

Bergh, A., Rooth, P., Widmark, A., and Damber, J. E. (1987). Treatment of rats with hCG induces inflammation-like changes in the testicular microcirculation. J. Reprod. Fert. 79:135–143.

Berne, R. M., and Levy, M. N. (1983). The microcirculation and lymphatics, in *Physiology* (Berne, R. M., and Levy, M. N., eds.) pp. 517–529, C. V. Mosby, St. Louis.

Bhasin, S., Krumman, L. A., Swerdloff, R. S., Morelos, B. S., Kim, W. H., diZerega, G. S., Ling, N., Esch, F., Shimasaki, S., and Toppari, J. (1989). Stage dependent expression of inhibin α and β-B subunits during the cycle of the rat seminiferous epithelium. Endocrinol. 124:987–991.

Brökelmann, J. (1963). Fine structure of germ cells and Sertoli cells during the cycle of the seminiferous epithelium in the rat. Z. Zellforsch. Mikrosk. Anat. 59:820–850.

Cheng, Y. C., Morris, P. L., Mathur, P. P., Marshall, A., Grima, J., Turner, T. T., Carreau, S., Tung, K.S.K., and Bardin, C. W. (1988). Identification of a factor in fractions of Sertoli cell-enriched culture medium containing CMB-21 that stimulates androgen production by Leydig cells, in *Nonsteroidal Gonadal Factors: Physiological Roles and Possibilities in Contraceptive Development* (Jones, H., ed.) pp. 192–201, Jones Institute Press, Norfolk.

Cheung, Y. M., Hwang, J. C., and Wong, P.Y.D. (1977). In vitro measurement of rate of fluid secretion in rat isolated seminiferous tubules: Effects of metabolic inhibitors and ions. J. Physiol. 269:1–15.

Christensen, A. K., Komorowski, T. E., Wilson, B., Ma, S. F., and Stevens, R. W. (1985). The distribution of serum albumin in the rat testis, studied by electron microscope immunocytochemistry on ultrathin frozen sections. Endocrinol. 116:1983–1996.

Cohen, J. P., Hoffer, A. P., and Rosen, S. (1976). Carbonic anhydrase localization in the epididymis and testis of the rat: histochemical and biochemical analysis. Biol. Reprod. 14:339–346.

Damber, J. E., Bergh, A., and Widmark, A. (1987a). Effect of an LHRH agonist on testicular microcirculation in hypophysectomized rats. Int. J. Androl. 10:785–791.

Damber, J. E., Bergh, A., and Widmark, A. (1987b). Testicular blood flow and microcirculation in rats after treatment with ethane dimethyl sulfonate. Biol. Reprod. 37:1291–1296.

Damber, J. E., Berg, A., Fagreell, B., Lindahl, O., and Rooth, P. (1986) Testicular microcirculation in the rat studied by video photometric capillaroscopy, fluorescence microscopy, and laser Doppler flowmetry. Acta. Physiol. Scand. 125:371–376.

DeKretser, D. M., and Kerr, J. B. (1988). The cytology of the testis, in *The Physiology of Reproduction* (Knobil, E., and Neil, J., eds.) Vol. 1, pp. 837–932, Raven, New York.

Desjardins, C. (1989). The microcirculation of the testis, in *Regulation of Testicular Function* (Ewing, L. L., and Robaire, B., eds.) pp. 243–249, N.Y. Academy of Science, New York.

Djakiew, D., and Dym, M. (1988). Pachytene spermatocyte proteins influence Sertoli cell function. Biol. Reprod. 39:1193–1206.

Dym, M., and Fawcett, D. W. (1970). The blood–testis barrier in the rat and the physiological compartmentation of the seminiferous epithelium. Biol. Reprod. 3:308–326.

Flickinger, C., and Fawcett, D. W. (1967). The junctional specializations of Sertoli cells in the seminiferous epithelium. Anat. Rec. 158:207–222.

Fritz, I. B., and Karmally, K. (1983). Hormonal influences on formation of plasminogen activator by cultured testis tubule segments at defined stages of the cycle of the seminiferous epithelium. Can. J. Biochem. Cell Biol. 61:553–560.

Fritz, I. B., Lyon, M. F., and Setchell, B. P. (1983). Evidence for a defective seminiferous tubule barrier in testes of Tfm and Sxr mice. J. Reprod. Fert. 67:359–363.

Galdieri, M., Monaco, L., and Stefanini, M. (1984). Secretion of androgen binding protein by sertoli cells is influenced by contact with germ cells. J. Androl. 5:409–415.

Gladwell, R. T. (1977). The effect of temperature on the potential difference and input resistance of rat seminiferous tubules. J. Physiol. (Lond) 268:111–121.

Griswold, M. D., Roberts, K., and Bishop, P. (1986). Purification and characterization of a sulfated glycoprotein secreted by Sertoli cells. Biochem. 25:7265–7270.

Hadley, M. A., and Dym, M. (1987). Immunocytochemistry of extracellular matrix in the lamina propria of the rat testis: Electron microscopic localization. Biol. Reprod. 37:1283–1289.

Hadley, M. A., Djakiew, D., Byers, S. W., and Dym, M. (1987). Polarized secretion of androgen-binding protein and transferrin by Sertoli cells grown in a bicameral cultured system. Endocrinol. 120:1097–1103.

Hadley, M. A., Byers, S. W., Suarez-Quian, C. A., Kleinman, H. K., and Dym, M. (1985). Extracellular matrix regulates Sertoli cell differentiation, testicular cord formation, and germ cell development in vitro. J. Cell Biol. 101:1511–1522.

Henning, R. D., and Young, J. A. (1971). Electrolyte transport in the seminiferous tubules of the rat studied by the stopped-flow microperfusion technique. Experientia 27:1037–1039.

Hinton, B. T. (1990). The testicular and epididymal luminal amino acid microenvironment in the rat. J. Androl. 11:498–505.

Hinton, B. T., and Keefer, D. A. (1983). Evidence for protein absorption from the lumen of the seminiferous tubule and rete testis of the rat testis. Cell Tissue Res. 230:367–375.

Hinton, B. T., and Howards, S. S. (1982). Rat testis and epididymis can transport [^{3}H]-3-O-methyl-D-glucose, [^{3}H]-inositol and [^{3}H]-α-aminoisobutyric acid across its epithelia in vivo. Biol. Reprod. 27:1181–1189.

Hinton, B. T., and Setchell, B. P. (1978). Fluid movement in the seminiferous tubules and epididymal duct of the rat. J. Physiol. 284:16P–17P.

Hoffer, A. P., and Hinton, B. T. (1984). Morphological evidence for a blood–epididymis barrier and the effects of gossypol on its integrity. Biol. Reprod. 30:991–1004.

Howards, S. S., Jessee, S. J., and Johnson, A. L. (1976). Micropuncture studies of the blood–seminiferous tubule barrier. Biol. Reprod. 14:264–269.

Hurt, G. S., Howards, S. S., and Turner, T. T. (1987). The effects of unilateral, experimental varicocele are not mediated through the ipsilateral testis. J. Anat. 8:403–408.

Janecki, A., Jakubowiak, A., and Steinberger, A. (1988). Effect of germ cells on vectoral secretion of androgen binding protein and transferrin by immature rat Sertoli cells in vitro. J. Androl. 9:126–132.

Jégou, B. (1991). Spermatids are regulators of Sertoli cell function. Ann. N.Y. Acad. Sci. 637:340–353.

Johnson, A. L., and Howards, S. S. (1976). Intratubular hydrostatic pressure in testis and epididymis before and after long-term vasectomy in the guinea pig. Biol. Reprod. 14:371–376.

Kato, M., Sung, W. S., Kato, K., and Goodman, D. S. (1985). Immunohistochemical studies on the localization of cellular retinol binding protein in rat testis and epididymis. Biol. Reprod. 32:173–190.

Kormano, M. (1967). Dye permeability and alkaline phosphatase activity of testicular capillaries in the postnatal rat. Histochemie 9:327–338.

Lacroix, M., Parvinen, M., and Fritz, I. B. (1981). Localization of testicular plasminogen activator in discrete portions (Stages VII and VIII) of the seminiferous tubule. Biol. Reprod. 25:143–146.

Levine, N., and Marsh, D. J. (1971). Micropuncture studies of the electrochemical aspects of fluid and electrolyte transport in individual seminiferous tubules, the epididymis and the vas deferens in rats. J. Physiol. (Lond) 213:557–570.

Li, L., Seddon, A. P., Meister, A., and Risely, M. S. (1989). Spermatogenic cell–somatic cell interactions are required for maintenance of spermatogenic cell glutathione. Biol. Reprod. 40:317–333.

Maquet, P. (1989). *Giovanni Alfonso Borelli (1608–1679). On the Movement of Animals.* Springer-Verlag, New York.

Martinez-Palomo, A., and Erlij, D. (1975). Structure of tight junctions in epithelia with different permeability. Proc. Natl. Acad. Sci. (USA) 72:4487–4491.

Mendis-Hadagana, S. M., Zirkin, B. R., and Ewing, L. L. (1988). Comparison of components of the testis interstitium with testosterone secretion in hamster, rat, and guinea-pig testis perfused in vitro. Am. J. Anat. 181:12–22.

Molenaar, R., de Rooij, D. G., Rommerts, F.F.G., Reivers, P. J., and van der Molen, H. (1985). Specific administration of ethane dimethane sulfonate. Biol. Reprod. 33:1213–1222.

Morales, C., Sylvester, S. R., and Griswold, M. D. (1987). Transport of iron and transferrin synthesis by the seminiferous epithelium of the rat in vivo. Biol. Reprod. 37:995–1005.

Muffly, K. E., Turner, T. T., Brown, M., and Hall, P. F. (1985). Content of K^+ and Na^+ in seminiferous tubule and rete testis fluids from Sertoli cell-enriched testes. Biol. Reprod. 33:1245–1251.

Nagano, T., and Suzuki, F. (1976). Freeze-fracture observations on the intercellular junctions of Sertoli cells and of Leydig cells in the human testis. Cell Tiss. Res. 166:37–48.

Nakanishi, T., Turner, R. J., and Burg, M. B. (1989). Osmoregulatory changes in myo-inositol transport by renal cells. Proc. Natl. Acad. Sci. (USA) 86:6002–6006.

Nicander, L. (1967). An electron microscopical study of cell contacts in the seminiferous tubules of some mammals. Z. Zellforsch. Mikrosk. Anat. 83:375–397.

O'Brien, D., Gabel, C. A., Welch, J. E., and Eddy, E. M. (1991). Mannose-6-phosphate receptors: Potential mediators of germ cell–Sertoli cell interactions. Ann. N.Y. Acad. Sci. 637:327–339.

Olson, G. E., and Hinton, B. T. (1985). Regional differences in luminal fluid polypeptides of the rat testis and epididymis revealed by two-dimensional gel electrophoresis. J. Androl. 6:20–34.

Parvinen, M., Nikula, H., and Huhtaniemi, I. (1984). Influence of rat seminiferous tubules on Leydig cell testosterone production in vitro. Mol. Cell Endocrinol. 37:331–336.

Perey, B., Clermont, Y., and LeBlond, C. P. (1961). The wave of the seminiferous epithelium in the rat. Am. J. Anat. 108:47–78.

Pickering, B. T., Birkett, S. D., Guldenaar, S.E.F., Nicholson, H. D., Worley, R.T.S., and Yavachev, L. (1989). Oxytocin in the testis: What, where and why? Ann. N.Y. Acad. Sci. 564:198–209.

Rivarola, M. A., Sanchez, P., and Saez, J. M. (1985). Stimulation of ribonucleic acid and deoxyribonucleic acid synthesis in spermatogenic cells by their co-culture with Sertoli cells. Endocrinol. 117:1796–1802.

Russell, L. D., Bartke, A., and Goh, J. C. (1989). Postnatal development of the Sertoli cell barrier, tubular lumen, and the cytoskeleton of Sertoli and myoid cells in the rat, and their relationship to tubular fluid secretion and flow. Am. J. Anat. 184:179–189.

Russell, L. D., and Peterson, R. N. (1985). Sertoli cell junctions: Morphological and functional correlates. Int. Rev. Cytol. 94:177–211.

Russell, L. D. (1978). The blood–testis barrier and its formation relative to spermatocyte maturation in the adult rat: A lanthanum tracer study. Anat. Rec. 190:99–112.

Russell, L. D. (1977). Movement of spermatocytes from the basal to the adluminal compartments of the rat testis. Am. J. Anat. 148:313–328.

Setchell, B. P. (1986). The movement of fluids and substances in the testis. Aust. J. Biol. Sci. 39:193–207.

Setchell, B. P. (1974). Secretions of the testis and epididymis. J. Reprod. Fertil. 36:165–177.

Setchell, B. P., Brooks, D. E. (1988). Anatomy, vasculature, innervation, and fluids of the male reproductive tract, in *The Physiology of Reproduction* (Knobil, E., and Neil, J., eds.) Vol. 1, pp. 753–836, Raven, New York.

Setchell, B. P., Pöllänen, P., and Zupp, J. L. (1988). Development of the blood–testis barrier and changes in vascular permeability at puberty in rats. Int. J. Androl. 11:225–233.

Setchell, B. P., and Rommerts, F.F.G. (1985). The importance of the Leydig cells in the vascular response to hCG in the rat testis. Int. J. Androl. 8:436–440.

Setchell, B. P., and Sharpe, R. M. (1981). Effect of injected human chorionic gonadotropin on capillary permeability, extracellular fluid volume, and the flow of lymph and blood in the testis of rats. J. Endocr. 91:245–254.

Setchell, B. P., Davies, R. V., Gladwell, R. T., Hinton, B. T., Main, S. J., Pilsworth, L., and Waites, G.M.H. (1978). The movement of fluid in the seminiferous tubules and rete testis. Anim. Biol. Anim. Bioch. Biophys. 18:623–632.

Setchell, B. P., and Waites, G.M.H. (1975). In *Handbook of Physiology, Section 7: Endocrinology* (Hamilton, D. W., and Greep, R. O., eds.) Vol. V. Male Reproductive System, pp. 143–172, American Physiological Society, Washington, D.C.

Setchell, B. P., Voglmayr, J. K., and Waites, G.M.H. (1969). A blood–testis barrier restricting passage from blood into rete testis fluid but not into lymph. J. Physiol. (Lond) 200:73–85.

Shabanowitz, R. B., and Kierszenbaum, A. L. (1986). Newly synthesized proteins in seminiferous intertubular and intratubular compartments of the rat testis. Biol. Reprod. 35:179–190.

Sharpe, R. M., Fraser, H. M., and Ratnasoariza, W. D. (1988). Assessment of the role of Leydig cell products other than testosterone in spermatogenesis and fertility in adult rats. Int. J. Androl. 11:507–523.

Sisk, C. L., and Desjardins, C. (1986). Pulsatile release of luteinizing hormone and testosterone in male ferrets. Endocrinol. 119:1195–1203.

Skinner, M. K., Fetherolf, P. M., and Anthony, C. T. (1988). Purification of a paracrine factor, P-Mod-S, produced by testicular peritubular cells that modulate Sertoli cell function. J. Biol. Chem. 263:2884–2890.

Skinner, M. K., Norton, J. N., Mullaney, B. P., Roselli, M., Whaley, P. D., and Anthony, C. T. (1991). Cell–cell interactions and the regulation of testis function. Am. N.Y. Acad. Sci. 637:354–363.

Sprando, R. L., Santulli, R., Awoniyi, C. A., Ewing, L. L., Zirkin, B. R. (1990). Does ethane 1,2-dimethane sulphonate (EDS) have a direct cytotoxic effect on the seminiferous epithelium of the rat testis? J. Androl. 11:344–352.

Steinberger, E. (1976). In *Male Infertility* (Cockett, A.T.K., and Urry, R. L., eds.) p. 12, Grune and Stratton, New York.

Susuki, F., and Nagano, T. (1986). Microvasculature of the human testis and excurrent duct system. Cell Tiss. Res. 243:78–89.

Sylvester, S. R., and Griswold, M. D. (1984). Localization of transferrin receptors in rat testis. Biol. Reprod. 31:195–203.

Sylvester, S. R., Skinner, M. K., and Griswold, M. D. (1984). A sulfated glycoprotein synthesized by Sertoli cells and by epididymal epithelial cells is a component of the sperm membrane. Biol. Reprod. 31:1087–1101.

Tuck, R. R., Setchell, B. P., Waites, G.M.H., and Young, J. A. (1970). The composition of fluid collected by micropuncture and catheterization from the seminiferous tubules and rete testis of rats. Pflugers Arch. 318:225–243.

Turek, F. W., and Van Cauter, E. (1988). Rhythms in Reproduction, in *The Physiology of Reproduction* (Knobil, E., and Neil, J., eds.) Vol. 2, pp. 1789–1830, Raven, New York.

Turner, T. T., and Howards, S. S. (1985). The tenacity of the blood-testis and blood-epididymal barriers, in *Male Fertility and its Regulation* (Lobl, T. J., and Hafez, E.S.E., eds.) pp. 383–393, MTP Press, Lancaster.

Turner, T. T., Jones, C. E., Howards, S. S., Ewing, L. L., Zegeye, B., and Gunsalus, G. L. (1984). On the androgen microenvironment of developing spermatozoa. Endocrinol. 115:1925–1932.

Ueda, H., Tres, L. L., and Kierszenbaum, A. L. (1988). Culture patterns and sorting of rat Sertoli cell secretory proteins. J. Cell Sci. 89:175–188.

Vihko, K. K., Toppari, J., and Parvinen, M. (1987). Stage-specific regulation of plasminogen activator secretion in the rat seminiferous epithelium. Endocrinol. 120:142–145.

Vitale, R., Fawcett, D. W., and Dym, M. (1973). The normal development of the blood–testis barrier and the effects of clomiphene and estrogen treatment. Anat. Rec. 176:333–344.

Waites, G.M.H., and Gladwell, R. T. (1982). Physiological significance of fluid secretion in the testis and blood–testis barrier. Physiol. Rev. 62:624–671.

Waites, G.M.H., and Einer-Jensen, N. (1974). Collection and analysis of rete testis fluid from Macaque monkeys. J. Reprod. Fertil. 41:505–508.

Weber, J. E., Turner, T. T., Tung, K.S.K., and Russell, L. D. (1988). Effects of cytochalasin D on the integrity of the Sertoli cell (blood–testis) barrier. Am. J. Anat. 182:130–147.

Widmark, A., Damber, J. E., and Bergh, A. (1986). Relationship between human chorionic gonadotropin induced changes in testicular microcirculation and the formation of testicular interstitial fluid. J. Endocrinol. 109:419–425.

Wright, W. (1988). Germ cell–Sertoli cell interactions: Analysis of the biosynthesis and secretion of cyclic protein-2. Dev. Biol. 130:45–56.

Wright, W. W., Parvinen, M., Musto, N. A., Gunsalus, G. L., Phillips, D. M., Mather, J. P., and Bardin, C. W. (1983). Identification of stage specific proteins synthesized by rat seminiferous tubules. Biol. Reprod. 29:257–270.

Yancey, P., Clark, M., Hand, S., Bowlus, R., and Somero, G. (1982). Living with water stress: Evolution of osmolyte systems. Science 217:1214–1222.

Yule, T. D., Montoya, G. D., Russell, L. D., Williams, T. M., and Tung, K.S.K. (1988). Autoantigenic germ cells exist outside the blood–testis barrier. J. Immunol. 141:1161–1167.

CHAPTER TWELVE

Spermatogonial Stem Cells

Marvin L. Meistrich and Maria E.A.B. van Beek

I. Introduction

Spermatogonial stem cells are responsible for the maintenance of spermatogenesis in the adult mammal. Knowledge of their regulation is important to understand the initiation of normal spermatogenesis, its recovery after toxic insults, and the modulation of fertility. Information regarding the identification of spermatogonial stem cells and their patterns of self-renewal is extensive and based largely on morphological studies. In contrast, very little of the biochemical and molecular characteristics of these stem cells is known. Unique gene transcripts and proteins, as have been identified in spermatocytes and spermatids (see Chapter 16), have not been described in spermatogonia. Although spermatogonial proliferation and differentiation must be regulated by paracrine and possibly endocrine factors, little is known about how such molecules are involved.

The intent of this chapter is to provide the necessary background to stimulate the application of new approaches in cell and molecular biology to studies of spermatogonial stem cells. First, we present general concepts of stem cell systems and then evaluate the current state of knowledge regarding the identity of the spermatogonial stem cell and its renewal. In addition, observations on conditions that modify spermatogonial proliferation and differentiation are presented because they provide clues as to how this process might be regulated. Finally, possible ways to characterize the stem cells biochemically as well as to identify molecules that regulate their proliferation and differentiation are discussed.

II. General Properties of Stem Cells

Differentiated cells, which are responsible for the function of any organ, are continually lost, yet the organ maintains its cell number and function. This is made possible by stem cells, which must produce both new stem cells and differentiating cells. Stem cells are generally described as undifferentiated, although they can give rise only to specific lineages of cells. The requirement for stem cells is quite apparent in renewing tissues, such as the hematopoietic system, the intestinal epithelium, the skin, and other squamous epithelia, and the testes, from which the mature cells are steadily eliminated and must be replenished.

Maintenance of the genetic integrity of stem cells is important to preserve the genetic quality and function of the differentiated cells. One way to accomplish this is by minimizing proliferation and, thus, the accumulation of replication errors in stem cells; this appears to be a feature of several of these stem cell systems (Lajtha, 1979; Potten, 1986).

A. Definition of Stem Cells

A stem cell is a cell or cell compartment that is a precursor of all the further differentiated cells of its lineage and is also capable of extended or indefinite production of more of its own kind (i.e., self-renewal). During the development of differentiated cells from stem cells, functional, morphological, and biochemical changes take place. The definition of a stem cell, however, is based on function. Although functional assays have enabled extensive characterization of the hematopoietic stem cell, they have not been as well developed in other tissues (e.g., the testes), and hence morphological studies have been a substitute for the identification of the stem cell.

Morphological criteria are ideally applied when the change in function (i.e., differentiation) and a change in morphology occur simultaneously. Functional differentiation, however, may occur without a corresponding change in morphology. Conversely, there may be minor morphological changes, such as those occurring during the cell cycle, that do not indicate functional changes. Despite these limitations, the following morphological definition will be useful: *Stem cells are the cells in the compartment that encompasses the self-renewing cells and those that are morphologically indistinguishable from them.*

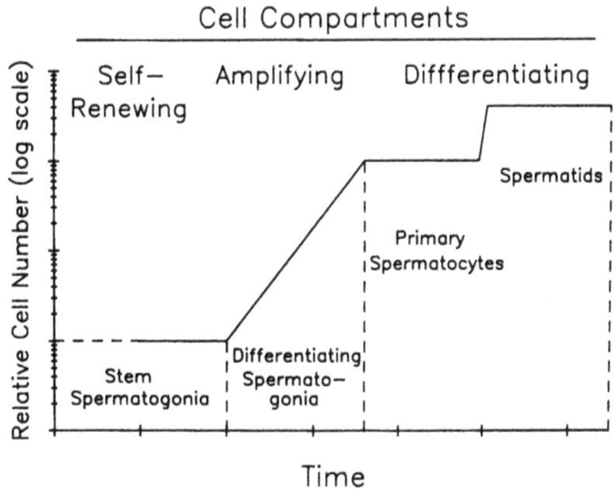

Figure 12–1. Schematic representation of the general compartments of cell populations in a hierarchical tissue containing stem cells. The cells in the spermatogenic sequence are presented as an example.

B. Hierarchical Levels in Renewing Tissues

The number of stem cells is relatively low in all systems that have been well characterized (Lajtha, 1979). Thus, between the stem cell compartment and the compartment of maturing differentiated cells, there must be an amplifying compartment in which numerous cell divisions occur, resulting in major increases in cell numbers. It is useful to describe the tissue in terms of three compartments (Fig. 12–1), although their exact boundaries may not be so clearly defined.

1. SELF-RENEWING COMPARTMENT

The role of the self-renewing compartment is dual: (1) to maintain stem cell numbers and (2) to produce cells that ultimately become differentiated cells. At some point a stem cell must become committed to produce either new stem cells or cells that will differentiate. Because of the limits of our testing systems, the commitment decision may not be apparent until later, when further differentiation takes place.

Often the outcome is first manifest at cell division. A stem cell could either produce two stem cells (self-renewing division) or two differentiated cells (differentiating division) (Fig. 12–2). The possibility also exists that a stem cell could divide and produce one stem cell and one differentiating cell (asymmetric division). The end result of two stem cells undergoing asymmetric divisions, however, is indistinguishable from one each undergoing a self-renewing and a differentiating division, respectively. Hence, in the following discussion it is only necessary to consider self-renewing and differentiating divisions.

For quantitative studies of stem cell renewal and its modification under experimental situations, the probability, P, that a stem cell will perform a self-renewing division is an important parameter. Alternatively, the stem cell could undergo a differentiating division with probability P_{diff}, or degenerate, before or at di-

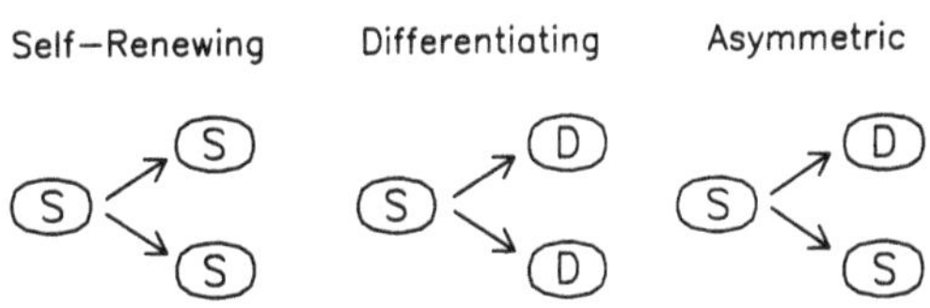

Figure 12–2. Schematic representation of three possible types of division a stem cell might undergo. S, stem cell; D, differentiating cell.

Table 12–1. Effects of Varying the Probability of Self-Renewal (*P*) and the Rate of Cell Division (D/C[a]) of Stem Cells on the Numbers of Stem Cells and the Production of Differentiated Cells

	$P = 0.5$ D/C = 1	$P = 0.5$ D/C = 3	$P = 0.75$ D/C = 1	$P = 0.75$ D/C = 3
Initial number of stem cells	40	40	40	40
Number of stem cells one cycle later[b]	40	40	60	135
Number of differentiating cells produced per cycle[b]	40	120	20	95

[a]Divisions per cycle represented by the number of divisions per unit of time, which may be the characteristic cycle time of the tissue.

[b]Characteristic cycle time of the tissue; in the testes it may be interpreted as the cycle of the seminiferous epithelium.

vision, with a probability P_{deg}. The sum of the probabilities for all these pathways a stem cell can take must be 1. Under steady-state conditions, stem cell numbers must be maintained; hence one self-renewing division, which generates one new stem cell, has to be performed for every stem cell that is lost to differentiation or degeneration. This criterion is satisfied when $P = 0.5$.

In addition to *P*, the rate of cell division is an important parameter of the stem cell compartment. The effect of varying *P* and the rate of stem cell division on the numbers of stem and differentiating cells produced is indicated in Table 12–1. The only way the number of stem cells can increase is if *P* is greater than 0.5. In that case, the rate of production of new stem cells is also affected by the rate of cell division. Note that the rate of production of differentiating cells is affected oppositely by *P* and by the rate of stem cell division.

2. AMPLIFYING COMPARTMENT

Once cells have committed to differentiate and have lost the potential to serve as stem cells, they are considered to be in the amplifying compartment. They divide several times within this compartment to produce a large number of differentiating cells from each stem cell division. In erythropoiesis an estimated 10 cell cycles result in the potential production of 1024 cells from a single precursor (Lajtha, 1979). The number of cell cycles that occur in the amplifying compartment can be flexible. After cell loss, the number of cell generations in this compartment can increase so that larger numbers of differentiated cells are produced (de Rooij et al., 1985).

3. DIFFERENTIATING COMPARTMENT

The third compartment is primarily devoted to cell differentiation. In the testes the differentiating compartment is quite distinct from the amplifying compartment, which includes up to type B spermatogonia. Spermatocytes, particularly those past the preleptotene stage, and spermatids constitute the differentiating compartment (Fig. 12–1).

C. *Hypotheses for Regulation of Stem Cell Kinetics and Renewal*

Regulatory processes controlling the proliferation, self-renewal, and differentiation of stem cells must allow for both the normal functioning of organs and recovery after damage. For a nongrowing tissue in the steady state, *P* must be 0.5 and the average proliferation rate of the stem cells must be constant.

If the number of stem cells is reduced, *P* will have to increase temporarily to regenerate stem cell number. This recovery may then be augmented by an increased proliferation of the stem cells. The production of differentiated cells can also be enhanced by increasing the proliferative rate of the stem cells (Table 12–1).

The various hypotheses for the regulation of stem cell renewal include the action of random processes, factors internal to the stem cell, and factors external to the stem cell. These hypotheses are not mutually exclusive. Under various circumstances the emphasis may shift from one regulatory mechanism to another.

In a large population of cells, purely random processes could maintain a stable population of both stem cells and differentiating cells. Other mechanisms, besides random processes, however, must exist to allow for recovery after damage, which involves alterations of both *P* and rate of stem cell proliferation, and also to correct for deviations of *P* from the steady-state value of 0.5.

Self-renewal could be regulated by factors internal to the stem cells. For example, molecules or cell organelles that are unequally divided at division may determine which daughter cell will remain as a stem cell and which will ultimately produce progeny that will differentiate. Aging of the stem cell, which might occur by accumulation of errors in the DNA, might also initiate the differentiation of a stem cell and result in a decrease in its self-renewal capacity (Harrison and Astle, 1982). Although internal factors could be modulated directly by cytotoxic treatment, the homeostatic processes that occur after cell depletion are more likely initiated by factors external to the stem cell.

Regulatory signals external to the stem cell could be transferred by paracrine factors and cell–cell contacts. The source of these signals could be the differentiated cells, whose number is usually reduced after damage, the other surviving stem cells, or the stromal supporting tissue. Signaling by the stromal tissue is more likely to be initiated by the loss of stem or differentiating cells than by a direct effect of the cytotoxic agent on the stroma.

D. Functional Assays for Stem Cells

Functional assays are important because the definition of a stem cell is primarily a functional one. Functional assays for stem cells may be transplantation or endogenous assays, and the end points used can be clonogenic or nonclonogenic.

Transplantation assays involve injecting a cell suspension containing stem cells into a recipient animal deficient in stem cells or plating the suspension in culture and evaluating colony formation or animal survival. This type of assay has been limited thus far to hematopoietic stem cells (Dexter, 1989).

Assays of other stem cells, such as those in the intestine, skin, and testes, can currently only be done endogenously (in their own tissue) (Withers et al., 1974). Endogenous assays for stem cells are usually performed after treatment with cytotoxic agents to kill most of the cells in the amplifying compartment and allow for the loss of the already differentiated cells, either by maturation or degeneration. In clonogenic assays, the tissue is examined for colonies of cells, each of which must have arisen from a single surviving stem cell. Therefore, such assays directly determine the numbers of surviving stem cells in the tissues. In nonclonogenic assays, the potential of the remaining stem cells to restore tissue function or cell number after cytotoxic damage is determined. The results thus measure a combination of the number of surviving stem cells and their subsequent function, which may be complicated by damage to the stromal tissue by the cytotoxic agent.

Comparison between functional assays and counts of cells with the morphological characteristics of stem cells is useful to validate morphological identification of stem cells. One must be aware, however, that some of the cells counted may not be functional stem cells.

E. Hematopoietic Stem Cells, A Model System

Much of our knowledge of stem cell systems is based on hematopoiesis, as these cells can be aspirated from bone marrow and assayed for function both in vivo and in vitro. The availability of such functional assays has allowed the study of growth and differentiation factors (interleukins and colony-stimulating factors) involved in the survival, proliferation, and development of the hematopoietic stem cells (Dexter, 1989), as well as the purification of these cells based on their surface antigens (Spangrude et al., 1988). The study of the hematopoietic stem cell, however, is more complex than that of the testicular stem cell. The former can produce cells of the monocytic, lymphocytic, erythroid, and granulocytic lineages, whereas the latter only produces cells of spermatogenic lineage.

III. Characterization of Type A Spermatogonia

Both morphological and functional assays have indicated that the stem cell for spermatogenesis in adult mammals is a subset of type A sper-

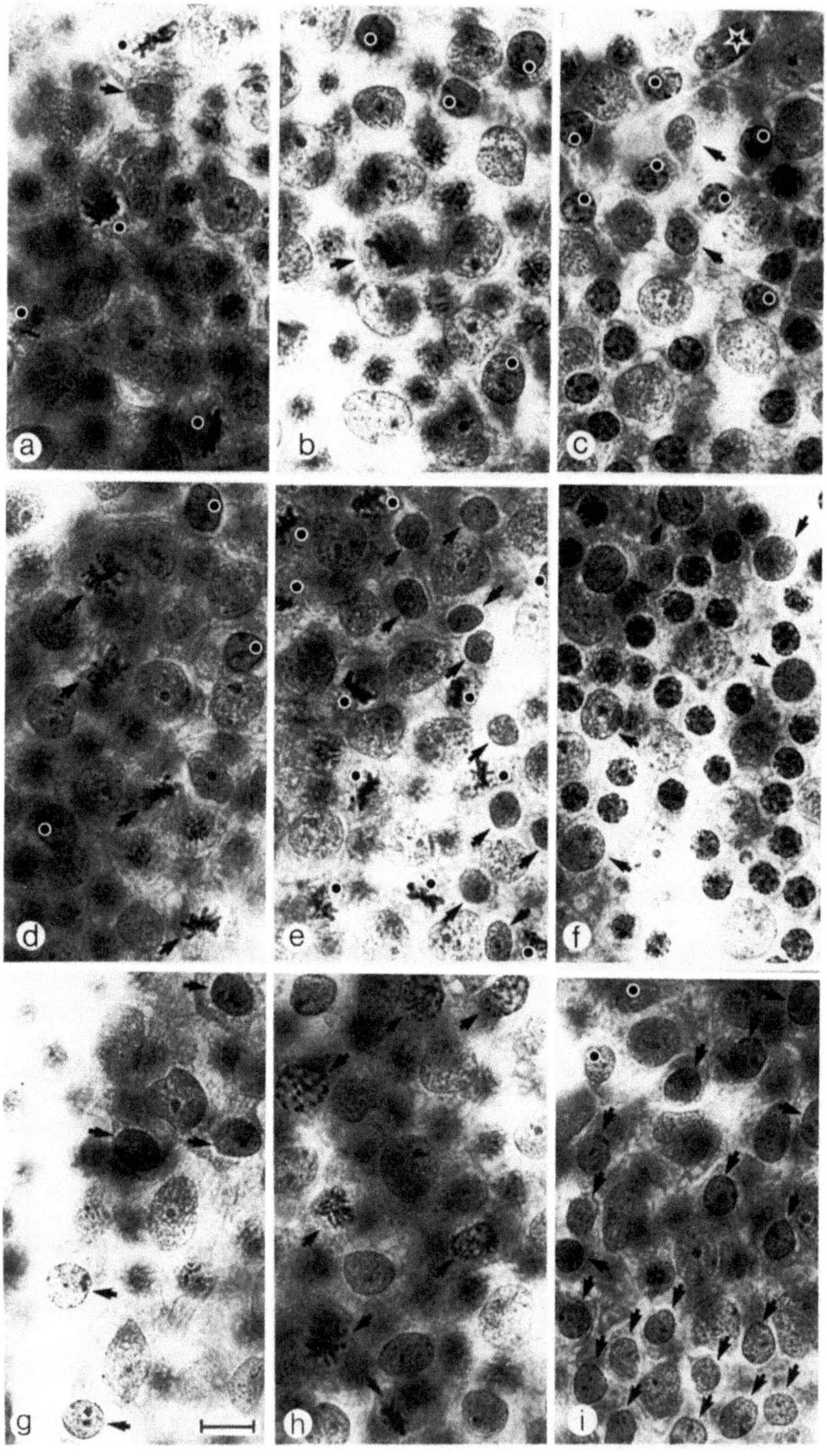

Figure 12–3. Photomicrographs of spermatogonia in whole-mounted seminiferous tubules of the rat. The cell types indicated by *arrows* are arranged in approximate order of development according to the model of Huckins (1971a). Numerous Sertoli cells with large pale nuclei and characteristic nucleoli (sometimes only visible in a different focal plane) are also seen. (**a**) Isolated undifferentiated type A spermatogonium, A_s *(arrow)* among dividing A_3 spermatogonia *(dot)* in stage XIV of the cycle of the seminiferous epithelium. (**b**) Isolated

matogonia. Therefore, further discussion of the stem cell compartment focuses on this category of cells. Type B, intermediate (In), and type A spermatogonia other than the stem cells are in the amplifying compartment and are described only for comparison with stem type A spermatogonia.

A. Nonprimates (Primarily Rodents)

Type A spermatogonia, including the stem cell, have been most extensively characterized in rodents, especially the rat, mouse, and Chinese hamster (reviewed in de Rooij, 1983; de Rooij et al., 1989). Studies have also been done in the ram (Lok et al., 1982; Hochereau-de Reviers, 1976), and to a lesser extent in the bull (Hochereau-de Reviers, 1976) and boar (Frankenhuis et al., 1982). The morphological types of A spermatogonia, their kinetics, and their mode of renewal are similar in at least five of these six species. In contrast, type A spermatogonia of primates show some different features and are discussed separately.

1. CELL MORPHOLOGY

Several types of spermatogonia are found along the basement membrane of seminiferous tubules. In general, the A spermatogonia have pale-stained nuclei containing granular or dustlike chromatin (Fig. 12–3). B spermatogonia (Fig. 12–3c) show distinct clumps or flakes of heterochromatin, just subjacent to the nuclear outline and a nucleolus located internally. These cells are the most differentiated spermatogonia and give rise to spermatocytes (Fig. 12–3f). In many species, a third type, the In spermatogonium (Fig. 12–3i), is recognized with characteristics between the A and B spermatogonia. These cell types represent a continuum of spermatogonial differentiation, and the distinctions between subsequent generations may be rather subtle.

There is some further heterogeneity among type A spermatogonia. In rats and mice these cells may be divided into undifferentiated and differentiating (A_1, A_2, A_3, and A_4) spermatogonia based on nuclear morphology and other factors. The undifferentiated spermatogonia generally have small nuclei with heterogeneous, finely granular chromatin and a small nucleolus, usually associated with a chromatin granule (Fig. 12–3a,c,e,i) (Huckins, 1971a). A_1 spermatogonia have larger nuclei with more finely dispersed chromatin and clearly visible nucleoli (Fig. 12–3f). During development from A_2 through A_3 (Fig. 12–3b) and A_4 (Fig. 12–3d,g) spermatogonia, flakes of more heavily stained chromatin attached to the nuclear outline become more prominent (Clermont and Bustos-Obregon, 1968) and the nuclear size becomes smaller (Huckins, 1971a), but is still larger than in the undifferentiated spermatogonia. Because nuclear size varies within the cell cycle, however, it is not a reliable criterion for identification.

The subclassification of type A cells requires more information besides their nuclear morphology. Their association with specific stages of the cycle of the seminiferous epithelium (see Chapter 8) is particularly helpful. In the rat, the differentiating type A spermatogonia just before stage IX are type A_1 spermatogonia, those between stages IX and XII are A_2, those be-

spermatogonium in mitosis *(arrow)*, identified as an A_s because it is isolated among interphase A_3 spermatogonia *(dots)*, stage XIV. **(c)** Pair of undifferentiated type A spermatogonia, A_{pr} *(arrows)* among B spermatogonia (some marked with *dots*), stage V. Note also an elongated A_s spermatogonium *(star)*. **(d)** A chain of four spermatogonia in mitosis *(arrows)*. These cells are identified as A_{al} because they are out of synchrony with the interphase type A_4 spermatogonia *(dots)* in this stage I tubule. **(e)** A chain of undifferentiated type A spermatogonia, A_{al} *(arrows)* among G_2 and mitotic intermediate spermatogonia *(dots)*, stage IV. The chain contains 16 cells (10 are shown). **(f)** A_1 spermatogonia *(arrows)* among preleptotene spermatocytes in a stage IX tubule. The A_1 cells are generally the largest of all the A spermatogonia. **(g)** A_4 spermatogonia *(arrows)* in a stage I tubule. At this stage they are larger and more have heterochromatin along the nuclear outline than do undifferentiated A spermatogonia. **(h)** A_4 spermatogonia in mitosis *(arrows)* at stage I. Note the wave of mitotic stages characteristic of the differentiated spermatogonia. **(i)** Intermediate spermatogonia *(arrows)* in interphase at stage II. Note that there is not much difference in morphology from the A_4 spermatogonia. A_{pr} spermatogonia are also present *(dots)* and are about the same size as the intermediates but lack the heterochromatin. All photographs ×760 magnification. Bar = 10 μm.

tween stages XII and XIV are A_3, and those between stage XIV and late stage I are A_4 (Fig. 12–4). Distinguishing these different types of spermatogonia without knowledge of the stage is unreliable.

Furthermore, the morphological distinction between some of these cells and the undifferentiated type A spermatogonia is subtle; even with knowledge of the stage, identification is difficult. For example, the same cell types in stages XIV–I of the rat have been reported to be type A_3 or A_4 spermatogonia by one laboratory (Bartmanska and Clermont, 1983) and undifferentiated spermatogonia by another (Huckins, 1971a). Other factors must also be considered for unequivocal subclassification of types of A spermatogonia; these factors include their topology within the epithelium and their kinetic behavior.

2. TOPOLOGY

In the tubular whole mount technique (Clermont and Bustos-Obregon, 1968) the spermatogonia are viewed through the upper surface of the basement membrane. This permits the identification of neighboring spermatogonia in all directions along the surface of the tubule.

Most of the type A spermatogonia display synchronous behavior over large areas of whole mounts. In a given area, large numbers of these cells may label with [^{3}H]thymidine, go into mitosis (Fig. 12–3h), or display morphological features typical of certain phases of the cell cycle (Fig. 12–3f; Huckins, 1971d). Furthermore, these events are associated with specific stages of the cycle of the seminiferous epithelium. The type A spermatogonia undergoing this synchronous behavior are the differentiating type A spermatogonia (A_1 through A_4). In the rat, their mitotic waves are observed in stages IX, XII, XIV, and at the end of stage I (Figs. 12–3h and 12–4).

The remainder of type A spermatogonia do not display such widespread synchrony. Cells with a particular morphology or in S-phase or mitosis are generally found isolated or in groups of 2 to 16 cells. These cells have the morphological characteristics consistent with undifferentiated spermatogonia. Their behavior is asynchronous, both with respect to the differentiating spermatogonia and to other groups of undifferentiated spermatogonia (Huckins, 1971a,b).

The cells within a group pass through their cycle synchronously (Huckins, 1971b) and, in some preparations, may be seen to be connected by intercellular bridges (Huckins, 1978a). Because it is not always possible to discern these bridges, a criterion based on the internuclear distance (less than 25 μm) may be

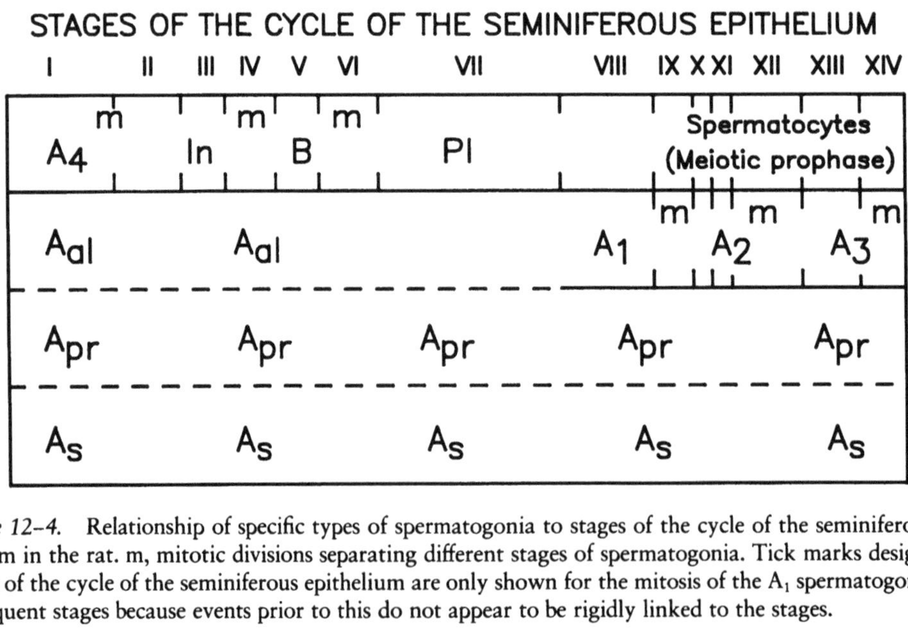

Figure 12–4. Relationship of specific types of spermatogonia to stages of the cycle of the seminiferous epithelium in the rat. m, mitotic divisions separating different stages of spermatogonia. Tick marks designating stages of the cycle of the seminiferous epithelium are only shown for the mitosis of the A_1 spermatogonia and subsequent stages because events prior to this do not appear to be rigidly linked to the stages.

applied to decide whether cells belong to the same group. When this criterion is applied, the numbers of cells in such groups are almost all represented by 2^n (Lok et al., 1982).

Isolated (i.e., with no morphologically similar cell within 25 μm), undifferentiated spermatogonia (Fig. 12–3a) have been designated A_s (Huckins, 1971a). The numbers of these cells (per Sertoli cell nucleus) do not vary significantly at different stages of the cycle of the seminiferous epithelium. This property is consistent with their being either a stem cell population with a *P* of 0.5, or a noncycling, reserve stem cell population.

Groups of two undifferentiated type A spermatogonia (Fig. 12–3c) have been designated A_{pr} (A_{paired}) (Huckins, 1971a). The numbers of these cells also do not vary significantly over the cycle of the seminiferous epithelium. This led to the proposal that A_s and A_{pr} were noncycling cells and were alternatively designated A_0 (Clermont and Bustos-Obregon, 1968).

Larger groups of apparently undifferentiated spermatogonia (Fig. 12–3e) have been designated A_{al} ($A_{aligned}$) (Huckins, 1971a). The type A_{al} spermatogonia are predominantly found in stages I through VII, with their numbers increasing fourfold over this span of stages. The stages at which the various types of undifferentiated spermatogonia are found are indicated in Figure 12–4.

Although Huckins (1971a) considered these aligned cells to be undifferentiated spermatogonia, it is difficult to distinguish them from type A_3 or A_4 spermatogonia on nuclear morphology alone (compare Figs. 12–3a,c,e with 12–3b,d,g). The lack of synchrony of aligned cells with the waves of labeling and mitoses of the differentiating spermatogonia (Figs. 12–3d,e) has been taken to imply that they are distinct cells (de Rooij, 1983). Others have argued, however, that the synchronous behavior of the differentiating spermatogonia may not be absolute and that these aligned cells may just be asynchronous differentiated type A spermatogonia (Bartmanska and Clermont, 1983). This question may be resolved with kinetic studies.

3. KINETIC STUDIES

Kinetic studies, which involve labeling cells with [^{3}H]thymidine, counting labeled cells and mitotic figures, and following the progression of the labeled cells with time, can further elucidate the nature and behavior of the type A_s, A_{pr}, and A_{al} spermatogonia. The most extensive studies have been done in the rat and in the Chinese hamster, although some studies have also been performed on the mouse (de Rooij, 1973) and ram (Lok et al., 1982). This discussion focuses on results in the rat; data obtained on other species are generally consistent with these results and are also applied to clarify observations made on the rat.

a. Labeling and Mitotic Indices

The proliferative activity of type A_s and A_{pr} (or alternatively A_0) spermatogonia has been investigated by examination of the labeling and mitotic indices of these cells. Some studies (Clermont and Hermo, 1975) could not conclusively demonstrate labeling of these cells; however, others did (Huckins, 1971b; Bartmanska and Clermont, 1983), reporting labeling indices of the A_s of about 25 percent in stages I and II, declining to 3 percent in stage V. In the Chinese hamster, the A_s cells were labeled throughout the cycle of the seminiferous epithelium (Lok and de Rooij, 1983), with labeling indices generally between 5 and 10 percent. Calculations based on these labeling indices and other cell kinetic parameters indicate that roughly 50 percent of type A_s spermatogonia actively cycle and that the A_s divide an average of twice during the cycle of the seminiferous epithelium. Thus, it appears that in rodents the A_s cells are indeed proliferating. Further support for this conclusion was provided by giving rats multiple injections of [^{3}H]thymidine over a 60-hour period; about 90 percent of the A_s cells in the stages studied (XIII–I) were labeled (Huckins, 1971c). Also the A_{pr} appear to have labeling indices similar to those of the A_s spermatogonia.

Measurements of the mitotic indices of type A_s (Fig. 12–3b) and A_{pr} spermatogonia in the rat also differ between laboratories. Clermont and Hermo (1975) report a mitotic index of 0.13 percent for these cells, whereas Huckins (1971a) reports between 0.14 and 3 percent depending on the stage of the cycle. Erickson and Hall (1983) obtained mitotic indices (including prophase through telophase) of 1.4 and 3.6 percent for the A_s and A_{pr}, respectively. Overall these data support the conclusion that type A_s and A_{pr} spermatogonia do proliferate.

The proliferative activity of type A_{al} spermatogonia is more strongly dependent than that of the A_s cells on the stage of the cycle of the seminiferous epitheium. In the rat, the labeling indices of the A_{pr} plus A_{al} cells ranges from 50 percent in stage I to 1 percent in stage V (Huckins, 1971b). Similarly in the Chinese hamster, the labeling indices of the A_{al} declined from about 20 percent in stages XI–II to 1 percent in stages V–VII (Lok and de Rooij, 1983). The labeling indices correlate with the stages at which the major increases in the numbers of undifferentiated spermatogonia occur (Huckins, 1971a; Lok et al., 1982).

b. Progression of Labeled Cells

Analysis of the progression of labeled cells can be used to further clarify the role of type A_s spermatogonia and the identity of type A_{al} spermatogonia. This has been done by the fraction of labeled mitoses (FLM) method, in which the fraction of mitotic figures that are labeled at different time intervals after injection of [^{3}H] thymidine is determined. In this method cells labeled in S-phase should first appear as mitoses after a time equal to the G_2 phase of the cell cycle and the FLM should reach a maximum when the cells labeled in S-phase pass through mitosis. Subsequently the FLM should decrease to zero as these mitoses represent cells that were in G_1 at the time of labeling. A second wave of labeled mitoses occurs in the daughter cells at about one cell cycle time after the peak of the initial wave. The heights of these peaks of the FLM curves are indicative of how synchronously the cells pass through their cycles. The data from FLM curves for the various types of spermatogonia in the rat are summarized in Table 12–2.

The observation of labeled mitoses of A_s cells confirms that they indeed proliferate. The lower values for the maximum FLM of the A_s cells, particularly in the second wave, indicate that these cells have much more variable cell cycle durations than type A_{al} or differentiating spermatogonia. Some A_s cells produce A_{pr} upon division; the cell cycle of the resulting A_{pr} is similar to that of the A_{al} but, as a second wave of labeled mitoses of A_s is not seen in the rat, it is not possible to evaluate the cycle time of the A_s cells.

These kinetic data can also resolve the question of whether the chains of type A spermatogonia dividing out of synchrony with the differentiating spermatogonia are actually undifferentiated A_{al} (Huckins, 1971a) or just asynchronous differentiating spermatogonia (Bartmanska and Clermont, 1983). The labeled A_{al} cells enter their first mitosis later than any

Table 12–2. Kinetic Properties of Various Types of Spermatogonia in the Rat as Deduced from Fraction of Labeled Mitoses (FLM) Curves

Cell types	Center of first wave (hr)	Maximum FLM (first wave)	Center of second wave (hr)	Maximum FLM (second wave)	Cell cycle[a] (hr)
A_s	30	0.82	No clear peak	0.15	Not evaluable[b]
A_{pr}[c]	—	—	≥81	0.60	≥51
A_{al}	25	0.97	77	0.90	55
A_2,A_3,A_4, In and B	15–20[d]	0.97	57–62[d]	0.96	42

[a]Calculated as the time interval between the half-maximum values on the leading edge of the FLM curves.

[b]A second peak in the FLM curves was observed in the Chinese hamster (Lok et al., 1983) with a maximum FLM of 0.25 and occurring at a time that would make the cell cycle time of these A_s cells equal to those of the A_{pr} and A_{al}.

[c]The first wave of labeled mitoses of A_{pr} is identical to A_{al}; in the second wave the A_{pr} would now be the A_{al} cells. The data for A_{pr} cells at the time of the second wave represent cells that were A_s at the time of labeling. Only lower limits could be evaluated because the experiment was carried out only until 81 hours, which appeared to be near the maximum FLM, but no further time points to demonstrate that it declined were taken.

[d]Range of values for various stages of differentiating spermatogonia. Individual types show high degrees of synchrony.

Data from Huckins, 1971b,d.

of the differentiating spermatogonia and have a cell cycle time of 55 hours, as opposed to 42 hours for the differentiating spermatogonia (Table 12–2). This discrepancy is not the result of the A_{al} having a high variability in their cell cycle parameters because the FLM reaches 0.90 or higher in their first two cycles. Thus, type A_{al} spermatogonia are not asynchronous differentiating spermatogonia, but have cell kinetic properties similar to those of type A_{pr} and some A_s spermatogonia, which is consistent with their identification as undifferentiated spermatogonia.

In both the rat (Huckins, 1971c) and Chinese hamster (Lok et al., 1984) labeled type A spermatogonia have been found in the testes at times after labeling equal to or greater than one cycle of the seminiferous epithelium (13 days in the rat and 17 days in the Chinese hamster). Actively proliferating cells would pass through multiple cell cycles during that period (Table 12–2) and the label will be so diluted that those cells will appear unlabeled. The only cells that would be labeled will be those that have not divided or have undergone only a few cell divisions during that period.

There are, however, some discrepancies. In the rat, type A_s spermatogonia that retained the label were only those that had been in stages XI–I at the time of labeling (Huckins, 1971c). In contrast, in the Chinese hamster the stage distribution of the labeling indices of the A_s one cycle after labeling, although maximal (11 to 15 percent) in stages IX–II and minimal (5 percent) in stages VII–IX, was identical to that observed at the time of labeling. Thus, all of the A_s cells seem to have the same probability, independent of the stage at which they are labeled, of producing daughter cells that remain as stem cells and have long cell cycle times. The type A_s spermatogonia appear to have a variable cell cycle duration, but the evidence for distinct subpopulations of short cycling A_s cells committed to differentiation and of long cycling cells that remain as stem cells is not conclusive.

B. Primates

Although the cell types of the seminiferous epithelium of primates appear similar to those of rodents and other nonprimates, there are differences, especially among type A spermatogonia. The numbers of type A spermatogonia, relative to Sertoli cells and other germ cells are much higher in primates than in rodents. Also primates have a unique subclass of type A spermatogonia, which has a darkly stained nucleus. These features require a separate discussion of type A spermatogonia in primates.

1. MORPHOLOGY

As in rodents, the nuclei of type A spermatogonia contain granular or dustlike chromatin. About half have pale-staining nuclei with granulated chromatin and are designated A_p (A_{pale}). The remainder, designated A_d (A_{dark}) contain homogeneous, finely granulated, deeply stained chromatin retracted from the nuclear membrane, and often show one or two prominent nucleoli attached to the nuclear membrane (Clermont, 1966; Clermont and Antar, 1973).

Both type A_d and A_p spermatogonia are observed at all stages of the cycle of the seminiferous epithelium in primates. In all primates the numbers of type A_d spermatogonia remain constant over the various stages of the cycle. In humans and some monkeys (e.g., *Cercopithecus aethiops,* Clermont, 1969) the number of A_p cells also stays constant, but in others (*Macaca arctoides,* Clermont and Antar, 1973; *Macaca fascicularis,* Fouquet and Dadoune, 1986) their number transiently doubles around stages VIII and IX (of 12 stages). Little change occurs in the nuclear morphology of the type A spermatogonia throughout the cycle of the seminiferous epithelium.

In monkeys, cells with the morphological characteristics of type B spermatogonia first appear in stage X at which time their numbers are equal to those of the type A_p spermatogonia of the previous stage (Clermont and Antar, 1973). The type B spermatogonia subsequently undergo mitoses and increase in numbers at specific stages, indicating four generations of these cells (Clermont, 1969). There is a progressive change in their morphology with the nuclei becoming smaller and showing increasing amounts of heterochromatin just subjacent to the nuclear membrane. The last mitotic division at stage VI results in the formation of preleptotene spermatocytes. In humans, type B spermatogonia are present in the equivalent stages of the cycle; however, their numbers in most stages are only twice the number of type A_p spermatogonia (Clermont, 1966), indicating only one or two generations of type B spermatogonia.

2. CELL TOPOLOGY

In whole-mounted tubules of monkeys and humans, type A_d and A_p spermatogonia are found in clusters in the epithelium (Clermont, 1966, 1969). Some areas contain only type A_d or A_p spermatogonia, whereas sometimes clusters of A_d and A_p cells are closely juxtaposed. Each cluster appears to contain only even numbers of cells with group sizes of two, four, and eight predominating, but groups of 6 and 12 are regularly observed as well. In the monkey, the mean number of cells in the clusters is the same for type A_d, A_p, and the newly formed B spermatogonia. During repopulation after irradiation, at which time the density of the spermatogonia is decreased, it was more evident that the number of either A_d or A_p cells in these clusters was either 1, 2, 4, 8, or 16 (van Alphen et al., 1988b).

3. KINETIC STUDIES

Injection of [^{3}H]thymidine produces labeling of type A_p but not A_d spermatogonia in the monkey. Labeling indices of type A_p spermatogonia are about 35 percent from stage VII to stage IX (Clermont, 1969), or to stage X (Clermont and Antar, 1973) depending on the species of monkey; no A_p cells are labeled in other stages. Differences in the span of stages at which type A_p spermatogonia are labeled correspond to whether these cells undergo one or two cell divisions.

Although type A_d spermatogonia are rarely (below 0.1 percent, Clermont and Antar, 1973) labeled within hours after injection of the label, they could become labeled within one cycle of the seminiferous epithelium (Fouquet and Dadoune, 1986). Such labeled A_d cells are in the same stage of the cycle of the seminiferous epithelium as the labeled A_p cells, but their labeling index is lower.

Mitotic divisions of type A spermatogonia in the monkey are observed either at stages VIII and X (Clermont and Antar, 1973) or only at stage X (Clermont, 1969) depending on the species. Based on labeling indices of cells just before these stages, the divisions must correspond to type A_p and not A_d spermatogonia.

The labeling of human spermatogonia has been studied only after incubation of testicular tissue in vitro with [^{3}H]thymidine (Chowdhury and Steinberger, 1977). Continuous labeling for up to 3 hours produced a labeling index of 20 percent for type A_p spermatogonia, but no A_d cells were labeled. After a chase in unlabeled medium, label disappeared from type A_p spermatogonia, but still no labeled A_d cells were found. Thus, in humans, as in monkeys, type A_p spermatogonia show limited proliferative activity, but the A_d cells show none.

IV. Models for Stem Cell Renewal in the Steady State

Models for stem cell renewal depend on the identification of the stem cell. Although the identity of the stem cells as type A spermatogonia is well accepted, different subpopulations among the A spermatogonia (Fig. 12–4) have been chosen as the stem cell. The three models proposed for stem cell renewal will be called the A_s model, the A_0/A_1 model, and the clone fragmentation model, based on the cells chosen as the stem cell or modes of generation of stem cells.

A. Description of the Models

1. THE A_S MODEL

In the A_s model (Fig. 12–5), the A_s (isolated undifferentiated type A spermatogonia) cells are considered to be the only population of stem cells. It was originally proposed for the rat (Huckins, 1971a), but it has also been applied to the mouse (de Rooij, 1973), Chinese hamster and ram (Lok et al., 1982), and boar (Frankenhuis et al., 1982).

When type A_s spermatogonia divide they can, with a probability P, produce two A_s cells or they may produce type A_{pr} spermatogonia, which remain connected to each other by a cytoplasmic bridge (Huckins, 1978a). When A_s cells divide into two A_s cells they usually migrate apart; if they do not migrate apart they are likely to be an A_{pr}. Because the nuclear morphology remains unchanged, the criterion for

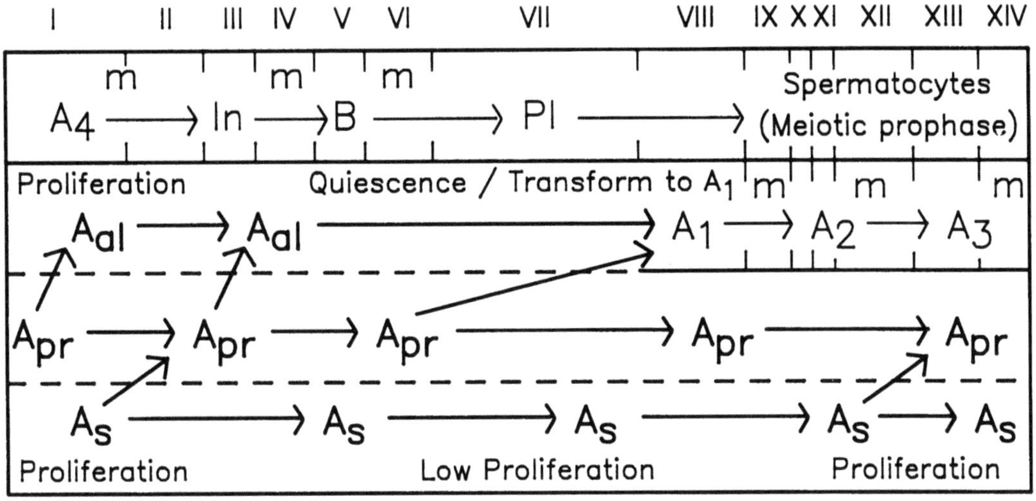

Figure 12–5. A_s model of stem cell renewal and spermatogonial differentiation in the rat. The undifferentiated spermatogonia and their pathways are displayed in bolder lines for emphasis. Arrows indicate pathways that a cell could follow by either remaining the same, undergoing mitosis, or morphological transformation. The A_s cells are considered to be the stem cells and the true A_{pr} and A_{al} are committed to the process of further differentiation.

differentiation is topological, based on either visualization of an intercellular bridge or, more commonly, the distance between the nuclei. The production of type A_{pr} spermatogonia is the first step toward differentiation. It is apparent from this criterion for differentiation that both members of the pair must be considered to be differentiating. Therefore, in this model of stem cell renewal only symmetric divisions can be observed.

Type A_{pr} spermatogonia that are formed divide again, preferentially during stages X–III (Lok and de Rooij, 1983), to produce groups of four type A_{al} spermatogonia, also connected to each other. The A_{al} cells proliferate (Fig. 12–3d), resulting in chains, usually consisting of either 4, 8, 16, or occasionally even 32 cells. Their proliferation declines during stages I–V of the cycle (Huckins, 1971a).

After they cease proliferating and before step VIII, most of the type A_{al} spermatogonia transform into type A_1 spermatogonia (compare Figs. 12–3e,f) without a cell division. In rodents, this morphological change may sometimes be subtle, but it is more distinct in the ram (Lok et al., 1982). Data in the ram confirm indirect observations on rodents, that A_s cells only rarely and A_{pr} cells only occasionally transform into type A_1 spermatogonia, and the probability for transformation of A_{al} cells increases dramatically with the size of the chain.

2. THE A_0/A_1 MODEL

The A_0/A_1 model is based on a slightly different identification scheme for type A spermatogonia in the rat from that presented in Figure 12–4. Type A_s and A_{pr} spermatogonia are designated A_0; the proliferating A_{al} cells were not recognized as a distinct class and were considered to be asynchronous A_3 or A_4 cells undergoing transformation into type A_1 spermatogonia.

According to this model, two compartments can serve as stem cells, both type A_0 spermatogonia and the compartment consisting of type A_1 through A_4 spermatogonia (Clermont and Bustos-Obregon, 1968). By virtue of their low labeling and mitotic indices in the unperturbed testis (Clermont and Hermo, 1975), type A_0 spermatogonia were considered to be nonproliferating and could not normally act as stem cells. After x-irradiation, however, their numbers and labeling indices indicated that they could function to repopulate the seminiferous epithelium, and hence they are referred to as a *reserve* stem cell population (Dym and Clermont. 1970).

The *renewing* stem cell compartment for normal spermatogenesis was assigned to type A_1 through A_4 spermatogonia. In this model, renewal is accomplished when type A_4 spermatogonia divide and produce some A_1 in addition to the more numerous In spermatogonia. This division usually occurs at the end of stage I of the cycle of the seminiferous epithelium, occasionally in stages II or III (Bartmanska and Clermont, 1983). Note the chains of type A spermatogonia produced by this division are considered to be type A_1 spermatogonia, instead of A_{al} as in the A_s model. These cells remain dormant until they divide into type A_2 spermatogonia in stage IX. Two subsequent divisions result in type A_3 and then A_4 spermatogonia again.

3. THE CLONE FRAGMENTATION MODEL

The clone fragmentation model (Erickson, 1981; Erickson and Hall, 1983) is a variation of the A_s model. In the A_s model, the A_s cells are the only stem cells and the production of an A_{pr} cell is a commitment toward differentiation and later yields A_{al}. Each group of A_{pr} or A_{al} cells is a *clone* because it encompasses all the progeny of one stem cell that underwent a differentiating division and, because of the cytoplasmic bridges, still forms one unit.

The clone fragmentation model proposes that these clones of A_{pr} and A_{al} cells, as well as the A_s, constitute the stem cell compartment. Because some type A_{al} spermatogonia, estimated to be 14 percent in the mouse (Erickson, 1981), do not transform into A_1 cells, these A_{al} cells could contribute to the production of undifferentiated or differentiated cells during the next cycle. Furthermore, clones of undifferentiated spermatogonia with cell numbers that deviate from the geometric series 2^n were found at an incidence of 12 percent in mice (Erickson, 1981). Clones with numbers other than 2^n should not exist according to the A_s model, and Erickson (1981) proposed that they arose by fragmentation of larger clones through breakage of intercellular bridges. Because the clones of undifferentiated type A spermatogonia can remain through multiple cycles, increase their size by mitotic division of cells, increase their number by fragmentation, and produce differentiating offspring by transformation into A_1 cells, they can function as stem cells. Several additional lines of reasoning were presented to support the need for such an alternative to the A_s model during recovery after irradiation; these are discussed in section V.B.

B. Evaluation of Models for Nonprimates

One source of discrepancy between the A_s and the A_0/A_1 model is whether the isolated and paired spermatogonia proliferate and divide in the normal seminiferous epithelium. Although some studies (Clermont and Hermo, 1975) failed to find significant labeling of these cells, numerous other reports (summarized in section III.A.3.) conclusively demonstrate that most of the A_s cells pass through an S-phase during the cycle of the seminiferous epithelium, supporting the A_s over the A_0/A_1 model.

The other aspect of this controversy is whether the mitotic indices of the A_s cells are sufficient to produce the numbers of differentiating type A_1 spermatogonia formed. In the rat, the number of A_1 spermatogonia formed per cycle is about eight times the numbers of A_s (Huckins, 1971a). Because the median size of the clones is also about eight cells, each A_s must produce one new A_s committed to differentiation to yield all of the A_1 cells formed. This could be achieved by one self-renewing division of A_s cells and one of the daughters undergoing a differentiating division per cycle. The time in mitosis (including prophase) is 2 hours (Huckins, 1971d) of a cycle of 310 hours; therefore, two divisions per cycle would represent a mitotic index of approximately two times (2 hours/310 hours), or 1.3 percent. This is consistent with values reported by Huckins (1971a) and Erickson and Hall (1983), but 10 times higher than the value reported by Clermont and Hermo (1975). Part of this discrepancy could be the result of scoring only metaphases, which represent only part of the time spent in mitosis. The low mitotic indices reported are, therefore, in the range expected for the A_s model.

The second major difference between the A_s and the A_0/A_1 model is whether there is a population of type A_{al} spermatogonia, distinct from the A_4, to serve as the source of type A_1 spermatogonia. Kinetic evidence (section

III.A.3.b.) confirmed the existence of a separate A_{al} population in stages where A_4 cells were also present.

The production of type A_1 from A_4 spermatogonia, as proposed in the A_0/A_1 model, represents dedifferentiation. The progression from type A_1 to A_4 spermatogonia is associated with a gradual morphological change involving the accumulation of heterochromatin subjacent to the nuclear membrane, a process that continues with the development to type In and B spermatogonia. Thus, reversion from type A_4 to A_1 spermatogonia requires a morphological dedifferentiation, which although not theoretically impossible, will require a more complex biochemical regulatory mechanism to explain. In other stem cell systems, the differentiation process is unidirectional (Hall and Watt, 1989), as in the A_s model.

The major difference between the A_s and the clone fragmentation model is whether type A_{al} spermatogonia are irreversibly committed to differentiation, or can self-renew by fragmentation. Fragmentation would give rise to clones of sizes other than 2^n, the frequency of which is a matter of disagreement. Although Erickson (1981) found that 12 percent of the clones of type A_{al} spermatogonia in the mouse are other than 2^n, van Beek and colleagues (1984) found less than 1 percent of such clones. These differences could arise from the criteria used to identify clones and the choice of stages at which to count clones.

To be members of the same clone, cells must be similar morphologically and within 25 μm of each other. Studies in the Chinese hamster using this distance criterion (Lok et al., 1982) showed that all type A_{al} spermatogonia were in groups of 2^n. Sometimes different-sized groups were found that could be identified as being combinations of A_s–A_{pr} or A_s–A_{al} because of differences in nuclear morphology. The 25-μm distance criterion for undifferentiated A_{al} spermatogonia does not apply to type A_1 spermatogonia because during the transformation of type A_{al} (Fig. 12–3e) into A_1 (Fig. 12–3f) spermatogonia in stages VI–VIII, the cells may move apart (Lok et al., 1982). Unfortunately, most of the observations leading to the clonal fragmentation model (Erickson 1981; Erickson and Hall, 1983) were performed in those stages.

Another phenomenon that may appear as clonal fragmentation is the delay in separation of two stem cells formed by a self-renewing division. In the Chinese hamster (Lok and de Rooij, 1983), two A_s cells formed by a self-renewing division may not move apart until their next division, resulting in about 20 percent of the apparent A_{pr} clones actually being false pairs. Although from a morphological point of view the separation of the false pairs may seem to be clonal fragmentation, from a functional point of view false pairs are consistent with the A_s model. Considering factors, such as delayed separation of A_s, chance association of two clones, viewing an entire clone, and small morphological differences apparently existing within one clone, the incidence of a few clones of size not equal to 2^n does not seem to be a compelling argument for clonal fragmentation.

The model that in our opinion explains the observations best is the A_s model. Although the A_{pr} (perhaps false pairs) may have some self-renewal capacity, they along with type A_{al}, A_1, A_2, A_3, and A_4 spermatogonia will be considered as part of the amplifying compartment. Further tests of this model using morphology alone will be difficult. The progression of cells should be investigated with molecular markers of spermatogonial differentiation, particularly ones that can be assessed in situ to take advantage of the extensive morphological studies.

C. Evaluation of Models for Primates

That two morphologically different classes of type A spermatogonia, A_d and A_p, are observed in primates, suggests the need for a different model for stem cell renewal. Clermont and Hermo (1976), drawing a parallel between primates and rodents, applied the A_0/A_1 model to primates, with the A_d in the role of the A_0 (reserve stem cells) and the A_p as the A_1 through A_4 (renewing stem cells). A scheme for stem cell renewal in the monkey *Macaca arctoides* (Clermont and Antar, 1973) is outlined in Figure 12–6, based on this analogy but incorporating some features of the A_s model.

The absence of any proliferation among type A_d spermatogonia is consistent with their role as reserve stem cells. A mechanism, however, is needed to explain the observation that type A_d spermatogonia are generally found in clusters (Clermont, 1969). De Rooij (1983) proposed

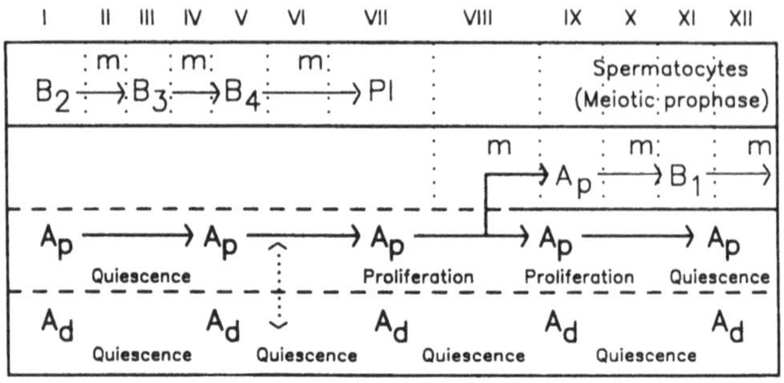

Figure 12–6. Model for stem cell renewal in the primate, *Macaca arctoides* (Clermont and Antar, 1973), showing features in common with the A_0/A_1 and the A_s models of stem cell renewal. The stem cells and their pathways are displayed in *bolder lines* for emphasis. The A_d spermatogonia appear to be noncycling, as proposed for the A_0 cells. The A_p spermatogonia encompass the class of undifferentiated A spermatogonia consisting of A_s, A_{pr}, and A_{al}. Some of the daughter A_p spermatogonia at stage IX as well as the B_1, B_2, B_3, and B_4 spermatogonia, correspond to the A_1 through B spermatogonia of the rat.

that, as the rate of proliferation of A spermatogonia in primates is lower than in rodents, some of them can enter an extended G_0 state, during which the nucleus becomes uniformly heterochromatic. Thus, the chain size and delayed labeling of the A_d spermatogonia could be a result of a transformation of A_p into A_d at low frequency. Conversely, after a cytotoxic injury, the A_d may transform into A_p and then proliferate (van Alphen et al., 1988a).

Type A_p spermatogonia are the only A spermatogonia that incorporate [^{3}H]thymidine and as such must serve as the renewing population of spermatogonia and the precursors of B spermatogonia. This is most apparent in several species of monkey in which a mitotic division resulting in an increase in the number of A_p spermatogonia (self-renewing division) is followed by another mitotis (differentiating division) that coincides with a decrease in A_p and the appearance of type B spermatogonia. This observation led to the proposal that at the second division, half of the A_p spermatogonia divide into B spermatogonia, whereas the other half remain as A_p spermatogonia and until the next cycle of the seminiferous epithelium during which they may again self-renew or differentiate.

The relationship of the self-renewal process to the topological arrangement of type A_p spermatogonia, which appears to be clonal with numbers of cells per chain generally given by 2^n, must be addressed. Assuming that clone fragmentation does not occur, the numbers of A_p spermatogonia per chain should double at each cycle. If, as in nonprimates, where the probability of A_{al} transforming into type A_1 spermatogonia increases with chain length, the probability of A_p differentiating into type B spermatogonia also increases with chain length, then the numbers of cells per chain will stabilize at a steady-state value.

An analogy may be made between type A_p spermatogonia and the A spermatogonia of rodents (de Rooij, 1983). In the example shown (Fig. 12–6), type A_p spermatogonia in stage IX that will produce B spermatogonia are analogous to the A_2 of rodents and the A_p that are precursors of these cells may be considered to be like the A_1. The rest of the A_p cells are like the A_s, A_{pr}, and A_{al}, except that they have a lower proliferative activity and the expansion of the undifferentiated A spermatogonia that occurs in each cycle in rodents does not occur in primates. Type B_1 through B_4 spermatogonia of monkeys seem to be analogous to the A_3 through B spermatogonia of rodents. Thus, some of the principles of the A_s model might also apply to type A_p spermatogonia.

It appears that in primates there is a dual stem cell compartment as in the A_0/A_1 model, but that the A_d and A_p cells can transform into each other rather than being distinct cell lineages. The self-renewal of the A_p spermatogonia appears to be analogous to that in the A_s model. Because most of the type A_p spermatogonia appear to be in clones of two, four, and eight, but isolated A_p cells may also exist, the stem cells could be isolated A_p spermatogonia. Preferential differentiation of the larger clones of A_p cells could be the mechanism for maintaining the clone size of these cells without clone fragmentation, although the latter cannot be completely ruled out at present.

V. Stem Cell Renewal After Cytotoxic Treatments

When the seminiferous epithelium is depleted by cytotoxic agents, restoration of spermatogenesis must occur from stem cells. Studies of this process can yield further insight into the identity of stem cells and feedback mechanisms regulating their proliferation and self-renewal.

The progeny of individual stem cells can best be observed in the depleted epithelium when lethally injured spermatogonia have been lost. All the progeny of a surviving stem cell constitute a single *colony*. Usually that stem cell will have undergone one or more self-renewing divisions to produce separate daughter A_s cells; each of these daughter stem cells, or its differentiating progeny connected by intercellular bridges, will be considered as a separate *clone* within a colony.

A. Assays for Stem Cell Survival

Measurements of stem cells after cytotoxic treatments have been done by both morphological and functional assays. Morphological assay of stem cell survival involves counting these cells, which is laborious and does not consider whether they are fully functional, or even reproductively viable. Hence, functional assays have an important role in evaluating stem cells after cytotoxic treatment and should be correlated with the identification of the surviving stem cells. As yet, functional assays of spermatogenic stem cells can only be done endogenously; both clonogenic and nonclonogenic assays (see section II.D.) can be used.

The simplest clonogenic assay for measuring stem cell survival is based on counts of repopulating and nonrepopulating seminiferous tubule cross sections at a given time after treatment with a high dose of a cytotoxic agent (Withers et al., 1974; Lu and Meistrich, 1979). Such observations are usually made when cells in colonies arising from surviving stem cells should have differentiated into midspermatids. Nonrepopulating tubules, containing only Sertoli cells, indicate death of all stem cells in sufficient proximity to the plane of section to produce repopulation by spermatogenic cells within the chosen period. Conversely, each repopulating cross section demonstrates the presence of at least one surviving stem cell close enough to the plane to produce differentiating cells in that section. In mice, the fraction of repopulated tubular cross sections is proportional to the number of colonies produced by surviving functional stem cells (de Ruiter-Bootsma et al., 1976). Alternatively, colonies can be quantified in tubular whole mounts. This can be done at shorter time intervals after irradiation, and allows a study of the effects of lower doses (van Beek et al., 1986a). A clonogenic assay appears useful for assessing spermatogonial survival in the Rhesus monkey (van Alphen et al., 1989a); however, it has not been as well validated as in rodents.

Nonclonogenic assays include counts of differentiated spermatogenic cells in histologic sections (van Alphen et al., 1989b), enzyme markers specific for differentiated germ cells, counts of late spermatid nuclei (sperm heads) in testicular homogenates (Meistrich et al., 1978), or determination of the time required to recover fertility (Cattanach, 1974). These assays are performed when the differentiated cells are the progeny of surviving stem cells. The validity of the nonclonogenic assays involving sperm head counts and duration of sterility is demonstrated in the mouse by the existence of relationships between them and the clonogenic assay, which hold irrespective of the cytotoxic agent used to kill stem cells (Meistrich, 1986).

B. Effects of Cytotoxic Treatments

Radiation has been the cytotoxic agent most widely used to investigate the behavior of stem spermatogonia. Direct observations on undifferentiated spermatogonia are available only for radiation and the cancer chemotherapeutic drugs, myleran (van Keulen and de Rooij, 1975) and cytosine arabinoside (de Rooij et al., 1985).

Because radiation and cancer chemotherapeutic drugs preferentially kill rapidly proliferating cells and have less toxicity toward nondividing cells, in self-renewing tissues they kill the cells in the amplifying compartment. In the testes these are the differentiating spermatogonia (Meistrich, 1986). Spermatogonial stem cells can also be killed by radiation and some chemotherapy agents, but they are not as sensitive as differentiating spermatogonia (Meistrich et al., 1982).

Cells lethally injured by these agents usually degenerate at division. Therefore, the rates of disappearance of such cells are related to their cell cycle times. The differentiating and the rapidly proliferating undifferentiated spermatogonia (mostly A_{al}) have cell cycles, in the rat, of about 2 days (Table 12–2). The differentiating spermatogonia do indeed disappear within a few days after cytotoxic treatment both in rodents and in primates. Type A_s spermatogonia, however, may divide at most a few times during a cycle of the seminiferous epithelium (Huckins, 1971c) and might not be lost for several weeks after the cytotoxic insult. In primates, because the proliferation rate of type A spermatogonia is lower than in rodents, this process may take even longer (Rowley et al., 1974; van Alphen et al., 1988a). Like the A_s cells of rodents, the disappearance of type A_p spermatogonia, which also divide only once or twice during a cycle, is gradual and requires several weeks. Type A_d spermatogonia, which normally show negligible proliferation, do not even begin to decline until 11 days after irradiation in the Rhesus monkey. This decline is associated with a transient rise in the number of A_p spermatogonia, suggesting that A_d are transforming into A_p. Two months or more are required for the A_d to reach their minimal values.

Spermatocytes and spermatids are generally not sensitive to these cytotoxic agents. After treatment they continue their differentiation process with normal kinetics, and are lost from the testes by maturation depletion because their numbers are not replenished in the absence of differentiating spermatogonia (Dym and Clermont, 1970; de Rooij et al., 1986).

C. Characterization and Surviving Stem Cells

Morphological studies have shown that after irradiation of rodent testes, the surviving spermatogonia, which appear to be responsible for recovery of spermatogenesis, are undifferentiated type A spermatogonia, mostly A_s and a few A_{pr} (Dym and Clermont, 1970). Spermatogenesis also recovers from type A_s spermatogonia after chemotherapeutic drug treatment (van Keulen and de Rooij, 1975; Meistrich et al., 1982). These results appear consistent with both the A_0/A_1 model and the A_s model. The ability to label these cells with [^{3}H]thymidine before irradiation (Huckins, 1978b; Huckins and Oakberg, 1978), however, is more consistent with the A_s model. Essentially all of the A_s cells that survive irradiation proliferate and give rise to additional stem and differentiating cells (van Beek et al., 1986a).

The clone fragmentation model, however, suggested that type A_{al} spermatogonia might also self-renew. This was based on observations during recovery from radiation, which revealed a greater number of mitoses in clones than in isolated or paired cells (Erickson, 1981) and increases in A_1 spermatogonia at times when the A_s cells were still decreasing (Erickson and Hall, 1983). Although these results were interpreted as evidence that type A_{al} spermatogonia were a self-renewing population, they could be explained in ways consistent with the A_s model. The increases in type A_1 spermatogonia and the high numbers of mitoses of A_{al}, despite decreases in numbers of A_s, could be a result of increasing the rate of division of the A_s cells (see Table 12–1) and increasing the number of amplifying divisions that the A_{al} spermatogonia undergo (see section V.D.1.). Indeed, the mitotic indices of the A_s do increase, resulting in a maintenance of the numbers of A_s mitoses despite a marked reduction in numbers of A_s cells (Erickson and Hall, 1983). In addition the number of A_s cells after irradiation may have been underestimated because radiation appears to inhibit the separation of undifferentiated A spermatogonia (van Beek et al., 1984).

In primates, the only germ cells surviving moderately high doses of radiation are some of the type A_d and A_p spermatogonia (Rowley et al., 1974; van Alphen et al., 1988a); hence, these must be the stem cells responsible for the recovery of spermatogenesis. Although the A_d are reserve stem cells, they must first transform into A_p before proliferating or differentiating. Because both A_d and A_p cells are present at all times after irradiation, it is not possible to determine the contribution of cells that were A_d versus A_p at the time of irradiation to the surviving stem cells that repopulate the epithelium.

Subpopulations of stem cells with differing radiosensitivities have been reported in the mouse using nonclonogenic functional assays (Cattanach, 1974; Lu et al., 1980). Causes for this variation have been investigated to identify whether different types of stem cells exist.

It has been suggested that the radioresistant stem cells are a long cycling subpopulation of type A_s spermatogonia (Huckins, 1978b). This was based on studies in the rat using [^{3}H]thymidine labeling 2 to 4 days before irradiation and the observation of numerous labeled A spermatogonia (largely A_s) at 7 days after irradiation. An alternate interpretation, however, that these are labeled by reutilization of label from dying differentiated spermatogonia, should be considered. Furthermore, although labeled type A_s spermatogonia did survive irradiation, it was not demonstrated that they *preferentially* survived.

Other causes for the variation in sensitivity have been described but these appear to be cyclical and not indicative of distinct subpopulations. A clonogenic assay revealed that in the mouse the radiation sensitivity of spermatogonial stem cells varied with the stage of the cycle of the seminiferous epithelium (van Beek et al., 1986a,b); the cells were most sensitive at stage VII and relatively resistant at stages IX–III (Table 12–3). The most radiosensitive stage is one of those at which the proliferation of the A_s cells is minimal and the radioresistant stages are those at which the proliferation is increasing and at a maximum. Studies using hydroxyurea, an S-phase specific cytotoxic drug, either shortly before or after irradiation (de Ruiter-Bootsma and Davids, 1981) or using [^{3}H]thymidine labeling at different times before irradiation (Oakberg, 1978), indicated that the phase of the cell cycle also affects radiosensitivity. Nearly all of the surviving stem cells were not in S-phase at the moment of irradiation; however, there were indications that the surviving stem cells had been in S-phase within 1 or 2 days before irradiation. Thus, it appears that stem cells are most resistant during the stages in which they are proliferatively active but they are sensitive during the S-phase of these cell cycles.

Table 12–3. Radiation Sensitivity and Proliferation of Rodent Stem Spermatogonia in Different Stages of the Cycle of the Seminiferous Epithelium

Stage of cycle	Stem cell survival after 100 cGy of neutron irradiation[a]	Labeling index of A_s spermatogonia[b]
XII–III	16%	+++
IV–VI	4%	++
VII	1%	+
IX–XI	27%	++

[a]Calculated from data in mouse (van Beek et al., 1986b).
[b]Estimated from data in Chinese hamster (Lok and de Rooij, 1983) and rat (Huckins, 1971b).

The clone fragmentation model suggests another source of heterogeneity of radiosensitivity among the spermatogonial stem cell population, the size of the clone. If this model is valid, the A_{al} and A_s cells could represent radiosensitive and radioresistant stem cell populations, respectively.

D. Stem Cell Renewal During Recovery

When stem cells are killed, complete recovery of spermatogenesis requires regeneration of stem cell numbers, in addition to the production of differentiated cells. The regulation of these processes, at least in rodents, indicates that feedback signals are operative on the stem cells. As indicated in Table 12–1, *P* must increase to regenerate stem cell numbers. Increases in the rate of stem cell division will also enhance the rates of both stem cell regeneration and the production of differentiated cells. These processes occur to different degrees and with different kinetics in rodents and primates.

1. RODENTS

Studies in the mouse using [^{3}H]thymidine injection 2 to 3.5 days after irradiation (Huckins and Oakberg, 1978) indicated that the surviving A_s cells became highly proliferative in all stages of the cycle. The sizes of colonies produced by these survivors demonstrated that spermatogonial stem cells undergo five cell divisions, with a cell cycle duration of 30 to 38 hours, during the first cycle of the seminiferous epithelium after irradiation (van Beek et al., 1990). Hydroxyurea treatments after irradiation showed that 90 percent of surviving stem cells enter S-phase within 72 hours after irradiation and the cells proliferate with a cell cycle time of 36 hours (Bootsma and Davids, 1988).

Experiments in the rat, however, yielded results with a different interpretation. When [^{3}H]thymidine was given before irradiation, labeled type A_s spermatogonia were observed up to 11 days after irradiation (Huckins, 1978b). This indicated that the cells surviving irradiation remained in a long cell cycle, as proliferation would have diluted the label. Although there may be interspecies differences, the problem of possible reutilization of label from dying spermatogonia should also be considered before accepting the above interpretation.

Analysis of the clonal arrangement of undifferentiated A spermatogonia within colonies in the irradiated epithelium provides information regarding the probabilities that stem cell divisions are self-renewing or differentiating (van Beek et al., 1990). The first division after cytotoxic treatment of mice is almost always a self-renewing one ($P = 1$). P declines markedly at the next division but remains at 0.7 or higher as late as the fifth postirradiation division, about 8 days later. Any value of P below 1.0 indicates that differentiating cells are being produced, and A_{pr} do appear at 3 days and A_{al} at 5 days after irradiation (Huckins and Oakberg, 1978).

The proliferative behavior of type A_{al} spermatogonia after cytotoxic treatments is also different from that in the control. Normally, the proliferation of these cells ceases by stage V in rodents. After local depletion of differentiating spermatogonia with the S-phase specific cytotoxic agent cytosine arabinoside, proliferation of A_{al} continues through stage VII, confined to the regions of tubules depleted of the differentiating spermatogonia (de Rooij et al., 1985). Similar results were also observed after myleran treatment, which kills all of the differentiating A spermatogonia (van Keulen and de Rooij, 1973). Despite the additional division, the A_{al} differentiate into new A_1 spermatogonia in stage IX.

A colony produced from a single stem cell can eventually encompass a long length of a seminiferous tubule. Colonies grow linearly at an average rate of 30 μm per day (van den Aardweg et al., 1983), until they reach a plateau (Meistrich et al., 1978) because of confluence of colonies or other causes. Because colony growth is a result of continued production of additional stem cells, the slow growth rate implies that, on the average, P is not much above 0.5. The linearity of growth is consistent with a model in which P returns to 0.5 in the interior of a colony, where recovery has occurred locally, but that P remains at some higher value at the edges.

2. PRIMATES

In contrast to rodents, in which recovery of the numbers of type A spermatogonia begins about one cycle of the seminiferous epithelium after cytotoxic treatment, in the Rhesus monkey the numbers of type A spermatogonia (both A_d and A_p) do not increase until after 44 days (van Alphen et al., 1988a,b), corresponding to about four cycles of the seminiferous epithelium. In humans, recovery of type A spermatogonial numbers does not occur until about 7 months (12 cycles) after irradiation (Paulsen, 1973; Rowley et al., 1974; Clifton and Bremner, 1983; Meistrich and van Beek, 1990).

During the first few months after irradiation, production of type B spermatogonia continues, indicating that some A_p cells must undergo differentiating divisions ($P < 1$). The relative reduction in type B spermatogonia and primary spermatocytes, however, is greater than that of the A_p spermatogonia. This inefficient production of differentiating cells along with the continued decrease in type A spermatogonia could be a result of several factors particular to primates. The A_d spermatogonia, which are in a long G_0 phase, likely survive until they are recruited to proliferate as A_p and die at their first mitosis or at subsequent mitoses as type B spermatogonia. It is also probable that in primates the processes that stimulate proliferation of the A_s and A_{al} in rodents on depletion of the differentiated cells do not occur and affect the A_p cells in the same time frame.

The recovery of spermatogonial numbers in the Rhesus monkey begins after 44 days postirradiation (van Alphen et al., 1988b). Thereafter the numbers of type A_d, A_p, and B spermatogonia increase in parallel. A_d spermatogonia are produced, presumably by arrest of some A_p spermatogonia in a G_0 state, when the density of spermatogonia in the epithelium is still quite low, indicating that cell density does not cause this transformation (van Alphen et al., 1988b). The increased production of type B and A_d spermatogonia, as well as increases in numbers of A_p, after 44 days indicates that, at 44 days, the A_p cells either increase their rate of cell division or decrease their degeneration and that P averages between 0.5 and 1.0.

Although repopulating colonies in rodents contain cells of all four generations, many of those in the Rhesus monkey are missing one or two generations (van Alphen et al., 1988b). Thus, it appears that during recovery from radiation damage in primates, in many repopulating colonies none of the stem spermatogonia undergo a differentiating division in some cycles of the seminiferous epithelium.

Some of the differences between rodents and primates in the behavior of stem spermatogonia after administration of cytotoxic drugs may be related to differences in the modes of stem cell renewal in the unperturbed testis. In addition, there might be species differences in how the testis responds to the insult by production of cell growth and differentiation factors, and in the response of spermatogonial stem cells to such factors.

VI. Assay Systems for Factors Affecting Stem Cells

To understand the phenomena described in the previous sections, the growth and differentiating factors that regulate stem spermatogonia must be identified. The problems associated with developing useful and convenient assay systems for such investigations are summarized here.

A. In vivo

In vivo assays of effects on stem cells should most nearly reflect the normal situation; however, such assays present numerous difficulties. The route of administration is one consideration. Potential growth and differentiation factors may be administered systemically (Clermont and Mauger, 1974), but large quantities would be required, the factors could be metabolized in other tissues, toxicity may be limiting, and the distribution might not mimic that of autocrine or paracrine factors. Alternatively, substances could be injected intratesticularly (van Dissel-Emiliani et al., 1989a), but their distribution within the testes may be uneven, complicating interpretation of any effects.

Furthermore, when adult animals with normal testes are used, the possibility that homeostatic mechanisms could counteract the effect of external factors must be considered. Alternatives are to use testes of immature or x-irradiated rats or mice (Clermont and Mauger, 1974), in which the number of stem cells is increasing ($P > 0.5$) and differentiating cells are being produced ($P < 1$). Hence, the effects of both stimulatory and inhibitory factors, for both proliferation and differentiation, may be studied. Furthermore, at the appropriate time after irradiation, many of the differentiating type A spermatogonia are eliminated, simplifying assessment of stem cells.

Evaluation of effects on stem cell proliferation requires careful cell counts in histological sections or, preferably, in whole mounts. Although cell counts alone are a measure of what the cells have done, autoradiographic analysis would more directly determine the entry of cells into S-phase and their progression. Although [^{3}H]thymidine incorporation into irradiated testes has been proposed as a more convenient end point than autoradiography of testicular sections (Thumann and Bustos-Obregon, 1978), if such studies are used to assess effects on stem cells, the only proliferating cells must be the undifferentiated type A spermatogonia (Cunningham and Huckins, 1979).

B. Tissue or Tubule Culture

An in vitro system for testing growth and differentiation factors would have the advantages of maintaining the factors in steady state, as well as obtaining higher concentrations and requiring smaller amounts, and of eliminating other molecules with competing or like activities through the use of defined media. Although spermatocytes and spermatids gradually disappear over 4 weeks in tissue culture, some spermatogonia, described as primitive type A cells, remain (Steinberger and Steinberger, 1966). This result indicates that the maintenance of production of differentiated cells requires the action of additional factors not in the medium.

Most studies have been done with shorter term cultures, in which some more differentiated elements remain. Such studies require discrimination between new production of differentiated cells and the maintenance of already existing cells. Methods for overcoming this problem include the use of 4-day-old rat testes (Steinberger and Steinberger, 1966), testes containing only type A spermatogonia from cryptorchid mice (Haneji et al., 1986), or segments of seminiferous tubules in defined stages of the spermatogenic cycle (Toppari and Parvinen, 1985).

Culture of cryptorchid mouse testes has proven useful for assay of factors involved in spermatogonial differentiation. Over a 9-day period in culture, differentiation of some type A spermatogonia into In and B spermatogonia occurs. The numbers of these differentiated cells and the proportion that incorporate [^{3}H]thymidine are dependent on addition of growth and differentiation factors to the medium (Nishimune et al., 1981; Haneji et al., 1984, 1986). In addition, the numbers of mitotic cells and remaining type A cells could also be determined. Discrimination between stem and differentiating type A spermatogonia, however, was not possible because the cells were counted in cross sections and, in the absence of spermatids, the stages of the cycle could not be determined. To overcome this difficulty, whole-mounted seminiferous tubules must be used and the topology of the type A spermatogonia examined.

Culturing isolated tubules could provide good material for this type of analysis, especially as they are ready for whole-mount preparations. The stages of the cycle of the seminiferous epithelium could be identified, first at the time of isolation by transillumination (Toppari and Parvinen, 1985) or by using animals with the seminiferous epithelium synchronized in only a few stages (Morales and Griswold, 1987), and again at the time of fixation. This would establish the spermatogonial types that should be present and their expected behavior between these two stages.

C. Cell Culture

In vitro culture of isolated spermatogonial stem cells would provide for direct assessment of the effects of growth and differentiation factors on these cells. As of yet, however, it has not been possible to isolate appreciably enriched populations of these cells from adult testes. Although type A spermatogonia can be separated from immature rodent testes (Bellve et al., 1977; Bucci et al., 1986), differentiating spermatogonia may still be mixed with the stem cells. Alternately, the use of fetal testes, which are enriched with the precursors of the spermatogonial stem cells (see Chapter 3), has been considered in the hope that these cells share some of the properties of spermatogonial stem cells. But no success with obtaining differentiation of primordial germ cells or with culturing type A spermatogonia has been reported, although culture of neonatal gonocytes is now possible (Orth and McGuinness, 1991).

VII. Regulators of Spermatogonial Proliferation and Differentiation

Regulation of stem spermatogonia may occur by modulation of any of the following parameters: their rate of proliferation, the probabilities that a cell division will be self-renewing or differentiating, and the probability that degeneration will occur. The committed but undifferentiated type A spermatogonia can be affected by changes in their production and degeneration rates, the numbers of divisions, and the probabilities that they will degenerate or transform into differentiating spermatogonia. The numbers of differentiating spermatogonia appear to be only affected by their production and degeneration rates, because their cell cycle times and differentiation pathway appear to be invariant in vivo. Thus, regulators of spermatogonial proliferation and differentiation can act at various specific points. Given the limitations of the assay systems, at present, we can only review observations of the effects of various agents on spermatogonial numbers.

A. Regulation by Cellular Interactions

The only cells in direct contact with the spermatogonia are the Sertoli cells. Hence, the effects of any other germ cells on the type A spermatogonia must occur through the Sertoli cells. Spermatogonia are also in contact with the basement membrane of the seminiferous tubules, and can, therefore, respond to paracrine factors secreted by myoid or Leydig cells.

1. DIFFERENTIATING GERM CELLS

Several lines of evidence show that the proliferation of undifferentiated spermatogonia can depend on the numbers of other spermatogonia. Elimination of differentiating spermatogonia in specific stages resulted in an additional division of type A_{al} spermatogonia in these stages (de Rooij et al., 1985). Within 3 days after irradiation in the mouse (Bootsma and Davids, 1988) and within 13 days in the rat (Huckins, 1978b), the proliferative activity of the A_s spermatogonia is enhanced. This result is most likely secondary to depletion of differentiating and A_{al} spermatogonia, rather than a direct effect of the cytotoxic agent. In addition, increased radiation sensitivity of stem cells is evident when the testis is exposed to a second dose of radiation 1 day after an initial dose (Meistrich et al., 1984); the 1-day interval corresponds to the time required for the degeneration of differentiating spermatogonia (Oakberg, 1959). This alteration in sensitivity may be related to the recruitment of the surviving stem cells into an active state of proliferation (de Ruiter-Bootsma and Davids, 1981) by changes in regulatory signals caused by the absence of the differentiating spermatogonia.

An interesting observation may be made by comparing the types of differentiating spermatogonia that are present and the proliferation of the chains of type A spermatogonia in different species. The stages of proliferation both of type A_{pr} and A_{al} spermatogonia in rodents (stages XI–III, Fig. 12–5) and of A_p in primates (stages VII–IX, Fig. 12–6) correspond to those in which type B spermatogonia are absent. This suggests another possible way in which specific stages of differentiating spermatogonia can affect the proliferation of undifferentiated spermatogonia and account for interspecies differences.

Although examples of negative feedback on proliferation of undifferentiated spermatogonia by the differentiating spermatogonia exist, it appears that the undifferentiated spermatogonia do not exert such an effect on each other. A study of the topographical arrangement of type A_s, A_{pr}, and A_{al} spermatogonia showed that clones of like size tended to cluster (de Rooij and Janssen, 1987), which is the opposite of what one would expect if the density of these cells exerted a negative feedback on the proliferation of others.

2. SERTOLI CELLS

Sertoli cells have large areas of contact with the spermatogonia. Although desmosome–gap junction complexes occur between Sertoli and various germ cells, they rarely appear to involve the type A spermatogonia (Russell, 1980). Hence, there is no evidence yet that direct cell–cell contacts have a role in regulation of stem cells.

It is more likely that Sertoli cells regulate spermatogonial proliferation by the secretion of paracrine factors. Sertoli cells secrete numerous molecules, including growth factors (see Chapter 9), and therefore, are excellent candidates for involvement in regulation of spermatogonial proliferation. The variation in the levels of some Sertoli cell secretory products with the cycle of the seminiferous epithelium (Bardin et al., 1988) could provide a mechanism to trigger stage-specific events in the proliferation and differentiation of spermatogonia. Because the Sertoli cells could respond to changes in the numbers and types of other germ cells or to external signals, they could have a role in the changes in spermatogonial behavior in response to cytotoxic or hormonal agents.

B. Regulation by Humoral Factors

The spermatogonia are located on the basal side of the Sertoli barrier and as such are exposed to paracrine factors secreted by Leydig, myoid, and other cells, as well as to systemic factors including hormones. Although spermatogenesis as a whole is regulated by steroid and gonadotropic hormones, the effects described are largely at postspermatogonial stages.

1. STEROID HORMONES

Testosterone is necessary for the maintenance of spermatogenesis, acting primarily on the Sertoli cell to effect efficient differentiation of spermatocytes and spermatids. Suppression of gonadotropins and intratesticular testosterone levels with silastic implants of testosterone (Huang and Nieschlag, 1986) did not significantly change the numbers of undifferentiated type A or A_1 spermatogonia, despite dramatic suppression of numbers of spermatids. Hypophysectomy did, however, result in quantitative changes in type A spermatogonial numbers. Four weeks after hypophysectomy there was a 40 percent reduction in the number of undifferentiated spermatogonia (largely A_{al}) in stages II–VI (Chowdhury, 1979), which was prevented by maintenance on testosterone. Twenty-one weeks after hypophysectomy there was a 60 percent reduction in type A_1 spermatogonia, which was in part reversed by testosterone replacement beginning at 8 weeks (Huang et al., 1987). Thus, in the rat, testosterone (and the gonadotropins) does not appear to be absolutely required for type A spermatogonial proliferation or differentiation. There are quantitative effects of testosterone on type A spermatogonia; however, the results could be explained by effects on the A_{al} rather than on the stem cells.

2. GONADOTROPINS

In rodents, follicle-stimulating hormone (FSH) does not appear to have any effect on the number of type A spermatogonia in vivo (Chemes et al., 1979). In vitro, however, FSH stimulates mitoses of mouse spermatogonia and the production of more type A, as well as In and B, spermatogonia (Haneji et al., 1984). Neither luteinizing hormone (LH) nor testosterone showed any stimulation (Haneji and Nishimune, 1982).

In monkeys, supraphysiological doses of FSH increase the numbers of type A_p, but not A_d, spermatogonia (van Alphen et al., 1988c). As the increase in the number of A_p spermatogonia was followed by a similar increase in the number of type B spermatogonia, it seems that proliferation of the A_p must have been increased (Table 12–1). It is not clear whether FSH acts directly on the spermatogonia or indirectly by the Sertoli cells, which are responsive to this hormone.

3. GROWTH FACTORS

To identify possible molecules that stimulate spermatogonial proliferation, the presence of growth factors in mammalian testes has been studied (reviewed in Bardin et al., 1988; Bellve and Zheng, 1989). Molecules related to growth factors, including transforming and insulin-like growth factors, have also been identified in Sertoli cells by cross reaction with antibodies and by Northern blotting with nucleic acid probes. Several proteins found in Sertoli cells, directly separated from testes or maintained in culture, or in Sertoli cell-conditioned medium have been identified as growth factors by their ability to stimulate growth of cultured cell lines or thymocytes, in the case of interleukin 1. Although these proteins are candidates for regulation of spermatogonial proliferation, they have been generally assayed with cultured cells, and not spermatogonia, as the target. In one study (van Dissel-Emiliani et al., 1989a), however, an increase in numbers of undifferentiated (probably A_{al}) spermatogonia was observed after intratesticular injection of a fraction from Sertoli cell-conditioned medium. Another study reported an increase in numbers of labeled cells (presumably spermatogonia) in the seminiferous epithelium of hypophysectomized rats after intratesticular injection of interleukin 1α (Pollanen et al., 1989). The need for future research in this area is apparent.

4. CHALONES

Before the identification of growth factors, studies of regulation of proliferation in renewing tissues focused on *chalones*. These were proposed to be tissue-specific inhibitors of proliferation of stem or other undifferentiated cells that were produced by (or in the presence of) the differentiated cells. The increases in proliferation of undifferentiated spermatogonia after cytotoxic damage (section VII.A.1.) are consistent with the release from a chalonelike feedback mechanism; however, if we assume paracrine action, there is no need for tissue specificity.

Clermont and Mauger (1974), using rats irradiated 11 days previously to deplete most of the differentiating spermatogonia, first demonstrated a chalone activity in the testes. When these rats were injected with a testicular extract, decreased labeling indices of type A spermatogonia, but not of In or B spermatogonia, were observed. Cunningham and Huckins (1979), however, followed similar protocols but found no effect. Other studies have also indicated the presence of chalone activities in testicular extracts; injection of such extracts reduced mitotic indices of type A spermatogonia in testes of irradiated rats (Irons and Clermont, 1979) and the numbers of A_{al} spermatogonia produced in normal mouse testes (de Rooij, 1980). We conclude that there appear to be inhibitors of type A spermatogonial proliferation in testicular extracts, but that none have been purified or identified.

Assays that are more convenient than quantitative counts of cells in autoradiograms after in vivo administration of fractionated extracts, appear to be required to monitor the purification of active factors in these crude preparations. Measurement of the amount of [^{3}H]thymidine incorporation after in vitro incubation of seminiferous tubules from irradiated rats with [^{3}H]thymidine and the extracts to be tested offers a simpler assay (Thumann and Bustos-Obregon, 1978). The possibility of incorporation of more label into surviving preleptotene spermatocytes (Cunningham and Huckins, 1979) or peritubular cells (Nishimune et al., 1981) than into type A spermatogonia, however, presents a serious problem.

5. VITAMIN A

Vitamin A (retinol) is essential for the maintenance of spermatogenesis, as well as for other epithelial cell functions and for growth. Retinoic acid relieves nearly all of the systemic effects of retinol deficiency but does not restore spermatogenesis (Coward et al., 1969). In vitamin A deficiency with no retinoic acid supplementation, the undifferentiated spermatogonia continue to proliferate; however, no differentiating A_1 spermatogonia are formed (de Rooij et al., 1989). When such deficient rats are given retinoic acid, differentiating spermatogonia and preleptotene spermatocytes, but no later stages of germinal cells, are found (Huang and Hembree, 1979; van Beek and Meistrich, 1992). These results suggest that retinoids (either retinoic acid or retinol) are required for transformation of type A_{al} into A_1 spermatogonia; however, the development of the young spermatocytes is dependent specifically on retinol.

Retinoids also promote the differentiation of type A spermatogonia in vitro. Increasing concentrations of retinol not only increase the numbers of type In and B spermatogonia and spermatogonial mitoses, but also decrease the numbers of type A spermatogonia; retinoic acid has similar but slightly weaker effects (Haneji et al., 1984, 1986). Thus it appears that maintenance and even division of some type A spermatogonia are possible in the absence of retinoids, but that differentiation requires retinoids. These results are consistent with the in vivo findings that retinoic acid can substitute for retinol to initiate further differentiation of type A_{al} spermatogonia.

The role of retinoids in regulation of spermatogenesis has further been demonstrated by observations that, after injection of a high dose of vitamin A into retinol-deficient animals, spermatogenesis is restored in a synchronous manner (Morales and Griswold, 1987). Only a few stages of the cycle of the seminiferous epithelium are present at any given time after the injection of retinol. It was proposed that synchronization of the epithelium occurred because A_{al} cells were blocked in a state just before their transformation into A_1 spermatogonia by retinol deficiency and the injection of retinol induced their coordinated release (de Rooij et al., 1989). This explanation has been questioned, based on the observation that synchronization also occurs in testes of rats maintained on retinoic acid, where this block is not present (van Beek and Meistrich, 1990).

6. INHIBIN

The homology between inhibin, which is synthesized by the Sertoli cell, and transforming growth factor-β suggests that it may function as a growth regulatory factor (see Chapter 10). Intratesticular injections of inhibin reduce the numbers of differentiating spermatogonia, apparently by degeneration, in the testes of mice and Chinese hamsters (van Dissel-Emiliani et al., 1989a). No effect of pure inhibin on the numbers of undifferentiated spermatogonia could be found.

VIII. Approaches to Molecular Characterization of Stem Cells

To understand how the various factors might regulate stem cell proliferation and differentiation, the molecular events occurring in these cells must be elucidated. Next, we discuss approaches that might be pursued and the problems that need to be overcome.

A. Purification of Spermatogonia

The characterization of spermatogonial stem cells would be greatly facilitated if these cells could be enriched and obtained in reasonable quantities. Because of the low numbers of spermatogonial stem cells in the adult, estimated to be 2×10^4 of 10^8 total cells in a mouse testis, it is unlikely that their purification can be achieved by sedimentation and density separation techniques. Even from 20-day-old rats, type A spermatogonial populations were enriched to only 50 percent, and likely were mostly differentiating type A cells (Bucci et al., 1986). Purification of stem spermatogonia will likely require development of specific antibodies and separation by fluorescence-activated cell sorting on a flow cytometer.

One way to obtain probes that might be specific for the stem spermatogonia of the adult testes is to separate their precursors and make probes for these cells. Some characteristics of the stem spermatogonia could be shared by their precursor cells, known as prospermatogonia, which include primordial germ cells, gonocytes, and T_2 prospermatogonia of neonatal testes (see Chapter 3). At 2 to 5 days after birth, these T_2 prospermatogonia have divided into cells that look like type A spermatogonia. Although Bellve and co-workers (1977) designated the type A spermatogonia in the 6-day-old mouse as primitive type A cells, Sutcliffe and Burgoyne (1989) indicate that at this time the population consists of both undifferentiated and differentiating (A_1 through A_4) spermatogonia.

Methods for purification of some of these prospermatogonia have been reported. Isolation of primordial germ cells is discussed elsewhere (see Chapter 3). Velocity sedimentation has been employed to separate gonocytes from 18-day-old fetal rats (van Dissel-Emiliani et al., 1989b), and type A spermatogonia from 6-day-old mice (Bellve et al., 1977). The latter population would more closely resemble the undifferentiated spermatogonia of the adult, but more differentiated type A cells may be included in this preparation.

B. Biochemical Characterization of Stem Spermatogonia

1. STATE OF THE GENOME

Cell differentiation is not only accompanied by changes in gene expression but may also involve changes in the genome itself. For example, the only truly totipotent stem cell is the embryonic stem cell. As differentiation proceeds the ability of the cell nucleus to substitute for that of the zygote decreases. Spermatogonial stem cells must be nearly totipotent (or capable of being reprogrammed to such a state) as they give rise to sperm that form the male pronucleus after fertilization. Although the male pronucleus is not completely totipotent because of imprinting, it only requires the presence of a female pronucleus to sustain normal development. It is not known whether imprinting of the male germ cell genome occurs before or after the stage of stem spermatogonia. The genomic state of the spermatogonial stem cell may also differ from that of the differentiating spermatogenic cells because the latter must activate genes required for sperm differentiation, whereas other genes involved in maintenance of the spermatogonial stem cell state may be turned off during differentiation by modifications of the DNA, such as methylation.

Because it is unlikely that stem spermatogonia can be isolated from the adult animal in sufficient purity and quantity for biochemical analyses, alternative approaches will be required for the study of their genomes. These would include in situ methods or other methods using DNA from isolation of only a few cells. Polymerase chain reaction (PCR) techniques allow the analysis of alterations in DNA sequence and methylation with extremely small amounts of DNA (Pfeifer et al., 1989).

2. GENE EXPRESSION IN SPERMATOGONIAL DIFFERENTIATION

In situ techniques will also probably have to be used to study gene expression in stem spermatogonia. Although in situ hybridization with nucleic acid probes has been used to identify gene transcripts in the later stages of spermatogenesis, it will be difficult to apply such techniques to spermatogonia for the following reasons: (1) the identification of the different types of A spermatogonia is difficult in sections; (2) in whole mounts, myoid and endothelial cells would affect hybridization to spermatogonia; (3) cell morphology is compromised by the mild fixation conditions necessary for hybridization and by the autoradiographic emulsion; and (4) the label will be largely over the cytoplasm, the boundaries of which are difficult to determine by light microscopy. Future directions for improvements include sensitive nonisotopic labeling techniques and the use of oligonucleotide probes to enhance penetration into fixed tissue.

When such techniques become available, proto-oncogenes are good candidates for examination, as are genes for the enzymes involved in DNA synthesis and for the proteins involved in mitosis. Proto-oncogenes are involved in activation of cells from a G_0 state and, therefore, may be initiators of stem cell proliferation. Using cell separation techniques, Wolfes and associates (1989) found the messenger RNAs (mRNAs) for the nuclear proto-oncogenes, c-*myc*, c-*fos*, and c-*jun*, as well as c-*raf* and the c-*ras* family, in A spermatogonia. At the protein level, c-*ras* but not c-*fos* was detected in the A spermatogonia. The levels of the nuclear proto-oncogene mRNAs peaked in the B spermatogonial fraction and those of the nonnuclear oncogenes at later stages. The nuclear proto-oncogenes, which may be candidates for altering programs of gene expression, are most predominant in the B spermatogonia and may have important functions in the differentiation of these cells.

Another approach will be to use genes that are known to be expressed in spermatogonia. Using separated cells, Thomas and colleagues (1989) showed that some cDNA probes obtained from later stage germ cells, hybridized to a mixed population of type A and B spermatogonia. For studies of stem spermatogonia, one could use a similar approach but starting with a cDNA library of type A spermatogonia.

3. PROTEINS OF SPERMATOGONIAL STEM CELLS

Identification of proteins characteristic of spermatogonial stem cells would be of great interest for several reasons. It would provide markers for better identification and possible sorting of these cells. Differentiation could be followed by the disappearance of a stem cell marker or the appearance of a differentiation marker. In addition, some of these markers may be related to cell proliferation and would help in understanding its regulation.

Limited information is available regarding proteins of even differentiated spermatogonia. Besides studies of the oncogenes, biochemical analyses have been limited primarily to histone variants. Among these variants, H2A.X and H1a are highly enriched in spermatogenic cells but are also found in somatic cells; TH3 is testis specific and is only found in the germinal cells. Cell separation studies have demonstrated high levels of H1a, H2A.X, and TH3 in a population of mostly differentiating type A and in B spermatogonia, as well as in later spermatogenic stages of the rat (Meistrich et al., 1985). In contrast, immunocytochemical analysis indicated that, in the mouse, levels of H1a and H2A.X were low in type A spermatogonia but high in early primary spermatocytes (Bhatnager, 1985; Rasheed et al., 1989). High levels of H1a were also found in type B spermatogonia, whereas high levels of H2A.X were found in gonocytes and prospermatogonia of the neonatal testis. The discrepancies between the biochemical and immunocytochemical assays could be a result of interspecies differences, the accessibility of the antigenic sites, or the qualitative nature of immunostaining. Thus, unusual histone variants appear to be present at high levels in precursors and progeny of spermatogonial stem cells and hence likely in the stem cells as well. They could provide markers for the germinal cells, although they are not likely to distinguish stem spermatogonia from other cells in their lineage.

Identification of surface molecules of stem spermatogonia would be valuable for cell sorting and for identification of surface receptors. Various polyclonal (Millette and Bellve, 1977) and monoclonal (Fenderson et al., 1984) antibodies to testicular cells have been prepared; however, none of those described to date appear to recognize type A spermatogonia. One possible approach is to isolate precursors of the stem spermatogonia and raise antibodies against surface markers, with the hope that some of the same epitopes will remain on the surface of the adult stem cells.

IX. Concluding Remarks

If there is little water in the stream, it is the fault, not of the channel, but of the source.
St. Jerome (c.342–420, translated by W. H. Fremantle)

The spermatogonial stem cells, as the cells responsible for the continued production of sperm throughout life, have an essential role in the reproductive process. Through elegant morphological studies, and related work on kinetics and recovery from cytotoxic insult, the stem cells of the spermatogenic system and their pathway of production of differentiated cells have been described. (Incidentally, spermatogonial stem cells have been better characterized morphologically than stem cells from any other renewing normal tissue or tumor system, and therefore, the results of studies on spermatogonial stem cells may be applied to these other systems.)

Maintenance of spermatogonial stem cell number and the regulated production of differentiating cells requires the action of growth and differentiation factors. Because there are many different control points in the type A spermatogonial compartment at which the yield of differentiating cells can be affected, thorough morphological evaluation based on the criteria and models discussed in earlier sections will be essential for the evaluation of where such factors may act. Although simpler assays should be developed to screen for factors affecting spermatogonial proliferation, their actions should be related to changes in counts or kinetics of specific cells.

Molecular characterization of this source, the spermatogonial stem cells, remains a formidable task. In situ methods will be required. Probes could be developed based on knowledge of growth regulation in other cell systems and from characterization of the immediate precursors and descendants of the stem cells.

Finally, manipulation of stem cell renewal and proliferation might eventually have practical implications. For example, it may be possible to increase sperm production in men who are oligospermic naturally or because of exposure to reproductive toxins.

ACKNOWLEDGMENTS

Support during the writing of this review and for research in the authors' laboratory was provided by Grants CA17364 and HD-16843 from the National Institutes of Health. The editorial assistance of Walter J. Pagel is appreciated.

REFERENCES

Bardin, C. W., Cheng, C. Y., Musto, N. A., and Gunsalus, G. L. (1988). The Sertoli cell, in *The Physiology of Reproduction* (Knobil, E., Neill, J., et al., eds.) pp. 933–974, Raven Press, New York.

Bartmanska, J., and Clermont, Y. (1983). Renewal of type A spermatogonia in adult rats. Cell Tissue Kinet. 16:135–143.

Bellve, A. R., and Zheng, W. (1989). Different pleiotropic actions of seminiferous growth factor (SGF), acidic fibroblast growth factor (aFGF), and basic fibroblast growth factor (bFGF). Ann. NY Acad. Sci. 564:116–131.

Bellve, A. R., Cavicchia, J. C., Millette, C. F., O'Brien, D. A., Bhatnagar, Y. M., and Dym, M. (1977). Spermatogenic cells of the prepuberal mouse. Isolation and morphological characterization. J. Cell. Biol. 74:68–85.

Bhatnagar, Y. M. (1985). Immunocytochemical localization of an H2A variant in the spermatogenic cells of the mouse. Biol. Reprod. 32:957–968.

Bootsma, A. L., and Davids, J.A.G. (1988). The cell cycle of spermatogonial colony forming stem cells in the CBA mouse after neutron irradiation. Cell Tissue Kinet. 21:105–113.

Bucci, L. R., Brock, W. A., Johnson, T. S., and Meistrich, M. L. (1986). Isolation and biochemical studies of enriched populations of spermatogonia and early primary spermatocytes from rat testes. Biol. Reprod. 34:195–206.

Cattanach, B. M. (1974). Spermatogonial stem cell killing in the mouse following single and fractionated x-ray doses, as assessed by length of sterile period. Mutation Res. 25:53–62.

Chemes, H. E., Dym, M., and Raj, H.G.M. (1979). The role of gonadotropins and testosterone on initiation of spermatogenesis in the immature rat. Biol. Reprod. 21:241–249.

Chowdhury, A. K. (1979). Dependence of testicular germ cells on hormones: A quantitative study in hypophy-

sectomized testosterone-treated rats. J. Endocrinol. 82:331–340.
Chowdhury, A. K., and Steinberger, E. (1977). In *The Testis in Normal and Infertile Men* (Troen, P., and Nankin, H. R., eds.) pp. 69–83, Raven Press, New York.
Clermont, Y. (1969). Two classes of spermatogonial stem cells in the monkey *(Cercopithecus aethiops)*. Am. J. Anat. 126:57–72.
Clermont, Y. (1966). Renewal of spermatogonia in man. Am. J. Anat. 118:509–424.
Clermont, Y., and Hermo, L. (1976). Spermatogonial stem cells and their behavior in the seminiferous epithelium of rats and monkeys, in *Stem Cells of Renewing Cell Populations* (Cairnie, A. B., Lala, P. K., and Osmond, D. G., eds.) pp. 273–286, Academic Press, New York.
Clermont, Y., and Hermo, L. (1975). Spermatogonial stem cells in the albino rat. Am. J. Anat. 142:159–176.
Clermont, Y., and Mauger, A. (1974). Existence of a spermatogonial chalone in the rat testis. Cell Tissue Kinet. 7:165–172.
Clermont, Y., and Antar, M. (1973). Duration of the cycle of the seminiferous epithelium and the spermatogonial renewal in the monkey *Macaca arctoides*. Am. J. Anat. 136:153–166.
Clermont, Y., and Bustos-Obregon, E. (1968). Re-examination of spermatogonial renewal in the rat by means of seminiferous tubules mounted "in toto." Am. J. Anat. 122:237–248.
Coward, W. A., Howell, J. M., Thompson, J. N., and Pitt, G.A.J. (1969). The retinol requirement of rats for spermatogenesis and vision. Br. J. Nutr. 23:619–626.
Clifton, D. K., and Bremner, W. J. (1983). The effects of testicular x-irradiation on spermatogenesis in man: A comparison with the mouse. J. Androl. 4:387–392.
Cunningham, G., and Huckins, C. (1979). Failure to identify a spermatogonial chalone in adult irradiated testes. Cell Tissue Kinet. 12:81–89.
de Rooij, D. G. (1983). Proliferation and differentiation of undifferentiated spermatogonia in the mammalian testis, in *Stem Cells, Their Identification and Characterization* (Potten, C. S., ed.) pp. 89–117, Churchill Livingstone, Edinburgh.
de Rooij, D. G. (1980). Effect of testicular extracts on proliferation of spermatogonia in the mouse. Virchows Arch. B Cell Pathol. 33:67–75.
de Rooij, D. G. (1973). Spermatogonial stem cell renewal in the mouse. I. Normal situation. Cell Tissue Kinet. 6:281–287.
de Rooij, D. G., van Dissel-Emiliani, F.M.F., and van Pelt, A.A.M. (1989). Regulation of spermatogonial proliferation. Ann. NY Acad. Sci. 564:140–153.
de Rooij, D. G., and Janssen, J. M. (1987). Regulation of the density of spermatogonia in the seminiferous epithelium of the Chinese hamster: I. Undifferentiated spermatogonia. Anat. Rec. 217:124–130.
de Rooij, D. G., van Alphen, M.M.A., and van de Kant, H.J.G. (1986). Duration of the cycle of the seminiferous epithelium and its stages in the rhesus monkey *(Macaca mulatta)*. Biol. Reprod. 35:587–591.
de Rooij, D. G., Lok, D., and Weenk, D. (1985). Feedback regulation of the proliferation of the undifferentiated spermatogonia in the Chinese hamster by the differentiating spermatogonia. Cell Tissue Kinet. 18:71–81.
de Ruiter-Bootsma, A. L., and Davids, J.A.G. (1981). Survival of spermatogonial stem cells in the CBA mouse after combined exposure to 1-MeV fission neutrons and hydroxyurea. Radiat. Res. 85:38–46.
de Ruiter-Bootsma, A. L., Kramer, M. F., de Rooij, D. G., and Davids, J.A.G. (1976). Response of stem cells in the mouse testis to fission neutrons of 1 MeV mean energy and 300 kV x rays. Methodology, dose-response studies, relative biological effectiveness. Radiat. Res. 67:56–68.
Dexter, T. M. (1989). Regulation of hemopoietic cell growth and development: Experimental and clinical studies. Leukemia 3:469–474.
Dym, M., and Clermont, Y. (1970). Role of spermatogonia in the repair of the seminiferous epithelium following x-irradiation of the rat testis. Am. J. Anat. 128:265–282.
Erickson, B. H. (1981). Survival and renewal of murine stem spermatogonia following ^{60}Co gamma radiation. Radiat. Res. 86:34–51.
Erickson, B. H., and Hall, G. G. (1983). Comparison of stem-spermatogonial renewal and mitotic activity in the gamma-irradiated mouse and rat. Mutation Res. 108:317–335.
Fenderson, B. A., O'Brien, D. A., Millette, C. F., and Eddy, E. M. (1984). Stage-specific expression of three cell surface carbohydrate antigens during murine spermatogenesis detected with monoclonal antibodies. Devel. Biol. 103:117–128.
Frankenhuis, M. T., Kramer, M. F., and de Rooij, D. G. (1982). Spermatogenesis in the boar. Veterinary Quarterly 4:57–61.
Fouquet, J. P., and Dadoune, J. P. (1986). Renewal of spermatogonia in the monkey *(Macaca fascicularis)*. Biol. Reprod. 35:199–207.
Hall, P. A., and Watt, F. M. (1989). Stem cells: The generation and maintenance of cellular diversity. Development 106:619–633.
Haneji, T., Koide, S. S., Nishimune, Y., and Oota, Y. (1986). Dibutyryl adenosine cyclic monophosphate regulates differentiation of type A spermatogonia with vitamin A in adult mouse cryptorchid testes in vitro. Endocrinology 119:2490–2496.
Haneji, T., Maekawa, M., and Nishimune, Y. (1984). Vitamin A and follicle-stimulating hormone synergistically induce differentiation of type A spermatogonia in adult mouse cryptorchid testes in vitro. Endocrinology 114:801–805.
Haneji, T., and Nishimune, Y. (1982). Hormones and the differentiation of type A spermatogonia in mouse cryptorchid testes incubated in vitro. J. Endocrinol. 94:43–50.
Harrison, D. E., and Astle, C. M. (1982). Loss of stem cell repopulating ability upon transplantation. Effects of donor age, cell number, and transplantation procedure. J. Exp. Med. 156:1767–1779.
Hochereau-de Reviers, M. T. (1976). Variation in the stock of testicular stem cells and in the yield of spermatogonial divisions in ram and bull testes. Andrologia 8:137–146.
Huang, H.F.S., Marshall, G.R., Rosenberg, R., and Nieschlag, E. (1987). Restoration of spermatogenesis by high levels of testosterone in hypophysectomized rats after long-term regression. Acta Endocr. 116:433–444.

Huang, H.F.S., and Nieschlag, E. (1986). Suppression of the intratesticular testosterone is associated with quantitative changes in spermatogonia populations in intact adult rats. Endocrinology 118:619–627.

Huang, H.F.S., and Hembree, W.C. (1979). Spermatogenic response to vitamin A in vitamin A deficient rats. Biol. Reprod. 21:891–904.

Huckins, C. (1978a). Spermatogonial intercellular bridges in whole-mounted seminiferous tubules from normal and irradiated rodent testes. Am. J. Anat. 153:97–122.

Huckins, C. (1978b). Behavior of stem cell spermatogonia in the adult rat irradiated testes. Biol. Reprod. 19:747–760.

Huckins, C. (1971a). The spermatogonial stem cell population in adult rats. I. Their morphology, proliferation and maturation. Anat. Rec. 169:533–558.

Huckins, C. (1971b). The spermatogonial stem cell population in adult rats. II. A radioautographic analysis of their cell cycle properties. Cell Tissue Kinet. 4:313–334.

Huckins, C. (1971c). The spermatogonial stem cell population in adult rats. III. Evidence for a long-cycling population. Cell Tissue Kinet. 4:335–349.

Huckins, C. (1971d). Cell cycle properties of differentiating spermatogonia in adult Sprague-Dawley rats. Cell Tissue Kinet. 4:139–154.

Huckins, C., and Oakberg, E. F. (1978). Morphological and quantitative analysis of spermatogonia in mouse testes using whole mounted seminiferous tubules. II. The irradiated testes. Anat. Rec. 192:529–542.

Irons, M. J., and Clermont, Y. (1979). Spermatogonial chalone(s): Effect on the phases of the cell cycle of type A spermatogonia in the rat. Cell Tissue Kinet. 12:425–433.

Lajtha, L. G. (1979). Stem cell concepts. Differentiation 14:23–34.

Lok, D., Jansen, M. T., and de Rooij, D. G. (1984). Spermatogonial multiplication in the Chinese hamster. IV. Search for long cycling stem cells. Cell Tissue Kinet. 17:135–143.

Lok, D., and de Rooij, D. G. (1983). Spermatogonial multiplication in the Chinese hamster. III. Labelling indices of undifferentiated spermatogonia throughout the cycle of the seminiferous epithelium. Cell Tissue Kinet. 16:31–40.

Lok, D., Jansen, M. T., and de Rooij, D. G. (1983). Spermatogonial multiplication in the Chinese hamster. II. Cell cycle properties of undifferentiated spermatogonia. Cell Tissue Kinet. 16:19–29.

Lok, D., Weenk, D., and de Rooij, D. G. (1982). Morphology, proliferation, and differentiation of undifferentiated spermatogonia in the Chinese hamster and the ram. Anat. Rec. 203:83–99.

Lu, C. C., Meistrich, M. L., and Thames, H. D. (1980). Survival of mouse testicular stem cells after gamma or neutron irradiation. Radiat. Res. 81:402–415.

Lu, C. C., and Meistrich, M. L. (1979). Cytotoxic effects of chemotherapeutic drugs on mouse testis cells. Cancer Res. 39:3575–3582.

Meistrich, M. L. (1986). Critical components of testicular function and sensitivity to disruption. Biol. Reprod. 34:17–28.

Meistrich, M. L., and van Beek, M.E.A.B. (1990). Radiation sensitivity of the human testis, in *Advances in Radiation Biology, Vol. 14, Relative Radiation Sensitivities of Human Organ Systems* (Lett, J. R., and Altman, K. I., eds.) pp. 227–268, Academic Press, New York.

Meistrich, M. L., Bucci, L. R., Trostle-Weige, P. K., and Brock, W. A. (1985). Histone variants in rat spermatogonia and primary spermatocytes. Devel. Biol. 112:230–240.

Meistrich, M. L., Finch, M., Lu, C. C., de Ruiter-Bootsma, A. L., de Rooij, D. G., and Davids, J.A.G. (1984). Strain differences in the response of mouse testicular stem cells to fractionated radiation. Radiat. Res. 97:478–487.

Meistrich, M. L., Finch, M., da Cunha, M. F., Hacker, U., and Au, W. W. (1982). Damaging effects of fourteen chemotherapeutic drugs on mouse testis cells. Cancer Res. 42:122–131.

Meistrich, M. L., Hunter, N. R., Suzuki, N., Trostle, P. K., and Withers, H. R. (1978). Gradual regeneration of mouse testicular stem cells after exposure to ionizing radiation. Radiat. Res. 74:349–362.

Millette, C. F., and Bellve, A. R. (1977). Temporal expression of membrane antigens during mouse spermatogenesis. J. Cell Biol. 74:86–97.

Morales, C., and Griswold, M. D. (1987). Retinol-induced stage synchronization in seminiferous tubules of the rat. Endocrinology 121:432–434.

Nishimune, Y., Haneji, T., and Aizawa, S. (1981). Differential effects of cyclic AMP on DNA synthesis by germ cells and non-germ cells in mouse testicular culture. J. Endocrinol. 89:257–262.

Oakberg, E. F. (1978). Differential spermatogonial stem-cell survival and mutation frequency. Mutation Res. 50:327–340.

Oakberg, E. F. (1959). Initial depletion and subsequent recovery of spermatogonia of the mouse after 20 R of gamma rays and 100, 300, and 600 R of x-rays. Radiat. Res. 11:700–719.

Orth, J. M., and McGuinness, M. P. (1991). Neonatal gonocytes co-cultured with Sertoli cells on a laminin-containing matrix resume mitosis and elongate. Endocrinology 129:1119–1121.

Paulsen, C. A. (1973). The study of irradiation effects on the human testis: including histologic, chromosomal and hormonal aspects, *Terminal Report, Atomic Energy Commission.* Contract No. AT(45-1-22-25).

Pfeifer, G. P., Steigerwald, S. D., Mueller, P. R., Wold, B., and Riggs, A. D. (1989). Science 246:810–813.

Pollanen, P., Soder, O., and Parvinen, M. (1989). Interleukin-1α stimulation of spermatogonial proliferation in vivo. Reprod. Fertil. Dev. 1:85–87.

Potten, C. S. (1986). Cell cycles in cell hierarchies. Int. J. Radiat. Biol. 49:257–278.

Rasheed, B.K.A., Whisenant, E. C., Ghai, R. D., Papaioannou, V. E., and Bhatnagar, Y. M. (1989). Biochemical and immunocytochemical analysis of a histone H1 variant from the mouse testis. J. Cell Sci. 94:61–71.

Rowley, M. J., Leach, D. R., Warner, G. A., and Heller, C. G. (1974). Effect of graded doses of ionizing radiation on the human testis. Radiat. Res. 59:665–678.

Russell, L. D. (1980). Sertoli–germ cell interrelations: A review. Gamete Res. 3:179–202.

Spangrude, G. J., Heimfeld, S., and Weissman, I. L. (1988).

Purification and characterization of mouse hematopoietic stem cells. Science 241:58–62.

Steinberger, A., and Steinberger, E. (1966). Stimulatory effect of vitamins and glutamine on the differentiation of germ cells in rat testes organ culture grown in chemically defined media. Exp. Cell Res. 44:429–435.

Sutcliffe, M. J., and Burgoyne, P. S. (1989). Analysis of the testes of H-Y negative XOSxr^b mice suggests that the spermatogenesis gene *(Spy)* acts during the differentiation of the A spermatogonia. Development 107:373–380.

Thomas, K, H., Wilkie, T. M., Tomashefsky, P., Bellve, A. R., and Simon, M. I. (1989). Differential gene expression during mouse spermatogenesis. Biol. Reprod. 41:729–739.

Thumann, A., and Bustos-Obregon, E. (1978). An "in vitro" system for the study of rat spermatogonial proliferative control. Andrologia 10:22–25.

Toppari, J., and Parvinen, M. (1985). In vitro differentiation of rat seminiferous tubular segments from defined stages of the epithelial cycle. Morphologic and immunolocalization analysis. J. Andrology 6:334–343.

van Alphen, M.M.A., van de Kant, H.J.G., and de Rooij, D.G. (1989a). Protection from radiation-induced damage of spermatogenesis in the Rhesus monkey *(Macaca mulatta)* by follicle-stimulating hormone. Cancer Res. 49:533–536.

van Alphen, M.M.A., van de Kant, H.J.G., and de Rooij, D. G. (1989b). Dose–response studies on the spermatogonial stem cells of the Rhesus monkey *(Macaca mulatta)* after x irradiation. Radiat. Res. 119:443–451.

van Alphen, M.M.A., van de Kant, H.J.G., and de Rooij, D. G. (1988a). Depletion of the spermatogonia from the seminiferous epithelium of the Rhesus monkey after x irradiation. Radiat. Res. 113:473–486.

van Alphen, M.M.A., van de Kant, H.J.G., and de Rooij, D. G. (1988b). Repopulation of the seminiferous epithelium of the Rhesus monkey after x irradiation. Radiat. Res. 113:487–500.

van Alphen, M.M.A., van de Kant, H.J.G., and de Rooij, D. G. (1988c). Follicle-stimulating hormone stimulates spermatogenesis in the adult monkey. Endocrinology 123:1449–1455.

van Beek, M.E.A.B., and Meistrich, M. L. (1990). A method for quantifying synchrony in testes of rats treated with vitamin A deprivation and readministration. Biol. Reprod., 42:424–431.

van Beek, M.E.A.B., Meistrich, M. L., and de Rooij, D. G. (1990). Probability of self-renewing divisions of spermatogonial stem cells in colonies, formed after fission neutron irradiation. Cell Tissue Kinet. 23:1–16.

van Beek, M.E.A.B., and Meistrich, M. L. (1992). Spermatogenesis in retinol-deficient rats maintained on retinoic acid. J. Reprod. Fert. 94:(in press).

van Beek, M.E.A.B., Davids, J.A.G., and de Rooij, D. G. (1986a). Nonrandom distribution of mouse spermatogonial stem cells surviving fission neutron irradiation. Radiat. Res. 107:11–23.

van Beek, M.E.A.B., Davids, J.A.G., and de Rooij, D. G. (1986b). Variation in the sensitivity of the mouse spermatogonial stem cell population to fission neutron irradiation during the cycle of the seminiferous epithelium. Radiat. Res. 108:282–295.

van Beek, M.E.A.B., van de Kant, H.J.G., Davids, J.A.G., and de Rooij, D. G. (1984). Response to fission neutron irradiation of spermatogonial stem cells in different stages of the cycle of the seminiferous epithelium. Radiat. Res. 97:556–569.

van den Aardweg, G.J.M.J., De Ruiter-Bootsma, A. L., Kramer, M. F., and Davids, J.A.G. (1983). Growth and differentiation of spermatogenetic colonies in the mouse testis after irradiation with fission neutrons. Radiat. Res. 94:447–463.

van Dissel-Emiliani, F.M.F., Grootenhuis, A. J., de Jong, F. H., and de Rooij, D. G. (1989a). Inhibin reduces spermatogonial numbers in testes of adult mice and Chinese hamster. Endocrinology 125:1898–1903.

van Dissel-Emiliani, F.M.F., de Rooij, D. G., and Meistrich, M. L. (1989b). Separation of rat gonocytes by velocity sedimentation at unit gravity. J. Reprod. Fert. 86:759–766.

van Keulen, C.J.G., and de Rooij, D. G. (1975). Spermatogenetic clones developing from repopulating stem cells surviving a high dose of an alkylating agent. I. First 15 days after injury. Cell Tissue Kinet. 8:543–551.

van Keulen, C.J.G., and de Rooij, D. G. (1973). Spermatogonial stem cell renewal in the mouse. II. After cell loss. Cell Tissue Kinet. 6:337–345.

Withers, H. R., Hunter, N., Barkley, H. T., Jr., and Reid, B. O. (1974). Radiation survival and regeneration characteristics of spermatogenic stem cells of mouse testis. Radiat. Res. 57:88–103.

Wolfes, H., Kogawa, K., Millette, C. F., and Cooper, G. M. (1989). Specific expression of nuclear proto-oncogenes before entry into meiotic prophase of spermatogenesis. Science 245:740–743.

CHAPTER THIRTEEN

The Process of Meiosis

Herbert Stern

I. Introductory Comments

A. The Gateway to Sexual Reproduction

An obvious and indispensable requirement for sexual reproduction is that the gametes involved, sperm or egg, contain a haploid number of chromosomes. This apparently simple requirement may be contrasted with the fact that mechanisms involved in fulfilling the requirement display considerable complexity. The latter can only be accounted for by features of chromosome behavior that are frequently disregarded in textbook descriptions of the process. From the standpoint of human reproduction it must be emphasized that the target haploid number must represent a fully balanced chromosome complement to provide normal progeny. The contribution by either gamete of even one additional chromosome to the fertilized egg leads most commonly to embryonic or fetal death, and the occasional fetuses that come to term are disabled to a greater or lesser degree depending on the chromosome involved in the resulting trisomy. Apart from the special case of sex chromosomes, no survivors are known among fetuses that have only one member of a homologous chromosome pair. Meiosis is the only divisional process capable of providing haploid cells for gamete formation and it may, therefore, be rightly considered as the gateway to sexual reproduction. It is the universal gateway among sexually reproducing organisms, be they fungi, flowering plants, or mammals.

B. Some Comparative Considerations

Meiosis is all too frequently described as a modified mitosis. It is a misleading description. The mitotic cycle cannot serve as a framework for analyzing the meiotic process. There are a number of similarities among the events involved in chromosome separation but, as will become evident, they constitute a small fraction of the events essential to meiosis. A useful introductory perspective to the complexity of meiotic organization is one that addresses the different processes by which its universal function is achieved. For mammals, in general, there are major differences in meiotic organization between females and males. In humans, female meiosis is initiated in the fetus during the fifth and sixth months of gestation, whereas in men the process is initiated at puberty. Moreover, in men, spermatocytes enter and complete meiosis without interruption. In women, oocytes enter meiosis while in the fetus and become arrested after completing the critical prophase stages. They then remain at the so-called *dictyate stage* until ovulation at which time they resume meiosis and proceed through the first metaphase but are arrested at the second metaphase. They complete meiosis only if they undergo fertilization. These interruptions could not occur in the absence of specific and separate regulatory controls. Another difference in pattern is evident at the disjunctional phase of meiosis. In men, all four haploid sets are retained for gamete formation; in women, three haploid sets are destroyed as polar bodies and only one is retained. This process of elimination in the oocyte requires a mechanism that regulates both the orientation of the spindle body and its peripheral location. For obvious reasons, such a mechanism is not required by spermatocytes.

Superimposed on the contrasts just described is one other comparison that bears directly on metabolic analyses of the meiotic process. Both spermatocytes and oocytes perform metabolic activities during meiosis that are related to gametogenesis but not to meiosis. Spermatocytes initiate metabolic activities during prophase that are addressed to spermiogenesis; these activities or related ones continue through the spermatid stage. Oocytes become very active metabolically when the "dormant" primordial follicle advances to the *primary* stage in preparation for fertilization events and for initial zygote development. Thus, there is a need to discriminate between those events that are accompaniments of meiosis and those that are directly required by the process itself. That need is not easily met but the issue will be given as close attention as possible in analyzing the meiotic phase of spermatogenesis.

C. Meiosis and Gametogenesis

The ultimate and exclusive target of germ line development is gamete formation, egg or sperm. Although meiosis is a prerequisite for gamete development it must be emphasized that spermatocyte meiosis and spermiogenesis are separate processes. Difficulties of analysis arise because, as already described, the two processes are not fully separated temporally. In the case of female meiosis, the insertion of events directed at embryonic needs into meiocyte metabolism is obvious by inspection. In the case of spermatogenesis, metabolic insertions are not at all obvious but, despite the limited information available, it is certain that some of the metabolic events during meiosis are related to postmeiotic processes. Studies of male meiotic metabolism in plants are simplified by the fact that the meiotic products enter an extended and metabolically active haploid stage during which the equivalent of spermiogenesis occurs. In considering this entire question it is important to bear in mind that whereas spermiogenesis represents a process of cellular differentiation, meiosis is almost exclusively a chromosomal process in which cytoplasmic activities are supportive but of secondary significance. Thus, in tracking meiotic events the issue at all times is how any one of them accounts for the unique behavior of the chromosomes.

II. Chromosomal Behavior During Meiosis: General Features

A. Outline of the Meiotic Process

This brief outline is intended to recall the elementary features of meiosis for those who have a somewhat hazy recollection of its nature. It should also serve to introduce some of the terminology that is used throughout the chapter.

Chromosomes of a meiocyte are replicated during the *premeiotic S-phase* and in this respect the meiocyte behaves much like a premitotic cell. There are, nevertheless, certain significant differences. As discussed in detail later, chromosome replication in premeiotic cells is not quite complete. Also, in contrast with mitotically dividing cells, the G2 phase is comparatively short and, in general, little attention is given to the duration of the G2 interval although some presumably G2 events are critical to meiosis. The first morphologically definable meiotic stage is leptotene; it is the first stage of the long prophase interval which terminates after pachytene at the diplotene–diakinesis stages. Leptotene and the next stage, zygotene, during which chromosome pairing occurs, extend for about 2 to 3 days in the mouse. These are followed by pachytene, which is the longest stage of meiotic prophase; it is estimated to extend for 4 to 5 days in the mouse spermatocyte. Remaining meiotic events require no more than 2 days. The significant feature of this schedule is the comparatively long duration of the prophase stages, one that contrasts sharply with that of a mitotic prophase, which may extend for 1 hour. The difference is readily accounted for by the numerous mechanisms required to effect the pairing and recombination of murine chromosomes.

B. Specific Contrasts between Meiosis and Mitosis

References are commonly made to the *mitotic cycle,* but meiosis can only be correctly referred to as a *process.* Daughter cells of a mitotic division can undergo mitosis but those of a meiotic division cannot repeat meiosis. In fungi and plants the haploid cells may undergo cycles of mitosis, but in animals the products of meiosis are normally subject to one of two fates, death or fertilization. In contemplating meiotic regulatory mechanisms it is important to keep in mind that for any cell line the regulatory program applies to a single and terminal cell generation. That program is repeated only after the zygote has differentiated and sets aside a new germ line. The contrast between mitotic and meiotic behavior is strikingly evident in the functioning of the mature testis. Spermatogonia in the basal regions of the seminiferous tubules undergo repeated mitoses, whereas spermatocytes, housed in the distinctive environment of the adluminal region, undergo a single meiotic division, the products of which differentiate into spermatozoa.

Certain details of the contrast in chromosomal behavior between meiotic and mitotic cells are useful to emphasize the distinctive aspects of meiotic chromosome activity. In mitosis, chromosome compaction through prophase to metaphase is a steady and uninterrupted process. In meiosis, it is a staged process such that each of the cytological stages of prophase is characterized by different degrees of compaction. Superimposed on the process of compaction is chromosome synapsis (pairing) which, as will be discussed, involves additional structural modifications of the chromosomes. As yet, there is no evidence that throws any light on the regulatory mechanisms governing the staged compaction of premetaphase meiotic chromosomes.

The respective behaviors of meiotic and mitotic kinetochores provide another contrast that begs for explanation. In mitosis, a signal occurs at the end of metaphase that causes the kinetochore to become functionally duplicated thus enabling sister chromatids to disjoin. In meiosis, the kinetochore remains single at the end of the first metaphase so that chromosome, rather than chromatid, separation occurs. Such an event also requires that the meiotic kinetochore be asymmetrically attached only to fibers leading to one spindle pole, in contrast to the mitotic kinetochore which is attached to fibers leading to both poles. A change in kinetochore behavior occurs at the end of the second meiotic metaphase when it mimics the mitotic pattern and the chromosomes disjoin equationally. Kinetochore behavior is clearly a critical component of the meiotic process. In the case of mitosis the issue of kinetochore behavior appears to be simple. One need only account for a mechanism by which kinetochores function as duplicated structures at the end of metaphase. A speculation no longer favored is that kinetochore DNA replication is delayed until the end of metaphase when its doubling would make an equational division possible. The weakness of that speculation lies in the absence of supportive evidence. The studies, currently accumulating, that are based on immunocytochemistry and electron microscopy point to a complex structure of kinetochore region with respect to DNA and protein components. Such evidence points strongly to a complex regulatory system, one that remains to be elucidated.

The evidence on kinetochore behavior is briefly summarized here and will not be discussed later. This treatment is accorded the meiotic kinetochore because, despite its importance, the information available offers little substantive understanding of meiotic disjunction. There is now general agreement not to use the terms kinetochore and centromere interchangeably. The centromere is the larger cytologically definable region that includes the kinetochore. The latter, identifiable by electron microscopy, is the site at which the spindle fibers attach to the chromosome. Its structure is that of a trilaminar plaque. The centromeric region in which it is found has been described as consisting of three adjacent domains (Earnshaw and Rattner, 1989). Next to the *kinetochore domain,* by this description, lies a *central domain,* and next to it is the *pairing domain.* Sister chromatids remain joined in the latter domain before anaphase in mitotic cells and beyond anaphase I and until anaphase II in meiotic cells. The model attached to the description is that the functional doubleness of the kinetochore depends on cleavage of the sister chromatid attachment in the pairing domain. Immunocytochemical analyses have identified six antigens that are unique to the centromere, but none of these can be assigned with certainty to the kinetochore. There is unambiguous evidence for specific DNA sequences being localized in the centromere, but except for *Saccharomyces cerevisiae*, the sequences have been found to be repetitive and in no case has a centromeric sequence been found to code for a protein. The relevant interactions that might account for apparent kinetochore behavior are believed to depend on specific DNA-protein binding. These interactions have not yet been identified for mitotic divisions, let alone for the more complex pattern of meiotic ones. The current studies of centromere behavior using immunochemical and molecular techniques are likely to disclose tangible mechanisms for both meiotic and mitotic systems.

C. Chromosome Disjunction and Recombination: Key Meiotic Events

Before discussing the detailed events underlying disjunction and recombination, it should be useful to make some general comments on the relevance of the two processes to fetal and postnatal health. The end point of meiosis in spermatogenesis is the formation of four spermatids, each appreciably smaller in volume than ¼ of the pachytene spermatocyte and destined to undergo further diminution in volume. Such reduction emphasizes the prime, but not exclusive, importance of the chromosomes in zygote formation. Two chromosomal properties are relevant to this consideration. One of these is the random assortment of maternally and paternally derived chromosomes in the haploid spermatids. Meiosis is the prime source of human diversity inasmuch as the probable occurrence of one particular assortment in a gamete, such as all maternally derived chromosomes, is $(\frac{1}{2})^{23}$. A second property is the stringent requirement for haploidy with respect to normalcy of the potential zygote. Abnormal chromosome constitutions account for 50 percent of all spontaneous abortions, and such constitutions are generally attributable to the oocyte (Hassold and Jacobs, 1984). This is not always the case. Triploidy occurs in 1 to 3 percent of all conceptions, very few of which survive to term and none of which survive long thereafter, but only 10 percent of these are due to the ovum. About 66 percent are attributable to dispermy and about 25 percent to diploid spermatozoa, the latter representing a failure in meiotic disjunction that is more than twice as frequent as the comparable event in the oocyte (Jacobs et al., 1978). Thirty-three percent of all XXY situations (Klinefelter's syndrome) are due to a failure of XY disjunction in the spermatocyte. About 20 to 25 percent of trisomy 21 (Down's syndrome), but only 7 percent of other trisomies, are paternal in origin. Unlike the situation in women, there is no significant increase in paternal chromosomal abnormalities with age.

Although meiosis in one or the other sex of a few species has been programmed to bypass crossing over, the event is essential to normal meiotic disjunction in all other species. This is so because the structure maintaining the paired chromosomes is transient. It does not persist after completion of pachytene and there is no mechanism to maintain pairing beyond pachytene if no crossover has occurred. Failure in crossing over is a major, but not exclusive, factor in the formation of aneuploid (absence of normal chromosome number) gametes.

None of the comments made thus far penetrates the meiotic mechanism. The comments should nevertheless provide a biological perspective on the role of spermatogenesis in the reproductive system that is superimposed on its serving as a source of spermatozoa for fertilization.

III. The Pattern of Metabolism in Meiotic Cells: An Outline

A. Experimental Considerations

Any meaningful presentation of meiotic metabolism must account for the coordination between the successive changes in chromosome behavior and the corresponding succession of changes in metabolic activities. To perform the needed analysis requires either an organism with synchronous meioses or a procedure for separating the different meiotic stages from a nonsynchronous population. It also requires adequate amounts for biochemical analyses. None of these needs is readily fulfilled. Naturally, synchronous systems that provide sufficient material for analysis are rare and only a few species in the lily family oblige. However, despite the widespread commercial interest in lilies, genetic information is totally lacking and there is little prospect for a change. Nevertheless, their natural synchrony, the 2 to 3 day duration of each of the prophase substages, and the feasibility of their being cultured in vitro through much of meiosis has made them a major source of biochemical information on the meiotic process. Fortunately, much effort has been put into spermatocyte fractionation and the "Staput" procedure does provide a useful but hardly perfect preparation of spermatocytes at different meiotic stages (Chandley et al., 1977). These are some of the factors that account for the relative paucity of meiotic research at the biochemical and molecular levels in higher organisms, mammals in particular (Stern, 1986).

The above deficiency contrasts markedly with the richness of information that is available from studies by light and electron microscopy. The still young but growing use of immunocytochemical procedures to analyze meiotic structures is providing an entry for molecular approaches as is the use of cDNA probes. Moreover, the molecular and genetic studies of meiosis in yeast will undoubtedly serve as a most fruitful path of entry into the molecular analyses of spermatocyte and oocyte meiosis. It is a little early for the contents of this chapter to reflect these anticipations. The presentation is based partly on direct and partly on indirect evidence; desirable coherence is achieved by substantial conjecture.

B. The Premeiotic Events

1. MEIOTIC INDUCTION

It is sufficient to focus on testicular activity alone to perceive the importance of identifying the factor responsible for initiating the meiotic process. The mitotic products of the spermatogonial B cell are obligatorily meiotic and are preceded by a long line of mitoses within the germ line. It follows that a highly specialized and tightly regulated mechanism must operate in the freshly formed spermatocytes, and probably also in the B cell, to cause so radical a shift in the chromosomal behavior. The most general comment that can be made on this issue at the nonmolecular level is that the conditions known to lead to an induction of meiosis vary widely. In many fungi, including yeast, effective starvation is the critical environmental signal. In many flowering plants, day length is the critical but temporally distant signal. In the fetal ovary it is attributed to a still unidentified *meiosis inducing substance,* one that, until the recent past, was frequently confused with the factor that induces *resumption* of meiosis in association with ovulation. There have been unestablished claims for identification of the inducing substance in testicular preparations. There may well be a universal molecular factor that is ultimately responsible for initiating meiosis in all organisms, but if so, it remains to be discovered.

Direct evidence for a factor regulating the induction of meiosis has been reported for *Saccharomyces cerevisiae* (Kassir et al., 1988). A gene, IME1, has been cloned and shown by various manipulations to induce meiosis in diploid yeast cells. Its transciption is stimulated by nitrogen starvation and it is not transcribed under most other conditions. There are difficulties, however, in applying the current information to meiotic induction in spermatocytes. The activity of IME1 is not only linked, at least under normal conditions, to starvation but it must also interact with the genes governing the mating type loci and and the meiotic repressor gene, RME1. These interactions are not fully clarified. Of more immediate relevance to spermatocyte behavior is the report by Wolfes and associates (1989) that the nuclear proto-oncogenes, c-*fos,* c-*jun, and* c-*myc,* are expressed at high levels in type B spermatogonia. The authors suggest that these probably function as transcription regulators in the B cells and thus lead to spermatocyte formation. The specific target(s) of regulation remains unknown but the recessive mutation, *ameiotic,* in maize, discovered by Rhoades (1956), may be relevant to the target in question. In later studies it was found that in homozygous mutants the normally premeiotic S-phase was followed by a mitosis, the daughter cells degenerating (Palmer, 1971). Unlike yeast, cells in higher eucaryotes cannot return to a mitotic population on reaching the developmental stage of meiotic initiation.

The latter observation points to some event related to or coincident with the premeiotic S-phase that is critical to meiotic induction. This does not imply an event sufficient for meiotic induction if only because, in animals, that event requires a prior differentiation of the germ line at an early embryonic stage. In the adult testis itself, spermatocyte meiosis must be preceded by a sequence of differentiated stages in the spermatogonial population. (Like the embryonic animal cells before differentiation, undifferentiated cells in the growing regions of mature higher plants retain the potential for germ line formation.) The distinctiveness of the premeiotic S-phase is already evident in its duration. The meiotic S-phase, compared with the mitotic one, has been found to be about five times as long in *Triturus,* two to three times as long in *Saccharomyces cerevisiae,* and about five times as long in *Lilium* (Stern, 1989). The reason for the attenuation of the premeiotic S-phase is not known but more will be said about it in discussing DNA replication. There is, however, an experimental event described in *Lilium* that may well have general relevance to one property of the premeiotic S-phase. Irreversible commitment to meiosis has been found to occur near the end of that phase. Lily meiocytes explanted from their natural environment into culture medium before S-phase completion revert to mitosis but enter meiosis if explanted after its completion (Stern, 1989). In general, cells that are disturbed during the period immediately following the premeiotic S-phase display various meiotic abnormalities, especially with respect to chromosome synapsis. It is certain that the interval surrounding S-phase termination includes a number of key events essential to the initiation and execution of meiosis.

C. Events in Meiotic Prophase

1. GENERAL

It is to be expected that the distinctive and unique changes in chromosome behavior as cells progress through meiosis would be accompanied by correspondingly distinctive changes in cell metabolism. Such changes are likely to be partially obscured by the housekeeping metabolic activities that are present throughout the process. Indeed, where techniques permit, relevant changes have been identified. There are two challenges associated with identifying meiosis-specific metabolic events. One is to distinguish between metabolic changes immediately related to chromosome behavior and those involved in postmeiotic events. The second challenge is to relate each specific metabolic change to a defined change in chromosome behavior. These challenges are not readily met, and although molecular techniques have provided and continue to provide data on molecular events during meiosis, we are still distant from a molecular characterization that will serve as an outline of the meiotic process.

2. RNA SYNTHESIS

The pioneering work on RNA synthesis during spermatogenesis was accomplished by Monesi and associates (1978). They made it clear that RNA was transcribed in both spermatocyte and spermatid, and they also found evidence for an unusual behavior of the transcripts. A significant fraction of pachytene transcripts were not released from nucleus to cytoplasm until diplotene (Monesi et al., 1978). The latter observation pointed to the involvement of meiotic RNA transcripts in postmeiotic events. An analysis of RNA transcription during spermatogenesis is to be found in Chapter 16. Two types of RNA transcripts have been described that probably have a role in either synapsis or recombination. Most of that information has been obtained from lily meiocytes and these can be more meaningfully discussed in the sections on synapsis and recombination.

One transcriptional event that is almost certainly relevant to meiosis is the differential behavior of the X chromosome in oocytes and spermatocytes. Autoradiographic studies have long made it evident that in oocytes both X chromosomes are transcriptionally active during meiotic prophase, whereas no more than one X chromosome is active in somatic cells of the female. The single X chromosome of the spermatocyte was thought to be transcriptionally inactive during much of the pachytene interval. This view has been modified in recent studies to indicate the occurrence of transcription in certain regions of the spermatocyte X chromosome (Chandley and McBeath, 1987). The cytogenetic evidence for sterility in most strains of male mice carrying balanced X autosome translocations has commonly been interpreted as being due to the activation of the X-chromosome segments by the translocated autosomes. This interpretation was recently tested by analyzing X-Y behavior in mice bearing a certain balanced autosome–autosome translocation that rendered them sterile (Richler et al., 1989). In such spermatocytes, the end of one member of the autosome–autosome pairing complex was associated with the X chromosome, made possible in those special cases by the incomplete pairing of the autosomes within the quadrivalent. (In meiocytes housing a reciprocal translocation between two different chromosomes, synapsis gives rise to a four-member pairing structure, the *quadrivalent.*) Two relevant observations were made on the consequences of this autosome X association. (1) The X-Y complex (see Fig. 13–3c), which normally becomes located at the periphery of the nucleus and thus spatially separated from the autosomes, remains intermingled with the autosomes. (2) At the gross level, there is no observable difference in transcriptional activity of the X chromosome between the sterile and fertile forms. The conclusion drawn by the investigators is that a separation of the X-Y complex from the autosomes is essential to fertility and that such separation probably involves the silencing of certain genes to permit spermatocyte meiosis to go to completion. Whether or not this conclusion is correct, it is likely that a selective transcriptional activity of the X chromosome plays a regulatory role in spermatocyte meiosis. Such selectivity probably occurs during the pachytene interval.

3. PROTEIN SYNTHESIS

Two classes of proteins are synthesized during meiotic prophase, those involved in the structural modifications of the chromosomes and those involved in the metabolic events underlying recombination. Many of these meiosis-specific syntheses are not readily apparent. If protein extracts are made from successive meiotic stages and compared by gel analysis, the similarities in gel pattern overwhelm the differences. The search for meiosis-specific proteins, those not present in cells other than meiocytes, is not highly rewarding. A murine gene transcript was recently described that is present in high levels at pachytene and is unique to spermatocytes except for residual low levels in spermatids. The sequence organization of the gene pointed to a regulatory function (Leroy et al., 1989). However, the transcripts were not detected in oocytes so that the gene's role is probably testis specific and might function in spermiogenesis. Many essential proteins, including meiosis-specific ones, are probably synthesized in premeiosis, but ongoing protein synthesis during meiosis is necessary for meiotic progress. Cells in meiotic prophase that are exposed to inhibitors of protein or RNA synthesis quickly become arrested (Parchman and Stern, 1969). As in the previous discussion of RNA synthesis, the specific information available on what appears to be meiosis-specific proteins is best considered in association with the relevant chromosomal events during prophase, chromosome synapsis, and recombination.

4. DNA REPLICATION

Among all the macromolecular syntheses occurring during meiotic prophase, that of DNA is easily the most distinctive. The pattern of DNA replication is tightly correlated with chromosome behavior, a correlation that makes it possible to relate the intervals of replication to specific aspects of chromosome behavior. It should prove helpful to introduce the topic by describing the profile of DNA replication from the premeiotic S-phase to the termination of meiotic prophase. An outline of events during that long interval is given in Figure 13–1. The distinctive feature of the DNA replication profile is the occurrence of three peaks. From a quantitative standpoint the major replication event occurs during the premeiotic S-phase, and the other two quantitatively minor replications occur during prophase. In the preceding discussion of premeiotic S-phase, only its extended duration relative to mitotic S-phases was considered. However, an important feature of that synthesis was not included. In liliaceous plants, it has been possible to demonstrate that premeiotic S-phase replication is only 99.8 to 99.9 percent complete. A group of DNA sequences, to be referred to as zygDNA (zygotene DNA), is blocked in replication. That replication is initiated in coordination with chromosome pairing at zygotene. Inhibition of that synthesis inhibits the process of synapsis (Stern and Hotta, 1987). Thus, there is not only a quantitative difference, but also a qualitative difference between premeiotic and premitotic DNA replication. Although no direct demonstration of zygDNA replication in mice has been possible, autoradiographic studies of mouse meiocytes

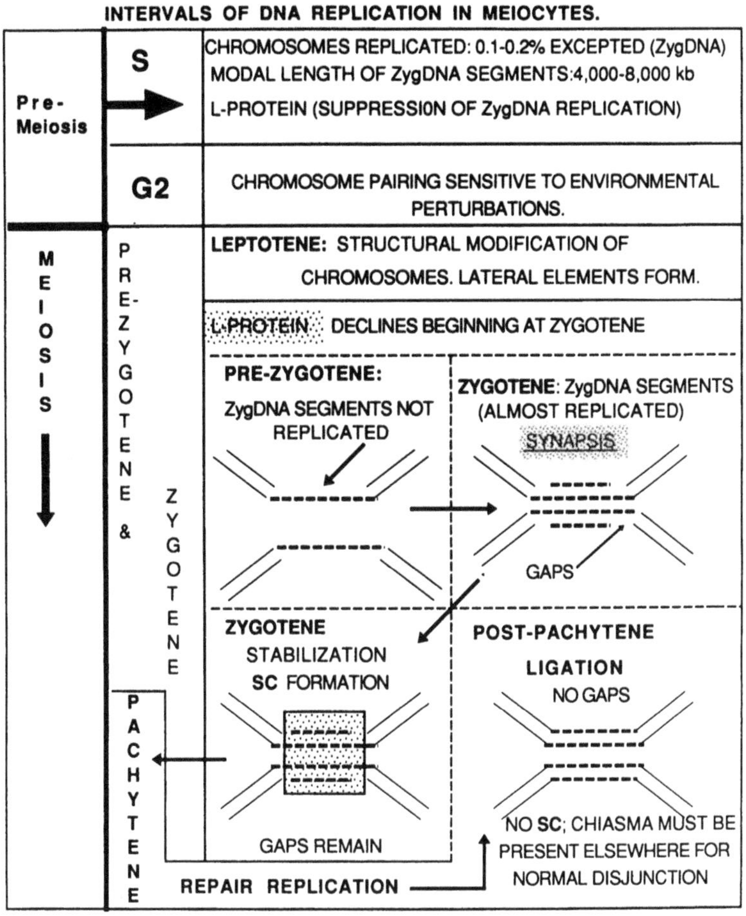

Figure 13–1. Successive intervals of DNA replication in meiotic cells. The figure outlines replication events beginning at the premeiotic S-phase and terminating after pachytene. The precise stage at which replication is completed has not been determined. zygDNA is the centerpiece of this figure inasmuch as its behavior has three critical stages: suppression of replication by L-protein during the S-phase and until zygotene. The zygDNA is incompletely replicated at the zygotene stage. The completion of zygDNA replication occurs after pachytene. Repair replication is the prominent feature of pachytene. The figure is based entirely on findings discussed in the text.

have revealed the occurrence of DNA replication during the zygotene phase (Moses et al., 1984). One other characteristic of zygDNA is noteworthy. The newly replicated zygDNA strands are not ligated to their neighbors until after the completion of pachytene (Hotta and Stern, 1976). The role of this delay in the meiotic process is unknown but rationalizing models are easily developed. It may be concluded that zygDNA sequences have a fundamental role in meiosis despite their being an infinitesimal fraction of the genome. Further details on the behavior of these sequences are given later.

The third interval of DNA replication occurs immediately after the completion of chromosome synapsis and its associated zygDNA synthesis. The DNA involved is referred to as PDNA (pachytene DNA) because of its occurrence during the pachytene stage, the stage at which crossing over is effected. In contrast with zygDNA, PDNA synthesis is not a consequence of delayed replication, although quantitatively it is of similar magnitude, about 0.2 percent of genomic DNA. It is a process of repair replication which correlates with the requirements for crossing over. PDNA replication has been demonstrated in mouse spermatocytes (Stubbs and Stern, 1986). The evidence is gratifying from the standpoint of geneticists because it helps lay to rest claims that crossing over occurs during the premeiotic S-phase. Additional evidence on this point is found in the section addressing the events associated with recombination.

This general survey of meiotic metabolism should make it evident that from the S-phase on there is a sequence of metabolic events that can be correlated with the sequence of events in chromosome behavior. As pointed out in the introductory remarks, these events have no counterparts in the mitotic cycle. The pairing process and the programmed recombination are unique features of meiosis. Moreover, each of these events involves several interacting mechanisms and thus accounts for the overall complexity of meiocyte metabolism. The challenge is to link the metabolic events to the mechanisms underlying the chromosomal ones. An attempt will be made to do so in the next two sections, but the linkages demonstrated will at best be thin.

IV. Chromosome Synapsis

A. General Aspects

In a small number of species among the Dipteran insects the chromosomes in all the somatic tissues are homologously paired. Such *somatic pairing* is structurally different from meiotic pairing, even if only the lack of synaptonemal complex is taken into account. These few cases are the only exception to the general rule that chromosome synapsis is a unique and universal feature of meiosis. Meiotic pairing is dependent on certain presynaptic processes that occur at leptotene and even earlier stages. Pairing initiation and completion are evident by light or electron microscopy at zygotene and, in most cases, the paired structures disassemble at the end of pachytene. The persistence of the paired state into metaphase occurs in species like the female silkworm as a mechanism for assuring normal chromosome disjunction in the absence of crossing over. Meiotic pairing is an elaborate but transient event; its termination is essential to the occurrence of disjunction just as its presence is essential to the occurrence of recombination.

It is a characteristic of meiotic pairing to involve only two chromosomes in any paired region although, as might be predicted, there are exceptions. Ordinarily, pairing in tetraploids, which have four homologues for each chromosome member, occurs by the formation of quadrivalents rather than bivalents because an individual chromosome may pair partly with one homologue and partly with another. Normal disjunction is not affected in such situations as long as homologous pairing is maintained. An essential property of synapsed chromosomes is the homology that governs paired regions, a condition that implies DNA sequence correspondence. Inversion of a chromosome segment and translocations between chromosomes are manifest in the geometry of the paired structures. There are genetic factors, at least in some species, that regulate the fidelity of homologous synapsis. The best example of this is the gene, Ph, in hexaploid wheat that prevents pairing between members of the three different but nearly homologous ("homoeologous") sets. The mechanism by which the gene effects its control is still unresolved (von Wettstein et al., 1984). Despite such regulatory factors there are situations in which the stringency in homologous pairing is violated. In an earlier discussion the effect of certain reciprocal autosomal translocations in mice on the meiotic behavior of the XY pair leading to male infertility was considered. The causal factor appears to be the high frequency with which the chromosome tip of one of the autosomes associates with the X chromosome. More extreme examples have been recorded in haploid plants, which have a nearly normal growth pattern (an apparently impossible condition in mammals), and in which nonhomologous chromosomes pair in the meiocytes (von Wettstein et al., 1984). Infertility is the expected consequence.

The initial challenge in approaching the phenomenon of synapsis lies in certain physical aspects of the process. The key consequence of meiotic pairing is an alignment of corresponding homologous regions between members of each chromosome pair. The action implies a correspondence in aligned DNA sequences. Conceiving a mechanism for such alignment presents a challenging problem. The problem arises in considering the structure of the chromosome at the time that pairing is initiated. Initiation occurs when the chromosomes are already partially compacted. The degree of compaction may be judged from the data assembled by von Wettstein and colleagues (1984) on the length of chromosomal DNA and the length of that chromosome in the paired configuration of pachytene. The ratio of chromosome/DNA lengths decreases with increasing DNA length. In *Saccharomyces cerevisiae* it is 0.3 percent, in lily it is 0.006 percent, and in one of the human chromosomes it is 0.01 percent. The chromosomes are somewhat less compacted at zygotene when pairing begins but hardly enough to alter the problem of getting DNA-specific alignment when the bulk of chromosomal DNA is not physically available for matching, a situation that is evident when examining electron microscope preparations (Figs. 13–2 and 13–3). Various models have been proposed to escape the dilemma but none of these has been generally accepted.

A set of histone analyses by Meistrich and co-workers, and also those of others (Meistrich and Brock, 1987) have an important, though not yet identifiable, bearing on one aspect of meiotic chromosome morphology. The histones are clearly important in governing the extent of chromosome compaction, either general or localized. Meiosis-specific variants of histones H1, H2A, H2B, and H3 have been reported. Each of these variants has a distinctive amino acid sequence and is coded for different gene loci. Their levels at the different stages of spermatogenesis reflect probable functions. The H1 variant is absent from spermatogonia and reaches its highest level in late pachytene when it is synthesized. The H2A variant is also ab-

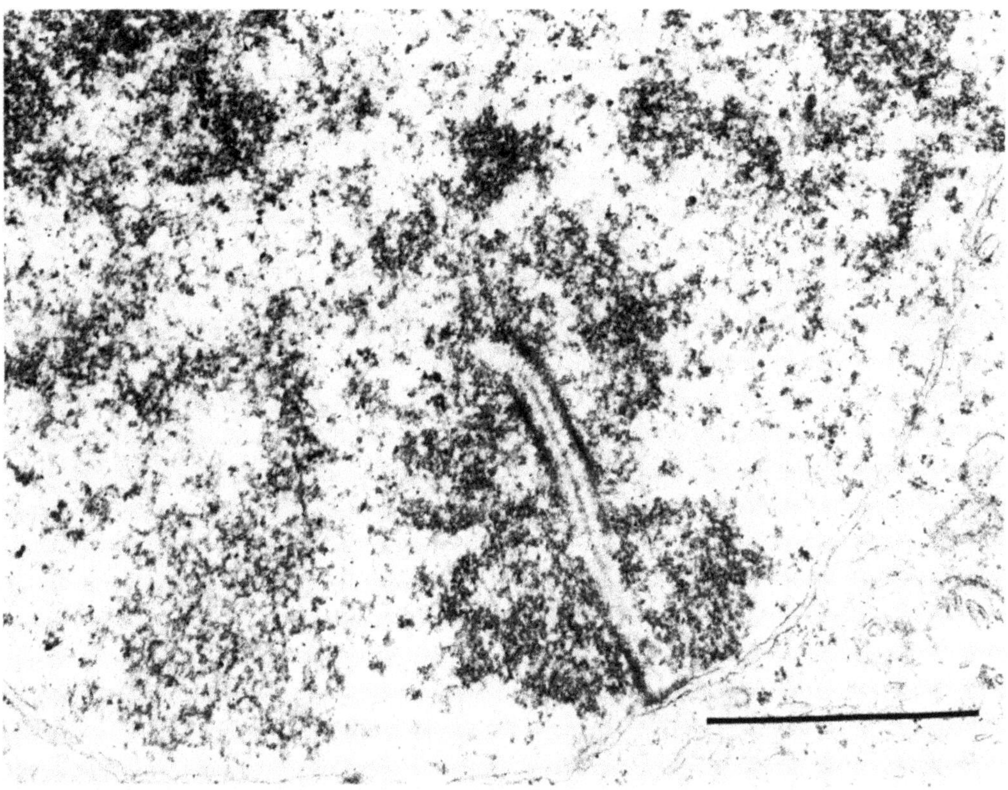

Figure 13–2. A transmission electron micrograph of an ultrathin section of a mouse spermatocyte at pachytene. The synaptonemal complex is attached to the nuclear membrane. The central element is evident in the central region, which is lined by the darkly stained lateral elements. The synaptonemal complex appears to be embedded in the chromatin simply because an exceedingly small fraction of the chromatin is in the paired region. Bar = 1 μm. (Reprinted with permission from Chandley et al., 1977.)

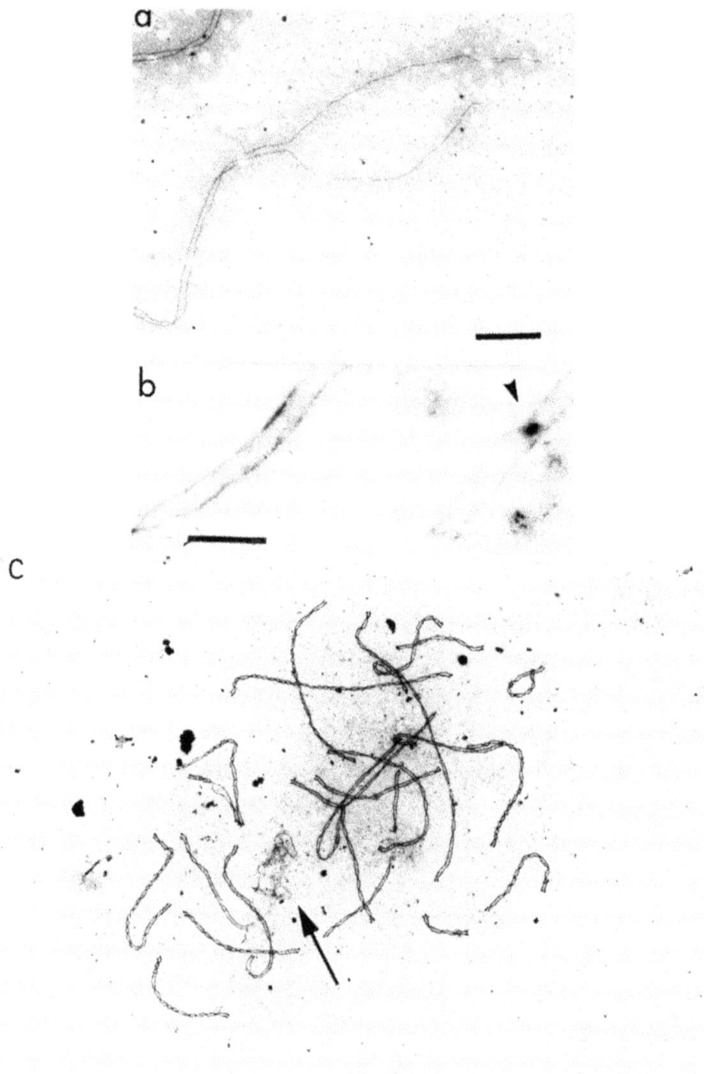

Figure 13–3. Segments of spread pachytene chromosomes. The lateral elements stain prominently in spreads. (**a** and **b**) Meiotic chromosomes of lily; courtesy of C. A. Hasenkampf (1989). Bar = 2 μm. In **a** part of the chromosomes is unpaired; the lightly stained material is chromatin. In **b** an *arrow* points to a recombination nodule. Its relevance is discussed in section V.A.2. (**c**) A spread of chromosomes from a human spermatocyte in late pachytene; courtesy of A. C. Chandley (1987). The lateral elements are prominent. The *arrow* points to an XY aggregate, a feature of male meiosis discussed in section III.C.2. Magnification record is unavailable.

sent, or nearly so, from spermatogonia but its level rises in the early prophase stages and remains so though the postmeiotic spermatid stage. The H2B variant's profile is similar to that of H2A except for a smaller but significant increase in level during prophase. The H3 variant is already present in spermatogonia in fairly high concentration and remains so through the prophase stages. The variants appear to replace the somatic histones and there is no evidence for a significant increase in the total histone complement. The specific action of these histone variants is a matter of speculation and, even at the speculative level, models are lacking. The relatively high levels of H2A and H2B at early meiotic prophase point to some role in pairing and/or recombination but, as yet, there is no demonstrated similarity in pattern across the phylogenetic spectrum to add strength to the speculation.

B. Ultrastructural Aspects of the Pairing Process

1. EXPERIMENTAL APPROACHES

The approaches have two principal targets, the process by which homologues are aligned and the mechanism by which the alignment is maintained. As already indicated, the first of these is remote from resolution, whereas the second of these has been effectively analyzed by electron microscopy. Serial sectioning for electron microscopy has made possible a three-dimensional representation of chromosomal configurations at successive prophase substages (Fig. 13–2). The later introduction of techniques for *chromosome spreads* greatly facilitated the analysis of paired chromosome behavior. This technique involves an abrupt release of chromosomes from an osmotically disrupted cell and a spreading of the released chromosomes over a liquid surface. Such spreads, which do not disrupt the paired condition, can be examined by light and electron microscopy (Fig. 13–3). The approach has been further broadened by the application of autoradiography and immunocytochemistry.

2. INITIATION OF PAIRING

Electron microscopic analysis has made it evident that a major structural change occurs in the chromosomes at the presynaptic stage of leptotene. Ribbonlike structures, 35 nm in width, are formed alongside the threadlike chromosomes at the presumed junction of sister chromatids. The junction is presumed because, unlike the prophase chromosomes in mitosis, there is no structural evidence of doubleness until after pachytene. The ribbonlike structures are referred to as lateral elements (LE) or, for functional descriptions, as axial elements because of their location at the presumed axis of the chromosome. The elements consist largely of protein along with some RNA (von Wettstein et al., 1984). Some unique polypeptides have been shown to be present in the LEs by immunochemical analysis, thus indicating that they are meiosis-specific proteins (Heyting et al., 1989).

Pairing itself begins at zygotene after the LEs have been formed. A controversy that still continues, but with lessening vigor, concerns the view that there is a prealignment event in the premeiotic cell generation (Del Fosse and Church, 1981). Although this view has the merit of allowing much more time for chromosome matching, it is not supported by electron microscopic observations of cells in leptotene or early zygotene. A more likely mechanism underlying the initiation of pairing involves the nuclear membrane near which or at which the telomeres of the chromosomes are clustered. Early paired telomeric segments are very frequently found in the membrane region, but the occurrence is not universal. The membrane site is still considered to be an important factor, but there are probably other paths of entry into homologous pairing. There are certainly many regions along a chromosome at which pairing is initiated. The simplistic model of a zipperlike pairing process originating at the telomere may be disregarded. There is no satisfactory model. What can be asserted is that at no time in the course of pairing has a molecularly intimate association, however brief, been observed. As different segments of chromosomes pair, their respective LEs remain separated by a central region, 100 nm wide, and as quickly as pairing occurs, the central region acquires a central element (CE), 30 nm wide. This completes the synaptic process.

3. PROPERTIES OF THE MEIOTIC BIVALENT

The three components just discussed, the LE, CE, and central region, constitute the synaptonemal complex (SC). It is a universal structure whose pattern and cross-sectional dimensions are very similar across the phylogenetic spectrum. The one variable is length and, as already mentioned, it is a function of chromosomal DNA content. Even superficial inspection of an electron micrograph is sufficient to indicate that the bulk of chromosomal DNA is excluded from the SC region. It reflects the problem of DNA sequence matching that was discussed earlier. Indeed, efforts to demonstrate the presence of some DNA strands cross the 150 to 200-nm spread of the SC have been unsuccessful. There are, nevertheless, intellectual difficulties in abandoning the role of the SC in chromosome alignment and some speculative models have assigned proteins the role of matching homologous chromosome segments. The molecular details of such models are not readily rationalized and the speculations have had a limited lifetime. The compelling evidence against a direct role of the SC in chromosome alignment is provided by EM studies in two types of cytogenetic situations. The meiocytes of certain hybrid plants display morphologically normal SC structures in paired nonhomologous segments and a similar situation prevails among the haploid plants referred to earlier (von Wettstein et al., 1984). A comparable behavior has been described for the phenomenon described as *synaptic adjustment,* a study that has centered largely on mouse spermatocytes (Moses et al., 1982). In strains carrying a well-characterized inverted region in one of the homologues and thus developing an inversion loop in the course of zygotene pairing, it was noted that in many cases the loop was not present at late pachytene stages. A normal SC was nevertheless present between the nonhomologously paired inverted and normal chromosome segments. This behavior has been attributed to the absence of a crossover in the affected region, which permits a release of tension by eliminating the inversion loop through synaptic adjustment. It is clear that bridging of paired chromosome segments by a morphologically normal SC neither implies nor requires homologous matching.

4. ROLE OF SYNAPTONEMAL COMPLEX IN STABILIZATION OF SYNAPSIS

The fact that both homologously and nonhomologously paired chromosome segments are bridged by SCs and that, in general, paired chromosome segments lacking SCs are not found in meiocytes, leads to the conclusion that pairing is stabilized by the SC. The mechanism of such stabilization is difficult to resolve, particularly with respect to the effective coordination between it and the alignment event that presumably requires DNA–DNA interactions just prior to SC formation. These interactions presumably require DNA sequence matching and, therefore, molecularly close apposition of DNA strands, but if so, the closeness is very quickly lost on formation of the SC. Nothing is known about the matching–stabilization sequence. It can only be said that when LEs achieve the required proximity they become linked by the formation of the CE and other components of the central region. That linkage is not necessarily triggered by homologous matching; the issue of accounting for the regularity of SC-stabilized homologous pairing is unresolved.

C. Metabolic Events Accompanying Synapsis

1. PRELIMINARY CONSIDERATIONS

Since its original application to meiotic chromosome pairing the term *synapsis* has been extended to cover the pairing of DNA strands in the course of recombination. Because of the relationship between chromosome pairing and recombination it is important to distinguish between these two very different processes. In the case of recombinational synapsis the process is easily described. It concerns the apposition of relatively short homologous DNA segments in molecular proximity to make possible the exchange of DNA strands. Recombinational synapsis must certainly occur during, or possibly prior to, the pachytene stage. Chromosome synapsis, which involves the full chromosome length, is a much more complex event. It concerns a positioning of chromosome pairs that reflects the distribution of DNA sequences, but it does not necessarily involve the regions undergoing recombinational synapsis. The latter is a necessary accompaniment rather than a component of chromosome synapsis, but information is sorely lacking on the relation between the two synaptic events. Moreover, as pointed out in the preceding section, there is no evidence for the presence of DNA spanning the SC between paired chromosomes and that adds to the difficulty in accounting for homologous chromosome alignment. Chromosomal synapsis is a precondition for recombinational synapsis, but it is not a direct part of the recombination process.

It may prove useful to propose an untested model that addresses the dilemma of achieving homologous synapsis between chromosomes that are so structured as to allow for molecular alignment of no more than minute amounts of DNA. The model proposed calls for numerous small stretches of appropriately located DNA that serve as the matching sites for chromosome synapsis. An obligatory requirement for such sites is that their respective locations at least roughly reflect the sequence composition of the comparatively very large stretches they each represent. Needless to say, there is no evidence for such sites but we propose that zygDNA sequences serve that role because in doing so the evidence discussed can be lodged in a functional if purely imaginative framework.

2. PROPERTIES OF zygDNA

If *microsporocytes* (the counterpart of spermatocytes in flowering plants) of *Trillium* or *Lilium* are pulse-labeled at successive intervals of prophase with a DNA precursor and the DNA is analyzed on a CsCl gradient, the fraction labeled during zygotene behaves like a GC-rich satellite (Hotta and Stern, 1971a). If hydroxyurea is added to the reaction mixture, labeling does not occur, which points to the synthesis not being a product of repair replication. If the zygotene cells are double-labled with BrdU and [^{3}H]deoxyadenosine followed by equilibrium centrifugation of the DNA in a CsCl gradient, the radiolabel is localized in the heavy component as would be expected if zygDNA is a product of semiconservative replication. When labeled material is hybridized to saturation with DNA preparations from successive stages extending from premeiosis to meiotic termination, the level of hybridization at premeiotic S-phase is half that at pachytene and postpachytene stages, the increase occuring during zygotene. The higher level is the same as that found in vegetative tissues. ZygDNA replication is clearly a delayed replication. Its coordination with chromosome synapsis makes it likely that a functional relationship exists between the two events.

The satellite-like behavior of zygDNA in CsCl gradients would point to its being a highly repeated sequence. Surprisingly, this is not at all the case. A Cot profile of the DNA (fraction of denatured DNA reassociated plotted against DNA concentration $\times$ time) indicates that the sequences are of low copy number (Hotta and Stern, 1975). The satellite behavior is due to yet another unusual phenomenon (see Fig. 13–1). Replication of zygDNA sequences at zygotene is incomplete and, as indicated in section III.C.4, the ends are not ligated to their neighbors until a stage following pachytene; that stage has not been precisely determined (Hotta and Stern, 1976). The sequences do not behave as a satellite in DNA preparations from premeiotic or posmeiotic cells. The delayed replication of zygDNA, which is presumed to be an essential part of the chromosome axis, may account for the apparent absence of doubleness in leptotene and zygotene chromosomes. Such incompleteness in replication may function as a physical connection between sister chromatids, a condition that is required to maintain the intactness of chiasmatic bivalents at diplotene when the SC is no longer present (Maguire, 1978). Before moving to the next topic it is of interest to point out that the absence of ligation through pachytene makes it possible to isolate zygDNA segments during that stage either by gentle mechanical shearing of the DNA or by treatment of nuclei with S1-nuclease.

3. zygDNA AND SYNAPSIS

Replication of zygDNA sequences occurs in coordination with other genomic sequences during premitotic S-phases (Hotta et al., 1984). It also occurs in premeiotic cells that have been caused to revert to mitosis. A meiosis-specific mechanism must, therefore, be present that is responsible for the suppression of zygDNA replication during the premeiotic S-phase. A mechanism has been identified by following the behavior of isolated meiotic nuclei under in vitro conditions (Hotta et al., 1984). Such behavior is revealing because under appropriate culture conditions the isolated nuclei mimic the in situ nuclei with respect to their pattern of DNA replication. Thus, nuclei isolated from cells in premeiotic S through leptotene stages do not replicate zygDNA sequences but do replicate other sequences. If, however, the nuclei are extracted with deoxycholate (DOC), they replicate zygDNA. The substance responsible for the observed suppression of replication is a protein, approximately 70 kd in molecular weight. It is referred to as the L-protein (leptotene protein) because of its role in suppressing zygDNA replication during the leptotene stage (see Fig. 13–1). It may be recalled that chromosomes acquire LEs during this stage but do not initiate synapsis until zygotene. Presumably, the action of that protein during leptotene is a critical factor in the initiation of synapsis. If DOC-extracted nuclei are treated with L-protein they lose the capacity to replicate zygDNA. The protein does not appear to have any significant effect on the replication of other DNA sequences.

If the level of L-protein is tracked through premeiosis and meiotic prophase its profile displays one expected feature and one that is unexpected. The expected feature is the gradual decrease of its level beginning at zygotene and reaching near absence at the beginning of the pachytene stage (see Fig. 13–7). The decrease is expected to account for the occurrence of zygDNA replication during the zygotene stage. No evidence is available for a mechanism that might account for the incompleteness of that replication. The unexpected feature is the time at which L-protein can be detected either by its activity or by gel analysis. The protein appears to be absent through almost all of the premeiotic S-phase. Its presence coincides approximately with the interval, close to S-phase termination, when the meiocytes can no longer be reverted to a mitotic division as described in section III.B.1. It is reasonable to suppose that the synthesis of L-protein is accompanied by an irreversible commitment to meiosis, but whether it is L-protein alone or a group of proteins that are so involved has not yet been determined. The time of commitment does not represent a sharp boundary between abnormal mitotic reversion and normal meiotic progression. Abnormalities in chromosome segregation are apparent in cells that are explanted before zygotene, the extent of abnormalities decreasing with the closeness of the explanted cells to zygotene. It is certain that the activities during the interval between S-phase and zygotene are essential to pairing (Ito and Stern, 1967). The sensitivity of prezygotene stages to subsequent meiotic upsets has been observed in a number of studies.

The late appearance of L-protein poses a question that remains unanswered. If the protein is not present until close to the end of the premeiotic S-phase, what mechanism suppresses its replication throughout much of the S-phase? Generally, late replicating sequences are the highly repeated ones and zygDNA sequences are not in that category. One might attribute the behavior to some unidentified cryptic factor and thus abandon the question or one might once again speculate. It is inviting to speculate that in all premitotic S-phases, zygDNA sequences or their equivalent mark the completion of the replication process and the resulting doubleness of mitotic prophase chromosomes. As indicated earlier, doubleness is not evident in meiotic prophase chromosomes. This implies a programming of the premeiotic S-phase to induce L-protein formation and probably that of others at the appropriate time, an implication that is consistent with the general property of an extended S-phase in meiotic as compared with mitotic cells. The speculation is testable but not yet tested.

The specificity of L-protein suppression of zygDNA replication requires a corresponding specificity of interaction between the protein and zygDNA sequences. This is indeed the case. The principal feature of that interaction is the comparatively strong selective binding of double-stranded (ds) zygDNA to L-protein, which has negligible binding affinity for the remainder of genomic dsDNA. The protein also binds to single-stranded (ss) DNA but without any sequence specificity. It is noteworthy that the protein has no binding affinity for mitochondrial and chloroplast DNA or for prokaryotic DNA (Y. Hotta, unpublished data). It does have an affinity for mammalian DNA, although the levels of binding are low, almost certainly because zygDNA presumably constitutes a very small fraction of mammalian genomes. The apparent universality in affinity for this protein among eucaryotic genomes lends support to the view that the protein is a critical component of the meiotic process.

The general binding of zygDNA sequences to L-protein invites the question of how a heterogeneous collection of low copy number sequences could display a common binding affinity for the protein. The answer is that the affinity lies in a component of zygDNA that constitutes about 1 percent of the total (Hotta et al., 1984). It is unknown whether the binding segment is comprised of one or several sequences. The difficulty in obtaining a clear answer derives from yet another property of the isolated zygDNA segments. Repeated efforts at cloning the segments as isolated have been unsuccessful. They have been cloned only after their ends were trimmed with nuclease and most of the clones obtained lacked binding regions for the L-protein. If it is assumed that such regions are located near the ends of the segments, it is evident the nuclease action is likely to remove them. This complicates answers to questions addressed to two other zygDNA properties that will now be considered.

Among the available plasmids with zygDNA insertions only two form binding complexes with L-protein. These, nevertheless, provide an unexpected but significant piece of information. No effect on plasmid structure occurs as a result of the binding. However, in the presence of ATP the supercoiled plasmids relax if bound to L-protein. Mixtures of L-protein, ATP, and plasmids without binding sites show no reaction. The relaxation is not due to topoisomerase activity. It is the product of a nick in one of the strands (Hotta et al., 1984). Whether the localized nicking is effected by the same polypeptide as the suppression of replication has not been determined, but the combination of binding to a specific subset of zygDNA sequences and the insertion of what appears to be a single nick is suggestive of a functional role in the pairing process. This consideration is returned to in the speculative paragraph on a molecular model for the synaptic process.

4. zygDNA ACTIVITIES IN THE COURSE OF SYNAPSIS

Two activities of these DNA sequences have already been described, semiconservative replication and stimulation of a particular nickase. To these must be added yet another significant aspect of zygDNA behavior, transcription. If poly(A)$^+$RNA extracts from cells in meiotic prophase are passed through a column of unfractionated zygDNA sequences bound to cellulose, an appreciable fraction of the RNA is retained and then released by formamide treatment. This fraction hybridizes specifically with certain sequences of zygDNA, which constitute about 1 to 2 percent of the total zygDNA population (Hotta et al., 1985a). Using zygDNA as a probe, these RNA sequences have not been detected in any extracts from nonmeiotic tissues. Moreover, the concentration profile of this RNA extends from midleptotene to the end of zygotene and is conveniently referred to as zygRNA (see Fig. 13–7). At its peak zygotene concentration it constitutes about 40 percent of the poly(A)$^+$RNA pool. That value drops quickly on either side of the peak. Its half-life seems to be shorter than that of the other transcripts, a conclusion based on the comparatively high degree of degradation in this fraction and its concentration profile. Whether the zygDNA sequences involved are transcribed simultaneously or in an ordered sequence has not been determined, but the latter seems more likely in view of the heterogeneity of the DNA sequences. There has been no demonstration of a translation of the zygRNA sequences.

Of immediate relevance to any discussion of spermatocyte behavior is the demonstration that zygRNA is also present during spermatogenesis although the timing of its presence cannot be as sharply defined as in lily microsporocytes. About 4 percent of the poly(A)$^+$RNA in extracts prepared from prophase spermatocytes is retained when passed through a cellulose-lily zygDNA column. Such RNA was not found in spermatocytes at premeiotic S-phase that were prepared from peripubertal mice. A more informative experiment associated with this finding was based on the evidence in lilies that newly replicated zygDNA segments remain unligated to their neighbors throughout pachytene and are, therefore, preferentially released when pachytene nuclei are treated with S1-nuclease (section IV.C.2.). If, as expected, zygDNA is present in mouse spermatocytes, treatment of pachytene nuclei with S1 nuclease should release a very small fraction of DNA that sediments less rapidly than the bulk of the genome and that can be detected by hybridizing with radiolabeled lily or mouse zygRNA. This has indeed been observed (Hotta et al., 1985a). Moreover, similar treatment of round spermatid or liver nuclei displayed hybridization only in the bulk fraction. These results point to a strong similarity in certain aspects of meiotic behavior between mice and lilies!

D. Speculation

None of the metabolic processes described can be linked demonstrably to the pairing process. Additional experiments centering on zygRNA were not particularly helpful. Microsporocytes of a lily hybrid that is defective in pairing have a low concentration of zygRNA. When normal meiocytes were treated with deoxyadenosine to inhibit zygDNA replication and thus synapsis, no apparent effect was observed with respect to zygRNA transcription. The latter cannot be considered a causal factor in pairing. It remains a highly coordinated accompaniment of synapsis.

A plausible scheme can be designed for the role of zygDNA. It begins with the assumption that the zygDNA segments are the sites serving to align chromosomes homologously in the first step of synapsis. Their low copy number and the limited evidence for the presence of zygDNA sequences in the region of the SC (Moses et al., 1984) are consistent with that role. If the assumption is correct, the delay in zygDNA replication can be rationalized by other assumptions, the first of which is that the presence of unreplicated zygDNA is essential for its role in the pairing process. This being so, the suppression of its replication until zygotene is understandable. The stabilization of pairing requires the presence of an SC, but that, in turn, requires the synthesis of essential components before initiation of the pairing process. The action of the L-protein in suppression of replication must be accompanied by that synthesis, hence the sensitivity of the prezygotene interval to various degrees of meiotic disruption. The nickase activity of the L-protein preparation is conveniently accounted for if it is assumed that its occurrence coincides with the initiation of synapsis among the various zygDNA segments (Fig. 13–4). If, as seems probable, the nick occurs at a specific site, an unwinding action at that site would make available hybridizable single-strand tails and, thus, provide segments for transient homologous alignment.

There is no certainty in these many speculations, but they should serve the heuristic purpose of illustrating the multiplicity of components that are likely to be involved in meiotic synapsis. Despite the attractiveness of simplistic diagrams that necessarily adorn textbooks, it needs to be emphasized that paired chromosomes are not mimics of paired complementary DNA strands. This latter thought serves as an appropriate entry into the equally complex phenomenon of meiotic recombination.

V. Processes in Meiotic Recombination

A. General

1. CHROMOSOMAL ASPECTS

Chromosome synapsis is an absolute prerequisite for meiotic recombination. Although recombination frequently occurs in somatic and vegetative cells where SCs do not form, there are no situations known in which meiocytes undergo normal recombination in the absence of synaptic structures. This relation is not tied to any evidence of a mechanism by which the SC functions directly in the recombinational process. One feature of all SCs examined (Fig. 13–3b) points to a structure that most probably serves as a site for recombination, the recombination nodule (RN). As yet, there is no information on its composition or on what aspect of recombination it functions despite its original

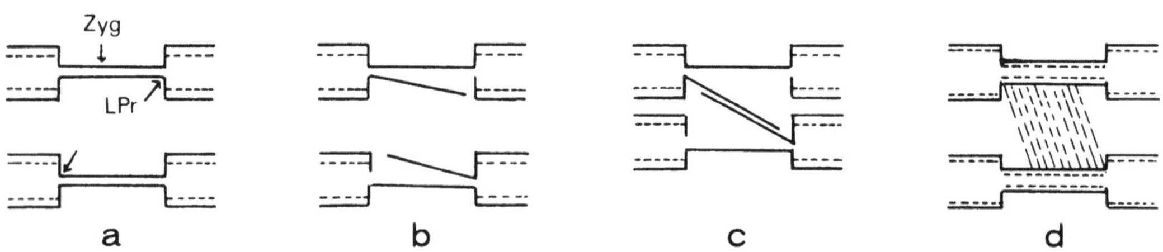

Figure 13–4. A model for the role of L-protein in making ssDNA tails available for homologous interaction and thus aligning the chromosomes for synapsis. A single and specific site of nicking in each of the zygDNA segments is required for the interaction to occur. In this model, the nicking occurs before replication. In some still undetermined way, replication follows alignment that is coordinated with stabilization by synaptonemal complex formation. The cluster of *dashed lines* represents the completed synaptonemal complex. (Reprinted with permission from Hotta et al., 1984.)

description about 15 years ago (Carpenter, 1975). Certain features of chromosomal recombination are, nevertheless, well established and are important in considering molecular mechanisms. Relative to the number of deoxynucleotides per diploid genome, the recombinational event is rare. In mouse spermatocytes the ratio is about three crossovers per 10^9 bases. Moreover, even at the gross level, the distribution is not random. Heterochromatin is virtually excluded from recombinational events even where, as in the Arabian oryx, it constitutes close to 50 percent of the genome (Stern and Hotta, 1980). From a positive standpoint, it needs to be emphasized that normal meioses require at least one crossover per chromosome pair and this is a regular occurrence in species where an average of one crossover per bivalent is the rule rather than the exception. In general, within a particular species, the number of crossovers per bivalent is related to chromosome length. The significant point of these comments is that the major feature of meiotic recombination is its *regularity* within each cell, the result of which is a high cellular frequency. Recombinational regularity, of course, is required for regularity of chromosome disjunction.

2. BIOCHEMICAL ASPECTS

The pachytene stage displays a distinctive pattern of biochemical activity, some of it shared with zygotene. A prominent feature of that pattern is a switch from the semiconservative DNA replication that is a characteristic of zygotene to a repair replication that is an exclusive characteristic of pachytene. The abrupt switch in DNA metabolism marks the beginning of crossing over, a long-recognized property of the pachytene stage. It is accompanied by the induction of nicking activity that, in turn, initiates repair replication (Howell and Stern, 1971). These metabolic events are partly the product of a freshly synthesized complement of proteins that are probably and, in some cases, certainly required for recombination. Several of these are meiosis specific, but many, particularly those involved in repair replication, are not novel components of the meiocytes. Specific distinctions will be indicated when the events are described. The synthesis of some meiosis-specific proteins is dependent on the prior occurrence of chromosome synapsis, whereas others are programmed to occur regardless of synapsis. As yet, no functional basis is known that might account for the difference.

It is relevant to indicate here that all the essential features of DNA behavior associated with recombination are shared by lily and mouse meiocytes, even though liliaceous material is more suitable for experimentation. In the case of zygDNA a direct demonstration of its delayed replication is very difficult, not only because it requires highly purified zygotene spermatocytes but also because the indirect evidence available indicates that, unlike lily, its buoyant density in CsCl solution is the same as that of bulk nuclear DNA (Stern and Hotta, 1985).

B. The Nicking Process

1. CHROMOSOMAL CHARACTERISTICS

Pulse labeling of pachytene cells with DNA precursors indicates the occurrence of DNA replication at that stage. Various lines of evidence point to its being a repair replication. It is resistant to inhibition by hydroxyurea; there are no changes in levels of hybridization saturation when pachytene-labeled sequences are tested against DNA samples from sequential meiotic stages. Pulse labeling of pachytene cells that had been labeled with BrdU during the S-phase displays an even distribution of label between the heavy and light strands (Hotta et al., 1977). Replication can be attributed to the action of an endonuclease that initiates nicking at the end of zygotene (Howell and Stern, 1971). The enzyme, when tested in vitro, acts only on dsDNA and the nicks are characterized by 3′P termini, a condition that requires a phosphatase to permit repair. The latter activity is present in the meiocytes. No information is available on how phosphatase activity might be regulated or whether such regulation is required.

The evidence for pachytene nicking was originally obtained by zonal sedimentation of native and denatured DNA samples from different meiotic stages and by comparing their respective profiles. No clear-cut differences are observed between the native DNA profiles, but those of ssDNA display a distinctive profile for pachytene samples. The profiles of all pachytene preparations, but none of the others, have two major components, one with a broad peak at about 100S to 105S, and the other with a slightly narrower spread at 60S to 63S. The first is common to preparations from all stages; the second is unique to pachytene. The latter is not present in achiasmatic preparations either from hybrids or from colchicine-treated cells, a recognized procedure for interfering with synapsis in plants but not in mammals. Moreover, DNA preparations of pachytene cells, labeled in situ in the case of mice and in vitro in the case of humans, display no labeling in the satellite components when centrifuged in $CsSO_4$ solutions. This is relevant to a property mentioned earlier, the selectivity of meiotic recombinational sites (Hotta and Stern, 1978a). In lilies, the 63S component that is a product of nicking accounts for about 50 percent of the nuclear DNA. In mouse spermatocytes, the slowly sedimenting fraction is also in the 63S range but it appears to represent a larger fraction of the total. The mean DNA length in these fractions is about 160kb, which, despite the size, accounts for an enormous excess of nicks relative to the number of crossovers.

Two inferences may be drawn from these observations. The first is that the regularity of crossing over is assured with respect to nicking by virtually saturating the system. The second inference bears on the unsettled debates as to whether nicks or double-strand cuts constitute the mechanism of choice for crossing over. It seems most unlikely that the highly organized program of nicking in both mouse and lily is irrelevant to recombination. This is not to dismiss the alternative mechanism, it is only to give recognition to nicking. The latter is certainly compatible with the Meselson–Radding model of molecular recombination (Radding, 1981).

2. THE REGULATION OF NICKING

The lily hybrid, "Black Beauty," is deficient in homologous pairing and, when pulse labeled during pachytene, displays little repair replication. As already indicated, the DNA isolated at pachytene contains only a small amount of the 63S component. When examined cytologically, one, but rarely two, bivalents are found. The cytological and biochemical data are compatible. Colchicine treatment of normally chiasmatic leptotene or preleptotene cells results in a similar achiasmatic situation. When such meiocytes are tested for meiotic endonuclease activity, its profile is indistinguishable from that of normal chiasmatic meiocytes (Hotta et al., 1979). If, as is possible, Black Beauty plants are treated to render the tetraploid, cytological examination of such meiocytes shows them to undergo normal synapsis, all bivalents having chiasmata. The cytological result is expected inasmuch as tetraploidy provides homologous pairing partners. It also indicates that the failure of pairing in the diploid hybrid is not due to some gene defect but only to an absence of pairing homology. Additional studies reveal that normal nicking and repair replication are restored. The conclusion is obvious: homologous pairing in some way determines whether or not the endonuclease can effect nicking.

3. CHROMATIN STRUCTURE REGULATES NICKING

If isolated nuclei from the achiasmatic hybrid are incubated in the presence of meiotic endonuclease, nicking does not occur. By contrast, isolated nuclei from normal meiocytes similarly treated do undergo nicking. It appears as though some structural properties of the achiasmatic nucleus block the effectiveness of endonuclease action. Direct evidence in support of this interpretation comes from experiments on chromatin digestion. The full set of studies is based on lily microsporocytes (Hotta and Stern, 1981); select but critical aspects of these studies have been performed with mouse spermatocytes (Stubbs and Stern, 1986).

Gel analysis of the DNA from deproteinized extracts of microsporocyte nuclei treated with micrococcal nuclease displays striking differences in label distribution with respect to meiotic stage when the gels are probed with DNA sequences labeled at pachytene. In pachytene preparations, the band representing the mononucleosomal DNA is exclusively labeled. In earlier stages label is more evenly distributed among the bands, indicating a relatively high sensitivity to micrococcal nuclease of chromatin sites that would have undergone repair replication at pachytene. At later stages, these sites become more resistant to digestion (Hotta et al., 1985b). At pachytene, the pachytene repair sites are more sensitive than the transcribed regions. When isolated nuclei are briefly digested with DNase II, a corresponding behavior is observed. In this case, it is possible to isolate and partially purify the chromatin fragments released by the treatment by adding Mg^{++} to the medium after the DNase incubation (Hotta and Stern, 1981). Such isolation made it possible to identify the components of these chromatin regions. Similar observations have been made with mouse spermatocytes (Stubbs and Stern, 1986). However, the DNase II technique is not as effective a procedure for probing the properties of PDNA, probably because of the relatively greater transcriptional activity in spermatocytes than in microsporocytes. Transcribed chromatin regions are also released by DNase II treatment, but the use of Mg^{++} to precipitate the transcriptionally active fragments is not sufficiently effective. Limited denaturation of the DNA followed by partial renaturation has been successfully applied to the spermatocytes. The technique makes evident the distinctiveness of the DNA sequences from cells that have been pulse labeled at pachytene (Stubbs and Stern, 1986).

The chromatin fragments subjected to zonal sedimentation have a sedimentation value of 11S. If late zygotene and early pachytene cells are incubated with radiolabeled precursors to DNA, RNA, and protein, all three macromolecules are present in the 11S fragments. Cot analysis of the DNA component yields a curve identical with that of the PDNA obtained by pulse labeling pachytene cells (see section III.C.4.). PDNA consists of middle repeat sequences, strikingly different from the zygDNA replicated at zygotene. Mouse PDNA has similar characteristics (Hotta et al., 1977). In mice, the isolated PDNA fragments range from 250 to 2500 bp (Stubbs and Stern, 1986). A similar situation occurs in lilies, but additional information is obtained by varying the length of DNasell treatment (Hotta and Stern, 1984). Shorter treatment time yielded labeled PDNA fragments that measured between 800 and 3000 bp. With longer treatment, the fragments were between 150 and 300 bp. Moreover, the two size classes reflected certain characteristics of PDNA organization. The smaller PDNA fragments are the virtually exclusive sites of repair replication activity. When the large fragments are treated with the exonuclease BAL-31, label is quickly removed leaving a relatively long unlabeled segment, one that lacks sequence homology with the shorter ones. The various lines of evidence point to PDNA consisting of two short labeled regions, referred to as PsnDNA (pachytene small nuclear), flanking an unlabeled middle segment (Fig. 13–5).

The RNA component of the 11S chromatin fragment will be referred to as PsnRNA because it is a small nuclear RNA, about 125 to 140 nucleotides in length; none of it is present in the cytoplasm. Approximately 97 percent of the RNA is localized in the chromatin. It hybridizes exclusively with PsnDNA, but none of the PsnRNA is present as an RNA/DNA hybrid in the chromatin. In vitro analyses of the hybrids formed by PsnRNA in the presence of excess PsnDNA indicate that they share the same sequence complexity. The conclusion that can be drawn in the case of lilies is that PsnRNA includes as many different sequence families as there are PsnDNA families, about 600 each repeated about 1000 times. Regardless of the complexity of this arrangement, the important fact to note is that the arrangement is specifically programmed for the pachytene stage and addressed to the recombinational events that occur then. The role of PsnDNA can be largely, if not precisely, accounted for. It must serve as the source of potential initiating sites for recombination, although no evidence is available as to how the thousands of PDNA sites relate to the 36 or less crossovers that occur in each meiotic cell. The role of PsnRNA is subject to more speculation and it is given due attention in the last component of this section. Before leaving the subject of PsnRNA two more properties need to be mentioned. The first is that the RNA is not only meiosis specific but that it is also an exclusive component of late zygotene and pachytene cells. This is made evident by gel analysis of RNA extracts from successive prophase stages using unfractionated PsnDNA as a hybridization probe (see Fig. 13–7). Clearly, it is transcribed to serve the process of recombination. The second property that deserves mention is the resistance of PsnRNA, when in the chromatin, to RNase action. The chromatin housing such resistance and also the PsnDNA and PsnRNA is conveniently referred to as Psn-chromatin. The resistance of PsnRNA to hy-

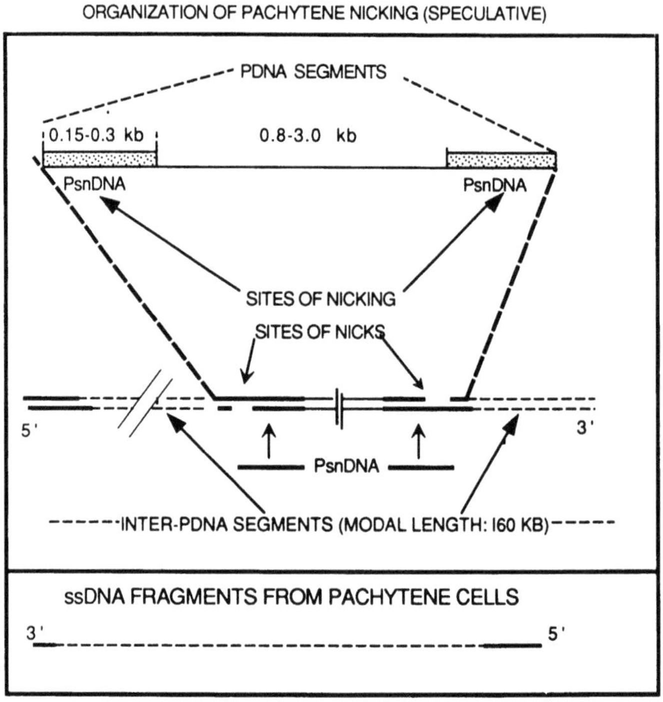

Figure 13–5. A diagram illustrating the way in which pachytene nicking is organized. A meiotic endonuclease is responsible for the nicking, which is initiated by a modification of the chromatin that renders the PDNA segments accessible to the enzyme. The active components within the segments are presumed to be located at its ends. They are referred to as PsnDNA because their sequences are homologous with families of small nuclear RNAs. Pachytene nicking is virtually confined to PsnDNA regions and the very large releasable ssDNA fragments from pachytene cells have PsnDNA sequences at their ends. The polarity of the nicking should be noted. (Based on findings of Hotta and Stern, 1984.)

drolysis cannot be attributed to hybrid formation with PsnDNA because such hybrids are not present. It would appear, rather, that the factor responsible is a protein.

An account of the proteins present in *Psn-chromatin* is, at best, tentative. The information available is limited and only some of the major characteristics can be mentioned (Stern, 1989). It is clear that there is little if any histone in Psn-chromatin. The isolated fragments dissociate readily to occur in the presence of 0.3 to 0.4M NaCl (Hotta and Stern, 1984). This is most unlikely to occur in the presence of histones and, indeed, gels of the protein components do not reveal any histones. The major component appears to be a single protein, not particularly basic, with a molecular weight in the range of 30 kd. It is referred to as *Psn-protein* because it has a high binding affinity for PsnRNA. It does bind nonspecifically to ssDNA as do all the meiotic proteins involved in some aspect of DNA metabolism, but it does not bind to dsPsnDNA. The protein protects PsnRNA from RNase action and this property almost certainly accounts for the ineffectiveness of RNase on the PsnRNA in the chromatin fragments, a property that is important in the isolation procedure. Gel analysis of the partially purified protein at successive meiotic stages indicates its virtually exclusive presence at the pachytene stage.

4. BEHAVIOR OF THE NICKED DNA REGIONS

A relevant question concerning DNA metabolism in pachytene cells is the extent to which the PsnDNA sites are nicked at pachytene (Fig. 13–5). A program in which only a fraction of the sites are subjected to nicking would point to a very different regulatory scheme from that in which all sites are affected. A direct test is possible because the 63S component of the ssDNA obtained from pachytene cells is a product of the programmed nicking and probing the sedimentation profile of total nuclear ssDNA from the cells with either PsnDNA or PsnRNA should reveal the distribution of the PsnDNA sequences. The experiments thus performed gave unambiguous results (Hotta and Stern, 1984). Almost all of the PsnDNA sequences are localized in the 63S component. Given the dynamic nature of the nicking process, the result permits the conclusion that all PsnDNA segments undergo nicking at pachytene. Because pachytene nicking is confined to the short PsnDNA segments, it is likely to be polarized. Speculations aside, there is direct evidence for polarity. PsnDNA sequences are at the termini of the 63S DNA fragments as demonstrated by their rapid removal upon treatment with an exonuclease. Moreover, superimposed on this latter characteristic is the evidence that most of the PsnDNA sequences are at the 5′ end of the 63S fragments. Digestion of the fragments with exonuclease-I that acts in a 3′–5′ direction results in a retention of the sequences until near completion of the entire fragment's hydrolysis. It is also possible to retrieve some components of the 63S fraction in duplex form and, as expected, the PsnDNA sequences are at the termini. These observations raise many unanswered questions but, even so, they reinforce the view that the process of pachytene nicking is highly regulated despite the general but unrefined genetic evidence that frequency of crossing over between two genes corresponds to the distance between them. This needs to be balanced by the equally general genetic evidence for *interference,* which refers to the fact that crossing over in adjacent regions of a chromosome is not independent but that a crossover in one region reduces the likelihood of a crossover in the adjacent one. The phenomenon is beyond the reach of current molecular approaches.

C. A Concluding Outline

The timing and siting of meiotic endonuclease action are determined by the programmed changes effected in chromatin structure. The nicking itself can be rationalized as the provision of a large excess of potential initiating sites for the ultimate event of a recombination. The nature of the mechanism that severely limits the number of crossovers while also assuring the occurrence of at least one per chromosome pair is unknown, but, if nicking serves as an example, that mechanism is not hindered by a failure of any of the earlier steps in the final set of events required for a crossover. This does not seem to be the case in *Saccharomyces cerevisiae* (Resnick, 1987), where the number of single-strand breaks is close to the number of recombinations, but the relationship may be affected by genome size. In higher eucaryotes there is an inverse relationship between genome size and number of recombinations per DNA unit. This relationship is illustrated in Figure 13–6, using lily, mouse, and *Drosophila* as examples.

These comments do not touch on the specific issues of the timing and siting of the chromatin changes. The issue of why there is an organization of specific nicking regions is understandably set aside. Even the simple issue of why heterochromatin lacks PsnDNA regions can only be explained by the fact that heterochromatic

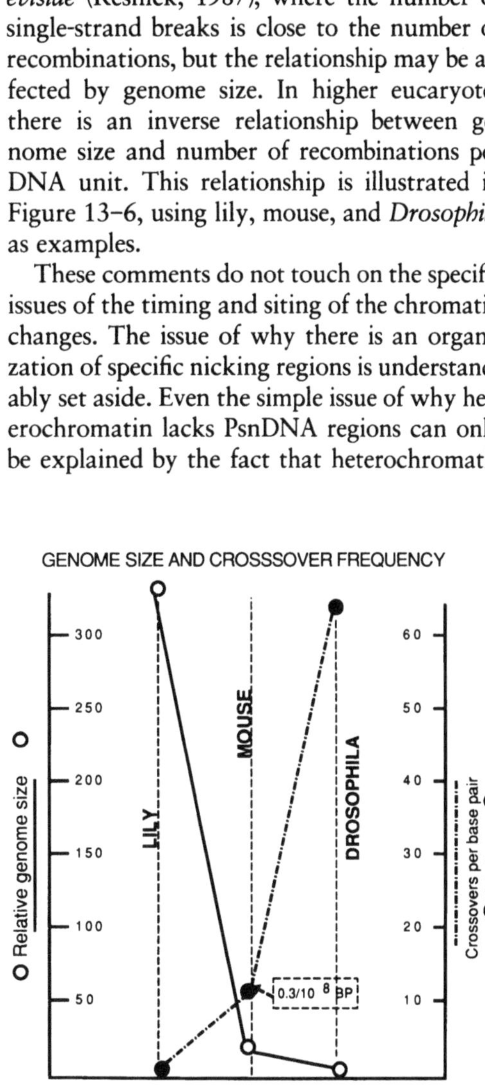

Figure 13–6. A comparison of genome size and crossover frequency per DNA content between three species comprising plant, insect, and mammal. The boxed insert contains the absolute frequency in the mouse. As indicated in the figure, all the comparisons are in relative terms.

DNA is not a recombinational region. The need to regulate the time of nicking is fairly evident. It must be coordinated with formation of synapsed chromosomes. The coordination is most likely effected by chromosomal action and it is not unreasonable to propose that an essential signal for that coordination is the transcription of PsnRNAs. Although one might suppose that each PsnDNA segment undergoes transcription, synchrony among the sites would probably be more readily achieved by some master sites of PsnRNA transcription. Each pachytene nucleus could thereby be flooded with PsnRNA molecules which could interact with their PsnDNA homologues. Such an action would cover the associated events of timing and siting. The complex of the PsnRNA with what has been called the Psn-protein could in some way lead to the replacement of the histones by that complex and render the PsnDNA open to the action of the meiotic endonuclease. Such a scenario is not without some experimental backing. In discussing the inaccessibility of meiotic endonuclease to PsnDNA sites in a nonpairing hybrid, the ineffectiveness of the enzyme was reported by incubating the isolated pachytene nuclei of the hybrid with meiotic endonuclease. Under those conditions no nicking occurred, a behavior that contrasted with that of normal chiasmatic pachytene nuclei, which displayed increased nicking. If, however, the hybrid nuclei were first incubated with PsnRNA, a protein fraction, and ATP, and then exposed to endonuclease, nicking did occur (Hotta and Stern, 1981). It did not occur if any one of the three components was omitted. Moreover, the nicking induced was in the PsnDNA regions. PsnRNA serves as a distinctive intranuclear messenger.

This ends a deficient account of intrachromosomal events. The final issue to be addressed is the role of nonchromosomal proteins in affecting chromosome behavior during the prophase interval of meiosis.

D. Proteins Affecting Chromosome Behavior at Prophase

1. GENERAL COMMENTS

There is no doubt that continued protein synthesis is a characteristic of meiotic prophase and that it is essential to the development of the meiotic process. This is readily demonstrated in lily meiocytes, which can be cultured in vitro and exposed to inhibitors or labeled precursors. In the rather distant past it was found that prophase cells exposed to [^{3}H]leucine displayed similar rates of incorporation into protein from leptotene through the remainder of meiotic prophase (Parchman and Stern, 1969). Moreover, exposure to cycloheximide (CHI) affected the rate of meiotic development and chromosome behavior depending on the concentration of CHI and the meiotic stage at which the drug was administered. Among the three concentrations tested at zygotene, 3.5×10^{-6}M, 1.8×10^{-6}M, and 3.5×10^{-7}M, the highest concentration caused a total meiotic arrest. The intermediate concentration caused the cells to be slowed in progress so that they required three times as long to reach the metaphase 1 stage (M1). Cells exposed to the lowest concentration required only 1.5 times as long to reach M1. Cells treated at leptotene or early zygotene were arrested at prophase. Those treated at late zygotene or early pachytene were rendered achiasmatic and displayed univalents rather than bivalents at M1. Those treated at late pachytene or diplotene were defective in disjunction. The action of the inhibitor is cumulative. Cells exposed to CHI for 24 hours and then returned to normal culture medium resumed normal meiotic development. The important consideration is the complexity and importance of biosynthesis during meiotic prophase. Whereas the synthesis of some proteins occurs during premeiosis in anticipation of meiotic requirements, other proteins must be synthesized in coordination with the processes they are addressed to. Comparable experiments with spermatocytes are unavailable, primarily because of the technical difficulties in their application.

Experiments with inhibitors, such as the ones described, demonstrate the need for a variety of proteins, but they provide no indications of functional specificity. The synaptonemal complex is an inviting target of protein analysis because of its role in pairing and recombination. Its protein components do have a functional specificity even if the precise nature of a particular protein function may be difficult to define. Several efforts have been made at isolating SCs in a pure state and using the preparations for protein analysis. The main analytical approach has been that of immunochemistry. The most elaborate studies have been reported by Heyting and colleagues (1989). They made electrophoretic gels of the protein components of purified rat SCs and also made monoclonal antibodies to many of the proteins. The antibodies were checked by immuno EM for the presence of the antigens in the SCs and those responding positively were checked by Western blotting for correspondence to the different gel bands. The results obtained thus far point to three polypeptides that are components of the LEs, one of which is also present in sperm heads. The other two are exclusive components of the LEs. The only additional piece of relevant information is that these proteins are products of de novo synthesis and are unique to meiotic chromosomes. A rigorous analysis of previously published immunocytochemical data on the presence of actin, myosin, or tubulin in the SC leads to the conclusion that the positive claims are highly doubtful (Dresser, 1987).

2. FUNCTIONAL HOUSEKEEPING PROTEINS

It should be useful to be able to discriminate between housekeeping and meiosis-specific enzymes because the latter point to the distinctive and unique features of the process. This, however, does not exclude the important fact that a number of enzymes which are active in mitotic cells play an important role in meiosis. The RAD 52 gene in *Saccharomyces cerevisiae* (Resnick, 1987) plays important roles in mitotic DNA repair and meiotic recombination. Comparable situations exist in higher eucaryotes with respect to proteins involved in DNA repair. Spermatocytes are capable of repairing DNA damage at almost all stages of meiosis, but there is a major difference between pachytene and all other stages with respect to DNA repair. It has been shown that pachytene spermatocytes have a much higher capacity to repair radiation or chemically induced DNA damage than those at earlier or later stages (Sega, 1974). Such high capacity is consistent with the evidence cited earlier for programmed nicking at the pachytene stage. There is no evidence for meiosis-specific repair enzymes, and although they may exist, it seems more likely that the increased capacity for repair is due to the level of one or more somatic enzymes that are present. Alternatively, regulatory genes could also be a causal factor and such a role has been proposed for the RAD 52 gene (Resnick, 1987).

Indirectly associated with intense DNA repair activity at pachytene is the behavior of DNA polymerases during meiotic prophase. In mouse spermatocytes there is a radical shift in activity levels of DNA polymerases α and β. DNA polymerase β virtually replaces the α form. Measurements made on early to late prophase spermatocytes indicate a β:α ratio of 50, a striking contrast with mouse liver polymerases in which the ratio is 0.8 (Sakaguchi et al., 1980). Despite the much higher spermatocyte ratio, the activity of liver DNA polymerase β is almost twice that of spermatocyte activity. There is no evidence about the role which DNA polymerase β might be playing in meiotic prophase. The likeliest interpretation is that this polymerase has a preferential role in DNA repair replication and that DNA polymerase α, which is known to have a major role in semiconservative replication, is no longer required. The fertilized egg provides ample polymerase for the earliest stages of development. The housekeeping enzyme, DNA polymerase β, would seem to have an important function in meiotic recombination.

Unlike the DNA polymerases, the activity profiles of other housekeeping enzymes in lily meiocytes that function in DNA metabolism are generally high during premeiotic interphase and remain so through pachytene (Stern and Hotta, 1983). These include polynucleotide kinase, polynucleotide ligase, topoisomerase II, and topoisomerase I. The latter activity is relatively low throughout the prophase interval. The meiotic role of topoisomerase II has not been adequately examined. It may well be a critical factor in removing interlocks among different chromosomes. Such interlocks occur frequently and must be resolved to avoid chromosome breakage during disjunction. The evidence that it is the most abundant nonhistone protein in isolated mitotic metaphase scaffolds (Boy de la Tour and Laemmli, 1988) and that it is also present in the cores of pachytene chromosomes (Moens and Earnshaw, in press) adds probability to the conjecture on its role in interlock removal. The evidence obtained on the few housekeeping enzymes tested makes it very likely that many more such enzymes are functionally involved in pachytene recombinational metabolism. A significant number of genetically and molecularly characterized DNA repair enzymes in yeast vegetative cells have been shown to be involved in meiotic recombination (Resnick, 1987); a similar situation must prevail among mammals.

3. MEIOSIS-SPECIFIC PROTEINS: HELICASES

One impressive and significant feature of metabolic organization at the zygotene and pachytene stages is the sharp appearance of a collection of proteins during zygotene (in some cases, at leptotene) that reaches a peak in early pachytene and returns to a low base level at its termination (Fig. 13–7). When tested, the proteins have been found to be pulse labeled during zygotene, thus indicating a coordination between synthesis and the rise in enzyme activity. The proteins studied can be related to synapsis or recombination, or both. They have their counterparts in somatic cells, but their properties are distinctive enough to justify their assignment to different genetic loci. Such assignment, it must be stressed, is without the benefit of genetic analysis.

Two of the proteins studied function in the winding or unwinding of DNA. Either process can be functional in synapsis and recombination. One of these, which is the first discussed, is an unwinding protein, *U-protein;* it has been studied only in lily microsporocytes (Hotta and Stern, 1978b). Its unwinding action is consistent with one of the many requirements for recombination. It unwinds 400 to 500 bp from a nicked site; the length unwound appears to be self-limiting. Duplex ends are also subject to unwinding but in this case, no more that 50 bp are unwound. The limiting mechanism is unknown. The process is ATP dependent, but the protein's ATPase activity is not tied to this unwinding action. It acts as an ATPase in the presence of ssDNA to which it readily binds. U-protein requires a 3′OH end to initiate unwinding so that dephosphorylation is required for the protein to act on the 3′P nicks introduced by the meiotic endonuclease. Binding between the U-protein and the site of unwinding is fairly stable. Once unwinding has been initiated the protein is not released even in the presence of a 200-fold excess of nicked duplex DNA. The stability may account for its limited unwinding action. Finally, the protein unwinds DNA regardless of source, but it has no effect on DNA–RNA hybrids. Its meiotic profile is sharp. Its concentration rises sharply at the beginning of zygotene, reaches a peak during late zygotene–early pachytene, and returns to a negligible level at the end of pachytene. No activity has been detected in vegetative cells.

The second helicase studied is more properly classified as a reassociation protein, *R-protein.* It has been studied in both lily microsporocytes and mammalian spermatocytes (Hotta and Stern, 1971b). Its meiotic profile is similar to that of the U-protein. Its activity begins to rise at mid-leptotene, reaches a peak at early pachytene, followed by a decline to a negligible level at the end of pachytene. The protein can unwind DNA in media lacking Mg^{++}, a condition in which the duplex is unstable. In presence of the cation, it is highly effective in catalyzing the renaturation of ssDNA. Sixty percent of denatured lily DNA, which is of exceedingly high sequence complexity, is converted to dsDNA in 2 hours at 25°C. The action does not require ATP. Its reassociation activity is effective with any ssDNA homologous mixture. Renaturation rate is dependent on sequence complexity. The protein has been found in rat, bull, and human spermatocytes, but not in mitotic cells of testicular tissue.

A distinctive property of the R-protein is its localization in nuclear lipoprotein, presumably as part of the nuclear membrane. It may be recalled that the L-protein is similarly localized. The relationship between these two proteins, if any, is appropriate for speculation, but if so an additional factor needs to be included. It has been shown (Hecht and Stern, 1971) that zygDNA is transiently associated with the nuclear membrane during its replication. There is little doubt that in some way the nuclear membrane is directly involved in a particular phase of synapsis, but the three physical associations do not throw light on a specific mechanism. As in the case of the U-protein, administering radiolabeled leucine to cultured leptotene–zygotene cells effects a radiolabeling of the R-protein. Moreover, labeling can also be effected by administering [^{32}P]phosphate. As will be seen,

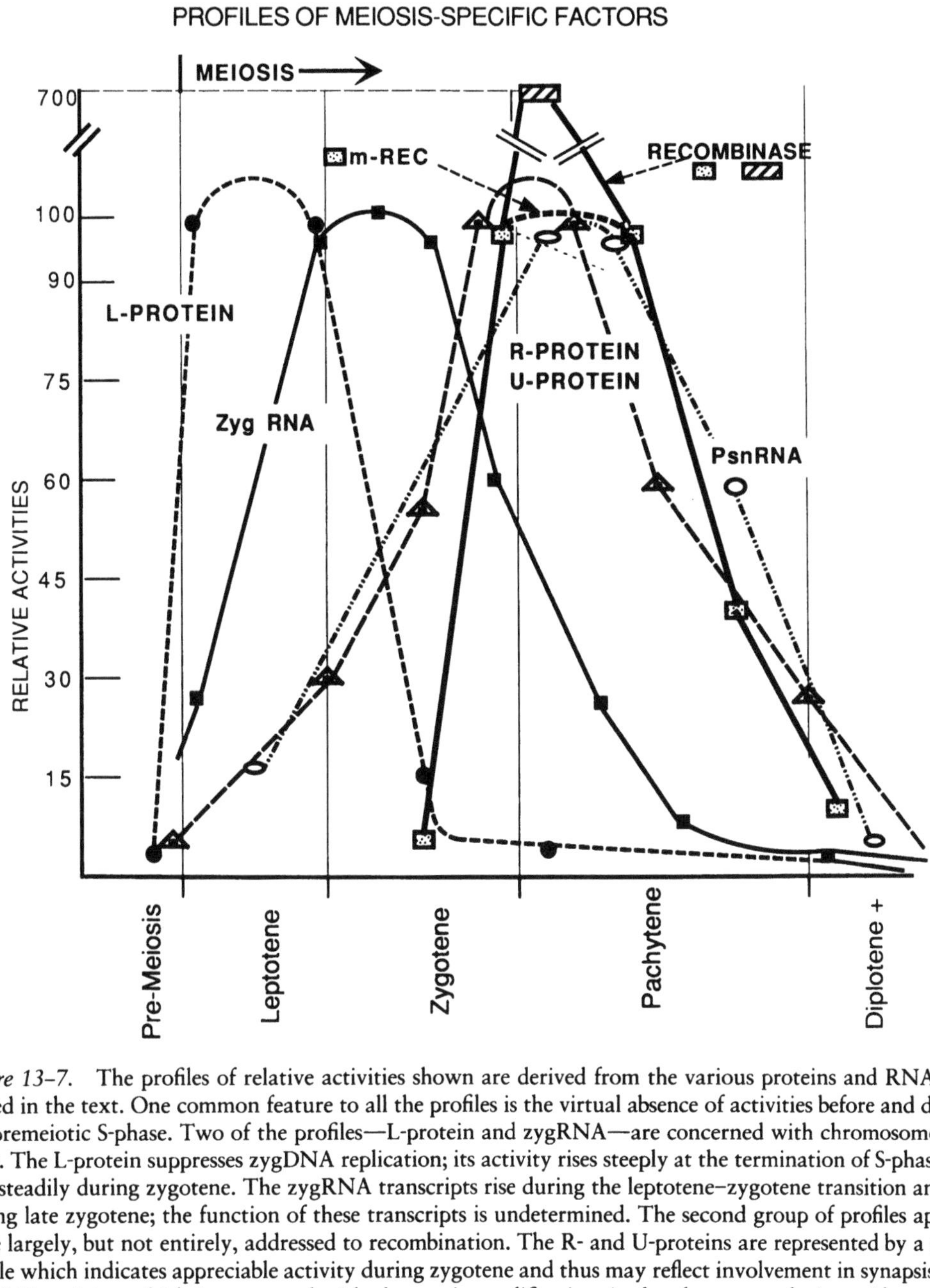

Figure 13–7. The profiles of relative activities shown are derived from the various proteins and RNAs discussed in the text. One common feature to all the profiles is the virtual absence of activities before and during the premeiotic S-phase. Two of the profiles—L-protein and zygRNA—are concerned with chromosome synapsis. The L-protein suppresses zygDNA replication; its activity rises steeply at the termination of S-phase and falls steadily during zygotene. The zygRNA transcripts rise during the leptotene–zygotene transition and fall during late zygotene; the function of these transcripts is undetermined. The second group of profiles appears to be largely, but not entirely, addressed to recombination. The R- and U-proteins are represented by a single profile which indicates appreciable activity during zygotene and thus may reflect involvement in synapsis. The PsnRNA activity, which is concerned with chromatin modification, is already present during mid-zygotene, presumably in anticipation of pachytene nicking. The m-rec and recombinase (in vitro) activities are represented by one curve except for the peak levels. The rise in recombinase activity is exponential, as shown in Figure 13.8 for mouse spermatocytes. The profile in this figure is based on lily microsporocytes because meiotic stage purification is much more satisfactory. Both recombinogenic proteins reach peak values at the transition from zygotene to pachytene. It must be emphasized that assignment of activities to intervals within stages or to borderlines between stages is looser than the diagram indicates.

such labeling is not due to phospholipid association but rather to a phosphorylation of the protein.

The capacity of R-protein to renature ssDNA depends on its phosphorylated state. Mather and Hotta (1977) found that the protein prepared from rat spermatocyte nuclei failed to bind and renature ssDNA if pretreated with cyclic adenosine monophosphate (cAMP) protein kinase. Treatment with alkaline phosphatase partially restored the binding. A more detailed study of phosphorylation was conducted with R-protein from lilies (Hotta and Stern, 1979). The ratio of protein bound to ssDNA versus that bound to dsDNA was 80 in the case of native R-protein, 1.09 in the case of phosphatase treated protein, and 40 in the case of the dephosphorylated protein treated with an endogenous protein kinase. Treatment of the native protein with a cAMP protein kinase increased the amount of bound phosphate, but the protein had negligible binding affinity for either ssDNA or dsDNA. The capacity of R-protein to reanneal ssDNA depends on its having a high binding affinity for ssDNA relative to that of dsDNA. Obviously, binding to ssDNA is a precondition of reassociation, and a lack of binding affinity for the reassociated duplex appears to be required for catalysis to proceed. The endogeneous protein kinase applied to the phosphatase-treated protein partially restores reassociation activity. This does not occur if R-protein is treated with cAMP protein kinase. Presumably the sites of phosphorylation need to be specific for functional activity. Indeed, only 1 to 2 phosphate residues have been estimated as bound to the native R-protein molecule.

The extent to which phosphorylation of R-protein regulates its activity in vivo is unknown. It is of interest, however, that phosphorylation controls yet another aspect of R-protein activity. In both mice and lilies R-protein interacts with DNA polymerase β and B, respectively (Sakaguchi et al., 1980). The protein stimulates the activity of DNA polymerase β in mice and its counterpart in lilies stimulates DNA polymerase B. It has no effect on polymerase α or A (lily). The murine R-protein has no effect on lily DNA polymerase B and the lily R-protein is equally ineffective on the activity of murine DNA polymerase β. In both species stimulation does not occur if the proteins are dephosphorylated or hyperphosphorylated; the capacity to reanneal has the same dependency on protein phosphorylation as the capacity to stimulate DNA polymerase β. Inasmuch as the two proteins have similar profiles of meiotic activity, it seems reasonable to conjecture that both are essential to the recombinational activities at pachytene. The conjecture offers no insights on the nature of the in vivo mechanism.

In lilies, but not in mice, it is possible to test for a regulatory relationship between the rise in R-protein and chromosome synapsis. As previously described, there is a lily hybrid that has limited homology for pairing so that there is very little effective pairing at pachytene. In tracing R-protein levels through meiotic prophase (Hotta et al., 1979), only a small increase in the proteins is detected during the zygotene and pachytene stages, an increase commensurate with the low level of chiasma formation. The characteristically high level of R-protein is restored along with chiasma formation in the tetraploids generated from the diploid hybrid. It is unlikely that the low protein level is the result of the absence of synapsed chromosomes because the R-protein normally begins to increase slightly before the initiation of synapsis. It would appear that the factors necessary to, but preceding, synapsis are the source of the defect. The L-protein, which is also localized in the nuclear lipoprotein fraction but increases in level prior to initiation of pairing, is only slightly reduced in the asynaptic hybrid. The marked reduction in R-protein under asynaptic conditions contrasts with the behavior of L-protein, which is the more directly associated with synapsis. The relationship between synapsis and certain meiosis-specific proteins is difficult to explain. Additional examples are provided by recombinases.

4. MEIOSIS-SPECIFIC PROTEINS: RECOMBINASES

The process of recombination, even at the molecular level, is far from being resolved. This holds for site-specific recombination and also for general recombination, the latter being the principal issue in meiotic behavior. Given this situation, there is an obvious difficulty in dealing with meiotic recombination from a molecular or biochemical standpoint. The ultimate set of events that underlies a reciprocal exchange of large or small fragments between two homologous DNA duplexes is clearly molecular in nature. But, this still unresolved set of events is embedded in processes that are much more remote from resolution. These processes have already been mentioned in discussing the cytogenetic aspects of chromosome behavior during meiotic prophase. The regularly fulfilled requirement of at least one crossover per chromosome bivalent, the phenomenon of positive interference that prohibits randomness of crossing over even at the gross chromosomal level, and the exclusion of recombination from heterochromatic regions are three examples pointing to the complex regulatory factors that govern the expression of the ultimate molecular events. Clearly, the general organization of recombinational activity within the meiocyte is a more immediate and meaningful target of study.

What follows in this section is an attempt to display whatever information is available concerning the general organization of recombination mechanisms in the meiotic cell. Several specific molecular components are discussed but no effort is made to present a coherent model of meiotic recombination. Such a model would be fanciful in the extreme and contribute little to a dissection of the process that would reveal the many serious gaps in our understanding. In effect, the presentation will not provide an attractive textbook picture.

An important aspect of the organization of recombinational mechanisms is inherent in the activity profiles of the proteins that are involved in the process. This has already been discussed for the helicases, which are probably components of recombinational mechanisms. Their properties, however, are also relevant to other processes, such as replication and transcription, and their functioning in the latter categories cannot be excluded. The distinctive feature of the recombinogenic proteins to be discussed is their direct involvement in recombination, and this feature is of special relevance to meiotic recombination. None of the proteins in question displays any significant level of activity until the meiocytes are already in zygotene (Fig. 13–7). This almost certainly lays to rest the periodic claims that recombination occurs during the premeiotic S-phase. The timing of the synthesis and appearance of the proteins coincides with cytogenetic evidence for recombination being a pachytene event. This does not appear to be precisely the case for yeast in which recombination overlaps the latter interval of premeiotic S (Resnick, 1987). Such overlaps are unlikely in the higher eucaryotes if only because of the cytogenetic evidence tying recombination to pachytene and the autoradiographic evidence excluding the S-phase from pachytene. Chromosome size may have an important bearing on the need to coordinate chromosome synapsis with recombination. Each of the chromosomes in *Saccharomyces cerevisiae* is smaller than the *Escherichia coli* chromosome. The structural differences between chromosome and recombinational synapsis in *E. coli* are less complex than those in mammalian meiocytes that must accommodate chromosomes of very much larger size. It is conceivable that in yeast meiocytes, the SC is indispensable for normal meiotic disjunction but somewhat dispensable for meiotic recombination.

The *E. coli* recA protein is, by far, the most thoroughly studied protein in the field of recombination. The genetic evidence for this role in recombination is overwhelming. The finding of a recA protein in meiotic cells that displays a typical zygotene–pachytene profile of activity is, therefore, of major interest (Hotta et al., 1985c). Essentially identical profiles have been found in mouse and lily. A similar, although not identical, enzyme has been identified in yeast meiocytes (Resnick, 1987). The recA-like proteins found in mouse and lily match those of *E. coli* with respect to the basic properties studied. Inasmuch as similar but not identical proteins have been found in meiotic and somatic tissues, the meiotic protein is referred to as *m-rec* and the somatic protein as *s-rec.*

Both m-rec and s-rec proteins bind to ssDNA and the binding is much increased in the presence of ATP. Under the latter conditions the proteins catalyze homologous reassociation. If the rates of reassociation catalyzed by mouse m-rec and s-rec are compared using ΦX174 as substrate, the m-rec rate is about 15 times as rapid as the s-rec rate. A more efficient catalysis by m-rec holds for all the properties tested. Except for the ATP requirement, m-rec catalyzes reassociation like R-protein. Roughly compared, the R-protein appears to be the more efficient. Under similar conditions, R-protein catalyzed 80 percent reassociation of ssT7DNA in 1 hour, whereas m-rec achieved only 40 percent reassociation. However, m-rec but not R-protein can couple the reassociation with other events required for recombination.

D-loop formation is one of the recombinogenic properties of recA. In the presence of ATP m-rec and s-rec promote the homologous insertion of ssDNA into a closed circular duplex. Because topological factors prohibit the penetrating strand from major displacement of the complementary strand, a D-loop configuration results. The joint molecules formed are stable when heated for 10 min at 65°C in presence of 0.1% sodium dodecylsulfate, 2 mM EDTA, and 0.15M NaCl. Under conditions identical to those required by recA, mouse and lily rec proteins catalyze *strand transfer,* the transfer of one strand from a linear duplex to a homologous circular ssDNA. The extent of strand transfer is very high, between 65 and 75 percent of the circular DNA becoming resistant to S1 nuclease. The formation of open circular duplexes is also evident from gel analysis of the reaction products. As in the case of D-loop formation, m-rec was found to be five to eight times more effective than s-rec in circular duplex formation. R-protein and U-protein appear to have no significant effect on the extent of strand transfer, unlike the effect of binding proteins on this activity of recA.

As indicated earlier for recombinogenic proteins in general, the m-rec protein displays a steep rise in activity at early zygotene, reaching the peak in late zygotene that persists through early pachytene. It then falls sharply so that cells at diplotene display no m-rec activity. The sharp rise coincides with the pairing process, but whether m-rec is involved in chromosome synapsis remains an open question. It is, nevertheless, significant that a decrease in synapsis is accompanied by corresponding reduction in m-rec activity. In this respect, m-rec and R-protein are similar in behavior. The most important feature of these observations is that in the metabolic organization of the meiocyte the activity of a major recombinogenic protein is centered on the interval of chromosome pairing. The activity of m-rec is not in evidence when the bulk of genomic DNA is replicated. That m-rec and s-rec are different or differently regulated proteins is discussed after a consideration of in vitro recombination.

A general but less specific analysis of recombinogenic protein activities during meiosis is to test for recombination in appropriately constituted cell extracts at successive meiotic stages. Insofar as the procedure yields demonstrable recombinational activities, it can serve as an indicator of any changes in levels of relevant proteins other than m-rec in the meiotic process. Neither recA nor m-rec has been found to generate recombinations under in vitro conditions. The method involves an incubation of two mutant plasmids in the cell extracts. Both plasmids contain the gene for resistance to ampicillin and each plasmid contains a distinctive mutation in the gene for resistance to tetracycline. Recombination between the two plasmids yields a plasmid that codes for resistance to tetracycline. Plasmids, incubated in cell extracts, are assayed for the latter property by transforming recA$^-$ *E. coli,* strain HB101, and measuring the resultant proportion of tetracycline/ampicillin-resistant colonies. The proportion in control incubations is 10^{-6}.

Extracts from mouse, lily, and yeast all have such in vitro recombinational activity (Hotta et al., 1985c). The structure of the plasmids has a significant effect on activity. Maximum rates of recombination are obtained when one plasmid is supercoiled and the other is a linear duplex. A slightly lower rate is obtained when both plasmids are supercoiled. No significant recombination occurs if one of the plasmids is single stranded, either circular or linear. Two relaxed circular plasmids also fail to undergo recombination. Clearly, one supercoiled duplex is essential to the success of the process in extracts of all three species. Moreover, their meiotic behavior with respect to recombination is approximately uniform. Cells of *Saccharomyces cerevisiae* exposed to sporulation medium begin to increase in activity about 3 hours after exposure and reach a peak at the sixth hour. The latter is about 70 times the rate in extracts from vegetative cells. Mouse spermatocytes reach a peak level at pachytene, which is 500 times the level in round spermatids and about 40 times the level in bone marrow extracts

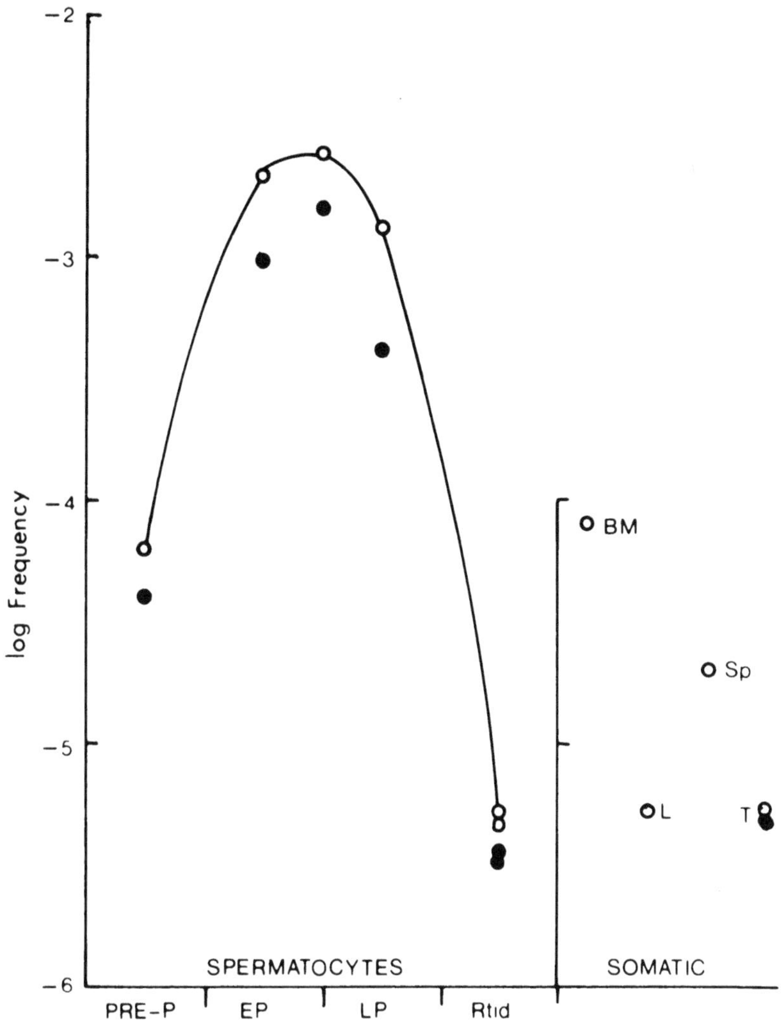

Figure 13–8. In vitro recombination activities as determined in mouse spermatocyte and somatic cell extracts. The somatic extracts were prepared from liver (L), thymus (T), spleen (Sp), and bone marrow (BM). The germline extracts assayed were from prepubertal spermatogonia and spermatocytes (Pre-P), spermatocytes in early prophase (EP), in late prophase (LP) and 90 percent round spermatids and about 10 percent premeiotic spermatocytes (Rtid). The *open circles* represent recombinations between linear and supercoiled plasmids; the *closed circles* represent recombination between two supercoiled plasmids. (From Hotta et al., 1985c.)

which are the highest found for somatic tissues (Fig. 13–8). Lily meiocytes, which can be more precisely staged than mouse spermatocytes, begin to increase at the beginning of zygotene, reach a peak representing a 700-fold increase at late zygotene–early pachytene, and drop to 1/1000 the activity at the termination of meiosis. The profiles of activity in mouse and lily meiocytes are the same as those of m-rec and thus reinforce the evidence that the timing of recombinational activity coincides with the achievement of chromosome pairing, which, in turn, coincides with the time of chromosome nicking and repair replication. Addition of m-rec to mouse extracts increases activity three-fold; U-protein addition generates a similar three-fold increase; U-protein and m-rec in combination yield the same increase (threefold) as each does independently. These increases are obviously small compared with those occurring during the leptotene to pachytene transition. In assessing the significance of these results it is important to bear in mind that although the extracts are required to effect the recombination observed in the transformed bacteria, information is lacking as to what steps in the process are effected by the extracts and which steps are provided by the bacteria. It is clear only that one or more essential and initial steps are present in the extracts; completion is effected by *E. coli.* These studies furnish pointers to the general way in which meiotic cells address the mechanism of recombination; they are distant from revealing specific mechanisms.

Although the rise and fall of in vitro recombinase activity is precisely coordinated with chromosome synapsis, its meiotic profile is unlike that of m-rec with respect to synapsis. The profile in lily meiocytes is indifferent to chromosome pairing. The achiasmatic hybrid displays the same sharp rise and fall in activity during the zygotene–pachytene interval as does the chiasmatic one. As mentioned earlier, such experimental tests are not possible in mice. However, an equivalent pattern of behavior is demonstrable in yeast. When yeast cells that are a/a or α/α in genetic constitution are transferred to sporulation medium they achieve peak in recombinase activity similar to that of the a/α diploids, although they do not enter meiosis. One or more of the initial steps in the recombinase activity are not regulated by chromosome synapsis, but we have no information on how many molecular pathways are available to the meiocyte at pachytene.

5. EFFECT OF TEMPERATURE ON TESTICULAR FUNCTION

The debilitating effect of elevated testicular temperature on fertility in humans and other mammals is well established. Analysis of the behavior of the recombinogenic proteins with respect to temperature serves two quite different purposes. On one hand, it provides information on the role, if any, that recombination may play in the temperature effect on fertility. On the other hand, it provides a relatively simple procedure to determine whether somatic and meiotic recombinase activities represent different proteins despite their similarities in behavior. The information provided by observations made is unambiguous. The temperature differential examined in mice is between 37 and 33°C, the former being optimal for all organs except the testes, which function optimally at 33°C. The rec proteins and those of the in vitro recombinase all display the corresponding optimal temperatures for activity.

The proteins m-rec and s-rec differ in temperature optima even with respect to ssDNA binding. The m-rec binds more strongly at 33°C than at 37°C, whereas the opposite is true for s-rec although their sedimentation values are measured at the same temperature. The m-rec displays three times the level of D-loop activity at 33°C than at 37°C. Strand transfer activity of m-rec at 33°C is five times that at 37°C (Fig. 13–9). No such differences have been observed for s-rec; activities at 37°C are slightly higher than those at 33°C. In all tests made, 37 to 39°C is the range for maximal activity. It is not known whether s-rec and m-rec are products of different genes or posttranslational modifications. The important point is that rec activity in spermatocytes is catalyzed by a different protein from that in somatic tissues. Although the assays for in vitro recombinase activity are more laborious than those for the rec protein, comparisons between somatic and meiotic cells yield the same contrast. Profiles of activities versus temperature reveal a peak activity for somatic preparations at 37 to 38°C and for meiotic ones at 33°C. Murine spermatocytes are programmed to function at the lower temperature and, presumably, so too are human spermatocytes.

A different approach to examine the effect of temperature on recombinational activity was made by Hotta and co-workers (1988) using surgical technique to translocate rat testes and their appendages to the abdominal cavity. Eight rats were so treated and an equal number were subjected to sham operations. At specific days after operation the testes were removed and extracts from spermatocytes prepared by the procedure used to assay for recombinase activity under standard conditions. Activities in the sham-operated rats remained unchanged. The cryptorchid testes dropped 50-fold in recombinase activity 24 hours after operation and by the fifth day it had dropped 350-fold. Beyond the fifth day there was no further decrease (Fig. 13–10). The result would be of little interest if the effect was general. This was not the case. The DNA-binding proteins underwent no change. In a separate study referred to

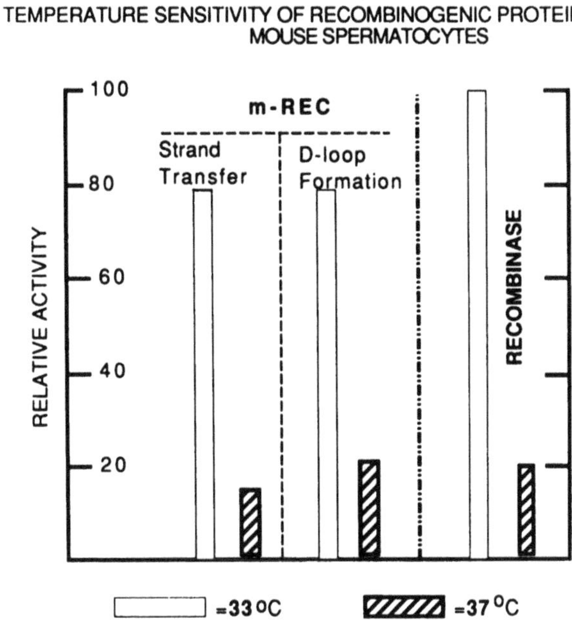

Figure 13–9. Activities of m-rec and recombinase (in vitro) at 37 and 33°C. The relative values are based on data in the publications of Hotta and co-workers (1985c, 1988). The optimal temperature for the somatic enzymes is, in all cases, about 37°C, slightly higher and appreciably lower activities occurring at 39 and 33°C, respectively.

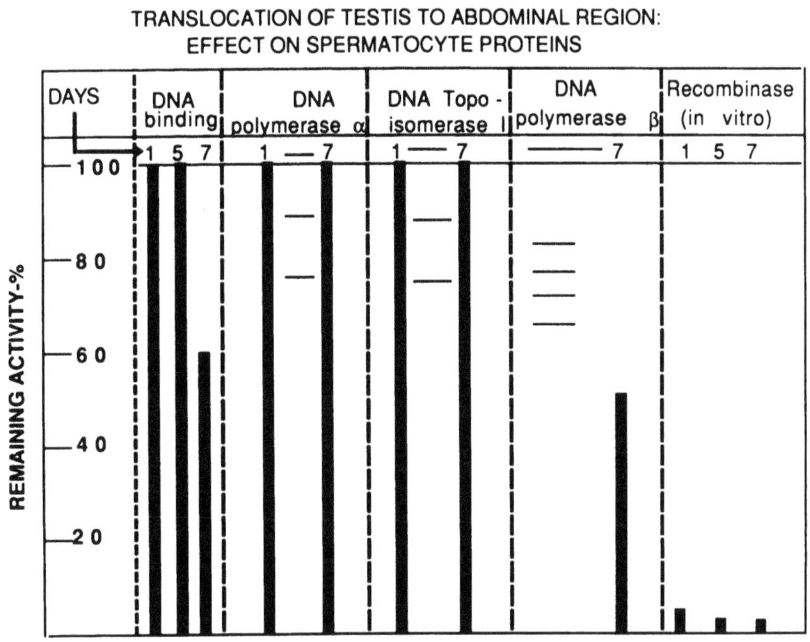

Figure 13–10. The diagram summarizes the experimental results reported by Hotta and co-workers (1988) in their attempt to determine the effect of exposing rat testes to abdominal temperature on the activities of various proteins in spermatocytes. It is evident that the recombinase activity drops precipitously within 24 hours following translocation. Presumably, the enzyme(s) is destroyed or inactivated as result of the temperature increase. No change was observed in these proteins after sham operations. It may be significant that DNA polymerase β, but not α, is affected by the temperature shift because the β enzyme is prominent in the spermatocytes.

in that report it was found that DNA polymerase α and topoisomerase I remained unchanged during the entire period. None of the latter enzymes displays marked activity during meiosis. On the other hand, DNA polymerase β, which is active during meiosis, was half as active after 7 days. These studies reflect an in vivo selective response by the testes to temperature. Among the various assays performed, the recombinase system would appear to be the most sensitive to temperature elevation. What remains undetermined is whether the loss in measured activity is due to a change in turnover rates or simply to protein inactivation and/or destruction, or both. It may, nevertheless, be concluded that one major temperature-sensitive component of the spermatogenic process is the enzymatic mechanism effecting recombination.

IV. Concluding Statement

A. Spermatocyte–Somatic Cell Interaction

The intimate association between spermatogenic and Sertoli cells has been fully discussed in Chapters 9 and 11. So too have the hormonal activities of the Leydig cells. In terms of what is known with respect to somatic cell contributions to the meiotic process, little can be added here. What needs to be emphasized is that despite the numerous studies and characterizations of Sertoli and Leydig cells, nothing tangible can be reported about the specific role which either of these cells plays in any particular phase of meiosis. Numerous efforts have been made to develop culture systems for spermatocytes that could match the performance of the culture systems now available for a large variety of somatic cells, but success has eluded those efforts. The evidence for the dependence of the meiocytes on the somatic cells is compelling but the acquisition of an understanding of that dependence may first require a much better understanding of the mechanisms underlying the meiotic process itself. Even the simple question about spermatocyte–spermatocyte interaction lacks a clear-cut answer. It is a characteristic of many gametogenic systems, animal and plant, that the germ cells develop as multinucleate groups, or *syncytia*. When meiosis is initiated, neighboring spermatocytes share relatively large intercytoplasmic passages. This undoubtedly contributes to developmental synchrony among neighbors, but such synchrony is unlikely to be a sufficient reason for so widespread a pattern of development. The phenomenon invites speculation.

A most intriguing set of experiments was performed by Lyon and colleagues (1975) to answer a basic question concerning the absence of spermatogenesis in individuals suffering from a mutant gene for the testosterone receptor, a condition called *testicular feminization.* Their aim was to establish whether the germ cells failed to develop because the cells were themselves unresponsive to testosterone or whether the failure occurred because the closely associated Sertoli cells were unresponsive. They approached the problem by constructing chimeric mice consisting of $X^{tfm}Y$ and XY genotypes, each of the strains carrying suitable genetic markers. Their starting population consisted of 57 chimeras from which they selected those males that were fertile and hence had a majority of XY somatic cells in the testis but also an appreciable proportion of $X^{tfm}Y$ cells in the germ line. When such males were mated with normal females a fraction of the female progeny were of the $X^{tfm}X$ genotype, but only if the $X^{tfm}Y$ germ cells could undergo spermatogenesis despite their lack of testosterone receptors. Those females could then be mated to XY males and half the XY progeny would be $X^{tfm}Y$ and female in phenotype. That result was indeed obtained and thus provided an unambiguous answer: dependency of spermatogenesis on testosterone derived from the Sertoli cell requirement for the hormone. The demonstration adds emphasis to the earlier statement that spermatocyte dependency on interactions with the somatic cells of the testes is unaccompanied by any understanding of the specific mechanisms underlying that dependency.

B. Perspective

This chapter has addressed three principal aspects of meiosis: its distinctive role in the reproductive system, the structural gymnastics which chromosomes undergo in the course of the process, and those features of metabolic organization that are essential, but not sufficient, to the fulfillment of that role. Although the chapter necessarily focuses on mammalian spermatocytes while stressing the universality of the process, it needs to be noted that more than one pathway is followed to achieve the same end. Such notation is important because the application of information obtained in studies of other species to the mammalian spermatocyte is unavoidable. Apart from light and electron microscopic studies, approaches to meiotic mechanisms in mammals have not yielded extensive information because of the limited experimental materials available. Some of the applications described here may turn out to be inappropriate despite the apparently compelling evidence based on the assumption that the major processes underlying meiosis are universal.

The distinctive role of meiosis in reproduction derives from two considerations, the requirement for gametes to be haploid and the genetic desirability of interchromosomal exchange and chromosomal reassortment in the course of achieving haploidy. Novel gene combinations in gametes means novel gene combinations in offspring, but the meiotic process is indifferent to the desirability of such novelty. The latter is a by-product of normal meiosis. In most cases, including all mammals, a failure in homologous interchromosomal exchange is hazardous to the achievement of haploidy. Many of the aneuploid zygotes that give rise to spontaneous abortions or to debilitated offspring are products of abnormal disjunction. The tight relationship between crossing over and disjunction was already recognized in the early genetic studies of *Drosophila,* but it is unfamiliar to many biologists.

Chromosomes must undergo homologous pairing for crossing over to occur, and crossing over must occur for the paired state to achieve its predetermined goal of balanced homologous reassortment. The events underlying that achievement involve a complex set of structural and compositional changes in the chromosomes. Quite apart from the changes that chromosomes undergo in transition from pachytene to the highly compacted metaphase state, the prepachytene changes are critical to chromosome synapsis and to recombination. The premeiotic interval, including at least part of the S-phase and extending to leptotene, encompasses critical events in chromosome organization. We have extremely limited information on the complement of proteins synthesized during this premeiotic period but we have indicated that these syntheses coincide with an irreversible commitment of cells to meiosis. The *L-protein* represents one of the syntheses. Less defined from a molecular standpoint is the formation of lateral elements which appear to consist of several proteins and which serve to provide for the stabilization of chromosome pairing once it has occurred. The mechanism by which homologous chromosomes are aligned remains a mystery. All that can be said is that the synaptonemal complex is not a source of alignment, and that if DNA matching is the source, only a small fraction of the DNA can be so involved. Some speculation on the subject has been provided. It should be evident that although we have extensive information at the level of microscopic analysis on the structural changes occurring during meiotic prophase, extremely little is known about the molecular aspects of those changes.

Little has been said about the process of chromosome disjunction, a process which has its counterpart in mitosis. Again, if the structural features of that process are put aside, there is little that can be discussed in molecular terms. It is the rich detail of chromosomal behavior during meiosis that makes the meiotic process an inviting target of research. It is the technical difficulties in obtaining convenient experimental material that discourages responses to the invitation. Nevertheless, the fact that meiosis is the gateway to sexual reproduction reassures the writer that sooner or later the invitation will be fully honored.

ACKNOWLEDGMENTS

The author wishes to express his thanks to N.I.C.H.D. whose research support made the writing of this chapter possible. He also gratefully acknowledges the continuous past support of NSF and the American Cancer Society. Together, these three institutions account for the numerous studies that have been described.

REFERENCES

Boy de la Tour, E., and Laemmli, U. K. (1988). The metaphase scaffold is helically folded: sister chromatids have predominantly opposite helical handedness. Cell 55:937–944.

Carpenter, A.T.C. (1975). Electron microscopy of meiosis in Drosophila melanogaster females: II. The recombination nodule—a recombination-associated structure at pachytene? Proc. Nat. Acad. Sci. (USA) 72:3186–3189.

Chandley, A. C. (1987). Meiotic analysis in germ cells of man and mouse, in *Mammalian development: a practical approach* (Monk, M., ed.) pp. 71–91, IRL Press, Eynsham.

Chandley, A. C., and McBeath, S. (1987). DNAse hypersensitive sites along the XY bivalent at meiosis in man include the X_pY_p pairing region. Cytogenet. Cell Genet. 44:22–31.

Chandley, A. C., Hotta, Y., and Stern, H. (1977). Biochemical analysis of meiosis in the male mouse. I. Separation and DNA labelling of specific spermatogenic stages. Chromosoma 62:243–253.

Del Fosse, F. E., and Church, K. (1981). Presynaptic chromosome behavior in Lillium. I. Centromere orientation and movement during premeiotic interphase (PMI) in *Lilium speciosum* var. *Rosemede*. Chromosoma 81:701–716.

Dresser, M. E. (1987). The synaptonemal complex and meiosis: an immunochemical approach, in *Meiosis* (Moens, P. B., ed.) pp. 245–274, Academic Press, Orlando.

Earnshaw, W. C., and Rattner, J. B. (1989). A map of the centromere (primary constriction) in vertebrate chromosomes at metaphase, in *Mechanisms of Chromosome Distribution and Aneuploidy* (Resnick, M. A., and Vig, B. K., eds.) pp. 33–42, Alan R. Lis Inc., New York.

Hasenkampf, C. A. (1989). A method for producing synaptonemal complex complements in lily and mouse. J. of Heredity 80:197–202.

Hassold, T. J., and Jacobs, P. A. (1984). Trisomy in man. Ann. Rev. Genet. 18:69–87.

Hecht, N. B., and Stern, H. (1971). A late replicating DNA complex from cells in meiotic prophase. Exp. Cell Res. 69:1–10.

Heyting, C., Dietrich, A. J., Moens, P. B., Dettmers, R. J., Offenberg, H. H., Redeker, E.J.W., and Vink, A.C.G. (1989). Synaptonemal Complex Proteins. Genome 31:81–87.

Hotta, Y., and Stern, H. (1971a). Analysis of DNA synthesis during meiotic prophase in *Lilium*. J. Mol. Biol. 55:337–355.

Hotta, Y., and Stern, H. (1971b). Meiotic protein in spermatocytes of mammals. Nature New Biol. 234:83–86.

Hotta, Y., and Stern, H. (1975). In *The Eukaryote Chromosome* (Peacock, W. J., Brock, R. D., eds.) pp. 283–300, The Australian National University Press, Canberra.

Hotta, Y., and Stern, H. (1976). Persistent discontinuities in late replicating DNA during meiosis in *Lilium*. Chromosoma 55:71–182.

Hotta, Y., and Stern, H. (1978a). Absence of satellite DNA synthesis during meiotic prophase in mouse and human spermatocytes. Chromosoma 69:323–330.

Hotta, Y., and Stern, H. (1978b). DNA unwinding protein from meiotic cells of *Lilium*. Biochemistry 17:1872–1880.

Hotta, Y., and Stern, H. (1979). The effect of dephosphorylation on the properties of a helix-destabilizing protein from meiotic cells and its partial reversal by a proteinkinase. Eur. J. Biochem. 95:31–38.

Hotta, Y., and Stern, H. (1981). Small nuclear RNA molecules that regulate nuclease accessibility in specific chromatin regions of meiotic cells. Cell 27:309–319.

Hotta, Y., and Stern, H. (1984). The organization of DNA segments undergoing repair synthesis during pachytene. Chromosoma 89:127–137.

Hotta, Y., Chandley, A. C., and Stern, H. (1977). Biochemical analysis of meiosis in the male mouse. II. DNA metabolism at pachytene. Chromosoma 62:255–268.

Hotta, Y., Tabata, S., and Stern, H. (1984). Replication and nicking of zygotene DNA sequences: control by a meiosis-specific protein. Chromosoma 90:243–253.

Hotta, Y., Bennett, M. D., Toledo, L. A., and Stern, H. (1979). Regulation of R-protein and endonuclease activities in meiocytes by homologous chromosome pairing. Chromosoma 72:191–201.

Hotta, Y., Tabata, S., Stubbs, L., and Stern, H. (1985a). Meiosis-specific transcripts of a DNA component replicated during chromosome pairing: homology across the phylogenetic spectrum. Cell 40:785–793.

Hotta, Y., de la Pena, A., Tabata, S., and Stern H. (1985b). Control of enzyme accessibility to specific DNA sequences during meiotic prophase by alterations in chromosome structure. Cytologia 50:611–620.

Hotta, Y., Tabata, S., Bouchard, R. A., Pinon, R., and Stern, H. (1985c). General recombination mechanisms in extracts of meiotic cells. Chromosoma 93:140–151.

Hotta, Y., Fujisawa, M., Tabata, S., Stern, H., and Yoshida, S. (1988). The effect of temperature on recombination activity in testes of rodents. Exp. Cell Res. 178:163–168.

Howell, S. H., and Stern, H. (1971). The appearance of DNA breakage and repair activities in the synchronous meiotic cycle of *Lilium*. J. Mol. Biol. 55:357–378.

Ito, M., and Stern, H. (1967). Studies of meiosis *in vitro* I. *In vitro* culture of meiotic cells. Devel. Biol. 16:36–53.

Jacobs, P. A., Angell, R. R., Buchanan, I. M., Hassold, T. J., Matsuyama, A. M., and Manuel, B. (1978). The origin of human triploids. Ann. Hum. Genet. London. 42:49–57.

Kassir, Y., Granot, D., and Simchen, G. (1988). IME1, a positive regulator gene of meiosis in S. cerevisiae. Cell 52:853–862.

Leroy, P., Alzari, P., Sassoon, D., Wolgemuth, D., and Fellous, M. (1989). The protein encoded by a murine male germ cell-specific transcript is a putative ATP-dependent RNA helicase. Cell 57:549–559.

Lyon, M. F., Glenister, P. H., and Lamoreux, M. L. (1975). Normal spermatozoa from androgen-resistant germ cells of chimaeric mice and the role of androgen in spermatogenesis. Nature 258:620–622.

Maguire, M. P. (1978). Evidence for separate genetic control of crossing over and chiasma maintenance in maize. Chromosoma 65:173–183.

Mather, J., and Hotta, Y. (1977). A phosphorylatable DNA-binding protein associated with a lipoprotein fraction from rat spermatocyte nuclei. Exp. Cell Res. 109:181–189.

Meistrich, M. L., and Brock, W. A. (1987). Proteins of the meiotic cell nucleus, in *Meiosis* (Moens, P. B., ed.) pp. 333–354, Academic Press, Orlando.

Moens, P. B., and Earnshaw, W. C. (in press). Anti-topoisomerase II recognizes meiotic chromosome cores. Chromosoma.

Monesi, V., Geremia, R., D'Agostino, A., and Biotani, C. (1978). Biochemistry of male germ cell differentiation in mammals. RNA synthesis in meiotic and postmeiotic cells. Current Topics in Develop. Biol. 12:11–36.

Moses, M. J., Dresser, M. E., and Poorman, P. A. (1984). Composition and role of the synaptonemal complex, in *Symp. Soc. Exp. Biol.* (Evans, C. W., and Dickinson, H. G., eds.) Vol. 38, pp. 245–270, The Company of Biologists Ltd., Cambridge.

Moses, M. J., Poorman, P. A., Roderick, T. H., and Davisson, M. T. (1982). Synaptonemal complex analysis of mouse chromosomal rearrangements. IV. Synapsis and synaptic adjustment in two paracentric inversions. Chromosoma 84:457–474.

Palmer, R. G. (1971). Cytological studies of ameiotic and normal maize with reference to premeiotic pairing. Chromosoma 35:233–246.

Parchman, L. G., and Stern, H. (1969). The inhibition of protein synthesis in meiotic cells and its effect on chromosome behavior. Chromosoma 26:298–311.

Radding, C. M. (1981). Recombination activities of *E. coli* recA protein. Cell 25:3–4.

Resnick, M. A. (1987). Investigating the genetic control of biochemical events in meiotic recombination, in *Meiosis* (Moens, P. B., ed.) pp. 157–210, Academic Press, Orlando.

Rhoades, M. M. (1956). Genetic control of chromosome behavior. Maize Genet. Coop. Newsletter 30:38–42.

Richler, C., Uliel, E., Rosenmann, A., and Wahrman, J. (1989). Chromosomally derived sterile mice have a 'fertile' active XY chromatin conformation but no XY body. Chromosoma 97:465–474.

Sakaguchi, K., Hotta, Y., and Stern, H. (1980). Chromatin-associated DNA polymerase activity in meiotic cells of lily and mouse; its stimulation by meiotic helix destabilizing protein. Cell Structure and Function 5:323–334.

Sega, G. A. (1974). Unscheduled DNA synthesis in the germ cells of male mice exposed in vivo to the chemical mutagen, ethylmethanesulfonate. Proc. Nat. Acad. Sci. (USA) 71:4955–4959.

Stern, H. (1986). Meiosis: some considerations. J. Cell Science. Suppl. 4, 29–43.

Stern, H. (1989). Meiosis, in *Chromosomes: eukaryotic, prokaryotic, and viral* (Adolph, K. W., ed.) Vol. II, pp. 3–37, CRC Press, Boca Raton.

Stern, H., and Hotta, Y. (1980). The organization of DNA metabolism during the recombination phase of meiosis with special reference to humans. Molec. Cell Biochem. 29:145–158.

Stern, H., and Hotta, Y. (1983). Meiotic aspects of chromosome organization, in *Stadler Genet. Symp.* (Gustafson, J. P., ed.) Vol. 15, pp. 25–41, University of Missouri, Columbia.

Stern, H., and Hotta, Y. (1985). Molecular biology of meiosis: synapsis-associated phenomena, in *Aneuploidy: etiology and mechanisms.* (Dellarco, V., Voytek, P. E., and Hollaender, A., eds.) pp. 305–316, Plenum, New York.

Stern, H., and Hotta, Y. (1987). The biochemistry of meiosis, in *Meiosis* (P. Moens, ed.) pp. 303–331, Academic Press, New York.

Stubbs, L., and Stern, H. (1986). DNA synthesis at selective sites during pachytene in mouse spermatocytes. Chromosoma 93:529–536.

von Wettstein, D., Rasmussen, S. W., and Holm, P. B. (1984). The synaptonemal complex in genetic segregation. Ann. Rev. Genet. 18:331–413.

Wolfes, H., Katsuhisa, K., Millette, C.F., and Cooper, G. M. (1989). Specific expression of nuclear proto-oncogenes before entry into meiotic prophase of spermatogenesis. Science 245:740–743.

CHAPTER FOURTEEN

Cell Biology of Mammalian Spermiogenesis

Yves Clermont, Richard Oko, and Louis Hermo

I. Spermiogenesis

A. Introduction

The spermatozoon is an unusual cell destined to deliver its nuclear DNA into a far-reaching target cell, the oocyte. It is a highly specialized cell with a head containing compact DNA in the nucleus and the acrosomic system provided with the enzymatic machinery necessary to penetrate the shell of the oocyte. It also has a tail made up of a contractile apparatus, the axoneme, in addition to cytoskeletal structures and mitochondria providing the energy necessary to maintain the tail's motion. This cell is the product of an extraordinarily intricate process of cell differentiation termed *spermiogenesis.* Spermiogenesis is the third and last phase of spermatogenesis taking place in the seminiferous epithelium of the testis. The objective of the present review is to examine the structural and biochemical features of spermiogenesis, beginning with a brief review of the structure of the developing spermatids and an examination of a number of general cytological features of spermiogenesis. Then, we examine systematically the changes of each organelle or cytoplasmic component of the spermatid during spermiogenesis. Although some of the features of spermiogenesis and the spermatozoon have been recently reviewed (de Kretser and Kerr, 1988; Eddy, 1988; Oko and Clermont, 1990), we include in this chapter several of our more recent findings on the rat.

B. Microanatomy of Spermatozoa

Mammalian spermatozoa are composed of a head and a tail. The *head* has four major components. The first is a condensed nucleus; the second is a perinuclear theca that has two regions, a subacrosomal region or perforatorium and a postacrosomal dense lamina or postacrosomal sheath (Figs. 14–1A,B). The perinuclear theca leaves a small area of the nucleus uncovered around the implantation fossa where the tail is attached; this theca-free area is referred to as the perifossal zone (Fig. 14–1B) (Lalli and Clermont, 1981). The acrosomic system, the third component, is a membrane-bound organelle that is tightly attached to the nucleus by means of the perinuclear theca (Fig. 14–1B). The acrosomic system has two distinct regions, the acrosome proper, which is a thickened portion over the dorsal surface of the nucleus, and the head cap, which is thinner and seen over the lateral and ventral surfaces of the nucleus (Fig. 14–1B). In some spatulate spermatozoa, this thinner head cap portion is called the equatorial segment (Fig. 14–1C) (Fawcett, 1975, 1979). Finally, the plasma membrane adheres closely to the head of the spermatozoon, with a thin film of cytoplasm between it and the outer acrosomal membrane. There is also a small accumulation of cytoplasmic matrix around the perifossal zone (Fig. 14–1B).

The heads of mammalian spermatozoa vary markedly in appearance owing to the species-

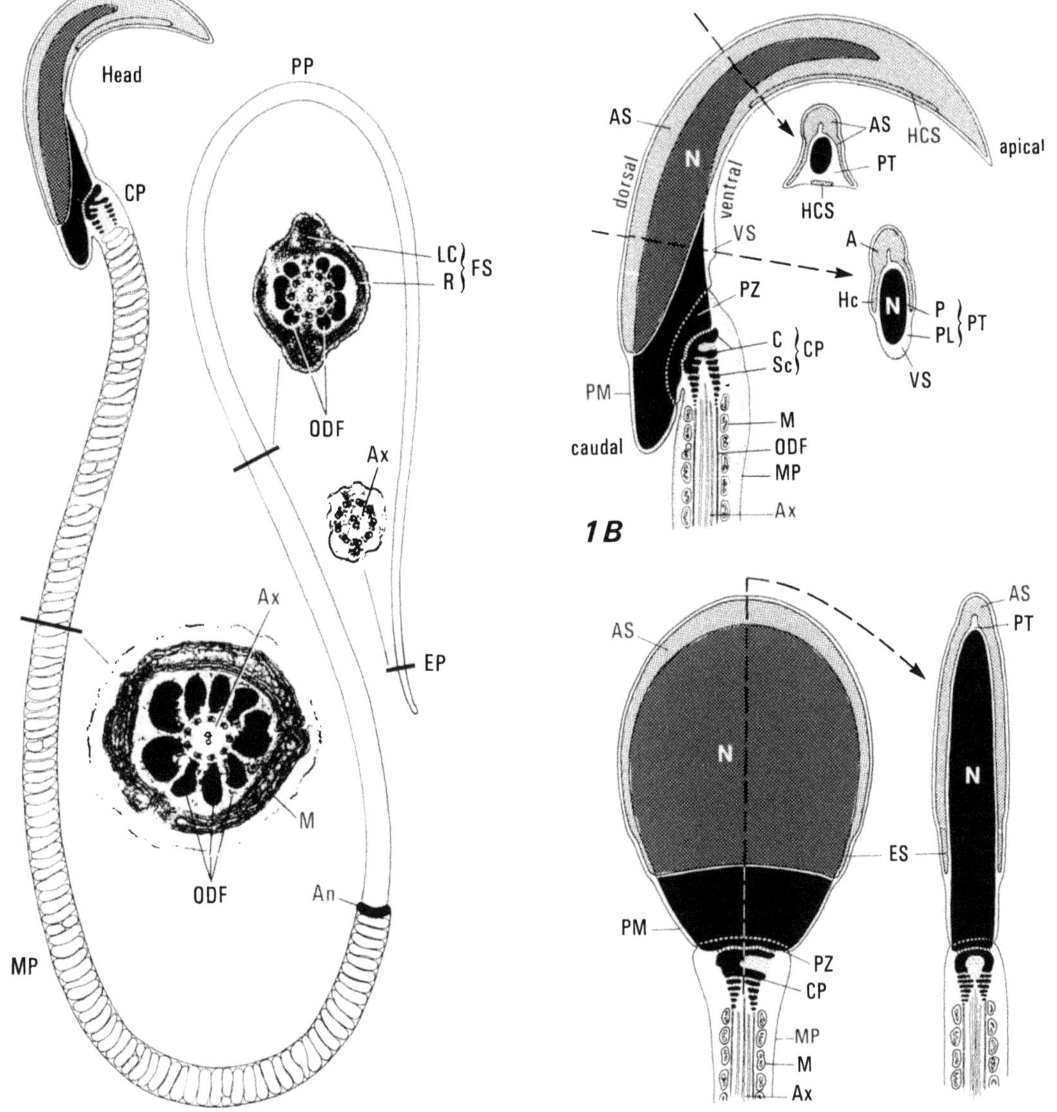

Figure 14–1. (**A**) A rat spermatozoon with its sickle-shaped head and its 190 μm long tail. (**B**) Some details of the falciform head of a rat spermatozoon. (**C**) A general representation of the spatulate head of mammalian spermatozoa. In **A** the four regions of the tail are indicated, that is, the neck or connecting piece (CP) joining the head to the tail, the middle piece (MP) showing the transversely oriented mitochondria, the principal piece (PP), and the end piece (EP). The annulus (An) is also indicated at the junction of the middle and principal pieces. Electron microscope photographs along the tail give the ultrastructural details of each region as seen in cross section. These show the axoneme (Ax), the outer dense fibers (ODF), the longitudinal columns (LC), and the transverse ribs (R) of the fibrous sheath (FS) and mitochondria (M). In **B** and **C** the various parts of the falciform and spatulate heads are shown in side and cross-sectional views. PM, plasma membrane; N, nucleus indicated in black; AS, acrosomic system indicated as *stippled area;* A, acrosome; HC, head cap; HCS, separated head cap segment; VS, ventral spur; *dotted white line* on the nucleus demarcates the area of the postacrosomal dense lamina (dorsally) from the perifossal zone (PZ) (ventrally). CP, connecting piece with its striated columns (Sc) and capitulum (C); MP, middle piece with its axoneme (Ax) associated with outer dense fibers (ODF) and mitochondria (M); P, perforatorium and postacrosomal dense lamina (PL) regions of the perinuclear theca (PT); ES, equatorial segment.

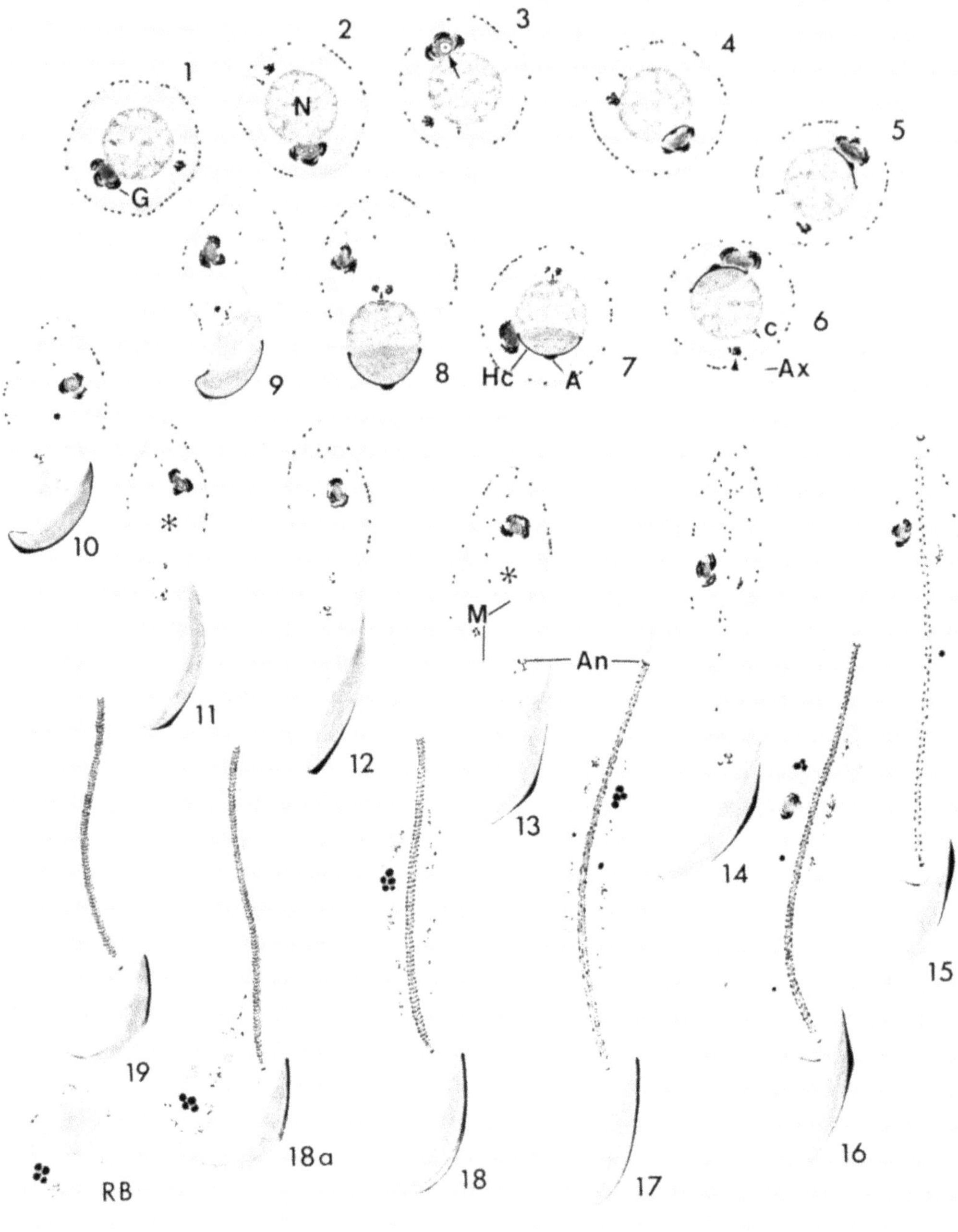

Figure 14–2. Drawing of the steps of spermiogenesis in the rat as seen with the light microscope in semithin sections of glutaraldehyde-fixed testes stained with toluidine blue. The numbers correspond to the steps of spermiogenesis as proposed in the classification of Leblond and Clermont (1952). During the first three steps or Golgi phase, the Golgi apparatus (G) elaborates proacrosomic granules that fuse to give rise to a single acrosomic granule *(arrow)* that becomes associated with the nucleus (N). During steps 4 to 7, or cap phase, the Golgi apparatus contributes to the formation of the acrosome proper (A) and head cap (Hc) portion of the acrosomic system. In these cells, the centrioles (c) and developing axoneme (Ax), the chromatoid body *(arrowhead),* and mitochondria (*small dots* along the plasma membrane) are also visible. During steps 8–14 or acrosome phase, the spermatid becomes polarized, the nucleus elongates and takes on a characteristic sickle shape; the bulk of cytoplasm forms the cytoplasmic lobe *(asterisks)* along the forming tail. The Golgi apparatus detaches from the acrosomic system, which undergoes marked structural changes at the surface of the nucleus.

specific shape of the nucleus. In mammals, spermatozoa generally fall into two main categories, the *spatulate* (Fig. 14–1C), which are more or less flattened and paddle-shaped, among which the pyriform or pear-shaped head of human spermatozoa is a variant, and the *falciform,* which are elongated, curved, and pointed, but present a wide variety of configurations with the sickle-shaped head of the rat spermatozoon being a well-known example (Fig. 14–1B). Some mammalian species have heads with extravagant shapes (e.g., bandicoot; Sapsford et al., 1969), but these will not be considered here as their description falls within the domain of comparative spermatology (see Fawcett, 1970, 1975, 1979; Koehler, 1970; Fawcett et al., 1971; Bernstein and Teichman, 1972; Baccetti and Afzelius, 1976).

The *tail* of mammalian spermatozoa varies considerably in length depending on the species (e.g., 190 μm long in the rat and only 60 μm long in humans). It consists of four distinct regions: the connecting or neck piece, the middle, the principal, and the end pieces (Fig. 14–1A). The *connecting or neck piece* is a short segment composed of an electron-dense material surrounding hollow cavities that in some cases contain remnants of the centrioles (Fawcett and Phillips, 1969b; Zamboni and Stefanini, 1970; Fawcett, 1975). It is composed of a capitulum and striated collar or columns (Fig. 14–1B) (Fawcett and Phillips, 1970). The capitulum is continuous distally with the striated collar, which in turn is continuous with the outer dense fibers at the junction of the connecting and middle pieces (Fig. 14–1B). The *middle piece* (Fig. 14–1A) includes: the axoneme composed of microtubules (nine peripheral doublets plus two central singlets); nine coarse outer dense fibers (ODFs), each with a characteristic droplike shape and each facing one of the nine microtubular doublets; mitochondria, elongated and crescentic in shape; and a thin film of cytoplasm delimited by the plasma membrane (Figs. 14–1A,B). The *principal piece* includes the axoneme and seven ODF showing a decreasing caliber facing microtubular doublets 1,2,4,5,6,7, and 9; ODF 3 and 8 are replaced by the longitudinal columns of the fibrous sheath (Fig. 14–1A). The fibrous sheath is a major cytoskeletal component of the principal piece made up of the longitudinal columns and transverse ribs (Fig. 14–1A). At the surface of the fibrous sheath there is a thin film of cytoplasm delimited by the plasma membrane (Fig. 14–1A). The end piece is composed of axonemal elements, which are disorganized distally, a film of cytoplasm, and the plasma membrane (Fig. 14–1A). The annulus, a plasma-membrane-associated cytoplasmic density is seen at the junction of the middle and principal pieces (Fig. 14–1A). In this review the plasma membrane of mammalian spermatozoa or spermatids is only mentioned in passing. This major component of germinal cells has been the subject of numerous studies that have been reviewed recently by Eddy (1988).

C. Steps of Spermiogenesis

Spermiogenesis is a continuous process stretching from the time of formation of the spermatid to the release of the fully formed spermatozoon. Evidently it is not sufficient to refer only to round and elongated spermatids when this multifaceted process of cell differentiation is analyzed. Conveniently spermiogenesis has been subdivided into the Golgi, cap, acrosome, and maturation phases, each being subdivided into steps on the basis of the changes observed in the nucleus and the developing acrosomic system (Leblond and Clermont, 1952; Clermont, 1954; Clermont and Leblond, 1955). Nineteen steps of spermiogenesis are defined in the rat (Fig. 14–2), 17 steps in the hamster (Clermont, 1954), 14 in the monkey (Clermont

---

The manchette (M) is visible as well as the annulus (An) seen next to the site of the implantation fossa of the tail. During the last steps 15–19 or maturation phase, the annulus (An) slides down the forming tail, the manchette disappears, and the mitochondria migrate toward the forming tail. The nucleus covered by the acrosomic system completes its evolution. The acrosomic system condenses and slides apically for a short distance. The underlying perforatorium develops at the tip of the condensed nucleus. The cytoplasmic lobe flows apically and the bulk of it separates from step 19 spermatid to form the residual body (RB). (Drawing from Clermont and Rambourg, 1978.)

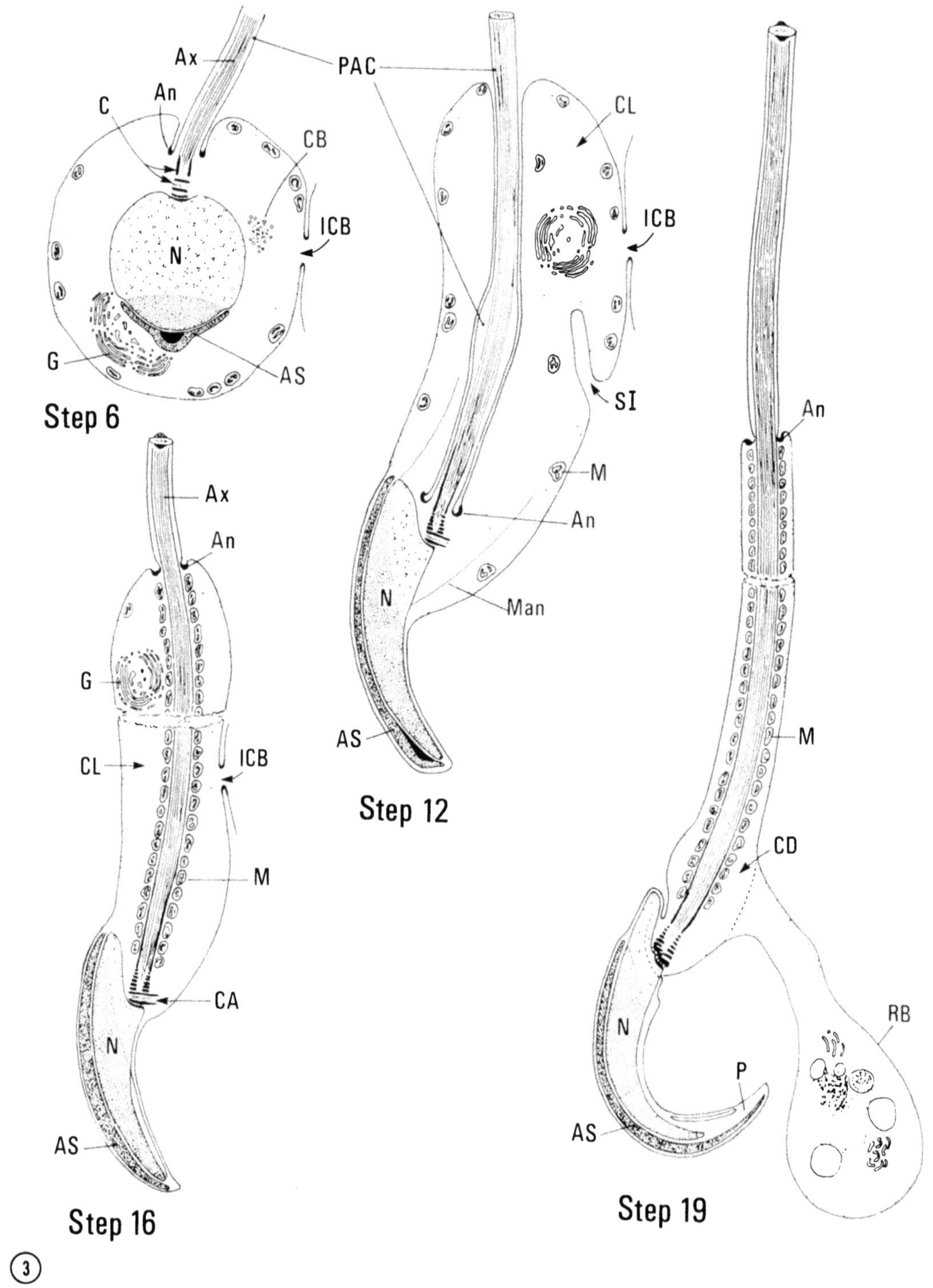

Figure 14–3. Diagram showing some of the main changes taking place during spermiogenesis of the rat as seen with the low power of the electron microscope. Four steps of spermiogenesis are illustrated: steps 6, 12, 16, and 19. These drawings show the transformation of the head components, that is, nucleus (N) and acrosomic system (AS), and cytoplasm that is spheroidal initially then becomes elongated and forms a drop-like lobe referred to as the cytoplasmic lobe (CL) along the forming tail. Toward the end of spermiogenesis, the

and Leblond, 1955), 6 in humans (Clermont, 1963), and so on. The classification proposed for the rat is now commonly used for both light and electron microscopic studies and serves as a system of reference in the present review (Fig. 14–2).

D. General Features of Spermiogenesis

Before entering the analysis of the genesis of the various components of the spermatozoon, it is useful to review some general features of the process of spermiogenesis.

1. In mammals, spermiogenesis is a long cell biological process. With the use of radioactive tracers and the radioautographic technique, it was possible to determine that spermiogenesis lasts approximately 20 days in the rat (Clermont et al., 1959) and 24 days in humans (Heller and Clermont, 1964), durations that are representative of the duration of spermiogenesis in most mammals (Clermont, 1972).
2. Spermiogenesis is a rigidly timed process and every step in the differentiation of the spermatid has a fixed duration within a given species (Clermont and Harvey, 1965). Furthermore, such durations have not been shown, as yet, to be extended or shortened by extracellular factors, such as, temperature, hormones, and growth factors. This feature needs to be remembered when regulatory mechanisms of spermatogenesis or spermiogenesis are investigated.
3. During the differentiation of the spermatid, several cellular systems are evolving simultaneously. Thus, for example, the endoplasmic reticulum, Golgi apparatus, mitochondria, centriolar–axonemal complex, as well as the nucleus itself, all simultaneously undergo structural and biochemical transformations, and these various processes have to be perfectly integrated and coordinated to yield a normal spermatozoon. In the following discussion, the evolution of the various organelles will be examined separately, but in reality these processes should be visualized in an integrated manner.
4. Spermiogenesis is a phenomenon that involves numerous spermatids all connected to each other by open intercellular bridges. This feature is not restricted to spermatids but also occurs in spermatogonia and spermatocytes (Fawcett et al., 1959; Dym and Fawcett, 1971; Weber and Russell, 1987) such that the eventual number of spermatids associated in such clones is remarkably large. This unique feature may contribute to the synchronization of spermatids into large groups of cells that are all at exactly the same step of spermiogenesis within any given segment of the seminiferous tubule.
5. The differentiation of spermatids is not structurally or functionally independent of their surroundings. Indeed, these cells develop in intimate relationship with Sertoli cells (review Bardin et al., 1988). In the present review a few examples of this close structural and functional association between spermatids and Sertoli cells are addressed.

---

bulk of the cytoplasm flows toward the head and detaches from the step 19 spermatid to form the residual body (RB), leaving behind a small cytoplasmic droplet (CD) around the neck region of the spermatozoon. In early spermatids, the centrioles (C), pulling the plasma membrane associated anlagen of the annulus (An) with them, move toward and become associated with the nuclear envelope (step 6). With the displacement of the cytoplasm caudally along the axoneme (Ax), there is the formation of an inner or proximal periaxonemal compartment (PAC) delimited by a double plasma membrane (step 12); the cytoplasm of the distal portion of the axoneme constituting the distal periaxonemal compartment is delimited by a single plasma membrane. With the migration of the annulus along the forming tail (compare step 12 with step 16), the narrow proximal periaxonemal compartment disappears; the mitochondria (M) now have a free access to the forming tail to which they bind to form the mitochondrial sheath that characterizes the middle piece of the tail. G, Golgi apparatus; CB, chromatoid body; ICB, open intercellular bridges that connect adjacent spermatids; Man, manchette (present only during the acrosome phase); CA, centriolar adjunct; SI, Sertoli invagination; P, perforatorium.

E. Cytoplasmic Compartments in Spermatids

The newly formed spermatid is spheroidal or polyhedral with a centrally located spherical nucleus. The spermatids maintain this shape during the Golgi (steps 1–3) and cap (steps 4–7) phases (Figs. 14–2 and 14–3). During these phases, the active growth of the axoneme is initiated by the distal centriole. The migration of the pair of centrioles toward the nucleus occurs at these phases and is accompanied by an invagination of the plasma membrane along the axoneme resulting in the formation of a distinct narrow cytoplasmic periaxonemal compartment (steps 6 and 12 of Fig. 14–3). This compartment can be further subdivided into two regions. The proximal periaxonemal compartment is the portion surrounded by the cytoplasmic lobe and consequently is demarcated by a double plasma membrane, its own as well as that of the cytoplasmic lobe (step 12, Fig. 14–3). The distal periaxonemal compartment is the portion of axonemal cytoplasm that is beyond the cytoplasmic lobe and consequently is delimited by only a single plasma membrane (step 12, Fig. 14–3).

In the acrosome phase (steps 8–14) the nucleus elongates, condenses, and loses its central position by shifting toward the cell surface. This shift in nuclear position results in the bulk of the cytoplasm flowing to the opposite pole of the cell where it takes on the shape of an elongated lobe that extends along the developing tail. Thus, in elongated spermatids, there exists only a thin rim of cytoplasm around the nucleus, whereas the bulk of the cytoplasm is within the cytoplasmic lobe that is situated beyond the nucleus and that demarcates the boundaries of the proximal periaxonemal compartment (step 12, Fig. 14–3).

At the onset (step 15) of the maturation phase (steps 15–19), as a result of the migration of the membrane-associated annulus along the axoneme, the narrow proximal periaxonemal compartment loses its demarcation (steps 16 and 19, Fig. 14–3). This situation now enables the bulk of cytoplasmic proteins to have direct access to the ODFs, which undergo their most rapid growth at this time along the axoneme and a direct contact of mitochondria with the ODFs along the middle piece of the tail, that is, between the connecting piece and the annulus (steps 16 and 19, Fig. 14–3). In steps 18 and 19 of the maturation phase, the cytoplasmic lobe migrates toward the head region of the spermatid and eventually detaches from the late step 19 spermatid to yield the residual body but leaves behind, attached to the neck region, a small mass of cytoplasm called the cytoplasmic droplet (step 19, Fig. 14–3) (Fawcett and Phillips, 1969a; Russell, 1984; Hermo et al., 1988). It should finally be noted that the intercellular bridges that connect spermatids to each other are maintained until spermiation when the residual cytoplasm separates from the spermatid.

II. Evolution of Cytoplasmic Elements

A. Precursors of Head Components

1. NUCLEUS AND NUCLEAR ENVELOPE

Modification of the Chromatin During Spermiogenesis

Immediately after its formation following the two meiotic divisions, the small spherical haploid nucleus of the spermatid shows a widely dispersed chromatin with few heterochromatin clumps and a small spherical nucleolus. The appearance of the chromatin does not change much during the Golgi and cap phases (steps 1–7). The first morphological evidence of chromatin condensation is observed in steps 8–10 when a few coarse clumps appear (Fig. 14–4). Thereafter, chromatin filaments form in step 11 and progressively thicken and aggregate into coarse granular masses throughout steps 12 and 13 (Fig. 14–5). The condensation of the chromatin is initiated at the apical end of the elongated nucleus, under the area occupied by the acrosomic system, and progresses caudally (Fig. 14–5). The coarse granular chromatin masses then coalesce gradually (Figs. 14–6 to 14–10) so that by step 18, the chromatin forms a solid mass with a few small cavities that later disappear to yield a solid homogeneous chromatin mass in late step 19 spermatids or spermatozoa (Fig. 14–11). Similar steps in the condensation of chromatin have been well described in the ram (Loir and Courtens, 1979; Loir and Lanneau, 1984).

In the spermatids of rats (Meistrich et al., 1978) and rams (Loir and Lanneau, 1984), chromatin condensation is accompanied by two major modifications in the basic nucleo-

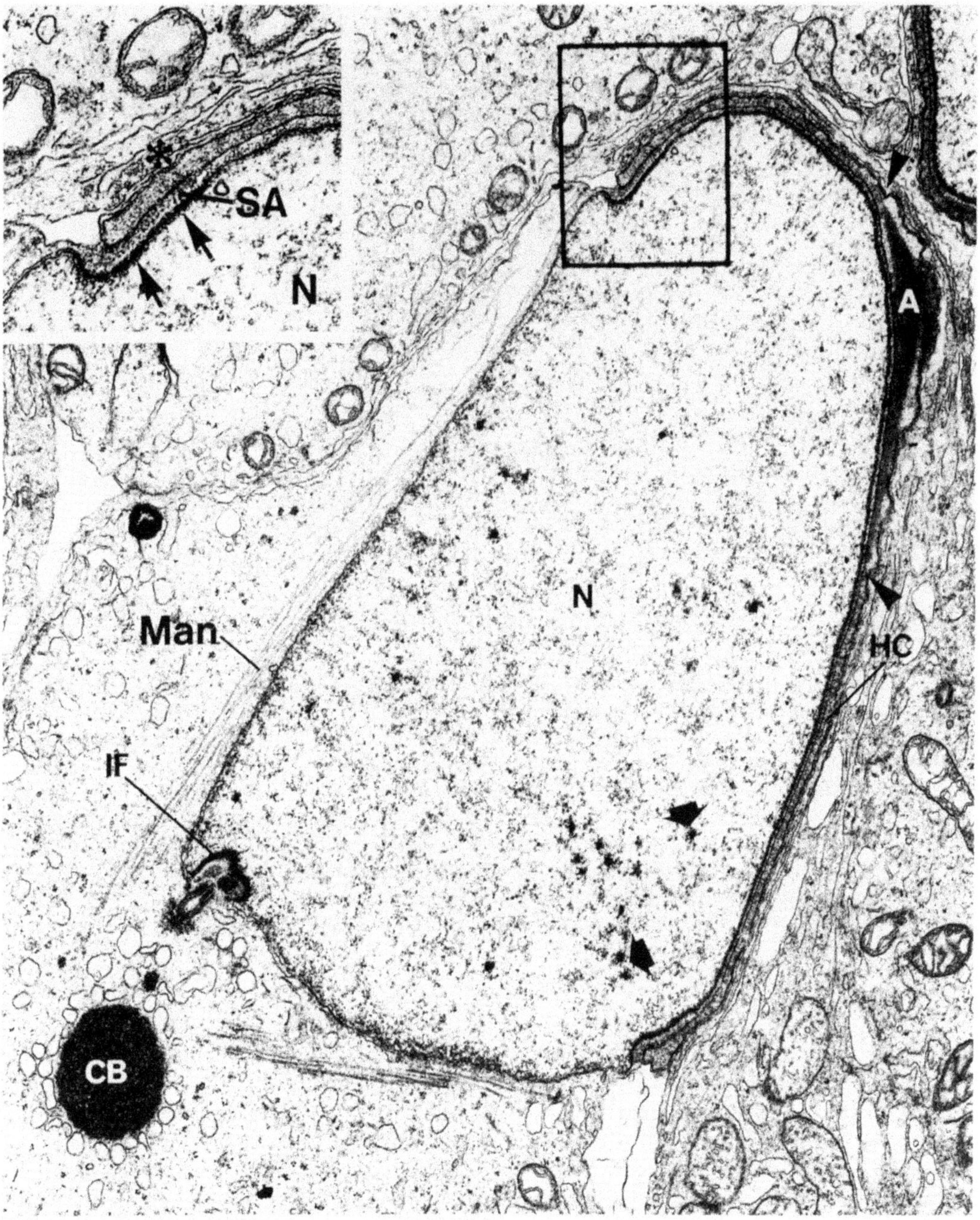

Figure 14–4. Longitudinal section through a step 10 spermatid. In the nucleus (N), which starts to elongate, the chromatin is dispersed except for some small granular masses *(arrows)*. The acrosomic system with the acrosome (A) (extending between the *two arrowheads*) and head cap (HC) portions cover tightly the apical extremity of the nucleus. At the other pole, the implantation fossa (IF) is visible. In the caudal cytoplasmic lobe (CL) a chromatoid body (CB) made up of a spherical mass of electron dense material surrounded by a corona of vesicles is visible. Microtubules of the manchette (Man) are also indicated. ×15,000. (Photographed by Dr. M. Lalli.) Inset: High power of boxed in area of Figure 14–4. In the subacrosomal space (SA) between the head cap and the nuclear envelope an electron dense granular material is present. Note layer of electron dense material on inner aspect of nuclear envelope *(arrows)*. *Asterisk,* ectoplasmic specialization of a Sertoli cell. ×30,000.

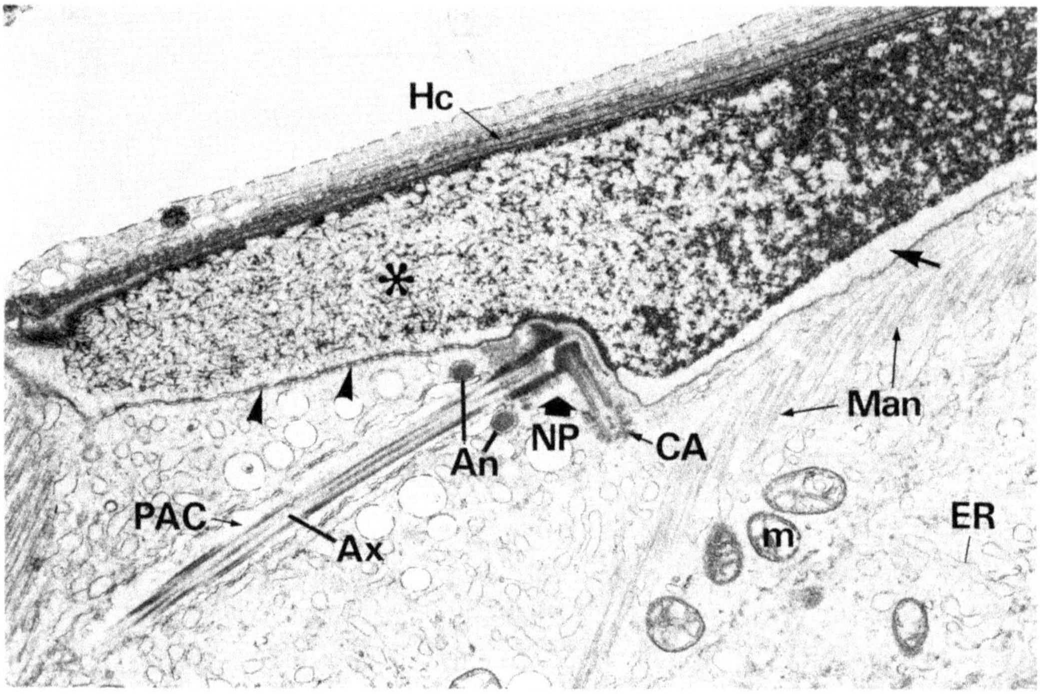

Figure 14–5. Longitudinal section through the caudal half of the head of a step 12 spermatid. The chromatin condensation is more advanced in the apical region of the nucleus than in the caudal region *(asterisk).* In the latter the chromatin has a coarsely filamentous texture, whereas apically such filaments aggregate in star-shaped masses that coalesce to form a dense felt-like chromatin mass that is retracted *(arrow)* from the nuclear envelope. The nuclear envelope at the caudal–ventral surface of the nucleus shows numerous nuclear pores *(arrowheads).* The early stage of formation of the neck piece (NP) inserted in the implantation fossa is visible. The centriolar adjunct (CA) with its extremity covered with a fuzzy material is indicated. The annulus (An) is associated with the plasma membrane, which at this step is invaginated along the axoneme (Ax); this area of cytoplasm around the axoneme is referred to as the proximal periaxonemal compartment (PAC). Man, microtubules of the manchette; ER, cisternae of endoplasmic reticulum; Hc, head cap portion of the acrosomic system; m, mitochondria. ×20,000. (Photographed by Dr. M. Lalli.)

protein composition of the nucleus. The first change involves the replacement of several testis-specific histones, which are present in both spermatocytes and round spermatids, by two (rat) or eight (ram) spermatid-specific basic proteins (spermatidal proteins). This replacement occurs gradually during the period of nuclear elongation and chromatin condensation, that is, during steps 9 to 12 in the rat and steps 8 to 11 in the ram. The second change, which is also gradual, involves the replacement of the spermatidal proteins just mentioned with one or more basic spermatozoal proteins, often referred to as protamines, which are particularly rich in arginine and cysteine. This second change in nucleoprotein composition occurs during steps 15 and 16 in the rat and during step 12 in the ram. In each case the nucleoproteins form S-S bonds between their cysteine residues, which probably contributes to the condensation and stabilization of the chromatin. The exact role of the nucleoproteins in the condensation of the chromatin of the spermatid's nucleus, however, is still unclear. One may wonder how the spermatidal nucleoproteins can be replaced by protamines considering that the chromatin is already highly condensed when this exchange takes place. The nature and modus operandi of the factor(s), presumably genetic in character, that determine the shape of the spermatozoon's nucleus remain unresolved questions.

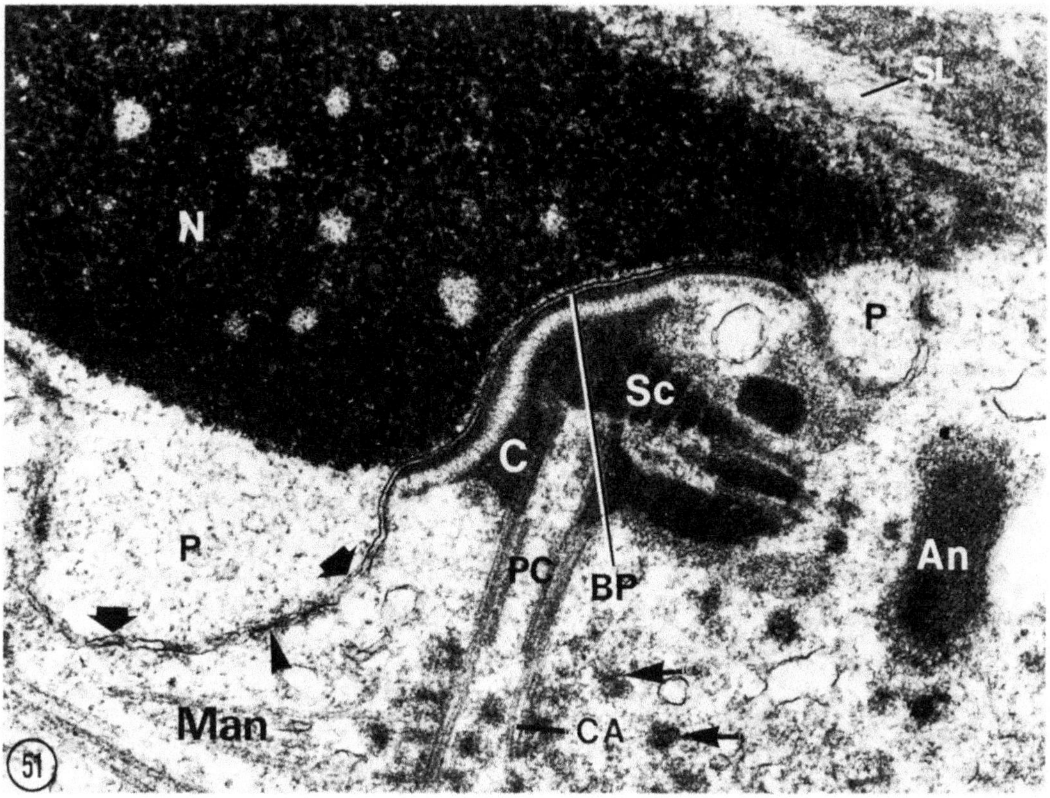

Figure 14–6. Longitudinal section through the caudal portion of the nucleus of an early step 15 spermatid in the region of the implantation fossa. In the nucleus (N), except in the pockets (P) demarcated by the redundant nuclear envelope *(arrows),* the condensed chromatin has a dense, coarsely granulated texture with few small areas of pale electron density. In the pockets the nuclear matrix shows a pale granular material. This redundant nuclear envelope shows nuclear pores *(arrowheads).* In the implantation fossa the following components of the connecting or neck piece can be identified: basal plate (BP), capitulum (C), subacrosomal lamina (SL), striated collar (Sc), and centriolar adjunct (CA) continuous with the proximal centriole (PC) located under the capitulum. AN, annulus cut longitudinally; Man, microtubules of manchette; *arrows,* small dense bodies around the tip of the centriolar adjunct. ×45,000. (Photographed by Dr. M. Lalli.)

Concerning the protamines, it is well established that they are synthesized toward the end of spermiogenesis (Hecht, 1987, 1988). However, messenger RNA (mRNA) synthesis stops early during spermiogenesis, that is, toward the end of the cap phase (Monesi, 1965; O'Brien and Bellvé, 1980). Thus, some long-lived mRNAs coding for the synthesis of protamines have to be stored in the spermatid's cytoplasm for an appreciable length of time before being activated and translated. The storage form of these mRNAs, as well as their location within the spermatid, has yet to be identified. Similar considerations may be made concerning cytoskeletal proteins (to be discussed later) synthesized late during spermiogenesis and deposited along the axoneme during the maturation phase (Oko and Clermont, 1989).

Nuclear Envelope

The nuclear envelope of the spermatid undergoes several regional modifications during spermiogenesis, some of which become permanent structural features of the spermatozoon.

In the subacrosomal domain, throughout steps 3–12, the inner nuclear membrane facing the nucleoplasm is distinctly associated with a thin layer of dense material (Fig. 14–4 inset and

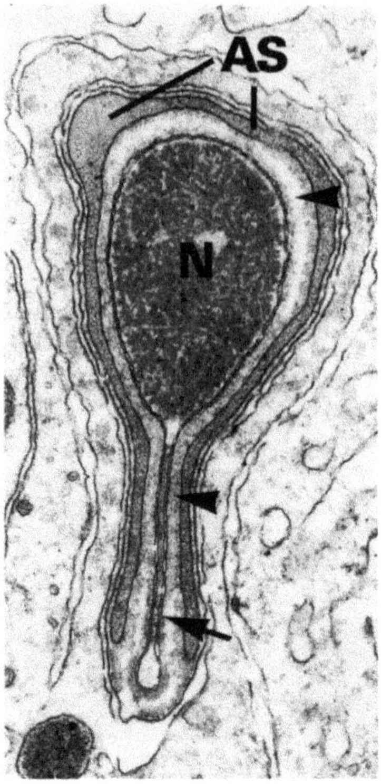

Figure 14–7. Cross section of the head of a spermatid. Photograph of a step 15 spermatid showing a flap of nuclear envelope *(arrow)* along the ventral aspect of the apical half of the nucleus; this superfluous nuclear envelope that forms as a result of the condensation of the chromatin will undergo resorption at step 17 of spermiogenesis. N, nucleus; AS, acrosomic system; subacrosomal space *(arrowheads)*. ×30,000. (Photographed by Dr. M. Lalli.)

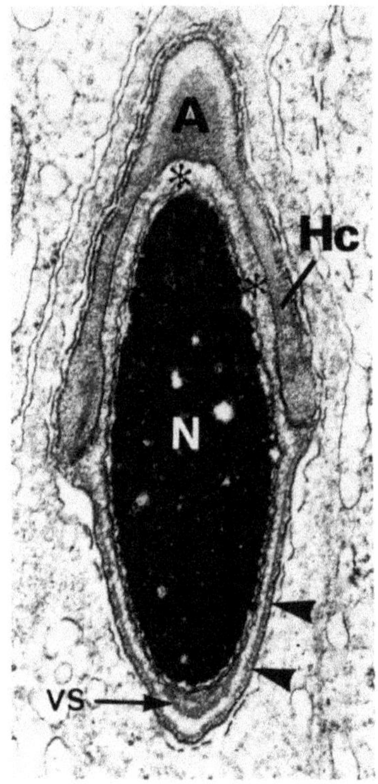

Figure 14–8. Cross section of the head of a step 17 spermatid showing the incompletely condensed chromatin of the nucleus (N) and the acrosomic system formed by the acrosome (A) and head cap (HC). On the ventral aspect of the head an outer layer of dense material is deposited along the plasma membrane *(arrowheads);* this is the initial stage of deposition of the postacrosomal dense lamina. An inner layer of material is seen, in the form of a crescent at the surface of the nuclear envelope and in the region of the ventral spur (VS). Between the acrosomic system and nucleus, the subacrosomal space *(asterisks)* contains a wispy material. ×30,000. (Photographed by Dr. M. Lalli.)

Fig. 14–16). Thereafter, the latter becomes separated from the condensed chromatin material by a thin pale layer of nucleoplasm (Figs. 14–8 and 14–9). In the implantation fossa facing the forming capitulum of the connecting piece of spermatids, beginning from step 8, there is a layer of dense material applied to the cytoplasmic face of the outer nuclear envelope that anchors the neck piece and is referred to as the basal plate (Figs. 14–4 and 14–6).

Throughout the cap and acrosome phases (Figs. 14–5 and 14–6), the pores of the nuclear envelope are either lost or shift down to the future site on the perifossal zone where at least some remain in the spermatozoon (Fig. 14–12). Exchanges between the nucleoplasm and the cytoplasm of the spermatid may take place at the level of these nuclear pores (Fawcett, 1975).

There is a marked rearrangement of two other distinct areas of the nuclear envelope in the rat with dissolution or resorption of portions of them. The first rearrangement is observed from steps 12 to 17 of spermiogenesis on the ventral aspect of the apical half of the nucleus. Concurrent with the condensation of the chromatin, a large ventral flap of superfluous nuclear envelope forms subjacent to the head

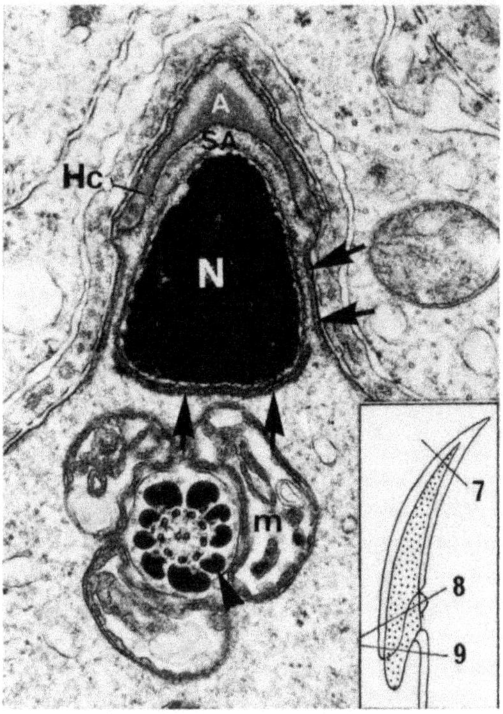

Figure 14–9. Cross section of the head of a step 17 spermatid showing the caudal extremity of the nucleus (N). The plane of section also goes through the tail with the axoneme surrounded by outer dense fibers *(arrowhead)* and mitochondria (m). The outer layer of the postacrosomal dense lamina is seen along the plasma membrane *(oblique arrows)* facing the lateral surfaces of the nucleus; the inner layer is seen on the nuclear envelope *(vertical arrows)*. Also labeled are the acrosome (A) and head cap (Hc) of the acrosomic system separated from the nucleus by the subacrosomal (SA) space. ×30,000. (Photographed by Dr. M. Lalli.) Inset: Section through the head for Figures 14–7 through 14–9 diagrammatically illustrated.

cap (Fig. 14–7). This ventral flap regresses during step 17 as the subacrosomal space enlarges and becomes filled with perinuclear material (compare Figs. 14–7 and 14–11; Lalli and Clermont, 1981). In spermatids of the hamster, a similar flap of superfluous nuclear envelope forms at the apical extremity of the condensing nucleus, but in this case the flap forms on the dorsal rather than the ventral aspect of the nucleus (Franklin and Fussell, 1972). Another segment of superfluous or redundant nuclear envelope is observed at the caudal extremity of the nucleus of maturation phase spermatids (Figs. 14–6 and 14–10D) and is observed not only in the rat but in most mammalian species (Franklin, 1968; Fawcett and Phillips, 1969b; Phillips, 1970). This segment of nuclear envelope, which shows numerous pores (Fig. 14–6), regresses at the very end of spermiogenesis, although part of it remains (Fig. 14–12) in the spermatozoa of most mammalian species (Zamboni and Stefanini, 1970).

2. PERINUCLEAR THECA

The perinuclear theca is a layer of dense cytoplasmic material that is present around the entire nucleus of all mammalian spermatozoa except in the area of the perifossal zone (Courtens et al., 1976; Lalli and Clermont, 1981). This layer shows two distinct regions, the subacrosomal region also known as the perforatorium, and the postacrosomal sheath also known as the postacrosomal dense lamina (Fig. 14–1B) (Fawcett, 1975; Lalli and Clermont, 1981). The perforatorium has been the subject of debate. Although it is poorly developed in spatulate spermatozoa, in humans in particular, it is a well-recognized entity in falciform spermatozoa where it was originally described and distinguished from the acrosomic system. In falciform spermatozoa, the perforatorium assumes the appearance of a pointed helmet that tightly fits the apical extremity of the nucleus (Figs. 14–11 and 14–11a,b). In certain rodents the perforatorium takes on a most curious shape with three pointed prongs (Breed, 1984). In spatulate spermatozoa, the perforatorium, although reduced in size, is present as a thin crescentic blade-like structure interposed between the acrosome and the apex of the nucleus (Courtens et al., 1976). In the rat, the perforatorium assumes a triangular shape with three prongs extending over the apical part of the nucleus, that is, one dorsal and two lateroventral (Fig. 14–11c). The dorsal prong extends further caudally than the other two (Fig. 14–11d). In the hamster there is a thickening of the perforatorium just under the caudal aspect of the acrosomic system that forms a beltlike layer around the nucleus (Yanagimachi and Noda, 1970).

The postacrosomal dense lamina begins where the acrosomic system ends and covers

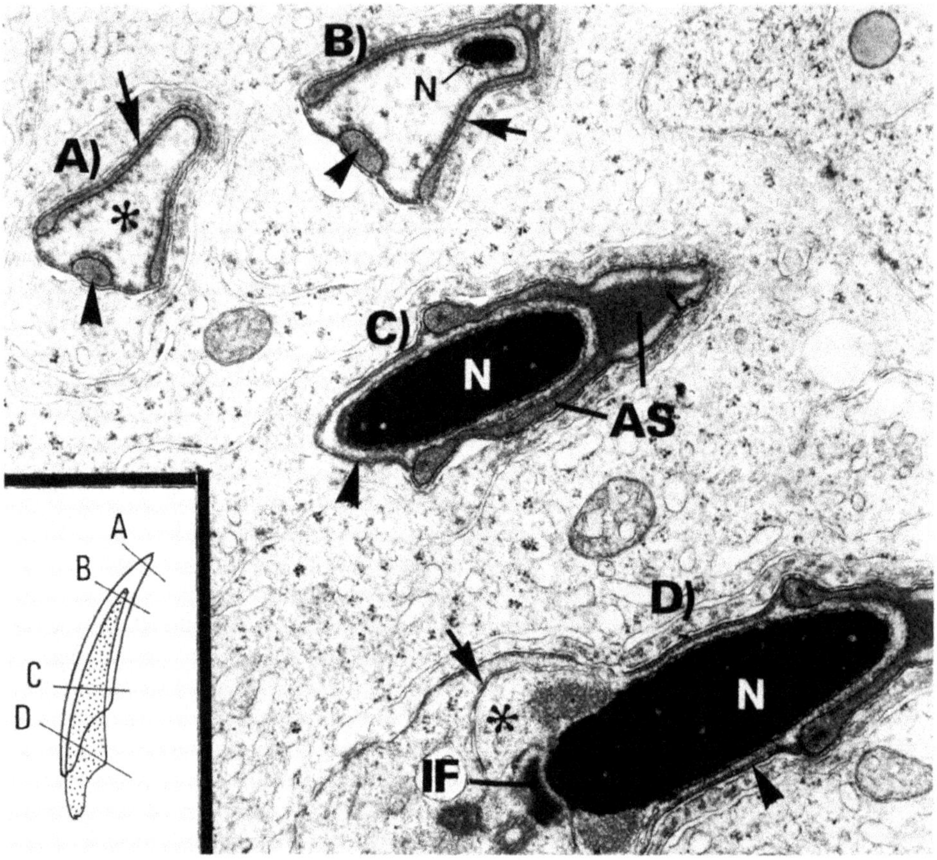

Figure 14–10. Photograph showing heads of step 17 spermatids cut at various levels labeled A, B, C, D; the plane of section for each is indicated on the small diagram. In A the plane of section misses the nucleus and shows the subacrosomal space filled with a wispy substance *(asterisk).* The acrosomic system itself shows a head cap portion *(arrow)* and a separated head cap segment *(arrowhead).* In B the plane of section goes through the tip of the nucleus (N). The acrosomic system shows the head cap *(arrow)* and separated head cap segment *(arrowhead).* In C the section goes through the nucleus (N) and the acrosomic system (AS). The forming postacrosomal dense lamina is seen along the plasma membrane *(arrowhead).* In D, the plane of section goes through the caudal extremity of the nucleus (N) covered by the acrosomic system (AS). The redundant nuclear envelope is visible *(arrow)* facing a nuclear matrix of low density *(asterisk).* IF, implantation fossa of the neck piece; *arrowhead,* postacrosomal dense lamina. ×20,000. (Photographed by Dr. M. Lalli.)

the caudal portion of the nucleus except for the perifossal zone (Figs. 14–8 and 14–11d). It is in fact composed of two distinct layers sandwiched between the plasma membrane and the nuclear envelope (Courtens et al., 1976; Lalli and Clermont, 1981). The layer related to the plasma membrane appears denser than the inner one associated with the nuclear envelope (Fig. 14–8). The inner layer presents a thickening along the ventral aspect of the nucleus of certain falciform spermatozoa referred to as the ventral spur (Figs. 14–1B, 14–8, and 14–11).

The formation of the postacrosomal dense lamina and perforatorium occurs at different times during spermiogenesis. The postacrosomal dense lamina appears first and is assembled as two discrete layers in the rat. An outer layer is deposited along the plasma membrane and an inner layer along the nuclear envelope (Figs. 14–8 and 14–9). The outer layer appears to be

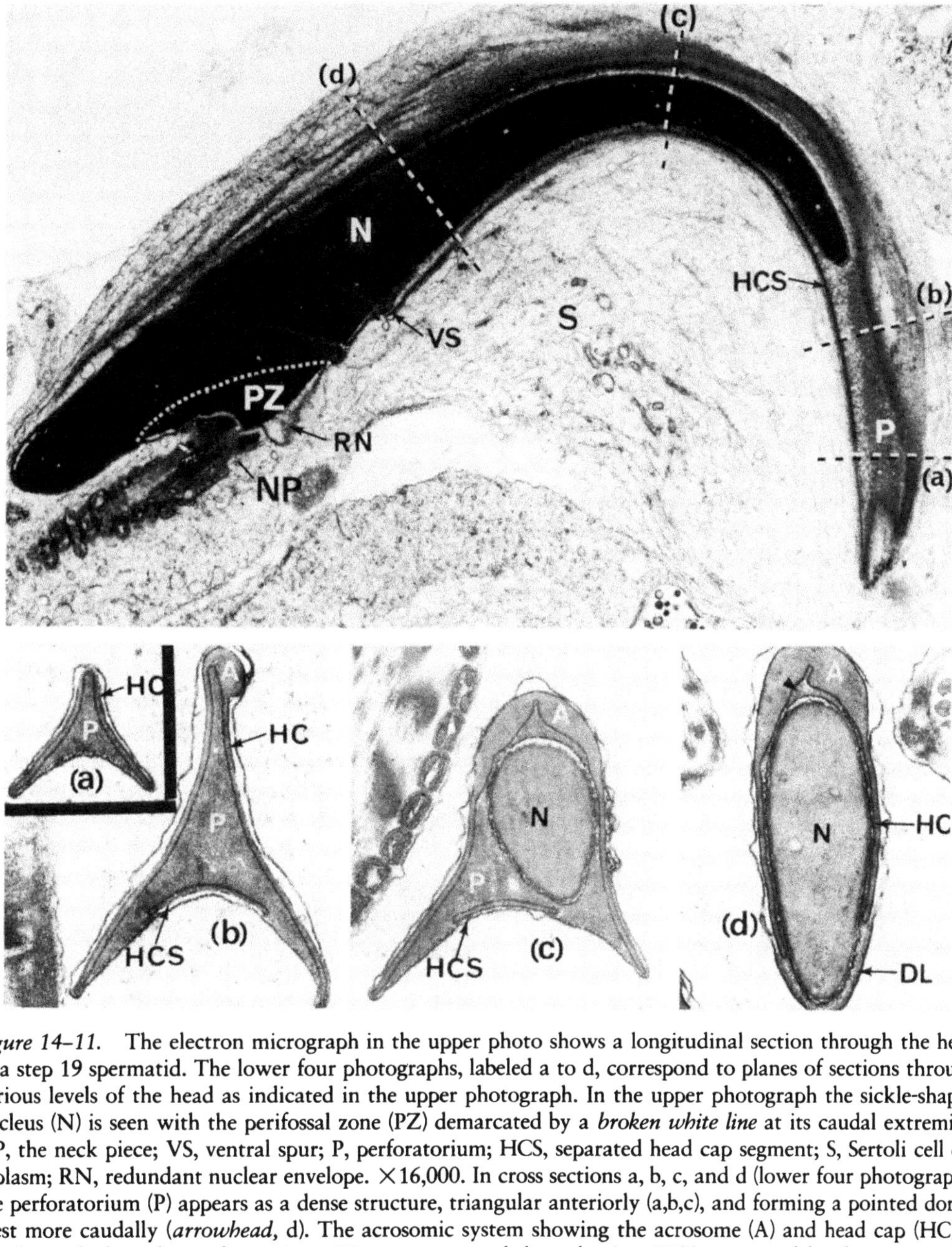

Figure 14–11. The electron micrograph in the upper photo shows a longitudinal section through the head of a step 19 spermatid. The lower four photographs, labeled a to d, correspond to planes of sections through various levels of the head as indicated in the upper photograph. In the upper photograph the sickle-shaped nucleus (N) is seen with the perifossal zone (PZ) demarcated by a *broken white line* at its caudal extremity. NP, the neck piece; VS, ventral spur; P, perforatorium; HCS, separated head cap segment; S, Sertoli cell cytoplasm; RN, redundant nuclear envelope. ×16,000. In cross sections a, b, c, and d (lower four photographs) the perforatorium (P) appears as a dense structure, triangular anteriorly (a,b,c), and forming a pointed dorsal crest more caudally (*arrowhead,* d). The acrosomic system showing the acrosome (A) and head cap (HC) is closely applied to the perforatorium. DL, postacrosomal dense lamina; HCS, separated head cap segment. ×35,000. (From Lalli and Clermont, 1981.)

continuous throughout its length, whereas the inner layer is segmented, being especially prominent at the caudal extremity of the nucleus (Fig. 14–9) and in the region of the ventral spur (Fig. 14–8). The assembly of the postacrosomal dense lamina proceeds caudally along the head beginning in step 11 and ending in step 16 of spermiogenesis. The perforatorium, on the other hand, condenses by step 19 from a wispy material of low electron density

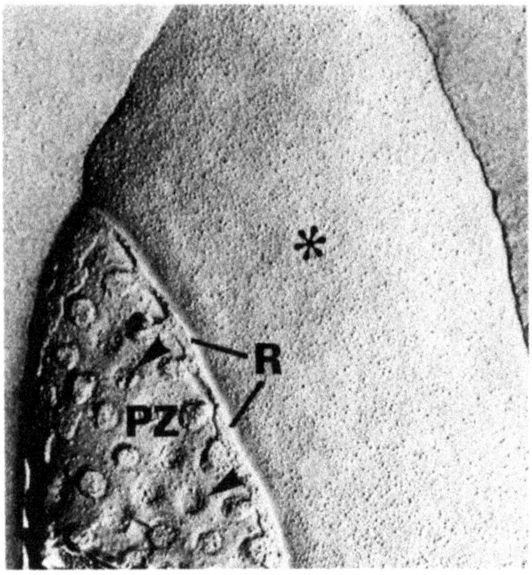

Figure 14–12. Photograph of the caudal and ventral extremity of the nucleus of a rat spermatozoon seen in a freeze-fracture preparation. The perifossal zone (PZ) with its numerous nuclear pores *(arrowheads)* is visible because the overlying plasma membrane was broken away during the preparation procedure. The limit of the perifossal zone is demarcated by a narrow depression or ring (R) at the surface of the head of the spermatozoon. At the level of the ring the plasma membrane establishes contact with the nuclear envelope. Above the ring, the plasma membrane *(asterisk)* is apposed to the postacrosomal dense lamina. ×45,000. (Photographed by Dr. M. Lalli.)

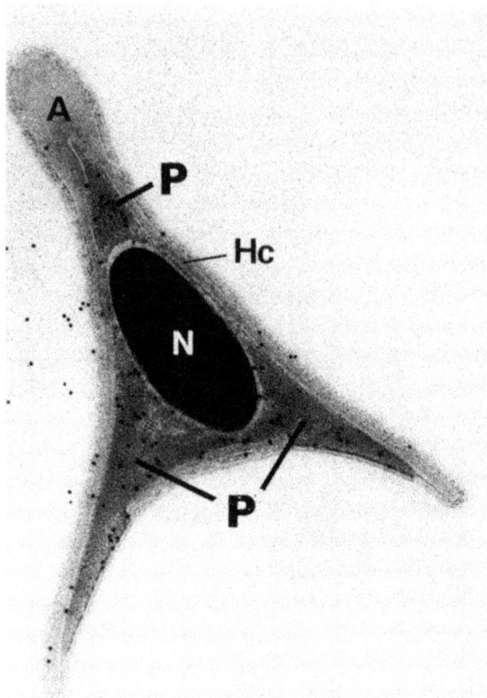

Figure 14–13. Electron micrograph of a cross section through the apical part of the head of a rat spermatozoon of material embedded in Lowicryl and immunogold labeled with antibodies raised against a whole isolated perforatorial fraction. Labeling occurs over the entire perforatorium (P), which has a triangular three-pronged shape in this region of the head. Although not shown, the postacrosomal dense lamina that begins where the acrosome ends and that is continuous with the perforatorium does not label. A, acrosome; Hc, head cap of acrosomic system; N, nucleus. ×53,000.

(Fig. 14–10) that accumulates in the subacrosomal space of earlier spermatids (Fig. 14–4) and results in the characteristic triangular shape of the definitive perforatorium (Figs. 14–11 and 14–11a–c).

An initial study on the isolated rat perforatorium by Olson and co-workers (1976a) indicated that this structure was composed of a single 13 kd polypeptide. A more recent study (Oko and Clermont, 1988) revealed that the perforatorium is composed of many polypeptides, six of which show prominence on gels (13, 13.4, 16, 33, 35, and 43 kd). Immunocytochemical localization at the electron microscopic level, with antibodies prepared against the whole perforatorial fraction, was found to be concentrated on the perforatorial portion of the perinuclear theca (Fig. 14–13) and also over the inner layer of the postacrosomal dense lamina at the level of the ventral spur. This observation clearly indicates a difference in protein composition between the perforatorium and the major portion of the postacrosomal dense lamina (Oko and Clermont, 1988). A more recent immunocytochemical investigation also revealed a differential distribution of polypeptides in the perforatorium that was dependent on the molecular weight of the polypeptides in question. The higher molecular weight polypeptides (34, 43, 57, and 63 kd) were found distributed over the whole perforatorium and the inner zone of the ventral spur, whereas lower molecular weight polypeptides (13, 13.4, and 16 kd) were restricted in their localization only to the thicker apical portion of the perforatorium, as well as the inner zone of the ventral spur (Oko et al., 1990). Another interesting finding made in the course of this study was that the antibodies against the high molecular weight polypeptides were also observed to be present over a thin layer of cytoplasm seen between the plasma membrane and the outer acrosomal membrane delimiting the head cap (Fig. 14–14). No immunolabeling, however, was obtained over the film of cytoplasm interposed between the plasma membrane and the outer acrosomal membrane delimiting the acrosome of the acrosomic system (Fig. 14–14) (Oko et al., 1990).

Recently, Oko and Clermont (1991) investigated by immunocytochemistry, at both light and electron microscopic levels, the distribution of antibodies raised against perforatorial polypeptides in spermatocytes and spermatids. It was observed that perforatorial proteins were ubiquitously distributed throughout the cytoplasm and the nucleus of both spermatocytes and early spermatids with the concentration being appreciably greater in the nucleus. With advancing steps of spermiogenesis (steps 11–13), the nuclear distribution of the perforatorial polypeptides gradually disappeared, whereas the immunoreactivity in the subacrosomal space slowly increased and reached its maximal intensity at step 19 of spermiogenesis. These results raise the interesting possibility of a nuclear source for some of the proteins that form the perinuclear theca.

One of the most striking features that characterize the head of spermatozoa is that both the inner acrosomal membrane and the plasma membrane are bound tightly to the perinuclear theca (Figs. 14–11a–d). It is apparent that if these membranes were not tightly bound to the surface of the spermatozoon's head, the hydrodynamic properties of this motile cell would be modified and the motion would be markedly modified or hindered. It is also feasible that the regional variations in composition of the perinuclear theca may reflect or determine the chemical or functional properties of adjacent membranes, the integrity of which may play a crucial role in the fertilization process. As well, the rigid, pointed, blade-like structure of the perforatorium in falciform and spatulate spermatozoa may have a mechanical role in traversing the zona pellucida of the oocyte after the acrosome reaction.

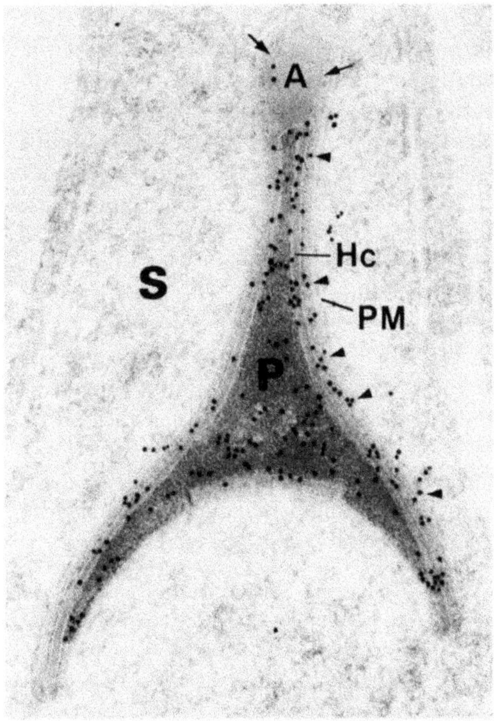

Figure 14–14. Electron micrograph of a cross section through the apical part of the head of a step 19 spermatid showing the immunogold labeling obtained with antibodies affinity purified from either the 34, 43, 57 or 63 kd perforatorial polypeptides. In this section the nucleus is not apparent (compare with Figure 14–13). Labeling occurs throughout the perforatorium (P), but is also found distributed over a thin layer of cytoplasm *(arrowheads)* seen between the plasma membrane (PM) and head cap (Hc) portion of the acrosomic system. No immunolabeling is found over the film of cytoplasm *(arrows)* interposed between the plasma membrane and the acrosome proper (A). Although not shown, antibodies raised against the lower molecular weight polypeptides (13, 13.4, and 16 kd) were restricted to the perforatorium only. S, Sertoli cell cytoplasm. ×60,000.

3. GOLGI APPARATUS AND ACROSOMIC SYSTEM

The Golgi apparatus is closely related to the developing acrosomic system during the Golgi and cap phases (Figs. 14–2 and 14–3). After its detachment from the acrosomic system at the beginning of the acrosome phase, the Golgi apparatus becomes lodged in the middle of the cytoplasmic lobe (Fig. 14–3) where it maintains its structural and functional activity (Clermont and Rambourg, 1978; Clermont and Tang, 1985). Because the role of the Golgi apparatus in the glycosylation and segregation of proteins is extensively investigated at present (reviews by Bennett, 1984, 1988; Farquhar, 1985; Rothman, 1985), it is of interest to examine its structural, cytochemical, and functional features and then examine its role in the formation of the acrosomic system. The functional significance of the relationship of the Golgi apparatus with multivesicular bodies and the chromatoid body will also be examined.

The Golgi Apparatus of Spermatids

At the time when cytologists were debating the existence of this organelle, many investigators working on spermiogenesis identified it correctly and demonstrated its role in the formation of the acrosomic system (review by Bennett, 1988). Using the periodic acid-Schiff technique for the detection of carbohydrates, Leblond and Clermont (1952), Clermont and Leblond (1955), and Clermont and associates (1955) demonstrated the presence of sugars in the Golgi apparatus and the developing acrosomic system. With the advent of the electron microscope, a role for this organelle in the formation of the acrosomic system was substantiated (Burgos and Fawcett, 1955; Clermont, 1956a; Sandoz, 1970).

The Golgi apparatus, whether it is hemi-

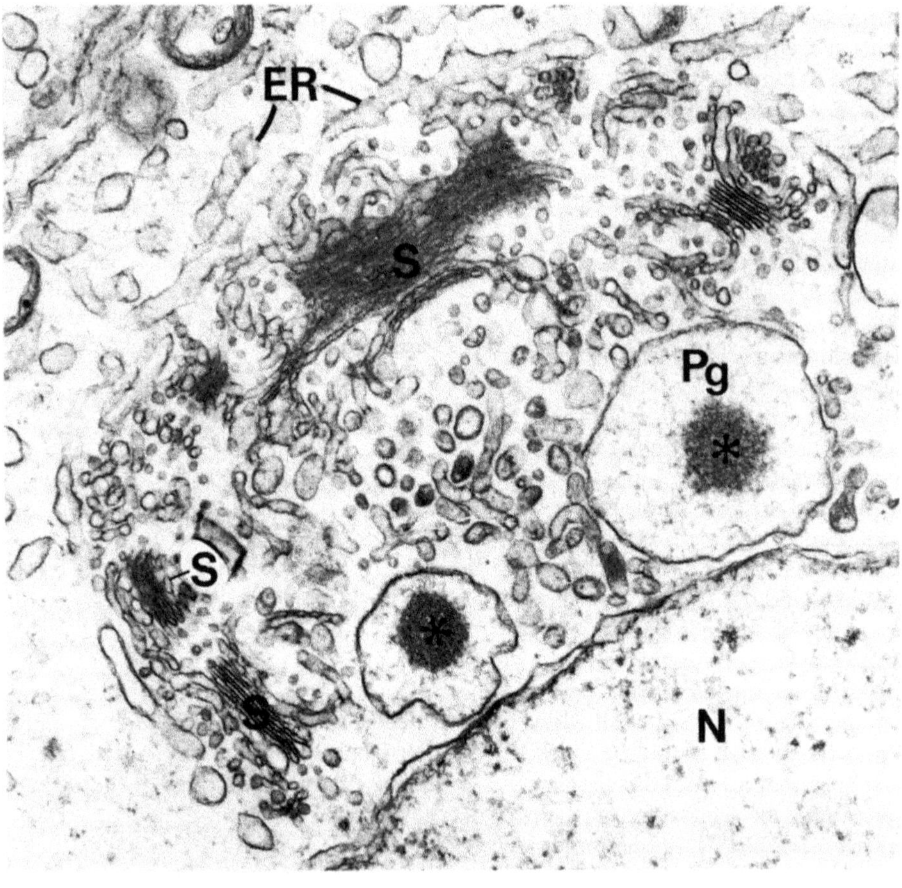

Figure 14–15. Photograph of the juxtanuclear Golgi apparatus of a step 2 spermatid. In the Golgi apparatus, short stacks of saccules (S) form the cortex. In the medulla two proacrosomic granules (Pg) with their characteristic dense core *(asterisks)* are seen. N, nucleus; ER, cisternae of endoplasmic reticulum. ×28,000.

spherical or spherical, shows two main regions: the cortex and the medulla. The cortex is composed mainly of interconnected stacks of saccules, and the medulla shows a loose aggregation of vesicular profiles, some showing a clathrin coat (Fig. 14–15 and see Fig. 14–17). A tridimensional analysis of the cortical region of the Golgi apparatus of steps 2–8 spermatids, using stereopairs, showed that the cortex was composed of saccular (compact) and intersaccular (noncompact) connecting regions (Hermo et al., 1980). In the saccular region, the stacks were composed of approximately seven to nine parallel saccules perforated with pores of various dimensions. These stacks showed a distinct *cis–trans* polarity. On the *cis*-side of the stack, three or four saccules were in places interrupted along their length by large perforations in register (see Fig. 14–17). These pan-shaped cavities, referred to as wells, contained small vesicles and their mouths were directed toward the *cis* or convex face of the stack and their

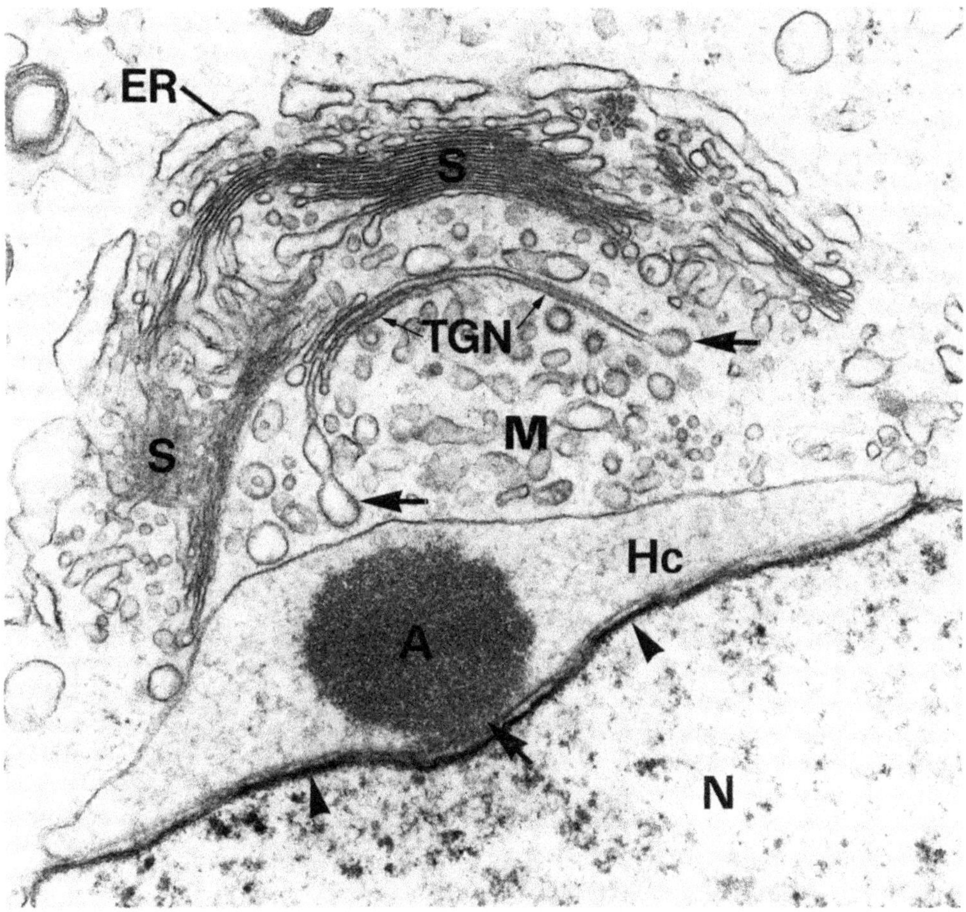

Figure 14–16. Photograph of a hemispherical Golgi apparatus and acrosomic system that are growing at the surface of the nucleus (N) of a step 5 spermatid. At this step of spermiogenesis the acrosome proper (A) is close to the inner acrosomal membrane *(oblique arrow)* but is still spherical. The head cap (Hc), containing a material of low density, is spread over the surface of the nucleus. Note the dense appearance of the nuclear envelope *(arrowheads)* underlying the acrosome. The Golgi apparatus itself shows a cortex composed of several interconnected stacks of saccules (S) and a medulla (M) containing several membranous vesicular and tubular profiles. Note the detached *trans* Golgi networks (TGN) on the *trans* aspect of the stack. One of these shows coated dilated edges *(horizontal arrows)*. Cisternae of the endoplasmic reticulum (ER) are closely applied to the *cis* aspect of the Golgi apparatus. ×36,000. (From Hermo et al., 1980.)

bottom closed by an underlying flattened saccule (see Fig. 14–17) (Hermo et al., 1980).

Overlying the first saccule of the stack that is usually dilated, there is a Golgi element that is osmiophilic and has the appearance of a network of anastomosed tubules, some of which extend in the direction of the cisternae of the endoplasmic reticulum (ER) that cap the Golgi apparatus (Figs. 14–15 to 14–17). On the *trans* side of the stack one to three saccules with a wider lumen show a "peeling-off" configuration (Figs. 14–15 and 14–17) or are separated from the stack (Figs. 14–16 and 14–17). The latter, referred to as the *trans* Golgi networks, have been shown to break down into tubules and vesicles (Thorne-Tjomsland et al., 1988). In the intersaccular regions, also referred to as the noncompact zone, numerous branching and anastomosing membranous tubules serve to connect the edge of one saccule of a stack with that of the corresponding saccule of an adjacent stack (Figs. 14–16 and 14–17). In these

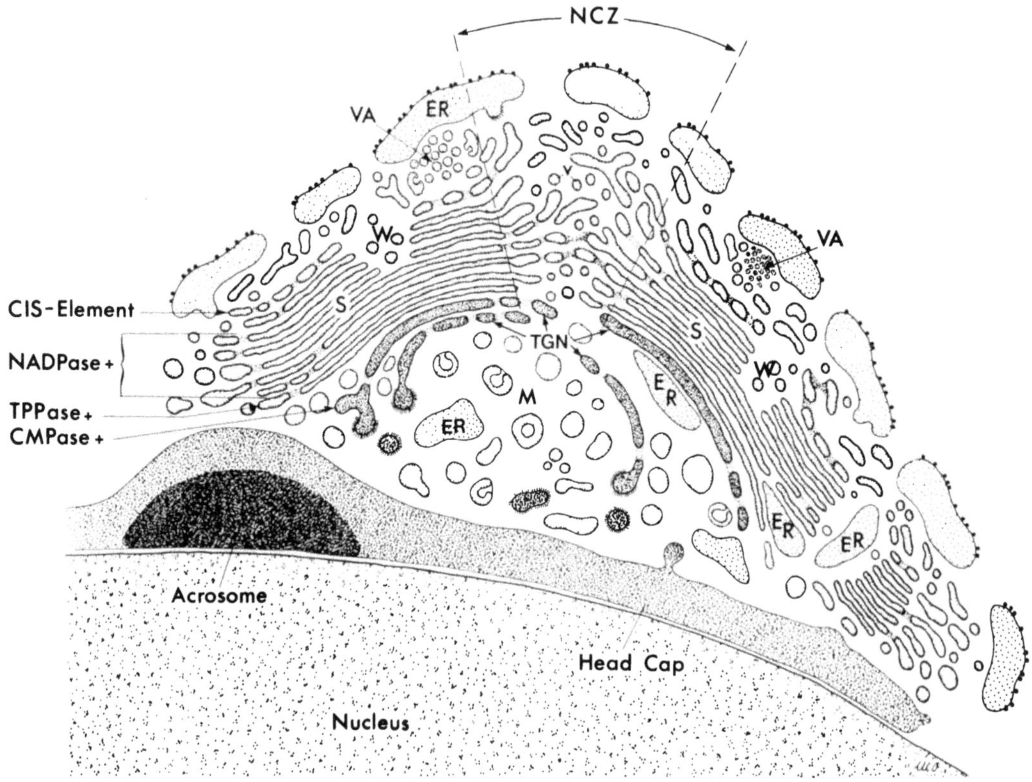

Figure 14–17. Diagram showing the structure of the hemispherical Golgi apparatus of a step 6 spermatid. The cortex of the Golgi apparatus is made up of compact zones formed of stacks of saccules (S) and noncompact zones (NCZ) containing dilated membranous tubules that serve to bridge together the saccules of adjacent stacks. The *cis* element is an anastomotic tubular network that overlies the first saccule *(not labeled).* The subjacent middle saccules are NADPase-positive, whereas one or two saccules on the *trans* face of the stack are TPPase-positive. The peeling-off or detached *trans* Golgi networks (TGN) on the *trans* face present a wider lumen and are CMPase-positive. The medulla (M) contains vesicles of different sizes, some coated, which show NADPase, TPPase, and CMPase activity. The acrosomic system (acrosome and head cap) also shows CMPase, NADPase and TPPase activity. Cisternae of endoplasmic reticulum (ER) are found overlying the *cis* aspect of the stack, between the saccules of the stack, in the noncompact zones, applied to the *trans* Golgi networks or in the medulla. Note aggregations of small vesicles (VA) inserted between the *cis* element and the overlying ER cisternae. Small vesicles are also seen in the wells (W) and at the edges of the saccules in the noncompact zones. (From Clermont et al., 1990.)

regions numerous small vesicles are seen at the edges of the saccules and between the tubules (Figs. 14–16 and 14–17).

Using various phosphatase markers, a characteristic distribution was noted in the cortex of the Golgi apparatus of steps 3–7 spermatids (Clermont et al., 1981; Tang et al., 1982; Thorne-Tjomsland et al., 1988; Smith et al., 1990) with nicotinamide adenine dinucleotide phosphatase (NADPase) being present in middle saccules, thiamine pyrophosphatase (TPPase) in one or two saccular elements on the *trans* side of the stack, and cytidine monophosphatase (CMPase) in the *trans* Golgi networks on the *trans* aspect of the stack (Fig. 14–17). Large vesicles in the medulla, as well as the acrosomic system's content, were strongly reactive with CMPase. The acrosomic system also showed a weaker reaction for NADPase and TPPase activity relative to that of CMPase, but one that was nevertheless significant compared with controls (Tang et al., 1982; Thorne-Tjomsland et al., 1988; Clermont et al., 1990). The proacrosomic granules were found to be strongly reactive for NADPase activity (Smith et al., 1990). The *cis*-osmiophilic element as well as the subjacent first dilated saccule were not found to contain any of these enzyme markers. Although the bridging tubules seen in the intersaccular regions showed the same cytochemical characteristics as the saccules to which they were connected, the small vesicles at the edges of the saccules, including those present in the wells, were negative to these cytochemical reactions. Thus, the Golgi apparatus has several cytochemically distinct compartments (Fig. 14–17).

During spermiogenesis the Golgi apparatus undergoes some major changes. In steps 1 and 2, the saccular components are small and the intersaccular regions predominate (Fig. 14–15). These stacks are involved in the formation of proacrosomic granules (Fig. 14–15). In steps 3 to 7, the cortex of the hemispherical Golgi apparatus gradually shows extensive saccular regions and less extensive intersaccular regions (Fig. 14–16). During these steps, the Golgi apparatus is actively involved in the formation of the acrosomic system. During steps 8–17, the cortex of the spherical Golgi apparatus is made up almost exclusively of stacks of saccules, the intersaccular noncompact regions being relatively less prominent (Fig. 14–18). In these steps, the CMPase-positive *trans* Golgi networks are either absent or considerably reduced in size (Tang et al., 1982; Thorne-Tjomsland et al., 1988; Clermont et al., 1990). The fate of these networks, as well as of the remaining Golgi elements beyond step 17 when stacks of saccules are no longer intact, has not yet been determined.

Functional Roles of the Golgi Apparatus

The main function of the Golgi apparatus is to glycosylate proteins that are delivered to it on its *cis*-face from the overlying ER cisternae (review Bennett, 1984, 1988). The presence of carbohydrates within the saccules, the *trans* Golgi networks as well as in the acrosomic system has been well-documented cytochemically (Susi et al., 1971). Injection of radioactive sugars used as tracers are concentrated first over the cortex of the Golgi apparatus, then over the medulla, and last over the developing acrosomic system (Sandoz, 1972; Clermont and Tang, 1985). Thus, the contribution of glycoproteins to the acrosomic system by the Golgi apparatus of spermatids postulated by Leblond and Clermont (1952) has been well documented. Furthermore, Clermont and Tang (1985) showed that the Golgi apparatus of acrosome and maturation phase spermatids, when the organelle is no longer associated with the acrosomic system, is still actively incorporating radioactive sugars and, therefore, is still involved in the biosynthesis of glycoproteins. The latter authors suggested that such glycoproteins are possibly destined to the plasma membrane that undergoes important biochemical modifications during spermiogenesis as was recently reviewed by Eddy (1988). Clermont and Tang (1985) also indicated that the medulla constitutes a distinct compartment of the Golgi apparatus where glycoproteins momentarily accumulate before being delivered, depending on the step of spermiogenesis, either to the acrosomic system or to the plasma membrane.

The role of the Golgi apparatus in the construction of the acrosomic system has been repeatedly documented through various electron microscopic studies (Susi et al., 1971; Mollenhauer et al., 1976) since the early study of Burgos and Fawcett (1955). It appears that glyco-

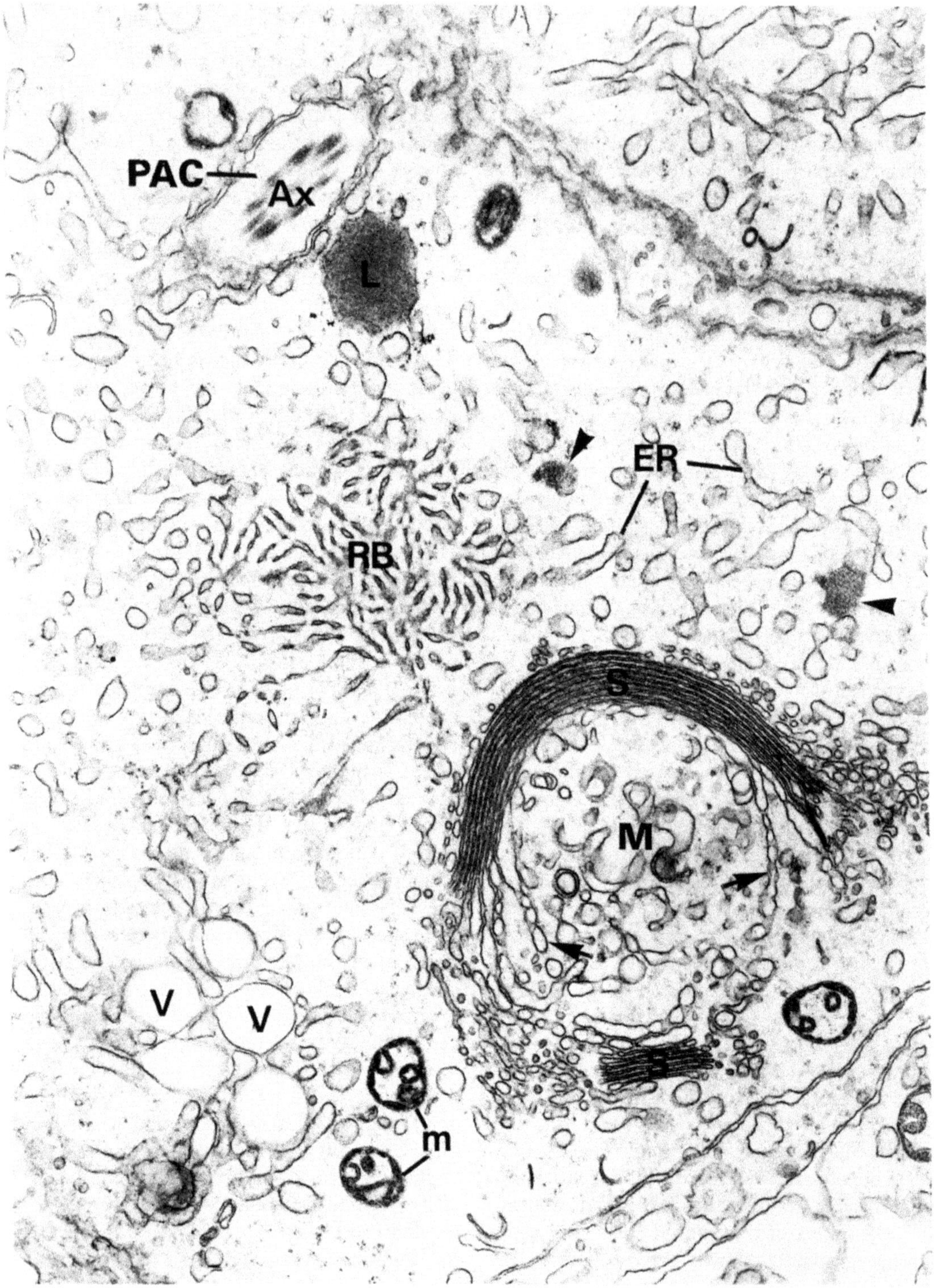

Figure 14–18. Photograph showing the Golgi apparatus of a step 12 spermatid after its separation from the acrosomic system. The spherical Golgi apparatus shows a cortex composed of large stacks of saccules (S) and a medulla (M) containing various irregular membranous profiles. The Golgi shows peeling-off *trans* elements *(arrows)*. In close proximity to the Golgi apparatus are numerous profiles of the endoplasmic reticulum (ER), some of which are arranged to form the radial body (RB). m, mitochondria with characteristic cristae; *arrowhead*, granulated bodies; V, large distended vesicles enveloped by ER cisternae; PAC, cross section of periaxonemal compartment; Ax, axoneme; L, lipid. ×40,000.

proteins are segregated and concentrated within the *trans* Golgi networks on the *trans* aspect of the stack; these elements break down into tubules and vesicles that then migrate toward and merge with the outer acrosomal membrane thereby emptying their content into the growing acrosomic system (acrosome and head cap) (Thorne-Tjomsland et al., 1988). The exact mechanism for the transport of glycosylated proteins from the *cis* to the *trans* aspect of the Golgi cortex, however, is still a problematic issue. Some authors have suggested an actual migration of saccules from the *cis* to the *trans* face of the Golgi stack (Susi et al., 1971; Mollenhauer et al., 1976); others have suggested that the transport of glycoproteins through the Golgi apparatus of various cell types takes place by means of small carrier vesicles that hop from one saccular element to the next (Farquhar, 1985; Rothman, 1985; Dunphy and Rothman, 1985).

The Golgi apparatus contributes glycoproteins by vesicular elements not only to the acrosomic system, but also to one or more multivesicular bodies that are seen in close relation with the Golgi apparatus during the earlier steps of spermiogenesis (Clermont and Tang, 1985). Such multivesicular bodies, which may be particularly numerous in the cytoplasm of some mammalian species (e.g., in bull; Barth and Oko, 1989), have a tendency to accumulate at the centriolar pole of the nucleus during steps 9–10 and eventually vanish. The presence of acid phosphatases in such multivesicular bodies indicates their possible lysosomal nature (Tang et al., 1982), but their function remains to be clarified. The Golgi apparatus also appears to contribute some CMPase and NADPase-positive vesicles to the chromatoid body when this particular structure is at proximity to the Golgi apparatus during the Golgi phase of spermiogenesis (Thorne-Tjomsland et al., 1988).

The *acrosomic system* is a membrane-bound organelle closely applied to the surface of the nucleus of the spermatid and the spermatozoon. Structurally and cytochemically it has two distinct regions, the acrosome and the head cap. The content of the acrosomic system, rich in glycoproteins (Allison and Hartree, 1970; Clermont et al., 1981), is composed of a number of hydrolytic enzymes including acid phosphatases (Thorne-Tjomsland et al., 1988). Phosphatase and lectin cytochemistry has shown a differential distribution of labeling in the acrosome and head cap (Thorne-Tjomsland et al., 1988; Bennett, 1988), thus emphasizing the existence of several compartments within the acrosomic system.

During spermiogenesis there are three distinct periods in the formation and evolution of the acrosomic system of the rat. During the Golgi phase, there is the formation of numerous proacrosomic granules on the *trans* aspect of the stacks that accumulate in the medullary region (Fig. 14–15). These membrane-bound granules show a dense core and a lighter cortex. The small proacrosomic granules fuse with each other to yield a single large acrosomic granule, that establishes a close relation with the nuclear envelope. During the cap phase, the spherical acrosomic granule enlarges by receiving glycoproteins from the Golgi apparatus by the fragmentation of the *trans* Golgi networks into vesicles and tubules accompanied by their fusion with the outer acrosomal membrane. As a result of the addition of membrane and content the developing acrosomic system grows and flattens over the surface of the nucleus. The dense acrosomic granule material applies itself to the inner acrosomal membrane and becomes hemispherical (Figs. 14–2, 14–3, and 14–17); this structure remains distinct throughout the rest of spermiogenesis and constitutes the acrosome proper. The rest of the acrosomic system contains a material of lower density and constitutes the head cap (Figs. 14–16 and 14–17). The head cap also remains distinct throughout spermiogenesis and in the fully formed spermatozoon.

The addition of material to the acrosomic system terminates at the end of the cap phase when the Golgi apparatus separates from it. In the rat, the structural transformation of the acrosomic system continues during the maturation phase, and although these structural changes merit attention, they are too complex to be dealt with in the present review, and hence the original publications should be consulted (Fawcett and Hollenberg, 1963; Lalli and Clermont, 1981). Last, some further changes may be observed in the acrosomic sytem of most mammalian spermatids after spermiation and as the spermatozoon travels along the epididymal canal. In general, there is a condensation of the acrosome proper (Bedford and Nicander, 1971; Bedford et al., 1973). This is obvious in some spatulate spermatozoa that have particularly large acrosomes, such as in the guinea pig (Fawcett and Hollenberg, 1963). In the rat, the final condensation of the acrosome is less striking but nevertheless evident.

4. ENDOPLASMIC RETICULUM AND ASSOCIATED CYTOPLASMIC DENSITIES

The *endoplasmic reticulum* (ER), a system of interconnected cisternae, undergoes extensive structural modifications during spermiogenesis, suggesting an active contribution of this organelle to the differentiation of the spermatid into a spermatozoon. The following description and findings on ER and ER-associated cytoplasmic bodies derive from a study using a special metallic impregnation technique (i.e., uranyl acetate, lead, and copper citrate) combined with a stereoscopic three-dimensional analysis at the electron microscopic level (Clermont and Rambourg, 1978). A more recent electron microscopic immunocytochemical investigation of ODF and fibrous sheath (FS) formation (Clermont et al., 1990) has also identified an ER-associated cytoplasmic body, the granulated body.

During the Golgi and cap phases of spermiogenesis, the ER is distributed throughout the cytoplasm as a three-dimensional network of spherical and distended tubular cisternae bridged by narrow connections. In addition, a loose network of tubular cisternae is located along the *cis* face of the Golgi apparatus (Fig. 14–16), as well as in the ectoplasm along the plasma membrane. Some ER cisternae are found to traverse the Golgi cortex at the level of the intersaccular regions, reach the medulla, and closely approximate themselves to the saccules on the *trans* aspect of the stacks or be inserted between the *trans* Golgi networks (Fig. 14–17) (Hermo et al., 1979). At the level of the intercellular bridges that connect spermatids to one another, the ER network extends across the bridge forming a continuum throughout all spermatids of one isogenic group (Clermont and Rambourg, 1978).

During the acrosome phase, the cytoplasm contains a more profuse ER network that shows the following modifications. (1) Along the inside and outside of the manchette, composed of microtubules, the cisternae form long tubes or plates, with lateral anastomoses, that run adjacent and parallel to these microtubules (Fig. 14–19). (2) Along the proximal periaxonemal compartment, the ER forms a "fenestrated sleeve" made up of a close network of tubular cisternae (Fig. 14–19); (3) similar networks are organized as "fenestrated spherules" enclosing large vesicles seen in the cytoplasmic lobe (to be discussed later). (4) In the midportion of the cytoplasmic lobe, at a short distance from the forming tail, the ER cisternae are continuous with a stack of annulate lamellae and an aggregate of collapsed cisternae arranged in the form of rosettes called the *radial body* (Figs. 14–18 to 14–20). Such a body is also seen in other species (Dym and Cavicchia, 1978). The radial body and annulate lamellae are always next to each other forming a radial body–annulate lamellae complex (Fig. 14–20).

During the maturation phase, the ER regresses by an apparent dissolution of their membranes at the level of the radial body–annulate lamellae complex with most of the ER cisternae disappearing at step 19; the radial body is the last element of the ER to regress (see Fig. 14–24). A few residual distended ER cisternae may be seen in the residual cytoplasm that detaches from the step 19 spermatid at the time of spermiation.

Some of the ER cisternae are associated with ribosomes. Ribosomes are also present free in the cytoplasm where they form polysomes that are particularly abundant during the acrosome phase (Fig. 14–18). Toward the end of spermiogenesis, ribosomes, some of which appear to dissociate from the ER membranes at the time of their dissolution, aggregate into clusters that in turn group to form a single large basophilic mass seen in the residual body, which is discarded from the late spermatid (Clermont, 1956b).

In the cytoplasm of acrosome phase spermatids there are a number of cytoplasmic densities that appear in relation to the cytoplasmic face of ER cisternae. Although the nature of some of these cytoplasmic densities remains mysterious, one of them, the granulated body, appears to have some relation with the formation of ODF as is seen later.

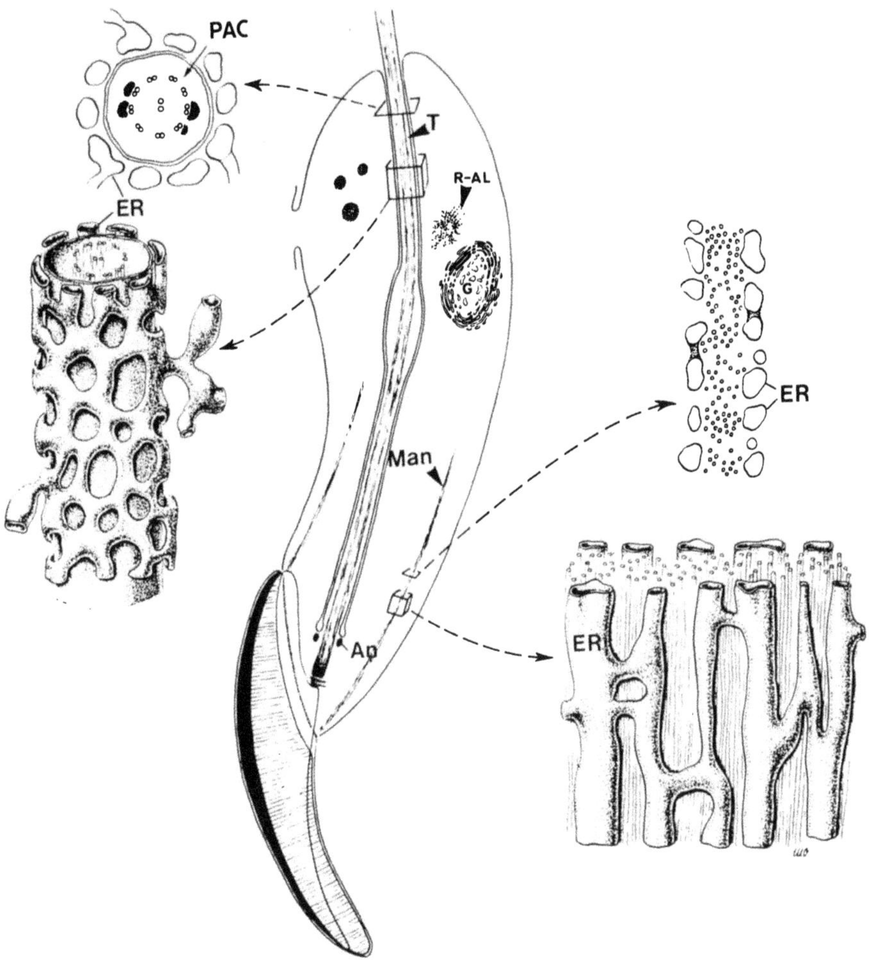

Figure 14–19. Diagrammatic representation of a step 14 spermatid showing two regional specializations of the endoplasmic reticulum (ER) as revealed from stereopairs of thick sections from Ur-Pb-Cu-stained testes. On the left a cross section of the tail (T) is depicted as seen in thin sections (above) and in a three-dimensional reconstruction from thick sections (below). In both drawings the axoneme is surrounded by a double plasma membrane and cisternal profiles of a "fenestrated ER sleeve." On the right, sections of a portion of the manchette (Man), composed of microtubules, are shown in association with cisternae of ER, as seen in thin sections (above) or as reconstructed from thick sections (below). In the latter case the roughly tubular ER cisternae form a double curtain with the microtubules located in between. The ER cisternae forming a fenestrated sleeve along the tail and the tubular ER cisternae along the manchette show numerous connections with the extensive network of ER cisternae distributed throughout the cytoplasm. G, Golgi apparatus; R–AL, radial body–annulate lamellae complex; An, annulus; PAC, proximal periaxonemal compartment. (From Clermont and Rambourg, 1978.)

Puffs and Granulated Bodies

In the cytoplasm of steps 8–9 spermatids, some "puffs" of a fuzzy fine filamentous material appear associated with the outer membrane (cytoplasmic face) of ER cisternae that are at close proximity to the radial body–annulate lamellae complex (Fig. 14–21). These ER associated puffs rapidly increase in number during steps 10–13, with a concomitant change in their texture from finely filamentous (Fig. 14–18) to granulated (Fig. 14–22). Such granulated bodies, composed of dense granules, 30 to 50 nm

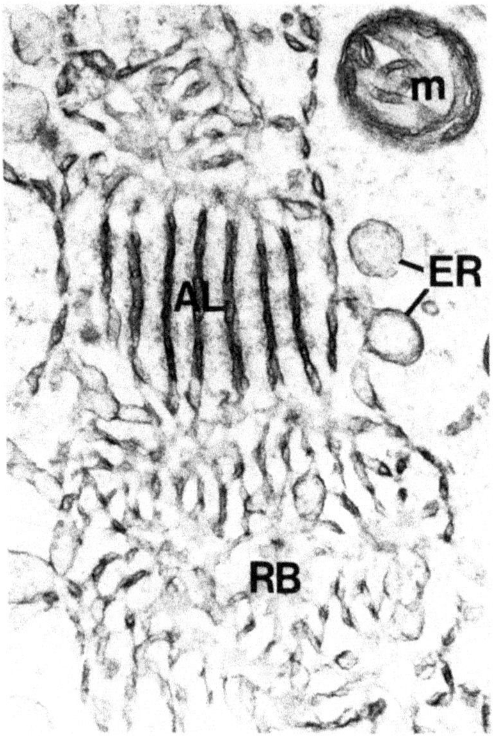

Figure 14–20. Photograph showing a differentiated portion of the endoplasmic reticulum (ER) cisternae of the spermatid's cytoplasm referred to as the radial body (RB)–annulate lamellae (AL) complex. This complex seen in the cytoplasmic lobe of spermatids during the acrosome and maturation phases of spermiogenesis is believed to be the site of resorption of ER membranes (see Clermont and Rambourg, 1978). ER, cisternae of endoplasmic reticulum; m, mitochondrion. ×40,000.

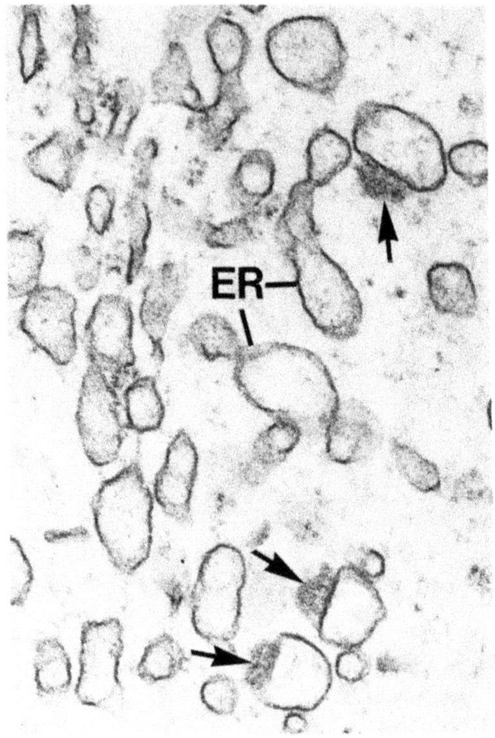

Figure 14–21. Photograph showing a small portion of a step 10 spermatid. At proximity to a stack of annulate lamellae *(not shown)* are numerous profiles of cisternae of endoplasmic reticulum (ER). Three of them show, on their cytoplasmic face, small puffs of a finely filamentous, fuzzy material *(arrows)*. At later steps of spermiogenesis, these puffs grow in size and transform into the granulated bodies shown in Figure 14–22. ×45,000.

in diameter, markedly increase in size and number during steps 14 to mid-18. They become irregular in shape and show at their periphery, small spherical masses or "satellites" composed of a fuzzy fine filamentous material that has a texture identical to that of the puffs (Fig. 14–22). At more advanced steps of spermiogenesis, that is, in late steps 18 and 19, the granulated bodies decrease in number, lose their satellites, and acquire a coarsely granulated appearance. One or two such coarsely granulated bodies are present in the residual cytoplasm discarded from the late spermatid. These granulated bodies are of interest because they appear to be composed of polypeptides destined for the formation of the ODF as is seen later.

Reticulated Body

In the cytoplasm of steps 14–16 spermatids another type of cytoplasmic density appears in association with a cluster of ER cisternae found at proximity to the radial body–annulate lamellae complex. This material, dense and finely filamentous, is organized as rod-like structures, 80 to 100 nm in diameter, deposited in parallel along the cytoplasmic face of ER cisternae. Coincident with the progressive dissolution of ER, this rod-like material separates from the ER and takes on the appearance of a network of

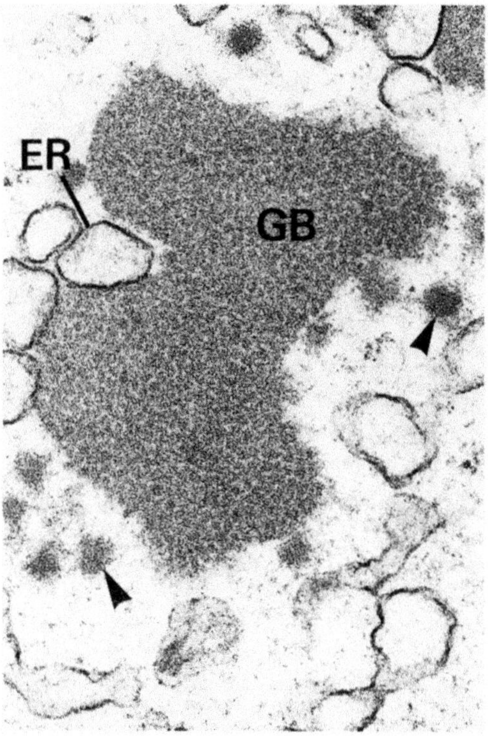

Figure 14–22. Photograph of a small portion of the cytoplasmic lobe of a step 14–15 spermatid showing a granulated body (GB) associated with cisternae of the endoplasmic reticulum (ER). Such bodies with a granular texture appear during the acrosome phase, become numerous during steps 15–17, and later decrease in number. They were found to contain outer dense fiber (ODF) proteins and were considered as storage sites of these cytoskeletal proteins. Note satellite bodies *(arrowheads),* similar in texture to the puffs, in vicinity of granulated bodies. ×50,000.

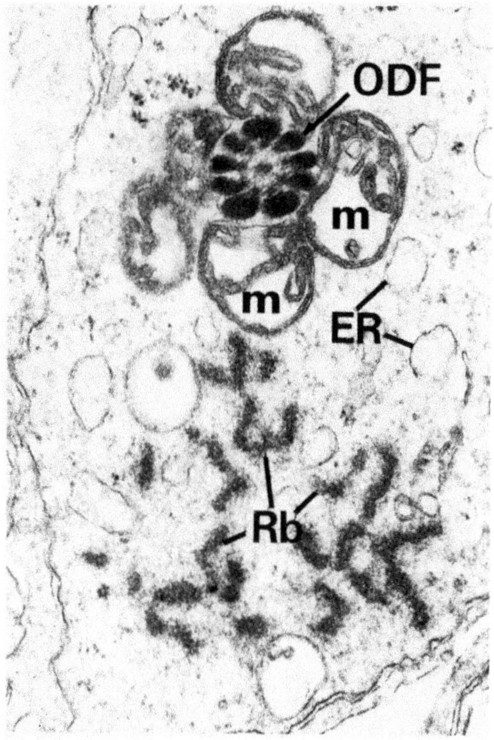

Figure 14–23. Photograph of a small portion of the cytoplasm along the tail of a step 17 spermatid. An electron dense material in the form of anastomosed cords is referred to as the reticulated body (Rb). There is only one such body per spermatid. ER, cisternae of endoplasmic reticulum; m, mitochondria next to outer dense fibers (ODF) of the forming tail. ×35,000.

anastomotic cords, termed the *reticulated body* (Fig. 14–23). There is only one such reticulated body in the cytoplasmic lobe of spermatids and it vanishes during step 18 (Clermont et al., 1990). The nature and function of this structure are unknown.

Radial Body Associated Densities

Toward the end of spermiogenesis, in steps 18 and 19, spherical masses of electron dense material appear and soon vanish in the middle of the regressing radial body (Fig. 14–24). The nature of this material is also unknown.

B. Precursors of Tail Components

1. CENTRIOLES AND NECK PIECE FORMATION

In the cytoplasm of newly formed spermatids there is a pair of centrioles, arranged at right angles to each other and located at a short distance from the plasma membrane (Fig. 14–3). During the Golgi phase, the centrioles migrate toward the nucleus pulling the plasma membrane with them. Simultaneously, one of the two centrioles initiates the formation of the axoneme. During the cap phase the centrioles become associated with the implantation fossa at the surface of the nucleus. In this fossa, the nuclear envelope is bound to the basal plate. Facing this plate, but separated from it by a nar-

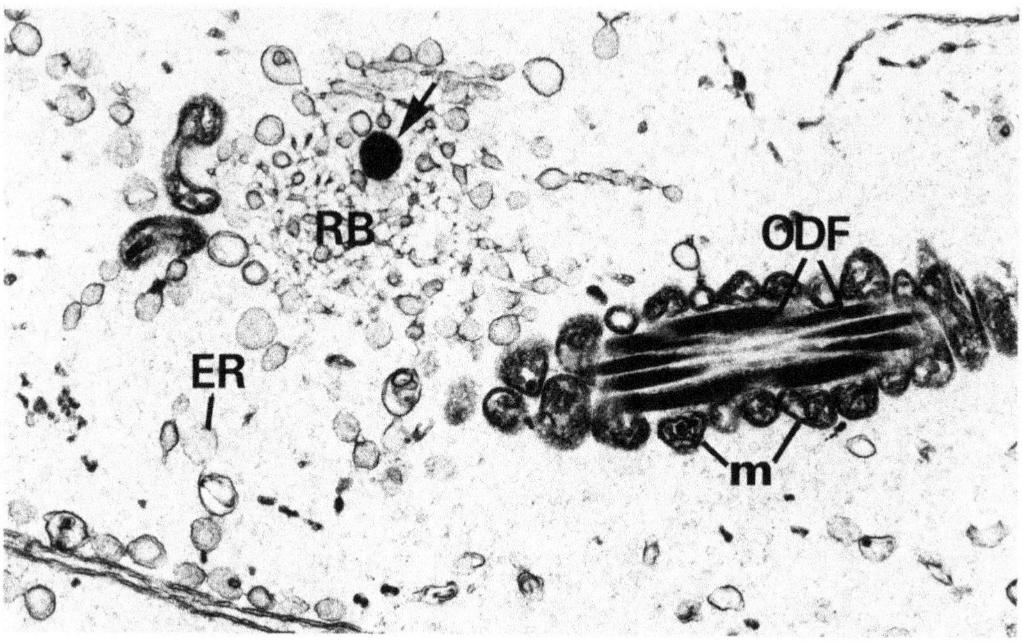

Figure 14–24. Photograph of a portion of a step 18 spermatid. Along the obliquely cut tail the mitochondria (m), disposed circularly, are elongated and condensed. The outer dense fibers (ODF) are seen here cut obliquely. At proximity there is a regressing radial body (RB) in the middle of which a spherical dense body *(arrow)* is visible. One or more such bodies are present within regressing radial bodies, but their composition and function are still unclear. In the cytoplasm of these late spermatids only few profiles of endoplasmic reticulum (ER) remain. ×35,000.

row gap, some electron dense material appears around and over the proximal centriole. This centriole forms the capitulum of the neck piece (Fawcett and Phillips, 1969b).

Throughout the acrosome phase, there is a progressive accumulation of electron dense material between the microtubular triplets and the external surface of both centrioles. This material becomes confluent and forms, in addition to a thicker and denser capitulum, the striated collar (Figs. 14–5 and 14–6). This collar was described as being formed of striated columns (Phillips, 1974; Fawcett, 1975), but in the rat no separate columns could be resolved and instead a striated collar, situated around the distal centriole, was observed (for a review, see Oko and Clermont, 1990).

During the cap phase, a peculiar structure known as the centriolar adjunct emerges from the growing tip of the proximal centriole and reaches its maximal length during the late acrosome phase (Figs. 14–5 and 14–6) (Fawcett and Phillips, 1969b; Irons, 1983). The wall of this adjunct, made up of modified microtubules, is an extension of the nine centriolar triplets. Present in most mammalian species it reaches considerable size in some animals (e.g., hamster; Phillips, 1974). Around the tip of the adjunct in the rat, at steps 12–14 spermatids, there are numerous small spherical bodies made up of a fine fuzzy material (Fig. 14–6). The composition and significance of this material are not known. In the middle of the maturation phase, the proximal centriole and adjunct undergo dissolution leaving an empty vault in the proximal portion of the connecting piece (Wooley and Fawcett, 1973; Irons, 1983). A partial, if not complete, dissolution of the microtubules of the distal centriole is also taking place, although in some species remnants of the distal centrioles may be seen in the connecting piece of mature spermatozoa (Zamboni and Stefanini, 1970).

In the rat, electron dense material that forms

the striated collar is continuous with the ODF that develop along the axoneme. Although the composition of the connecting piece has not yet been clarified, recent immunocytochemical results by Oko (1988) and Clermont and colleagues (1990) indicate that this structure, including the basal plate, capitulum, and striated collar, contains proteins that are also present in the ODF. A biochemical resemblance between these two cytoskeletal structures was also emphasized by Calvin (1976) who showed that both are rich in disulfide bonds.

2. AXONEME AND MANCHETTE

Axoneme

The axoneme of spermatids, the growth of which is triggered by the distal centrioles, elongates as in any growing cilia by the addition of α and β tubulins at the distal ends of the microtubular singlets and doublets. In the rat, the full growth of the axoneme, which reaches a length of 150 μm, appears to be completed by the end of the cap phase of spermiogenesis (Irons and Clermont, 1982a). The detailed structure and composition of the axoneme, because of its complexity, are not discussed in the present review; it has been well covered elsewhere (Fawcett, 1975; Link, 1979; Oko and Clermont, 1990).

Manchette

The manchette or caudal tube composed of microtubules characterizes the spermatid during the acrosome phase (Fig. 14–2). This is a curtain-like structure that has a certain rigidity and extends from just below the acrosomic system into the cytoplasmic lobe (step 12 of Fig. 14–3). It has a general flattened, truncated conical shape. The manchette is inserted on a ringlike structure or perinuclear insertion ring that surrounds the nucleus and is associated with the plasma membrane at a short distance from the caudal edge of the head cap (step 12 of Fig. 14–3). The manchette thus surrounds the implantation fossa, the developing neck piece, and part of the axoneme ensheathed by the proximal periaxonemal cytoplasm (Figs. 14–4 and 14–19). The wall of this manchette, which reaches its maximal size during step 14, is composed of numerous microtubules running parallel to each other and linked by lateral fine filamentous bridges (MacKinnon et al., 1973). These microtubules were strongly labeled in a recent electron microscopic immunocytochemical study with antibodies raised against detyrosinated microtubules (Hermo et al., 1991).

Certainly one striking feature is the sudden disruption and dissolution of the manchette during step 15 of spermiogenesis simultaneous with the rapid migration of the annulus along the axoneme (step 16 of Fig. 14–3). The microtubules break down as a result of the depolymerization of tubulins, and no remnant of the manchette can be seen in late spermatids or spermatozoa. The perinuclear insertion ring persists for a while and finally undergoes complete dissolution (Lalli and Clermont, 1981).

Several hypotheses have been put forward to explain the manchette's existence. Cole and colleagues (1988) suggest that the manchette is involved in nuclear shaping on the basis that improper positioning of the manchette is associated with an abnormal sperm head or nuclear shaping. Another possibility is that the manchette creates cytoplasmic currents that may serve to rearrange organelles (e.g., mitochondria) in the cytoplasmic lobe. Cytoplasmic motions are especially noticeable in living spermatids during the acrosome phase (Clermont, 1967) at the time the manchette prevails. Finally, it is feasible that unidentified microtubular translocator proteins are present on the manchette that would serve to transport proteins (e.g., cytoskeletal proteins) up or down its length. A precedent for such a transport along microtubules was established by Vale and associates (1985a,b).

3. CHROMATOID BODY, ANNULUS

The chromatoid body and annulus are two structures that are distinct morphologically but become momentarily associated during spermiogenesis.

The *chromatoid body* is seen in the cytoplasm of rat spermatids during the Golgi, cap, and early acrosome phases. The chromatoid body in early steps of spermiogenesis is a small, irregular, spongy mass made up of an electron dense granular material associated with several small vesicular profiles (Fig. 14–25) that contain carbohydrates (Susi and Clermont, 1970). It is frequently seen at proximity to the Golgi apparatus (Parvinen and Jokelainen, 1974),

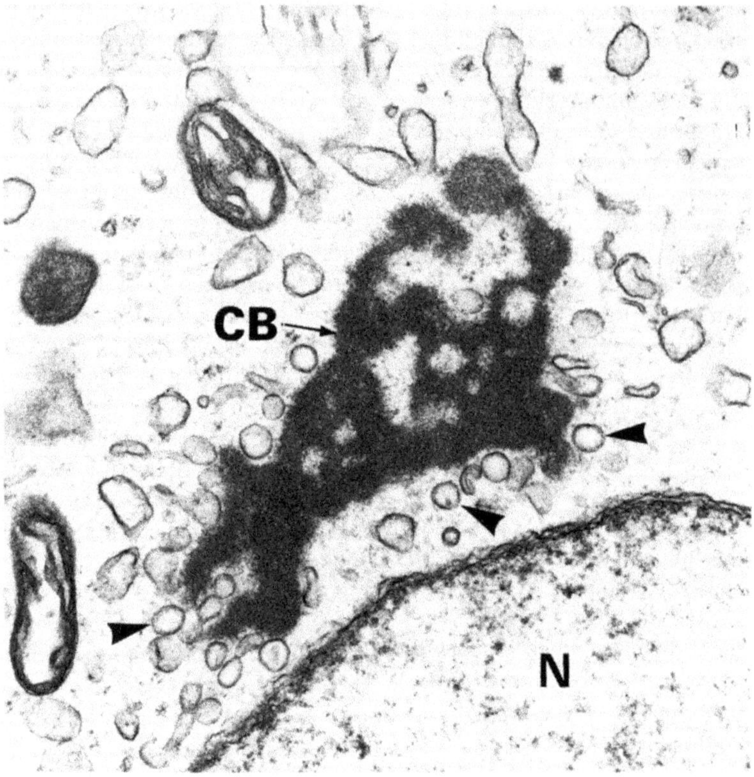

Figure 14–25. Next to the nucleus (N) of an early step 5 spermatid there is a chromatoid body (CB) made up of a fine electron dense granulofilamentous material having a characteristic spongy appearance. Numerous small vesicular profiles *(arrowheads),* some of Golgi origin and containing phosphatases (Thorne-Tjomsland et al., 1988), are seen associated with the chromatoid body. ×30,000.

which appears to contribute some small photophatase-containing vesicles to this body (Thorne-Tjomsland et al., 1988). Toward the end of the cap phase the chromatoid body moves to the centriolar pole of the nucleus (step 6 of Fig. 14–3), changes in shape, and forms a crescentic structure around the distal centriole and proximal part of the axoneme. It appears to contribute material to the annulus (Fawcett et al., 1970; Susi and Clermont, 1970) after which it transforms into a dense spherical body surrounded by distended vesicles (Fig. 14–4). Thereafter, the chromatoid body disappears from the cytoplasm of the elongated spermatids.

The composition of the chromatoid body is still unknown. A light microscopic study indicated that it contained RNA (Daoust and Clermont, 1955), but this has been questioned (Eddy, 1970), although ribosomal particles are associated with it. Some of its vesicular components were found to contain phosphatases (Thorne-Tjomsland et al., 1988).

The *annulus* is a ring-like structure composed of an electron dense finely filamentous material that is closely associated with the plasma membrane and is seen in fully formed spermatozoa at the junction of the middle and principal piece of the tail (Figs. 14–1A, 14–3, 14–5, and 14–26). The early stage in formation of the annulus can already be detected in Golgi phase spermatids at the level of the plasma membrane that surrounds the developing axoneme next to the distal centriole (step 6 in Fig. 14–3). At this point the plasma membrane is invaginated slightly and it is at the very bottom of the invaginated plasma membrane that the annulus appears first as an anlagen and later by

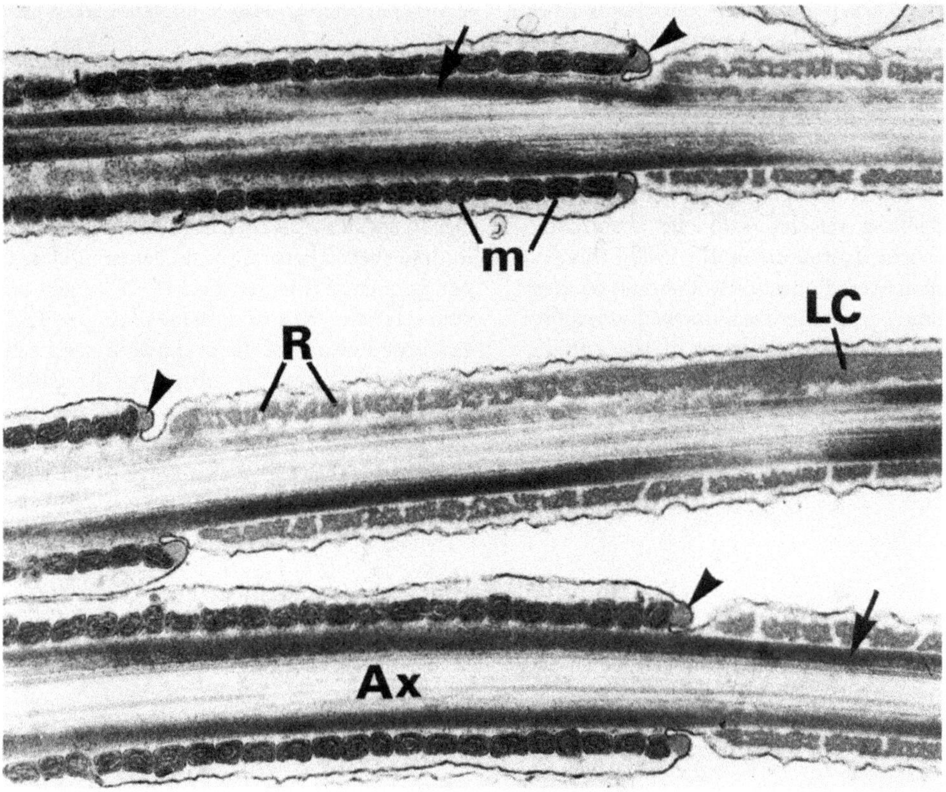

Figure 14–26. Longitudinal sections of the tails of three spermatozoa at the junction of the middle piece *(at left)* and principal piece *(at right)*. The annulus *(arrowheads)*, associated with a small infolding of the plasma membrane, is visible in each tail. The markedly condensed mitochondria (m) of the middle piece are visible as small squarish profiles. The transverse ribs (R) of the fibrous sheath separated from the plasma membrane are seen in cross section. In the tail located in the center of the photograph some of the ribs are replaced by a section through the longitudinal column (LC). The microtubules of the axoneme (Ax) as well as views of the outer dense fibers *(arrows)* can also be identified. ×30,000.

the addition of electron dense material to this anlagen (Fawcett et al., 1970). This primitive annulus is seemingly attached to the edges of the centriolar structure, as the pair of centrioles migrates toward the implantation fossa it pulls with it the plasma membrane and associated annulus (step 12 in Fig. 14–3). During the late cap phase when the chromatoid body is already surrounding the centrioles, a portion of the chromatoid body material is seemingly added to the annulus on its cytoplasmic face (Fawcett et al., 1970; Susi and Clermont, 1970; Phillips, 1974). Thus, this ring has two regions, the dense and finely filamentous annulus proper and a layer of granular material of chromatoid body origin. This condition persists throughout the acrosome phase until step 15, when the annulus rapidly slides down along the axoneme (step 16 in Fig. 14–3). At this time the chromatoid body's portion of the ring apparently undergoes partial dissolution, whereafter only a residue of it persists (Phillips, 1974); none of it is seen in the spermatozoon.

The significance of these complex changes undertaken by the annulus and the chromatoid body during spermatogenesis remains mysterious. It has been suggested that the chromatoid body may serve as a carrier of information in the form of RNA (Hecht, 1987). It is true that in certain species the chromatoid body establishes a close association with the nuclear pores of early spermatids (Fawcett et al., 1970).

Whether or not the nucleus contributes RNA to it at this point is not known. Fawcett and associates (1970) suggested that the chromatoid body may originate from a material associated with mitochondria of spermatocytes, on the basis of a similarity in density of both structures. As to the association of some CMPase and NADPase vesicles with the chromatoid body (Thorne-Tjomsland et al., 1988), this also raised unanswered questions. It should be mentioned finally that there is a marked variability in size and internal structure of the annulus from one species to another (Fawcett and Phillips, 1970), but within a given species, its structural characteristics are constant from one given spermatid to another.

4. OUTER DENSE FIBERS, FIBROUS SHEATH

Outer Dense Fibers

The ODFs, a major cytoskeletal element of the spermatozoon's tail, are present in all mammalian spermatozoa although they show marked variations in size and shape from species to species (Fawcett, 1970, 1975). Along the middle piece there are nine coarse ODFs, one per axonemal doublet (Fig. 14–27). Each has a characteristic size and shape (Fig. 14–1A), is composed of a medulla and a thin cortex, and has been shown to present a periodic substructure (Fawcett, 1975; Olson and Sammons, 1980). The ODFs decrease in diameter from the proximal to the distal extremity of the middle

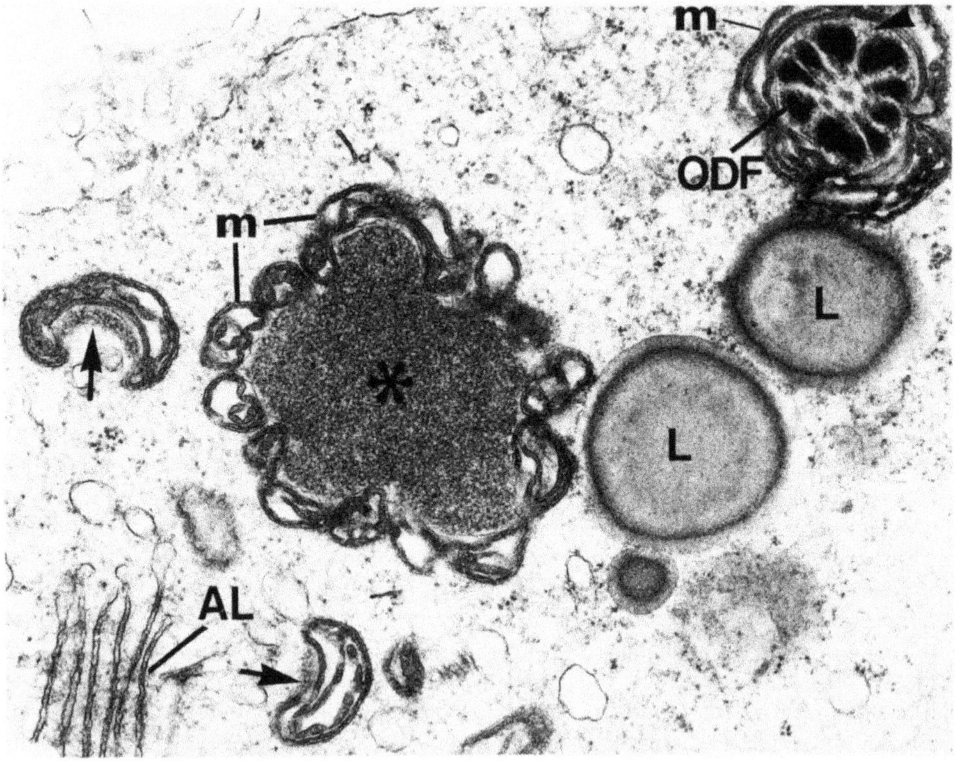

Figure 14–27. Photograph showing in the cytoplasm of a step 17 spermatid a large mass of granular material *(asterisk)* to which crescentic mitochondria (m) are bound. A similar granular material is seen in the concavity of free condensed crescentic mitochondria *(arrows)*. A material of similar appearance *(arrowhead)* is seen between the outer dense fibers (ODF) and the crescentic mitochondria (m) forming the middle piece of the tail. The granular material appears to serve as a binding material that attaches mitochondria to ODF. AL, annulate lamellae; L, lipid droplets. ×35,000.

piece. In the principal piece ODFs are also present, but ODF numbers three and eight are replaced by the longitudinal columns of the fibrous sheath (Fig. 14–1A). The remaining ODFs decrease in diameter distally and disappear in the end piece of the tail (Fig. 14–1A). Polyacrylamide gel analyses suggest that there are five or six major ODF polypeptides ranging from 11,500 to 87,000 daltons as well as several minor ones (reviewed by Eddy, 1988; Oko and Clermont, 1990). Compositional analysis has shown that the major rat ODF polypeptides have a high content of lysine, arginine, cysteine, aspartate, serine, leucine, and proline, and a large amount of phosphate bound as phosphoserine (Olson and Sammons, 1980;

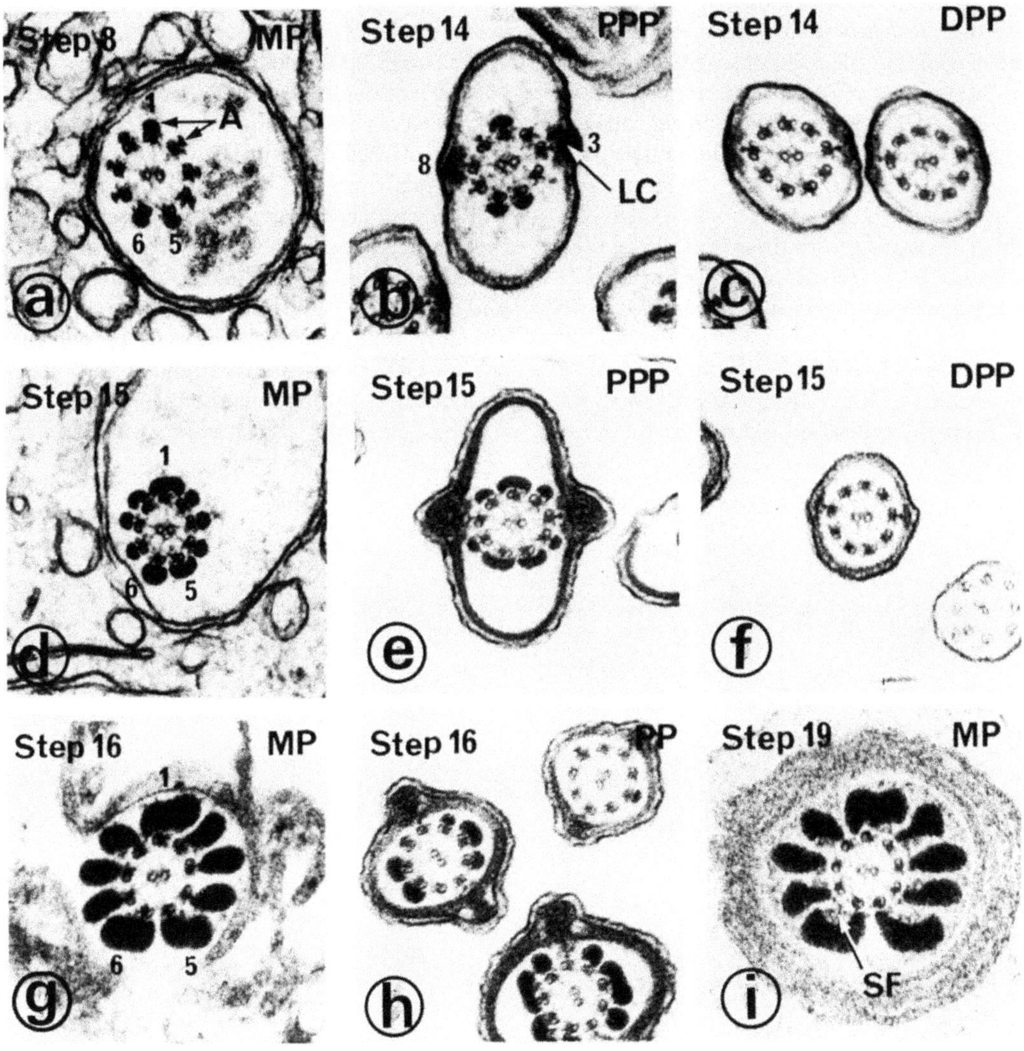

Figure 14–28. Cross sections through the forming tail of rat spermatids. The steps of spermiogenesis are indicated as well as the portion of the tail sectioned, such as middle piece (MP), proximal principal piece (PPP), and distal principal piece (DPP). These photographs show the stages of development of the outer dense fibers along the outer doublets of the axoneme. Outer dense fibers 1, 5, and 6 are indicated in **a**, **d**, and **g**. As seen in sections **b** and **c** of the principal piece, the longitudinal columns (LC) of the fibrous sheath occupy positions 3 and 8 around the axoneme. Indicated in **a**, the anlagen (A) of the outer dense fibers. SF, satellite fibers. ×40,000. (From Irons and Clermont, 1982a.)

Vera et al., 1984). Disulfide bonds cross-linking proteins are present and increase during epididymal transit (Calvin and Bedford, 1971). Various studies indicate that ODF proteins may comprise a unique family of proteins not found in mammalian somatic cells.

The morphogenesis of ODFs has been investigated by electron microscopic studies in the rat by Irons and Clermont (1982a) and in a few other species (Sapsford et al., 1969; de Kretser, 1969). Outer dense fiber anlagen first appears along the most proximal portion of the axoneme in step 8 in the form of small projections on the external face of each microtubular doublet (Fig. 14–28a). From steps 9–14 the anlagen gradually increases in length along the axoneme in a proximal-to-distal direction (Fig. 28a–c). During these steps very small ODFs develop in association with these projections (Fig. 14–28a,b). In a second phase (steps 15–17), these rudimentary fibers increase markedly in diameter with the most dramatic growth occurring in step 16 (Fig. 14–28 compare d and e with h and i). This transformation, which results from the deposition of an electron dense material along these rudimentary fibers coincides with a period of most rapid incorporation of [^{3}H]proline- and [^{3}H]cysteine-containing proteins into these fibers (Irons and Clermont, 1982a). In a final phase (steps 17–19), the fibers continue to enlarge assuming their definitive size only during the last step of spermiogenesis (Fig. 14–26).

In recent light and electron microscopic immunocytochemical studies using affinity purified antibodies prepared against specific ODF proteins, the observations on the assembly of ODFs were confirmed (Oko and Clermont, 1989; Clermont et al., 1990). In addition, these studies were able to distinguish the time of formation of ODF proteins in the cytoplasm of spermatids from the time of their assembly. Outer dense fiber proteins were present in the cytoplasm of spermatids from steps 9–19, but reached a peak of synthesis from step 16 to mid-step 18 in parallel with the time of peak assembly along the axoneme. The immunolabels for ODF proteins were found to be distributed diffusely throughout the cytoplasmic matrix but especially concentrated over specific

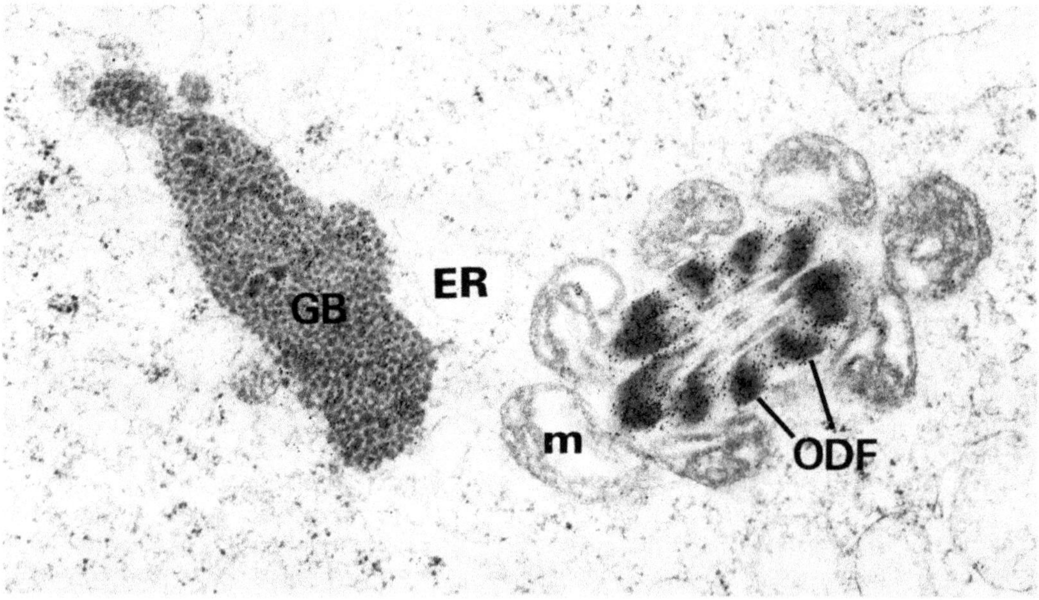

Figure 14–29. Electron micrograph of a section of material embedded in Epon through the cytoplasm of a step 17 spermatid, immunogold labeled with antibodies affinity purified against individual outer dense fiber (ODF) polypeptides. Both the ODF and the granulated body (GB) are strongly labeled. A relatively light diffuse labeling is found in the cytoplasmic matrix. m, mitochondria; ER, endoplasmic reticulum. ×42,500.

granulated bodies, which as previously mentioned are associated with ER (Figs. 14–29 and 14–30). Considering these observations we suggest that ODF polypeptides, presumably biosynthesized like any other cytoskeletal proteins on polysomes, can either be targeted directly to the growing ODFs or stored temporarily in these granulated bodies. During the last few steps of spermiogenesis, when ODF protein synthesis ceases, we envision that ODF proteins within granulated bodies are released and delivered to the surface of the still growing ODFs. As a consequence there would be a progressive resorption of the granulated bodies in these last few steps of spermiogenesis that does appear to be the case (Clermont et al., 1990). If granulated bodies indeed serve as transitory storage stations for ODF proteins, then regulatory mechanisms would have to be positioned to account for the binding and concentration of ODF proteins in granulated bodies and later for their release into the cytoplasmic matrix.

The Fibrous Sheath

The fibrous sheath (FS) is a complex structure composed of longitudinal columns that run lengthwise opposite microtubular doublets 3 and 8 and numerous transverse ribs that bridge the longitudinal columns (Phillips, 1974; Fawcett, 1975; Oko and Clermont, 1990) (Fig. 14–1A). Its appearance is variable from one species to another (Fawcett, 1975). Originally described as being composed of a single polypeptide of 80 kd (Olson et al., 1976b), a more recent investigation revealed at least 18 distinct polypeptide bands in sodium dodecylsulfate-polyacrylamide gels of FS fractions (Oko, 1988). Of these bands the 75, 27.5, and 14.4 kd polypeptides were more prominent.

The morphogenesis of the FS in the rat sper-

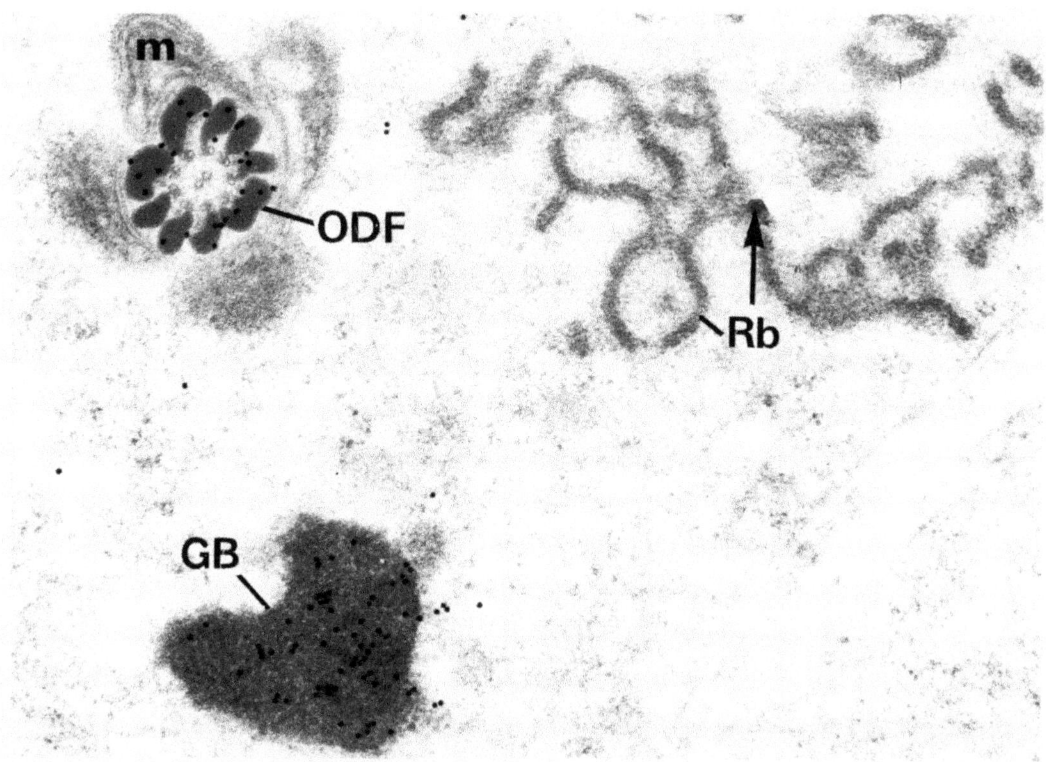

Figure 14–30. Electron micrograph of a Lowicryl section through the cytoplasm of a step 17 spermatid, immunogold labeled with antibodies affinity purified against individual outer dense fiber (ODF) polypeptides. The labeling is confined to the ODF and the granulated body (GB). No labeling is found over the reticulated body (Rb). m, mitochondria. ×48,750.

matid has been studied by electron microscopic radioautography after injection of [^{3}H]proline (Irons and Clermont, 1982b), and although it was briefly alluded to in a number of mammalian species it was analyzed in detail in only one other species, the marsupial bandicoot rat (Sapsford et al., 1969). The development of the FS takes place in a distal-to-proximal direction, starting at the end of the principal piece (Fig. 14–31). The anlagen of the longitudinal columns is initially deposited along doublets three and eight in the distal portion of the principal piece in steps 2–10 (Fig. 14–31). The longitudinal columns are gradually laid down on the anlagen in a proximal direction ending at the level of the annulus in step 17 (Fig. 14–31). The ribs on the other hand are deposited on anlagen, which is present on the plasma membrane along the length of the principal piece during steps 11–17 (Fig. 14–31). The observations of Irons and Clermont (1982b) indicate that the columns and ribs are assembled independently of each other except for a period of 2.5 days between steps 12 and 14. In steps 15–17 there is a convergence of the ribs and a further deposition of FS proteins to form definitive ribs and longitudinal columns that then separate from the plasma membrane (Fig. 14–26). The

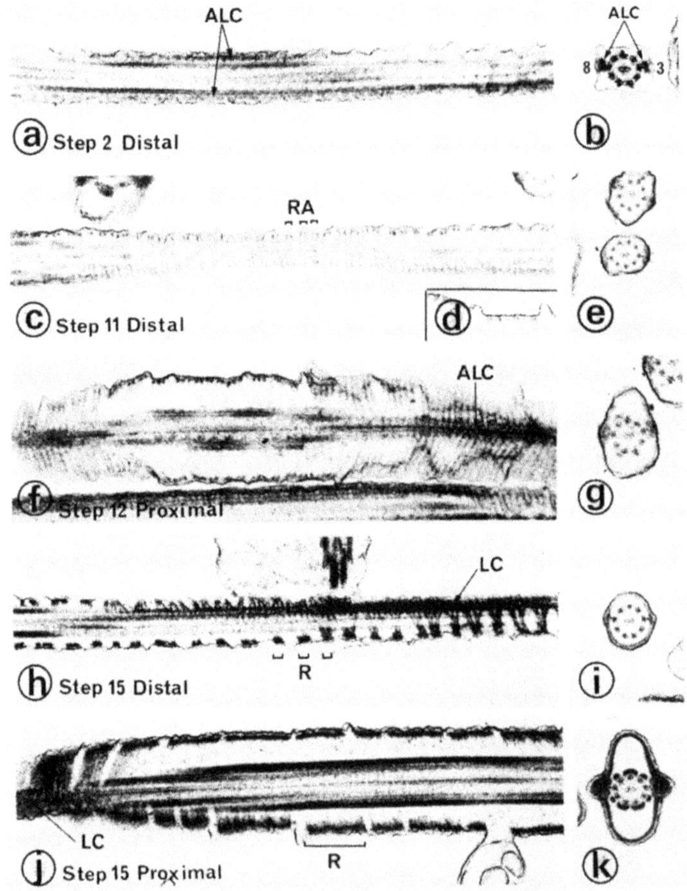

Figure 14–31. Longitudinal and cross sections through the distal and proximal regions of the forming tail of rat spermatids at various steps of spermiogenesis. The photographs show the formation of the fibrous sheath. ALC, anlagen of the longitudinal columns; RA, anlagen of the ribs appearing in close relation to the plasma membrane; R, ribs forming at the surface of the rib anlagen; LC, longitudinal column. ×33,000. (From Irons and Clermont, 1982b.)

assembly of the FS is thus a long multistep process that extends from steps 2 to 17 of spermiogenesis.

In the rat, the rib anlagen appears to originate from bundles of proteinaceous filamentous material seen along the axoneme in the distal periaxonemal compartment. A similar material was described in the periaxonemal cytoplasm of bandicoot spermatids (Sapsford et al., 1969). In humans and rabbits, a fibrous sheath precursor material was also described in the periaxonemal compartment, although in these instances it assumed the configuration of microtubules (de Kretser, 1969; Phillips, 1974; Wartenberg and Holstein, 1975).

The results of recent light and electron microscopic immunocytochemical studies (Oko and Clermont, 1989; Clermont et al., 1990) generally verified the earlier observations made on the assembly of the FS (Irons and Clermont, 1982b). In addition, these immunocytochemical observations demonstrated that peak production of FS proteins is reached in step 15 of spermiogenesis and that FS proteins are diffusely spread throughout the cytoplasmic matrix of spermatids from steps 9–18 (Fig. 14-32). Interestingly the antibodies raised against the FS proteins, of which there were several, did not react with the FS rib anlagen or with the filamentous material seen in the periaxonemal compartment, which is presumably FS precursor material (Clermont et al., 1990). Thus, the anlagen that may serve as a trigger for FS assembly appears to be composed of proteins distinct from those that compose the definitive FS. In contrast to ODF proteins, there appears to

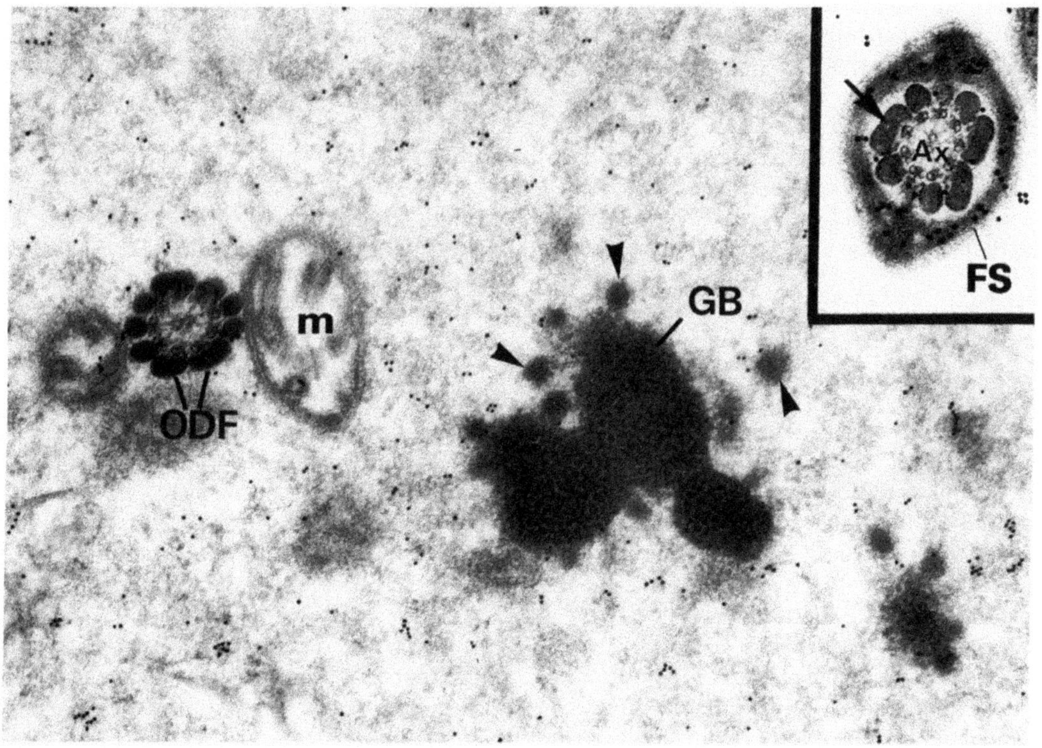

Figure 14–32. Electron micrograph of a Lowicryl section through the cytoplasm of a step 15 spermatid, immunogold labeled with antibodies affinity purified against individual fibrous sheath polypeptides. The gold labeling is distributed evenly throughout the cytoplasmic matrix. It is not found over the outer dense fibers (ODF) or the granulated body (GB). (Compare with Figs. 14-29 and 14-30.) m, mitochondria; *arrowheads,* satellite bodies. ×55,000. Inset: Depicts immunogold labeling with anti-fibrous sheath antibodies over the fibrous sheath (FS) of the principal piece. No labeling is found over the outer dense fibers *(arrows)* or the axoneme (Ax). ×62,500.

be no storage loci for FS proteins in the spermatid's cytoplasm during the formation of the fibrous sheath.

As to the functions of ODFs and FS, although they obviously modulate the type of motion of the axoneme, their exact role in this process remains unknown. Similarly the molecular events that regulate the biosynthesis of ODF and FS proteins as well as their deposition along the axoneme constitute an open field of investigation.

5. MITOCHONDRIA

The structure and arrangement of mitochondria along the middle piece of the tail of fully formed spermatozoa soon attracted the attention of cytologists when the electron microscope became available. It was discovered that condensed, crescentic mitochondria are arranged side by side and transversely along the cytoskeletal components of the tail. They are not arranged horizontally, however, but at a small angle such that the mitochondria, disposed tip to tip, are helically arranged along the middle piece (Fawcett, 1975; Phillips, 1977). The genesis of this arrangement of modified mitochondria was studied by André (1962) in several rodents and was recently reexamined closely in the mouse by Otani and co-workers (1988).

The mitochondria of the newly formed spermatid are peculiar in that their intracristal space is distended with an electron lucent fluid, whereas the mitochondrial matrix is condensed and reduced in volume (Figs. 14–5 and 14–9). Thus, there is a reversal of contrast, with the intracristal space being more important volumetrically than the matrical space. Before their elongation and helical rearrangement in the maturation phase, the spermatid's mitochondria are spheroidal and distributed at the periphery of the cytoplasm (Fig. 14–3).

During the acrosome phase, as the cytoplasm extends along the axoneme, the mitochondria do not change much in structure, although their number appears to increase considerably. Evidence for the synthesis of mitochondrial DNA has been presented by Hecht and Liem (1984); however, morphological evidence of division of this organelle has not been documented.

At step 15 of the maturation phase the annulus descends down the axoneme. As a consequence, the proximal periaxonemal compartment disappears (step 16 in Fig. 14–3) allowing mitochondria to have direct access to the rapidly growing ODFs. The mitochondria, by unexplained mechanisms migrate from the peripheral cytoplasm, contact the ODFs, and rapidly dispose themselves side by side along the future middle piece (Figs. 14–9 and 14–23). At this point they have a hemispherical shape and their surface, apposed to the ODFs, is often scalloped. During steps 17 and 18 the mitochondria elongate and present a crescentic shape (Fig. 14–1A). This change in shape is accompanied by a loss of electron lucent fluid from the intracristal space, the volume of which now approximates the volume of the matrix (Figs. 14–1A, 14–24, and 14–26). Many of the mitochondria do not take part in forming the middle piece of the tail, although they do undergo identical structural changes as those comprising the mitochondrial sheath. These superfluous mitochondria are discarded within the residual body at the end of spermiogenesis.

Mitochondria Associated Granular Material

In steps 9–15 spermatids, there are several large-sized spherical bodies composed of a finely granular material appearing in the middle of the cytoplasmic lobe surrounded by ER cisternae. Some of the superfluous mitochondria become disposed around the periphery of these spherical bodies beginning at step 16 (Fig. 14–27). During subsequent steps they condense, elongate, and take on a crescentic shape, in synchrony with all the other mitochondria. At this time, the mitochondria that are aligned around the ODFs accumulate a granular material on their concave surface (Fig. 14–27). This material is similar to the material found in the mitochondrial-associated granular body and appears to originate from it (Fig. 14–27) (Hermo and Clermont, 1980; Oko and Hrudka, 1982, 1984). It is retained in the spermatozoa (Olson et al., 1987; Barth and Oko, 1989) and most likely binds the mitochondrial sheath to the ODFs. Friend and Heuser (1981) have shown a stepladder arrangement of particle triplets on the concave aspect of the outer mitochondrial membrane that may correspond to this granular material.

III. Spermatid–Sertoli Cell Structural Relations

One of the most striking histological features of the seminiferous epithelium is the clustering of elongated spermatids within the apical cytoplasm of Sertoli cells. Although the ultrastructural characteristics of the Sertoli cell ectoplasm facing spermatids have been well described (Russell, 1984), there is still no explanation for the migration of spermatids in and out of these deep crypts. One possibility could be attributed to the numerous microtubules present in the Sertoli cell cytoplasm (Amlani and Vogl, 1988). These microtubules in an electron microscopic immunocytochemical study have recently been shown to be strongly labeled with antibodies raised against tyrosinated microtubules but not to those of detyrosinated and acetylated microtubules (Hermo et al., 1991). Nevertheless, this intimate relation of spermatids with Sertoli cells from steps 8–9 until spermiation favors exchanges of different substances between these two classes of cells.

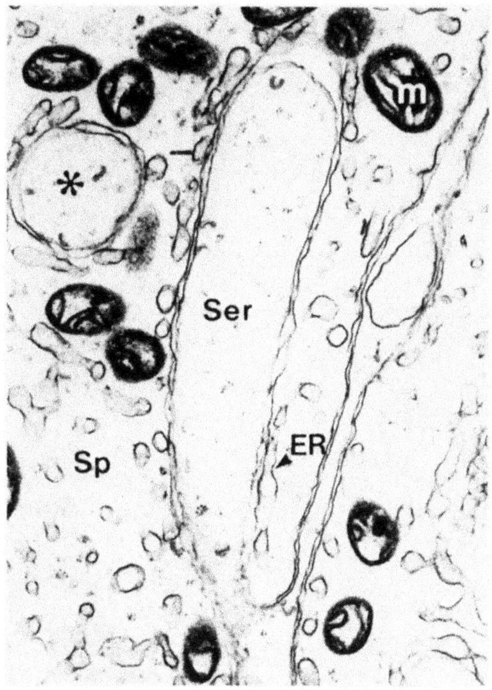

Figure 14–33. Section of a small portion of a cytoplasmic lobe of a step 13 spermatid showing a longitudinally cut, digit-like Sertoli cell process (Ser) invading a spermatid (Sp). One such process is also seen in cross section *(asterisk)*. Numerous profiles of cisternae of endoplasmic reticulum (ER) are aligned along the plasma membrane of the spermatid enclosing the Sertoli cell process. m, mitochondria. ×15,000.

A. Invagination of Sertoli Cell Processes Into the Spermatid Cytoplasm

During the last two phases of spermiogenesis, two types of Sertoli cell processes invaginate the cytoplasmic lobe of spermatids (Morales and Clermont, 1982). Some, observed in steps 11–15, are associated with a tight network of ER cisternae located within the spermatid's cytoplasm subjacent to the plasma membrane facing the Sertoli cell process, whereas others seen from step 11 until the end of spermiogenesis are not (Figs. 14–33 to 14–35). After their formation, the ER-associated Sertoli cell processes are cleaved into double-walled vesicles (Figs. 14–33 to 14–35). Concomitantly, the enclosed Sertoli cell cytoplasm undergoes autolysis. This is followed by the dissolution of the Sertoli cell plasma membrane leaving within the spermatid's cytoplasm many empty-looking vacuoles delimited merely by the plasma membrane of the spermatid (Fig. 14–35). Thereafter, in steps 15 and 16, these vacuoles cluster, collapse, and give rise to a single compact mass of interlaced ER cisternae and small residual vacuoles. With the dissolution of ER cisternae in step 18, this mass finally disperses (Morales and Clermont, 1982).

These observations suggested that the ER-associated Sertoli cell processes may constitute a pathway for the delivery of small molecules from Sertoli cells to spermatids, resulting from the lytic breakdown of the Sertoli cell cytoplasm (Morales and Clermont, 1982). In contrast, the Sertoli cell processes not associated with a network of ER cisternae do not show cleavage or autolysis and appear to be involved mainly in holding the elongating spermatids in

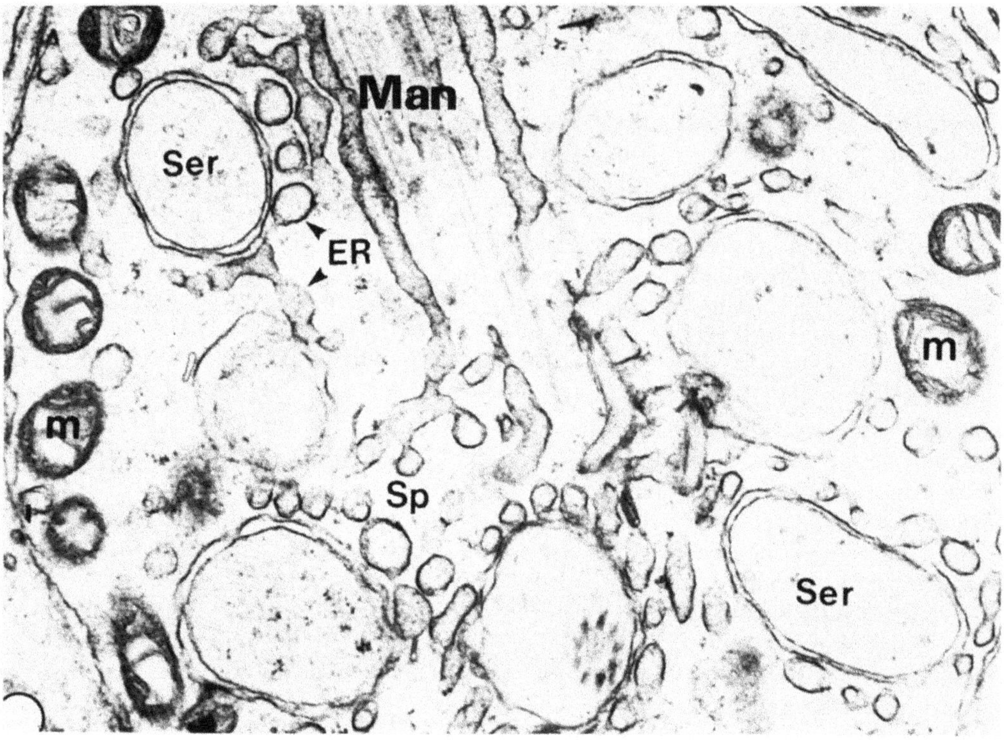

Figure 14–34. Cross section of a cytoplasmic lobe of a step 12 spermatid (Sp) showing several Sertoli cell processes (Ser) in cross section. These Sertoli cell invaginations are demarcated by a double membrane, the inner membrane being the membrane of the Sertoli cell, the outer membrane being the membrane of the spermatid. The latter membrane is associated with cisternae of endoplasmic reticulum (ER). m, mitochondria; Man, manchette. ×18,000. (From Morales and Clermont, 1982.)

place. Such Sertoli cell invaginations of spermatids have been observed in a number of mammalian species (reviewed in Morales and Clermont, 1982) but their evolution during spermiogenesis has been fully analyzed in the rat (Morales and Clermont, 1982) and more recently in the ground squirrel (Vogl et al., 1985).

B. Spermatid Invagination Into Sertoli Cells: Tubulobulbar Complexes

Near the end of spermiogenesis, the late spermatids remain attached to the superficial layer of the seminiferous epithelium for a period of 3 or 4 days. The sickle-shaped heads of the spermatids are embedded in a discoid apical process of Sertoli cell cytoplasm that is connected to the rest of the cell by a narrow stalk (Russell and Clermont, 1976). In the concavity of the head several long (2 to 3 μm), very narrow tubular projections of the spermatid's plasma membrane invaginate into the discoid mass of Sertoli cell cytoplasm. These tubular processes terminate by a bulbous swelling that may measure up to 1 μm in diameter (Fig. 14–36). Along the tubular part, the plasma membrane of the Sertoli cell is closely apposed to the spermatid's membrane, the intercellular space being only 6 to 8 nm wide. In the Sertoli cytoplasm immediately surrounding the tubular portion there is an accumulation of filamentous material, whereas next to the bulbous extremity there are, at short distance, smooth surfaced ER cisternae. This structure, originally described in the rat and termed the tubulobulbar complex (Russell and Clermont, 1976) was also found in a number of mammalian species (Russell, 1984) and appears to be of

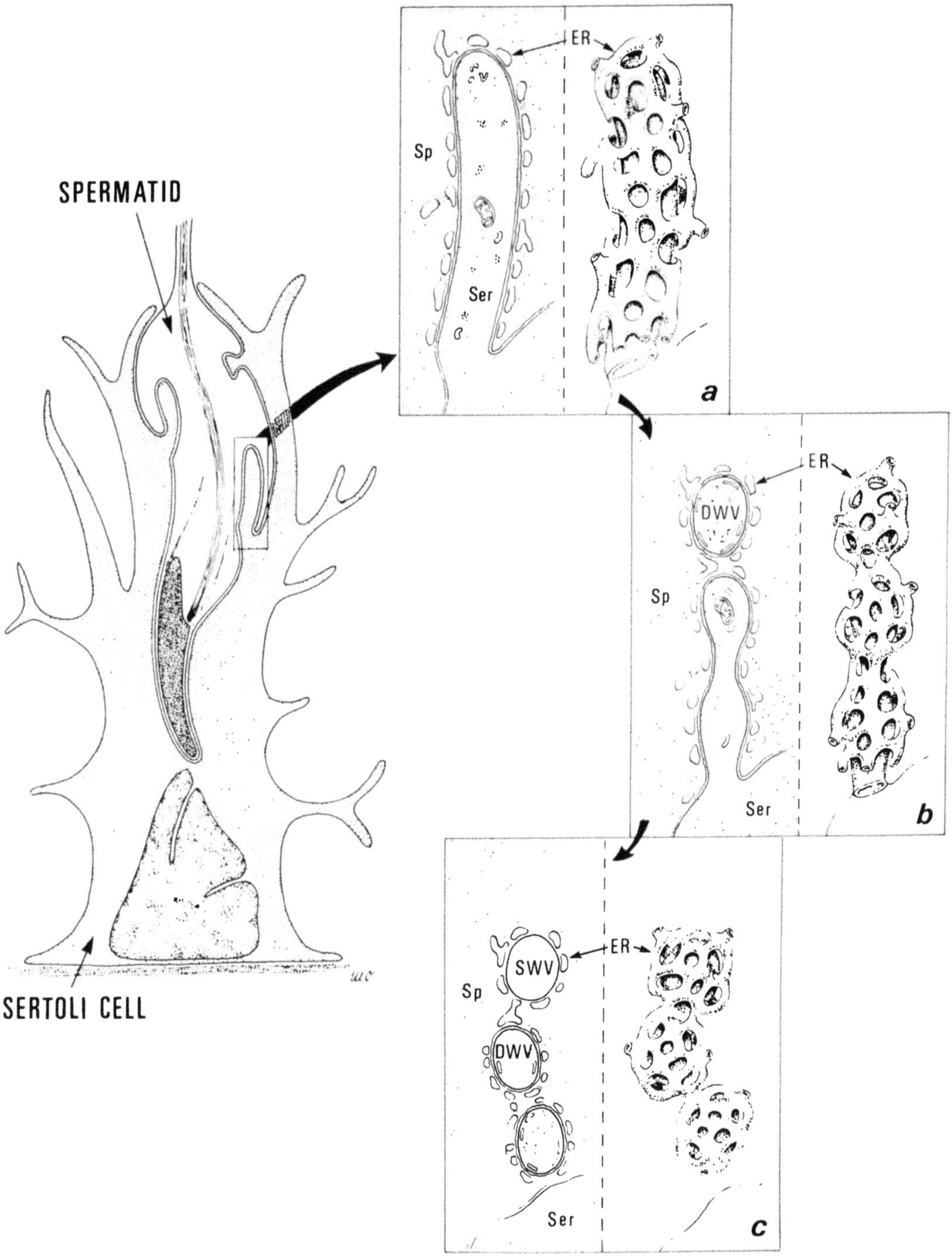

Figure 14–35. Diagram showing a step 15 spermatid inserted into the cytoplasm of a Sertoli cell as well as processes of the latter invading the cytoplasm of the spermatid. On the right, in boxes **a**, **b**, and **c**, pairs of drawings show diagrammatically the evolution of the Sertoli cell processes (Ser) after they invade the spermatid (Sp). In each box, the drawing on the left shows the invaginating Sertoli cell process or its derivatives, such as double-walled vesicles (DWV) or single-walled vacuoles (SWV), as seen in thin sections. The drawing on the right represents the same structures as seen in three dimensions to illustrate the relationship of cisternae of endoplasmic reticulum (ER) of the spermatid with that of the Sertoli cell invaginations. The fragmentation and dissolution of Sertoli cell cytoplasmic processes invaginating the spermatids is considered to be a possible route of passage of molecules from the Sertoli cell to the spermatids. (From Morales and Clermont, 1982.)

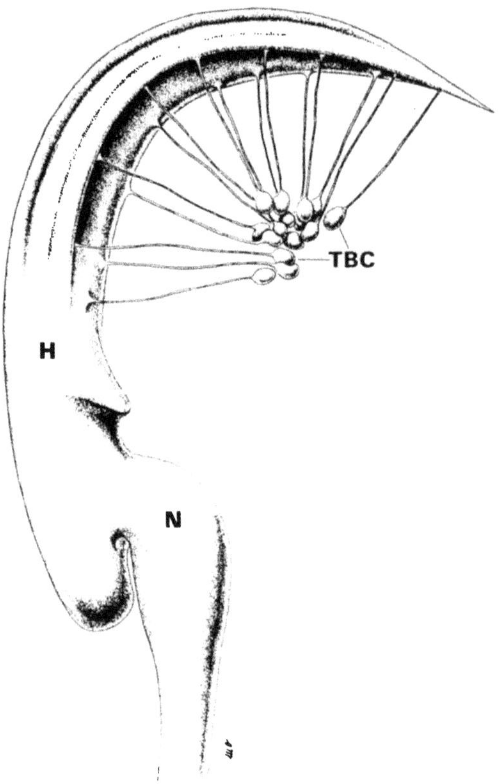

Figure 14–36. Three-dimensional drawing of the external form of the head (H) and neck (N) piece region of a step 19 spermatid. This drawing shows only the superficial view of the spermatid when the head is still encapsulated by an apical process of the Sertoli cell (not shown). Note several long tubulobulbar processes originating from the plasma membrane of the spermatid that would project into the Sertoli cell cytoplasm (not shown). These arise from the ventral aspect of the sickle-shaped head and end in a central cluster of bulbous dilatations. Such tubulobulbar complexes (TBC) are lost at the time of spermiation as the head of the spermatid disengages from the Sertoli cell cytoplasm. (From Russell and Clermont, 1976.)

general occurrence but with some variation in appearance. These tubulobulbar complexes, of which there are up to 24 per rat spermatid, appear as these cells complete their migration at the apex of the Sertoli cells. They disappear just before the release of the spermatids into the lumen of the seminiferous tubule as a result of the clipping off of the spermatid's membrane at the base of the tubulobulbar complex next to the nucleus. Morphological evidence suggests that the tubulobulbar complexes serve as anchoring devices that retain the spermatids at the surface of the seminiferous epithelium (Russell and Clermont, 1976). Russell (1979) has suggested that water molecules may flow from the spermatid to the Sertoli cell at the level of these peculiar processes of the spermatid's plasma membrane.

C. The Elimination of Residual Cytoplasm of Spermatids by Sertoli Cells

There is also an interaction between the Sertoli cell and the spermatid at the very end of spermiogenesis that leads to the liberation of the mature step 19 spermatid's superfluous cytoplasm. This was well described in the early electron microscopic studies of Fawcett and Hollenberg (1963) and reviewed by Russell (1984). It appears that a flap of Sertoli cell cytoplasm associates with the cytoplasmic lobe and embraces it. As a result, when the heads of late spermatids move toward the tubular lumen, the cytoplasmic lobe stays put. This cytoplasmic spheroidal mass contains unused mitochondria, lipid droplets, some vesicular profiles, granulated bodies, and clustered ribosomes and is referred to as the residual body. It detaches from the neck region of the step 19 spermatid, leaving behind a small mass of cytoplasm referred to as the cytoplasmic droplet containing some short flattened membranous vesicule profiles showing galactosyl transferase activity (Oko et al., 1989). The cytoplasmic droplet flows along the middle piece of the tail of the spermatozoon during its passage through the epididymis and is eventually discarded in the corpus and the cauda epididymidis (Hermo et al., 1988). The residual body itself is phagocytosed by the Sertoli cell and lyzed, after fusion with secondary lysosomes (Morales et al., 1985). Thus, the Sertoli cell not only contributes to holding the spermatids in place throughout spermiogenesis but also provides sustenance. After spermiation it contributes actively and efficiently to the elimination of the cytoplasmic residues discarded by the late spermatid.

General Conclusion

The present review, using mainly the rat spermatid as a model, stresses the extraordinary complexity of the transformation of the nucleus and organelles of newly formed spermatids as they metamorphose into spermatozoa. Interestingly enough, with the exception of mitochondria, none of the standard cytoplasmic organelles (e.g., ER, Golgi, centrioles) constitute components of the mature spermatozoa. Instead what we observe in the spermatozoon are derivatives of these organelles, that is, acrosomic system elaborated by the ER–Golgi system; the axoneme derived from one of the two centrioles; and peculiar cytoskeletal elements of the spermatozoon, such as, the perinuclear theca in the head and the ODFs, FS, and annulus in the tail, which are structures that appear during spermiogenesis as a result of an active biosynthetic activity, presumably taking place at the level of polysomes. In addition, the "originality" of this process of cell differentiation is emphasized by features for which there is no equivalent in any other cell type. One of these processes concerns the nucleus that, as a result of an exquisitely controlled mechanism of chromatin condensation and nuclear envelope disposal, takes a compacted and species-specific microanatomy. The close association of a Golgi-derived lysosome-like structure, the acrosomic system, with the spermatid's nucleus and similarly the association of the pair of centrioles with the nuclear envelope during the early steps of spermiogenesis followed by the formation of a structurally complex neck piece are rather unique cases of nuclear and cytoplasmic organelle associations. The structural modifications of mitochondria as well as their displacement toward and disposition along the middle piece of the forming tail of the spermatozoon have no equivalent in any other cell type. Last, the evolution and/or formation of cytoskeletal components, such as the manchette, annulus, ODFs, FS, and perinuclear theca, are all spermatid-specific characteristics. The remarkable association of the transforming spermatids with the Sertoli cell, of which only a few facets were presented here, should also be considered as peculiar examples of intercellular interactions. All cytological events taking place during spermiogenesis presuppose a multitude of regulatory mechanisms, biochemical and physicochemical in nature, about which very little information is presently available and that constitutes an open and vast field of research for cell and molecular biologists.

ACKNOWLEDGMENTS

The work of the authors presented in this chapter has been supported by grants from the Medical Research Council of Canada. The authors gratefully acknowledge the help of Dr. Michael Lalli who provided several electron micrographs to illustrate the text. The drawings were prepared by Mrs. Margo Oeltschner.

REFERENCES

Allison, A. C., and Hartree, F. (1970). Lysosomal enzymes in the acrosome and their possible role in fertilization. J. Reprod. Fertil. 21:501–515.

Amlani, S., and Vogl, A. W. (1988). Changes in the distribution of microtubules and intermediate filaments in mammalian Sertoli cells during spermatogenesis. Anat. Rec. 220:143–160.

André, J. (1962). Contribution à la connaissance du chondriome: Étude de ses modifications ultrastructurales pendant la spermiogénèse. J. Ultrastruct. Res. (Suppl.) 3:1–185.

Baccetti, B., and Afzelius, B. A. (1976). The biology of the sperm cell, in *Monographs in Developmental Biology* (Wolsky, A., ed.) pp. 101–106, S. Karger, Basel, München, Paris, London, New York, Sydney.

Bardin, C. W., Cheng, C. Y., Mustro, N. A., and Gunsalus, G. L. (1988). The Sertoli cell, in *The Physiology of Reproduction* (Knobil, E., Neill, J. D., et al., eds.) pp. 933–998, Raven, New York.

Barth, A. D., and Oko, R. J. (1989). Abnormal morphology of bovine spermatozoa, pp. 1–285, Iowa State University Press, Ames, Iowa.

Bedford, J. M., and Nicander, L. (1971). Ultrastructural changes in the acrosome and sperm membranes during maturation of spermatozoa in the testis and epididymis of the rabbit and monkey. J. Anat. 108:527–543.

Bedford, J. M., Calvin, H., and Cooper, G. W. (1973). The maturation of spermatozoa in the human epididymis. J. Reprod. Fertil. [Suppl. 18]: 199–213.

Bennett, G. (1988). Radioautographic and cytochemical studies on the synthesis and intracellular transport of glycoproteins. Acta Histochemica, Suppl.-B and XXXVI, S. 9–49.

Bennett, G. (1984). Role of the Golgi complex in the secretory process, in *Cell Biology of the Secretory Process* (Cantin, M., ed.) pp. 102–147, Karger, Basel.

Bernstein, M. H., and Teichman, R. J. (1972). Regional differentiation in the heads of spermatozoa of rabbit, man and bull. Am. J. Anat. 133:165–178.

Breed, W. C. (1984). Sperm head structure in the *Hydromyinae* (Rodentia: Muridae): A further evolutionary development of the subacrosomal space in mammals. Gamete Res. 10:31–44.

Burgos, M., and Fawcett, D. W. (1955). Studies on the fine structure of the mammalian testis. I. Differentiation of the spermatid in the cat *(Felis domestica)*. J. Biophys. Biochem. Cytol. 1:287–300.

Calvin, H. I. (1976). Isolation and subfractionation of mammalian sperm heads and tails, in *Methods in Cell Biology,* Vol. 13 (Prescott, D. M., ed.) pp. 85–104, Academic Press, New York.

Calvin, H. I., and Bedford, J. M. (1971). Formation of disulphide bonds in the nucleus and accessory structures of mammalian spermatozoa during maturation in the epididymis. J. Reprod. Fert. (Suppl.) 13:65–75.

Clermont, Y. (1972). Kinetics of spermatogenesis in mammals: Seminiferous epithelium cycle and spermatogonial renewal. Physiol. Rev. 52:198–236.

Clermont, Y. (1967). Cinétique de la spermatogénèse chez les mammifères. Arch. d'Anat. Micr. et Morph. Expér. 56 (Suppl. 3–4):7–60.

Clermont, Y. (1963). The cycle of the seminiferous epithelium in man. Am. J. Anat. 112:35–52.

Clermont, Y. (1956a). The Golgi zone of the rat spermatid and its role in the formation of cytoplasmic vesicles. J. Biophys. Biochem. Cytol. 2:119–122.

Clermont, Y. (1956b). The submicroscopic structure responsible for the cytoplasmic basophilia of the rat spermatids. Exp. Cell Res. 11:214–217.

Clermont, Y. (1954). Cycle de l'épithélium séminal et mode de renouvellement des spermatogonies chez le hamster. Rev. Canad. Biol. 13:208–245.

Clermont, Y., Oko, R., and Hermo, L. (1990). Immunocytochemical localization of proteins utilized in the formation of outer dense fibers and fibrous sheath in rat spermatids: An electron microscopic study. Anat. Rec. 227:447–457.

Clermont, Y., Hermo, L., Rambourg, A., and Thorne-Tjomsland, G. (1990). Structural and cytochemical characteristics of the Golgi apparatus during the formation of the acrosomic system in rat spermatids, in *Cellular and Molecular Events in Spermiogenesis* (Hamilton, D., and Waites, G.M.H., eds.) Cambridge University Press, New York, pp. 17–29.

Clermont, Y., and Tang, X. M. (1985). Glycoprotein synthesis in the Golgi apparatus of spermatids during spermiogenesis of the rat. Anat. Rec. 213:33–43.

Clermont, Y., Lalli, M. F., and Rambourg, A. (1981). Ultrastructural localization of nicotinamide adenine dinucleotide phosphatase (NADPase), thiamine pyrophosphatase (TPPase) and cytidine monophosphatase (CMPase) in the Golgi apparatus of early spermatids in the rat. Anat. Rec. 201:613–622.

Clermont, Y., Rambourg, A. (1978). Evolution of the endoplasmic reticulum during rat spermiogenesis. Am. J. Anat. 151:191–212.

Clermont, Y., and Harvey, S. C. (1965). The duration of the cycle of the seminiferous epithelium of normal, hypophysectomized and hypophysectomized-hormone treated albino rats. Endocrinol. 76:80–89.

Clermont, Y., Leblond, C. P., and Messier, B. (1959). Durée du cycle de l'épithélium séminal du rat. Arch. Anat. Micr. 48 bis:37–56.

Clermont, Y., and Leblond, C. P. (1955). Spermiogenesis of man, monkey, ram and other mammals as shown by the "periodic acid-Schiff" technique. Am. J. Anat. 96:229–254.

Clermont, Y., Glegg, R. E., and Leblond, C. P. (1955). Presence of carbohydrates in the acrosome of the guinea pig spermatozoon. Exp. Cell Res. 8:453–458.

Cole, A., Meistrich, M. L., and Trostle-Weige, P. K. (1988). Nuclear and manchette development of normal and *azh/azh* mutant mice. Biol. Reprod. 38:385–401.

Courtens, J. L., Courot, M., and Fléchon, J. E. (1976). The perinuclear substance of boar, bull, ram and rabbit spermatozoa. J. Ultrastruct. Res. 57:54–64.

Daoust, R., and Clermont, Y. (1955). Distribution of nucleic acids in germ cells during the cycle of the seminiferous epithelium in the rat. Am. J. Anat. 96:255–284.

de Kretser, D. M. (1969). Ultrastructural features of human spermiogenesis. Z. Zellforsch. 98:577–605.

de Kretser, D. M., and Kerr, J. B. (1988). The cytology of the testis, in *The Physiology of Reproduction* (Knobil, E., Neill, J. D., et al., eds.) pp. 837–932, Raven Press, New York.

Dunphy, W. G., and Rothman, J. E. (1985). Compartmental organization of the Golgi stack. Cell 42:13–21.

Dym, M., and Cavicchia, J. C. (1978). Functional morphology of the testis. Biol. Reprod. 18:1–15.

Dym, M., and Fawcett, D. W. (1971). Further observations on the numbers of spermatogonia, spermatocytes, and spermatids connected by intercellular bridges in the mammalian testis. Biol. Reprod. 4:195–215.

Eddy, E. M. (1988). The spermatozoon, in *The Physiology of Reproduction* (Knobil, E., Neill, J. D., et al., eds.) pp. 27–68, Raven Press, New York.

Eddy, E. M. (1970). Cytochemical observations on the chromatoid body of the male germ cell. Biol. Reprod. 2:114–128.

Farquhar, M. G. (1985). Progress in unraveling pathways of Golgi traffic. Annu. Rev. Cell Biol. 1:447–488.

Fawcett, D. W. (1979). Comparative aspects of the organization of the testis and spermatogenesis, in *Animal Models for Research on Contraception and Fertility* (Alexander, N. J., ed.) pp. 84–104, Harper & Row, New York.

Fawcett, D. W. (1975). The mammalian spermatozoon. Dev. Biol. 44:394–436.

Fawcett, D. W. (1970). A comparative view of sperm ultrastructure. Biol. Reprod. 2 (Suppl. 2):90–127.

Fawcett, D. W., Anderson, W. A., and Phillips, D. M. (1971). Morphogenetic factors influencing the shape of the sperm head. Dev. Biol. 26:220–251.

Fawcett, D. W., and Phillips, D. M. (1970). Recent observations on the ultrastructure and development of the mammalian spermatozoon, in *Comparative Spermatology* (Baccetti, B., ed.) pp. 13–28, Academic Press, New York.

Fawcett, D. W., Eddy, E. M., and Phillips, D. M. (1970). Observations of the fine structure and relationships of the chromatoid body in mammalian spermatogenesis. Biol. Reprod. 2:129–153.

Fawcett, D. W., and Phillips, D. M. (1969a). Observations on the release of spermatozoa and on changes in the head during passage through the epididymis. J. Reprod. Fert. 6 (Suppl.):405–418.

Fawcett, D. W., and Phillips, D. M. (1969b). The fine structure and development of the neck region of the mammalian spermatozoon. Anat. Rec. 165:153–184.

Fawcett, D. W., and Hollenberg, R. D. (1963). Changes in the acrosome of guinea-pig spermatozoa during passage through the epididymis. Z. Zellforsch. Mikrosk. Anat. 60:276–292.

Fawcett, D. W., Ito, S., and Slautterback, D. (1959). The occurrence of intercellular bridges of groups of cells exhibiting synchronous differentiation. J. Biophys. Biochem. Cytol. 5:453–460.

Franklin, L. E. (1968). Formation of the redundant nuclear envelope in monkey spermatids. Anat. Rec. 161:149–162.

Franklin, L. E., and Fussell, E. N. (1972). Evolution of the apical body in golden hamster spermatids with some reference to primates. Biol. Reprod. 7:194–206.

Friend, D. S., and Heuser, J. E. (1981). Orderly particle arrays on the mitochondrial outer membrane in rapidly-frozen sperm. Anat. Rec. 199:159–175.

Hecht, N. B. (1988). Post-meiotic gene expression during spermiogenesis, in *Meiotic Inhibition: Molecular Control of Meiosis* (Hàse Hihc, F., ed.) pp. 291–313, Alan R. Liss, N.Y.

Hecht, N. B. (1987). Gene expression during spermiogenesis. Ann. NY Acad. Sci. 513:90–101.

Hecht, N. B., and Liem, H. (1984). Mitochondrial DNA is synthesized during meiosis and early spermiogenesis in the mouse. Exp. Cell. Res. 154:293–297.

Heller, C. G., and Clermont, Y. (1964). Kinetics of the germinal epithelium in man. Rec. Prog. Hormone Res. 20:545–575.

Hermo, L., Oko, R., and Hecht, N. B. (1991). Differential post-translational modifications of microtubules in cells of the seminiferous epithelium of the rat: a light and electron microscope immunocytochemical study. Anat. Rec. 229:31–50.

Hermo, L., Dworkin, J., and Oko, R. (1988). Role of epithelial clear cells of the rat epididymis in the disposal of the contents of cytoplasmic droplets detached from spermatozoa. Am. J. Anat. 183:107–124.

Hermo, L., and Clermont, Y. (1980). Possible source of material binding mitochondria to the flagellum of spermatids during spermiogenesis of the rat. (Abstr.) Anat. Rec. 196:77.

Hermo, L., Rambourg, A., and Clermont, Y. (1980). Three-dimensional architecture of the cortical region of the Golgi apparatus of rat spermatids. Am. J. Anat. 157:357–373.

Hermo, L., Clermont, Y., and Rambourg, A. (1979). Endoplasmic reticulum–Golgi apparatus relationships in the rat spermatid. Anat. Rec. 193:243–256.

Irons, M. J. (1983). Synthesis and assembly of connecting-piece proteins as revealed by radioautography. J. Ultrastruct. Res. 82:27–34.

Irons, M. J., and Clermont, Y. (1982a). Formation of the outer dense fibers during spermiogenesis in the rat. Am. J. Anat. 202:463–471.

Irons, M. J., and Clermont, Y. (1982b). Formation of the fibrous sheath during spermiogenesis in the rat. Am. J. Anat. 165:121–130.

Koehler, J. K. (1970). A freeze-etching study of rabbit spermatozoa with particular reference to head structure. J. Ultrastruct. Res. 33:598–614.

Lalli, M. F., and Clermont, Y. (1981). Structural changes of the head components of the rat spermatid during late spermiogenesis. Am. J. Anat. 160:419–434.

Leblond, C. P., and Clermont, Y. (1952). Spermiogenesis of rat, hamster and guinea pig as revealed by the "periodic acid-fuchsin sulfurous acid" technique. Am. J. Anat. 90:167–215.

Linck, R. W. (1979). Advances in the ultrastructural analysis of the sperm flagellar axoneme, in *The Spermatozoon* (Fawcett, D. W., and Bedford, J. M., eds.) pp. 99–115, Urban and Schwarzenberg, Baltimore, Munich.

Loir, M., and Lanneau, M. (1984). Structural function of the basic nuclear proteins in ram spermatids. J. Ultrastruct. Res. 86:262–276.

Loir, M., and Courtens, J. L. (1979). Nuclear reorganization in ram spermatids. J. Ultrastruct. Res. 67:309–324.

MacKinnon, E. A., Abraham, P. J., and Svatek, A. (1973). Long link induction between microtubules of the manchettes in intermediate stages of spermiogenesis. Z. Zellforsch. Mikrosk. anat. 136:447–460.

Meistrich, M. L., Brock, W. A., Grimes, S. R., Platz, R. D., and Hnilica, L. S. (1978). Nuclear protein transitions during spermatogenesis. Fed. Proc. 37:2522–2525.

Mollenhauer, H. H., Hass, B. S., and Morré, D. J. (1976). Membrane transformations in Golgi apparatus of rat spermatids. A role for thick cisternae and two classes of coated vesicles in acrosome formation. J. Microsc. Biol. Cell 27:33–36.

Monesi, V. (1965). Synthetic activity during spermatogenesis in the mouse. RNA and protein synthesis. Exp. Cell Res. 39:197–224.

Morales, C., Clermont, Y., and Hermo, L. (1985). Nature and function of endocytosis in Sertoli cells of the rat. Am. J. Anat. 173:203–217.

Morales, C., and Clermont, Y. (1982). Evolution of Sertoli cell processes invading the cytoplasm of rat spermatids. Anat. Rec. 203:233–244.

O'Brien, D. A., and Bellvé, A. R. (1980). Protein constituents of the mouse spermatozoon. II. Temporal synthesis during spermiogenesis. Dev. Biol. 75:405–418.

Oko, R. (1988). Comparative analysis of proteins from the fibrous sheath and outer dense fibers of rat spermatozoa. Biol. Reprod. 39:169–182.

Oko, R., and Clermont, Y. (1991). Origin and distribution of perforatorial proteins during spermatogenesis of the rat. Anat. Rec. 230:489–501.

Oko, R., and Clermont, Y. (1990). Mammalian spermatozoa: Structure and assembly of the tail, in *Controls of Sperm Motility* (Gagnon, C., ed.) pp. 3–29, CRC Press, Boca Raton, FL.

Oko, R., Moussakova, L., and Clermont, Y. (1990). Regional differences in composition of the perforatorium and of the outer periacrosomal layer of rat spermatozoa as revealed by immunocytochemistry. Am. J Anat. 188:64–73.

Oko, R., and Clermont, Y. (1989). Immunocytochemical

analysis of the formation of the fibrous sheath and outer dense fibers in the seminiferous epithelium of the rat. Anat. Rec. 225:46–55.

Oko, R., Chan, P.T.K., and Hermo, L. (1989). Galactosyl transferase activity within lamellar elements of cytoplasmic droplets of rat epididymal spermatozoa. J. Cell Biol. 109:250A.

Oko, R., and Clermont, Y. (1988). Isolation, structure and protein composition of the perforatorium of rat spermatozoa. Biol. Reprod. 39:673–687.

Oko, R., and Hrudka, F. (1984). Gossypol-induced early and delayed effects in the seminiferous epithelium of the adult rat. Contraceptive Delivery Systems 5:335–356.

Oko, R., and Hrudka, F. (1982). Segmental aplasia of the mitochondrial sheath and sequelae induced by gossypol in rat spermatozoa. Biol. Reprod. 26:183–195.

Olson, G. E., Winfrey, V. P., and Flaherty, S. P. (1987). Cytoskeletal assemblies of mammalian spermatozoa, in *Cell Biology of the Testis and Epididymis* (Orgebin-Crist, M.-C., and Danzo, B. J., eds.) Ann. NY Acad. Sciences 513:222–246.

Olson, G. E., and Sammons, D. W. (1980). Structural chemistry of outer dense fibers of rat sperm. Biol. Reprod. 22:319–332.

Olson, G. E., Hamilton, D. W., and Fawcett, D. W. (1976a). Isolation and characterization of the perforatorium of rat spermatozoa. J. Reprod. Fert. 47:1–9.

Olson, G. E., Hamilton, D. W., and Fawcett, D. W. (1976b). Isolation and characterization of the fibrous sheath of rat epididymal spermatozoa. Biol. Reprod. 14:517–530.

Otani, H., Tanaka, O., Kasai, K.-I., and Yoshioka, T. (1988). Development of mitochondrial helical sheath in the middle piece of the mouse spermatid tail: Regular dispositions and synchronized changes. Anat. Rec. 222:26–33.

Parvinen, M., and Jokelainen, P. T. (1974). Rapid movements of the chromatoid body in living early spermatids of the rat. Biol. Reprod. 11:85–92.

Phillips, D. M. (1977). Mitochondrial disposition in mammalian spermatozoa. J. Ultrastruct. Res. 58:144–154.

Phillips, D. M. (1974). *Spermiogenesis.* Academic Press, New York, London.

Phillips, D. M. (1970). Ultrastructure of spermatozoa of the woolly opossum, *Caluromys philander.* J. Ultrastruct. Res. 33:381–397.

Rothman, J. E. (1985). The compartmental organization of the Golgi apparatus. Sci. Am. 84–95.

Russell, L. D. (1984). Spermiation. The sperm release process: Ultrastructural observations and unresolved problems, in *Electron Microscopy in Biology and Medicine: Ultrastructure of Reproduction* (Van Blerkom, J., and Motta, P. M., eds.) pp. 46–66, Martinus Nijhoff, Boston.

Russell, L. D. (1979). Spermatid–Sertoli tubulobulbar complexes as devices for elimination of cytoplasm from head region of late spermatids of the rat. Anat. Rec. 194:233–246.

Russell, L. D., and Clermont, Y. (1976). Anchoring device between Sertoli cells and late spermatids in rat seminiferous tubules. Anat. Rec. 185:259–278.

Sandoz, D. (1972). Etude autoradiographique de l'incorporation in vitro de galactose-^{3}H dans les spermatides de souris. J. Micros. 15:403–408.

Sandoz, D. (1970). Evolution des ultrastructures au cours de la formation de l'acrosome du spermatozoide chez la souris. J. Micros. 9:535–558.

Sapsford, C. S., Rae, C. A., and Cleland, K. W. (1969). Ultrastructural studies on maturing spermatids and on Sertoli cells in the bandicoot *Parameles nasuta* (Marsupalia). Austr. J. Zool. 17:195–292.

Smith, C. E., Hermo, L., Fazel, A., Lalli, M. F., and Bergeron, J.J.M. (1990). Ultrastructural distribution of NADPase within the Golgi apparatus and lysosomes of mammalian cells. Progress Histochem. Cytochem. 21/3, pp 1–124.

Susi, F. R., Leblond, C. P., and Clermont, Y. (1971). Changes in the Golgi apparatus during spermiogenesis in the rat. Am. J. Anat. 130:251–267.

Susi, F. R., and Clermont, Y. (1970). Fine structural modifications of the rat chromatoid body during spermiogenesis. Am. J. Anat. 129:177–192.

Tang, X. M., Lalli, M. F., and Clermont, Y. (1982). Cytochemical study of the Golgi apparatus of the spermatid during spermiogenesis in the rat. Am. J. Anat. 163:283–294.

Thorne-Tjomsland, G., Clermont, Y., and Hermo, L. (1988). Contribution of the Golgi apparatus components to the formation of the acrosomic system and chromatoid body in rat spermatids. Anat. Rec. 221:591–598.

Vale, R. D., Reese, T. S., and Sheetz, M. P. (1985a). Identification of a novel force generating protein, kinesin, involved in microtubule-based motility. Cell 42:39–50.

Vale, R. D., Schnapp, B. J., Mitchisou, T., Steuer, E., Reese, T. S., and Sheetz, M. P. (1985b). Different axoplasmic protein generate movement in opposite directions along microtubules in vitro. Cell 43:623–632.

Vera, J. C., Brito, M., Zuvic, T., and Burzio, L. O. (1984). Polypeptide composition of rat sperm outer dense fibers. J. Biol. Chem. 259:5970–5977.

Vogl, A. W., Soucy, L. J., and Foo, V. (1985). Ultrastructure of Sertoli cells penetrating processes found in germ cells of the golden-mantled ground squirrel *(Spermophilus lateralis).* Am. J. Anat. 172:75–86.

Wartenberg, H., and Holstein, A. F. (1975). Morphology of the "spindle-shaped body" in the developing tail of human spermatids. Cell Tissue Res. 159:435–443.

Weber, J. E., and Russell, L. D. (1987). A study of intercellular bridges during spermatogenesis in the rat. Am. J. Anat. 180:1–24.

Wooley, D. M., and Fawcett, D. W. (1973). The degeneration and disappearance of the centrioles during the development of the rat spermatozoon. Anat. Rec. 177:289–302.

Yanagimachi, R., and Noda, Y. D. (1970). Fine structure of the hamster sperm head. Am. J. Anat. 128:367–388.

Zamboni, L., and Stefanini, M. (1970). The fine structure of the neck of mammal spermatozoa. Anat. Rec. 169:155–172.

CHAPTER FIFTEEN

Cellular Interactions in the Seminiferous Epithelium

William W. Wright

I. Introduction

Cell–cell interactions are essential to the development, maturation, and mature function of the seminiferous tubule, the site of spermatogenesis. These interactions occur between Sertoli cells and the underlying peritubular myoid cells, between adjacent Sertoli cells, between Sertoli cells and their surrounding spermatogenic cells, and are manifest in the morphology of the seminiferous epithelium. For example, interactions between Sertoli cells and the underlying peritubular myoid cells are evident in their coordinate deposition of the basement membrane of the seminiferous epithelium (Skinner et al., 1985). Interactions between adjacent Sertoli cells are obvious in their formation of both tight and gap junctions (Gilula et al., 1976). Finally, interactions between Sertoli cells and germ cells are obvious in the desmosomes, ectoplasmic specializations, and the tubulobulbar complexes formed between these cells (Russell et al., 1983). These extensive cell–cell interactions are also manifest in the highly ordered spatial arrangement of germ cells within the seminiferous tubule. Particular types of germ cells reside next to specific areas of Sertoli cell plasma membrane: spermatogonia and early preleptotene spermatocytes are found between the basement membrane of the seminiferous tubule and the basal plasma membrane of Sertoli cells. Spermatocytes, once they cross the blood–testis barrier, reside next to the basolateral surfaces of Sertoli cells just above the blood–testis barrier, whereas spermatids are adjacent to the more apical plasma membrane domains of the Sertoli cell (Weber et al., 1983). This segregation of the particular types of germ cells along different domains of the Sertoli cell plasma membrane has led to the hypothesis that specific and different interactions between Sertoli cells and germ cells occur along the different domains of the Sertoli cell (Wright, 1988).

The purpose of this review is to examine how the Sertoli cells, the germ cells, and the myoid cells interact to form a functional seminiferous epithelium. Most of the literature that is cited comes from studies of the rat or cells isolated from that species. Four themes run throughout this review. The first is that the importance of specific interactions between the cells of the seminiferous epithelium changes as the gonad develops, matures, and expresses its adult function. The second theme is that all of the basic types of cellular interactions are extant in this tissue. Thus, we examine stromal–epithelial (i.e., myoid–Sertoli) interactions, cell adhesion and junction formation, metabolic cooperation, and the secretion of growth and paracrine factors. The third theme is that in the adult animal, Sertoli cells appear to interact simultaneously with a number of different types of germ cells and these interactions may have multiple consequences for the interacting cells. The fourth theme is that cell–cell interactions stimulate different intracellular pathways than those stimulated by follicle-stimulating hormone (FSH), the principal polypeptide hormone involved in regulating the differentiation and maturation of the seminiferous epithelium.

II. The Changing Role of Cell–Cell Interactions During Development and Maturation of the Seminiferous Epithelium

The first theme of this review is that the importance of specific cellular interactions in the seminiferous epithelium varies as a function of the state of its maturation. This can be appreciated by dividing the life history of this tissue into three epochs: (1) fetal and neonatal development, (2) prepubertal maturation, and (3) mature function. What follows is a brief outline of some of the major attributes of these three epochs in the rat testes. We refer to the following outline in subsequent discussions of specific cellular interactions in an attempt to place these interactions in the context of the life history of the tissue.

A. The Development of the Fetal and Neonatal Seminiferous Epithelium (from 13.5 days postconception to 4 to 5 days postpartum in the rat)

The earliest sign of the differentiation of seminiferous tubules in the indifferent embryonic rat gonad is the development of adherent junctions between adjacent fetal Sertoli cells, and the onset of the formation of a basement membrane along the basal surface of the testis cords, the anlage of the seminiferous epithelium (Pelliniemi and Lauteala, 1981). Cord formation is very rapid in the rat as it occurs from 13.5 to 14.5 days postconception (pc). During this time, there is a disappearance of fibronectin from within the presumptive testicular cords and the deposition of laminin, fibronectin, and types IV and V collagen and heparin sulfate proteoglycan in the basal lamina (Pelliniemi et al., 1987). Taketo and colleagues (1984) demonstrated that indifferent gonads of 11- and 12-day pc male mouse embryos would differentiate into testes in vitro. They noted, however, that this differentiation could be inhibited by increasing internal cyclic adenosine monophosphate (cAMP) pools in the tissue, and that a component of this inhibition was the destruction of the basement membrane of the testis cords. These effects of cAMP were limited to a defined period of development because by 14 days pc, when formation of the testis cords was complete, experimentally increased cAMP pools in the tissue were without discernible effect. Thus, during a brief period of testicular development, there appears to be a link between intercellular cAMP levels, deposition of basement membrane, and the formation of testis cords.

After the formation of the testis cords there is increased proliferation of Sertoli cells. In the rat fetus, this proliferation is maximal by 20 days pc, and then begins to decrease until, at 22 days postpartum (pp), replication of Sertoli cells has ceased (Orth, 1982). It is noteworthy that during late fetal life, there are dramatic changes in the replication of the germ cells. At 16 days pc, prespermatogonia are dividing; however, this division has stopped by 17 days, and is not resumed until 4 to 5 days pp (Orth, 1982). This difference between the replication of prespermatogonia and Sertoli cells suggests that these two cell types may be targets of different growth factors. Indeed, as will be seen subsequently, a number of different growth factors have been identified in the testes.

B. The Maturation of the Prepubertal Seminiferous Epithelium (from 5 days to 45 days postpartum in the rat)

Prepubertal maturation of the rat seminiferous epithelium begins with resumption of the division of prespermatogonia and ends with production of the first cohort of sperm. Prespermatogonia, which are quiescent from day 17 pc to approximately day 4 pp, face one of two fates: degeneration or replication (Clermont and Perey, 1957). The progeny of the cells that successfully enter mitosis become typical type A spermatogonia and are committed to the process of spermatogenesis. In contrast to what is noted for prespermatogonia, at 5 days pp, there is still extensive replication of Sertoli cells. By 22 days, however, replication of Sertoli cells has ceased and concomitant with the decrease in Sertoli cell replication is a loss of gap junctions and the formation of tight junctions between adjacent Sertoli cells (Orth, 1982; Gilula et al., 1976). From 20 to 45 days of age, there are marked changes in the size and morphology of the Sertoli cell, an increase in the diameter and length of the seminiferous tubule and, thus, a requirement for a continuous deposition of the basement membrane around the growing tubule by Sertoli and myoid cells. Coincident with final growth of the Sertoli cell population is the morphogenesis of the peritubular myoid cell. In newborn mice, the cells surrounding the seminiferous tubule are fibroblastic in appearance, but by 22 days, these cells have differentiated into peritubular myoid cells through a testosterone-dependent process (Bressler and Ross, 1972). Taken together, the changes in the maturing seminiferous epithelium demonstrate a continued interaction of Sertoli and myoid cells in the deposition of the basement membrane and suggest that growth or paracrine factors are involved in the final rounds of replication of these somatic cells as well as the replication and differentiation of the prespermatogonia.

C. Functions of the Mature Epithelium

By the end of puberty (45 days in the rat), most of the growth of the testes has occurred and spermatogenesis is fully established. Coincident with this final maturation are three major physiological changes in the rat seminiferous tubule. First, with establishment of full tubular diameter and length, there appears to be a marked reduction in the deposition in basement membrane around the tubule as electron microscopic immunocytochemical analysis detects only small amounts of basement membrane components within the myoid or Sertoli cells (Hadley and Dym, 1987). Second, there is a fundamental change in the hormonal regulation of spermatogenesis. In immature rats, it is well established that both FSH and testosterone are required for completion of the first round of spermatogenesis, whereas testosterone is sufficient to insure sperm production in the adult animal (Chemes et al., 1979; Dym et al., 1979; and see chapter 8). The loss of the requirement for FSH occurs concomitantly with the loss of responsiveness of the Sertoli cell to FSH and is due, in part, to an alteration of the N_s GTP binding protein or to a decoupling of this protein from the adenylate cyclase associated with the FSH receptor (Heindel, 1988). Sometime during this period of decreasing FSH responsiveness of Sertoli cells there is an increase in the synthesis and secretion of particular Sertoli cell products at specific stages of the cycle of the seminiferous epithelium (see section VII. A., stage-specific protein synthesis). Thus, maturation of the seminiferous epithelium appears to be associated with a decreased responsiveness of Sertoli cells to FSH and increased responsiveness of Sertoli cells to germ cells.

III. Interactions Between Myoid and Sertoli Cells

The remainder of this chapter examines specific types of cellular interactions in the seminiferous epithelium and begins with a review of interactions between peritubular myoid and Sertoli cells. Interactions between myoid and Sertoli cells fall into two categories: interactions via the deposition of the basement membrane of the seminiferous tubule and secretion of paracrine factors by myoid cells that influence the differentiated function of Sertoli cells. The first part of this section is a discussion of the structure and source of the basement membrane of the seminiferous tubule. The second part examines the biological effect of this basement membrane on Sertoli cell structure and function. The third part reviews data on the secretion by myoid cells of a paracrine factor, P-mod-S, that stimulates specific differentiated functions of cultured Sertoli cells.

A. Structure and Source of the Basement Membrane of the Seminiferous Tubule

The basement membrane of the rat seminiferous tubule is part of a two-component lamina propria. The first component resides between the Sertoli and peritubular myoid cells; the second component resides between the myoid cells and the lymphatic endothelial cells (Fig. 15–1). Hadley and Dym (1987) defined the basement membrane as that portion of the first component of the lamina propria that is immediately beneath the Sertoli cell. Using electron micro-

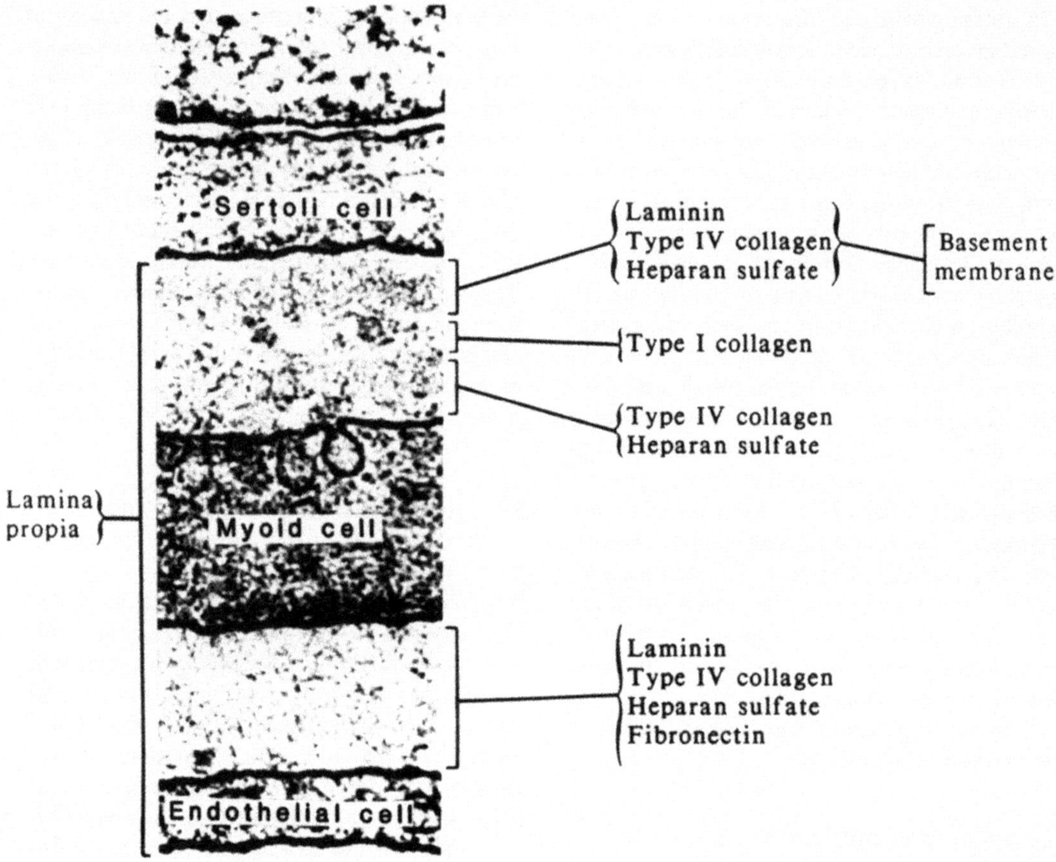

Figure 15–1. Photographic montage of the lamina propria of the seminiferous tubule that summarizes the distribution of extracellular matrix molecules as determined by electron microscopic immunocytochemistry (×100,000). (Reprinted with permission from Hadley and Dym, 1987.)

scopic immunocytochemistry, these authors demonstrated that the basement membrane contains laminin, type IV collagen, and heparin sulfate proteoglycan. Beneath this is a layer of type I collagen and adjacent to the myoid cell, a layer containing a mixture of type IV collagen and heparin sulfate proteoglycan. Between the myoid and the endothelial cells were identified laminin, type IV collagen, heparin sulfate proteoglycan, and fibronectin. Thus, the specific topological distribution of the molecules of the lamina propria suggests that this structure has polarity and raises the possibility that this polarity is biologically important to the development and function of the seminiferous tubule. Studies by Enders and co-workers (1986) support this hypothesis. These authors isolated the bilaminar lamina propria of the seminiferous tubule in the form of cell-free sleeves and then examined the binding of Sertoli and other cells to these sleeves. Sertoli cells were found to bind only to the luminal side of the tubule (i.e., to the basement membrane), whereas a cell line, murine 3T3 cells, binds both the luminal and exterior surface. Thus, although both sides of the lamina propria contain molecules to which cells can bind (for example, type IV collagen), the Sertoli cell must express receptors for basement membrane molecules that are accessible only on the basement membrane.

Investigations of the source of the basement membrane constituents of the rat seminiferous tubule have used Sertoli cells isolated from rats of different ages and cultured under a variety of conditions. As a result of the differences in the experimental conditions of the various studies, there is only a partial consensus about the source of the various basement membrane components. Borland and associates (1986) isolated Sertoli cells from 15-day-old rats and cultured them on plastic dishes in medium supplemented with insulin, transferrin, retinoic acid, epidermal growth factor, and ascorbic acid. They examined the basement membrane components in the cells by light and electron microscopic immunocytochemistry and characterized the basement membrane components deposited by the cells by immunoprecipitation and Western blotting. This analysis identified type IV collagen, fibronectin, laminin, and heparin sulfate proteoglycan as being products of the cultured Sertoli cells. Skinner and colleagues (1985) isolated Sertoli cells from 20-day-old rats and cultured them on plastic dishes in medium supplemented with FSH, insulin, retinol, and testosterone. Immunocytochemical analysis revealed type IV collagen and laminin, but not fibronectin or type I collagen within the cells. In contrast, peritubular myoid cells isolated from the same age rats were found to contain type I and IV collagens as well as fibronectin. Hadley and associates (1985) also failed to demonstrate type I collagen in Sertoli cells cultured on plastic, but noted a significant secretion of this protein when the cells were cultured on a reconstituted basement membrane prepared from Engelbreth-Holm-Swarm (EHS) sarcoma tumors.

Although there is some disagreement about the source of some of the molecules in the basement membrane, Skinner and co-workers (1985) provide clear evidence that the Sertoli and myoid cells cooperate in the deposition of basement membrane. They noted that when these two cell types were co-cultured, there was a marked increase in the amount of fibronectin, types I and IV collagen, and laminin deposited as a basement membrane, a membrane that was shown in other studies to separate the myoid and Sertoli cells in co-culture and thus be reminiscent of the basement membrane of the seminiferous tubule (Tung and Fritz, 1987). Additional studies indicate that the interactions between the Sertoli and myoid cells leading to the deposition of this basement membrane in vitro are disrupted by adding cAMP to the co-cultures. As we already noted, cAMP has also been shown to inhibit the deposition of the basement membrane around the fetal testis cords (Taketo et al., 1984) and thus, there are similar effects of cAMP on the deposition of basement membrane by fetal testis cords and by prepubertal Sertoli cells and myoid cells in co-culture.

B. The Biological Effect of the Basement Membrane on Sertoli Cell Structure and Function

There is general agreement that basement membrane extracts (biomatrix) promote the polarized morphology of cultured immature Sertoli cells (Tung and Fritz, 1984; Hadley et al., 1985). Rat Sertoli cells, which are isolated from 4- to 20-day-old rats and cultured on plastic dishes, form a low squamous epithelium, whereas the same cells, cultured on a biomatrix prepared either from seminiferous tubules or EHS sarcoma, are highly polarized and form basolateral tight junctions similar to those observed in vivo. This biomatrix also inhibits replication of Sertoli cells isolated from 4- to 10-day-old rats (Tung and Fritz, 1984; Hadley et al., 1985). There is mounting evidence that biomatrix promotes not only the polarized morphology of Sertoli cells, but also the secretion of specific Sertoli cell products, and interactions between Sertoli cells and germ cells. First we consider the effect of biomatrix on the secretion of specific proteins. Hadley and associates (1985) isolated Sertoli cells from 10-day-old rats and cultured them on plastic or one of two biomatrices (EHS sarcoma or Sertoli–myoid biomatrix) in the presence of FSH, insulin, transferrin, vitamins E and A, hydrocortisone, epidermal growth factor, testosterone, and estradiol. The cells cultured on both biomatrices were found to secrete significantly more androgen-binding protein (ABP) and transferrin than cells cultured on the plastic dishes. In agreement with these observations, Sertoli cells from 19-day-old rats have been shown to secrete more transferrin when they are cultured without hormone supplements on a biomatrix deposited by Sertoli–myoid cell co-cultures than when they are cultured directly on plastic (Swinnen et al., 1990). When Anthony and Skinner (1989) cultured Sertoli cells from 20-day-old rats in FSH, insulin, retinol, and testosterone, however, they noted no effect of biomatrix from EHS sarcoma tumors on either ABP or transferrin secretion. Differences between the findings of Anthony and Skinner (1989) and other investigators raise the possibility that the effect of biomatrix on the differentiated function of Sertoli cells may be modulated by the source of the biomatrix, by hormones, growth factors or vitamins, or by the age of the Sertoli cells.

Biomatrix has also been shown to promote germ cell–Sertoli cell interactions. Tung and Fritz (1984) noted an enhanced survival of germ cells when the Sertoli cells were cultured on Sertoli–myoid biomatrix. Hadley and colleagues (1985) discovered that Sertoli cells cultured *within* EHS biomatrix formed tubes with basal and adluminal compartments. They also noted that spermatogonia adhering to the Sertoli cells migrated into the adluminal compartment of the cord and entered meiosis. It is noteworthy that such in vitro differentiation of germ cells has not been documented with Sertoli cells cultured *on* biomatrix. In fact, the only other documentation of substantial in vitro differentiation of rat germ cells has occurred with cultures of intact seminiferous tubules (Parvinen et al., 1983). Thus, the three-dimensional structure of the seminiferous tubule, which is dictated in part by the basement membrane, may be essential for creating the environment required for germ cell differentiation.

C. Paracrine Interactions Between Myoid Cells and Sertoli Cells

Recent evidence indicates that, in addition to synthesizing part of the basement membrane of the seminiferous tubule, myoid cells also secrete an androgen-regulated paracrine factor, P-mod-S, that regulates Sertoli cell function (Skinner et al., 1988). P-mod-S, which has been isolated as two forms (P-mod-S A and P-mod-S B), causes a rapid increase in cyclic guanosine monophosphate (cGMP) levels of Sertoli cells, increases steady-state levels of transferrin and ABP in RNAs in these cells, and increases their secretion of transferrin and ABP (Fig. 15–2) (Skinner and Fritz, 1985; Norton and Skinner, 1989). In addition, P-mod-S synergizes with a combination of FSH, insulin, retinol, and testosterone to increase transferrin secretion (Fig. 15–2). The fact that P-mod-S stimulates intracellular cGMP content and FSH stimulates cAMP content indicates that these two factors stimulate different intracellular pathways in the Sertoli cell.

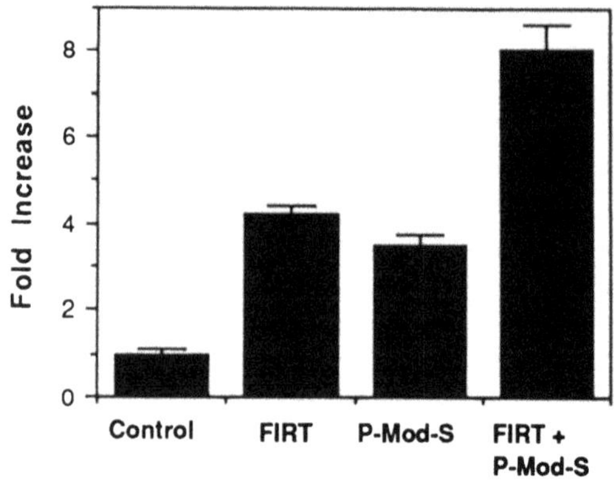

Figure 15–2. The effect of FSH, insulin, retinol, and testosterone (FIRT), P-mod-S A, and FIRT plus P-mod-S A on the secretion of transferrin by Sertoli cells in culture. Sertoli cells were isolated from 20-day-old rats and cultured for 5 days in Ham's F-12 medium (control), or medium supplemented with FIRT and/or P-mod-S A. Medium and supplements were changed at day 2 of culture and cells continued in culture for an additional 3 days. Transferrin concentration in the medium collected at day 5 was measured by radioimmunoassay and data (mean $\pm$ SEM) expressed as fold increase in transferrin secretion by Sertoli cells cultured in absence (control) or presence of FIRT and/or P-mod-S A. Transferrin production by control cells was defined as onefold. (Data are from Norton and Skinner, 1989.)

IV. Adhesion and Intercellular Junction Formation in the Seminiferous Tubule

There are two characteristics of the mature rat seminiferous epithelium that illustrate the importance of cell adhesion and intercellular junction formation in the life history of this tissue. The first characteristic is the sequential distribution of the stages of the cycle of the seminiferous epithelium along the length of a tubule, referred to as the spermatogenic wave. For example, a segment of a rat seminiferous tubule at stage V of the cycle is always next to segments of tubule at stage IV or VI of the cycle. This linear organization of the stages along a tubule must be established during the time when prespermatogonia resume cell division, differentiate into spermatogonia, and become committed to spermatogenesis. Such organization may result from the radial propagation of developmental signals along the tubule of the newborn animal, possibly by the passage of cyclic nucleotides or specific ions through the gap junctions that couple adjacent, immature Sertoli cells (Gilula et al., 1976; Eusebi et al., 1983). The second characteristic was already noted in the introduction to this chapter, that specific types of germ cells are located next to particular plasma membrane domains of the Sertoli cell. As these germ cells mature, they must, therefore, move within the epithelium to assume a new position relative to the Sertoli cell. This movement of the germ cells requires that these cells disrupt their old connections with the Sertoli cell, move to the new domain of the Sertoli cell plasma membrane, adhere to that domain, and then establish junctions with the Sertoli cell to stabilize the physical relationship between the two cell types.

The following discussion of cell adhesion and junction formation in the seminiferous tubule is divided into four sections. First we examine general mechanisms and characteristics of such interactions in other biological systems. We do this because we know more specific details of interactions in these other systems and they may provide insight into how cell adhesion promotes germ cell development. Second, we discuss intercellular junctions between adjacent Sertoli cells. Third, we examine adhesion of germ cells to Sertoli cells and the formation of specific junctions between these cells. Fourth, we examine evidence suggesting that the importance of adhesion to germ cell development differs between species.

A. General Mechanisms and Consequences of Cell Adhesion

Cells adhere to one another by specific cell adhesion molecules. It has been found that the state of development of a cell may affect the level of expression and the biochemical modification of these cell surface molecules (Tosnye et al., 1986; Rutishauser et al., 1988). For example, nerve cell adhesion molecule (N-cam) is initially expressed at very high levels on dividing myotubes. With the development of mature synapses between nerve and muscle, however, there is a marked decrease in the amount of N-cam on the muscle (Tosney et al., 1986). Chemical modifications of the N-cam molecule during development affect its ability to promote cellular interactions. For example, the N-cam of embryonic chick (E7) brain cells has a high polysialic acid content that inhibits the binding of one cell to another. With maturation of these cells, however, there is a decrease in the sialic acid content of N-cam that enables the molecule to promote adhesion of adjacent cells (Rutishauser et al., 1988). Four consequences of cell adhesion have been identified. (1) Cell adhesion promotes cell survival (Edelman, 1984). (2) Adhesion can promote the differentiated function of the adherent cells (Rutishauser et al., 1988). (3) In many instances, cell adhesion appears to be followed by the establishment of specific intercellular junctions; for example, adhesion of neuroepithelial cells by N-cam is a prerequisite for the establishment of gap junctions (Rutishauser et al., 1988). In addition, it has been suggested that in the small intestine, another cell adhesion molecule, uvomorulin, establishes the cell–cell interactions required for assembly of tight junctions (Gumbiner and Simons, 1986). (4) Cell adhesion can cause reorganization of the cellular cytoskeleton. In their analysis of the adherens junction-specific cell adhesion molecule (A-cam) of cultured lens epithelial cells, Volk and Gieger (1986) noted that this molecule co-localized in the plasma membrane with the cytoskeletal elements vinculin and actin. In addition, they noted that when cell adhesion was inhibited by monovalent anti A-cam IgG, there was marked disorganization of the microfilaments of the cells. Taken together, these observations indicate that cell adhesion can be a regulated process that has profound effects on a number of different aspects of cell structure and function.

B. Adhesion and Junction Formation Between Sertoli Cells

We noted in our discussion of the effect of biomatrix on Sertoli cell structure and function that cells cultured within a biomatrix will migrate within the gel and form tubule-like structures (Hadley et al., 1985). There are two implications of this experiment. First, it indicates that dispersed cells have the ability to adhere to one another and, once they adhere, to form intercellular junctions. Second, in agreement with studies of adhesion of other cell types (Rutishauser et al., 1988), it indicates that basement membrane components can modulate cell adhesion and junction formation.

In vivo, types and organization of junctions between adjacent Sertoli cells change during the maturation of the testes (Gilula et al., 1976). In testes of 10-day-old rats, typical gap junctions are observed between adjacent Sertoli cells. Electrophysiological studies have shown that these gap junctions are functional, electrically coupling adjacent cells and, thus providing the opportunity for the passage of small biologically active molecules, such as cAMP, between these cells (Eusebi et al., 1983). By 15 days, the first components of the Sertoli–Sertoli tight junctions are observed along with the gap junctions. Freeze fracture studies, however, demonstrate that the intramembrane particles forming these junctions are not aligned and are thus unable to form a functional Sertoli–Sertoli tight junction (Gilula et al., 1976). By 20 days, however, the numbers of these intramembrane particles have increased markedly and formed the linear rows that provide the morphological basis for the blood–testis barrier (Gilula et al., 1976) (Fig. 15–3A). Simultaneous with the development of the Sertoli–Sertoli occluding junctions, there is a decrease in the numbers of gap junctions. Thus, in the adult, there is a marked reduction in the interaction of adjacent Sertoli cells by gap junctions.

The Sertoli–Sertoli occluding junction of the mature testis is comprised of 50 parallel rows of particles, involving over 4 percent of the total surface area of the Sertoli cell (Gilula et al., 1976; Weber et al., 1983). Despite the fact that this is an extremely tight junction, it is cyclically modified as small segments of each junction open and reclose to allow the passage of preleptotene spermatocytes from the basal to the adluminal compartment of the tubule (Russell, 1977). Recent evidence suggests that movement of the germ cells through this junction occurs because at stages VII–VIII, the Sertoli cells secrete a protease, plasminogen activator, that reduces the integrity of these tight junctions (Vihko et al., 1984; Allenberg and Fritz, 1989) (see section VII. A on stage specific biosynthesis of Sertoli cell proteins).

C. Adhesion of Germ Cells to Sertoli Cells and the Formation of Junctions

A number of in vitro studies have demonstrated the ability of isolated pachytene spermatocytes to adhere to and then form desmosome-like junctions with cultured Sertoli cells (D'Agostino and Stefanini, 1987; Enders and Millette, 1988). Two types of experiments indicate that the adhesion of pachytene spermatocytes to Sertoli cells is mediated by a glycoprotein on the cell surface of pachytene spermatocytes (D'Agostino and Stefanini, 1987). First, antiserum raised against germ cell surface antigens inhibited, in a dose-dependent manner, the binding of pachytene spermatocytes to Sertoli cells. Second, one of the pachytene spermatocyte cell surface proteins recognized by this polyvalent antiserum was isolated by sodium dodecylsulfate (SDS) gel electrophoresis and electroelution and the isolated protein was shown to inhibit adherence of germ cells to Sertoli cells. Two pieces of evidence suggest that binding of pachytene spermatocytes to Sertoli cells is specific. First, neither red blood

Figure 15–3 (**A**) Tight junctions between adjacent Sertoli cells. This picture shows tight junctions between a number of mature Sertoli cells that have been cross sectioned. *Arrowheads* point to two of the many sites of fusion of plasma membranes of adjacent Sertoli cells. *Triangle* points to cisternae of endoplasmic reticulum. *Arrow* points to actin filament bundles between the Sertoli cell plasma membrane and the endoplasmic reticulum. ×75,000. (**B**) Ectoplasmic specialization of a Sertoli cell that is adjacent to a step 9 rat spermatid. SC, Sertoli cell cytoplasm; SN, spermatid nucleus; AC, spermatid acrosome. *Triangle* points to cisternae of endoplasmic reticulum that is part of the ectoplasmic specialization. *Arrow* points to actin filament bundles between the endoplasmic reticulum and the Sertoli cell plasma membrane. Sertoli cell plasma membrane is indicated by (**) and spermatid plasma membrane is indicated by (*). ×70,000.

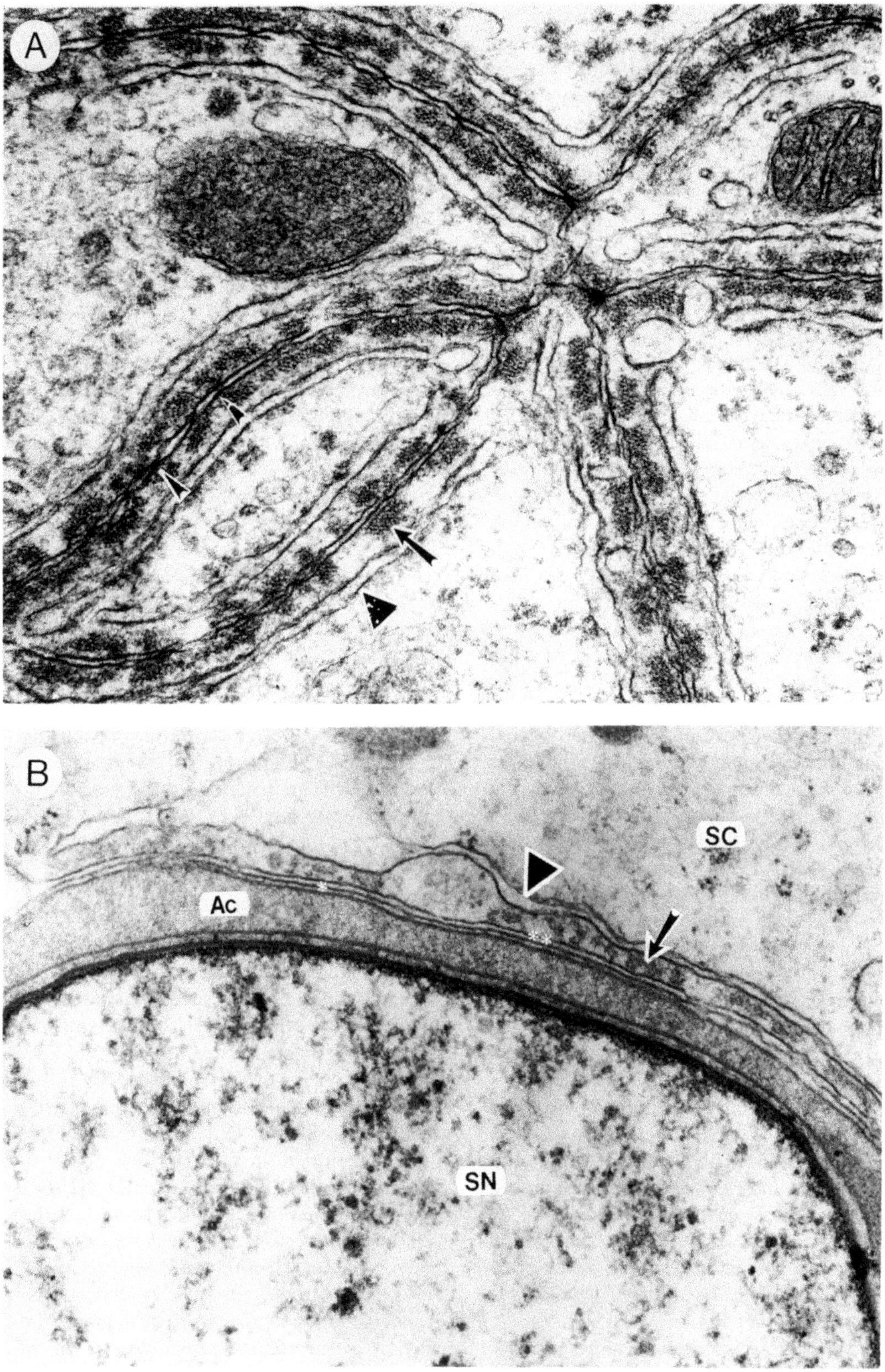
A
B
SC
Ac
SN

cells nor spleen cells bind Sertoli cells (DePhilip and Danahey, 1987), and germ cells do not bind another somatic cell, 3T3 fibroblasts (Enders and Millette, 1988). Second, pachytene spermatocytes bind Sertoli cells with more avidity than round spermatids (Enders and Millette, 1988). Binding of germ cells to Sertoli cells requires a metabolically active Sertoli cell, as this binding is inhibited by cold and formaldehyde fixation of the Sertoli cell (DePhilip and Danahey, 1987) and suggests that Sertoli cells must aggregate their adhesion molecules at the point of contact with the germ cell. In contrast, formaldehyde-fixed germ cells bind Sertoli cells, suggesting that it is sufficient that the germ cells have their adhesion molecules in a topological orientation allowing access by the Sertoli cell. Taken together, these data suggest that germ cells and Sertoli cells have distinct adhesion molecules that allow specific binding of these two cell types to one another.

Electron microscopic analysis has documented that types of junctions between rat Sertoli cells and germ cells change in vivo as the germ cells mature and move adjacent to different membrane domains of the Sertoli cell (Russell et al., 1983). Spermatogonia and spermatocytes are attached to Sertoli cells only through desmosome/gap junctions (analyses have not been conducted at sufficiently high magnification to distinguish between these two types of junctions). With the development of germ cells, there is a decrease in the number of desmosome/gap junctions between germ cells and Sertoli cells until, in the compacted spermatids, these types of junctions are no longer observed. In vivo, round spermatids and Sertoli cells develop an area of tight, trypsin-sensitive adhesion that is associated with thin filaments that span the plasma membrane of the two cells (Russell et al., 1988). This area of adhesion is associated with a cytoskeletal complex in the Sertoli cells called the ectoplasmic specialization (Fig. 15–3B). The ectoplasmic specialization persists almost to the end of spermatogenesis when it is removed by the action of the tubulobulbar complex in preparation for release of these spermatids into the lumen of the seminiferous tubule (see Russell et al., 1988, and references therein).

The principal intercellular feature of the ectoplasmic specialization is an area of densely packed actin filaments that is attached both to the area of the plasma membrane at the point of adhesion and to cisterna of endoplasmic reticulum deeper within the cell (Russell et al., 1988; Vogel, 1988). The importance of the actin filaments to the adhesive process was demonstrated by Russell and colleagues (1988) when they observed that cytochalasin D-induced depolymerizion of actin filaments in the Sertoli cell resulted in a loss of ectoplasmic specialization. This loss caused separation of compacted spermatids from Sertoli cells and random orientation of step 8 spermatids, which are normally pointed toward the basement membrane of the seminiferous tubule.

In examining the relationship of the ectoplasmic specialization to the microtubules of the Sertoli cell, Vogel (1988) has noted that once spermatids elongate and occupy the crypts within the Sertoli cell (at stages II–V of the cycle of the seminiferous epithelium), there develop abundant microtubules in the Sertoli cell that surround the ectoplasmic specialization. He suggests that the endoplasmic reticulum associated with the ectoplasmic specialization rides along these microtubules and allows for the movement of the spermatids toward the lumen of the tubule at stages VI and VII. This suggestion is consistent with our previous discussion that cell adhesion molecules and cytoskeleton often are structurally and functionally interactive (see section IV. A.).

D. The Importance of Cell Adhesion to Spermatogenesis Differs in Different Species

The extent and importance of cell adhesion in the testes differs between species. In the rat, the importance of cell adhesion to successful spermatogenesis is documented not only by the evidence discussed previously, but also by the observation that when segments of seminiferous tubules were placed in culture, the main requirement for in vitro progression of the tubules from stage XII to stage V of the cycle was maintenance of normal contact between Sertoli cells and germ cells (Parvinen et al., 1983). Disruption of this adhesion was found to lead to the death of the germ cells. In contrast, for *Xenopus,* adhesion appears to be much less important to successful spermatogenesis. In this species, elongated spermatids are the only germ cells that make extensive physical contact with

Sertoli cells (Reed and Stanley, 1982). This in vivo situation may therefore explain the observation of Risley (1983) that isolated leptotene spermatocytes cultured in the presence of 10 percent fetal calf serum differentiate to elongate spermatids. It is noteworthy, in light of Reed and Stanley's morphological observations, that isolated elongate *Xenopus* spermatids do not form sperm in vitro (Risley, 1983). Sperm are formed in vitro, however, when explants of *Xenopus* testes are placed in culture (Risley et al., 1987), which suggests that adhesion of compacted spermatids to Sertoli cells is required for completion of spermatogenesis in *Xenopus*.

V. Metabolic Cooperation Between Sertoli Cells and Germ Cells

The metabolic effects of Sertoli cells on germ cells have been reviewed recently and will only be outlined here (Grootegoed et al., 1989). An essential metabolic requirement of the germ cells that is met by the Sertoli cells is a source of exogenous lactate and pyruvate. Data outlined in that review and in other publications demonstrate that this requirement by the germ cells is met by the Sertoli cells, which convert glucose to lactate and pyruvate via glycolysis. Apparently most of the pyruvate or lactate is not metabolized further in Sertoli cells by the tricarboxylic acid cycle; instead these two metabolites are secreted from the cells. Because of this secretion, it has been suggested that adenosine triphosphate (ATP) requirements of the Sertoli cell are primarily met by metabolism of lipids and amino acids (Grootegoed et al., 1986).

The requirement for exogenous lactate and pyruvate by germ cells stems in part from their inability to metabolize glucose efficiently and because in vivo, there is very little glucose for the germ cells to metabolize. Studies of the carbohydrate content of rete testis fluid indicate that this fluid contains very little glucose; however, its lactate concentration is equivalent to serum (Setchell and Waites, 1975; Grootegoed et al., 1989). In vitro studies demonstrate that lactate and pyruvate are essential to germ cell survival. Isolated spermatids and spermatocytes that are cultured in glucose suffer an immediate depletion of ATP and die. In contrast, endogenous lactate stimulates oxygen consumption and supports ATP production, as well as RNA and protein synthesis (Jutte et al., 1981; Grootegoed, 1989).

Another reason that germ cells require high concentrations of lactate is that these cells contain high amounts of the enzyme, lactate dehydrogenase-X (LDH-X) which converts pyruvate to lactate. This conversion reduces the amount of pyruvate that enters the tricarboxylic acid cycle via the action of the enzyme pyruvate dehydrogenase (Grootegoed et al., 1986). The high concentrations of lactate in testicular fluids, however, may inhibit conversion of endogenous pyruvate to lactate by LDH-X (Grootegoed et al., 1984), and in addition, serve as an energy source for germ cells by its oxidation back to pyruvate by LDH-X (Grootegoed et al., 1984).

Sertoli cells also regulate glutathionine synthesis by germ cells. This tripeptide is of interest because it has diverse biological functions, including maintenance of thiol groups and protection against peroxides (Li et al., 1989). Grootegoed and colleagues (1989) note that glutathione content of germ cells in seminiferous tubules is maintained over a 48-hour culture period because of de novo glutathione synthesis by these cells. Isolated germ cells, however, which have been cultured for 48 hours have 50 percent less glutathione than freshly isolated cells. Studies with isolated frog spermatogenic cells in culture (Li et al., 1989) demonstrate that the incorporation of exogenous [^{35}S]cysteine into intercellular glutathione decreases with increasing time in culture. In contrast, there is no significant effect of time in culture on glutathione efflux out of the cells. Thus, in agreement with Grootegoed and associates (1989), these authors conclude that Sertoli cells either stimulate the importation of cysteine into the germ cells or increase the enzyme activities required for glutathione synthesis.

VI. Growth Factors in the Seminiferous Epithelium

The life history of the seminiferous epithelium suggests that specific growth factors regulate the replication and differentiation of particular testicular cell types. A few examples of these events are: First, in the newborn rat (1 to 4 days), Sertoli cells, but not prespermatogonia, proliferate (Orth, 1982). This suggests that either Sertoli cell division is specifically stimulated or prespermatogonial replication is specifically inhibited. Second, in prepubertal animals, once prespermatogonia give rise to spermatogonia, the new cells are competent to begin the precisely timed set of events that comprise spermatogonial replication, meiosis, and spermiogenesis (Clermont and Perey, 1957). This progression of cells through spermatogenesis is maintained throughout the life of a fertile animal and it has been suggested that the temporal precision of spermatogenesis results from the action of testicular growth factors on the germ cells (Bellve and Zheng, 1989a). Third, in the mature rat, the rate of replication of undifferentiated spermatogonia increases when the number of differentiated spermatogonia decreases (de Rooij et al., 1985).

Our discussion of growth factors in the testes begins with a brief, general review of how the growth factors regulate division and differentiation of cells. After this review, we focus on a subset of the growth factors that have been identified in the testes to illustrate the importance of growth factors to the replication and differentiation of the cells of the seminiferous epithelium. The growth factors we discuss are fibroblast growth factor, somatomedin C, testis-specific growth factor, transforming growth factor β, inhibin, and activin. A detailed review of the identity and biological properties of growth factors in the testis can be found in Bellve and Zheng (1989a,b).

A. Growth Factors and Cell Replication and Differentiation

A great deal of what is known about the regulation of the cell cycle by growth factors has been learned using the murine 3T3 cell line. These studies have demonstrated that growth factors regulate the entry and progression of cells through the G_1 phase of the cell cycle (Fig. 15–4). Replicating cells may enter G_1 immediately upon completing mitosis; quiescent cells at G_0 can be rendered competent to enter G_1 by the action of growth factors, such as fibroblast growth factor and platelet-derived growth factor, which are called competence factors (Gospodarowicz et al., 1987; Pardee, 1989). Entry into G_1 by 3T3 cells is dependent on the progression factors, somatomedin C (SmC) and epidermal growth factor while progression through G_1 requires the continued presence of SmC (Pardee, 1989). Once cells have completed G_1 and entered S phase, they continue through S, G_2, and M without additional stimulation. This restriction of growth factor action to G_0 and G_1 of the cell cycle is important in considering the effects of growth factors on spermatogenic cells, as only spermatogonia and preleptotene spermatocytes undergo DNA replication and thus are the only spermatogenic cells where growth factors can act to stimulate cell proliferation. Growth factor action is not restricted to cell replication, however, for growth factors and hormones have been shown to regulate the exiting of the cells from the cell cycle for the purposes of cell differentiation (Fig. 15–4). Thus, because there is considerable

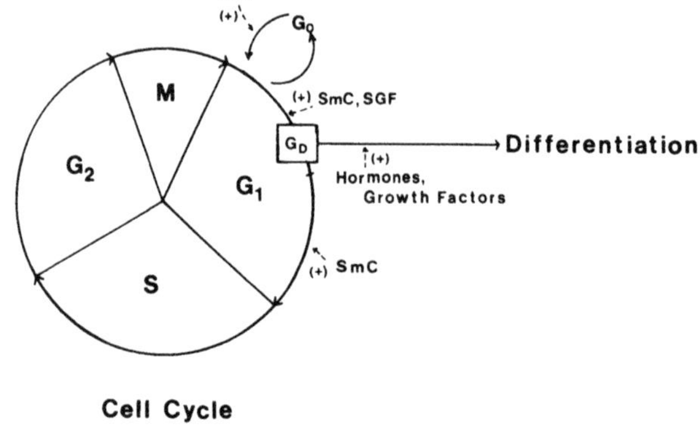

Figure 15–4. The cell cycle and the role of three growth factors identified in the testes in the regulation of this cycle. FGF, basic fibroblast growth factor; SmC, somatomedin C; SGF, seminiferous growth factor; M, mitosis; G_1, period between mitosis and DNA synthesis; S, period of DNA synthesis; G_2, period between DNA synthesis and mitosis; G_D, point of exit from the cycle leading to cellular differentiation; G_0, period when cells are out of the cell cycle. Cells at G_0 regain the ability to reenter the cell cycle upon stimulation by competence factors such as FGF.

cell differentiation during meiosis and spermiogenesis, it is possible that germ cells later than preleptotene spermatocytes also are targets for growth factors.

B. Basic Fibroblast Growth Factor

Basic fibroblast growth factor (bFGF), with a truncated amino terminus, has been isolated from mature bovine testes and shown to have a biological activity equivalent to full-length basic fibroblast growth factor (Ueno et al., 1987). In studies of Sertoli cells isolated from 20 to 26-day-old pigs, Jaillard and co-workers (1987) demonstrated that bFGF was more potent than SmC in stimulating cell replication and, in addition, synergized with SmC to increase DNA synthesis by these Sertoli cells (Fig. 15–5). Synergism between bFGF and SmC is consistent with the observation (discussed earlier) that these two growth factors can act at different parts of the cell cycle. Basic fibroblast growth factor has also been shown to stimulate specific differentiated functions of porcine Sertoli cells and of the murine Sertoli cell-derived TM4 cell line. When added to porcine Sertoli cell cultures, bFGF stimulates FSH receptor number, increases cAMP content of FSH-treated cells, and promotes plasminogen activator secretion by cells cultured in the presence or absence of FSH (Jaillard et al., 1987). In addition, bFGF has been shown to stimulate the secretion of sulfated glycoprotein-1 by the murine Sertoli cell-derived TM4 cell line (Bellve and Zheng, 1989b). Taken together, these data suggest two in vivo functions for bFGF in the testis: (1) stimulating cell division by immature Sertoli cells by making them competent to enter the cell cycle and (2) regulating the differentiated function of Sertoli cells.

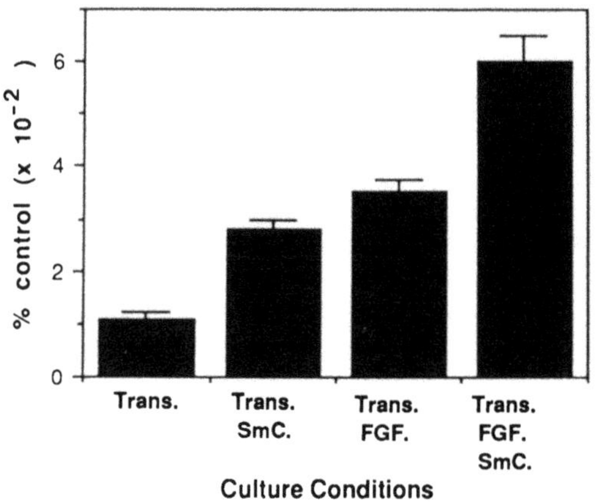

Figure 15–5. Effect of several growth factors on DNA synthesis by immature porcine Sertoli cells cultured for 3 days in 1:1 mixture of Ham's F-12:Dulbecco's Modified Eagles Medium supplemented with transferrin, insulin, and vitamin A. These cells were then subcultured in the same medium for 24 hours and medium replaced with fresh medium without supplements (control) or medium supplemented with transferrin (Trans.), transferrin and somatomedin C (Trans. SmC.), transferrin and fibroblast growth factor (Trans. FGF.), or transferrin and fibroblast growth factor plus somatomedin C (Trans. FGF. SmC.). After 2 days of culture under these conditions, cells were incubated for 18 hours in 1 μCi/mL [^{3}H]thymidine. Data (mean $\pm$ SEM) for each group are expressed as percent of [^{3}H]thymidine incorporation measured in control cells. (Data are from Jaillard et al., 1987.)

C. Somatomedin C

The gene for SmC has been shown to be expressed in rat testes by cloning and sequencing a cDNA prepared from testis mRNA (Casella et al., 1987). Both immature rat and porcine Sertoli cells secrete SmC in vitro and this secretion by porcine cells is stimulated by bFGF and epidermal growth factor (Smith et al., 1987; Chatelin et al., 1987). Immature rat Sertoli cells have been shown to contain approximately 2000 receptors/cell for SmC and data suggest that these numbers increase dramatically with age (Oonk and Grootegoed, 1988). Because Sertoli cells are both a source and a target for SmC, it is of interest that SmC has been shown to stimulate DNA synthesis by immature porcine Sertoli cells (Fig. 15–5), and total protein synthesis and lactate secretion by immature rat Sertoli cells (Jaillard et al., 1987; Borland et al., 1988). Thus, SmC may be an autocrine regulator of Sertoli cell replication and metabolism.

The presence of SmC receptors on germ cells is controversial. Oonk and Grootegoed (1988) attempted to measure SmC receptors on isolated rat pachytene spermatocytes and concluded from their results that germ cells lack a cell surface SmC receptor. Immunocytochemical studies of SmC receptors in human testes, however, have led to the opposite conclusion.

Use of a monoclonal antibody to localize immunocytochemically SmC receptors in biopsy sections of human testes demonstrated a consistent and high concentration of these receptors in secondary spermatocytes and early spermatids (Vanelli et al., 1988). In contrast, immunolocalization of SmC on cell surfaces of rat germ cell–Sertoli cell co-cultures identified SmC bound to a subset of pachytene spermatocytes (Tres et al., 1986).

D. Seminiferous Growth Factor

Seminiferous growth factor (SGF), an acidic polypeptide with a molecular weight of 15,700 has been isolated from bovine and murine testes and data suggest that it originates from Sertoli cells (Bellve and Zhuang, 1989b). Seminiferous growth factor is a progression factor for 3T3 murine fibroblasts (see Fig. 15–4) and stimulates proliferation of immature murine Sertoli cells, the murine Leydig cell-derived line TM3, and the murine Sertoli cell-derived line TM4 (Bellve and Zhuang, 1989a,b). This ability to stimulate the two cell lines is not observed with two other growth factors of similar molecular size as SGF, bFGF, and acidic FGF (Bellve and Zhuang, 1989b), indicating that SGF is different from these two fibroblast growth factors. Seminiferous growth factor does share one attribute with bFGF, however, the ability to stimulate specific functions of TM4 cells. When these cells are cultured in the presence of SGF, they markedly increase their secretion of SGP-1 and of a number of other proteins (Bellve and Zhuang, 1989b). Thus, SGF may regulate both the growth and differentiated function of Sertoli cells in vivo.

E. Inhibin, Activin, and Transforming growth factor (TGF-β)

Inhibin, activin, and TGF-β, members of the TGF-β gene family, have been shown to have diverse and sometimes opposite effects on the growth and differentiated function of a number of cell types (Roberts et al., 1988). Subunits for mRNAs of inhibin, activin, and TGF-β have all been identified in the seminiferous epithelium, which suggests that these molecules are involved in cellular interactions within the seminiferous epithelium and between this tissue and the interstitial cells (Roberts et al., 1989; Skinner and Moses, 1989). Inhibin, composed of two subunits (α and β_A or β_B) is secreted by Sertoli cells (Morris et al., 1988). The observation of relatively equal amounts of mRNA for inhibin α and β_A subunits in Sertoli cells suggests that activin (composed of two inhibin β_A subunits) is also secreted by this cell (Roberts et al., 1989). TGF-β, composed of two identical subunits that exhibit sequence homology with the β subunits of inhibin, is secreted by both Sertoli and myoid cells (Skinner and Moses, 1989).

The effects of activin and inhibin on Sertoli and myoid cell replication have yet to be evaluated. Two studies with other cell types, however, indicate how these two factors can have opposing effects on cells and, therefore, suggests that their effects on these two testicular cell types may be complex. First, studies with Chinese hamster ovary (CHO) cells demonstrate that a dose-dependent inhibition of DNA synthesis by activin is partially blocked by inhibin (Gonzalez-Machon and Vale, 1989). Second, studies with human K562 erythroleukemic cell line demonstrate that activin stimulates differentiation of these cells and inhibin antagonizes this effect of activin (Yu et al., 1987).

Inhibin is one of the few gonadal peptides that have been implicated as having a direct effect on germ cells. Van Dissel-Emiliani and colleagues (1989) demonstrated that when Chinese hamster testes (one testis/animal) were injected once a day for 4 days with 1512 units of a partially purified inhibin preparation, there was a significant reduction in numbers of A_3, intermediate, B_1 and B_2 spermatogonia to 90, 87, 66, and 93 percent of numbers found in the controlateral, noninjected testis. Significant reductions in numbers of intermediate and B_1 spermatogonia were also noted when testes were injected once a day for 4 days with 750 units of biochemically homogeneous inhibin. The authors concluded that they noted a significant effect of partially purified inhibin, but not of purified inhibin, on A_3 and B_2 spermatogonia, because they injected more inhibin activity with the partially purified preparation.

TGF-β has not been shown to have a significant effect in vitro on either the replication or function of Sertoli cells isolated from 20-day-old rats (Skinner and Moses, 1989). Neither is there an effect of TGF-β on numbers of peritubular myoid cells. TGF-β, however, was shown to have a marked effect on the in vitro morphology of myoid cells and myoid–Sertoli co-cultures, as in both cases it caused marked aggregation of the cells (Skinner and Moses, 1989). As, with a number of different cell types, TGF-β significantly stimulates the secretion of extracellular matrix proteins and enhances the response of cells to these matrix proteins (Roberts et al., 1988), aggregation of myoid cells and myoid–Sertoli co-cultures in response to TGF-β may result from the ability of this growth factor to promote basement membrane deposition and cell–substrate interactions.

VII. Effects of Germ Cells on Sertoli Cells

In mammals, specific types of spermatogonia, spermatocytes, and spermatids are found together within each segment of the seminiferous tubule and define the stages of the cycle of the seminiferous epithelium (Leblond and Clermont, 1952). The germ cells in these cellular associations develop synchronously and progress in unison within a defined period through this cycle (Clermont, 1972). This coordinated maturation of the germ cells suggests the presence of a central organizer within the seminiferous tubule. Based on our previous discussion of the morphological interaction of Sertoli cells and germ cells, it is reasonable to propose that the Sertoli cell plays a fundamental role in coordinating the synchronous development of the different types of germ cells found within a segment of tubule. For Sertoli cells to play this role, however, they would have to recognize and respond to the maturation of the germ cells. This last section of our review, therefore, examines data that demonstrate that germ cells regulate Sertoli cell function. This section is divided into two parts. First, we examine how the synthesis of specific Sertoli cell products varies with the stages of the cycle of the seminiferous epithelium. This discussion emphasizes studies on the synthesis of two different proteins that exhibit dramatic but different cycles of synthesis. These proteins are cyclic protein-2 (CP-2) and plasminogen activator (Wright et al., 1983; Vihko et al., 1984). The second part of this section examines how co-culturing germ cells with Sertoli cells changes specific functions of Sertoli cells.

A. Stage-specific Protein Synthesis by Sertoli Cells

The first demonstration that the function of rat Sertoli cells changes with progression of the cycle of the seminiferous epithelium resulted from the development of procedures to isolate living seminiferous tubules at defined stages of the cycle (Parvinen and Ruokonen, 1982). Using these techniques, Parvinen and his colleagues measured stage-specific changes in the synthesis, secretion, and mRNA content for specific Sertoli cell products. Subsequently, Griswold and co-corkers employed in situ and solution hybridization to analyze the relative amounts of specific mRNAs in Sertoli cells at different stages of the cycle (Morales et al., 1987). The proteins that have been shown to exhibit the greatest stage-specific variations in synthesis are cyclic protein-2 (CP-2) (Wright, 1988) and plasminogen activator (Lacroix et al., 1981; Vihko et al., 1987) (Fig. 15–6). Cyclic protein-2, recently shown to be the proenzyme form of Cathepson-L (Erickson-Lawrence et al., 1991), is maximally synthesized and secreted in vitro at stage VI and the first half of stage VII of the cycle. The synthesis of CP-2 decreases sharply thereafter; at stages VIII and XII, its synthesis is 14 and 5 percent, respectively, of that observed at stage VI (Wright, 1988). In contrast, the secretion of urokinase-type plasminogen activator activity is low at stage VI, increases markedly by stage VII, remains elevated during stage VIII, and then decreases to its lowest level at stages IX–XI (Lacroix et al., 1981; Vihko et al., 1987). Precise estimates of the stage-specific rates of synthesis of this protein have not been made, but Northern blot analysis of the stage-specific changes in urokinase-type plasminogen activator mRNA at different stages of the cycle revealed stage-specific changes that were similar to that observed with measurements of secreted plasminogen activator (Vihko et al., 1989). Transferrin and sulfated glycoprotein-2 (SGP-2) mRNA contents have also been shown to vary in a stage-specific manner although to a lesser extent than is observed with CP-2 and plasmino-

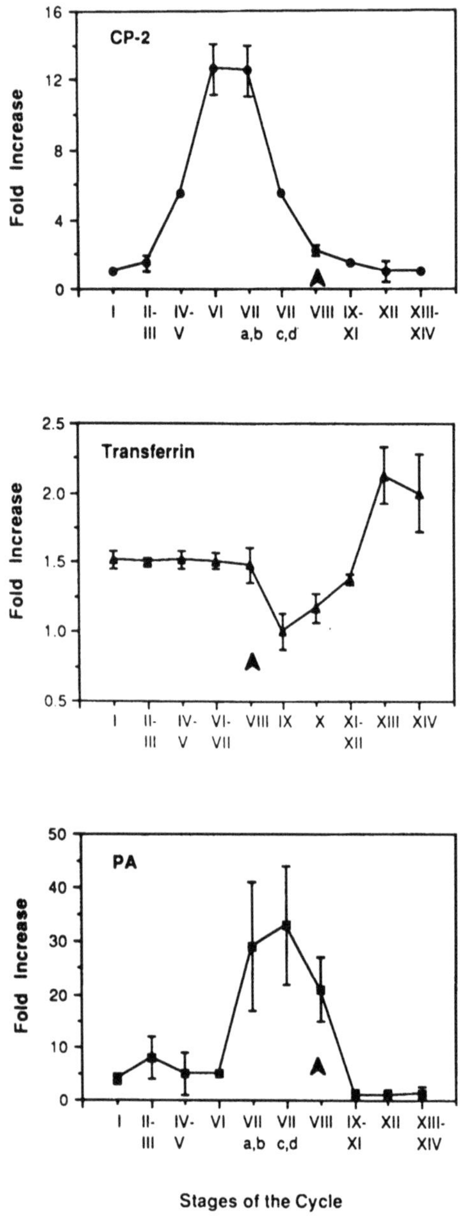

Figure 15–6. Synthesis of cyclic protein-2 (CP-2), plasminogen activator (PA), and transferrin by mature rat Sertoli cells within seminiferous tubules at defined stages of the cycle of the seminiferous epithelium. All data (mean ± SEM) are expressed as fold increase over the stage of minimal synthesis of each protein, where minimal synthesis is defined as 1. CP-2 synthesis was measured by incubation of seminiferous tubules at defined stages of the cycle of the seminiferous epithelium with [^{35}S]methionine and measurement of incorporation of the radiolabeled amino acid into CP-2. PA synthesis was assayed by incubation of seminiferous tubules at defined stages of the cycle for 20 hours and measurement of plasminogen activator activity in culture medium. Transferrin synthesis was assessed by analysis of the amount of transferrin mRNA at each stage of the cycle by in situ cytohybridization. Note that the scales for the abscissas of graphs for CP-2 and PA differ for abscissa for graph of transferrin. *Arrow* on each graph points to data from stage VIII of the cycle. (Data are from Wright et al., 1983; Wright, 1988; Vihko et al., 1987; Morales et al., 1987.)

gen activator (Morales et al., 1987). Transferrin mRNA is highest at stages XIII–I and decreases about threefold by stages VII–VIII (Fig. 15–6); SGP-2 mRNA is highest at stage VII–VIII and decreases 22 percent by stage IX–XII. It should be noted, however, that the synthesis of most Sertoli cell proteins appears not to vary with progression of the cycle. For example, the amount of SGP-1 mRNA is constant from one stage to the next (Morales et al., 1989). In addition, two-dimensional gel analysis of the synthesis of intracellular proteins revealed that almost all of these proteins were synthesized at a constant rate from one stage to the next (Wright et al., 1983).

Studies of CP-2 conducted in our laboratory suggest a number of important characteristics about the cellular interactions in the seminiferous tubule. First, although CP-2 is rapidly secreted by cultured Sertoli cells, it exits seminiferous tubules very slowly (Wright, 1988). Because immunocytochemical data failed to indicate that CP-2 was nonspecifically adsorbed to germ cells, we concluded that the secreted CP-2 was trapped in microenvironments within the seminiferous tubule and we proposed that substantial time was required for the protein to diffuse out of these environments (Wright, 1988; Wright et al., 1989; Zabludoff et al., 1990a). In light of our previous discussion that there are specific domains of Sertoli cell plasma membrane to which particular germ cells bind, we suggest that different Sertoli cell products may be targeted at different types of germ cells and, thus, enter different microenvironments of the seminiferous tubule (Wright et al., 1989). Second, Northern blot and immunocytochemical analyses demonstrated that in vivo, significant amounts of CP-2 mRNA and protein were only present in Sertoli cells within stage VI and VIIa,b tubules. Thus, the stage-specific synthesis of CP-2 that was noted in vitro also occurs in vivo (Zabludoff et al., 1990a; Erickson-Lawrence et al., 1991). Third, analysis of the effect of testicular maturation on CP-2 synthesis indicates that substantial amounts of CP-2 are synthesized only after 35 days of age and that the greatest increase in CP-2 synthesis occurs between 35 and 45 days (Zabludoff et al., 1990b) (Fig. 15–7). Additional studies demonstrate, however, that when Sertoli cells from mature and immature animals are separated from germ cells and the Sertoli cells placed in culture, the rate of synthesis of CP-2 by Sertoli cells from two different age rats is identical (Wright et al., 1989). Taken together, these observations suggest that from 35 to 45 days, there is either a quantitative change in the amount of stimulus coming from each germ cell and/or a marked increase in the sensitivity of the Sertoli cell to this stimulus. Finally, comparison of the hormonal regulation of CP-2 and transferrin secretion by mature Sertoli cells in culture demonstrated that transferrin, but not CP-2 secretion was hormonally regulated (Wright et al., 1989). This finding, coupled to the observation that CP-2 and transferrin exhibit different stage-specific patterns of synthesis, suggests that germ cells and hormones have different effects on the synthesis of these two proteins.

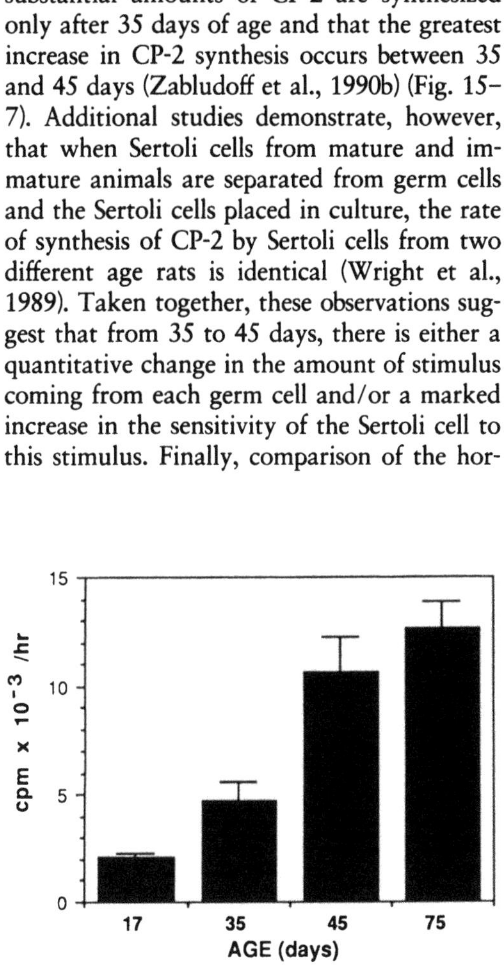

Figure 15–7. Effect of testicular maturation on the synthesis of cyclic protein-2 by seminiferous tubules. Seminiferous tubules (20 cm) from all stages of the cycle were isolated from testes of rats of 17, 35, 45, and 75 days of age, incubated for 1 hour in [^{35}S]methionine and CP-2 synthesis measured by immunoprecipitation. Data (mean ± SEM) are expressed as cpm of [^{35}S]methionine incorporated into CP-2 by 20 cm of tubules. (Data are from Zabludoff et al., 1990b.)

Additional facets of the regulation and function of stage-specific protein synthesis are revealed by studies of urokinase-type plasminogen activator. In studying the effect of irradiation-induced maturation depletion on the stage-specific secretion of plasminogen activator activity, Vihko and colleagues (1984) noted that this secretion was minimal at all stages of the cycle when preleptotene spermatocytes were lost from the epithelium, but that the stage-specific secretion resumed with reappearance of the preleptotene spermatocytes. Thus, Vihko and associates (1984) concluded that preleptotene spermatocytes regulate plasminogen activator secretion. Studies of the hormonal regulation of plasminogen activator mRNA, however, demonstrated that mRNA for urokinase-type plasminogen activator was unaffected by FSH (Vihko et al., 1989). This finding, along with studies of P-mod-S effects on Sertoli cells, is consistent with the hypothesis, stated earlier, that hormones and cell–cell interaction stimulate Sertoli cells through different intracellular pathways. Finally, studies by Ailenberg and Fritz (1989) provide insight into the biological significance of the stimulation of plasminogen activator secretion by preleptotene spermatocytes. These investigators studied the effect of plasminogen activator on the permeability of the barrier created by immature Sertoli cells cultured on a two-chamber assembly. This in vitro permeability barrier is created by the same tight junctions that in vivo form the blood–testis barrier. They noted that there was an inverse relationship between the amount of plasminogen activator activity in the culture medium and the tightness of the permeability barrier created by the Sertoli cells. Thus, the stage-specific secretion of plasminogen activator, which is regulated by preleptotene spermatocytes, may facilitate the sequential opening and reclosing of small segments of the blood–testis barrier to allow the translocation of these spermatocytes from the basal to the adluminal compartment of the seminiferous tubule.

B. Effect on Sertoli Cell Function of Co-culturing Germ Cells with Sertoli Cells

The effect of germ cells on Sertoli cell function also has been studied by co-culturing germ cells with Sertoli cells. La Magueresse and Jegou (1988) examined how age, hormones, and germ cells interacted to regulate Sertoli cell function. They noted that when adherent germ cells were removed from the cultured Sertoli cells by hypotonic shock, there was a significant decrease in ABP secretion. In addition, they noted that when a crude germ cell suspension was added back to hypotonically shocked Sertoli cells there was a significant increase in ABP secretion. This stimulation was achieved when purified round spermatids or pachytene spermatocytes were added to Sertoli cells from 20- and 45-day-old rats. When Sertoli cells from 10-day-old rats were examined, however, they noted much less stimulation of ABP secretion by crude germ cells or purified spermatocytes or spermatids. This observation, therefore, agrees with the previous suggestion that age modulates the response of the Sertoli cell to germ cells.

La Magueresse and Jegou (1988) also observed that germ cells eliminated the response to FSH of Sertoli cells from 45-day-old rats. These authors noted that FSH stimulated ABP secretion by Sertoli cells from 10, 20, and 45-day-old rats when these cells were cultured without germ cells. In contrast, FSH stimulated ABP secretion by Sertoli cells isolated from 10- and 20-day-old, but not from 45-day-old, rats when these cells were co-cultured with germ cells. The addition of dbcAMP to Sertoli–germ cell co-cultures, however, increased ABP secretion with Sertoli cells isolated from all three ages. This observation, therefore, suggests that stimulation of 45-day-old Sertoli cells by germ cells leads to a desensitization of the cells to FSH, possibly by decoupling the adenylate cyclase from the FSH receptor.

Djakiew and Dym (1988) studied the effect of pachytene spermatocytes on the secretion of transferrin by Sertoli cells cultured in two-compartment culture chambers. In their studies they noted that proteins released by cultured pachytene spermatocytes caused a dose-response stimulation of transferrin secretion. Part of this increased transferrin secretion was shown to be due to an overall stimulation of protein synthesis by Sertoli cells; although transferrin secretion increased threefold in response to the highest dose of pachytene spermatocyte proteins, total protein secretion increased twofold. Castellon and colleagues (1989) also studied the effect of secreted factors from germ cells on Sertoli cell function and compared their effect with the effect of adhesion of the germ cells to the Sertoli cells (Fig. 15–8). To test the effect of secreted factors, they cultured germ cells on a permeable membrane suspended in the medium above the cultured Sertoli cells. When they isolated Sertoli cells from 18- and 26-day-old rats, they noted that Sertoli cells in direct contact with germ cells secreted more ABP and transferrin than Sertoli cells cultured beneath the germ cells in the upper chamber (Fig. 15–8). Sertoli cells cultured alone, however, secreted significantly less ABP and transferrin than Sertoli cells co-cultured with germ cells under both conditions. Thus, germ cells seem to stimulate Sertoli cells by at least two different mechanisms; one requires adhesion of the germ cell to the Sertoli cell, and the other is mediated by soluble factors released from the germ cells.

Two additional studies shed light on Sertoli cell intracellular pathways, which are stimulated by germ cells and provide additional evidence that cell–cell interactions stimulate different intracellular pathways in Sertoli cells than do hormones. Ireland and Welch (1987) noted that when Sertoli cells were co-cultured with germ cells or cultured in germ cell-conditioned medium, there was an immediate phosphorylation of two Sertoli cell proteins. Additional data indicated that the phosphorylation of one protein occurred by the action of a calmodulin-dependent protein kinase, whereas the other protein was phosphorylated by a phospholipid-dependent protein kinase C. Activation of proteinase C by germ cells is noteworthy because Monaco and Conti (1987) demonstrate that phorbol ester-induced activation of protein kinase C decreases the ability of FSH to stimulate adenylate cyclase of Sertoli

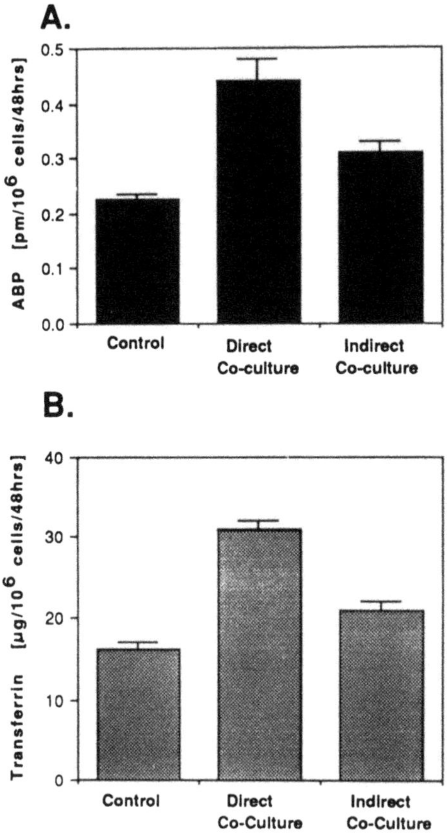

Figure 15–8. Effect of direct and indirect Sertoli cell–germ cell co-culture on ABP (**A**) and transferrin (**B**) secretion by Sertoli cells isolated from 26-day-old rats. Sertoli cells were cultured for 2 days in a 1:1 mixture of Ham's F-12:Dulbecco's Modified Eagles Medium supplemented with insulin, EGF, vitamins A and E, progesterone and hydrocortisone. On day 2, adherent germ cells were removed from the Sertoli cell cultures. Eight to ten hours after hypotonic treatment of the Sertoli cell cultures, germ cells were isolated from 26-day-old rat testes and these germ cells were added directly on top of the Sertoli cells (direct co-cultures) or placed on a permeable membrane suspended above the Sertoli cells (indirect co-cultures). Germ cells were not added to control cultures of Sertoli cells. After an additional 48 hours, medium was collected from the control Sertoli cell cultures and Sertoli–germ cell co-cultures. ABP and transferrin concentrations in the medium were measured by radioimmunoassays and data (mean ± SEM) expressed as amounts of the two proteins released into the medium by 10^6 Sertoli cells in 48 hours. (Data are from Castellon et al., 1989.)

cells. Thus, stimulation of Sertoli cell protein kinase C activity by germ cells may explain why germ cells cause a desensitization of 45-day-old Sertoli cells to FSH (Le Magueresse and Jegou, 1988).

VIII. Summary

This review has demonstrated that the Sertoli cells, spermatogenic cells, and myoid cells interact by a variety of mechanisms and that these interactions are essential for the development, maturation, and mature function of the seminiferous epithelium. In the introduction of this review, we set forth themes that deserve repeating. First, we postulated that specific interactions are more or less important during development and maturation of the seminiferous epithelium. For example, the interaction of Sertoli cells and myoid cells in the deposition of the basement membrane was shown to be an essential event in fetal development. In contrast, there appears to be minimal synthesis of basement membrane components in the mature animal. In addition, we have reviewed data that suggest that interactions between Sertoli cells and germ cells responsible for the stage-specific synthesis of CP-2 are not established until late in pubertal development of the testes. Some interactions between germ cells and Sertoli cells, however, may be essential for most of the life history of the seminiferous epithelium. Thus, the secretion of lactate and pyruvate by Sertoli cells may be required as soon as prespermatogonia give rise to type A spermatogonia and begin the process of spermatogenesis. Second, we hypothesized that cell–cell interactions act through a number of different pathways and have multiple consequences for the interacting cells. As just reviewed, we noted that germ cells stimulate phosphorylation of specific Sertoli cell intracellular proteins. This stimulation appears to increase the secretion of both ABP and transferrin. Third, we suggested that cell–cell interactions stimulated different intracellular pathways than do hormones. Studies of the effects on Sertoli cells of germ cells and of myoid cells support this hypothesis. Fourth, we hypothesized that during testicular maturation, interactions between germ cells and Sertoli cells become more important to the regulation of the Sertoli cell than the interaction of this somatic cell with FSH. Thus, we noted that Sertoli cells from 45-day-old rats were most sensitive to stimulation by germ cells. In contrast, when these were co-cultured with germ cells, they failed to respond to FSH.

Obviously many questions are left unanswered in this review. We know nothing about the effect of Sertoli cells on gene expression by germ cells. In addition, we do not know whether the effect of germ cells on Sertoli cell function is the result of increased gene transcription by the Sertoli cells and/or stabilization of specific mRNAs. Answers to these and many other questions await the results of future experiments.

ACKNOWLEDGMENTS

Research in the author's laboratory is supported by National Institutes of Health grants HD 17989, HD 21648, and AG 08321. I thank Dr. David Phillips, of The Population Council, for the electron micrograph of Sertoli cell tight junctions and Dr. Robert Sprando of The Division of Reproductive Biology, Johns Hopkins University School of Hygiene and Public Health for the electron micrograph of the ectoplasmic specialization of Sertoli cells.

REFERENCES

Ailenberg, M., and Fritz, I. B. (1989). Influences of follicle-stimulating hormone, proteases, and anti-proteases on permeability of the barrier generated by Sertoli cells in a two-chambered assembly. Endocrinology 124:1399–1407.

Anthony, C. T., and Skinner, M. K. (1989). Actions of extracellular matrix on Sertoli cell morphology and function. Biol. Reprod. 40:691–702.

Bellve, A. R., and Zheng, W. (1989a). Growth factors as autocrine and paracrine modulators of male gonadal functions. J. Reprod. Fert. 85:771–793.

Bellve, A. R., and Zheng, W. (1989b). Different pleiotypic actions of seminiferous growth factor (SGF), acidic fibroblast growth factor (aFGF), and basic fibroblast growth factor (bFGF). Annals N.Y. Acad. Sci. 564:116–131.

Borland, K., Muffly, K. E., and Hall, P. F. (1986). Production of components of extracellular matrix by cultured Sertoli cells. Biol. Reprod. 35:997–1008.

Bressler, R. S., and Ross, M. H. (1972). Differentiation of peritubular myoid cells of the testis: Effects of intratesticular implantation of newborn mouse testes into normal and hypophysectomized adults. Biol. Reprod. 6:148–159.

Casella, S. J., Smith, E. P., van Wyk, J. J., Joseph, D. R., Hynes, M. A., Hoyt, E. C., and Lund, P. K. (1987). Isolation of rat testis cDNAs encoding an insulin-like growth factor I precursor. DNA 6:325–330.

Castellon, E., Janecki, A., and Steinberger, A. (1989). Influence of germ cells on Sertoli cell secretory activity in direct and indirect co-culture with Sertoli cells from rats of different ages. Mol. Cell. Endocrinol. 64:169–178.

Chatelain, P. G., Naville, D., and Saez, J. M. (1987). Somatomedin-C/insulin-like growth factor 1-like material secreted by porcine Sertoli cells in vitro: Characterization and regulation. Biochem. Biophys. Res. Comm. 146:1009–1017.

Chemes, H. E., Dym, M., and Raj, H.G.M. (1979). Hormonal regulation of Sertoli cell differentiation. Biol. Reprod. 21:251–262.

Clermont, Y. (1972). Kinetics of spermatogenesis in mammals: Seminiferous epithelium cycle and spermatogonial renewal. Physiol. Rev. 52:198–263.

Clermont, Y., and Perey, B. (1957). Quantitative study of the cell population of the seminiferous tubules in immature rats. Am. J. Anat. 100:241–260.

D'Agostino, A., and Stefanini, M. (1987). A rat spermatocyte surface protein is involved in adhesion of pachytene spermatocytes to Sertoli cells in vitro. Mol. Cell. Biology 7:1250–1255.

DePhillip R. M., and Danahey, D. G. (1987). Germ cell binding to rat Sertoli cells in vitro. Biol. Reprod. 37:1271–1282.

de Rooij, D. G., Lok, D., and Weenk, D. (1985). Feedback regulation of the proliferation of the undifferentiated spermatogonia in the Chinese hamster by the differentiating spermatogonia. Cell Tissue Kinet. 18:71–81.

Djakiew, D., and Dym, M. (1988). Pachytene spermatocyte proteins influence Sertoli cell function. Biol. Reprod. 39:1193–1205.

Dym, M., Raj, H.G.M., Lin, Y. C., Chemes, H. E., Kotite, N. J., Nayfeh, S. N., and French, F. S. (1979). Is FSH required for maintenance of spermatogenesis in adult rats? J. Reprod. Fert. (Suppl.) 26:175–181.

Edelman, G. M. (1984). Modulation of cell adhesion during induction, histogenesis, and perinatal development of the nervous system. Annu. Rev. Neurosci. 7:339–377.

Enders, G. C., and Millette, C. F. (1988). Pachytene spermatocyte and round spermatid binding to Sertoli cells in vitro. J. Cell Science 90:105–114.

Enders, G. C., Henson, J. H., and Millette, C. F. (1986). Sertoli cell binding to isolated testicular basement membrane. J. Cell Biol. 103:1109–1119.

Erickson-Lawrence, M., Zabludoff, S., and Wright, W. W. (1991). Cyclic protein-2, a secretory product of rat sertoli cells, is the proenzyme form of cathepsin L. Mol. Endocrinol. 5:1789–1798.

Eusebi, F., Ziparo, E., Fratamico, G., Russo, M. A., and Stefanini, M. (1983). Intercellular communication in rat seminiferous tubules. Develop. Biol. 100:249–255.

Gilula, N. B., Fawcett, D. W., and Aoki, A. (1976). The Sertoli cell occluding junctions and gap junctions in mature and developing mammalian testis. Develop. Biol. 50:112–168.

Gospodarowicz, D. N., Ferrara, N., Scheigerer, L., and Neufeld, G. (1987). Structural characterization and biological functions of fibroblast growth factor. Endocrine Rev. 8:95–114.

Grootegoed, J. A., Oonk, R. B., Jansen, R., and van der Molen, H. J. (1986). Metabolism of radiolabelled energy-yielding substrates by rat Sertoli cells. J. Reprod. Fert. 77:109–118.

Grootegoed, J. A., Jansen, R., and van der Molen, H. J. (1984). The role of glucose, pyruvate and lactate in ATP production by rat spermatocytes and spermatids. Biochem. Biophys. Acta. 767:248–256.

Gonzalez-Manchon, C., and Vale, W. (1989). Activin-A, inhibin, and transforming growth factor-β modulate growth of two gonadal cell lines. Endocrinology 125:1666–1672.

Groogetoed, J. A., Boer, P.J.D., and Mackenbach, P. (1989). Sertoli cell–germ cell communication. Annals NY Acad. Sci. 564:232–242.

Gumbiner, B., and Simons, K. (1986). A function assay for proteins involved in establishing an epithelial occluding barrier: Identification of a uvomorulin-like polypeptide. J. Cell Biol. 102:457–468.

Hadley, M. A., and Dym, M. (1987). Immunocytochemistry of extracellular matrix in the lamina propria of the rat testis: Electron microscopic localization. Biol. Reprod. 37:1283–1289.

Hadley, M. A., Byers, S. W., Suarez-Quian, C. A., Kleinman, H. K., and Dym, M. (1985). Extracellular matrix regulates Sertoli cell differentiation, testicular cord formation, and germ cell development in vitro. J. Cell Biol. 101:1511–1522.

Heindel, J. J. (1988). Age-related and testicular regression-induced changes in adenosine 3', 5'-monophosphate responses in cultured hamster Sertoli cells. Endocrinology 122:475–481.

Ireland, M. E., and Welch, M. J. (1987). Germ cell stimulation of Sertoli cell protein phosphorylation. Endocrinology 120:1317–1326.

Jaillard, C., Chatelain, P. G., and J. M. Saez. (1987). In vitro regulation of pig Sertoli cell growth and function: Effects of fibroblast growth factor and somatomedin-C. Biol. Reprod. 37:665–674.

Jutte, H.P.M., Grootegoed, J.A., Rommerts, F.F.G., and van der Molen, H. J. (1981). Exogenous lactate is essential for metabolic activities in isolated rat spermatocytes and spermatids. J. Reprod. Fert. 62:399–405.

Lacroix, M., Parvinen, M., and Fritz, I. B. (1981). Localization of testicular plasminogen activator in discrete portions (stages VII and VIII) of the seminiferous tubule. Biol. Reprod. 25:143–146.

La Magueresse, B., and Jegou, B. (1988). In vitro effects of germ cells on the secretory activity of Sertoli cells recovered from rats of different ages. Endocrinology 22:1672–1680.

Leblond, C. P., and Clermont, Y. (1952). Definition of the stages of the cycle of the seminiferous epithelium in the rat. Ann. NY Acad. Sci. 55:548–573.

Li, L., Seddon, A. P., Meister, A., and Risley, M. S. (1989). Spermatogenic cell–somatic cell interactions are required for maintenance of spermatogenic cell glutathione. Biol. Reprod. 40:317–331.

Monaco, L., and Conti, M. (1987). Inhibition by phorbol esters and other tumor promoters of the response of the Sertoli cell to FSH: Evidence for dual site of action. Mole. Cell. Endocrinol. 49:227–236.

Morales, C. R., Alcivar, A. A., Hecht, N. B., and Griswold, M. D. (1989). Specific mRNAs in Sertoli and germinal cells of testes from stage synchronized rats. Mol. Endocrinol. 3:725–733.

Morales, C., Hugly, S., and Griswold, M. (1987). Stage-dependent levels of specific mRNA transcripts in Sertoli cells. Biol. Reprod. 36:1035–1046.

Morris, P. L., Vale, W. W., Cappel, S., and Bardin, C. W. (1988). Inhibin production by primary Sertoli cell-enriched cultures: Regulation by follicle-stimulating hormone, androgens, and epidermal growth factor. Endocrinology 122:717–725.

Norton, J. N., and Skinner, M. K. (1989). Regulation of Sertoli cell function and differentiation through the actions of a testicular paracrine factor P-mod-S. Endocrinology 124:2711–2719.

Oonk, R. B., and Grootegoed, J. A. (1988). Insulin-like growth factor I (IGF-I) receptors on Sertoli cells from immature rats and age-dependent testicular binding of IGF-1 and insulin. Mol. Cell Endocrinol. 55:33–43.

Orth, J. M. (1982). Proliferation of Sertoli cells in fetal and postnatal rats: A quantitative autoradiographic study. Anat. Rec. 203:485–492.

Pardee, A. B. (1989). G_1 events and regulation of cell proliferation. Science 246:603–609.

Parvinen, M., Wright, W. W., Phillips, D. M., Mather, J. P., Musto, N. A., and Bardin, C. W. (1983). Rat spermatogenesis in vitro: Completion of meiosis and early spermiogenesis. Endocrinology 112:1150–1152.

Parvinen, M., and Ruokinen, A. (1982). Endogenous steroids in rat seminiferous tubules. Comparison of different spermatogenic stages isolated by transillumination-assisted microdissection. J. Androl. 3:211–220.

Pelliniemi, L. J., and Lauteala, L. (1981). Development of sexual dimorphism of the embryonic gonad. Human Genetics 58:64–67.

Pelliniemi, L. J., Paranko, J., Grund, S. K., Fröjdman, K., Foidart, J. -M., and Lakkala-Pananko, T. (1987). Extracellular matrix in testicular differentiation. Annals NY Acad. Sci. 513:405–416.

Reed, S. C., and Stanley, H. P. (1972). Fine structure of spermatogenesis in the South African clawed toad *Xenopus laevis* Daudin. Ultrastructural Res. 41:277–295.

Risley, M. S. (1983). Spermatogenic cell differentiation in vitro. Gamete Research 4:331–346.

Risley, M. S., Miller, A., and Bumcrot, D. A. (1987). In vitro maintenance of spermatogenesis in *Xenopus laevis* testis explants cultured in serum-free media. Biol. Reprod. 36:985–997.

Roberts, V., Meunier, H., Sawchenko, P. E., and Vale, W. (1989). Differential production and regulation of inhibin subunits in rat testicular cell types. Endocrinology 125:2350–2359.

Roberts, A. B., Flanders, K. C., Kondaiah, P., Thompson, N. L., van Obberghen-Schilling, E., Wakefield, L., Rossi, P., de Crombrugghe, B., Heine, U., and Sporn, M. B. (1988). Transforming growth factor β: Biochemistry and roles in embryogenesis, tissue repair and remodeling and carcinogenesis. Recent Prog. Hormone Res. 44:157–193.

Russell, L. D. (1977). Movement of spermatocytes from the basal to the adluminal compartment of the rat testis. Am. J. Anat. 148:313–328.

Russell, L. D., Goh, J. C., Rashed, R.M.A., and Vogel, A. W. (1988). The consequences of actin disruption at Sertoli cell ectoplasmic specialization sites facing spermatids after in vivo exposure of rat testis to cytochalasin D. Biol. Reprod. 39:105–118.

Russell, L. D., Taloon-Doran, M., Weber, J. E., Wong, V., and Peterson, R. N. (1983). Three dimensional reconstruction of a rat stage V Sertoli cell: III. A study of specific cellular relationships. Am. J. Anatomy 167:181–192.

Rutishauser, U., Acheson, A., Hall, A. K., Mann, D. M., and Sunshine, J. (1988). The neural cell adhesion molecule (NCAM) as a regulator of cell–cell interactions. Science 240:53–57.

Skinner, M. K., and Moses, H. L. (1989). Transforming growth factor β gene expression and action in the seminiferous tubule: Peritubular cell-Sertoli cell interactions. Mol. Endocrinol. 4:625–634.

Skinner, M. K., Fetterolf, P. M., and Anthony, C. T. (1988). Purification of paracrine factor, P-Mod-S, produced by testicular peritubular cells that modulate Sertoli cell function. J. Biol. Chem. 263:2884–2890.

Skinner, M. K., and Fritz, I. B. (1985). Testicular peritubular cells secrete a protein under androgen control that modulates Sertoli cell functions. Proc. Natl. Acad. Sci. 82:114–118.

Skinner, M. K., Tung, P. S., and Fritz, I. B. (1985). Cooperativity between Sertoli cells and testicular peritubular cells in the production and deposition of the extracellular matrix components. J. Cell Biol. 100:1941–1947.

Smith, E. P., Svoboda, M. E., van Wyk, J. J., Kierszenbaum, A. L., and Tres, L. L. (1987). Partial characterization of a somatomedin-like peptide from the medium of cultured rat Sertoli cells. Endocrinology 120:186–193.

Swinnen, K., Cailleau, J., Heyns, W., and Verhoeven, G. (1990). Prostatic stromal cells and testicular peritubular cells produce similar paracrine mediators of androgen action. Endocrinology 126:142–150.

Taketo, T., Thau, R. B., Adeyemo, O., and Koide, S. (1984). Influence of adenosine 3':5'-cyclic monophosphate analogues on testicular organization of fetal mouse gonads in vitro. Biol. Reprod. 30:189–198.

Tosney, K. W., Watanabe, M., Landmesser, L., and Rutishauser, U. (1986). The distribution of NCAM in the chick hindlimb during axon outgrowth and synaptogenesis. Dev. Biol. 114:437–452.

Tres, L. L., Smith, E. P., van Wyk, J. J., and Kierszenbaum, A. L. (1986). Immunoreactive sites and accumulation of somatomedin-C in rat Sertoli-Spermatogenic cell co-cultures. Exp. Cell Res. 162:33–50.

Tung, P. S., and Fritz, I. B. (1987). Morphological restructuring and formation of basement membranes by Sertoli cells and testis peritubular cells in co-culture: Inhibition of the morphogenetic cascade by cyclic AMP derivatives and by blocking direct cell contact. Dev. Biol. 120:139–153.

Tung, P. S., and Fritz, I. B. (1984). Extracellular matrix promotes rat Sertoli cell histotypic expression in vitro. Biol. Reprod. 30:213–229.

Ueno, N., Baird, A., Esch, F., Ling, N., and Guillemin, R. (1987). Isolation and partial characterization of basic fibroblast growth factor from bovine testis. Mol. Cell. Endocrinol. 49:189–194.

van Dissel-Emiliani, F.M.F., Grootenhuis, A. J., de Jong, F. H., and de Rooij, D. G. (1989). Inhibin reduces spermatogonial numbers in testes of adult mice and Chinese hamsters. Endocrinology 125:1899–1903.

Vanelli, B. G., Barni, T., Orlando, C., Natali, A., Serio, M., and Balboni, G. C. (1988). Insulin-like growth factor-1 (IGF-1) and IGF-1 receptor in human testis: An immunohistochemical study. Fert. Steril. 49:666–669.

Vihko, K. K., Penttila, T-L, Parvinen, M., and Belin, D. (1989). Regulation of urokinase- and tissue-type plasminogen activator gene expression in the rat seminiferous epithelium. Mol. Endocrinol. 3:52–259.

Vihko, K. K., Toppari, J., and Parvinen, M. (1987). Stage-specific regulation of plasminogen activator secretion in the rat seminiferous epithelium. Endocrinology 120:142–145.

Vihko, K. K., Suominen, J.J.O., and Parvinen, M. (1984). Cellular regulation of plasminogen activator secretion during spermatogenesis. Biol. Reprod. 31:383–389.

Vogel, A. W. (1988). Changes in the distribution of microtubules in rat Sertoli cells during spermatogenesis. Anat. Rec. 222:34–41.

Volk, T., and Geiger, B. (1986). A-Cam: A 135-kd receptor of intercellular adherens junctions. I. Immunoelectron microscopic localization and biochemical studies. J. Cell Biol. 103:1441–1450.

Weber, J. E., Russell, L. D., Wong, V., and Peterson, R. N. (1983). Three-dimensional reconstruction of a rat stage V Sertoli cell: II. Morphometry of Sertoli–Sertoli and Sertoli–germ cell relationships. Am. J. Anat. 167:163–179.

Wright, W. W (1988). Germ cell–Sertoli cell interactions: Analysis of the biosynthesis and secretion of cyclic protein-2. Dev. Biol. 130:45–56.

Wright, W. W., Zabludoff, S. D., Erickson-Lawrence, M., and Karzai, A. W. (1989). Germ cell–Sertoli cell interactions. Studies of cyclic protein-2 in the seminiferous tubule. Annals NY Acad. Sci. 564:173–185.

Wright, W. W., Parvinen, M., Musto, N. A., Gunsalus, G. L., Philips, D. M., Mather, J. P., and Bardin, C. W. (1983). Identification of stage-specific proteins synthesized by rat seminiferous tubules. Biol. Reprod. 29:257–270.

Yu, J., Shao, L., Lemas, V., Yu, A. L., Vaughan, J., Rivier, J., and Vale, W. (1987). Importance of FSH-releasing

protein and inhibin in erythrodifferentiation. Nature 330:765–767.
Zabludoff, S., Erickson-Lawrence, M., and Wright, W. W. (1990a). Sertoli cell, proximal convoluted tubules in the kidney and neurons in the brain contain Cyclic protein-2. Biol. Reprod. 43:15–24.
Zabludoff, S., Karzai, A. W., and Wright, W. W. (1990b). Germ cell–Sertoli cell interactions: The effect of testicular maturation on the synthesis of cyclic protein-2. Biol. Reprod. 43:25–33.

CHAPTER SIXTEEN

Gene Expression During Male Germ Cell Development

Norman B. Hecht

I. The Testis as a Model System to Study Cellular Differentiation

A. Introduction

The spermatozoon, a genetically quiescent, hydrodynamically shaped, and self-propelled cell, is a marvel of differentiation. To produce a spermatozoon, a testicular stem cell enters a differentiative pathway that ensures both genetic diversity of each gamete and a reduction in chromosome number to a haploid level. Major cytological and structural changes occur as the male germ cell passes through meiosis and spermiogenesis en route to the creation of the uniquely shaped spermatozoon. The cellular changes during spermatogenesis are likely controlled by many of the common regulatory mechanisms for differentiating somatic eukaryotic cells and thus provide an excellent system to study general transcriptional and translational regulatory mechanisms of mammalian cells. In addition, the many unique testicular isozymes and structural proteins suggest that the testis should be an excellent source of organ or cell-specific factors that modulate gene expression.

Well-established procedures to isolate populations of testicular cell types allow the investigator to dissect biochemical events during spermatogenesis by examining gene expression at specific "windows" of male germ cell differentiation during spermatogenesis. With this approach, regulatory mechanisms for gene expression programmed for initiation during meiosis, for example, can be compared with the expression of another set of genes activated during spermiogenesis. The existence of structures unique to germ cells, such as the synaptonemal complex in meiotic cells and the manchette in postmeiotic cells, will provide many novel proteins and genes to examine. The natural cessation of transcription during spermiogenesis further allows an examination of translational regulatory mechanisms for numerous sperm proteins that are synthesized toward the end of spermatogenesis.

This chapter will not present an exhaustive review of the literature on gene expression in the testis but instead attempts to identify approaches and pose research problems emphasizing the differentiating germ cell. It focuses primarily on state of the art molecular biology approaches to study male germ cell differentiation. The important roles that growth factors, hormones, proto-oncogenes, and the somatic cells of the testis play in germ cell development are only discussed in passing as other chapters in this book address these topics. For additional information on the biology, genetics, and molecular biology of spermatogenesis, the reader is directed to excellent recent reviews from the laboratories of Balhorn (1989), Bellvé (1979, 1983), Dixon (1985), Eddy (1988), Goldberg (1977), Griswold (1986, 1988), Handel (1987), Hecht (1986, 1989), Kistler (1989), Meistrich (1989), Mezquita (1985), Olds-Clarke (1988),

Poccia (1986), Willison (1987), Wolgemuth (1989), and their colleagues.

1. STEM CELL RENEWAL AND DIFFERENTIATION

For spermatogenesis to proceed normally throughout the life of an animal, stem cells must be renewed constantly and a subpopulation of the cells must continually enter the differentiative pathway that leads to the formation of spermatozoa (Fig. 16–1). Unfortunately, we have little understanding of the signals or genetic programs that regulate these critical events in the early stages of spermatogenesis. From cytological studies, we do know that primitive spermatogonia are produced from gonocytes that have migrated to a location adjacent to the basal lamina of the seminiferous tubule during fetal development (reviewed in Bellvé, 1979). The primitive spermatogonia serve as progenitors of the definitive spermatogonia that produce the primitive type A and presumptive type A spermatogonia. Type A, intermediate, and type B spermatogonia appear sequentially in the testes of species, such as the mouse, 7 to 10 days after birth.

A primary limitation to biochemical studies of spermatogonial differentiation has been the difficulty in obtaining sufficient numbers of highly purified spermatogonia. It is now possible to use cell separation procedures, however, to obtain enriched populations of type A and type B spermatogonia from the testes of prepuberal animals such as 8-day-old mice (Romrell et al., 1976). When combined with signal-enhancing procedures, such as the polymerase chain reaction, analysis of small samples of DNA or RNA from limited numbers of cells can be undertaken (Saiki et al., 1988). A short-term research goal that needs to be pursued is the identification of marker molecules specific to spermatogonia and the substages of stem cell renewal. Longer term efforts should be directed toward an understanding of the trigger events that shift spermatogonia from a cycling to a differentiating destiny. These endeavors go far beyond their importance to spermatogenesis as cellular proliferation from stem cells is a common theme in differentiation. Advances in understanding other systems, such as the hematopoietic system, should be carefully scrutinized to identify unifying themes that will aid in defining the mechanisms of stem cell differentiation in the testis.

Recently, Beamer and colleagues described a mouse mutant, juvenile spermatogonial depletion (jsd), whereby homozygous mutant mice 3 to 10 weeks of age undergo one wave of spermatogenesis but fail to maintain mitotic proliferation of type A spermatogonia (Beamer et al., 1988). Vitamin A supplementation by retinol did not remedy the mutation and analysis of serum hormones showed normal testosterone levels at all ages, suggesting that the testes of the jsd/jsd mutant may be deficient in proliferative signals, perhaps from the somatic Sertoli cells. Creative utilization of mutants, such as the juvenile spermatogonial depletion mouse, could provide insight into the signals required for renewal or differentiation of specific populations of spermatogonia. A more detailed discussion of spermatogonial stem cell proliferation is presented in Chapter 12.

2. MEIOSIS

Meiosis, the interval when synapsis develops between homologous chromosomes allowing genetic recombination to occur, is central to the development of both the male and female germ cells. In the male mouse, meiosis requires 11 to 12 days of the approximately 34-day germ cell cycle. Meiotic prophase commences with the preleptotene primary spermatocyte, the cell type in which the last semiconservative DNA replication of the differentiating male germ cell occurs. Chromosome condensation commences concomitantly with the translocation of the leptotene and zygotene stage spermatocytes from the basal membrane region of the seminiferous tubule toward the adluminal (central) compartment of the seminiferous epithelium. The cellular movement and chromosome pairing between homologues is completed in the pachytene spermatocyte. As the chromosomes condense, axial elements become visible between the two sister chromatids of each pairing chromosome. The addition of a central element produces the synaptonemal complex, a structure believed to be required for effective synapsis in eukaryotes. Although its molecular structure and precise functions have not yet been delineated, the synaptonemal complex facilitates the recombination process. After meiotic prophase, the male gamete divides twice,

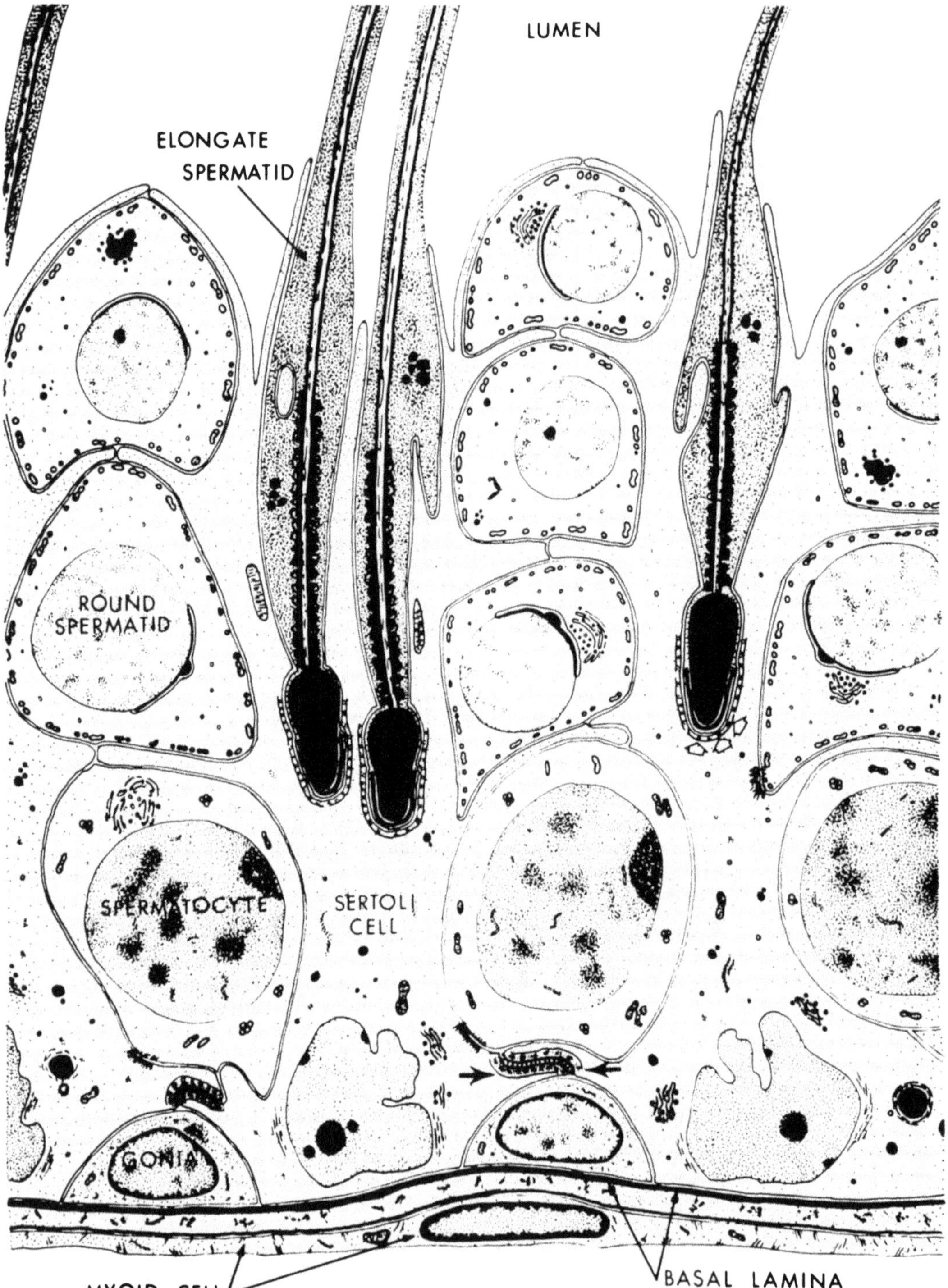

Figure 16–1. Spermatogenesis. The events of spermatogenesis commence at the basement membrane (gonia) and proceed toward the lumen. (Kindly provided by Dr. L. Russell, Southern Illinois University School of Medicine. Reprinted with permission from *Reproductive and Developmental Toxicity of Metals,* edited by Clarkson et al., Plenum Press.)

producing the briefly present diploid secondary spermatocyte and then the haploid round spermatid. Chapter 13 in this volume focuses on the events of meiosis.

3. SPERMIOGENESIS

The differentiation of the round spermatid to the spermatozoon occurs during spermiogenesis, an interval of about 14 days in the mouse that is subdivided into 16 steps. Despite the cessation of RNA synthesis around step 8 of spermiogenesis (reviewed in Kierszenbaum and Tres, 1978), major cellular architectural changes occur throughout this interval. The round spermatid undergoes a series of shape changes ultimately producing the species-specific shaped spermatozoon. During spermiogenesis an acrosome with novel enzymatic components develops in the nucleus and, in mammals, the chromatin proteins are replaced by protamines producing a tightly compacted DNA protamine complex with extensive disulfide bridge crosslinking. The reasons and mechanism whereby the heads of spermatozoa have species-specific shapes are not understood. The shaping of the sperm nucleus is not controlled by the protamines as only a very limited number (one to three variants) of mammalian protamines are ubiquitously shared among mammalian spermatozoa. In addition to the changes in the nucleus during spermiogenesis, the axoneme and tail of the spermatozoon are also formed. This cellular reshaping requires a major reorganization of the gamete including a substantial reduction in mitochondrial number. Ultimately, most of the spermatid's cytoplasm is lost as the residual body is pinched off. Lacking cytoplasmic ribosomes, only the mitochondria of the spermatozoon are capable of protein synthesis.

II. Novel Testicular Proteins

A. Testicular Isozymes

One especially fascinating aspect of gene expression in male germ cells is the existence of a large number of novel proteins and unique isozyme variants in the testis (Table 16–1). Testicular isozyme variants have been well characterized for lactate dehydrogenase, cytochrome c, and phosphoglycerate kinase (reviewed in Goldberg, 1977). Additional variants for enzymes, such as enolase (Edwards and Grootegoed, 1983), alcohol dehydrogenase (Keung, 1988), diaphorase, hexokinase, carbonic anhydrase, and phosphoglycerate mutase (Fundele et al., 1987), await further study.

Of all isozymes, the isozymic variants of lactate dehydrogenase (LDH) have been the most extensively studied. In somatic tissues five distinct isozymic forms of LDH have been found. These five variants result from different mixtures of polypeptides A and B, coded from two different genes. Although these variants are generally ubiquitously expressed, they often show tissue-specific patterns of expression. A third gene C codes for LDH C4, the isozyme only found in the germ cells of the testis (Goldberg, 1977; Millan et al., 1987; Takaro and Li, 1989). Physiological and biochemical analyses of LDH C4 indicate that, compared with LDH A and B, the testicular variant is more sensitive to pyruvate inhibition and less sensitive to lactate inhibition. LDH C4 oxidizes lactate more readily than LDH A or B and would, therefore, make more reduced coenzyme available for processing in the mitochondrial electron transport system of the sperm. Goldberg and colleagues have speculated that this preference for lactate may be the reason LDH C4 is used in spermatozoa, because high concentrations of lactate are present in oviductal fluids (Goldberg, 1977). Other properties of LDH C4 that could influence its selective value in the testis are a low enzyme turnover number, a broad substrate and coenzyme specificity in certain species, a resistance to thermal denaturation in certain species, and a possible association with the mitochondria of the spermatozoa. Considering that these properties are variable among the large number of species containing LDH C4, however, the reason LDH C4 is found in testes is not obvious. Moreover, the reported association of LDH C4 and mitochondria (Montamat and Blanco, 1976) needs confirmation. The presence of LDH in mitochondria may be due to adventitious binding as the protein appears to be freely distributed in the cytoplasm of the spermatozoon. Despite this uncertainty, the unique sequence and cellular location of LDH C4 have been successfully exploited to derive sperm-specific peptides used in an immunological approach for contraception (reviewed in Goldberg, 1987).

Table 16–1. Representative Novel Proteins Expressed in the Testis

Protein	Reference
Cytochrome c_t	reviewed in Goldberg, 1977
Lactate dehydrogenase C4	reviewed in Goldberg, 1977
Phosphoglycerate kinase 2	reviewed in Goldberg, 1977
Testicular histone variants	reviewed in Meistrich, 1989
Transition protein 1	Heidaran and Kistler, 1987a, b; Kleene et al., 1988
Transition protein 2	Kleene and Flynn, 1987; Luerssen et al., 1989
Protamine 1	Kleene et al., 1985
Protamine 2	Yelick et al., 1987
t-complex polypeptide 1 (TCP-1)	Silver, 1985; Willison et al., 1986, 1989
Testicular α tubulin	Distel et al., 1984; Hecht et al., 1988
Heat shock proteins	reviewed in Wolgemuth et al., 1988; Allen et al., 1988
Acrosin	reviewed in Goldberg, 1977
Fibrous sheath proteins	Fenderson et al., 1988; Oko, 1988
Androgen binding protein	Joseph et al., 1988
Hyaluronidase	reviewed in Goldberg, 1977; Harrison and Gaunt, 1988
Outer dense fibers	Vera et al., 1984; Oko, 1988
Alcohol dehydrogenase	Keung, 1988
ATP-dependent RNA helicase[a]	Leroy et al., 1989
Regulatory subunit for cAMP-dependent protein kinase	Sandberg et al., 1987; Oyen et al., 1988
Thyroid hormone receptor	Benbrock and Pfahl, 1987
Paracrine factor (P-mod-S)	Skinner et al., 1988
Calmodulin-dependent phosphodiesterase	Rossi et al., 1988
Perinuclear theca proteins	Longo et al., 1987
Enolase	Edwards and Grootegoed, 1983
Phosphoglycerate mutase	Fundele et al., 1987
Clone 49HT8	Seboun et al., 1990
t complex-testis expressed 1	Lader et al., 1989
cAMP phosphodiesterases	Swinnen et al., 1989
Angiotensin converting enzyme	Roy et al., 1988; Ehlers et al., 1989
Calspermin	Ono and Means, 1990

The above list indicates some of the proteins believed to be unique to or predominantly expressed in the mammalian testis. The references listed are not exhaustive and are intended only to provide the reader with a source to additional references.
[a]Testis-specific expression of this protein has not been definitively established yet.

Cytochrome c is a small heme-containing protein present in all mitochondria. In mammals, two functionally similar but distinct cytochromes c have been detected (reviewed in Goldberg, 1977). One protein, cytochrome c_s, is present in all somatic cells, whereas cytochrome c_t is a testis-specific isozyme. Cytochrome c_t has been detected by indirect immunofluorescence in primary spermatocytes and in cell types comprising the later stages of spermatogenesis (Goldberg et al., 1977). Spermatogonia, Sertoli cells, and interstitial cells contain cytochrome c_s. Cytochrome c_t messenger RNAs (mRNAs) have been detected in both pre-meiotic and meiotic cells (Hake et al., 1990). The rat testicular isozyme is encoded by a single copy gene of approximately 7 kb with three introns and a single polyadenylation site. The functional mouse cytochrome c_s gene has three functionally equivalent polyadenylation sites producing mRNAs of approximately 1.3, 1.1, and 0.8 kb. In the rat, cytochrome c_t differs from cytochrome c_s in 15 of 104 amino acids (Virbasius and Scarpulla, 1988). Although the cytochrome c_s proteins of mouse and rat are identical, the mouse and rat cytochrome c_t differ by four amino acids, suggesting that cytochrome c_t is evolving more rapidly than cytochrome c_s.

Phosphoglycerate kinase (PGK) is a third well-studied isozyme with a testis-specific variant (Vande Berg et al., 1973). PGK is a metabolic enzyme functioning in the Embden-Meyerhof pathway that converts glucose or fructose to pyruvate. In mammals, the somatic variant, PGK 1, is present on the X chromosome, whereas the testicular form, PGK 2, is on an autosome. The functional human gene for PGK 2 lacks introns and contains remnants of a poly $(A)^+$ sequence and adjacent direct repeat sequences suggesting that this variant arose by reverse transcriptase-mediated processing of a transcript from a separate gene, presumably PGK 1 (McCarrey and Thomas, 1987). It has been suggested that PGK 2 is an evolutionary compensatory response to the inactivation in early stages of spermatogenesis of the X-linked PGK 1 gene. Although a reasonable explanation for the existence of PGK 2, this hypothesis cannot explain other testicular isozymes such as LDH, which are not X chromosome-linked (Edwards et al., 1987). Additional novel testicular proteins are listed in Table 16–1.

Although puzzling, the phenomenon of multiple testicular isozymes is not unique to either the testis or isozymes. An analogous situation exists with the multiple isotypes of structural proteins such as the tubulins. In general, mammalian cells express a wide variety of tubulin isotypes (reviewed in Cleveland and Sullivan, 1985). In experiments using sera specific to β tubulin isotypes, no segregation of β tubulin isotypes in microtubules was detected suggesting no functional differences among the isotypes (Lewis et al., 1987). Both interphase cytoskeletal and mitotic spindle microtubules are mixed copolymers of all expressed β tubulin isotypes. The lack of subcellular subunit sorting makes it difficult to explain the absolute interspecies conservation of many mammalian tubulin isotypes.

Despite much biochemical analysis, the reasons the testis contains unique isozymes are not known. It is possible some testicular isozymes may be required to perform male germ cell specific physiological functions that the somatic forms would not perform optimally. It is equally likely that testicular isozymes have been selected because the variant genes contain essential testicular regulatory control sequences absent in the regulatory regions of the genes of the somatic variants. To determine the function of protein variants, it is necessary to remove or selectively inactivate specific genes. This can be accomplished by homologous recombinational procedures where altered or deleted genes are substituted for the wild type gene (Smithies et al., 1985). Advances in targeting an altered gene to a specific site on a chromosome will provide a means to determine whether, for instance, spermatogenesis could proceed normally without LDH C4. Alternatively, the transgenic antisense approach described later in this chapter could be used to inactivate the function of a specific isozyme. Such studies would determine whether specific testicular isozymes are essential for spermatogenesis because they optimize a particular enzyme reaction or they simply represent a residue of evolution.

B. Additional Unique Testicular Proteins

In addition to isozymes, the testis contains many other novel enzymes and structural proteins (Table 16–1). Some, such as acrosin and the protamines, are unique to the testis. Others, such as the meiotic histones, are enriched in the testis. Most of these proteins are required for specific functions and structures of the testis and their expression is temporally controlled. In the nucleus, the somatic histones of the male germ cell are replaced during meiosis by several histones unique to or enriched in the testis (reviewed in Meistrich, 1989). As spermiogenesis proceeds, an additional histone variant is expressed at the time the transition proteins begin to replace the histones (Moss et al., 1989). At the end of spermiogenesis, the transition proteins are replaced by the protamines that further compact the sperm DNA. In addition to the chromatin reorganization of the sperm nucleus, the acrosome, a modified lysosome essential for fertilization, also develops. The acrosome contains novel enzymes such as acrosin (Adham et al., 1989) and hyaluronidase and specific membrane proteins involved in sperm–egg interaction (Leyton and Saling, 1989). A group of insoluble basic proteins that contribute to the formation of the perinuclear theca, a structure involved in nuclear shape changes, have also been described (Longo et al., 1987).

The synthesis of the axoneme and tail of the spermatozoon during spermiogenesis requires the formation of an elaborate microtubular structure associated with numerous unique fibrous sheath and sperm tail proteins. A better understanding of the role played by the cytoskeleton of the germ cell as the maturing gamete changes shape is needed.

In addition to the large number of novel testicular isozymes and structural proteins, a variety of proto-oncogenes are expressed in the testes (reviewed in Propst et al., 1988). Some, such as c-*raf*, three members of the *ras* family, and c-*abl* are expressed throughout spermatogenesis. Other nuclear proto-oncogenes, including c-*fos*, c-*jun*, *jun* B, and c-*myc*, are expressed in early stages of germ cell differentiation with highest levels in type B spermatogonia (Wolfes et al., 1989; Alcivar et al., 1990b). The proto-oncogenes, c-*mos*, *pim*-1, and *int*-1, are predominantly expressed in postmeiotic testicular cells.

Recently, a developmentally regulated protein with high homology to the murine translation initiation factor, elF-4A, and to the *Drosophila* protein, vasa, was identified from a testis cDNA library (Leroy et al., 1989). It is expressed only in the testis with high levels of transcript being present in meiotic and postmeiotic germ cells. From its sequence, a putative mononucleotide binding region and DNA/RNA domains have been deduced suggesting an adenosine triphosphate (ATP)-dependent RNA helicase function for this male germ cell-specific protein.

Heat shock and stress proteins are widely expressed in prokaryotes and eukaryotes (Wolgemuth et al., 1989). In the testis, in addition to some of the ubiquitous HSP70 proteins, a unique member of the heat shock gene family has been found (Allen et al., 1988; Zakeri et al., 1988a, b; Maekawa et al., 1989). The gene contains extensive similarity to a heat shock-inducible HSP70 gene within the coding region but differs in the 5′ and 3′ untranslated regions. This gene, expressed during meiotic prophase, also differs from another member of this family that is expressed in postmeiotic male germ cells.

Studies of the pathways regulating cyclic adenosine monophosphate (cAMP) in spermatogenesis and in sperm have also begun to identify unique testicular regulatory subunits of a cAMP-dependent protein kinase (Oyen et al., 1988), a testis-specific phosphodiesterase (Rossi et al., 1988), and a family of genes homologous to the dunce encoded phosphodiesterase of *Drosophila* (Swinnen et al., 1989).

Many unique components of the sperm nucleus, midpiece, and tail remain to be identified. A more complete understanding of the components of the spermatozoon is likely to provide an explanation for many sperm defects of animals and humans, as well as provide novel targets for contraceptive efforts.

C. Spermiogenesis: A Window to Investigate Temporal Gene Expression

In germ cells of the mammalian testis, RNA synthesis has been detected in spermatogonia, primary spermatocytes, and in spermatids (reviewed in Monesi, 1978; Hecht, 1986). Based primarily on nuclear grain counts of [^{3}H]uridine incorporation from autoradiographic studies, the vast majority of testicular RNA synthesis is believed to occur during meiosis with a reduced level of RNA synthesis during the early stages of spermiogenesis. Analyses of round spermatid RNAs indicate that after a brief labeling interval with [^{3}H]uridine, poly (A)$^+$ RNA was synthesized. Quantitation of RNA isolated from populations of meiotic and postmeiotic cells reveals that, after normalization for DNA content, the ratio of cytoplasmic RNA to DNA is constant in pachytene spermatocytes and round spermatids. Moreover, round spermatids contain twice as much poly (A)$^+$ RNA per microgram of RNA as pachytene spermatocytes (Kleene et al., 1983). Although some of this mRNA may be stored from earlier stages, the absence of functional nucleoli in round spermatids suggests that mRNAs will constitute the majority of postmeiotic RNA synthesis.

To establish whether postmeiotic RNA synthesis simply represents a continuation of genes transcribed during meiosis or contains, at least in part, a group of novel transcripts, spermatid expressed genes have been sought by differential screening of cDNA libraries. Summarizing work from several laboratories, proteins such as the transition proteins, protamines, a regulatory subunit for a protein kinase, and certain proto-oncogenes, have been established to be transcribed first in mammalian round spermatids (Table 16–2). The postmeiotic expression of these gene products has been demonstrated by Northern blot analysis of RNA preparations from: (1) successive stages of developing prepuberal testes (in the mouse, for instance, the first wave of spermatogenesis initiates soon after birth leading to the presence of meiotic cells in the testes of 2-week-old animals and sperm formation within 5 to 6 weeks); (2) isolated populations of premeiotic, meiotic, and postmeiotic cell types obtained by separation of dissociated cells by unit gravity sedimentation in bovine serum albumin gradients, Percoll density gradient centrifugation, or elutriation; or (3) in situ hybridization. The polymerase chain reaction procedure can now be used to define precisely when an mRNA can first be detected. Because of the ability of this technique to detect very small numbers of molecules, however, care must be used to distinguish between "promiscuous" and true gene expression (Chelly et al., 1989). Accepting these caveats, the existence of a group of genes programmed for expression solely during spermiogenesis has been established.

The genetic evidence for transmission ratio distortion in t-haplotypes, a phenomenon where male mice heterozygous for a complete t-haplotype transmit the t-haplotype chromosome at a frequency far above the expected mendelian segregation ratio, can be explained by haploid gene expression (Silver, 1985). The differential expression and partitioning of the gene products from each germ cell during spermiogenesis would produce populations of unique spermatozoa. Recent transgenic studies, however, have demonstrated movement of transcripts through the intracellular bridges connecting germ cells thereby producing spermatozoa that are phenotypically equivalent (Braun et al., 1989a) (see section C.5.). Whether these results for a transgene can be extrapolated to all endogenous genes is not known.

Transcription of a group of genes during spermiogenesis provides an excellent entree to examine programmed gene expression in the testis (Table 16–2). The regulated transcription of a novel testicular gene at a specific interval of spermatogenesis is presumably dependent on specific regulatory signals operating in the testis. In addition to testis-specific genes, other genes coding for proteins, such as the smooth muscle γ actin (SMGA), are also postmeiotically expressed during spermatogenesis (Kim et al., 1989). The fact that the SMGA is also expressed in the intestine and the uterus suggests it should contain regulatory signals recognized by additional tissues.

Similar comparisons need to be undertaken with genes expressed during meiotic and premeiotic intervals of spermatogenesis (Tables 16–2 and 16–3). The existence of meiotically expressed genes, such as PGK 2 (Erickson et al., 1980; Gold et al., 1983) and LDH C4 (Fujimoto et al., 1988), provides a second group of temporally regulated testis-specific genes. Although currently few marker proteins for spermatogonia are known, this approach can be extended in the future when unique spermatogonial gene products have been identified.

III. Control Mechanisms

A. At the Cellular Level

Although it is likely that germ cells regulate limited aspects of their developmental destiny, the overall differentiation of male germ cells is heavily dependent on their interactions with somatic cells of the testis. Some of these cellular interactions are mediated at the hormonal level through the effects of molecules such as the androgens, follicle-stimulating hormone (FSH), luteinizing hormone (LH), and prolactin on somatic cells and also by autocrine and paracrine actions of growth factors (Bellvé and Zheng, 1989). Molecules modulating steroidogenesis in Leydig cells and influencing peritubular and Sertoli cell function also play critical roles in regulating germ cell development. Especially important to later stages of germ cell differentiation are interactions between the Sertoli cell and germ cell compartments whereby nutritional and regulatory factors shuttle between the somatic and germ cells. Co-culture of germ cells and Sertoli cells in bicameral chambers has revealed that the addition of pachytene spermatocyte-conditioned medium to a Sertoli cell culture stimulates the synthesis and secretion of total protein, transferrin, and two peptides

Table 16–2. Representative Genes Predominantly Expressed at Specific Intervals of Spermatogenesis and/or In Specific Cell Types of the Testis

Genes expressed	Reference
Premeiotic stages	
c-*myc*	Wolfes et al., 1989
c-*fos*	Wolfes et al., 1989
c-*jun*	Wolfes et al., 1989
jun B	Alcivar et al., 1990
c-*kit*	Ponzetto, Geremia personal communications
histones	reviewed in Meistrich, 1989
lactate dehydrogenase A	Li et al., 1989
lactate dehydrogenase B	Alcivar et al., 1990
cytochrome c_s	Hake et al., 1990
TH3	Meistrich and Brock, 1987
Meiotic stages	
Lactate dehydrogenase C4	reviewed in Goldberg, 1977; Wieben, 1981; Fujimoto et al., 1988; Jen et al., 1989
Phosphoglycerate kinase 2	McCarrey and Thomas, 1987
Cytochrome c_t	Hake et al., 1990
Histone variants	reviewed in Meistrich, 1989
Heat shock proteins	reviewed in Wolgemuth et al., 1988; Allen et al., 1988.
Mak	Matsushime et al., 1990
NASP	Welch et al., 1990
Calmodulin	Slaughter and Means, 1989
Calspermin	Ono et al., 1990
Acrosin	Klemm et al., 1990
Postmeiotic stages	
c-*mos*	Goldman et al., 1987; Mutter and Wolgemuth, 1987; Propst et al., 1987, 1988
int-1	Shackleford and Varmus, 1987
pim-1	Sorrentino et al., 1988
Protamine 1	Kleene et al., 1985; Klemm et al., 1989
Protamine 2	Yelick et al., 1987
Regulatory subunit type II protein kinase	Oyen et al., 1988
Smooth muscle γ actin	Kim et al., 1989
Tubulin variants	Distel et al., 1984; Villasante et al., 1986; Hecht et al., 1988
t complex polypeptide 1 (TCP1)	Silver, 1985; Willison et al., 1986, 1989
Transition protein 1	Heidaran and Kistler, 1987a, b; Kleene et al., 1988
Transition protein 2	Kleene and Flynn, 1987; Leuerssen et al., 1989
c-*abl* (mRNA size variant only)	Ponzetto and Wolgemuth, 1985
MEA	Lau et al., 1989
SP-10	Wright et al., 1990
G3PD	Schatte et al., 1990
PH-20	Lathrop et al., 1989
fer T	Fischman et al., 1990
β1,4 gal Tase	Shaper et al., 1990
cAMP PDE	Swinnen et al., 1989
Leydig cell	
Pro-opiomelanocortin	Gizang-Ginsberg and Wolgemuth, 1985
Renin	Deschepper et al., 1986
Angiotensin II receptors	Millan and Aguilera, 1988

Genes expressed	Reference
Sertoli cell	
Follicle-stimulating hormone receptor	Griswold et al., 1986
Androgen-binding protein	Joseph et al., 1985
Ceruloplasmin	Griswold et al., 1986
Transferrin	Griswold et al., 1986
Sulfated glycoproteins	Griswold et al., 1986
Inhibin	Mason et al., 1985
Mullerian-inhibiting substance	Cate et al., 1986

The above list indicates some of the testicular gene products shown to be temporally expressed. These data are generally based on an RNA detection level provided by Northern blots and the predominant cell type or stage of expression is indicated.

(Djakiew and Dym, 1988). Conditioned media from 3T3 cells and round spermatids did not stimulate the Sertoli cells. The recent data suggesting that antibodies to rat Sertoli cell secretory proteins recognize antigenic sites in developing spermatids and spermatozoa suggest structural proteins could also be transported (Kierszenbaum et al., 1988). Chapters 7, 8, 9, and 15 provide a detailed discussion of cell–cell interactions and the cell biology of several of the cell types essential for germ cell differentiation.

B. At the DNA Level

The patterns of DNA synthesis in premeiotic and meiotic testicular cells differ considerably, being both predictable and novel. In the sexually mature testis, normal cell cycle-regulated DNA synthesis occurs continuously in cells such as the spermatogonia, whereas cell types such as the Sertoli cells show little, if any, DNA synthetic activity after puberty. Other cell types, such as pachytene spermatocytes, do not replicate any DNA but do synthesize small amounts of DNA. This synthesis is presumably repair DNA synthesis, a result of recombination. In the preleptotene spermatocyte, the final semiconservative DNA replication in the cycle of male germ cell differentiation occurs. These cells have an unusually long S phase period compared with normal replicating cells. This phenomenon is not unique to mammals but has also been observed in amphibians, such

Table 16–3. Representative Genes That Are Widely Expressed in the Testis

Genes	Reference
DNA polymerase β	Alcivar et al., 1992
Ornithine decarboxylase	Alcivar et al., 1989
α Tubulin	reviewed in Hecht, 1986
β Tubulin	reviewed in Hecht, 1986
β Actin	Waters et al., 1985
γ Actin	Waters et al., 1985
Genes encoded by mitochondrial genome	Alcivar et al., 1989
c-*raf*	reviewed in Propst et al., 1988
jun D	Alcivar et al., 1991
c-*abl*	reviewed in Propst et al., 1988
Plasminogen activator	Huarte et al., 1987
Nerve growth factor[a]	Ayer-LeLievre et al., 1988
Insulinlike growth factor I precursor[a]	Casella et al., 1987
Cholecystokinin	Persson et al., 1989

[a]Temporal expression not established.

as *Triturus,* where the increased length of the preleptotene DNA replication period is believed to be due to a reduction in the number of initiation sites for replication in the chromosomal DNA. Although the physiological significance of this extended S phase is not known, it may be a result of chromosomal changes that reprogram the germ cell nucleus as it prepares to enter meiotic prophase. Moreover, in *Lilium* it has been shown that a small percentage (approximately 0.3 percent) of the genome is not replicated in the preleptotene stage of meiotic prophase but is synthesized later in the zygotene stage of meiotic prophase. This late DNA synthesis is believed essential for chromosome pairing between homologues. Chapter 13 discusses the multiple DNA synthetic intervals of meiosis in more detail.

1. REPLICATION ENZYMES

Eukaryotic DNA replication is dependent on several replicative enzymes and a host of additional proteins essential for DNA strand unwinding, single strand DNA stabilization, synthesis, initiation, and termination. Although comparative studies with different cell types of the testis offer an excellent system to relate the enzymes and factors required for semiconservative DNA replication to those essential for DNA recombinational repair, this avenue of research has not been explored leaving us with only a rudimentary knowledge of these events in the testis (reviewed in Hecht, 1986). We do know that high levels of DNA polymerase α, an enzyme responsible for DNA elongation synthesis, have been found in spermatogonia, preleptotene spermatocytes, and in pachytene spermatocytes (Grippo et al., 1978; Hecht et al., 1978). DNA polymerase β, an enzyme believed to play an important role in DNA repair, is found at high levels in both meiotic and postmeiotic cells. DNA polymerase γ, an enzyme involved in mitochondrial DNA synthesis in somatic cells, has been isolated from spermatozoa. DNA polymerase γ is presumably involved in mitochondrial DNA replication throughout spermatogenesis.

Additional components essential for DNA synthesis in the testis that need investigation include the primase that synthesizes RNA primer and the cellular factors involved in DNA unwinding, ligating, and proofreading functions. Moreover, our lack of knowledge relating the important cyclin proteins, topoisomerases, and DNA polymerases δ and ϵ to the DNA replicative and repair events of the testis needs to be remedied.

2. CHROMATIN

The chromatin compositions of mammalian testicular cells are among the most varied of any tissue. In addition to cell types containing the usual nucleosomal somatic histones, certain testicular cells also contain numerous testis-specific or testis-enriched histones, several high mobility group (HMG) proteins, a group of transition proteins, and protamines (Bhatnagar et al., 1983, reviewed by Balhorn, 1989). The chromatin of the germ cells undergoes a major shift in function during mid-spermiogenesis when the condensational events of the spermatid nucleus terminate transcription. This cessation of RNA synthesis in the male germ cell nucleus is maintained until after fertilization when the protamines of the male pronucleus are replaced by histones. Although dogma argues that the termination of transcription after steps 8 to 9 of spermiogenesis is absolute, mouse sperm nuclei contain RNA polymerase II (Hecht and Williams, 1978) and sperm nuclei of certain species contain residual amounts of histones (Balhorn, 1989). The nuclei of human spermatozoa, for example, contain up to 10 percent histone. A determination whether the histones of spermatozoa are localized to specific genes or chromosomal locations would help define whether certain genes could be more accessible for transcription in the sperm nucleus or soon after fertilization (Gatewood et al., 1987). Data are also badly needed on the mechanism of chromosome condensation during spermiogenesis. For instance, do chromosome condensation and compaction always follow a specific chromosome order? Are there "late" expressed genes? Do specific chromosomes always occupy specific sites in the sperm nucleus? Is there a specific chromosome order of protamine displacement after fertilization?

The mechanism by which transcription terminates in the testis also merits investigation. Transcription ceases in mouse spermatids long before the protamines or even the transition proteins are synthesized indicating that other factors must mediate these changes. Although the loss of nucleosome structure is temporally correlated with termination of transcription, how these structural changes in chromatin lead to transcription termination needs to be determined.

In somatic cells the chromatin structure of certain expressed and nonexpressed genes has been shown to differ with regard to susceptibility to DNase I digestion, solubility, composition, and amount of H1 histones and other postsynthetically modified histones, and DNA methylation. In the testis, chromatin differences have been reported for the trout protamine genes (Nickel and Davie, 1989) and among genes that are constitutively expressed, tissue specific, and inactive in chicken sperm chromatin (Groudine and Conklin, 1985). An increased database relating chromatin structure and active and inactive genes in the mammalian testis is needed.

3. DNA METHYLATION

In mammalian DNA the CpG dinucleotide sequence is sometimes methylated at the deoxycytosine residue. Changes in methylation state of genes have been proposed to modulate chromatin structure and function and to alter the transcriptional state of genes between active and inactive forms (reviewed in Cedar, 1988). Although reduced methylation at specific sites of a gene is often correlated with active transcription of the gene, there are numerous exceptions and it is apparent that methylation is only one of many interacting regulatory elements that control transcription.

In general, tissue-specific genes of somatic cells, such as β globin, ovalbumin, and vitellogenin, are methylated at more CpG sites in spermatozoal DNA than in their tissue of expression (Groudine and Conklin, 1985). Moreover, DNA from the testes of 18-day-old prepuberal mice or from adult mice have identical methylation states for the three somatic tissue-specific genes, β globin, pancreatic amylase, and a histocompatibility gene, and for the testis-specific gene, LDH C4 (Rahe et al., 1983). This suggests these genes undergo no methylation changes in meiotic or postmeiotic cells. In contrast, several satellite DNA sequences are less methylated in the DNA of spermatozoa than in somatic DNAs (Ponzetto-Zimmerman and Wolgemuth, 1984). This undermethylation of satellite DNAs appears to occur very early in spermatogenesis as primitive type A, type A, and type B spermatogonia show identical undermethylation states. In the rooster, meiotic and postmeiotic cells contain about 30 percent less methylcytosines than premeiotic cells, indicating a genome-wide reduction of DNA methylation as spermatogenesis proceeds (Rocamora and Mezquita, 1989). Studies analyzing three postmeiotically expressed testis-specific genes, transition protein 1 and protamines 1 and 2, reveal methylation changes during germ cell differentiation in the mouse testis (Trasler et al., 1990). The changes are gene-specific, with 5′ flanking CpG sites of transition protein 1 becoming less methylated, whereas equivalent CpG sites for the protamines 1 and 2 become more methylated as spermatogenesis proceeds. Analysis of DNA prepared from isolated populations of male germ cells indicates that these methylation changes occur in spermatogonia and in meiotic prophase. Although methylation changes are generally believed to be coupled to DNA replication, both de novo methylation and demethylation events occur for these three genes during meiotic prophase. Perhaps the binding of specific proteins may be an initial event underlying such methylation changes (Meehan et al., 1989). The increase or decrease in methylation suggests that for TP1 and the protamines, the methylation or demethylation of DNA at the specific sites examined is not directly correlated with gene activation.

Differential methylation also provides a mechanism to imprint regions of a genome in the germ line (Reik et al., 1987; Sapienza et al., 1987; Swain et al., 1987). In studies of the in vivo effect of an activated *c-myc* oncogene, a transgenic mouse strain was produced with an unexpected inheritance pattern (Swain et al., 1987). Progeny inheriting the transgene from the male parent expressed the transgene as expected in the heart, whereas progeny inheriting the transgene from the female parent did not express the transgene. This pattern of expression precisely correlated with the imprinted methylation state of the transgene in all tissues. The methylation of the transgene occurred in passage through the female parent and was lost in the germ cells of the male. These studies suggest that at least for specific genes and/or specific chromosomal locations, DNA methylation plays an important additional role in gene expression.

C. At the RNA Level

The initiation of RNA synthesis is an important control point for regulating gene expression in eukaryotic cells. Profound developmental changes are controlled at the level of differential transcription. In recent years much progress has been made in defining the *cis* acting DNA elements of genes and the protein factors that recognize them. As a result, many sequences have been found that serve as either enhancers (stimulators) or silencers (repressors) for transcription, and a growing number of transcription factors have been characterized (Mitchell and Tjian, 1989). In the testis, differential transcription is likely to play an important role in temporal gene expression as specific groups of mRNAs are differentially transcribed in premeiotic, meiotic, and postmeiotic cells. As more of the regulatory elements of the genes for testicular proteins are identified and sequenced, comparisons of the *cis* acting sequences and the either ubiquitous or testis-specific factors that interact with them can be made. The start of such an analysis is presented below.

1. *cis* ACTING DNA ELEMENTS AND *trans* ACTING FACTORS

At this time (late 1989), promoter sequences have been defined for only a few testicular genes including phosphoglycerate kinase 2 (McCarrey and Thomas, 1987), protamines 1 and 2 (Johnson et al., 1988; Krawetz et al., 1988), transition protein 1 (Heidaran et al., 1989), a testis-specific H2B histone (Kim et al., 1987), and a homeobox gene, Hox 1.4 (Wolgemuth et al., 1989). Preliminary sequence comparisons looking for general consensus regulatory elements among these genes have revealed few common sequences with the exception of the two protamine genes. In the mouse the genes for protamine 1 and 2 (mP1 and mP2) are single copy genes closely linked on chromosome 16. They code for two distinct, small, arginine-rich proteins. Despite their frequent use of arginine codons, the coding regions of the mP1 and mP2 genes differ considerably in sequence, in fact, arg complement DNAs (cDNAs) for mP1 and mP2 do not cross-hybridize. There are, however, several common sequences present in their 5′ flanking regions and a strongly conserved sequence adjacent to the polyadenylation site in the 3′ untranslated region of the mP1 and mP2 mRNAs (Fig. 16–2). This latter sequence is present in every protamine gene or cDNA sequenced to date.

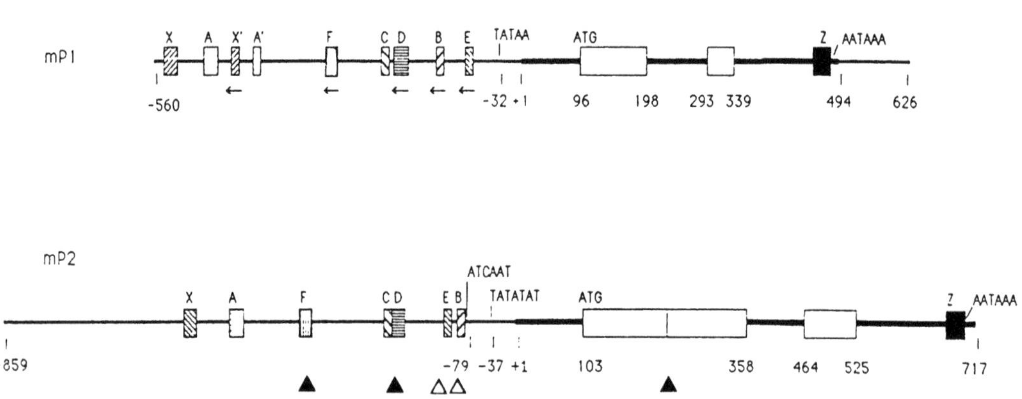

Figure 16–2. Structure of the mouse protamine 1 and 2 genes. *Open boxes* indicate the amino acid coding regions. *Heavy lines* indicate the transcribed regions. *Hatched* and *shaded boxes* represent sequences that are found in both genes in similar positions. An *arrow* under the box indicates that the sequence is present on the complementary DNA strand. Positions indicated in this figure are relative to the mRNA cap sites. The *vertical solid bar* in the mP2 coding region shows the processing site for the mature mouse protamine 2. (Reprinted with permission from Johnson et al., 1988.)

From the small number of genomic sequences known for testis-specific genes, it is impossible to draw any conclusions about testicular or stage-specific consensus regulatory sequences. It is perhaps not surprising that the conserved sequences in the 5′ flanking regions found in protamine 1 and 2 are not present in genes like PGK-2, the testicular H2B histone, and Hox 1.4, which are transcribed during meiosis. There is, however, also a lack of most of the protamine conserved sequences between transition protein 1 and the two protamines, three postmeiotically expressed proteins. This may indicate that: (1) it is simplistic to assume that identical regulatory sequences exist for coordinately regulated genes; (2) *cis* acting sequences are exceptionally fine tuned and specific for each gene; (3) molecular identity may exist in a form involving similar conformation, not necessarily similar sequence; or (4) different sequences may be recognized by the same factors that control temporal expression. It is also possible that TP1 and the protamines are not under the same temporal control. Recent in situ hybridization studies in mouse and rat indicate that TP1 mRNA can be detected one step of spermiogenesis before the protamine mRNAs suggesting that TP1 initiates transcription before the protamines (Mali et al., 1989).

Many other controlling factors in addition to the interaction of *trans* acting factors and *cis* acting DNA elements regulate transcription. For instance, the rat testicular H2A and H2B histones share a common 5′ upstream sequence of approximately 200 nucleotides (Hwang et al., 1990). This region contains S phase-specific transcription elements common to DNA replication-dependent histones despite the fact that the testicular H2A and H2B histones are DNA replication independent and are believed to be specific to pachytene spermatocytes. Clearly, the state of methylation and perhaps the existence of chromatin domains for active and inactive genes are also likely to modulate the expression of these testicular histone genes.

Numerous protein factors that activate transcriptional initiation of RNA polymerase II have been described (reviewed in Mitchell and Tjian, 1989). Many of these DNA-binding proteins function as sequence-specific and cell-specific transcription factors. It is likely cell-specific testicular transcription factors will be found. Sequence knowledge of additional testis-specific genes will provide a growing database to identify stage-specific factors and sequences that could temporally regulate the expression of genes in the testis. The in vitro transcription studies and transgenic mouse studies described below represent a start at understanding the molecular basis for transcription regulation in the testis.

2. TESTICULAR GENE TRANSCRIPTION IN VITRO

Regulation of transcription in developmental programs of higher eukaryotic cells is mediated by the frequency and sites of synthesis initiation of mRNAs. Factors that interact with specific DNA sequences in gene promoters control this process. In vitro transcription systems, whereby protein extracts are prepared from isolated nuclei, provide a useful methodology to identify tissue-specific and cell-type specific factors that modulate the initiation of transcription by RNA polymerase II. The lack of a permanent spermatogenic germ cell line and the short life span (a few days) of dissociated testicular cells in culture make it difficult to develop transfection assays with male germ cells to define critical regulatory sequences of testicular genes. Analyzing genes in protein extracts from premeiotic, meiotic, and postmeiotic intervals by in vitro transcription has been used as an alternate procedure to dissect transcriptional control during spermatogenesis.

Several genes have been analyzed in transcriptionally active extracts prepared from mouse and rat testes (Bunick et al., 1990a, b). In contrast to in vitro transcription systems from somatic tissues such as liver, brain, and spleen, in vitro transcription in a testicular extract has a unique temperature optimum. Maximal transcription in adult testis extracts occurs at 20°C, whereas somatic in vitro systems favor 30°C. Extracts prepared from 16-day-old testes, a stage in prepuberal development where meiosis has been initiated but no postmeiotic cells have yet differentiated, show the 30°C transcription optimum of somatic cells suggesting that the factor(s) responsible for the lowered temperature optimum is present in later stage germ cells. The "somatic-like" temperature optimum of the 16-day testis extract suggests the synthesis or modification of an existing factor(s) after meiosis. Whether such a factor is unique to postmeiotic cells of the testis remains to be determined. The change in in vitro transcription temperature optima in extracts prepared from prepuberal and sexually mature testes parallels in vivo temperature sensitivity of developing male germ cells and may be a contributory factor to the lowered temperature requirement for spermatogenesis.

A second novel feature of the rodent testicular in vitro transcription system is the favored in vitro transcription of a testis expressed gene promoter, the mouse protamine 2 promoter, over somatic cell-specific promoters. In contrast to the protamine 2 gene, both γ-fibrinogen, a liver-specific gene, and a fetal ∈ globin gene transcribe better in a Hela cell extract than in a testis extract.

To further define why the protamine 2 gene seems to have a greater promoter effectiveness in the testicular extract, a deletion analysis of the protamine 2 promoter was undertaken. Removing several upstream elements of mP2 sequentially, a positive promoter element that increased transcription was identified at −170 to −82 nucleotides, whereas further upstream a larger region of negative influence was observed. This negative regulatory region was only active in a heterologous extract prepared from Hela cells suggesting that transcription of the mP2 gene may involve a release from repression in the sexually mature testis but not in somatic cells. The factors that recognize the negative and positive elements of the protamine 2 gene remain to be defined.

3. AN EVOLUTIONARY APPROACH TO UNDERSTAND THE REGULATION OF PROTAMINE TRANSCRIPTION: WHY DO MOUSE SPERMATOZOA AND RAT SPERMATOZOA DIFFER IN PROTAMINE CONTENT?

A comparison of protamine gene expression in the rat and mouse provides a useful model to understand gene regulation in the testis. Unlike the mouse where both protamine 1 and 2 are found in spermatozoa (in a ratio of 1:2), rat spermatozoa only contain protamine 1. The absence of protamine 2 in rat spermatozoa appears to be due to a differential expression of the protamine 2 gene in mouse and rat because the rat genome contains a gene coding for protamine 2 (Tanhauser and Hecht, 1989) and low levels of rat protamine 2 mRNA are found on polysomes (Bower et al., 1987). Isolation and sequence analysis of the rat protamine 2 (rP2) gene reveals the coding regions of the mouse and rat protamine 2 genes share over 90 percent identity (Fig. 16–3). Notable differences are detected, however, at several sites in the 5′ flanking regions of the mouse and rat protamine 2 genes.

Comparing the upstream sequences of mouse and rP2 genes, there are several deletions and numerous base pair changes between the genes. To determine whether such sequence differences could be contributing to the low level of rP2 mRNA expression, promoter strengths were analyzed in testicular in vitro transcription systems. When equimolar amounts of equivalent rat and mouse protamine 2 (mP2) promoters are competitively transcribed in vitro, the rP2 promoter transcribes at only about 30 percent of the efficiency of the mP2 promoter, suggesting that under in vitro assay conditions the gene sequence differences do reduce the ability of the rat gene to be transcribed (Bunick et al., 1990b). The specific sequences responsible for these differences need to be determined. Preliminary results from in vitro protein binding studies indicate that at least one protein–DNA complex is formed with the mP2 positive regulatory region that is not formed with the rP2 promotor. Equal levels of in vitro transcription of the mouse or rP2 promoters in either mouse or rat testicular extracts suggest that both rat and mouse testes contain the *trans* acting factors essential for protamine 2 in vitro transcription. Although the in vitro and in vivo data both show reduced levels of rP2 mRNA, one must be cautious when extrapolating in vitro data to in vivo observations because mRNA turnover does not occur under the in vitro assay conditions. These data, however, make it tempting to postulate that the reduced level of rP2 mRNA is at least partially the result of reduced transcription of the gene in vivo.

Although present at a low level, the rP2 mRNA on polysomes predicts that low levels of the protamine 2 protein should be synthesized. The absence of any mature rP2 in spermatozoa may be the result of a few critical amino acid differences in the coding regions of the mouse and rP2 genes. In the mouse, the first 43 amino acids of protamine 2 are cleaved from the initial translation product in several steps in the nucleus of the maturing spermatid (Balhorn et al., 1984). One of the amino acid differences between the mouse and rP2 gene sequences is at amino acid position 43 of the mouse (Fig. 16–3). This site, where the histidine of the mouse is replaced by glutamate in the rat, is the amino terminus of the mature mouse protamine 2. The detection of low levels of an rP2 precursor in an extract of total sperm basic nuclear proteins from epididymal spermatozoa (Bunick et al., 1990a,b) suggests the rP2 precursor is synthesized, but at a very low level, results consistent with the low level of rP2 polysomal mRNA detected (Bower et al., 1987). The absence of any mature protamine 2 in ejaculated rat spermatozoa suggests that the precursor is not processed. How a putative mutation at one or more processing sites results in the loss of the rP2 precursor from the spermatozoa is not known.

4. TESTICULAR TRANSCRIPTION IN CULTURED CELLS

Much of our general understanding of *cis* acting regulatory elements of eukaryotic genes has been derived from transient assays of DNA constructs that are transfected into cultured cells. Using such assays, constructs containing deletions, additions, and sequence perturbations can be quickly assayed for relative transcriptional abilities and regulatory sequences can be identified. Relevant to the genes discussed in this chapter, this approach has been well utilized with the rat somatic cytochrome c gene (Evans and Scarpulla, 1988, 1989). Using transfection assays in kidney cell lines, three major regions of the rat cytochrome c_s gene that are required for maximal transcription have been defined. One region is an intragenic controlling element in the 5′ end of the first intron of cytochrome c_s, whereas the other two regions are present in the 5′ upstream promoter element. Mutations in each of the regulatory regions diminish cytochrome c promoter activity and eliminate the specific binding of cognate nuclear factors including nuclear respiratory factor 1 and transcriptional activators ATF and Sp1. The ATF recognition sequence in the upstream promoter is identical to a c-AMP response element and confers a cAMP response to the gene.

Similar studies have been conducted with testicular cell types that can be maintained in culture such as the Sertoli cell. Day and Lee (1989) have expressed a gene for a stress inducible protein in a Sertoli cell line. Glucose-regulated proteins (GRP) are members of a family of stress-inducible proteins that share homology with the 70,000 dalton heat shock protein, HSP70. Using a rat promoter for a GRP78 gene in a mouse Sertoli cell line, testosterone has been shown to serve as a transcription inducer (Day and Lee, 1989). Moreover, cellular factors present in Sertoli cells under either noninduced or calcium ionophore induced conditions, bind specifically to a region of the GRP78 promoter previously shown to be essential for basal level expression and induction of transcription. Such studies will help identify the multiple factors and/or the posttranslational modifications of existing factors that regulate GRP78 transcription.

Unfortunately, a similar approach cannot be followed by those wishing to examine the regulatory sequences of genes expressed in germ cells. Despite the availability of excellent cell separation procedures for germ cells (Romrell et al., 1976; Meistrich et al., 1981), the dissociated germ cells (before or after separation) only survive in culture for a few days suggesting they require additional metabolic components and/or cellular interactions. Efforts to co-culture germ cells with Sertoli cells offer promise as a means to improve the viability of germ cells in culture (Stefanini et al., 1980; Djakiew and Dym, 1988). The research group that establishes the culture conditions to maintain primary cultures of male germ cells or develops a procedure to transform male germ cells would earn the eternal gratitude of many investigators studying the regulation of gene expression during spermatogenesis.

5. TESTICULAR TRANSCRIPTION IN VIVO

The introduction of foreign genes into the germ line of an organism provides a powerful tool to modify its genetic endowment. Applications of transgenic technology are manyfold, ranging from the analysis of sequence elements

```
                                                 -400
GCCAGTGCTAGGGCTGACAGCACAGACTGGGCTAAGAGTTGGTAGAGTGCTTCCTCGAATGCCCA
    *  *  **  **   *   **  *  **** ********        **************
TAATGCACATGGAGTGGGTGGGAAGCTTTTGCTAGGAGTTGGTGAGTGCTCTCCTCGAATGCCCA

ACCAAC---------------------------------------GCATGGCTGGTAGCAAACGT
*****                                        ******  *    ** * **
ACCAAGGCCATGCATGGGCTGCATCTCCAACACTGCATGTCCAGGGCATGGAAGCACACATATGT

       -350
GCAATTCCAGCCCTTGG-AGGTGAAGAGACCAGGGTTAGAAGTTCAA------------------
*  ****** *** *** **** *** ********************
G--ATTCCAACCCCTGGGAGGTAAAGGGACCAGGGTTAGAAGTTCAAGGTCATTCTTCATTGTGT

             -300
--GGCTGTCTCACATTAAATAAGTCAGCATGCTACAGGACCTCAAGGCAAGACGAGTAACTTGGA
  ***  ********* **** ***********        **** ****** ***********
GAGGCCATCTCACATTCAATA-GTCAGCATGCT--------TCAAAGCAAGATGAGTAACTTGGC

                                                -200
CCCTACTCCCAATCTGCCAGCCAGTGCTGCAAACTCTGTGTTGCCCTCACAGAGGAGACTGGGCA
*****             ******* ******** *****  ************* *********
CCCTA-------------AGCCAGTCCTGCAAACCCTGTGCCGCCCTCACAGAGGGGACTGGGCA

                                           -150
GGGTGGGAACAATCAATCAGGGGTGGGCCGACAGGTCACAGTGGGGTTCACCTTTATATATGAGC
************************************************ ****************
GGGTGGGAACAATCAATCAGGGGTGGGCCGACAGGTCACAGTGGGGTTTACCTTTATATATGAGC

                            -100
CCTCTGAGAGCCCCCCACACACACACCAGACCATCATCACCACCAAGAGCAGGTGGGCAGGCTAC
**************  ****      *************************************
CCTCTGAGAGCCCCAAACAC------CAGACCATCATCACCACCAAGAGCAGGTGGGCAGGCTTT

              -50
CGTCCCTCCTCTTCCAATCCAGGTCAGCTGCAGCCTCAACCCAGAACCTCCTGACCTCCTGGCAC
*********** *************************** ************** **********
CGTCCCTCCTCCTCCAATCCAGGTCAGCTGCAGCCTCAATCCAGAACCTCCTGATCTCCTGGCAC

                                                  Gln
T ATG GTT CGC TAC CGA ATG AGG AGC CCC AGT GAG GGT CAG CAC CAG GGG
  *** *** *** *** *** *** *** *** *** *** *** *** * * *** *** ***
C ATG GTT CGC TAC CGA ATG AGG AGC CCC AGT GAG GGT CCG CAC CAG GGG
 +1   Val Arg Tyr Arg Met Arg Ser Pro Ser Glu Gly Pro His Gln Gly

  50                                                      Glu
  CCT GGG CAA GAC CAT GAG CGC GAG GAG CAG GGG CAG GGG CAA GAG CTG
  *** **  *** *** *** **  *** *** *** *** *** *** *** *** * * ***
  CCT GGA CAA GAC CAT GAA CGC GAG GAG CAG GGG CAG GGG CAA GGG CTG
  Pro Gly Gln Asp His Glu Arg Glu Glu Gln Gly Gln Gly Gln Gly Leu

    100                                       Glu
  AGC CCA GAG CGC GTG GAG GAC TAT GGG AGG ACA GAA AGG GGC CAC CAC
  *** *** *** *** **  *** *** *** *** *** ***  *  *** *** *** ***
  AGC CCA GAG CGC GTA GAG GAC TAT GGG AGG ACA CAC AGG GGC CAC CAC
  Ser Pro Glu Arg Val Glu Asp Tyr Gly Arg Thr His Arg Gly His His
                                                 ↑
```

```
        150
CAC --- AGA CAC AGG CGC TGC --- --- AAG AGG CTT CAC AGG ATC CAC
***     *** *** *** *** ***         *** * * **  **  *** *** ***
CAC CAC AGA CAC AGG CGC TGC TCT CGT AAG AAG CTA CAT AGG ATC CAC
His His Arg His Arg Arg Cys Ser Arg Lys Arg Leu His Arg Ile His

                       200
AAG AGG CGC CGG TCA TGC AGA AGG CGG AGG AGA CAC TCC TGC CGC CAC
*** *** **  *** *** *** *** *** *** *** *** *** *** *** *** ***
AAG AGG CGT CGG TCA TGC AGA AGG CGG AGG AGA CAC TCC TGC CGC CAC
Lys Arg Arg Arg Ser Cys Arg Arg Arg Arg Arg His Ser Cys Arg His

                          250
AGG AGG CGG CAT CGC AGA G taagcacccagtagccaagtcccccgctacctctgct
*** *** *** *** *** *** * *********   **** *  ***  ** **** ****
AGC AGG CGG CAT CGC AGA G taagcaccccacagccgaccccctggccacctgtgct
Arg Arg Arg His Arg Arg

          300                                                350
gctgcccacgaaggcagggtgcccctctccaaaccctgttgccttccgggcagcctcgcaaacct
*****                   ***  ** ******** ******** * *     *******
gctgctg---------------ccccatctaaaccctgctgccttccagacgtaga--caaacct

                                                      Ser
gaactttctctacccag  GC TGC AGA AGA TCC CGA AGG AGG AGG AGC TGC
* ***** ** *    ***  ** *** *** *** *** *** *** *** *** **  ***
ggactttcctttgtacag  GC TGC AGA AGA TCC CGA AGG AGG AGG AGA TGC
                   Gly Cys Arg Arg Ser Arg Arg Arg Arg Arg Cys

                        Trp     Tyr Tyr             450
AGG TGC AGG AAA TGC AGG TGG CAC TAT TAT TAA GCCTCCCCAGGTCGATCCAT
*** *** *** *** **  ***  ** ***  **   *  *** *********** *  *****
AGG TGC AGG AAA TGT AGG AGG CAC CAT CAC TAA GCCTCCCCAGGCCTGTCCAT
Arg Cys Arg Lys Cys Arg Arg His His His

                                                    500
TCTGCCTGGAGCTAAGGAAGTCACCTGCCTCAGGAA--GTCACCTGCCCAAGCAAAGTCATGTA
************ *********** ****  *****  *****************  ******
TCTGCCTGGAGCCAAGGAAGTCACTTGCCCAAGGAATAGTCACCTGCCCAAGCAACATCATGTG

                                      550
AGGCCACAACACCATTCCATGTTGATATCTGAGCCCTGAGCTGCCAAGGAGCCACGAAATCTAA
******** ************* *** ******************************* **** *
AGGCCACACCACCATTCCATGTCGATGTCTGAGCCCTGAGCTGCCAAGGAGCCACGAGATCTGA

                  600
GTAATGAGCAAAGCCACCTGCTAAATAAAGCTT
*** ***************** ***********
GTACTGAGCAAAGCCACCTGCCAAATAAAGCTT
```

Figure 16–3. Sequence comparison of the rat and mouse protamine 2 genes. Top line: rat sequence; bottom line: mouse sequence. The CAAT and TATA boxes are *underlined.* +1 indicates the start of transcription. The *arrow* indicates the processing site for the mature form of the mouse protamine 2. *Asterisks* indicate identity and *dashes* indicate deletions. At sites where the protein sequences between the rat and the mouse genes differ, the rat amino acid is written above its sequence. (Modified from Bunick et al., 1990.)

essential for gene regulation in a tissue to functional analysis of gene products through insertional mutagenesis or endogenous overexpressed or underexpressed molecules. Using the transgenic mouse model, the elegant work from Richard Palmiter and colleagues has answered important biological questions in spermatogenesis as well as partially delineated the transcriptional and translational regulatory signals of the protamine 1 gene (Peschon et al., 1987; Braun et al., 1989a,b).

Although haploid gene expression in the mammalian testis has been well established, the physiological significance of haploid gene expression is debatable. The transmission ratio distortion of t-haplotypes in mice, where a particular t-haplotype is transmitted at frequencies far above the predicted mendelian inheritance, argues that postmeiotic expression could confer unique properties on individual spermatozoa (Silver, 1985). Other have proposed that the 1 μm in diameter intracellular bridges between germ cells allow movement of the allelic gene products between spermatids. Recently, Braun and co-workers (1989a) have used the transgenic mouse model to examine this question.

Producing transgenic mice with a construct containing the mouse protamine 1 promoter and a human growth hormone reporter gene, Braun and colleagues used in situ hybridization to localize transcripts from the transgene. They found in both homozygous and hemizygous progeny, over 90 percent of the late spermatids contained human growth hormone mRNAs. These results indicate that the human growth hormone mRNA readily diffuses through the intracellular bridges, an event that negates specific cellular localization for each specific gene allele expressed in postmeiotic cells. If the human growth hormone mRNA is representative of the endogenous testicular mRNAs expressed in the spermatids, one would predict that all (most?) gene products are shared among spermatids. Whether exceptions to this mechanism lead to the non-mendelian inheritance of genes such as those of the t-haplotypes needs to be established.

Insight into essential *cis* acting elements of testis genes has also been provided by transgenic mice studies. In experiments where the mouse protamine 1 promoter has been used, a region no greater than and perhaps less than −465 nucleotide to the start of transcription has been shown to be sufficient for tissue and cell-type specific transcription (Peschon et al., 1987; Braun et al., 1989a,b). Similar studies examining the promoter of mouse protamine 2 have shown that a 859-base-pair sequence contains all the information needed for tissue-specific protamine expression of the transgene (Stewart et al., 1988). It is likely that smaller 5′ flanking regions for both protamine 1 and 2 will be sufficient to provide the elements needed for postmeiotic gene expression in the testis. Surprisingly, transgenes with the protamine 1 promoter also are expressed in the heart and temporal bone of mice (Behringer et al., 1988). Transgenic mice were generated from a chimeric gene composed of the SV40 T-antigen gene attached to the 5′ and 3′ flanking sequence of mP1. The transgene was expressed in round spermatids but no testicular pathology was observed. The transgene, however, was also expressed in the heart where it produced rhabdomyosarcomas in the right atrium. Bilateral osteosarcomas were also found within the petrous portion of the temporal bone. The expression of mP1 constructs in the heart and inner ear of mice is intriguing and may be the result of an aberrant cellular response to the particular construct used. Such ectopic gene expression should be explored further as the unexpected expression of a testis promoter in the heart and ear may provide important insight into common enhancer elements or *trans* acting factors.

Transgenic mice also provide a system to assay the function of cellular products by either their overexpression or inhibition of expression. Recently, a testis-isolated promoter for a homeobox gene, Hox 1.4, which shows its highest level of expression in adult animals in developing male germ cells, has been used to overexpress its product (Wolgemuth et al., 1989). A construct containing approximately 10 kb of sequence of the homeodomain of Hox 1.4 and about 400 base pairs of the 3′ untranslated region of the SV40 T-antigen coding region and poly (A) addition site was used to make transgenic mice. Although expression of the transgene was lower in the testis than the endogenous gene in two lineages, the construct exhibited the appropriate tissue specificity of expression. Most importantly, transcript levels were elevated in the embryonic gut, a tissue that normally expresses this gene. The overexpression of the homeobox gene product in the gut resulted in the inherited abnormal phenotype known as congenital megacolon, a condition where animals have difficulty extruding feces through the most distal portion of their colon. Future studies where testicular genes are either overexpressed or ectopically expressed need to be conducted with testicular isozymes, structural protein isotypes, and proto-oncogenes.

In an alternative to overexpression of a gene product, the expression of specific genes can be repressed in prokaryotes or eukaryotes by the formation of antisense RNA (RNA complementary in sequence to regions of an mRNA) (reviewed in Melton, 1988). The antisense RNA to a specific gene prevents utilization of the mRNA by an unknown mechanism. The expression of antisense mRNAs targeted to specific testicular isozymic variants in transgenic mice would offer an exciting means to inactivate genes. The paucity of publications using this procedure, however, suggests its successful application may only be limited to special situations. A notable success is the recent conversion of phenotypically normal mice heterozygous for the shiverer mutation to the mutant shiverer phenotype (Katsuki et al., 1988). Antisense RNA was produced to the myelin basic protein (MBP) in transgenic mice using about 1.3 kb of the mouse MBP gene and a portion of the rabbit β globin gene carrying the mouse MBP cDNA in the antisense orientation. Depending on the amount of MBP produced, different gradations of the shiverer phenotype are observed. It is likely that this study was successful because it was possible to detect an alteration in phenotype after a reduction in MBP mRNA level rather than the complete inactivation of all MBP mRNA. Because for many genes, especially enzymes, a very low level of product is still sufficient for normal function in the progeny, this approach may have only limited use.

D. Regulation by Alternative Transcriptional Initiation and mRNA Processing in the Testis

Many testicular transcripts of novel size have been detected. Among the proto-oncogenes, the sizes of testicular mRNAs for c-*mos*, c-*abl*, and *pim*-1 differ from corresponding mRNAs in other tissues (reviewed in Propst et al., 1988). Differences in c-*mos* mRNA size are even seen between male and female germ cells with a c-*mos* transcript of 1.7 kb in testis and a 1.4 kb c-*mos* mRNA in ovary. Although both mRNAs have the same open reading frame and thus code for the same protein, their 5′ untranslated regions differ in length. These differences are likely due to the use of alternate sites for transcription initiation. Such differences in mRNA length may allow differential translational regulation for c-*mos* in male and female germ cells.

Size differences are also seen for pro-opiomelanocortin (POMC) and proenkephalin mRNAs (Gizang-Ginsberg and Wolgemuth, 1985; Kilpatrick et al., 1987). POMC exists as a heterogenous mRNA of about 800 nucleotides in the testis and 1.5 kb in the pituitary, whereas a 1.7-kb proenkephalin mRNA is present in spermatogenic cells and a 1.45-kb mRNA in somatic cells. A smaller testicular prodynorphin mRNA is also expressed in Sertoli cells (Garrett et al., 1989).

The c-*abl* proto-oncogene expresses mRNAs of about 7kb, 6kb, and 4.7 kb in the testis (reviewed in Wolgemuth et al., 1989). Although the two larger size mRNAs are also present in somatic tissues, the 4.7-kb mRNA is only found in postmeiotic testicular cells. A detailed characterization of the testicular c-*abl* mouse mRNAs has revealed that the 4.7-kb postmeiotic mRNA arises from the use of an alternative polyadenylation site of a transcript initiated from the same promoter as the 6-kb c-*mos* mRNA (Meijer et al., 1987). Although the use of this alternative polyadenylation site deletes over 1 kb from the 3′ untranslated region of the mRNA, the same protein is coded for by the two mRNAs.

Although the use of different transcription initiation sites and alternative polyadenylation sites could modulate the translational efficiency of c-*mos* and c-*abl* transcripts, the mRNAs produced still code for testicular proteins identical to the somatic proteins. This may not be true for all c-*mos* and c-*abl* testicular transcripts as additional short mRNAs that would produce truncated proteins have been identified for these two proto-oncogenes (reviewed in Propst et al., 1988).

Posttranscriptional processing is not restricted to proto-oncogenes and opioid peptide precursors. The cytochromes also undergo extensive postranscriptional processing. The cytochrome c_s gene has three functionally equivalent sites for polyadenylation in all organs (Virbasius and Scarpulla, 1988). In contrast, the single copy rat cytochrome c_t gene has a single polyadenylation site. In addition to multiple start sites for transcription, the heterogeneous mouse cytochrome c_t mRNAs also contain variable poly (A) lengths (Hake et al., 1990). The physiological function(s), if any, of such heterogeneous mRNAs is not known.

In avian round spermatids, histone mRNAs that are polyadenylated 26 to 28 nucleotides 3′ to the usual poly(A)$^-$ terminus are synthesized (Challoner et al., 1989). This alternate processing pathway in postmeiotic testicular cells disregards the usual hairpin structures and purine-rich elements of replication-dependent histones of somatic cells.

E. Translation

In most mammalian tissues, gene expression is believed to be primarily regulated at the level of transcription with occasional modulation at the translational level. This is also likely to be true in the premeiotic and meiotic stages of spermatogenesis. Chromatin changes in the developing male gamete, however, lead to loss of nucleosomes with a concomitant cessation of RNA synthesis during mid-spermiogenesis. By necessity, many of the proteins of the spermatozoon must be either synthesized before mid-spermiogenesis and stored or their transcripts must be stored and translationally regulated. Although it is likely both mechanisms are used, studies with three DNA-binding proteins, transition protein 1 (TP1), and the two protamines (P1 and P2) indicate that translational regulation during spermiogenesis plays a primary role in their expression.

1. RNA STORAGE

In rodents, the transition proteins are synthesized during mid-spermiogenesis while the protamines are synthesized at the termination of spermatogenesis (steps 12 to 15 in the mouse). Both TP1 and the protamines are synthesized after transcription has terminated, requiring in the mouse that TP1 mRNA be stored approximately 3 days and the protamine mRNAs for about 7 days. During these intervals the already processed mRNAs (the TP1 and protamine genes each contain one intron) are stored in the cytoplasm of the spermatid. The mTP1 and mP1 and mP2 mRNAs are not detectable in pachytene spermatocytes but can be detected first in round spermatids where they have sizes of about 600, 580, and 730 nucleotides, respectively (Kleene et al., 1984; Heidaran and Kistler, 1987a, b, 1988; Yelick et al., 1989). In elongating spermatids where the proteins are synthesized, the mRNAs shorten with the TP1 mRNAs ranging from 480 to 600 nucleotides, whereas mP1 and mP2 mRNAs are about 450 and 600 nucleotides, respectively. Fractionation of postmitochondrial supernatants by sucrose gradients reveals the longer mRNAs are in the nonpolysomal fraction, whereas the shortened mRNAs are present on polysomes. Temporally, the shortened mRNAs are first detected at the time the proteins are synthesized. Studies using RNase H and thermal chromatography of mRNAs with poly (U) sepharose have demonstrated that the shortening of the mRNAs is the result of a partial deadenylation to molecules containing about 30 poly (A)s.

2. POLYADENYLATION CHANGES

The shortening of the polysomal mRNAs of TP1 and the two protamines is remarkable but not unique for rodents or for these particular genes. A similar shortening of protamine polysomal mRNAs has been noted in trout (reviewed in Dixon et al., 1985), hamster, rat (Bower et al., 1987), and rooster (Oliva et al., 1988). Other genes including transition protein 2 (Kleene and Flynn, 1987) and a seleno protein of the sperm mitochondrial capsule (HEM 1050) (Kleene, 1989; Kleene et al., 1990) also show similar length reductions that are correlated with translation. The physiological significance of these changes in mRNA length is not known as well as whether the length reductions are a result of mRNA translation or a requirement for translation. In the testis, shortened polysomal mRNAs have only been reported for postmeiotically expressed genes. Reductions in poly(A) length are, however, a general phenomenon of eukaryotic mRNA metabolism (Bernstein and Ross, 1989).

Reductions in mRNA length at the time of translation do not occur for all testicular genes. Ornithine decarboxylase (ODCase) and cytochrome c_t, two genes that are transcribed before spermatogenesis, do not show shortened mRNAs during spermiogenesis (Alcivar et al., 1990; Hake et al., 1990). In fact, cell separation studies suggest that both of the ODCase mRNAs are longer in round spermatids than in pachytene spermatocytes as a result of an increased level of adenylation. Because the ODCase protein is believed to be synthesized in both meiotic and postmeiotic testicular cells, the reason for the increased polyadenylation is not obvious. It may increase the stability of the mRNA as has been reported for vasopressin and human growth hormone. Changes in poly(A)$^+$ RNA length for LDH C4 mRNA have also been noted (Fujimoto et al., 1988).

3. ROLE OF THE 3′ UNTRANSLATED REGION IN REGULATING TEMPORAL EXPRESSION

Analysis of two mP1 chimeric gene constructs in transgenic mice has recently established that the time of translation of the mouse protamine 1 mRNA is under the control of the 3′ untranslated region (UTR) of the mP1 mRNA (Braun et al., 1989b). Two fusion genes were constructed. The first construct contained 4.1 kb of 5′ flanking sequence of the mP1 gene, 91 base pairs of the 5′ UTR of mP1, the human growth hormone (hGH) structural gene, and 156 nucleotides of the 3′ UTR of mP1. The second construct was identical except the mP1 3′ UTR was replaced with the normal 112 nucleotide 3′ UTR of hGH. Immunocytochemical analysis of testes from transgenic mice produced from these constructs revealed hGH was first detected in elongating spermatids at about stage 12 when the construct contained the 3′ UTR of mP1. In contrast, the testes of transgenic animals produced with the construct containing the hGH 3′ UTR, synthesized hGH in round spermatids, at the same time that mP1 mRNA is first detected in the testes and many days before mP1 is translated. The synthesis of hGH in round spermatids rather than in elongating spermatids also resulted in a different subcellular localization of the protein. The early synthesized hGH was localized in the acrosome of the spermatid, whereas the hGH protein synthesized in elongating spermatids accumulated intracellularly. The hGH mRNA with the 3′ UTR of hGH also appeared to have a shorter stability than the hGH mRNA with the 3′ UTR of protamine 1. These studies demonstrated the importance of the 3′ UTR of the mP1 for temporal translational control.

To define the precise regions of the protamine 3′ UTR that help confer temporal translation regulation, specific protein binding to the 3′ UTR of protamine 2 has been examined (Kwon and Hecht, 1991). Using RNA gel retardation and UV crosslinking, two essential *cis* elements that bind protein have been identified between nucleotides +537 and +572 of the mP2 mRNA. A protein of about 18 kD was shown to bind to one of the elements. This protein is present in cytoplasmic extracts prepared from populations of round spermatids (where protamine mRNA is stored), but not in elongated spermatid extracts (where protamine is synthesized). Moreover, protein binding to mRNA constructs containing the protamine 2 3′ UTR, but not the human growth hormone 3′ UTR, represses mRNA translation in a cell-free translation system suggesting mRNA–protein interactions are likely to regulate translation during spermiogenesis.

F. Post-translational Regulation

The regulated synthesis of a protein only represents the start of its cellular regulation. Although the importance of post-translational modifications of a protein is often overlooked by molecular biologists, such changes create innumerable regulatory possibilities. There are two general classes of protein modification. A protein may be modified irreversibly by a mechanism, such as cleavage, or reversibly with enzymes that add and remove a component to the original primary amino acid sequence. In the testes, acrosin and protamine 2 are examples of irreversible protein modifications, whereas proteins such as the protein kinases undergo reversible phosphorylations.

The enzyme acrosin is synthesized as a precursor molecule that must be sequentially cleaved to become functional (Arboleda and Gerton, 1988). The recent cloning of a cDNA for boar preproacrosin has revealed a sequence that codes for 416 amino acids with a hydrophobic leader sequence and the light and heavy chain of proacrosin (Adham et al., 1989). A proline-rich domain in the C-terminal region may be a cellular binding site for the enzyme.

Protamine 2 is a structural protein that also requires proteolytic processing. In the mouse and probably other species whose sperm contain protamine 2, a precursor molecule of nearly double the size of the protamine 2 found in spermatozoa, is synthesized (Yelick et al., 1987). It is processed stepwise in the nucleus of the elongating spermatid. Because no posttranslational cleavage is required for any protamine 1, it is not obvious why protamine 2 is synthesized as a precursor. One can speculate that the N-terminus may be initially essential for proper interaction with the P1 protamine and for correct deposition onto the DNA.

Proteins that are reversibly modified undergo charge modifications that allow them to move to specific intracellular or extracellular locations. Enzyme activities can be stimulated or repressed during stages of the cell cycle or states of differentiation by phosphorylations or dephosphorylations, an increasingly important factor in the modulation of DNA and RNA binding proteins. In the mammalian testis the growing numbers of serine, threonine, and tyrosine protein kinases encoded by proto-oncogenes suggest these modifications will play important regulatory functions during spermatogenesis (reviewed in Wolgemuth et al., 1989).

IV. The Mitochondrion in the Testis

In the mammalian testis, mitochondria exhibit several distinct morphologies as germ cell differentiation proceeds. Spermatogonia and somatic testicular cells contain the traditional "cigar-shaped" mitochondria found in most somatic tissues. During meiosis, mitochondria with diffuse and vacuolated matrices begin to replace the "somatic" mitochondria. By the beginning of spermiogenesis, the "somatic" mitochondria have been totally replaced by the "germ cell" mitochondria. The "germ cell" mitochondria ultimately account for over 80 percent of the total mitochondria in the testes of sexually mature rats. During meiosis, the mitochondria aggregate near the nucleus. Subsequently during spermiogenesis, the "germ cell" mitochondria orient under the plasma membrane and are replaced by the crescent-shaped mitochondria of the spermatozoon. Each spermatozoal midpiece contains approximately 75 mitochondria, structurally unique organelles that retain their α helical shape even after release by sonication. Analysis of the densities and protein composition of the different classes of testicular mitochondria reveals that the mitochondria from spermatozoa are more dense and contain distinct polypeptides absent in other testicular mitochondria. One of these proteins is a selenium containing protein of 17 to 20 kd that maintains and stabilizes the crescent shape of the outer mitochondrial membrane (reviewed in Hecht and Kennington, 1983). A second mitochondrial protein is the unique cytochrome c_t, in the rat differing in 15 of 104 amino acids from the somatic cytochrome c. Cytochrome c_t totally replaces cytochrome c_s in the mitochondria of spermatozoa.

In vivo labeling studies with [^{35}S]methionine indicate that many of the mitochondrial proteins are synthesized over prolonged periods of meiosis and spermiogenesis suggesting that the spermatozoal mitochondria are produced by sequential substitution of new proteins into existing mitochondria rather than the abrupt appearance of a new class of mitochondria during spermiogenesis.

A. *Mitochondrial DNA of the Testis*

Breeding studies with many different vertebrates have demonstrated that progeny always inherit their mitochondria from their mother. The maternal inheritance of mitochondria raises the question "Does the zygote select against the paternal mitochondria and if so, how?" The apparent loss of the paternal mitochondria may be the result of a dilution effect, as less than 10^2 sperm mitochondria enter an egg that contains approximately 10^5 mitochondria. Alternatively, the paternal mitochondria may be "debilitated" or programmed for nonproliferation. Although the current biochemical data are incomplete, the genetic evidence for maternal inheritance suggests that no paternal mitochondria survive in progeny.

Restriction enzyme digestion analysis of mitochondrial DNA (mtDNA) from meiotic and postmeiotic mouse male germ cells (including spermatozoa) indicates there are no gross deletions in the mitochondrial genome during spermatogenesis (Hecht et al., 1984). Between meiosis and the end of spermatogenesis, however, the number of mitochondrial genomes per haploid nuclear genome decreases 8- to 10-fold with spermatozoa containing on average one mitochondrial genome per mitochondrion (most somatic cell types contain multiple copies of mtDNA in each mitochondrion). Despite the fact that there is such a substantial decrease in the total amount of mtDNA as spermiogenesis advances, mtDNA is synthesized during meiosis and early spermiogenesis. Whether this mtDNA replication simply represents normal replacement synthesis or could be a means whereby the spermatozoal mtDNA becomes selectively marked needs to be determined. As found for all other mammalian mitochondrial genomes, there is no evidence for DNA methylation of testicular, spermatozoal, or oocyte mitochondria indicating that the maternal inheritance of mitochondria cannot be mediated by a differential state of methylation of the maternal and paternal mitochondrial DNA as has been suggested for the chloroplast DNA of *Chlamydomonas.*

B. Mitochondrial Gene Expression in the Testis

The mitochondrial genome of the mouse and other mammals is a double-stranded, circular DNA of about 16 kb that codes for two ribosomal RNAs, 22 transfer RNAs, and 13 polypeptides. From sequence analysis of the mitochondrial genome, subunits I, II, and III of cytochrome c oxidase, subunits 6 and 8 of ATPase, cytochrome b, and seven subunits of NADH dehydrogenase have been shown to be mitochondrially encoded (Bibb et al., 1981). The many additional proteins essential to produce a functional mitochondrion are coded for by nuclear genes and are synthesized in the cytoplasm of the cell. Many of these proteins are synthesized as precursors with leader sequences that facilitate transport into the mitochondria. RNA can also be transported into mitochondria. One nuclear RNA transcript, the RNA component of the mitochondrial RNA-processing endoribonuclease, appears to be transported across mitochondrial membranes into mitochondria (Chang and Clayton, 1989). To date, no nuclear encoded mRNAs have been detected in mitochondria. The different structural proteins found in mitochondria of different organs are the result of differential gene expression of nuclear-encoded mitochondrial proteins.

Although much is known about somatic mitochondrial transcription and translation, little is known about these events in germ cells and, specifically, in spermatozoa. Several laboratories have reported that the mitochondria of bovine, rat, and hamster epididymal spermatozoa synthesize RNA (reviewed in Alcivar et al., 1989). However, the radiolabeled 23S and 16S "sperm mitochondrial" RNAs they identified co-migrated with bacterial rRNAs making bacterial contamination seem likely. A report of 28S and 18S radiolabeled RNAs synthesized when mouse epididymal spermatozoa were incubated with [^{3}H]uridine also is suspect as the RNAs co-migrate with eukaryotic ribosomal RNAs and no cytoplasmic ribosomes have been detected in spermatozoa.

To better define the capabilities of testicular mitochondria for protein synthesis, the relative levels of RNA transcripts for several mitochondrial genes in the differentiating testicular mitochondria were examined by Northern blot analysis with specific probes (Alcivar et al., 1989). The levels of RNA for the two mitochondrial ribosomal RNAs and for the mitochondrial mRNAs coding for cytochrome oxidase subunits I and II, cytochrome b, ATPase subunits 6 and 8, and subunit I of NADH dehydrogenase showed no major qualitative or quantitative differences in total RNA extracts prepared from prepuberal and sexually mature testes. When total RNA from individual cell types was examined, reduced amounts of mitochondrial compared with cytoplasmic mRNAs were seen in enriched preparations of RNA from pachytene spermatocytes, round spermatids, and residual bodies. These reductions parallel the reduced amount of mitochondrial DNA/cell found in germ cells from later stages of spermatogenesis.

To see if a programmed deficiency of mitochondrial gene expression in spermatozoa could be the reason for maternal inheritance, RNA was isolated from epididymal mouse spermatozoa. Hybridization analysis indicated that normal transcripts were present for several mitochondrial mRNAs and the two mitochondrial ribosomal RNAs in RNA preparations from the mitochondria of epididymal spermatozoa suggesting the paternal mitochondria are likely to be translationally competent. The reason mitochondria are solely inherited from the mother remains unknown.

V. Concluding Questions

1. What is the molecular trigger that induces spermatogonia to enter the pathway of germ cell differentiation rather than continue in the stem cell renewal pathway?
2. What are the mechanisms controlling chromosome pairing and genetic recombination during meiosis?
3. Are the testicular isozymes essential for spermatogenesis?
4. What information passes from the germ cell to the Sertoli cell? From the Sertoli cell to the germ cell?
5. What is the mechanism that causes the decline of transcription during spermiogenesis?
6. Why do many genes expressed in the testis use transcription start or processing sites that differ from those used in other tissues?
7. Does the mammalian spermatozoon provide any essential epigenetic components to the egg at fertilization?
8. How is chromatin compaction regulated; that is, are the chromosomes condensed and later decondensed following fertilization in a specific temporal and spatial order?
9. How are the stored mRNAs induced to be translated?
10. Why have protamines evolved in mammals when histones work very well in lower eukaryotes? Why do multiple protamine types exist?

ACKNOWLEDGMENTS

The research presented in this manuscript from the author's laboratory was supported by NIH and EPA grants. The author thanks his laboratory research colleagues for their critical readings and many useful comments on this chapter and Valerie Ricciardone for her outstanding secretarial assistance. The author is especially indebted to Professor C.R. Woese who introduced him to molecular biology and Professor H. Stern who introduced him to the wonders of germ cells.

REFERENCES

Adham, I. M., Klemm, U., Maier, W.-M., Hoyer-Fender, S., Tsaousidou, S., and Engel, W. (1989). Molecular cloning of preproacrosin and analysis of its expression pattern in spermatogenesis. Eur. J. Biochem. 182:563–568.

Alcivar, A. A., Hake, L. E., and Hecht, N. B. (1992). DNA polymerase β and poly(ADP)ribose polymerase are differentially expressed during development of male germ cells. Biol. Reprod. 46:201–207.

Alcivar, A. A., Hake, L. E., Hardy, M., and Hecht, N. B. (1990). Increased levels of junB and c-jun mRNAs in male germ cells following testicular cell dissociation: Maximal stimulation in prepuberal animals. J. Biol. Chem. 265:20160–20165.

Alcivar, A. A., Hake, L. E., Millette, C. F., Trasler, J. M., and Hecht, N. B. (1989). Mitochondrial gene expression in male germ cells of the mouse. Dev. Biol. 135:263–271.

Alcivar, A. A., Hake, L. E., Mali, P., Kaipia, A., Parvinen, M., and Hecht, N. B. (1990). Developmental and differential expression of the ornithine decarboxylase gene in rodent testis. Biol. Reprod. 41:1133–1142.

Alcivar, A. A., Hake, L. E., Kwon, Y. K., Goggins, R., and Hecht, N. B. (1991). JunD mRNA expression differs from c-jun and junB mRNA expression during male germinal cell differentiation. Mol. Reprod. Devel. 30:187–193.

Alcivar, A. A., Trasler, J. M., Hake, L. E., Salchi-Ashtiane, K., Goldberg, E., and Hecht, N. B. (1991). DNA methylation and expression of the genes coding for lactate dehydrogenases A and C during rodent spermatogenesis. Biol. Reprod. 44:527–535.

Allen, R. L., O'Brien, D. A., Jones, C. C., Rockett, D. L., and Eddy, E. M. (1988). Expression of heat shock proteins by isolated mouse spermatogenic cells. Mol. Cell. Biol. 8:3260–3266.

Amat, J A., Fields, K. L., and Schubart, U. K. (1990). Stage-specific expression of phosphoprotein p19 during spermatogenesis in the rat. Mol. Reprod. Devel. 26:383–390.

Arboleda, C. E., and Gerton, G. L. (1988). Proacrosin/acrosin during guinea pig spermatogenesis. Dev. Biol. 125:217–225.

Ariel, M., McCarrey, J., and Cedar, H. (1991). Methylation patterns of testis-specific genes. Proc. Nat. Acad. Sci. USA 88:2317–2321.

Ashman, J. B., Hall, E. S., Eveleth, and Boekelheide, K. (1992). Tau, the neuronal heat-stable microtubule-associated protein, is also present in the cross-linked microtubule network of the testicular spermatid manchette. Biol. Reprod. 46:120–129.

Ayer-LeLievre, C., Olson, L., Ebendal, T., Hallböök, F., and Persson, H. (1988). Nerve growth factor mRNA and protein in the testis and epididymis of mouse and rat. Proc. Natl. Acad. Sci. USA 85:2628–2632.

Baarends, W. M., Themmen, A. P. N., Blok, L. J., Mackenbach, P., Brinkmann, A. O., Meijer, D., Faber, P. W., Trapman, J., and Grootegoed, J. A. (1990). The rat androgen receptor gene promoter. Mol. Cell. Endo. 74:75–84.

Balhorn, R. (1989). Mammalian protamines: Structure and molecular interactions, in *Molecular Biology of Chromosome Function* (Adolph, K. W., ed.) pp. 366–395, Springer Verlag, New York.

Balhorn, R., Weston, S., Thomas, C., and Wyrobek, A. (1984). DNA packaging in mouse spermatids. Synthesis of protamine variants and four transition proteins. Exp. Cell Res. 150:298–308.

Beamer, W. G., Cunliffe-Beamer, T. L., Shultz, K. L., Langley, S. H., and Roderick, T. H. (1988). Juvenile spermatogonial depletion *(jsd):* A genetic defect of germ cell proliferation of male mice. Biol. Reprod. 38:899–908.

Behringer, R. S., Peschon, J. J., Messing, A., Gartside, C. L., Hauschka, S. D., Palmiter, R. D., and Brinster, R. L. (1988). Heart and bone tumors in transgenic mice. Proc. Natl. Acad. Sci. USA 85:2648–2652.

Bellvé, A. R. (1979). The molecular biology of mammalian spermatogenesis, in *Oxford Reviews of Reproductive Biology* (Finn, C. A., ed.) pp. 151–261, Oxford University Press, London and New York.

Bellvé, A. R., and Zheng, W. (1989). Growth factors as autocrine and paracrine modulators of male gonadal functions. J. Reprod. Fert. 85:771–793.

Bellvé, A. R., and O'Brien, D. A. (1983). The mammalian spermatozoon: structure and temporal assembly, in *Mechanism and Control of Animal Fertilization* (Hartman, J. F., ed.) pp. 55–137, Academic Press, London.

Benbrook, D., and Pfahl, M. (1987). A novel thyroid hormone receptor encoded by a cDNA clone from a human testis library. Science 238:788–791.

Bernstein, P., and Ross, J. (1989). Poly(A), poly(A) binding protein and the regulation of mRNA stability. T.I.B.S. 14:373–377.

Bhatnagar, Y. M., McCullar, K., Faulkner, R. D., and Ghai, R. D. (1983). Biochemical and immunological characterization of a histone variant associated with spermatogenesis in the mouse. Biochim. Biophys. Acta 760:25–33.

Bibb, M. J., van Etten, R. A., Wright, C. T., Walberg, M. W., and Clayton, D. A. (1981). Sequence and gene organization of mouse mitochondrial DNA. Cell 26:167–180.

Bower, P. A., Yelick, P. C., and Hecht, N. B. (1987). Both protamine 1 and 2 genes are expressed in the mouse, hamster, and rat. Biol. Reprod. 37:479–488.

Braun, R. E., Behringer, R. R., Peschon, J. J., Brinster, R. L., and Palmiter, R. D. (1989a). Genetically haploid spermatids are phenotypically diploid. Nature 337:373–376.

Braun, R. E., Peschon, J. J., Behringer, R. R., Brinster, R. L., and Palmiter, R. D. (1989b). Protamine 3′-untranslated sequences regulate temporal translational control and subcellular localization of growth hormone in spermatids of transgenic mice. Genes & Development 3:793–802.

Braun, R. E., Lo, D., Pinkert, C. A., Widera, G., Flavell, R. A., Palmiter, R. D., and Brinster, R. L. (1990). Infertility in male transgenic mice: Disruption of sperm development by HSV-tk expression in postmeiotic germ cells. Biol. Repro. 43:684–693.

Bunick, D., Johnson, P., Johnson, T. R., and Hecht, N. B. (1990a). Transcription of the testis-specific protamine 2 gene in a homologous *in vitro* transcription system. Proc. Nat. Acad. Sci. USA 87:891–895.

Bunick, D., Balhorn, R., Stanker, L. H., and Hecht, N. B. (1990b). Expression of the rat protamine 2 gene is suppressed at the level of transcription and translation. Exp. Cell Res. 188:147–152.

Burfeind, P., and Hoyer-Fender, S. (1991). Sequence and development expression of a mRNA encoding a putative protein of rat sperm outer dense fibers. Dev. Biol. 148:195–204.

Caldwell, K. A., and Handel, M. (1991). Protamine transcript sharing among postmeiotic spermatids. Proc. Nat. Acad. Sci. USA 88:2407–2411.

Casella, S. J., Smith, E. P., Van Wyk, J. J., Joseph, D. R., Hynes, M. A., Hoyt, E. C., and Lund, P. K. (1987). Isolation of rat testis cDNAs encoding an insulin-like growth factor I precursor. DNA 6:325–330.

Cate, R. L., Mattaliano, R. J., Hession, C., Tizard, R., Farber, N. M., Cheung, A., Ninfa, E. G., Frey, A. Z., Gash, D. J., Chow, E. P., Fisher, R. A., Bertonis, J. M., Torres, G., Wallner, B. P., Ramachandran, K. L., Ragin, R. C., Manganaro, T. F., MacLaughlin, D. T., and Donahoe, P. K. (1986). Isolation of the bovine and human genes for Mullerian inhibiting substance and expression of the human gene in animal cells. Cell 45:685–698.

Cedar, H. (1988). DNA methylation and gene activity. Cell 53:3–4.

Challoner, P. B., Moss, S. B., and Groudine, M. (1989). Expression of replication-dependent histone genes in avian spermatids involves an alternate pathway of mRNA 3′-end formation. Mol. Cell Biol. 9:902–913.

Chang, D. D., and Clayton, D. A. (1989). Mouse RNase MRP RNA is encoded by a nuclear gene and contains a decamer sequence complementary to a conserved region of mitochondrial RNA substrate. Cell 56:131–139.

Chelly, J., Concordet, J-P., Kaplan, J-C., and Kahn, A. (1989). Illegitimate transcription: Transcription of any gene in any cell type. Proc. Natl. Acad. Sci. USA 86:2617–2621.

Cleveland, D. W., and Sullivan, K. F. (1985). Molecular biology and genetics of tubulin. Ann. Rev. Biochem. 54:331–365.

Cole, I. D., and Kistler, W. S. (1987). Nuclear transition protein 2 (TP2) of mammalian spermatids has a very basic carboxyl terminal domain. Biochem. Biophys. Res. Commun. 147:437–442.

Cole, K. D., Kandaia, J. C., and Kistler, W. S. (1986). Isolation of the gene for the testis-specific H1 histone variant H1t. J. Biol. Chem. 261:7178–7183.

Dahl, H.-H. M., Brown, R. M., Hutchison, W. M., Maragos, C., and Brown, G. K. (1990). A testis-specific form of the human pyruvate dehydrogenase E1α subunit is coded for by an intronless gene on chromosome 4. Genomics 8:225–232.

Day, A. R., and Lee, A. S. (1989). Transcriptional regulation of the gene encoding the 78-kD glucose-regulated protein GRP78 in mouse Sertoli cells: Binding of specific factor(s) to the GRP78 promoter. DNA 8:301–310.

De, S. K., Enders, G. C., and Andrews, G. K. (1991). High levels of metallothionein messenger RNAs in male germ cells of the adult mouse. Mol. Endo. 5:628–636.

Deschepper, C. F., Mellon, S. H., Cumin, F., Baxter, J. D., and Ganong, W. F. (1986). Analysis by immunocytochemistry and *in situ* hybridization of renin and its mRNA in kidney, testis, adrenal, and pituitary of the rat. Proc. Natl. Acad. Sci. USA 83:7552–7556.

Distel, R. J., Kleene, K. C., and Hecht, N. B. (1984). Haploid expression of a mouse testis α tubulin gene. Science 224:68–70.

Dixon, G. H., Aiken, J. M., Jankowsky, J. M., McKenzie, D. I., Moir, R., and States, J. C. (1985). Organization and evolution of the protamine genes of salmonid fishes, in *Chromosomal Proteins and Gene Expression* (Reeck, G., Goodwin, G., and Puigdomenech, P., eds.) pp. 287–314, Plenum Press, New York.

Djakiew, D., and Dym, M. (1988). Pachytene spermatocyte proteins influence Sertoli cell function. Biol. Reprod. 39:1193–1205.

Dudley, K., Potter, J., Lyon, M. F., and Willison, K. R. (1984). Analysis of male sterile mutations in the mouse using haploid-stage expressed cDNA probes. Nucleic Acids Res. 12:4281–4293.

Eddy, E. M. (1988). The spermatozoon, in *The Physiology of Reproduction* (Knobil, E., and Neill, J., eds.) pp. 27–68, Raven Press, New York.

Edwards, Y. H., and Grootegoed, J. A. (1983). A sperm-specific enolase. J. Reprod. Fert. 68:305–310.

Edwards, Y. H., Povey, S., LeVan, K. M., Driscoll, C. E., Millan, J. L., and Goldberg, E. (1987). Locus determining the human sperm-specific lactate dehydrogenase, LDH-C, is syntenic with LDH-A. Dev. Gen. 8:219–232.

Ehlers, M.R.W., Fox, E. A., Strydom, D. J., and Riordan, J. F. (1989). Molecular cloning of human testicular angiotensin-converting enzyme: The testis isozyme is identical to the C-terminal half of endothelial angiotensin-converting enzyme. Proc. Natl. Acad. Sci. USA 86:7741–7745.

Elsevier, S. M., Noiran, J., and Carre-Eusebe, D. (1991). Processing of the precursor of protamine P2 in mouse: Identification of intermediates by their insolubility in the presence of sodium dodecyl sulfate. Eur. J. Biochem. 196:167–175.

Erickson, R. P., Kramer, J. M., Rittenhouse, J., and Salkeld, A. (1980). Quantitation of mRNAs during mouse spermatogenesis: Protamine-like histone and phosphoglycerate kinase 2 mRNAs increase after meiosis. Proc. Natl. Acad. Sci. USA 77:6086–6090.

Evans, M. J., and Scarpulla, R. C. (1989). Interaction of nuclear factors with multiple sites in the somatic cytochrome c promoter. J. Biol. Chem. 264:14361–14368.

Evans, M. J., and Scarpulla, R. C. (1988). Both upstream and intron sequence elements are required for elevated expression of the rat somatic cytochrome c gene in COS-1 cells. Mol. Cell Biol. 8:35–41.

Fenderson, B. A., Toshimori, K., Muller, C. H., Lane, T. F., and Eddy, E. M. (1988). Identification of a protein in the fibrous sheath of the sperm flagellum. Biol. Reprod. 38:345–357.

Fischman, K., Edmun, J. C., Shackleford, G. M., Turner, J. A., Rutter, W. J., and Nir, U. (1990). A murine *fer* testis-specific transcript (fe T) encodes a truncated fer protein. Mol. Cell. Biol. 10:146–153.

Fujimoto, H., Erickson, R. P., and Toné, S. (1988). Changes in polyadenylation of lactate dehydrogenase-X mRNA during spermatogenesis in mice. Mol. Reprod. Dev. 1:27–34.

Fundele, R., Winking, H., Illmensee, K., and Jägerbauer, E. (1987). Developmental activation of phosphoglycerate mutase-2 in the testis of the mouse. Dev. Biol. 124:562–566.

Garrett, J. E., Collard, M. W., and Douglass, J. O. (1989). Translational control of germ cell-expressed mRNA imposed by alternative splicing: Opioid peptide gene expression in rat testis. Mol. Cell. Biol. 9:4381–4389.

Gatewood, J. M., Cook, G. R., Balhorn, R., Bradbury, E. M., and Schmid, C. W. (1987). Sequence-specific packaging of DNA in human sperm chromatin. Science 236:962–964.

Gizang-Ginsberg, E., and Wolgemuth, D. J. (1985). Localization of mRNAs in mouse testes by *in situ* hybridization: Distribution of α tubulin and developmental stage specificity of pro-opimelanocortin transcripts. Dev. Biol. 111:293–305.

Gold, B., Fujimoto, H., Kramer, J. M., Erickson, R. P., Hecht, N. B. (1983). Haploid accumulation and translational control of phosphoglycerate kinase-2 messenger RNA during mouse spermatogenesis. Dev. Biol. 98:392–399.

Goldberg, E. (1987). In *Contraception Research for Today and the Nineties: Progress in Birth Control Vaccines* (Talwar, G. P., ed.) pp. 277–283, Springer-Verlag, New York.

Goldberg, E. (1977). In *Isozymes: Current Topics in Biological Research* (Rattazzi, M. C., Scandalios, J. S., and Litt, G. S., eds.) pp. 79–124, Alan R. Liss, New York.

Goldberg, E., Sberna, D., Wheat, T. E., Urbanski, G. J., and Margoliash, E. (1977). Cytochrome c: Immunofluorescent localization of the testis-specific form. Science 196:1010–1012.

Goldman, D. S., Kiessling, A., Millette, C. F., and Cooper, G. M. (1987). Expression of c-*mos* RNA in germ cells of male and female mice. Proc. Natl. Acad. Sci. USA 84:4509–4513.

Goldsborough, A., Ashworth, A., and Willison, K. (1990). Cloning and sequencing of POU-boxes expressed in mouse testis. Nucleic Acids Res. 18:1634.

Goto, M., Tamura, T., Mikoshiba, K., Masamune, Y., and Nakanishi, Y. (1991). Transcription inhibition of the somatic-type phosphoglycerate kinase 1 gene *in vitro* by a testis-specific factor that recognizes a sequence similar to the binding site for Ets oncoproteins. Nucleic Acids Res. 19:3959–3963.

Grippo, P., Geremia, R., Locorotondo, G., and Monesi, V. (1978). DNA-dependent DNA polymerase species in male germ cells of the mouse. Cell Diff. 7:237–248.

Griswold, M. D., Collard, M., Hugly, S., and Huggenvik, J. (1986). The use of specific cDNAs to assay Sertoli cell functions, in *Molecular and Cellular Aspects of Reproduction* (Dhindsa, D. S., and Bahl, O. P., eds.) pp. 301–317, Plenum Press, New York.

Griswold, M. D., Morales, C., and Sylvester, S. R. (1988). Molecular biology of the Sertoli cell, in *Oxford Reviews of Reproductive Biology,* pp. 124–161, Oxford University Press, London and New York.

Groudine, M., and Conklin, K. F. (1985). Chromatin structure and *de novo* methylation of sperm DNA: Implications for activation of the paternal genome. Science 228:1061–1068.

Hake, L., Alcivar, A. A., and Hecht, N. B. (1990). Changes in mRNA length accompany translational regulation of the somatic and testis-specific cytochrome c genes during spermatogenesis in the mouse. Devel. 110:249–257.

Handel, M. (1987). Genetic control of spermatogenesis in mice, in *Results and Problems in Cell Differentiation in Spermatogenesis: Genetic Aspects* (Hennig, W., ed.) pp. 1–62, Springer-Verlag, Berlin.

Harrison, R.A.P., and Gaunt, S. J. (1988). Multiple forms of ram and bull sperm hyaluronidase revealed by using monoclonal antibodies. J. Reprod. Fert. 82:777–785.

Hecht, N. B. (1989a). Mammalian protamines and their expression, in *Histones and Other Basic Nuclear Proteins* (Hnilica, L., Stein, G., and Stein, J., eds.) pp. 347–373, CRC Press, Boca Raton, Florida.

Hecht, N. B. (1989b). Molecular biology of structural chromosomal proteins of the mammalian testis, in *Molecular Biology of Chromosome Function* (Adolph, K. W., ed.) pp. 396–420, Springer Verlag, New York.

Hecht, N. B., Distel, R. J., Yelick, P. C., Driscoll, C. E., Goldberg, E., and Tung, K.S.K. (1988). Localization of a novel testicular α tubulin antigenic epitope that is not detectable in brain. Mol. Cell Biol. 8:996–1000.

Hecht, N. B. (1986). Regulation of gene expression during mammalian spermatogenesis, in *Experimental Approaches to Mammalian Embryonic Development* (Rossant, J., and Pedersen, R. A., eds.) pp. 151–193, Cambridge University Press, New York.

Hecht, N. B., Liem, H., Kleene, K. C., Distel, R. J., and Ho, S. M. (1984). Maternal inheritance of the mouse mitochondrial genome is not mediated by a loss or gross alteration of the paternal mitochondrial DNA or by methylation of the oocyte mitochondrial DNA. Dev. Biol. 102:452–461.

Hecht, N. B., and Kennington, E. (1983). Synthesis of mitochondrial proteins from mouse epididymal spermatozoa during spermatogenesis, in *The Sperm Cell* (André, J., ed.) pp. 419–422, Martinus Nijhoff Publishers, The Hague.

Hecht, N. B., and Williams, J. L. (1978). Synthesis of RNA by separated heads and tails from bovine spermatozoa. Biol. Reprod. 19:573–579.

Hecht, N. B., Farrell, D., and Williams, J. L. (1978). DNA polymerases in mouse spermatogenic cells separated by sedimentation velocity. Biochim. Biophys. Acta 561:358–368.

Heidaran, M. A., Kozak, C. A., and Kistler, W. S. (1989). Nucleotide sequence of the *Stp-1* gene coding for rat spermatid nuclear transition protein 1 (TP1): Homology with protamine P1 and assignment of the mouse *Stp-1* gene to chromosome 1. Gene 75:39–46.

Heidaran, M. A., Showman, R. M., and Kistler, S. W. (1988). A cytochemical study of the transcriptional and translational regulation of nuclear transition protein 1 (TP1), a major chromosomal protein of mammalian spermatids. J. Cell Biol. 106:1427–1433.

Heidaran, M. A., and Kistler, W. S. (1987a). Isolation of a cDNA clone for transition protein 1 (TP1), a major chromosomal protein of mammalian spermatids. Gene 54:281–284.

Heidaran, M. A., and Kistler, W. S. (1987b). Transcriptional and translational regulation of the message for Transition Protein 1, a major chromosomal protein of mammalian spermatids. J. Biol. Chem. 262:13309–13315.

Höög, C., Schalling, M., Grunder-Brundell, E., and Daneholt, B. (1991). Analysis of a murine male germ cell-specific transcript that encodes a putative zinc finger protein. Mol. Reprod. Devel. 30:173–181.

Howard, T. E., Shai, S-Y., Langford, K. G., Martin, B. M., and Bernstein, K. E. (1990). Transcription of testicular angiotensin-converting enzyme (ACE) is initiated within the 12th intron of the somatic ACE gene. Mol. Cell. Biol. 10:4294–4302.

Huarte, J., Belin, D., Bosco, D., Sappino, A.-P., and Vassalli, J.-D. (1987). Plasminogen activator and mouse spermatozoa: Urokinase synthesis in the male genital tract and binding of the enzyme to the sperm cell surface. J. Cell Biol. 104:1281–1289.

Huh, N-E., Hwang, I., Lim, K., You, K-H., and Chae, C-B. (1991). Presence of a bi-directional S phase-specific transcription regulatory element in the promoter shared by testis-specific TH2A and TH2B histone genes. Nucleic Acids Res. 19:93–100.

Hwang, I., Lim, K., and Chae, C-B. (1989). S phase-specific transcription regulatory elements are present in a replication-independent testis-specific H2B histone gene. Mol. Cell Biol. 9:1005–1013.

Ibáñez, C. F., Pelto-Huikko, M., Söder, O., Ritzèn, E. M., Hersh, L. B., Hökfelt, T., and Persson, H. (1991). Expression of choline acetyltransferase mRNA in spermatogenic cells results in an accumulation of the enzyme in the postacrosomal region of mature spermatozoa. Proc. Natl. Acad. Sci. USA 88:3676–3680.

Jen, J., Deschepper, C. F., Shackleford, G. M., Lee, C.Y.G., and Lau, Y.-F. (1990). Stage-specific expression of the lactate dehydrogenase-x gene in adult and developing mouse testes. Mol. Reprod. Devel. 25:14–21.

Johnson, P. A., Bunick, D., and Hecht, N. B. (1991). Protein binding regions in the testis-specific mouse protamine 2 gene. Biol. Reprod. 44:127–134.

Johnson, P. A., Peschon, J. J., Yelick, P. C., Palmiter, R. D., and Hecht, N. B. (1988). Sequence homologies in the mouse protamine 1 and 2 genes. Biochim. Biophys. Acta 950:45–53.

Joseph, D. R., Hall, S. H., Conti, M., and French, F. S. (1988). The gene structure of rat androgen-binding protein: Identification of potential regulatory deoxyribonucleic acid elements of a follicle-stimulating hormone-regulated protein. Mol. Endocrinol. 2:3–13.

Joseph, D. R., Hall, S. H., and French, F. S. (1985). Identification of complementary DNA clones that encode rat androgen binding protein. J. Androl. 6:392–395.

Kasahara, M., Gutknecht, J., Brew, K., Spurr, N., and Goodfellow, P. N. (1989). Cloning and mapping of a testis-specific gene with sequence similarity to a sperm-coating glycoprotein gene. Genomics 5:527–534.

Kashiwabara, S-I., Baba, T., Takada, M., Watanabe, K., Yano, Y., and Arai, Y. (1990). Primary structure of mouse proacrosin deduced from the cDNA sequence and its gene expression during spermatogenesis. J. Biochem. 108:785–791.

Katsuki, M., Sato, M., Kimura, M., Yokoyama, M., Kobayashi, K., and Nomura, T. (1988). Conversion of normal behavior to shiverer by myelin basic protein antisense cDNA in transgenic mice. Science 241:593–595.

Keung, W.-M. (1988). A genuine organ specific alcohol dehydrogenase from hamster testes: Isolation, characterization and developmental changes. Biochem. Biophys. Res. Comm. 156:38–45.

Kierszenbaum, A. L., Ueda, H., Ping, L., Abdullah, M., and Tres, L. L. (1988). Antibodies to rat Sertoli cell secretory proteins recognize antigenic sites in acrosome and tail of developing spermatids and sperm. J. Cell Sci. 91:145–153.

Kierszenbaum, A. L., and Tres, L. L. (1978). RNA transcription and chromatin structure during meiotic and postmeiotic stages of spermatogenesis. Fed. Proc. 37:2512–2516.

Kilpatrick, D. L., Borland, K., and Jin, D. F. (1987). Differential expression of opioid peptide genes by testicular germ cells and somatic cells. Proc. Natl. Acad. Sci. USA 84:5695–5699.

Kilpatrick, D. L., Zinn, S. A., Fitzgerald, M., Higuchi, H., Sabol, S. L., and Meyerhardt, J. (1990). Transcription of the rat and mouse proenkephalin genes is initiated at distinct sites in spermatogenic and somatic cells. Mol. Cell. Biol. 10:3717–3726.

Kim, E., Waters, S. H., Hake, L. E., and Hecht, N. B. (1989). Identification and developmental expression of a smooth muscle γ actin in post-meiotic male germ cells of mice. Mol. Cell Biol. 9:1875–1881.

Kim, Y.-J., Hwang, I., Tres, L. L., Kierszenbaum, A. L., and Chae, C.-B. (1987). Molecular cloning and differential expression of somatic and testis-specific H2b histone genes during rat spermatogenesis. Dev. Biol. 124:23–34.

Kistler, W. S. (1989). Structures of testis-specific histones, spermatid transition proteins and their genes in mammals, in *Histones and Other Basic Nuclear Proteins* (Hnilica, L., Stein, G., and Stein, J., eds.) pp. 331–346, CRC Press, Boca Raton, FL.

Kleene, K. C. (1989). Poly (A) shortening accompanies the activation of translation of five mRNAs during spermiogenesis in the mouse. Development 106:367–373.

Kleene, K. C., Smith, J., Bozorgzadeh, A., Harris, M., Hahn, L., Karimpour, I., and Gerstel, J. (1990). Sequence and developmental expression of the mRNA encoding the seleno-protein of the sperm mitochondrial capsule in the mouse. Dev. Biol. 137:395–402.

Kleene, K. C., Bozorgzadeh, A., Flynn, J. F., Yelick, P. C., and Hecht, N. B. (1988). Nucleotide sequence of a cDNA encoding mouse transition protein 1. Biochim. Biophys. Acta 950:215–220.

Kleene, K. C., and Flynn, J. F. (1987). Characterization of a cDNA clone encoding a basic protein, TP2, involved in chromatin condensation in the mouse. J. Biol. Chem. 262:17272–17277.

Kleene, K. C., Distel, R. J., and Hecht, N. B. (1985). The nucleotide sequence of a cDNA clone encoding mouse protamine 1. Biochem. 24:719–722.

Kleene, K. C., Distel, R. J., and Hecht, N. B. (1984). Translational regulation and coordinate deadenylation of a haploid mRNA during spermiogenesis in the mouse. Dev. Biol. 105:71–79.

Kleene, K. C., Distel, R. J., and Hecht, N. B. (1983). cDNA clones encoding cytoplasmic poly(A)$^+$ RNAs which first appear at detectable levels in haploid phases of spermatogenesis in the mouse. Dev. Biol. 98:455–464.

Klemm, U., Lee, C-H., Burfeind, P., Hake, S., and Engel, W. (1989). Nucleotide sequence of a cDNA encoding rat protamine and the haploid expression of the gene during rat spermatogenesis. Biol. Chem. Hoppe-Seyler 370:293–301.

Klemm, U., Maier, W-M., Tsaousidou, S., Adham, I. M., Willison, K., and Engel, W. (1990). Mouse preproacrosin: cDNA sequence, primary structure and postmeiotic expression in spermatogenesis. Differentiation 42:160–166.

Kramer, J. M., and Erickson, R. P. (1981). Developmental programs of PGK-1 and PGK-2 isozymes in spermatogenic cells of the mouse: Specific activities and rates of synthesis. Dev. Biol. 87:37–45.

Krawetz, S. A., Connor, W., and Dixon, G. H. (1988). Bovine protamine genes contain a single intron. The structure of two alleles. J. Biol. Chem. 263:321–326.

Krawetz, S. A., Herfort, M. H., Hamerton, J. L., Pon, R. T., and Dixon, G. H. (1989). Chromosomal localization and structure of the human P1 protamine gene. Genomics 5:639–645.

Kremling, H., Luerssen, H., Adham, I. M., Klemm, U., Tsaousidou, S., and Engel, W. (1989). Nucleotide sequences and expression of cDNA clones for boar and bull transition protein 1 and its evolutionary conservation in mammals. Differentiation 40:184–190.

Kwon, Y. K., and Hecht, N. B. (1991). Cytoplasmic protein binding to highly conserved sequences in the 3' untranslated region of mouse protamine 2 mRNA, a translationally regulated gene of male germ cells. Proc. Natl. Acad. Sci. USA 88:3584–3588.

Lader, E., Ha, H-S., O'Neill, M., Artzt, K., and Bennett, D. (1989). *tctex-1:* A candidate gene family for a mouse *t* complex sterility locus. Cell 58:969–979.

Langford, K. G., Shai, S-Y., Howard, T. E., Kovac, M. J., Overbeek, P. A., and Bernstein, K. E. (1991). Transgenic mice demonstrate a testis-specific promoter for angiotensin-converting enzyme. J. Biol. Chem. 266:15559–15562.

Lathrop, W. F., Carmichael, E. R., Myles, D. G., and Primakoff, P. (1989). Cloning and sequencing of the gene for the PH-20 protein of guinea pig sperm. J. Cell Biol. 109:124.

Lathrop, W. F., Carmichael, E. P., Myles, D. G., and Primakoff, P. (1990). cDNA cloning reveals the molecular structure of a sperm surface protein, PH-20, involved in sperm–egg adhesion and the wide distribution of its gene among mammals. J. Cell Biol. 111:2939–2949.

Lattion, A.-L., Soubrier, F., Allegrini, J., Hubert, C., Corvol, P., and Alhenc-Gelas, F. (1989). The testicular transcript of the angiotensin I-converting enzyme encodes for the ancestral, non-duplicated form of the enzyme. FEBS Lett. 252:99–104.

Lau, Y.-C.C., Chan, K., Sparkes, R. (1989). Male enhanced antigen gene is phylogenetically conserved and expressed at late stages of spermatogenesis. Proc. Nat. Acad. Sci. USA 86:8462–8466.

Lee, S-J. (1990). Expression of HSP86 in male germ cells. Mol. Cell. Biol. 10:3239–3242.

Leroy, P., Alzari, P., Sassoon, D., Wolgemuth, D., and Fellous, M. (1989). The protein encoded by a murine male germ cell-specific transcript is a putative ATP-dependent RNA helicase. Cell 57:549–559.

Leyton, L., and Saling, P. (1989). 95 kd sperm proteins bind ZP3 and serve as tyrosine kinase substrates in response to zona binding. Cell 57:1123–1130.

Leuerssen, H., Maier, W. M., Hoyer-Fender, S., and Engel, W. (1989). The nucleotide sequence of rat transition protein 2 (TP2) cDNA. Nucleic Acids Res. 17:3585.

Lewis, S. A., Gu, W., and Cowan, N. J. (1987). Free intermingling of mammalian β-tubulin isotypes among functionally distinct microtubules. Cell 49:539–548.

Li, S.S.-L., O'Brien, D. A., How, E. W., Versola, J., Rockett, D. L., Eddy, E. M. (1989). Differential activity and synthesis of lactate dehydrogenase isozymes A (muscle), B (heart), and C (testis) in mouse spermatogenic cells. Biol. Reprod. 40:173–180.

Longo, F. J., Krohne, G., and Franke, W. W. (1987). Basic proteins of the perinuclear theca of mammalian spermatozoa and spermatids: A novel class of cytoskeletal elements. J. Cell Biol. 105:1105–1120.

McCarrey, J. R., and Thomas, K. (1987). Human testis-specific PGK gene lacks introns and possesses characteristics of a processed gene. Nature 326:501–505.

MacGregor, G. R., Russell, L. D., VanBeek, M. E. A. B., Hanten, G. R., Kovac, M. J., Kozak, C. A., Meistrich, M. L., and Overbeek, P. A. (1990). Symplastic spermatids (sys): A recessive insertional mutation in mice causing a defect in spermatogenesis. Proc. Natl. Acad. Sci. USA 87:5016–5020.

Maekawa, M., O'Brien, D. A., Allen, R. L., and Eddy, E. M. (1989). Heat-shock cognate protein (hsc71) and related proteins in mouse spermatogenic cells. Biol. Reprod. 40:843–852.

Mali, P., Kaipia, A., Kangasniemi, M., Toppari, J., Sandberg, M., Hecht, N. B., and Parvinen, M. (1989). Stage-specific expression of nucleoprotein mRNAs during rat and mouse spermiogenesis. Reprod. Fert. Devel. 1:369–382.

Mason, A. J., Hayflick, J. S., Ling, N., Esch, F., Ueno, N., Ying, S. Y., Guillemin, R., Niall, H., and Seeburg, P. H. (1985). Complementary DNA sequences of ovarian follicular fluid inhibin show precursor structure and homology with transforming growth factor. Nature 318:659–663.

Matsushime, H., Jinno, A., Takagi, N., and Shibuya, M. (1990). A novel mammalian protein kinase gene (mak) is highly expressed in testicular germ cells at and after meiosis. Mol. Cell. Biol. 10:2261–2268.

Mayerhofer, A., Russell, L. D., Grothe, C., Rudolf, M., and Gratzl, M. (1991). Presence and localization of a 30-kDa basic fibroblast growth factor-like protein in rodent testes. Endocrinology 129:921–927.

Mazarakis, N. D., Nelki, D., Lyon, M. F., Ruddy, S., Evans, E. P., Freemont, P., and Dudley, K. (1991). Isolation and characterisation of a testis-expressed developmentally regulated gene from the distal inversion of the mouse t-complex. Development 111:561–571.

Meehan, R. R., Lewis, J. D., McKay, S., Kleiner, E. L., and Bird, A. P. (1989). Identification of a mammalian protein that binds specifically to DNA containing methylated CpGs. Cell 58:499–507.

Meijer, D., Hermans, A., von Lindern, M., van Agthoven, T., de Klein, A., Mackenbach, P., Grootegoed, A., Talarico, D., Della Valle, G., and Grosveld, G. (1987). Molecular characterization of the testis specific *c-abl* mRNA in mouse. EMBO J. 6:4041–4048.

Meistrich, M. L. (1989). In *Histones and Other Basic Nuclear Proteins* (Hnilica, L., Stein, G., and Stein, J., eds.) pp. 165–182, CRC Press, Boca Raton, FL.

Meistrich, M. L., and Brock, W. A. (1987). Proteins of the meiotic cell nucleus, in *Meiosis* (Moens, P. B., ed.) p. 333, Academic Press, Orlando, FL.

Meistrich, M. L., Longtin, J., Brock, W. A., Grimes, S. J., and Mace, M. L. (1981). Purification of rat spermatogenic cells and preliminary biochemical analysis of these cells. Biol. Reprod. 25:1065–1077.

Melton, D. A. (1988). *Antisense RNA and DNA*. Cold Spring Harbor Laboratory, New York.

Merlino, G. T., Stahle, C., Jhappan, C., Linton, R., Mahon, K. A., and Willingham, M. C. (1991). Inactivation of a sperm motility gene by insertion of an epidermal growth factor receptor transgene whose product is overexpressed and compartmentalized during spermatogenesis. Genes & Dev. 5:1395–1406.

Meyerhof, W., Müller-Brechlin, R., and Richter, D. (1991). Molecular cloning of a novel putative G-protein coupled receptor expressed during rat spermiogenesis. FEBS Lett. 284:155–160.

Mezquita, C. (1985). Chromatin proteins and chromatin structure in spermatogenesis, in *Chromosomal Proteins and Gene Expression* (Reeck, G., Goodwin, G., and Puigdomenech, P., eds.) pp. 315–332, Plenum Press, New York.

Millan, M. A., and Aguilera, G. (1988). Angiotensin II receptors in testes. Endo. 122:1984–1990.

Millan, J. L., Driscoll, C. E., LeVan, K. M., and Goldberg, E. (1987). Epitopes of human testis-specific lactate dehydrogenase deduced from a cDNA sequence. Proc. Natl. Acad. Sci. USA 84:5311–5315.

Mitchell, P. J., and Tjian, R. (1989). Transcriptional regulation in mammalian cells by sequence-specific-DNA binding proteins. Science 245:371–378.

Monesi, V. (1978). Biochemistry of male germ cell differentiation in mammals: RNA synthesis in meiotic and post-meiotic cells. Curr. Top. Dev. Biol. 12:11–36.

Montamat, E. E., and Blanco, A. (1976). Subcellular distribution of the lactate dehydrogenase isozyme specific for testis and sperm. Exp. Cell Res. 103:241–245.

Morales, C. R., Kwon, Y. K., and Hecht, N. B. (1991). Cytoplasmic localization during storage and translation of the mRNAs of transition protein 1 and protamine 1, two translationally regulated transcripts of the mammalian testis. J. Cell Sci. 100:119–131.

Moss, S. B., Challoner, P. B., and Groudine, M. (1989). Expression of a novel histone 2B during mouse spermatogenesis. Dev. Biol. 133:83–92.

Mutter, G. L., and Wolgemuth, D. J. (1987). The proto-oncogene *c-mos* is expressed with distinct developmental patterns in the male and female germ cell lines. Proc. Nat. Acad. Sci. USA 84:5301–5305.

Nagamine, C. M., Chan, K., Hake, L. E., Lau, Y.-F. C. (1990). The two candidate testis-determining Y genes (*Zfy-1* and *Zfy-2*) are differentially expressed in fetal and adult mouse tissues. Genes & Dev. 4:63–74.

Nelki, D., Dudley, K., Cunningham, P., and Akhavan, M. (1990). Cloning and sequencing of a zinc finger cDNA expressed in mouse testis. Nucleic Acids Res. 18:3655.

Nickel, B. E., and Davie, J. R. (1989). The protamine gene

chromatin in developing trout testis exists in an altered state. Biochim. Biophys. Acta. 1007:23–29.

Oko, R. (1988). Comparative analysis of proteins from the fibrous sheath and outer dense fibers of rat spermatozoa. Biol. Reprod. 39:169–182.

Oko, R., and Clermont, Y. (1989). Light microscopic immunocytochemical study of fibrous sheath and outer dense fiber formation in the rat spermatid. Anat. Rec. 225:46–55.

Olds-Clarke, P. (1988). Genetic analysis of sperm function in fertilization. Gam. Res. 20:241–264.

Oliva, R., Goren, R., and Dixon, G. H. (1989). Quail (*Coturnix japonica*) protamine, full-length cDNA sequence, and the function and evolution of vertebrate protamines. J. Biol. Chem. 264:17627–17630.

Oliva, R., Mezquita, J., Mezquita, C., and Dixon, G. H. (1988). Haploid expression of the rooster protamine mRNA in the postmeiotic stages of spermatogenesis. Dev. Biol. 125:332–340.

Ono, T., and Means, A. R. (1990). Calspermin is a testis specific calmodulin-binding protein closely related to Ca^{2+}/Calmodulin-dependent protein kinases. Adv. Exp. Med. Biol. 255:263.

Oyen, O., Scott, J. D., Cadd, G. G., McKnight, G. S., Krebs, E. G., Hansson, V., and Jahnsen, T. (1988). A unique mRNA species for a regulatory subunity of cAMP-dependent protein kinase is specifically induced in haploid germ cells. FEBS Letters 229:391–394.

Passananti, C., Felsani, A., Caruso, M., and Amati, P. (1989). Mouse genes coding for "zinc-finger"-containing proteins: Characterization and expression in differentiated cells. Proc. Natl. Acad. Sci. USA 86:9417–9421.

Paulssen, R. H., Paulssen, E. J., Gordeladze, J. O., Hansson, V., and Haugen, T. B. (1991). Cell-specific expression of guanine nucleotide-binding proteins in rat testicular cells. Biol. Repro. 45:566–571.

Persson, H., Rehfeld, J. F., Ericsson, A., Schalling, M., Pelto-Huikko, M., and Hokfelt, T. (1989). Transient expression of the cholecystokinin gene in male germ cells and accumulation of the peptide in the acrosomal granule: Possible role of cholecystokinin in fertilization. Proc. Natl. Acad. Sci. USA 86:6166–6170.

Persson, H., Ayer-Le Lievre, C., Soder, O., Villar, M. J., Metsis, M., Olson, L., Ritzen, M., and Hokfelt, T. (1990). Expression of β-nerve growth factor receptor mRNA in Sertoli cells downregulated by testosterone. Science 247:704–707.

Peschon, J. J., Behringer, R. R., Brinster, R. L., and Palmiter, R. D. (1987). Spermatid-specific expression of protamine 1 in transgenic mice. Proc. Natl. Acad. Sci. USA 84:5316–5319.

Poccia, D. (1986). Remodeling of nucleoproteins during gametogenesis, fertilization, and early development. Int. Rev. Cytol. 105:1–65.

Ponzetto-Zimmerman, C., and Wolgemuth, D. J. (1984). Methylation of satellite sequences in mouse spermatogonia and somatic DNAs. Nucleic Acids Res. 12:2807–2822.

Ponzetto, C. P., and Wolgemuth, D. J. (1985). Haploid expression of a unique *c-abl* transcript in the mouse male germ line. Mol. Cell Biol. 5:1791–1794.

Propst, F., Rosenberg, M. P., Iyer, A., Kaul, K., and VandeWoude, G. (1987). *c-mos* proto-oncogene RNA transcripts in mouse tissues: Structural features, developmental regulation, and localization in specific cell types. Mol. Cell Biol. 7:1629–1637.

Propst, F., Rosenberg, M. P., and Vande Woude, G. (1988). Proto-oncogene expression in germ cell development. Trends Genetics 4:183–187.

Racamora, N., and Mezquita, C. (1989). Chicken spermatogenesis is accompanied by a genomic-wide loss of DNA methylation. FEBS Lett. 247:415–418.

Rahe, B., Erickson, R. P., and Quinto, M. (1983). Methylation of unique sequence DNA during spermatogenesis in mice. Nucleic Acids Res. 11:7947–7959.

Rasheed, B. K. A., Whisenant, E. C., Ghai, R. D., Papaioannou, V. E., and Bhatnagar, Y. M. (1989). Biochemical and immunocytochemical analysis of a histone H1 variant from the mouse testis. J. Cell Sci. 94:61–71.

Reik, W., Collick, A., Norris, M. L., Barton, S. C., and Surani, A. (1987). Genomic imprinting determines methylation of parental alleles in transgenic mice. Nature 328:248–251.

Roberts, V., Meunier, H., Sawchenko, P. E., and Vale, W. (1989). Differential production and regulation of inhibin subunits in rat testicular cell types. Endocrinology 125:2350–2359.

Robertson, N. G., Pomponio, R. J., Mutter, G. L., and Morton, C. C. (1991). Testis-specific expression of the human MYCL2 gene. Nucleic Acids Res. 19:3129–3137.

Robinson, M. O., McCarrey, J. R., and Simon, M. I. (1989). Transcriptional regulatory regions of testis-specific *PGK2* defined in transgenic mice. Proc. Natl. Acad. Sci. USA 86:8437–8441.

Romrell, L. J., Bellvé, A. R., and Fawcett, D. W. (1976). Separation of mouse spermatogenic cells by sedimentation velocity: A morphological characterization. Dev. Biol. 49:119–131.

Rossi, P., Giorgi, M., Geremia, R., and Kincaid, R. L. (1988). Testis-specific calmodulin-dependent phosphodiesterase. J. Biol. Chem. 263:15521–15527.

Rossi, P., Albanesi, Ch., Grimaldi, P., and Germeia, R. (1991). Expression of the mRNA for the ligand of C-kit in mouse Sertoli cells. Biochem. Biophys. Res. Comm. 176:910–914.

Rotem, R., Paz, G. F., Homonnai, Z. T., Kalina, M., and Naor, Z. (1990). Protein kinase C is present in human sperm: Possible role in flagellar motility. Proc. Natl. Acad. Sci. USA 87:7305–7308.

Roy, S. N., Kusari, J., Soffer, R. L., Lai, C. Y., and Sen, G. C. (1988). Isolation of cDNA clones of rabbit angiotensin converting enzyme: Identification of two distinct mRNAs for the pulmonary and the testicular isozymes. Biochem. Biophys. Res. Comm. 155:678–684.

Russell, L. D., Saxena, N. K., and Turner, T. T. (1989). Cytoskeletal involvement in spermiation and sperm transport. Tissue Cell 21:361–379.

Saffer, J. D., Jackson, S. P., and Annarella, M. B. (1991). Developmental expression of Sp1 in the mouse. Mol. Cell. Biol. 11:2189–2199.

Saiki, R. K., Gelfand, D. H., Stoffel, S., Scharf, S. J., Higuchi, R., Horn, G. T., Mullis, K. B., and Erlich, H. A. (1988). Primer-directed enzymatic amplification of DNA with a thermostabile DNA polymerase. Science 239:487–491.

Sakata, J., Abe, Y., and Uyeda, K. (1991). Molecular cloning of the DNA and expression and characterization of rat testes fructose-6-phosphate, 2-kinase:fructose-2, 6-bisphosphatase. J. Biol. Chem. 266:15764–15770.

Sandberg, M., Tasken, K., Oyen, O., Hansson, V., and Jahnsen, T. (1987). Molecular cloning, cDNA structure and deduced amino acid sequence for a type I regulatory subunit of cAMP-dependent protein kinase from human testis. Biochem. Biophys. Res. Comm. 149:939–945.

Sapienza, C., Peterson, A. C., Rossant, J., and Balling, R. (1987). Degree of methylation of transgene is dependent on gamete of origin. Nature 328:251–254.

Schatte, E. C., Welch, J. E., and Eddy, E. M. (1989). Isolation and characterization of a testis-specific cDNA encoding a glyceraldehyde 3-phosphate dehydrogenase-like protein. J. Cell Biol. 109:66a.

Seboun, E., Mattei, M. G., Fellous, M., and Bishop, C. E. (1990). Isolation of an autosomally located testis specific transcript with a human Y-derived DNA probe. EMBO J., in press.

Shackleford, G. M., and Varmus, H. E. (1987). Expression of the proto-oncogene int-1 is restricted to post-meiotic male germ cells and the neural tube of mid-gestational embryos. Cell 50:89–95.

Shaper, N. L., Wright, W. W., and Shaper, J. H. (1990). Murine β1,4-galactosyltransferase: Both the amounts and structure of the mRNA are regulated during spermatogenesis. Proc. Natl. Acad. Sci. USA 87:791–795.

Silver, L. M. (1985). Mouse t haplotypes. Ann. Rev. Genetics 19:179–208.

Skinner, M. K., Fetterolf, P. M., and Anthony, C. T. (1988). Purification of a paracrine factor, P-mod-S, produced by testicular peritubular cells that modulates Sertoli cell function. J. Biol. Chem. 263:2884–2890.

Slaughter, G. R., and Means, A. R. (1989). Analysis of expression of multiple genes encoding calmodulin during spermatogenesis. Mol. Endocrinol. 3:1569.

Smithies, O., Gregg, R. G., Boggs, S. S., Koralewski, M. A., and Kucherlapati, R. S. (1985). Insertion of DNA sequences into the human chromosomal β globin locus by homologous recombination. Nature 317:230–234.

Sorrentino, V., McKinney, M. D., Giorgi, M., Geremia, R., and Fleissner, E. (1988). Expression of cellular proto-oncogenes in the mouse male germ line: A distinctive 2.4 kilobase pim-1 transcript is expressed in haploid postmeiotic cells. Proc. Natl. Acad. Sci. USA 85:2191–2195.

Stallard, B. J., and Griswold, M. D. (1990). Germ cell regulation of Sertoli cell transferrin mRNA levels. Mol. Endo. 90:393–401.

Stallard, B. J., Collard, M. W., and Griswold, M. D. (1991). A transferrinlike (hemiferrin) mRNA is expressed in the germ cells of rat testis. Mol. Cell. Biol. 11:1448–1453.

Stefanini, M., Ziparo, E., Palombi, F., Galdieri, M., Geremia, R., and Russo, M. A. (1980). Evidence for *in vitro* interaction between Sertoli cells and germ cells in the rat, in *Animal Models of Reproduction* (Serio, M., and Martini, L., eds.) pp. 149–157, Raven Press, New York.

Stewart, T. A., Hecht, N. B., Hollinghead, P. G., Johnson, P. A., Leong, J.A.C., and Pitts, S. L. (1988). Haploid-specific transcription of protamine-*myc* and protamine-Tag fusion genes in transgenic mice. Mol. Cell Biol. 8:1748–1755.

Swain, J. L., Stewart, T. A., and Leder, P. (1987). Parental legacy determines methylation and expression of an autosomal transgene: A molecular mechanism for parental imprinting. Cell 50:719–727.

Swinnen, J. V., Joseph, D. R., and Conti, M. (1989). The mRNA encoding a high-affinity cAMP phosphodiesterase is regulated by hormones and cAMP. Proc. Natl. Acad. Sci. USA 86:8197–8201.

Swinnen, J. V., Joseph, D. R., and Conti, M. (1989). Molecular cloning of rat homologues of the *Drosophila melanogaster* dunce cAMP phosphodiesterase: Evidence for a family of genes. Proc. Natl. Acad. Sci. USA 86:5325–5329.

Takano, T., and Li, S.S.-L. (1989). Human testicular lactate dehydrogenase-c gene is interrupted by six introns at positions homologous to those of LDH-A (muscle) and LDH-B (heart) genes. Biochem. Biophys. Res. Comm. 159:579–583.

Tanhauser, S. M., and Hecht, N. B. (1989). Nucleotide sequence of the rat protamine 2 gene. Nucleic Acids Res. 17:4395.

Teruya, J. H., Kutsunai, S. Y., Spear, D. H., Edwards, P. A., and Clarke, C. F. (1990). Testis-specific transcription initiation sites of rat farnesyl pyrophosphate synthetase mRNA. Mol. Cell. Biol. 10:2315–2326.

Teruya, J. H., Salido, E. C., Edwards, P. A., and Clarke, C. F. (1991). Testis-specific transcripts of rat farnesyl pyrophosphate synthetase are developmentally regulated and localized to haploid germ cells. Biol. Reprod. 44:663–671.

Thomas, K. H., Wilkie, T. M., Tomashefsky, P., Bellvé, A. R., and Simon, M. I. (1989). Differential gene expression during mouse spermatogenesis. Biol. Reprod. 41:729–739.

Thomas, K., Del Mazo, J., Eversole, P., Bellvé, A., Hiraoka, Y., Li, S.S.L., and Simon, M. (1990). Developmental regulation of expression of the lactate dehydrogenase (LDH) multigene family during mouse spermatogenesis. Development 109:483–493.

Trasler, J. M., Johnson, P. A., Hake, L. E., Alcivar, A. A., Millette, C. F., and Hecht, N. B. (1990). DNA methylation and demethylation of post-meiotically expressed genes during meiotic prophase in the mouse testis. Mol. Cell. Biol. 10:1828–1834.

Vande Berg, J. L., Cooper, D. W., and Close, P. J. (1973). Mammalian testis phosphoglycerate kinase. Nature New Biol. 243:48–50.

van der Hoorn, F. A., Tarnasky, H. A., and Nordeen, S. K. (1990). A new rat gene RT7 is specifically expressed during spermatogenesis. Devel. Biol. 142:147–154.

Vera, J. C., Brito, M., Zuvic, T., and Burzio, L. O. (1984). Polypeptide composition of rat sperm outer dense fibers. J. Biol. Chem. 259:5970–5977.

Villasante, A., Wang, D., Dobner, P., Dolph, P., Lewis, S. A., and Cowan, N. J. (1986). Six mouse α-tubulin mRNAs encode five distinct isotypes: Testis-specific expression of two sister genes. Mol. Cell Biol. 6:2409–2419.

Virbasius, J. V., and Scarpulla, R. C. (1988). Structure and expression of rodent genes encoding the testis-specific cytochrome c. Differences in gene structure and evolution between somatic and testicular variants. J. Biol. Chem. 263:6791–6796.

Watanabe, K., Baba, T., Kashiwabara, S-I., Okamoto, A.,

and Arai, Y. (1991). Structure and organization of the mouse acrosin gene. J. Biochem. 109:828–833.

Waters, S. H., Distel, R. J., and Hecht, N. B. (1985). Mouse testes contain two size classes of actin mRNA that are differentially expressed during spermatogenesis. Mol. Cell Biol. 5:1649–1654.

Welch, J. E., Zimmerman, L. J., Joseph, D. R., and O'Rand, M. G. (1990). Characterization of a nuclear autoantigenic sperm protein. I. Complete sequence and homology with the *Xenopus* protein, N1/N2. Biol. Reprod. 43:559–568.

Wieben, E. D. (1981). Regulation of the synthesis of lactate dehydrogenase-X during spermatogenesis in the mouse. J. Cell. Biol. 88:492–498.

Willison, K., Lewis, V., Zuckerman, K. S., Cordell, J., Dean, C., Miller, K., Lyon, M. F., and Marsh, M. (1989). The t complex polypeptide 1 (TCP-1) is associated with the cytoplasmic aspect of Golgi membranes. Cell 57:621–632.

Willison, K., and Ashworth, A. (1987). Mammalian spermatogenic gene expression. Trends Genetics 3:351–355.

Willison, K. R., Dudley, K., and Potter, J. (1986). Molecular cloning and sequence analysis of a haploid expressed gene encoding t complex polypeptide 1. Cell 44:727–738.

Wolfe, S. A., and Grimes, S. R. (1991). Protein–DNA interactions within the rat histone H4t promoter. J. Biol. Chem. 266:6637–6643.

Wolfes, H., Kogawa, K., Millette, C. F., and Cooper, G. M. (1989). Specific expression of nuclear proto-oncogenes before entry into meiotic prophase of spermatogenesis. Science 245:740–743.

Wolgemuth, D. J., Behringer, R. R., Mostoller, M. P., Brinster, R. L., and Palmiter, R. D. (1989). Transgenic mice overexpressing the mouse homeobox-containing gene *Hox-1.4* exhibit abnormal gut development. Nature 337:464–467.

Wolgemuth, D. J., Gizang-Ginsberg, E., Ponzetto, C., and Zakeri, Z. F. (1988). Genetic control of germ cell function: developmentally regulated gene expression during gametogenesis, in *Molecular Biology of Fertilization* (Schatten, G., and Schatten, H., eds.) pp. 235–258, Academic Press, New York.

Wright, R. M., John, E., Klotz, K., Flickinger, C. J., and Herr, J. C. (1990). Cloning and sequencing of cDNAs coding for the human intraacrosomal antigen SP-10. Biol. Reprod. 42:693–701.

Yan, Y. C., Bai, Y., Wang, L., Miao, S., and Koide, S. S. (1990). Characterization of cDNA encoding a human sperm membrane protein related to A4 amyloid protein. Proc. Nat. Acad. Sci. USA 87:2405.

Yelick, P. C., Balhorn, R., Johnson, P. A., Corzett, M., Mazrimas, J. A., Kleene, K. C., and Hecht, N. B. (1987). Mouse protamine 2 is synthesized as a precursor whereas mouse protamine 1 is not. Mol. Cell Biol. 7:2173–2179.

Yelick, P. C., Kwon, Y. K., Flynn, J. F., Borzorgzadeh, A., Kleene, K. C., and Hecht, N. B. (1989). Mouse transition protein 1 is translationally regulated during the postmeiotic stages of spermatogenesis. Mol. Rep. Dev. 1:193–200.

Zabludoff, S. D., Karzai, A. W., and Wright, W. W. (1990). Germ cell–Sertoli cell interactions: The effect of testicular maturation on the synthesis of cyclic protein-2 by rat Sertoli cells. Biol. Repro. 43:25–33.

Zakeri, Z. F., Ponzetto, C., and Wolgemuth, D. J. (1988a). Translational regulation of the novel haploid-specific transcripts for the *c-abl* proto-oncogene and a member of the 70 kDa heat shock gene family in the male germ line. Dev. Biol. 125:417–422.

Zakeri, Z. F., Wolgemuth, D. J., and Hunt, C. R. (1988b). Identification and sequence analysis of a new member of the mouse *HSP70* gene family and characterization of its unique cellular and developmental pattern of expression in the male germ line. Mol. Cell. Bio. 8:2925–2932.

Zinn, S. A., Ebert, K. M., Mehta, N. D., Joshi, J., and Kilpatrick, D. L. (1991). Selective transcription of rat proenkephalin fusion genes from the spermatogenic cell-specific promoter in testis of transgenic mice. J. Biol. Chem. 266:23850–23855.

ADDENDUM

The following papers appeared since the completion of this chapter.

Fischman, K., Edman, J. C., Shackleford, G. M., Turner, J. A., Rutter, W. J., and Nir, U. (1990). A murine *fer* testis-specific transcript *(ferT)* encodes a truncated fer protein. Mol. Cell. Biol. 10:146–153.

Jen, J., Deschepper, C. F., Shackleford, G. M., Lee, C.Y.G., and Lau, Y.-F. C. (1990). Stage-specific expression of the lactate dehydrogenase-X gene in adult and developing mouse testes. Mol. Reprod. Dev. 25:14–21.

Kasahara, M., Gutknecht, J., Brew, K., Spurr, N., and Goodfellow, P. N. (1989). Cloning and mapping of a testis-specific gene with sequence similarity to a sperm-coating glycoprotein gene. Genomics 5:527–534.

Krawetz, S. A., Herfort, M. H., Hamerton, J. L., Pon, R. T., and Dixon, G. H. (1989). Chromosomal localization and structure of the human P1 protamine gene. Genomics 5:639–645.

Kremling, H., Luerssen, H., Adham, I. M., Klemm, U., Tsaousidou, S., and Engel, W. (1989). Nucleotide sequences and expression of cDNA clones for boar and bull transition protein 1 and its evolutionary conservation in mammals. Differentiation 40:184–190.

Lattion, A.-L., Soubrier, F., Allegrini, J., Hubert, C., Corvol, P., and Alhenc-Gelas, F. (1989). The testicular transcript of the angiotensin I-converting enzyme encodes for the ancestral, non-duplicated form of the enzyme. FEBS Lett. 252:99–104.

Yan Cheng, C., Grima, J., Stahler, M. S., Lockshin, R. A., and Bardin, C. W. (1989). Testins are structurally related Sertoli cell proteins whose secretion is tightly coupled to the presence of germ cells. J. Biol. Chem. 264:21386–21393.

CHAPTER SEVENTEEN

Patterns of Expression and Potential Functions of Cellular Oncogenes and Heat Shock Protein Genes During Germ Cell Development

Debra J. Wolgemuth, Carol M. Gruppi, Vida Vambutas, and Axel G. Wadewitz

I. Introduction

Major developmental processes are involved in gametogenesis, including mitotic and meiotic division, genetic recombination, and morphological differentiation (reviewed by Bellve, 1979). The expression of specific genes coincident with these events suggests that they play key roles in these processes. Spermatogenesis, in particular, provides a useful model system to identify the genetic basis of cellular differentiation within a defined cell lineage. Several features of spermatogenesis in mouse facilitate the study of lineage-specific gene expression in a mammalian system. Because there is a gradual temporal progression of spermatogenic cell types (Nebel et al., 1961), cells at later stages can be eliminated by using animals at specific stages of fetal or postnatal development. Also, enriched populations of germ cell types can be prepared readily (e.g., Wolgemuth et al., 1985). Several sterile male mouse mutants, in which germ cell development is arrested at different stages, are also known (Searle, 1982). The use of transgenic animals to elucidate gene function should also become extremely valuable as stage-specific promoters could be used either to overexpress or interfere with the function of specific genes at precise developmental stages (Hanahan, 1988; Jaenisch, 1988).

Elucidating the genetic program that governs mammalian germ cell differentiation has been made feasible by recent advances in molecular biology. Several strategies can be used to identify and characterize specific genes involved with the processes of spermatogenic differentiation. The strategies can be outlined as follows:

1. Examine genes expressed in other tissues but that exhibit unique patterns of expression in germ cells, especially those with potential function as proto-oncogenes, growth factors and their receptors, and cell-cycle genes
2. Identify genes expressed only in germ cells by subtractive hybridization screening or by enhancer-trap and gene-trap cloning
3. Use conserved sequences from genes with known development- or differentiation-regulating function in other systems to identify genes with potential regulatory function during gametogenesis

Each of the above approaches has certain advantages and certain inherent limitations. The first strategy usually permits immediate assessment of the pattern of expression and potential function because the genes are by definition already identified. The two classes of genes reviewed in this chapter were identified using this first approach.

II. Cellular Oncogenes

A. Introduction and Rationale for Studying Cellular Oncogene Expression and Function During Germ Cell Development

The genomes of several RNA tumor viruses contain genes, the so-called oncogenes, that cause the host cell to undergo transformation (reviewed in Bishop, 1983; Herrlich and Ponta, 1989). These viral oncogenes frequently have homologues in cellular DNA, known as proto-oncogenes or cellular oncogenes. Why then are viral oncogenes implicated in the production of transformed phenotypes? Differences in coding potential between the viral and cellular oncogenes can result in the altered oncoproteins having increased activity compared with that of the proto-oncogenes. Frequently viral oncogenes produce proteins, containing, as in retroviruses, either gag or env sequences attached to a partial proto-oncogene sequence lacking the normal amino terminus. Highly elevated levels of expression can occur as a result of the insertion of viral promoters next to cellular genes.

Cellular oncogenes are thought to play critical roles in normal growth and development. For example, in the mouse Harvey c-*ras* is expressed at all stages of embryogenesis and at high levels in extraembryonic tissues (Muller et al., 1982). The cellular oncogene c-*fos* is expressed abundantly in the mouse amnion and placenta (Muller et al., 1983), whereas c-*myc*, c-*erb A*, c-*src*, and c-*sis* are expressed at low levels in the embryo (Slamon and Cline, 1984).

Several cellular oncogenes have been shown to encode for growth factors or their receptors, genes clearly implicated in developmental pathways. The cellular oncogene c-*sis* encodes the B chain of the platelet-derived growth factor and c-*erb B* and c-*fms* encode the epidermal growth factor and colony stimulating factor-1 receptors, respectively (Waterfield et al., 1983; Downward et al., 1984; Scherr et al., 1985). Additional evidence for the importance of cellular oncogene function during development comes from recent studies in *Drosophila* where mutations in the homologues of mammalian cellular oncogenes have been shown to affect development. For example, a mutation in the *int*-1 homologue results in the *wingless* phenotype (Rijsewijk et al., 1987), whereas mutation of the c-*rel* homologue produces the *dorsal* phenotype (Steward, 1987).

Most proto-oncogenes can be divided into seven general groups: (1) growth factor receptor tyrosine protein kinases, which become activated on binding their respective ligands; (2) plasma membrane-associated protein kinases, also tyrosine-specific protein kinases, which are anchored to the membrane by myristylation; (3) cytoplasmic serine- and threonine-specific protein kinases; (4) plasma membrane-associated GTP-binding proteins; (5) nuclear proto-oncogenes; (6) proto-oncogenes encoding growth factors; and (7) tumor suppressor genes, whose wild type function is antiproliferative. Another type of cellular oncogene is represented by *int*-1. Although the *int*-1 product appears to be a secretory protein, its function remains unknown (Shackleford and Varmus, 1987).

A subset of the cellular oncogenes exhibit the interesting properties (for the purposes of this review) either of being expressed at high levels, of being expressed in unique, developmentally regulated patterns, and/or of yielding novel-sized transcripts in reproductive tissues. With the exception of the cellular oncogenes encoding growth factors, members of all classes are known to be expressed in both testis and ovary. Table 17–1 lists several members of each group that are expressed in germ cell-containing tissues. Currently 10 cellular oncogenes have been shown to be expressed specifically in germ cells and it would not be surprising if additional ones will be identified as new proto-oncogenes are discovered and RNA from reproductive tissues is examined for their expression.

B. Cellular Oncogenes Expressed in Germ Cells

1. c-*abl*

The cellular oncogene c-*abl* is expressed ubiquitiously in all murine tissues and cell lines ex-

Table 17-1. Examples of Proto-Oncogenes Expressed in Germ Cell-Containing Tissues

Gene	Size	Site of expression	Function	Reference
A. Growth factor receptor protein kinases				
c-elk	? kd	Adult testis, ovary	tyr kinase(?)	12
c-fms	140 kd	Adult testis	tyr kinase	14
c-kit	145 kd	Adult testis, ovary	tyr kinase	5,16
c-ret	? kd	Adult testis	tyr kinase(?)	22
c-trkB	145 kd	Adult testis, ovary	tyr kinase(?)	9
B. Plasma membrane-associated protein kinases				
c-flk	? kd	Adult testis, ovary	tyr kinase(?)	12
c-src	60 kd	Adult testis	tyr kinase	4
c-yes	62 kd	Oocytes	tyr kinase	21
C. Cytoplasmic protein kinases				
c-mos	43 kd	Spermatids, oocytes	ser/thr kinase	6,15
c-pim	33 kd	Early spermatids, adult ovary	ser/thr kinase	20
c-raf	74 kd	Meiotic spermatocytes, adult ovary	ser/thr kinase	20,24
D. Plasma membrane-associated GTP-binding proteins				
H-*ras*	21 kd	Male germ cells, oocytes	GTPase	11,24
K-*ras*	21 kd	Male germ cells, oocytes	GTPase	11,24
N-*ras*	21 kd	Male germ cells, oocytes	GTPase	11,24
E. Nuclear proto-oncogenes				
c-abl[a]	150 kd	Male germ cells, oocytes	tyr kinase	8,17
c-erb A	55 kd	Testis	thyroid hormone receptor	1
c-ets-1	56 kd	Immature testis	DNA binding protein(?)	3
c-ets-2	56 kd	Immature testis	DNA binding protein(?)	3
c-fos	62 kd	Spermatogonia	transcription activator	24
c-jun	39 kd	Spermatogonia	transcription activator	7,24
c-myc	62 kd	Premeiotic and meiotic male germ cells, oocytes	transcription activator	22,24
F. Tumor suppressor genes				
RB	~110 kd	Adult testis, ovary	nuclear phosphoprotein	2,10
p53	53 kd	Adult testis, ovary	nuclear phosphoprotein	18
G. Other proto-oncogenes				
int-1	36–44 kd	Spermatids	secretory glycoprotein (?)	19
c-rel	68 kd	Adult testis, ovary	?	13

[a]*c-abl* may also be cytoplasmic and membrane associated.

References: (1) Benbrook and Pfahl, 1987; (2) Bernards et al., 1989; (3) Bhat et al., 1987; (4) Boulter and Wagner, 1988; (5) Geissler et al., 1988; (6) Goldman et al., 1987; (7) Hirai et al., 1989; (8) Iwaoki, Y., Mutter, G. L., and Wolgemuth, D. J., unpublished observations; (9) Klein et al., 1989; (10) Lee et al., 1987; (11) Leon et al., 1987; (12) Letwin et al., 1988; (13) Moore and Bose, 1989; (14) Muller et al., 1983; (15) Mutter and Wolgemuth, 1987; (16) Nocka et al., 1989; (17) Ponzetto and Wolgemuth, 1985; (18) Rogel et al., 1985; (19) Shackleford and Varmus, 1987; (20) Sorrentino et al., 1988; (21) Steele et al., 1989; (22) Tahira et al., 1988; (23) Taylor et al., 1986; (24) Wolfes et al., 1989.

amined, yielding transcripts of 6.2 and 8.0 kb in size (e.g., Wang and Baltimore, 1983). An interesting exception to this pattern is observed in the adult mouse testes, where high levels of a novel 4.7 kb transcript are found (Muller et al., 1982). The expression of this novel transcript is regulated developmentally and restricted to haploid germ cells (Ponzetto and Wolgemuth, 1985).

The cellular oncogene c-*abl* is present as a single copy gene in the mouse genome and spans over 100 kb (Wang and Baltimore, 1983; Bernards et al., 1988). Molecular cloning and Northern blot hybridization analysis have shown that at least five distinct messenger RNAs (mRNAs) are produced (Ben-Neriah et al., 1986; Oppi et al., 1987; Meijer et al., 1987). Mouse lymphoid cells have four types of c-*abl* transcripts that differ in their 5′ sequences, encoding unique N-terminal regions of 20 to 45 amino acids. The two major mRNAs are initiated by separate promoters, whereas the two minor transcripts result from alternative splicing (Bernards et al., 1988).

The testis-specific transcript has been reported to have a 5′ sequence identical to the 6.2-kb somatic transcript (Oppi et al., 1987; Meijer et al., 1987), although the 3′ end lacks approximately 1 kb of 3′ untranslated sequences (Duggal et al., 1987; Meijer et al., 1987). The predicted amino acid sequence from complement DNA (cDNA) cloning (Oppi et al., 1987; Meijer et al., 1987), as well as our studies on c-*abl* proteins (Ponzetto et al., 1987), indicates that the 4.7-kb transcript produces a protein of about 150 kd, the same as that resulting from the somatic 6.2-kb transcript.

The c-*abl* protein encoded by the 8.0 kb somatic transcript (type IV, Ben-Neriah et al., 1986) initiates with the sequence Met-Gly-Gln, which is the myristylation site in both v-*abl* and c-*scr* proteins, suggesting that the protein may be anchored in the plasma membrane (Ben-Neriah et al., 1986). Recently, however, van Etten and colleagues (1989) found that a large fraction of type IV c-*abl* protein overexpressed in nontransformed 3T3 fibroblasts is localized in the nucleus. Analysis of the amino acid sequence indicates that the type IV c-*abl* protein has a nuclear localization signal similar to SV40 large T antigen (van Etten et al., 1989). Because the coding sequences of all c-*abl* transcripts are identical except for the 5′ region sequences, it is likely that the other c-*abl* isoforms may also function in the nucleus.

Immunoprecipitation and an in vitro kinase assay were used by our laboratory to demonstrate the presence of c-*abl* protein throughout the spermatogenic cell lineage (Ponzetto et al., 1989). As expected, each cell type examined contained the 150-kd c-*abl* tyrosine protein kinase. In addition, a phosphoprotein of about 74 kd (P74) was found in immunoprecipitates of extracts of postmeiotic spermatids and residual bodies. Double immunoprecipitation, using secondary site-specific c-*abl* antibodies to reprecipitate c-*abl* protein after heat denaturation in the presence of sodium dodecylsulfate and β-mercaptoethanol, demonstrated that P74 is not a truncated form or fragment of c-*abl*. Two-dimensional phosphopeptide mapping also indicated that the c-*abl* protein and P74 were distinct. The presence of P74 in anti-*abl* immunoprecipitates is most likely due to co-precipitation by virtue of its close association with the c-*abl* protein kinase.

Interestingly, phosphoamino acid analysis of the c-*abl* and P74 proteins indicates that both contain predominately phosphoserine, with approximately equimolar amounts of phosphotyrosine and phosphothreonine (Wadewitz and Wolgemuth, 1990). Because c-*abl* kinase is tyrosine specific, an additional serine/threonine protein kinase must be present in the immunoprecipitates, possibly the co-precipitated P74. Such a result is intriguing in light of the recent report that the platelet-derived growth factor (PDGF) receptor, a tyrosine protein kinase, co-precipitates with the c-*raf* serine/threonine protein kinase (Morrison et al., 1989).

2. c-*mos*

The cellular oncogene c-*mos* encodes a cytoplasmic serine/threonine protein kinase that, in the mouse, is expressed most abundantly in adult gonads, and at very low levels in adult epididymis and other somatic tissues (Mutter and Wolgemuth, 1987; Goldman et al., 1987; Propst et al., 1987). Very low levels of c-*mos* are also expressed in day 19 embryos (Propst et al., 1987). Tissue-specific variations in the size of c-*mos* transcripts appear to be due to the use of different promoters (Propst et al., 1987). In testis, c-*mos* is expressed as a 1.7-kb transcript almost exclusively in germ cells and most abundantly in early spermatids.

Studies of the function of c-*mos* during spermatogenesis are lacking, in marked contrast with the considerable amount of information concerning the possible function of c-*mos* in the vertebrate oocyte. Murine ovarian c-*mos* mRNA levels exceed testicular levels by several-fold (Goldman et al., 1987; Mutter and Wolgemuth, 1987). Northern blot and in situ hybridization analyses indicate an absence of c-*mos* RNA in primordial or resting oocytes (Mutter and Wolgemuth, 1987; Keshet et al., 1987). Transcripts of 1.4 kb in length begin to accumulate in the cytoplasm of growing oocytes, increasing 36- to 90-fold as the oocyte increases in volume (Mutter and Wolgemuth, 1987; Goldman et al., 1987).

In murine oocytes, germinal vesicle breakdown (GVBD) occurs within 4 hours of the resumption of meiosis and meiosis I is complete by 6 to 8 hours. The mature oocyte will arrest again at metaphase II until fertilization, at which time meiosis II is completed. The levels of c-*mos* transcripts appear to be stable as the oocytes resume meiosis, then drop precipitously at the metaphase II stage (Mutter et al., 1988). No c-*mos* transcripts were detected in two-cell or early preimplantation embryos by in situ hybridization analysis (Mutter et al., 1988).

Mouse oocytes undergoing meiotic maturation, as well as fully grown, IBMX-arrested oocytes, and metaphase II oocytes all contained the c-*mos* protein product, $pp39^{mos}$, whereas growing oocytes did not (Paules et al., 1989). (IBMX, a phosphodiesterase inhibitor, was used to maintain meiotic arrest in the fully grown murine oocytes.) Thus, although c-*mos* transcripts accumulate in the growing oocyte, they are not translated until maturation.

The function of c-*mos* in murine oocytes was investigated by injection of c-*mos* antisense oligodeoxynucleotides (oligos) (O'Keefe et al., 1989). Oocytes injected with c-*mos* oligos failed to arrest at metaphase II, and often cleaved into two cells. It was concluded that loss of c-*mos* function leads to premature reformation of a nucleus, essentially permitting ensuing DNA synthesis and cell cleavage. These results also implied that the c-*mos* product is needed for oocyte maturation.

In most species, fully grown oocytes are arrested at the first meiotic prophase (Masui and Clark, 1979). In *Xenopus*, this arrest is released by progesterone and is followed by the appearance of the maturation promoting factor (MPF), which causes GVBD. Progesterone-induced oocyte maturation also results in the appearance of the c-*mos* product, $pp39^{mos}$ (Sagata et al., 1989a). The increase in c-*mos* product appears to involve translation of c-*mos* mRNAs that had accumulated during the growth phase (Sagata et al., 1989a). These observations further support a role for c-*mos* during oocyte maturation, here correlated with progesterone-induced maturation and the activity of MPF (Sagata et al., 1988; Watanabe et al., 1989).

Early attempts to immunoprecipitate $pp39^{mos}$ from *Xenopus* oocyte homogenates were unsuccessful, most likely owing to its rapid turnover or to the low concentrations in the cytosol (<0.001 percent of the total soluble protein; Watanabe et al., 1989). The $pp39^{mos}$ degradation in homogenates was prevented by addition of EGTA, a calcium chelator (Watanabe et al., 1989). Eggs activated by treatment with the calcium ionophore A23187 or by fertilization similarly lacked $pp39^{mos}$ (Watanabe et al., 1989). The mechanisms responsible for the $pp39^{mos}$ turnover have not been elucidated, but the influx of Ca^{2+} during fertilization may activate a specific protease. Calpain, a Ca^{2+}-dependent cysteine protease has been recently implicated in selective $pp39^{mos}$ degradation after fertilization or activation (Watanabe et al., 1989).

Microinjection of cytoplasm from mature oocytes into blastomeres of two-cell embryos induces cleavage arrest at metaphase, apparently due to a cytostatic factor (CSF) (Masui and Markert, 1971). When 5′ capped c-*mos* RNA (which directs the synthesis of $pp39^{mos}$) was injected into one of the blastomeres of the two-cell embryo, all of the blastomeres ceased to cleave within 30 to 90 minutes after injection (Sagata et al., 1989b; Watanabe et al., 1989). Injection of c-*mos* RNA preannealed with antisense oligos did not induce cleavage arrest nor did c-*mos* RNA lacking the ATP binding domain. It has been suggested that the c-*mos* product may be the CSF responsible for arrest of metaphase II oocytes. Fertilization generates a signal to degrade the c-*mos* product, thereby removing the block to division of the zygote.

3. c-*raf*-1

The cellular oncogene c-*raf*-1, also a member of the cytoplasmic serine/threonine protein kinase family, is expressed throughout spermatogenesis as a 3.1-kb transcript. The C-*raf*-1 mRNAs are found in type A and B spermatogonia through the round spermatid stage, but not in residual bodies. The highest levels are observed at the pachytene spermatocyte stage (Wolfes et al., 1989). Adult ovary also abundantly expresses a 3.1-kb transcript (Sorrentino et al., 1988).

The potential role of c-*raf* during germ cell development is particularly intriguing as c-*raf* kinase associates with the ligand-activated PDGF growth factor receptor, a tyrosine protein kinase. Morrison and co-workers (1989) have shown that baculoviral-expressed PDGF receptor kinase and c-*raf* kinase associate in vitro. After co-incubation, c-*raf* kinase activity is stimulated four to sixfold. Similarly, cells transformed with other tyrosine kinases, such as v-*src*, v-Ha-*ras*, or v-*fms*, show elevated c-*raf* kinase activity (Roberts et al., 1988). Thus, c-*raf* may play a critical role in the signal transduction of growth factor receptor kinases expressed during spermatogenesis.

4. *pim*-1

In the mouse, *pim*-1, another member of the serine/threonine protein kinase family, is expressed predominantly in thymus and spleen as a 2.8-kb transcript. The testis contains an additional abundant transcript of 2.4 kb in size. The 2.4-kb testis-specific transcript is expressed only in postmeiotic spermatids. Adult mouse ovary expresses only the 2.8-kb transcript (Sorrentino et al., 1988).

5. c-*ras*

The mammalian *ras* family has three members designated H-*ras*, K-*ras*, and *N*-*ras*, originally isolated from rat Harvey and Kirsten sarcomas, and a human neuroblastoma cell line, respectively (Barbacid, 1987). All three code for very similar 21 kd proteins that can bind and hydrolyze GTP. The expression of the *ras* gene family has been shown to vary with tissue type and during development, and to exhibit a unique pattern of expression in the testis.

The H-*ras* transcripts are present in spermatogonia, meiotic spermatocytes, and in postmeiotic round spermatids and residual bodies (Wolfes et al., 1989). The K-*ras* gene is transcribed at low levels in spermatogonia, at highest levels in pachytene spermatocytes; it decreases sharply by the round spermatid stage, and is lacking in residual bodies (Wolfes et al., 1989). The H-*ras* and K-*ras* germ cell transcripts are the same size as the somatic transcripts (Leon et al., 1987). The N-*ras* transcription during spermatogenesis is more complex. A 5.0-kb transcript is found in all male germ cells, and three additional transcripts (3.5, 1.9, and 1.4 kb) are detectable in meiotic prophase and postmeiotic germ cells. The 3.5- and 1.9-kb transcripts appear to be male germ cell specific (Leon et al., 1987; Wolfes et al., 1989). All three *ras* members are also expressed in adult ovary, with N-*ras* having the highest expression (Leon et al., 1987).

Interestingly, other members of the *ras* family are also expressed at high levels in testis and ovary. The highest expression of *Ral* A is found in the testis. The *Rho* 12, another *ras*-related gene, is transcribed at high levels in both ovary and testis, compared with the low levels expressed in all somatic tissues examined (Olofsson et al., 1988).

Like c-*mos*, in *Xenopus* oocytes injection of c-*ras* protein also induces maturation (Birchmeier et al., 1985; Allende et al., 1988). Injection of monoclonal antibodies that inactivate c-*ras* inhibit maturation induced by insulin but not by progesterone (Deshpande et al., 1987; Korn et al., 1987). The c-*ras* product may be functioning upstream of c-*mos*, however, since interfering with c-*mos* function by injecting antisense oligos blocked $p21^{ras}$ induction of oocyte maturation (Barrett et al., 1990).

6. THE NUCLEAR PHOTO-ONCOGENES c-*fos*, c-*myc*, AND c-*jun*

Wolfes and associates (1989) have shown that three nuclear proto-oncogenes, c-*fos*, c-*jun*, and c-*myc*, are expressed at high levels in type B spermatogonia. The cellular oncogenes c-*myc* and c-*jun* were also detected at high levels in preleptotene spermatocytes. The c-*myc* protein, however, has only been detected in nuclei of spermatogonia (Koji et al., 1988). Spermatogenic cell types at later stages of development express c-*jun* and c-*fos* at levels approximately tenfold lower and their transcripts become undetectable in haploid cells. Detectable levels of c-*fos* protein are found only in type B spermatogonia (Wolfes et al., 1989). Germ cell transcripts of these three cellular oncogenes appear to be the same size as those of somatic cells (Gonda and Metcalf, 1984; Rauscher et al., 1988), in contrast to the observation of germ cell-specific sized transcripts for c-*abl,* c-*mos,* and so forth. Furthermore, the higher levels of c-*myc,* c-*fos,* and c-*jun* in mitotic stages of germ cell differentiation are very different from the pattern observed for c-*abl* and c-*mos,* for example, and suggest that these nuclear proto-oncogenes may function in regulating mitotic cell division rather than in postmeiotic spermatogenic events.

The cellular oncogene c-*jun* may also play an important role during oogenesis as even higher levels of c-*jun* transcripts are found in the ovary (Hirai et al., 1989). Interestingly, Hirai and colleagues (1989) found that the two other members of the *jun* family, *jun* B and *jun* D, are also expressed in germ-cell containing tissues. Very low levels of *jun* D were found in mature testes and moderate expression was observed in ovaries. The *jun* B is also expressed in testes and ovaries, at levels approximately 5 to 10 times higher than *jun* D (Hirai et al., 1989).

In early oogenesis high levels of c-*myc* mRNA accumulate (Taylor et al., 1986; Mechali et al., 1988), reaching a maximum during midoogenesis. Mature oocytes contain 5×10^6 c-*myc* mRNA molecules, whereas somatic cells typically contain only 10 to 30 copies (Mechali et al., 1988). In situ hybridization experiments indicate that c-*myc* mRNA is evenly distributed in the cytoplasm of the oocyte. The accumulation of c-*myc* in the nonmitotically dividing oocyte is in marked contrast to its pattern of accumulation in the male germ line, where it accumulates at highest levels in type B spermatogonia (Wolfes et al., 1989).

Recently, c-*fos* and c-*jun* proteins were shown to form a complex that binds to AP-1 transcriptional control elements (Curran and Franza, 1988; Vogt and Tjian, 1988; Rauscher et al., 1988). Lech and colleagues (1988) have shown that c-*myc* protein also functions as a transcriptional activator when expressed as a fusion protein in yeast. The simultaneous transcription of these nuclear proto-oncogenes in type B spermatogonia suggests that they mediate changes in gene expression during this stage of differentiation.

7. *int*-1

Among cellular oncogenes, *int*-1 is unusual in that the protein product appears to be a secreted glycoprotein of approximately 36 to 44 kd (Papkoff et al., 1987). Unlike most proto-oncogenes, which are often expressed in several cell types and tissues, the expression of *int*-1 is restricted to the neural tube of midgestational embryos and postmeiotic male germ cells (Shackleford and Varmus, 1987). Murine *int*-1 transcripts of 2.6 kb are expressed abundantly in round spermatids, with sharply lower levels in elongating spermatids. Pachytene spermatocytes contain little or no *int*-1 transcripts. The *int*-1 transcripts in the embryonic central nervous system are also 2.6 kb in length (Shackleford and Varmus, 1987).

The restricted expression pattern of *int*-1 has striking parallels with that of the homeobox-containing gene, *Hox*-1.4. As discussed later, *Hox*-1.4 is expressed almost exclusively in pachytene spermatocytes, postmeiotic spermatids and the embryonic nervous system (Wolgemuth et al., 1986, 1987, 1989b; Rubin et al., 1986; Duboule et al., 1986). It is curious that there are also some recessive neurological mutations, such as the murine mutants *quaking* and *wobbler,* which also affect germ cell differentiation (Searle, 1982).

C. Cellular Oncogenes Expressed in Germ Cells in Patterns as Yet Undefined With Respect to Development

1. c-*kit*

The c-*kit* cellular oncogene, a putative receptor tyrosine protein kinase, was shown recently to be the product of the dominant *white spotted* (*W*) locus in the mouse (Geissler et al., 1988; Nocka et al., 1989). The pleiotrophic phenotypes of *W* mutants include white coat color, sterility in both sexes, and anemia, and have been attributed to the failure of stem cells to migrate or proliferate during development (Mintz and Russell, 1957). Mapping experiments reported by Geissler and colleagues (1988) demonstrate that the c-*kit* structural gene maps within the w^{19H} deletion that covers the *W* locus. To support the hypothesis that DNA rearrangements of the c-*kit* locus are associated with *W* mutations, these authors investigated two other *W* mutants. They reported that c-*kit* gene is disrupted in spontaneous alleles, W^{44} and W^{X}. In W^{44} DNA encoding amino acids 240 to 342 of c-*kit* is disrupted and in W^{X}, the disruption occurs between amino acids 342 to 791 (Geissler et al., 1989).

The *Steel* (*Sl*) gene has been identified as the ligand for the c-*kit* receptor (Copeland et al., 1990; Flanagan and Leder, 1990; Huang et al., 1990; Williams et al., 1990; Zsebo et al., 1990). *Sl* is expressed abundantly in the stromal cells of the mesoderm along the migratory pathway of the primordial germ cells during the same period of fetal development in which the germ cells themselves express the c-*kit* gene; *Sl* is also expressed in the genital ridge (Matsui et al., 1990). Recent studies on *Sl* expression in adult mouse tissues revealed that in adult gonads, *Sl* is expressed in follicular cells of the ovary and in Sertoli cells of the testis (Motro et al., 1991; Tajima et al., 1991).

2. TUMOR SUPPRESSOR GENES (*RB* AND *P53*)

In addition to the proto-oncogenes involved in the regulated activation of cell proliferation, another class of genes, the tumor suppressor genes, normally function to inhibit proliferation. Loss of function mutations of these genes result in deregulation of the cell cycle or a block to the cell cycle exit (Cooper and Whyte, 1989). The best characterized member of this group is the retinoblastoma susceptibility gene (*RB*). In the mouse, *RB* has been shown to be expressed as a 4.7-kb transcript in all adult tissues examined (Bernards et al., 1989). In addition to the normal 4.7-kb transcript seen in somatic tissues, adult testes also express a 2.8-kb species, the appearance of which coincides with the development of postmeiotic spermatids (Bernards et al., 1989). The adult ovary expresses only the larger transcript (Lee et al., 1987).

The product of *p53*, originally discovered as complexing with the SV40 large T antigen (Lane and Crawford, 1979), has been detected in numerous actively proliferating nontransformed cells but is undetectable or present at low levels in resting cells. Both adult ovary and testis express relatively high levels of *p53*, compared with most somatic tissues (Rogel et al., 1985).

Impairment of normal tumor suppressor gene function is increasingly being associated with a wide variety of carcinomas. Interestingly, Saksela and associates (1989) failed to detect any *RB* expression in a human seminoma, a testicular germ-cell tumor. In addition, the c-*myc* homologue, N-*myc*, was found to be expressed at a much higher level than normal. Thus, *myc* activation and decreased *RB* expression may synergize the oncogenic potential of each other, as has been suggested for other tumor types (Saksela et al., 1989).

3. *flk* AND *elk*

Letwin and co-workers (1988) screened a rat brain lambda gt11 cDNA expression library with antiphosphotyrosine antibodies and identified two clones. One of the two, *flk*, shared homology to the *fps/fes* proto-oncogene and was found to be expressed widely, but most abundantly in adult testes. Testes also contained two *flk* transcript sizes not seen in somatic tissues. The other clone, *elk*, shared homology with the *eph* proto-oncogene and demonstrated limited tissue-specific expression, being abundant in the brain and present at low levels in the testes. A similar experimental approach using testis cDNA libraries would likely reveal additional unknown tyrosine protein kinases.

D. Expression of Cellular Oncogenes in Nongerminal Compartments of Testes

Proto-oncogene expression in nongerminal cells of the testes has been documented much less extensively than expression in germ cells, and the observations are based primarily on in vitro systems. Follicle-stimulating hormone (FSH)-induced expression of c-*fos* in primary Sertoli cell cultures was observed by Hall and colleagues (1988). Increased transcription of the c-*fos* gene was likely mediated by cyclic adenosine monophosphate (cAMP) as FSH is known to stimulate cAMP levels in target cells (Parvinen, 1982). Lin and colleagues (1988) evaluated the effect of in vivo administration of human chorionic gonadotropin (hCG) on nuclear proto-oncogene expression in Leydig cells of rat testes. Rats were treated with hCG and Leydig cells were isolated and purified 1 to 24 hours later. Although hCG caused a transient expression of c-*myc* mRNA, c-*fos* mRNA levels were unaffected.

Low levels of expression of several proto-oncogenes with protein kinase activity, which are known to be expressed in the germ line, may also be expressed in the surrounding somatic cells of the gonads. The two c-*abl* somatic transcripts of approximately 6 and 8 kb are also expressed in the somatic cells of the testes (Goldman et al., 1987; Propst et al., 1988). Low levels of c-*mos* transcripts were also detected in the sterile *W* mouse mutants and in sex-reversed mice that lack germ cells, as well as in mutants in which spermatogenesis is arrested at the pachytene or diplotene stages (Propst et al., 1988). Although the low levels of c-*mos* expression may be due to the presence of a few germ cells in the mutant gonads, it is equally possible that the somatic compartment of the male gonad does contribute c-*mos* mRNAs.

III. Heat Shock Protein Genes

A. Definition and Classification of Heat Shock Proteins

Prokaryotic and eukaryotic organisms respond to exogenous stress by synthesizing a distinct set of proteins. This cellular reflex, the *heat shock response,* was first recognized during studies on the temperature sensitivity of *Drosophila* embryos (Ritossa, 1964). It was observed that the puffing pattern of the polytene chromosomes was altered with a change in temperature from 25° to 30°C. This alteration in transcriptional activity was associated with the synthesis of a discrete set of proteins, the heat shock proteins (hsp) (Tissieres et al., 1974). The heat shock response was found to be present in a wide variety of organisms, including bacteria, plants, and animals, and to be induced by exposure to a wide variety of conditions or agents such as anoxia, heavy metals, and viral infection (reviewed in Craig, 1985; Lindquist, 1986). The evolutionary conservation of the cellular response to stress suggests that hsp play a critical role in the maintenance of cellular homeostasis.

The expression of hsp genes and their corresponding proteins is not limited to a stress response, because the genes for hsp and hsp-related family members are also induced during normal development and growth. As is the case for the bona fide heat shock response, developmental induction of the hsp can be found in a wide variety of organisms including flies, frogs, mice, and humans (Craig, 1985).

The hsp synthesized by most species examined to date can be divided into three families based on molecular weight (Pardue, 1988): (1) the hsp20 family, with hsp in the molecular weight range of 15 to 30 kd; (2) the hsp70 family, which includes proteins that range in size from 65 to 75 kd; and (3) the hsp90 family, with proteins in the range of 80 to 90 kd (Pelham, 1985; Pardue, 1988). The apparent molecular weights of the hsp vary within a family between species and even among family members of the same species. For example, *Drosophila* has seven major hsp genes (hsp83, hsp70, hsp68, hsp27, hsp26, hsp23, and hsp22). The hsp90 family contains hsp83 as its only member; the hsp70 family contains hsp70 and hsp68 as members, and the hsp20 family contains hsp27, hsp26, hsp23, and hsp22. In contrast, in mouse both the hsp70 and hsp90 families are multigene families. The murine hsp70 family appears to contain at least five genes (Lowe et al., 1986), whereas the murine hsp90 family contains eight genes (Moore et al., 1989). Sequence and protein data suggest that only two of the genes in the mouse hsp90 family encode functional proteins; the remaining genes are apparently pseudogenes (Moore et al., 1989).

Although the molecular weight within a hsp family varies, the hsp all exhibit a high degree of conservation at both the nucleotide and amino acid level. The hsp70 from *Escherichia coli,* yeast, *Drosophila,* and humans exhibit greater than 80 percent homology at the amino acid level (Schlessinger, 1986). Mouse hsp84 has 81 percent amino acid homology and 74 percent nucleotide sequence homology to *Drosophila* hsp83 (Moore et al., 1987).

Homology among the heat shock families is not limited to the coding region of the hsp genes, but also extends to the 5′ regulatory region. Consensus sequences 5′ to the promoter are required for proper stress and/or developmental induction of the hsp genes. This consensus sequence (5′CTnGAAnnTTCnAG3′) is termed a heat shock element (HSE) and has been shown to be present in multiple copies in the 5′ region of hsp genes from several species (Pelham, 1985; Lindquist, 1986).

Deletion analysis of the 5′ regulatory region of the hsp26 gene of *Drosophila* has shown that different regulatory elements are required for its proper developmental and heat induction (Cohen and Meselson, 1985). Sequences upstream of −341 are required for ovarian expression of hsp26 but not for heat-induced expression. Regions between −351 and −53 are required for heat-induced expression but not for ovarian expression. Hsp26 expression in *Drosophila* spermatocytes also requires 5′ regulatory elements (Glaser et al., 1986). A hsp 26-lacZ fusion gene was introduced into the germ line of *Drosophila* by P-element mediated transformation. The construct contained 2 kb of the 5′ flanking region of hsp26. Expression of β-galactosidase was observed in nurse cells and spermatocytes; this pattern is identical to endogenous hsp26 expression. A second construct containing 278 bp of 5′ flanking region expressed β-galactosidase only in spermatocytes. These results imply that distinct 5′ regulatory elements are involved in the expression of hsp26 in the germ line of *Drosophila*.

Analysis of transgenic mice carrying a mouse hsp68-lacZ construct also demonstrated that there are distinct 5′ regions responsible for proper heat and developmental induction of the mouse gene as well (Kothary et al., 1989). Approximately 800 bp of the 5′ flanking region of hsp68, a region known to contain five HSE, was sufficient for proper heat induction of the hsp68 gene but was not sufficient for developmental induction (Kothary et al., 1989).

B. Functions of hsp

Although there is much information on the structure and expression of hsp, the exact function of these proteins remains elusive. It is generally assumed that hsp expression "rescues" the cell from damage caused by exogenous stress. Speculation on the functions of hsp during stress and developmental induction has been based primarily on the subcellular location of the hsp and their association with other proteins.

One postulated function of hsp in stressed and nonstressed cells may involve mediating protein–protein interactions. Hsp are induced by injection of denatured proteins into nonstressed *Xenopus* cells (Ananthan et al., 1986). Hsp90 is associated with the inactive form of several steroid receptors (Joab et al., 1984; Catelli et al., 1985; Sanchez et al., 1985). It is thought that dissociation of the steroid receptor from hsp90 allows it to convert to its active form (discussed in Pratt, 1987; Beato, 1989). Hsp90 has also been found to be associated with the *src* tyrosine kinase as it is transported from the nuclear to plasma membrane (Brugge, 1986). It is postulated that binding of hsp90 to the hydrophobic protein keeps the kinase soluble until it reaches the plasma membrane (Pelham, 1985, 1986), essentially acting as a "molecular chaperone" (Pelham, 1988). Hsp in general appear to facilitate movements of proteins within the cell and prevent protein complexes from forming insoluble aggregates.

Hsp may also function in the assembly and disassembly of macromolecular structures (Pelham, 1986). An hsp70 protein appears to function as an uncoating ATPase in the disassembly of clathrin trimers in coated vesicles (Chappell et al., 1986). In vitro studies have shown hsp90 crosslinks actin fibrils in a calmodulin-dependent manner (Koyasu et al., 1986; Nishida et al., 1986). These observations further support the role of hsp as mediators of protein–protein interaction in the cell.

Another class of stress-induced proteins are the glucose-regulated proteins (GRP). GRPs are related to hsp but are induced by glucose deprivation of cells, rather than by heat shock (reviewed by Pelham, 1986). GRPs are also localized in the endoplasmic reticulum of nonstressed cells where they may also mediate protein–protein interaction, as has been proposed for the other hsp (Pelham, 1986).

C. Rationale for Examining hsp Expression and Function in Germ Cells

Hsp have been shown to be expressed during development of the germ cell lineage in a wide variety of organisms. In yeast, hsp26 and hsp24 are induced strongly during meiosis and acrospore formation (Kurtz et al., 1986). Hsp83, hsp28, and hsp26 are produced in the ovarian nurse cells of *Drosophila* and subsequently transported to the oocyte (Zimmerman et al., 1983). Hsp26 is expressed in the male germ cell lineage in *Drosophila* as well, specifically in spermatocytes (Glaser et al., 1986). In the germ cell lineage of mouse, both the 70-kd and 89-kd hsp have been shown to be expressed in preovulatory oocytes as well as in granulosa cells (Curci et al., 1987).

Two members of the hsp70 gene family have been shown to be expressed in a developmentally regulated pattern in germ cells of mouse testes (Zakeri and Wolgemuth, 1987; Zakeri et al., 1988). The expression of hsp genes in germ cells likely extends to other mammalian species, as members of the hsp70 and hsp20 families have been shown to be expressed in human testes (Zakeri and Wolgemuth, 1987; C. M. Gruppi and D. J. Wolgemuth, unpublished observations).

The conservation of hsp expression during oogenesis and spermatogenesis in evolutionarily diverse organisms suggests a fundamental role for hsp during germ cell development. The effect of hyperthermia on male germ cells was recognized at the time of Hippocrates (cited in Rock and Robinson, 1965). Histological studies in the early part of this century demonstrated that it is the germinal compartment of testes that is damaged by heat (Moore and Chase, 1923; Young, 1927). In addition, a differential sensitivity to heat was observed within the germ cell population. Spermatogonia and spermatozoa are the most resistant, whereas primary spermatocytes and early spermatids are the most sensitive (Young, 1927).

D. Heat Shock Gene Expression During Mammalian Development

1. EMBRYOGENESIS

Mammalian embryos, like those of flies, sea urchins, and frogs, express hsp during normal development. A member of the hsp70 gene family is among the first products of the mouse zygotic genome and is expressed at high levels in the early mouse embryo (Bensaude et al., 1983; Morange et al., 1984). Two electrophoretically distinct hsp70 proteins are synthesized in the mouse preimplantation embryo and in undifferentiated embryonal carcinoma (EC) cells (Bensaude and Morange, 1983; Barnier et al., 1987). Upon in vitro differentiation of EC cells the levels of the two hsp70 proteins decrease, suggesting that expression of the hsp70 proteins corresponds to developmental cues (Levine et al., 1984; Barnier et al., 1987). Developmental expression of hsp70 family members was observed in placenta starting at day 8.5 and in yolk sac starting at day 11.5 (Kothary et al., 1987). No expression of this hsp70 family member was detected in embryonic tissues until day 15.5 (Kothary et al., 1987). Members of the hsp90 gene are also expressed in the developing mouse conceptus. Transcripts from two distinct hsp90 genes are seen at readily detectable and relatively constant levels in the embryonic and extra-embryonic compartments throughout midgestation (Gruppi et al., 1991).

It has been demonstrated that mammalian embryos incur developmental abnormalities when subjected to hyperthermia. High temperatures increase the incidence of abnormal cleavage in preimplantation embryos, causing retarded fetal development (Bellve, 1972; Bellve, 1973; Ulberg et al., 1973) and birth defects (Webster et al., 1984; Germain et al., 1985). As in other developing systems, heat induction of hsp is stage dependent in the mammalian embryo. Before the blastocyst stage, no induction of hsp in response to heat shock is observed (Morange et al., 1984). After the blastocyst stage, a strong induction of the heat inducible hsp70 family member can be detected (Morange et al., 1984). The mechanism underlying the existence of this refractory period for heat induction of hsp is unknown.

2. GERM CELL DIFFERENTIATION

Hsp genes are expressed in a precise temporal pattern in the developing mammalian germ line. Two hsp70 family members exhibit unique patterns of expression during male germ cell differentiation in the mouse. An abundant 2.7-kb hsp70 transcript is detected in adult mouse testes in the absence of exogenous stress (Zakeri and Wolgemuth, 1987). This hsp70 transcript is unique to testes and distinct from the hsp70 transcripts of 3.5 kb and 2.4 kb seen in somatic tissues (Zakeri and Wolgemuth, 1987). The testicular transcript, designated as HSP 70.1, is restricted to the germinal rather than the somatic compartment of the testes. The HSP 70.1 transcript first appears during the haploid stages of spermatogenesis and is stable throughout spermiogenesis (Zakeri and Wolgemuth, 1987).

A second hsp70 family member, designated as HSP 70.2, is also expressed at high levels in mouse testes (Zakeri et al., 1988). The expression of HSP 70.2 is restricted to the germinal compartment of testes, and like HSP 70.1 is 2.7 kb in size (Zakeri et al., 1988). The developmental pattern of expression of HSP 70.2 is distinct from that of HSP 70.1 in the germ cells of testes, as HSP 70.2 is expressed primarily in meiotic prophase cells (Zakeri et al., 1988).

Members of the hsp90 family also exhibit cellular specificity of expression in the murine testis. Two distinct hsp90 transcripts, approximately 3.2 kb and 2.9 kb in size, are detected in mouse testes (Gruppi et al., 1991). Based on the size of the putative corresponding proteins, the 3.2-kb transcript is designated as being the product of the *HSP86* gene, and the 2.9-kb transcript is likely to be the product of the *HSP84* gene.

Hsp 86 is more abundant in testes than is hsp84. Hsp86 is expressed primarily in the germinal compartment of testes, particularly in germ cells in meiotic prophase, whereas hsp84 is expressed primarily in the somatic compartment of testes and at extremely low levels in the germinal compartment of testes. Both hsp90 family transcripts were expressed in testes of other mammals, suggesting an important role for hsp90 in testicular function (Gruppi et al., 1991).

Studies of the expression of hsp at the protein level in the male germ line have focused on the hsp70 family. Low levels of a 72-kd and a 73-kd hsp are detected in the mouse testes (Zakeri et al., 1990). A unique isoform of approximately 73 kd, designated 73T, is expressed in the germinal compartment of the testes. The 73T is present in meiotic prophase and postmeiotic germ cells and is not induced by heat shock. The 72-kd hsp is induced only slightly in the adult mouse testes. In contrast, induction of the 72-kd protein is substantial in the testes of germ-cell-deficient mice. These data suggest that it is the somatic compartment, not the germinal compartment of the testes, that is responsible for the production of bona fide hsp70 in response to heat shock.

There are alternative observations that suggest that the level of 72-kd protein is increased by heat shock in meiotic and postmeiotic germ cells (Allen et al., 1988). Evidence from studies on mouse oocytes, however, reinforces the concept that mammalian germ cells are refractory to heat shock induction of hsp (Curci et al., 1987). Unstressed mouse oocytes and granulosa cells actively synthesize hsp70 and hsp89. Upon heat shock the murine oocytes continue to synthesize the 70-kd and 89-kd proteins at the same level. Heat-shocked granulosa cells synthesize de novo the 33-kd and 68-kd hsp and exhibit increased expression of the 70-kd protein.

E. Possible Functions of Heat Shock Genes in Male Germ Cell Differentiation

One can hypothesize several potential functions for hsp during mammalian spermatogenesis because expression of members of the hsp90 and hsp70 correlates with discrete points in testicular maturation and with particular cell types. It has been speculated that hsp are involved in the assembly and disassembly of macromolecular structures. These observations are significant with respect to spermatogenesis because of the major morphogenetic changes that occur during germ cell differentiation. Major protein structures, like the synaptonemal complex, are assembled and disassembled during meiotic prophase. During spermiogenesis, structures including the acrosome and the flagellum are formed.

As previously mentioned, it has also been postulated that hsp mediate protein–protein interaction. Particularly relevant to spermatogenesis is the association of hsp90 with steroid receptors. Hsp90 has been shown to be associated with various steroid receptors in the non-DNA binding state; disassociation of hsp90 from the receptor converts it to its DNA-binding state (Pratt, 1987; Beato, 1989). Because the testis is both a steroid hormone producing and responsive tissue, the hsp90 family may function as a binding protein for various steroid receptors. Given the role of retinoic acid and its derivatives in the maintenance and progression of spermatogenesis (Wolbach and Howe, 1925), it will be of interest to determine if hsp90 proteins are also associated with the receptors for these steroid hormone family members (Petrovich et al., 1987; Giguere et al., 1987).

IV. Strategies to Identify Function of Cellular Oncogenes and Heat Shock Genes During Germ Cell Differentiation

Genetic approaches have proved very powerful in elucidating the control of development in many systems. Several of these strategies can be used to study gene function during spermatogenesis. One approach would involve altering the expression of the gene in question during germ cell differentiation in a negative or positive manner in transgenic mice. Negative interference, accomplished by mutating the endogenous gene by homologous recombination or by overproducing an antisense orientation of the gene, would produce a loss of function mutation. Positive interference could be accomplished by ectopic or elevated levels of expression of a fully functional gene, under the direction of the endogenous or a heterologous promoter, yielding a gain of function phenotype.

The classic approach of generating mutations, which is now feasible in mice because of recent advances in the use of targeting recombination in embryonic stem cells (Robertson et al., 1986; Thomas and Capecchi, 1987; Frohman and Martin, 1989), involves disrupting the endogenous gene and examining the resulting phenotype in the heterozygous or homozygous state. As most cellular oncogenes or hsp genes are also expressed in various embryonic and adult tissues, a concern is that completely disrupting the function of these genes might result in embryonic or neonatal lethality. Recently, Schwartzberg and colleagues (1989) reported the disruption of the c-*abl* proto-oncogene and the generation of mice carrying the disrupted gene in the germ cell line. It will be most interesting to examine the resulting phenotype in these animals, especially with respect to reproductive function. The fact that the heterozygous animals are fertile implies that, at very least, mice need only one functional c-*abl* gene in the germ line.

The usefulness of producing gain of function mutations is supported by our recent experiments in which we manipulated the expression of *Hox*-1.4 in transgenic mice and produced an interesting developmental abnormality (Wolgemuth et al., 1989a). Elevated levels of *Hox*-1.4 in the mesodermal germ layer of midgestation embryos correlate with the development of congenital megacolon in neonatal and adult animals, apparently due to abnormal differentiation of the enteric nervous system. Whether the abnormality arises in the migrating neural crest cells that differentiate into the enteric ganglia or to the microenvironment of the gut into which the cells migrate remains to be determined.

Nonetheless, the experiment underscores the usefulness of manipulating the expression of particular genes in a specific tissue or cell type. For example, the presence of high levels of uniquely sized c-*abl* and c-*mos* transcripts in haploid spermatogenic cells suggests that these genes may be important for spermiogenesis or later gamete function. Therefore, it would be of interest to use spermatogenic-stage-specific promoters to elevate levels of their protein products at the appropriate stages or to produce such proteins at a developmental stage in which they are not expressed normally. The former experiment might be accomplished by using the protamine 1 promoter to drive expression (Peschon et al., 1987), whereas the phosphoglycerate kinase 2 promoter (Robinson et al., 1989) might be useful to achieve essentially premature expression of c-*abl* or c-*mos*.

The fact that such experiments can be proposed is due to a synthesis of the results obtained using the molecular approaches to identify genes involved in gametogenesis that were outlined at the beginning of this chapter. Protamine 1 was isolated using the second strategy, that of selecting for genes expressed only in the developing germ cells (Kleene et al., 1985). Various applications of the subtractive hybridization strategies are currently in use to identify genes expressed specifically in each spermatogenic cell type (e.g., Thomas et al., 1989). Such genes might then provide the appropriate promoters to manipulate expression throughout spermatogenesis. A potentially more powerful method of identifying germ cell stage-specific genes involves the use of gene trap or enhancer trap approaches (Gossler et al., 1989). Here, a readily expressed reporter gene, such as bacterial β-galactosidase, is inserted into constructs designed such that the expression of β-galactosidase in the resulting transgenic animals requires insertion of the construct within a gene (gene trap) or in the vicinity of the 5′ regulatory region (enhancer trap) of a gene. The screening method here involves selecting among the transgenic animals for those expressing the inserted β-galactosidase gene in a developmentally regulated pattern within the male (or female) germ line. Unlike subtractive hybridization, this technique may be used to identify genes expressed at low levels but that may be critical in more regulatory functions.

Our laboratory also has used the third strategy (outlined at the beginning of this chapter) to isolate genes with potential development-regulating function. This approach uses genes with known development-regulating function in simpler systems to screen for related genes in the genomes of more complex organisms. For example, we used a gene containing a homeobox, a region of 61 amino acids, which is conserved among several *Drosophila* genes that are known to be important in regulating development and differentiation. Observations of the expression of homeobox-containing genes during oogenesis in lower vertebrates and in mammalian embryonal carcinoma cell lines of germ cell origin prompted us to consider if genes containing homeobox sequences are expressed during mammalian male germ cell development (discussed in Wolgemuth et al., 1986). To test this hypothesis, a mouse testis cDNA library was screened with cloned homeobox-containing sequences from the *Antennapedia* gene of *Drosophila* under conditions of reduced stringency (Wolgemuth et al., 1986). We isolated a cDNA and subsequently the mouse homeobox-containing gene *Hox*-1.4, and observed interesting specificity of expression of this gene both in the adult male germ line and during embryogenesis. A complete discussion of *Hox*-1.4 expression is beyond the scope of the present review but may be found in several recent publications (e.g., Wolgemuth et al., 1986, 1987, 1989a, b). Although *Hox*-1.4 is the only mouse homeobox-containing gene to date identified specifically because of its expression in testes, other mouse genes containing homeo domains are expressed in testes (discussed in Wolgemuth et al., 1989b). These include other members of the *Hox*-1 cluster and the homologues of *Hox*-1.4, *Hox*-5.1, and *Hox*-2.6.

Finally, the strategy of using conserved domains to screen for related genes in more complex genomes has been used with other DNA binding motifs, including zinc finger domains and the paired box domain. The *Drosophila kruppel* gene, which contains zinc fingers, was used to isolate several mouse genes (e.g., Chowdhury et al., 1987). At least one of these genes, *mkr5*, is known to be expressed in testis and ovary, although the precise cell type has not been determined (Chowdury et al., submitted). Similarly, the paired box-domains of the *Drosophila paired* and *gooseberry* genes have been used to identify mouse genes that bear striking homology to the *Drosophila* genes both within the paired box and elsewhere in the coding regions (Deutsch et al., 1988; Burri et al., 1989). The usefulness of this approach is only beginning to be realized in the effort to unravel the complex genetic network responsible for the development and differentiation of specific lineages within higher organisms, including differentiation of the germ cell line.

ACKNOWLEDGMENTS

We thank Dr. Martin Winer for critical reading of this manuscript. This work was supported in parts by grants from the National Institutes of Health, HD05077 and HD07093, and from NASA, NAGW 3-385 NGT-50315.

REFERENCES

Allen, R. L., O'Brien, D. A., and Eddy, E. M. (1988). A novel hsp 70-like protein (P70) is present in mouse spermatogenic cells. Mol. Cell. Biol. 8:828–832.

Allende, C., Hinrichs, M. V., Santos, E., and Allende, J. E. (1988). Oncogenic *ras* protein induces meiotic maturation of amphibian oocytes in the presence of protein synthesis inhibitors. FEBS Lett. 234:426–430.

Ananthan, J., Goldberg, A. L., and Voellmy, R. (1986). Abnormal proteins serve as eukaryotic stress signals and trigger the activation of heat shock genes. Science 232:522–524.

Barbacid, M. (1987). *Ras* proteins. Ann. Rev. Biochem. 56:779–827.

Barnier, J. V., Bensaude, O., Morange, M., and Babinet, C. (1987). Mouse 89 kD heat shock protein. Two polypeptides with distinct developmental induction. Exp. Cell. Res. 170:186–194.

Barrett, C. B., Schroetke, R. M., Van der Hoorn, F. A., Nordeen, S., and Maller, J. L. (1990). Ha-*ras*$^{val\text{-}12,thr\text{-}59}$ activates S6 kinase and p34^{cdc2} kinase in *Xenopus* oocytes: Evidence for c-*mos*^{xe}-dependent and -independent pathways. Mol. Cell. Biol. 10:310–315.

Beato, M. (1989). Gene regulation by steroid hormones. Cell 56:335–344.

Bellve, A. R. (1979). The molecular biology of mammalian spermatogenesis, in *Oxford Reviews in Reproductive Biology,* Vol. 1 (Finn, C. A., ed.) pp. 159–261, Clarendon Press, Oxford, England.

Bellve, A. R. (1973). Development of mouse embryos with abnormalities induced by parental heat stress. J. Reprod. Fert. 35:393–403.

Bellve, A. R. (1972). Viability and survival of mouse embryos following parental exposure to high temperature. J. Reprod. Fert. 30:71–81.

Benbrook, D., and Pfahl, M. (1987). A novel thyroid hormone receptor encoded by a cDNA clone from a human testis library. Science 238:788–791.

Ben-Neriah, Y., Bernards, A., Paskind, M., Daley, G. Q., and Baltimore, D. (1986). Alternative 5′ exons in c-*abl* mRNA. Cell 44:577–586.

Bensaude, O., Babinet, C., Morange, M., and Jacob, F. (1983). Heat shock proteins, first major products of zygotic gene activity in mouse embryo. Nature 305:331–333.

Bensaude, O., and Morange, M. (1983). Spontaneous high expression of heat-shock proteins in mouse embryonal carcinoma cells and ectoderm from day 8 mouse embryo. EMBO J. 2:173–177.

Bernards, A., Paskind, M., and Baltimore, D. (1988). Four murine c-*abl* mRNAs arise by usage of two transcriptional promoters and alternative splicing. Oncogene 2:297–304.

Bernards, R., Schackleford, G. M., Gerber, M. R., Horowitz, J. M., Friend, S. H., Schartl, M., Bogenmann, E., Rapaport, J. M., McGee, T., Dryja, T. P., and Weinberg, R. A. (1989). Structure and expression of the murine retinoblastoma gene and characterization of its encoded protein. Proc. Natl. Acad. Sci. USA 86:6474–6478.

Bhat, N. K., Fisher, R. J., Fujiwara, S., Ascione, R., and Papas, T. S. (1987). Temporal and tissue-specific expression of mouse *ets* genes. Proc. Natl. Acad. Sci. USA 84:3161–3165.

Birchmeier, C., Broek, D., and Wigler, M. (1985). *Ras* proteins can induce meiosis in *Xenopus* oocytes. Cell 43:615–621.

Bishop, J. M. (1983). Cellular oncogenes and retroviruses. Ann. Rev. Biochem. 52:301–354.

Boulter, C. A., and Wagner, E. F. (1988). Expression of c-*src* and c-*abl* in embryonal carcinoma cells and adult mouse tissues. Exp. Cell Res. 179:214–221.

Brugge, J. S. (1986). Interaction of the Rous sarcoma virus protein with the cellular proteins pp50 and pp90. Curr. Topics Microbiol. Immun. 123:1–22.

Burri, M., Tromvoukis, Y., Bopp, D., Frigerio, G., and Noll, M. (1989). Conservation of the paired domain in metazoans and its structure in three isolated human genes. EMBO J. 8:1183–1190.

Catelli, M. G., Binart, N., Jung-Testas, I., Renoir, J. M., Baulieu, E. E., Feramisco, J. R., and Welch, W. J. (1985). The common 90-kd protein component of non-transformed '8S' steroid receptors is a heat-shock protein. EMBO J. 4:3131–3135.

Chappell, T. G., Welch, W. J., Schlossman, D. M., Palter, K. B., Schlessinger, M. J., and Rothman, J. E. (1986). Uncoating ATPase is a member of the 70 kilodalton family of stress proteins. Cell 45:3–13.

Chowdhury, K., Deutsch, U., and Gruss, P. (1987). A multigene family encoding several "finger" structures is present and differentially active in mammalian genomes. Cell 48:771–778.

Cohen, R. S., and Meselson, M. (1985). Separate regulatory elements for the heat-inducible and ovarian expression of the *Drosophila hsp26* gene. Cell 43:737–746.

Cooper, J. A., and Whyte, P. (1989). RB and the cell cycle: Entrance or exit. Cell 58:1009–1011.

Copeland, N. G., Gilbert, D. J., Cho, B. C., Donovan, P. J., Jenkins, N. A., Cosman, D., Anderson, D., Lynman, S. D., and Williams, D. E. (1990). Mast cell growth factor maps near the steel locus on mouse chromosome 10 and is deleted in a number of steel alleles. Cell 63:175–183.

Craig, E. A. (1985). The heat shock response. Crit. Rev. Biochem. 18:239–280.

Curci, A., Bevilacqua, A., and Mangia, F. (1987). Lack of heat-shock response in preovulatory mouse oocytes. Dev. Biol. 123:154–160.

Curran, T., and Franza, B. R. (1988). *Fos* and *jun:* The AP-1 connection. Cell 55:395–397.

Deshpande, A. T., and Kung, H.-F. (1987). Insulin induction of *Xenopus* oocyte maturation is inhibited by monoclonal antibody against p21 proteins. Mol. Cell. Biol. 7:1285–1288.

Deutsch, U., Dressler, G. R., and Gruss, P. (1988). *Pax* 1, a member of a paired box homologous murine gene family, is expressed in segmented structures during development. Cell 53:617–625.

Downward, J., Yarden, Y., Mayes, E., Screace, G., Totty, N., Stockwell, P., Ullrich, A., Schlessinger, J., and Waterfield, M. D. (1984). Close similarity of epidermal growth factor receptor and v-*erb*-B oncogene protein sequences. Nature 307:521–527.

Duboule, D., Baron, A., Mahl, P., and Galliot, B. (1986). A new homeo box is present in overlapping cosmid clones which define the mouse *Hox-1* locus. EMBO J. 5:1973–1980.

Duggal, R. N., Zakeri, Z. F., Ponzetto, C., and Wolgemuth, D. J. (1987). Differential expression of the c-*abl* proto-oncogene and the homeobox-containing gene *Hox-1.4* during mouse spermatogenesis, in *Cell Biology of the Testis and Epididymis,* Vol. 513 (Orgebin-Crist, M. C., ed.) pp. 112–127, Annals of the New York Academy of Sciences, New York.

Flanagan, J. G., and Leder, P. (1990). The kit ligand: A cell surface molecule altered in steel mutant fibroblasts. Cell 63:185–194.

Frohman, M. A., and Martin, G. R. (1989). Cut, paste, and save: New approaches to altering specific genes in mice. Cell 56:145–147.

Geissler, E. N., Ryan, M. A., and Housman, D. E. (1988). The dominant-white spotting (W) locus of the mouse encodes the c-*kit* proto-oncogene. Cell 55:185–192.

Germain, M. A., Webster, W. S., and Edwards, M. J. (1985). Hyperthermia as a teratogen: Parameters determining hyperthermia-induced head defects in the rat. Teratology 31:265–272.

Giguere, V., Ong, E. S., Segui, P., and Evans, R. M. (1987). Identification of a receptor for the morphogen retinoic acid. Nature 330:624–629.

Glaser, R. L., Wolfner, M. F., and Lis, T. (1986). Spatial and temporal pattern of hsp26 during normal development. EMBO J. 5:747–754.

Goldman, D. S., Kisseling, A. A., Millette, C. F., and Cooper, G. M. (1987). Expression of c-*mos* RNA in germ cells of male and female mice. Proc. Natl. Acad. Sci. USA 84:4509–4513.

Gonda, T. J., and Metcalf, D. (1984). Expression of *myb, myc,* and *fos* proto-oncogenes during the differentiation of a murine myeloid leukemia. Nature 310:249–251.

Gossler, A., Joyner, A. L., Rossant, J., and Skarnes, W. C. (1989). Mouse embryonic stem cells and reporter constructs to detect developmentally regulated genes. Science 244:463–465.

Gruppi, C. M., Zakeri, Z. F., and Wolgemuth, D. J. (1991). Stage and lineage-related expression of two HSP90 transcripts during mouse germ cell differentiation and embryogenesis. Mol. Reprod. & Develop. (in press).

Hall, S. H., Joseph, D. R., French, F. S., and Conti, M. (1988). Follicle-stimulating hormone induces transient expression of the proto-oncogene c-*fos* in primary Sertoli cell cultures. Mol. Endocrin. 2:55–61.

Hanahan, D. (1988). Dissecting multistep tumorgenesis in transgenic mice. Ann. Rev. Genet. 22:479–519.

Huang, E., Nocka, K., Beier, D. R., Chu, T.-Y, Buck, J., Lahm, H.-W., Wellner, D., Leder, P., and Besmer, P. (1990). The hematopoietic growth factor KL is encoded by the *Sl* locus and is the ligand of the c-*kit* receptor, the gene product of the *W* locus. Cell 63:225–233.

Herrlich, P., and Ponta, H. (1989). 'Nuclear' oncogenes convert extracellular stimuli into changes in the genetic program. Trends Genet. 5:112–116.

Hirai, S.-I., Ryseck, R.-P., Mechta, F., Bravo, R., and Yaniv, M. (1989). Characterization of *junD:* A new member of the *jun* proto-oncogene family. EMBO J. 8:1433–1439.

Jaenisch, R. (1988). Transgenic animals. Science 240:1468–1474.

Joab, I., Radanyi, C., Renoir, M., Buchou, T., Catelli, M.-G., Binart, N., Mester, J., and Baulieu E.-E. (1984). Common non-hormone binding component-transformed chick oviduct receptors of four steroid hormones. Nature 308:850–853.

Keshet, E., Rosenberg, M. P., Mercer, J. A., Propst, F., Vande Woude, G. F., Jenkins, N. A., and Copeland, N. G. (1987). Developmental regulation of ovarian-specific *mos* expression. Oncogene 2:235–240.

Kleene, K., Distel, R. J., and Hecht, N. B. (1985). Nucleotide sequence of a cDNA clone encoding mouse protamine 1. Biochemistry 24:719–722.

Klein, R., Parada, L. F., Coulier, F., and Barbacid, M. (1989). *trk*B, a novel tyrosine protein kinase receptor expressed during mouse neural development. EMBO J. 8:3701–3709.

Koji, T., Izumi, S., Tanno, M., Moriuchi, T., and Nakane, P. K. (1988). Localization in situ of c-*myc* mRNA and c-*myc* protein in adult mouse testis. Histochem. J. 20:551–557.

Korn, L. J., Siebel, C. W., McCormick, F., and Roth, R. A. (1987). *Ras* p21 as a potential mediator of insulin action in *Xenopus* oocytes. Science 236:840–842.

Kothary, R., Clapoff, S., Darling, S., Perry, M. D., Moran, L. A., and Rossant, J. (1989). Inducible expression of an hsp68–lacZ hybrid gene in transgenic mice. Development 105:707–714.

Kothary, R., Clapoff, S., Darling, S., Perry, M. D., Moran, L. A., and Rossant, J. (1987). Cell-lineage specific expression of the mouse hsp68 gene during embryogenesis. Dev. Biol. 121:342–348.

Koyasu, S., Nishida, E., Kadowaki, T., Matsuzaki, F., Iida, K., Harada, F., Kasuga, H., Sakai, H., and Yahara, I. (1986). Two mammalian heat shock proteins, HSP 90 and HSP 100, are actin-binding proteins. Proc. Natl. Acad. Sci. USA 83:8054–8058.

Kurtz, S., Rossi, J., Petko, L., and Lindquist, S. (1986). An ancient developmental induction: Heat-shock proteins induced in sporulation and oogenesis. Science 231:1154–1156.

Lane, D. P., and Crawford, L. V. (1979). T antigen is bound to a host protein in SV40-transformed cells. Nature 278:261–263.

Lech, K., Anderson, K., and Brent, R. (1988). DNA bound *fos* protein activates transcription in yeast. Cell 46:603–611.

Lee, W.-H., Bookstein, R., Hong, F., Young, L.-J., Shew, J.-Y., and Lee, E. Y.-H. P. (1987). Human retinoblastoma susceptibility gene: Cloning, identification, and sequence. Science 235:1394–1399.

Leon, J., Guerrero, I., and Pellicer, A. (1987). Differential expression of the *ras* gene family in mice. Mol. Cell. Biol. 7:1535–1540.

Letwin, K., Yee, S.-P., and Pawson, T. (1988). Novel protein-tyrosine kinase cDNAs related to *fps/fes* and *eph* cloned using anti-phosphotyrosine antibody. Oncogene 3:621–627.

Levine, R. A., La Rosa, G. J., and Gudas, L. J. (1984). Isolation of cDNA clones for genes exhibiting reduced expression after differentiation of murine teratocarcinoma stem cells. Mol. Cell. Biol. 4:2142–2150.

Lin, T., Blaisdell, J., Barbour, K. W., and Thompson, E. A. (1988). Transient activation of c-*myc* proto-oncogene expression in Leydig cells by human chorionic gonadotropin. Biochem. Biophys. Res. Comm. 157:121–126.

Lindquist, S. (1986). The heat shock response. Ann. Rev. Biochem. 55:1151–1191.

Lowe, D. G., and Moran, L. A. (1986). Molecular cloning and analysis of DNA complementary to three mouse Mr = 68,000 heat shock protein mRNAs. J. Biol Chem. 261:2102–2120.

Masui, Y., and Clarke, H. J. (1979). Oocyte maturation. Int. Rev. Cytology 57:185–193.

Masui, Y., and Markert, C. L. (1971). Cytoplasmic control of nuclear behavior during meiotic maturation of frog oocytes. J. Exp. Zool. 177:129–146.

Matsui, Y., Zsebo, K. M., and Hogan, B.L.M. (1990). Embryonic expression of a haematopoietic growth factor by the *Sl* locus and the ligand for *c-kit.* Nature 347:667–669.

Mechali, M., Gusse, M., Vriz, S., Taylor, M., Andeol, Y., Moreau, J., Hourdry, J., Leibovici, M., Brulfert, A., Almouzni, G., and Mousseron-Grall, S. (1988). Protooncogenes and embryonic development. Biochimie 70:895–899.

Meijer, D., Hermans, A., von Lindern, M., van Agthoven, T., de Klein, A., Mackenbach, P., Grootegoed, A., Talarico, D., Della Valle, G., and Grosveld, G. (1987). Molecular characterization of the testis specific c-*abl* mRNA in mouse. EMBO J. 6:4041–4048.

Mintz, B., and Russell, E. S. (1957). Gene-induced embryological modifications of primordial germ cells in the mouse. J. Exp. Zool. 134:207–237.

Moore, B. E., and Bose, H. R. (1989). Expression of the c-*rel* and c-*myc* proto-oncogenes in avian tissues. Oncogene 4:845–852.

Moore, C. R., and Chase, D. (1923) Heat application and testicular differentiation. Anat. Rec. 26:344–345.

Moore, S. K., Kozak, C., Robinson, E. A., Ullrich, S. J., and Appella, E. (1989). Murine 86-and 84-kDa heat shock proteins, cDNA sequences, chromosomes assignments, and evolutionary origins. J. Biol. Chem. 264:5343–5351.

Moore, S. K., Kozak, C., Robinson, E. A., Ullrich, S. J., and Appella, E. (1987). Cloning and nucleotide sequence of the murine hsp 84 cDNA and chromosome assignments of related sequences. Gene 56:29–40.

Morange, M., Diu, A., Bensaude, O., and Babinet, C. (1984). Altered expression of heat shock proteins in embryonal carcinoma cells and mouse early embryonic cells. Mol. Cell. Biol. 4:730–735.

Morrison, D. K., Kaplan, D. R., Escobedo, J. A., Rapp, U. R., Roberts, T. M., and Williams, L. T. (1989). Direct activation of the serine/threonine kinase activity of *raf*-1 through tyrosine phosphorylation by the PDGF beta-receptor. Cell 58:649–657.

Motro, B., Van Der Kooy, D., Rossant, J., Reith, A., and Bernstein, A. (1991). Contiguous patterns of c-*kit* and *steel* expression: Analysis of mutations at the W and Sl loci. Development 113:1207–1221.

Muller, R., Slamon, D. J., Adamson, E. D., Tremblay, J. M., Muller, D., Cline, M. J., and Verma, I. M. (1983). Transcription of c-onc genes c-ras^{ki} and c-*fms* during mouse development. Mol. Cell. Biol. 3:1062–1069.

Muller, R., Slamon, D. J., Tremblay, J. M., Cline, M. J., and Verma, I. (1982). Differential expression of cellular oncogenes during pre- and post-natal development in the mouse. Nature 299:640–644.

Mutter, G. L., Grills, G. S., and Wolgemuth, D. J. (1988). Evidence for the involvement of the proto-oncogene c-*mos* in mammalian meiotic maturation and possibly very early embryogenesis. EMBO J. 7:683–389.

Mutter, G. L., and Wolgemuth, D. J. (1987). Distinct developmental patterns of c-*mos* protooncogene expression in female and male germ cells. Proc. Natl. Acad. Sci. USA 84:5301–5305.

Nebel, B. R., Amarose, A. P., and Hackett, E. M. (1961). Calendar of gametogenic development in the prepuberal male mouse. Science 134:832–833.

Nishida, E., Koyasu, S., Sakai, H., and Yaharas, I. (1986). Calmodulin-regulated binding of the 90-kDa heat shock protein to actin filaments. J. Biol. Chem. 261:16003–16036.

O'Keefe, S. J., Wolfes, H., Kiessling, A. A., and Cooper, G. M. (1989). Microinjection of antisense c-*mos* oligonucleotides prevents meiosis II in the maturing mouse egg. Proc. Natl. Acad. Sci. USA 86:7038–7042.

Olofsson, B., Chardin, P., Touchot, N., Zahraoul, A., and Tavitian, A. (1988). Expression of the *ras*-related *ra1A, rho12* and *rab* genes in adult mouse tissues. Oncogene 3:231–234.

Oppi, C., Shore, S. K., and Reddy, E. P. (1987). Nucleotide sequence of testis-derived c-*abl* cDNAs: Implications for testis-specific transcription and *abl* oncogene activation. Proc. Natl. Acad. Sci. USA 84:8200–8204.

Papkoff, J., Brown, A.M.C., Varmus, H. E. (1987). The *int*-1 proto-oncogene products are glycoproteins that appear to enter the secretory pathway. Mol. Cell. Biol. 7:3978–3984.

Pardue, M. L. (1988). The heat shock response in biology and human disease: A meeting review. Genes Develop. 2:783–785.

Parvinen, M. (1982). Regulation of the seminiferous epithelium. Endocrine Rev. 3:404–417.

Paules, R. S., Buccione, R., Moschel, R. C., Vande Woude, G. F., and Eppig, J. (1989). Mouse *mos* proto-oncogene product is present and functions during oogenesis. Proc. Natl. Acad. Sci. USA 86:5395–5399.

Pelham, H. (1988). Coming in from the cold. Nature 332:776–777.

Pelham, H. (1986). Speculations on the functions of the major heat shock and glucose-regulated proteins. Cell 46:959–961.

Pelham, H. (1985). Activation of heat shock genes in eukaryotes. Trends Genet. 1:31–35.

Peschon, J. J., Behringer, R. R., Brinster, R. L., and Palmiter, R. D. (1987). Spermatid-specific expression of protamine 1 in transgenic mice. Proc. Natl. Acad. Sci. 84:5316–5319.

Petrovich, M., Brand, N. J., Krout, A., and Chambon, P. (1987). A human retinoic acid receptor which belongs to the family of nuclear receptors. Nature 330:444–450.

Ponzetto, C., Wadewitz, A. G., Pendergast, A. M., Witte, O. N., and Wolgemuth, D. J. (1989). $P150^{c-abl}$ is detected in mouse male germ cells by an in vitro kinase assay and is associated with stage-specific phosphoproteins in haploid cells. Oncogene 4:685–690.

Ponzetto, C., and Wolgemuth, D. J. (1985). Haploid expression of a unique c-*abl* transcript in the mouse male germ line. Mol. Cell. Biol. 5:1791–1794.

Pratt, W. B. (1987). Transformation of glucocorticoid and progesterone receptors to the DNA-binding state. J. Cell. Biochem. 35:51–68.

Propst, F., Rosenberg, M. P., Oskarsson, M. K., Russell, L. B., Nguyen-Huu, M. C., Nadeau, J., Jenkins, N. A., Copeland, N. G., and Vande Woude, G. F. (1988). Genetic analysis and developmental regulation of testis-specific RNA expression of *mos, abl,* actin and *Hox-1.4.* Oncogene 2:227–233.

Propst, F., Rosenberg, M. P., Iyer, A., Kaul, K., and Vande

Woude, G. F. (1987). *C-mos* proto-oncogene RNA transcripts in mouse tissues: Structural features, developmental regulation, and localization in specific cell types. Mol. Cell. Biol. 7:1629–1637.

Rauscher, F. J., Cohen, D. R., Curran, T., Bos, T. J., Vogt, P. K., Bohmann, D., Tjian, R., and Franza, B. R. (1988). *Fos*-associated protein p39 is the product of the *jun* proto-oncogene. Science 240:1010–1016.

Rijsewijk, F., Schuermann, M., Wagenaar, E., Parren, P., Weigel, D., and Nusse, R. (1987). The *Drosophila* homolog of the mouse mammary oncogene *int*-1 is identical to the polarity gene *wingless*. Cell 50:649–657.

Ritossa, F. M. (1964). Behavior of RNA and DNA synthesis at the puff level in salivary gland chromosomes of *Drosophila*. Exp. Cell Res. 36:515–523.

Roberts, T. M., Kaplan, D., Morgan, W., Keller, T., Mamon, H., Piwnica-Worms, H., Druker, B., Cohen, B., Schaffhausen, B., Whitman, M., Cantley, L., Rapp, U., and Morrison, D. (1988). Tyrosine phosphorylation in signal transduction. Cold Spring Harbor Symp. Quant. Biol. 53:161–172.

Robertson, E. J., Bradley, A., Kuehn, M., and Evans, M. (1986). Germ-line transmission of genes introduced into cultured pluripotential cells by retroviral vectors. Nature 323:445–447.

Robinson, M. O., McCarrey, J. R., and Simon, M. I. (1989). Transcriptional regulatory regions of testis-specific PKG2 defined in transgenic mice. Proc. Natl. Acad. Sci. USA 86:8437–8441.

Rock, J., and Robinson, D. (1965). Effect of induced intrascrotal hyperthermia on testicular function in man. Am. J. Obstet. Gynecol. 93:793–801.

Rogel, A., Popliker, M., Webb, C. G., and Oren, M. (1985). P53 cellular tumor antigen: Analysis of mRNA levels in adult tissues, embryos, and tumors. Mol. Cell. Biol. 5:2851–2855.

Rubin, M. R., Toth, L. E., Patel, M. D., D'Eustachio, P., and Nguyen-Huu, M. C. (1986). A mouse homeo box gene is expressed in spermatocytes and embryos. Science 233:663–667.

Sagata, N., Daar, I., Oskarsson, M., Showalter, S. D., and Vande Woude, G. F. (1989a). The product of the *mos* proto-oncogene as a candidate "initiator" for oocyte maturation. Science 245:643–645.

Sagata, N., Watanabe, N., Vande Woude, G. F., and Ikawa, Y. (1989b). The *c-mos* proto-oncogene product is a cytostatic factor responsible for meiotic arrest in vertebrate eggs. Nature 342:512–518.

Sagata, N., Oskarsson, M., Copeland, T., Brumbaugh, J., Showalter, S. D., and Vande Woude, G. F. (1988). Function of *c-mos* proto-oncogene product in meiotic maturation in *Xenopus* oocytes. Nature 335:519–525.

Saksela, K., Makela, T. P., and Alitalo, K. (1989). Oncogene expression in small-cell lung cancer cell lines and a testicular germ-cell tumor: activation of the N-*myc* gene and decreased RB mRNA. Int. J. Cancer 44:182–185.

Sanchez, E. R., Toft, D. O., Schlessinger, M. L., and Pratt, W. B. (1985). Evidence that the 90-kDa phosphoprotein associated with untransformed L-cell glucocorticoid receptor is a murine heat shock protein. J. Biol. Chem. 260:12398–12401.

Scherr, C. J., Rettenmier, C. W., Sacca, R., Roussel, M. F., Look, A. T., and Stanley, E. R. (1985). The *c-fms* proto-oncogene product is related to the receptor for the mononuclear phagocyte growth factor, CSF-1. Cell 41:665–676.

Schlessinger, M. L. (1986). Heat shock proteins: The search for functions. J. Cell Biol. 103:321–325.

Schwartzberg, P. L., Goff, S. P., and Robertson, E. J. (1989). Germ-line transmission of a *c-abl* mutation produced by targeted gene disruption in ES cells. Science 246:799–803.

Searle, A. G. (1982). The genetics of sterility in the mouse, in *Genetic Control of Gamete Production and Function* (Crosignani, P. G., Rubin, B. L., and Fraccaro, M., eds.) pp. 93–114, Academic Press, New York.

Shackleford, G. M., and Varmus, H. E. (1987). Expression of the proto-oncogene *int*-1 is restricted to postmeiotic male germ cells and the neural tube of mid-gestational embryos. Cell 50:89–95.

Slamon, D. J., and Cline, M. J. (1984). Expression of cellular oncogenes during embryonic and fetal development of the mouse. Proc. Natl. Acad. Sci. USA 81:7141–7145.

Sorrentino, V., McKinney, M. D., Giorgi, M., Geremia, R., and Fleissner, E. (1988). Expression of cellular proto-oncogenes in the mouse male germ line: A distinctive 2.4-kilobase *pim*-1 transcript is expressed in haploid post-meiotic cells. Proc. Natl. Acad. Sci. USA 85:2191–2195.

Steele, R. E., Irwin, M. Y., Knudsen, C. L., Collett, J. W., and Fero, J. B. (1989). The *yes* proto-oncogene is present in amphibians and contributes to the maternal RNA in the oocyte. Oncogene Res. 1:223–233.

Steward, R. (1987). *Dorsal*, an embryonic polarity gene in *Drosophila*, is homologous to the vertebrate proto oncogene, *c-rel*. Science 238:692–694.

Tahira, T., Ishizaka, Y., Sugimura, T., and Nagao, M. (1988). Expression of proto-*ret* mRNA in embryonic and adult rat tissues. Biochem. Biophys. Res. Comm. 153:1290–1295.

Tajima, Y., Onoue, H., Kitamura, Y., and Nishimune, Y. (1991). Biologically active kit ligand growth factor is produced by mouse Sertoli cells and is defective in Sl^d mutant mice. Development 113:1031–1035.

Taylor, M. V., Gusse, M., Evan, G. I., Dathan, N., and Mechali, M. (1986). *Xenopus myc* proto-oncogene during development: Expression as a stable maternal mRNA uncoupled from cell division. EMBO J. 5:3563–3570.

Thomas, K. H., Wilkie, T. M., Tomashefsky, P., Bellve, A. R., and Simon, M. I. (1989). Differential gene expression during mouse spermatogenesis. Biol. Reprod. 41:729–739.

Thomas, K. R., and Capecchi, M. R. (1987). Site-directed mutagenesis by gene targeting in mouse embryo-derived stem cells. Cell 51:503–512.

Tissieres, A., Mitchell, H. K., and Tracy, U. M. (1974). Protein synthesis in salivary glands of *Drosophila melanogaster:* Relation to chromosome puffs. J. Mol. Biol. 84:389–398.

Ulberg, L. C., and Sheean, L. A. (1973). Early development of mammalian embryos in elevated ambient temperatures. J. Reprod. Fert. 19(Suppl):155–161.

Van Etten, R. A., Jackson, P., and Baltimore, D. (1989). The mouse type IV *c-abl* gene product is a nuclear protein, and activation of transforming ability is associated with cytoplasmic localization. Cell 58:669–678.

Vogt, P. K., and Tjian, R. (1988). *Jun:* A transcriptional regulator turned oncogenic. Oncogene 3:3–7.

Wadewitz, A. G., and Wolgemuth, D. J. (1990). Characteristics of the expression of the c-*abl* and c-*mos* protooncogenes in mouse germ cells, in *Growth Factors in Fertility Regulation* (Haseltine, F. P., and Findlay, J., eds.) pp. 35–49, Cambridge University Press, Cambridge.

Wang, J., and Baltimore, D. (1983). Cellular RNA homologous to the Abelson murine leukemia virus transforming gene: Expression and relationship to the viral sequence. Mol. Cell. Biol. 3:773–779.

Watanabe, N., Vande Woude, G. F., Ikawa, Y., and Sagata, N. (1989). Specific proteolysis of the c-*mos* proto-oncogene product by calpain on fertilization of *Xenopus* eggs. Nature 342:505–511.

Waterfield, M. D., Scrace, G. T., Whittle, N., Stroobant, P., Johnsson, A., Wasteson, A., Westermack, B., Heldin, C. H., Huang, J. S., and Deuel, T. F. (1983). Platelet-derived growth factor is structurally related to the putative transforming protein $p28^{sis}$ of simian sarcoma virus. Nature 304:35–39.

Webster, W. S., and Edwards, M. J. (1984). Hyperthermia and the induction of neural tube defects in mice. Teratology 29:417–425.

Williams, D. E., Eisenman, J., Baird, A., Rauch, C., Van Ness, K., March, C. J., Park, L. S., Martin, U., Mochizuki, D. Y., Boswell, H. S., Burgess, G. S., Cosman, D., and Lyman, S. D. (1990). Identification of a ligand for the c-*kit* proto-oncogene. Cell 63:167–174.

Wolbach, S. B., and Howe, P. R. (1925). Tissue changes following deprivation of fat soluble vitamin A. J. Exp. Med. 42:753–777.

Wolfes, H., Kogawa, K., Millette, C. F., and Cooper, G. M. (1989). Specific expression of nuclear proto-oncogenes before entry into meiotic prophase of spermatogenesis. Science 245:740–743.

Wolgemuth, D. J., Behringer, R. R., Mostoller, M. P., Brinster, R. L., and Palmiter, R. D. (1989a). Transgenic mice overexpressing the mouse homeobox-containing gene *Hox1.4* exhibit abnormal gut development. Nature 337:464–467.

Wolgemuth, D. J., Viviano, C. M., and Watrin, F. (1989b). Expression of homeobox-containing genes in the male mammalian germ line, in *Development and Function of the Reproductive Organs,* Vol VIII (Joss, N., ed.) pp. 37–48, Ares Serono Symposia Press, Rome, Italy.

Wolgemuth, D. J., Viviano, C. M., Gizang-Ginsberg, E., Frohman, M. A., Joyner, A. L., and Martin, G. R. (1987). Differential expression of the mouse homeobox-containing gene Hox-1.4 during male germ cell differentiation and embryonic development. Proc. Natl. Acad. Sci. USA 84:5813–5817.

Wolgemuth, D. J., Engelmeyer, E., Duggal, R. N., Gizang-Ginsberg, E. E., Mutter, G. L., Ponzetto, C., Vivano, C., and Zakeri, Z. F. (1986). Isolation of a mouse cDNA coding for a developmentally regulated, testis-specific transcript containing homeo box homology. EMBO J. 5:1229–1235.

Wolgemuth, D. J., Gizang-Ginsberg, E., Engelmyer, E., Gavin, B. J., and Ponzetto, C. (1985). Separation of mouse testis cells on a Celsep™ apparatus and their usefulness as a source of high molecular weight DNA or RNA. Gam. Res. 12:1–10.

Young, W. C. (1927). The influence of high temperature on the guinea-pig testis. J. Exp. Zoo. 49:459–499.

Zakeri, A. F., and Wolgemuth, D. J. (1987). Developmental stage specific expression of the hsp70 gene family during differentiation of the mammalian male germ line. Mol. Cell. Biol. 7:1791–1796.

Zakeri, A. F., Wolgemuth, D. J., Hunt, C. R. (1988). Identification and sequence analysis of a new member of the mouse hsp70 gene family and characterization of its unique cellular and developmental pattern of expression in the male germ line. Mol. Cell. Biol. 8:2925–2932.

Zakeri, Z. F., Welch, W. J., and Wolgemuth, D. J. (1990). Characterization and inducibility of hsp70 proteins in the male mouse germ line. J. Cell. Bio. 111:1785–1792.

Zimmerman, J. L., Petri, W., and Meselson, M. (1983). Accumulation of a specific subset of *D. melanogaster* heat shock mRNAs in normal development without heat shock. Cell 32:1161–1170.

Zsebo, K. M., Williams, D. A., Geissler, E. N., Broudy, V. C., Martin, F. H., Atkins, H. L., Hsu, R.-Y., Birkett, N. C., Okino, K. H., Murdock, D. C., Jacobsen, F. W., Langley, K. E., Smith, K. A., Takeishi, T., Cattanach, B. M., Galli, S. J., and Suggs, S. V. (1990). Stem cell factor is encoded at the *Sl* locus of the mouse and is the ligand for the c-*kit* tyrosine kinase receptor. Cell 63: 213–224.

CHAPTER EIGHTEEN

Sperm Cell Surface Proteins of Testicular Origin: Expression and Localization in the Testis and Beyond

Diana G. Myles

I. Introduction

Spermatogenesis is an example of cell differentiation that involves dramatic alterations in all aspects of the cell. In early stages, the germ cell divides mitotically to bring about a total reorganization of the genome. During the later stage of spermatogenesis (spermiogenesis) there is a complete transformation of the spermatid to make it capable of carrying out the vital function of conveying the male genome to the egg. The sperm surface is an integral part of this cellular differentiation. Throughout spermatogenesis in the testes and also in subsequent steps of sperm maturation in the male and female reproductive tracts, the cell surface is being transformed, both in its overall composition and in the localization of surface molecules.

Spermatogenic cells are subjected to continual change in their extracellular interactions. The earliest cells (mitotic spermatogonia and preleptotene spermatocytes) are found at the periphery of the seminiferous tubule. The spermatogonia are in contact with the basal lamina and slightly flattened against it, presumably through cell adhesions. When the preleptotene primary spermatocytes are formed as the result of the last mitotic division of the spermatogonia, they move away from the basal lamina toward the lumen of the seminiferous tubule, become more spherical in appearance, and are more deeply engulfed in crypts in the Sertoli cells. The germ cells display a continually changing pattern of junctions and other surface associations with the Sertoli cell plasma membrane (Russell and Peterson, 1985). Eventually, cell adhesions between Sertoli cells and germ cells are broken (except in the region of the residual body) when the sperm are released into the lumen of the seminiferous tubule. At this point the sperm is no longer part of a tissue but becomes an isolated cell. Until it reaches the egg, its main interactions are with the ever-changing fluid environment and epithelial cell surfaces of the male and female reproductive tracts. In some species the sperm cells are associated with each other during part of their passage through the male tract. For example, in guinea pig, they become stacked in rouleaux and display a specialized association between the two plasma membranes of adjacent sperm in the stack (Fawcett, 1975).

No more vital cell–cell interaction happens in the entire cycle of the organism than that which occurs when the sperm reaches the egg. The surface of the mammalian sperm is the initial effector of fertilization. It is the sperm's surface that contains the information that allows it to recognize and bind to the extracellular egg coat, the zona pellucida. At least in some cases, sperm surface molecules also transduce a signal from the egg to bring about the exocytotic fusion of the acrosome membrane with the plasma membrane during the acrosome reaction. Furthermore, surface enzymes have been implicated in penetration through the zona and it is the surface of the sperm, along with the egg plasma membrane, that is responsible for first the binding and then the fusion of male and female gametes.

The purpose of this chapter is to discuss recent developments in the study of protein composition and organization of the differentiating sperm surface. This is not meant to be a comprehensive review. Rather, references are chosen as illustrative of the phenomena being discussed. The examples are mostly limited to those proteins that have at least been tentatively identified and for which there is good evidence of a surface location.

During spermatogenesis, the composition of the germ cell surface is altered as newly synthesized sperm-specific proteins are incorporated into the sperm membrane. After the sperm leave the testes, new proteins may be added to the surface from the fluids of the male reproductive tract. In addition, both in the testes and post-testicularly, proteins already associated with the surface may be lost or altered by enzymatic action. A striking feature of the sperm surface is its topographical organization. Using a variety of probes, it has been shown that the sperm surface is arranged into distinct molecular domains (Eddy, 1988). Organization of surface proteins into domains begins in the testes, and continues post-testicularly. It is generally thought that this heterogeneous distribution reflects a segregation of function into different surface regions. For example, it has been observed, both in the case of zona binding and fusion with the egg plasma membrane, that distinct regions of the head surface are involved (Yanagimachi, 1988). Changes in localization are important because they bring about changes in the composition of the surface within a specific domain. By controlling the composition of the surface domains, through rearrangement of surface molecules, the cell is in fact controlling the composition of its functional surface units. Furthermore, by altering patterns of localization, surface functions may be regulated by the changing interactions between molecules within a specific region.

This chapter concentrates on sperm surface proteins that are a part of the sperm when it leaves the testes. Modifications to these testicular proteins are traced, including changes in their surface expression, localization, and structure. For the purpose of this review, the cell surface is considered to be all the cell components that are accessible to the external environment so that they can either carry out a surface function or be recognized by a surface probe (Fig. 18–1). Potential components of the sperm surface can be divided into two classes: (1) those that are integral to the membrane and are partially incorporated into the lipid bilayer and (2) peripheral components that are completely external to the bilayer (Fig. 18–1). Integral components of the membrane are lipids and proteins that are not removed from the surface except by disruption of the membrane (e.g., by detergents). Integral membrane proteins include transmembrane proteins that have one or more hydrophobic membrane spanning regions and, therefore, contact the cytoplasm as well as being exposed on the external surface of the cell to the extracellular environment. Integral membrane proteins can also be anchored in only the outer leaflet of the lipid bilayer by glycosylphosphatidylinositol (GPI) (Ferguson and Williams, 1988). Both types of proteins are commonly glycoproteins or proteoglycans with a more or less extensive carbohydrate moiety in the extracellular domain. Carbohydrate groups may also be associated with glycolipids that are a part of the outer lipid leaflet. Although it is theoretically possible for integral membrane proteins to be inserted into the membrane from external sources, current understanding of membrane biogenesis suggests that both transmembrane and GPI-anchored proteins are synthesized in spermatogenic cells, in conjunction with the endoplasmic reticulum. They are then transported to the membrane surface in membrane vesicles. Peripheral components of the membrane might include adsorbed molecules (glycoproteins and proteoglycans) or soluble ligands that are associated with the surface by binding to surface receptors. The source of these molecules would be either secretions of the sperm itself or secretions of Sertoli cells or any of the tissues that contribute to the fluids of the male and female reproductive tracts.

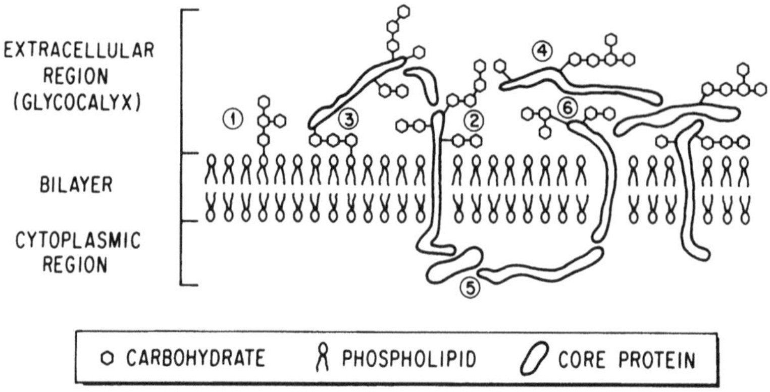

Figure 18–1. Diagrammatic representation of the cell surface and the plasma membrane. The surface includes the extracellular regions of carbohydrate-rich glycocalyx and the hydrophilic regions of the membrane bilayer that are available for surface functions and are accessible to surface probes. The intrinsic molecules of the bilayer are lipids, glycolipids (1), transmembrane glycoproteins (2), and glycosylphosphatidylinositol (GPI)-anchored proteins (3). Extrinsic surface molecules include adsorbed glycoproteins (4) and cytoplasmic proteins (5). In some cases molecules that extend into the extracellular region may be covert (6) and are, therefore, not considered here as part of the surface. Proteoglycans may be associated with the membrane in any of the configurations shown for glycoproteins.

In many cases a carbohydrate-rich glycocalyx or cell coat (Fig. 18–1) has been described around sperm cells that can be visualized with an electron microscope in thin sections stained to enhance carbohydrate groups and in surface replicas (e.g., Friend and Fawcett, 1974; Suzuki and Nagano, 1980; Friend, 1982). Carbohydrate groups in this layer include the carbohydrate regions of integral glycoproteins, proteoglycans, and glycolipids, as well as peripheral carbohydrate-containing molecules.

II. Surface Composition During Spermatogenesis

There are stage-specific surface proteins that first appear at each of the three stages of spermatogenesis: mitotic proliferation, meiosis, and spermiogenesis. The occurrence of stage-specific proteins is not surprising. Functional maturation of sperm cells capable of fertilizing eggs would predict surface changes as would alterations in cell–substrate and cell–cell interactions as the cells proceed through differentiation.

The method used to identify stage-specific markers varies with different investigations and, therefore, gives us different information about the surface. Some studies have identified the stage at which synthesis of a particular protein can be detected. In other studies only specific regions of the protein molecule (e.g., antigenic epitopes or glycosyl groups) are identified and, therefore, do not distinguish between stage-specific protein modifications and new surface proteins. Many investigations have focused on identifying germ cell-specific (differentiation) antigens for at least two different reasons. First, they are likely candidates in germ cell functions, including fertilization. Second, the unique germ cell surface antigens that first appear at puberty are not present when immunological tolerance to self is induced, and are consequently recognized as foreign. The mechanism that prevents them from causing an autoimmune response is therefore of interest.

Surface changes may occur by the incorporation of new proteins into the cell surface. In addition to proteins that are synthesized by the developing germ cells themselves, there is evidence that secretions from the surrounding Sertoli cells can become associated with the sperm cell surface (Griswold, 1988).

A. Stage-Specific Proteins

1. MITOTIC STAGE OF SPERMATOGENESIS

Germ cells of the mitotic stage are formed in a compartment of the testis (basal compartment), separated from later stage cells that develop in the adluminal compartment. These separate compartments are formed by Sertoli cells, columnar epithelial cells that span the seminiferous tubule from basal lamina to lumen. Sertoli cells are in close contact with the developing germ cells and also form tight junctions with each other, creating the blood–testis barrier and dividing the testis into two compartments. Cells outside this barrier, in the basal compartment, are exposed to all the components of the blood and, therefore, the immune system, but inside the barrier, antigens may at least be partially sequestered from the immune system (Mahi-Brown et al., 1988). The blood–testis barrier is thought to allow the existence of late arising germ cell-specific proteins that would otherwise be autoantigenic. Therefore, it has been thought that cells outside the blood–testis barrier (i.e., in the basal compartment) would not express autoantigens. Recent evidence, however, indicates that mitotic spermatogonia and preleptotene spermatocytes can express autoantigens that are accessible to circulating autoantibodies; therefore, mechanisms other than sequestration behind the blood–testis barrier must be available to suppress an immune response in the testes (Mahi-Brown et al., 1988; Yule et al., 1988; and see Chapter 19).

A few surface proteins have been identified on premeiotic germ cells in the mouse. Galactosyltransferase (GalTase) can be detected on the surface of mouse germ cells by immunofluorescence as early as the spermatogonial stage (Scully et al., 1987). The XT-2 protein of mouse is present on spermatogonia and stages through pachytene and then becomes undetectable and is not found on mature sperm (Bechtol et al., 1979). Also, high molecular weight ($>1 \times 10^6$) glycoconjugates, identified by Western blots of testis extracts after gel electrophoresis, have been detected on preleptotene spermatocytes after sialidase treatment with the testis-specific anticarbohydrate monoclonal antibodies (mAbs) C6 and A5 (Fenderson et al., 1984).

2. MEIOTIC STAGE OF SPERMATOGENESIS

Progression of cells into meiosis is accompanied by a shift in the surface protein composition. Included in this developmental stage are all the cells from the leptotene stage of prophase I up to the round spermatid stage. Many studies of the surface have used purified populations of primary spermatocytes at pachytene stage: these cells are abundant (this is a long-lasting stage), and are easy to separate and identify. Of course, many differences in cell surface composition may exist between different cell types within the meiotic stages of spermatogenesis.

In the early stages of meiosis, cells migrate away from the basement membrane and become sequestered inside the blood–testis barrier. Because these cells are within the blood–testis barrier they are thought to be immunologically privileged and, therefore, are particular candidates for having autoantigenic sperm-specific proteins on their surfaces.

Several surface proteins or epitopes have been reported to be absent from spermatogonia but present on meiotic cells. They include, in the mouse, XT-1 antigen (Bechtol, 1984) and three cell surface carbohydrates recognized by three mAbs, J1, C6, A5 (Fenderson et al., 1984); in the rat, a 28kd antigen (Gaunt, 1982) and a galactosyl receptor (Abdullah and Kierszenbaum, 1989); and in the rabbit, RSA-1 (O'Rand and Romrell, 1981) (Table 18–1). At least one major surface polypeptide (p151/6.0) has been found to be synthesized by mouse pachytene spermatocytes, but not by round spermatids even though it is still present in the round spermatid membrane (Millette et al., 1987). This is an example of differential synthesis between meiotic germ cells (pachytenes) and postmeiotic round spermatids (Table 18–1).

3. SPERMIOGENESIS

Further changes occur in the cell surface as spermatocytes finish meiosis, progress into the round spermatid stage, and then become transformed, during spermiogenesis, into a polarized sperm cell. One area of surface change is in the interactions with surrounding Sertoli cells. For example, pachytene spermatocytes apparently have a greater binding affinity for Sertoli cells than round spermatids (Ziparo et al., 1980; Enders and Millette, 1988) and a specific candidate for stage-specific adhesion between pachytenes and Sertoli cells has been identified (D'Agostino and Stefanini, 1987; D'Agostino and Stefanini, 1990). Furthermore, adhering junctions between round spermatids and Sertoli cells are rarely seen and, as spermatids start to elongate, all conventional junctions are lost (Russell and Peterson 1985). Trypsin-sensitive material, appearing between elongating sper-

Table 18–1. Surface Proteins Reported on Meiotic Spermatocytes

Species	Probe	Stages	Identify (mol wt of protein)	Reference
Mouse	PAb enzymatic activity	spermatogonia-cauda sperm	galactosyl transferase	Sculley et al., 1987
	MAb	spermatogonia-pachytene	XT-2 (n.d.)	Bechtol et al., 1979
	MAb	leptotene-condensing spermatid	XT-1 (n.d.)	Bechtol, 1984
	MAb	pachytene-round spermatid	J1 ($>1 \times 10^6$) C6 ($>1 \times 10^6$)	Fenderson et al., 1984
	MAb	pachytene-cauda sperm	28 kd	Gaunt, 1982
	plasma membrane prep	?-pachytene	90 kd (p.I. 5.6) 56 kd (p.I. 6.0)	Millette and Moulding, 1981
	PAb	?-pachytene	57 kd (p.I. 6.4)	O'Brien and Millette, 1986
	PAb	?-pachytene-spermatid	65 kd (p.I. 6.1) 151 kd (p.I. 6.0) 51 kd	O'Brien and Millette, 1986
	ConA	?-pachytene	73 kd (p.I. 5.7)	Millette and Scott, 1984
Rat	PAb	leptotene-cauda sperm	54 kd	Abdullah and Kierszenbaum, 1989
	PAb	pachytene-early spermatids	n.d.	Tung and Fritz, 1978
Rabbit	PAb	pachytene-ejaculated sperm	at least 9 polypeptides	O'Rand and Romrell, 1980; Romrell et al., 1982
	PAb	pachytene-ejaculated sperm	RSA-1 (13 kd)	O'Rand and Romrell, 1981

n.d., not determined.

matids and Sertoli cells, however, has been described in surface regions where there is apparent adherence between the two cell types (ectoplasmic specializations) (Russell and Peterson, 1985). In the final stages of spermiogenesis this intercellular region appears to lose its adhesive properties and to be replaced by adhesive zones between the anterior head plasma membrane of the sperm and the Sertoli cell plasma membrane (tubulobulbar complexes) (Russell and Peterson, 1985). There is also a specialized zone that forms between the residual body of the sperm and the Sertoli cell that may lead to the eventual retention of the residual body when sperm are released.

These dramatic changes in cell–cell adhesion alone would predict that there might be multiple changes in cell surface composition. In addition, this is the final chance for proteins, synthesized by the sperm itself, to be incorporated directly into the plasma membrane. During the early stages of spermiogenesis, protein synthesis is still occurring. Protein synthesis during spermiogenesis can be translated either from messenger RNA (mRNA) that was transcribed at earlier (diploid) stages in spermatogenesis or from newly transcribed mRNA (haploid gene expression) (Hecht, 1988; see Chapter 16).

A number of surface proteins present on round spermatids have been identified. Some of these surface proteins are also present on earlier stages and are included in Table 18–1, but others are first found on round spermatids (Table 18–2). These include proteins that have been

Table 18–2. Surface Proteins First Detected on Round Spermatids

Species	Probe	Stages	Identify (mol wt of protein)	Reference
Mouse	Plasma membrane prep	R.S. - ?	90–95 kd (p.I. 5.9) ~88 kd (p.I. 5.5) 58 kd (p.I. ~ 6.2)	Millette and Moulding, 1981
	ConA	R.S. - ?	84 kd (p.I. 6.3)	Millette and Scott, 1984
	PAb	R.S. - ?	40 kd (p.I. 4.8) 46 kd (p.I. 4.8)	O'Brien and Millette, 1986
Rabbit	PAb	R.S.-ejaculated sperm	31 kd 34 kd 103 kd	O'Rand and Romrell, 1980 Romrell et al., 1982
Guinea pig	PAb	R.S.-testicular sperm	n.d.	Tung et al., 1979
	MAb sperm	R.S.-cauda	PH-20 (64 kd)	Phelps and Myles, 1987

R.S., round spermatid; n.d., not determined.

detected by two-dimensional gel electrophoresis of purified plasma membranes (Millette and Moulding, 1981), some of which were identified as glycoproteins (Millette and Scott, 1984). Some germ cell surface autoantigens, identified by complex antisera, are first detected on round spermatids in rabbit and guinea pig (O'Rand and Romrell, 1980; Tung et al., 1979) and the PH-20 protein, an autoantigen (Primakoff et al., 1988), is first detected on round spermatids in guinea pig (Phelps and Myles, 1987; Table 18–2).

In some cases, proteins have been reported to be on the surface of testicular sperm but not on earlier germ cell stages (e.g., Klint et al., 1987; Koyama et al., 1988). This could result from failure to detect a protein that is associated with the cell surface at earlier stages either because of technical problems or because the region of the molecule recognized by the probe is covert or has not yet been created. New addition of protein to the surface during the final stages of spermiogenesis would not be expected to include protein newly synthesized by the sperm, but it could be protein that is stored in some internal pool and then becomes incorporated into the late stage spermatid surface. It was found that a sperm-specific surface antigen could be detected in early spermatids (postmeiotic), but that it appeared to be intracellular at that stage (Klint et al., 1987). Also, proteins secreted by Sertoli cells might become part of the sperm surface at these late stages. The secreted Sertoli cell protein of rat, SGP-2, was found on the heads and tails of late spermatids and residual bodies in testicular sections. Spermatogonia, spermatocytes, and early spermatids showed no significant staining (Sylvester et al., 1984).

B. Alterations in Existing Surface Proteins During Spermatogenesis

Changes in surface composition may, in many cases, result from synthesis of new membrane proteins that are inserted into the germ cell membrane at different stages. Some changes in surface composition, however, may result from modifications of already existing proteins. There is evidence for at least three mechanisms for change of surface proteins during spermatogenesis: (1) elimination of the protein, (2) change in glycosylation of the glycoproteins, and (3) limited proteolytic processing. Of course, other types of modifications may occur. Depending on the type of surface probe used, modifications of surface proteins could result in either the new detection or loss in detection of a surface protein that is continuously a part of the surface.

Once present on the surface, antigens or at least the expression of a particular epitope, may be lost. The mAb (A5) that recognizes a surface carbohydrate on mouse meiotic cells and round spermatids no longer recognizes that epitope on epididymal sperm (Fenderson et al., 1984) and, therefore, the expression of this epitope must be lost during later stages in spermatogenesis or epididymal passage. Elimination of certain antigens or partial populations of these antigens may occur by their segregation into the residual body so they will be left behind when the sperm is released into the lumen of the seminiferous tubule. A polyclonal antiserum, generated against spermatogonia of mice and then extensively absorbed with somatic cells, reveals a class of antigens that are on germ cells through the round spermatid stage and then become partitioned into the residual body and, as a result, are lost from testicular sperm (Millette and Bellve, 1980). The segregation of surface sites into the residual body might occur because they are no longer required in sperm cell function or because they are involved in the adhesion between the residual body and the Sertoli cell.

Both glycosylation and deglycosylation of surface molecules have been reported during spermatogenesis. Depending on the probe used this could alter the detectability of the protein. Changes in glycosylation could also affect protein function. Gerton and Millette (1986) compared fucosylation patterns on two-dimensional gels of glycoproteins during cell culture of isolated pachytenes and round spermatids. One of the proteins that was fucosylated by both cell types migrates at the position corresponding to the protein p151 whose synthesis apparently terminates after meiosis. This would indicate that there may be fucosylation of this protein after its synthesis and insertion into the surface (Gerton and Millette, 1986). There is also evidence that fucosyltransferase may be on the surface of developing germ cells of mouse and it could conceivably add fucose residues to other surface proteins (Millette et al., 1987).

A potential example of surface degylcosylation has also been observed on mouse spermatogenic cells. As mentioned previously, two mAbs, C6 and A5, were shown to label pachytenes and round spermatids, but earlier germ cell stages (through zygotene spermatocytes) were not labeled. Earlier stages could, however, be labeled after treatment with sialidase (Tables 18–1 and 18–2). A possible interpretation of these results is that terminal sialic acid substitutions are masking or altering the determinants recognized by these mAbs so that they are not available on earlier cell stages, and that these antigens are desialylated during the pachytene stage (Fenderson et al., 1984).

Proteases have been found in the testes, including angiotensin-converting enzyme (Hohlbrugger et al., 1982; Vanha-Perttula et al., 1985) and plasminogen activator (see LaCroix et al., 1981). Plasminogen activator is secreted by Sertoli cells (Ailenberg and Fritz, 1988) at specific stages in the spermatogenic cycle (LaCroix et al., 1981). An inhibitor of plasminogen activator activity in the testis has also been found (Hettle et al., 1988). It has been suggested that plasminogen activator may be affecting intercellular adhesion during translocation of germ cells from the basal to the adluminal compartment and during sperm release (see Hettle et al., 1988). Proteolysis of surface proteins could lead to the elimination of all or most of a protein molecule from the surface. Another possible action of proteases on surface proteins is limited proteolysis, which is a common method of activating inactive proteins (Neurath, 1989). The surface protein PH-30 of guinea pig sperm appears to be proteolytically processed during spermatogenesis. The mature form of PH-30 on cauda epididymal sperm is composed of two subunits, α ($\sim$ 45 kd) and β ($\sim$ 25 to 28 kd). Immunoblots of testicular cells have revealed a precursor form of the α subunit ($\sim$ 105 kd) that is processed during spermatogenesis to the 45 kd form (Blobel et al., 1991). This change may occur after protein insertion into the plasma membrane; the addition of exogenous proteinase K to spermatogenic cells results in cleavage of the 105 kd precursor of the α subunit (Blobel, C. P., and White, J. M., personal communication).

III. Post-testicular Modification of Surface Composition

When sperm leave the testes they are not fully functional. Full fertilization capacity is only gained after epididymal passage, capacitation, and the acrosome reaction. Two functional changes have been described after the sperm have passed through the epididymis: changes in the capacity for forward motility and fertilization (for review see Orgebin-Crist and Olson, 1984). If testicular sperm are diluted into a buffered solution, they show very restricted motility and no directed forward motility. At some time between the caput and the cauda, depending on the species, they have acquired this ability. Attempts to fertilize eggs with testicular sperm have not been successful. Recent studies of washed caput sperm of mouse have shown that they are capable of binding to the zona pellucida (McLaughlin and Shur, 1987); however, passage through the epididymis is thought to be required to attain full fertilizing ability. Even after sperm are ejaculated or removed from the cauda epididymis, they must still be capacitated and acrosome react before fertilization (Yanagimachi, 1988). Surface changes are thought to be important in the acquisition of sperm functions at each of these maturational steps and have been described for each stage (for review, see Yanagimachi, 1988).

Post-testicular modifications apparently take place in the absence of any significant synthesis of surface proteins by the sperm themselves. When sperm are released into the lumen of the seminiferous tubule, ribosomes and endoplasmic reticulum, required for cytoplasmic synthesis of membrane proteins, are left behind in the residual body. Theoretically, the small amount of cytoplasm left behind in the cytoplasmic droplet could be a source of either synthesis or some storage of surface molecules. There is, however, only evidence of protein synthesis by the mitochondria of sperm cells (Bragg and Handel, 1979).

A major source of potential change in composition is the addition of surface molecules from the surrounding fluids through which the sperm pass in the male and female reproductive tracts. In particular, the addition of molecules from the epididymal fluid (see Orgebin-Crist and Olson, 1984; Eddy et al., 1985) or from the seminal fluid has been discussed (Eddy, 1988). Because this review is focused on changes that occur to proteins that are associated with the sperm when they leave the testes, newly added proteins are considered here only as far as they affect access of probes to testicular sperm proteins.

There are also many possibilities for modification of testicular sperm surface proteins during post-testicular maturation. Potentially surface molecules can be made inaccessible to probes by the addition of new molecules on top of them or by changes in their conformation. Others may be modified by enzymes present in the surrounding fluids, on the surface of cells lining the male/female tracts, or enzymes that are carried along with the sperm either internally (e.g., in the cytoplasmic droplet) or on the surface.

A. Changes in the Detection of Surface Proteins

Functionally, only those molecules or regions of molecules that are exposed on the sperm surface can be considered part of the surface. Changes in binding of a given probe to the surface could result from structural change in a limited region of the molecule or from simply covering (addition of new surface molecules) or uncovering of surface groups (Fig. 18–1). The sperm surface may be viewed best as a layered structure with change occurring at several levels and the final surface depending on all those levels of change.

1. NEW SURFACE EXPRESSION

The ability to detect surface changes and to distinguish between the addition of new molecules and the alteration of old ones will depend on the probe used. For example, the PH-30 protein cannot be detected with two mAbs (PH-1 and PH-30) on sperm from the testis or early segments of the epididymis (Primakoff et al., 1987). If testicular sperm are stained with a variety of polyclonal antibodies (pAbs) raised against the purified PH-30 protein, however, they are able to stain PH-30 on testicular sperm (Phelps et al., 1990). The existence of PH-30 on testicular sperm can be confirmed by surface labeling sperm with ^{125}I and immunoprecipitating labeled PH-30 from detergent extracts of testicular sperm with a pAb (Phelps et al., 1990).

These results illustrate that the inability to detect a protein with a single surface probe cannot be interpreted as the absence of that protein from a particular cell type. Rather, specific probes may be detecting differences in either access of the probe to the region of the molecule that it recognizes or a change in the molecule that alters the ability of the protein to be recognized. Such changes may be the most important aspects of surface changes in terms of function.

2. LOSS OF SURFACE EXPRESSION

Actual loss of a protein might occur by different mechanisms. The external hydrophilic domain of proteins could be potentially removed from the surface by proteolysis (e.g., Thorley-Lawson et al., 1986). The GPI-anchored proteins could be released by the action of phospholipases and in fact, an apparent reduction in the amount of the GPI-anchored protein, PH-20, has been measured after epididymal passage (Phelps, B. M., and Myles, D. G., unpublished observation). Surface proteins could be completely eliminated by sorting them into the membrane surrounding the cytoplasmic droplet that is eventually shed during epididymal passage. Peripheral proteins that are more loosely associated with the surface may also be lost gradually by changes in the fluid conditions that allow them to be released. Experiments showing that mouse caput sperm can bind to the zona pellucida after washing demonstrate that molecules responsible for mature sperm function may be found on the surface of sperm before epididymal passage, but be inhibited by loosely associated surface components (McLaughlin and Shur, 1987). Previous studies had shown that GalTase on uncapacitated cauda epididymal sperm is loaded with high molecular weight polylactosaminyl substrates that block the GalTase activity in zona binding and that these substrates are released during capacitation in vitro (Shur and Hall, 1982a).

In contrast to actual loss, the apparent loss of surface molecules could result from loss of access to the protein by whatever probe is being used due to modification of the surface. These changes could include the rearrangement of existing surface components or the addition of new surface components.

B. *Modification of Surface Expression by Protein Structural Changes*

1. CHANGES IN GLYCOSYLATION OF SURFACE PROTEINS

Many investigators have suggested the possibility that sperm surface glycoproteins could undergo modification of their carbohydrate groups during epididymal passage and there is circumstantial evidence to support the plausibility of this phenomenon. However, no single convincing example of surface glycoprotein modification has emerged.

Some of the factors that have led to the interest in carbohydrate modification during epididymal transit include the following: mouse sperm have on their surface a galactosyl transferase and a fucosyl transferase (Shur and Hall, 1982b; Ram et al., 1989) and these ecto-enzymes could glycosylate other sperm surface proteins, if appropriate sugar donors were available in the epididymal fluid. Hamilton (1980, 1981) reported the presence of a soluble galactosyl transferase in rat rete testis and epididymal fluid and also the presence of an α-lactalbumin-like activity in rat epididymal fluid. Focusing on glycosidases rather than glycosyl transferases, Killian and collaborators studied levels of various glycosidases in sperm removed from different epididymal regions and various glycosidases in cell types from epididymal tissue (Chapman and Killian, 1984; Hall and Killian, 1987). The relevance of these intracellular glycosidase levels to potential glycosidase action on the sperm surface is unclear. More suggestive was the study of Skudlarek and Orgebin-Crist (1986), who showed that rat epithelial epididymal cells secreted two glycosidases, β-galactosidase and β-glucuronidase. However, the physiological substrate(s) of these secreted glycosidases is(are) unknown (Skudlarek and Orgebin-Crist, 1986). None of these investigations showed that the enzymes studied act on sperm surface carbohydrate moieties.

Some studies have focused on particular sperm surface components and examined the possibility that these undergo carbohydrate modification during epididymal transit. Cossu and Boitani (1984) found that mouse spermatogenic cells synthesize surface lactosaminoglycans that are unusual in that they are not fucosylated. They showed that surface lactosaminoglycans on mouse epididymal sperm are fucosylated and demonstrated a fucosyl transferase activity in epididymal homogenates. The fucosylated lactosaminoglycans on epididymal sperm, however, may not be the same lactosaminoglycans present on spermatogenic cells as they only have in common similiar behavior on a gel filtration column (Cossu and Boitani, 1984). Furthermore, the fucosyl transferase activity was detected in homogenates of epididymal tissue; therefore, evidence is lacking that it is secreted into epididymal fluid.

An important candidate for carbohydrate modification is the ~26-kd protein identified on the surface of rat sperm from the cauda. This protein has been reported to have molecular weights in the range of 24 to 37 kd because of variable behavior on sodium dodecylsulfate-polyacrylamide gel electrophoresis (SDS-PAGE) (Zeheb and Orr, 1984; Hamilton et al., 1986). Most recently the ~26-kd protein has been suggested as an identifying term (Moore et al., 1989). The ~26-kd protein originally aroused interest because it can be detected by labeling with the galactose oxidase/sodium borohydride technique on cauda sperm but not on caput sperm (Olson and Hamilton, 1978; Zeheb and Orr, 1984; Hamilton et al., 1986). Despite considerable study, the cell type (spermatogenic or epididymal) in which the ~26-kd protein is synthesized remains unknown (Moore et al., 1989). Thus, at least three possibilities remain to be distinguished in the study of this protein. (1) The ~26-kd protein may be synthesized by the epididymal epithelium, secreted and then bound to sperm in the cauda as an integral membrane protein; this would be striking as the protein is anchored in the membrane by linkage to glycosyl phosphatidylinositol (Moore et al., 1989). (2) The ~26-kd protein may be present on testicular sperm, but the carbohydrate that can be labeled is masked and then revealed by other (noncarbohydrate-related) surface modifications occurring during transit. (3) The protein may be present on testicular sperm and a modification to a carbohydrate moiety on it may occur during transit (so that it can be labeled by galactose oxidase) on sperm that have reached the cauda.

2. LIMITED PROTEOLYSIS OF SURFACE PROTEINS

It has been suggested that proteolytic processing may be a form of post-translational modification of the sperm surface during epididymal passage (e.g., Eddy et al., 1985) and there is evidence for proteolytic enzymes in the epididymis (Hohlbrugger et al., 1982; Vanha-Perttula et al., 1985; Huarte et al., 1987). Proteolytic processing has also been suggested to play a role in surface modification of sperm during capacitation (Talbot and Chacon, 1981).

A decrease in the apparent molecular weight (from SDS-PAGE) of four surface antigens of mouse sperm recognized by the mAbs, M42, M5, W33, and W108, after passage through the epididymis has been described (Lakoski et al., 1989). The post-translational modification of the M42 antigen was of particular interest as this mAb interferes with the ability of sperm to undergo a zona-induced acrosome reaction and sperm are not fully capable of these physiological acrosome reactions until they reach the cauda (Lakoski et al., 1988).

In guinea pig, there is evidence that proteolysis of the sperm surface may be a prominent aspect of epididymal maturation (Phelps et al., 1990; Blobel et al., 1990). Brief trypsin treatment of testicular sperm resulted in the mimicking of several aspects of epididymal maturation. Two surface antigens, PH-20 and PH-30, undergo changes in molecular weight that can be mimicked by limited proteolysis in vitro. Furthermore, in vitro proteolytic treatment of testicular sperm produces an overall gel pattern of detergent extracts of surface-labeled sperm that are altered from the testicular pattern to resemble the cauda pattern (Phelps et al., 1990). Other alterations of the sperm surface that occur during epididymal passage, including the change in surface topography (see section IV.A.2.) and the appearance of surface epitopes, can also be induced by treatment of testicular sperm with trypsin. As an example of epitope appearance, trypsin treatment in vitro, and by implication proteolytic processing in vivo, results in the ability of two mAbs to stain PH-30 protein on testicular sperm. Because the protein is known to be present on testicular sperm by studies with pAbs (see section III. A.-1.), these results indicate that two new epitopes are either revealed or created by proteolysis. This is of particular interest because antibody inhibition studies indicate that the epitope recognized by the PH-30 mAb is required for sperm–egg plasma membrane fusion (Primakoff et al., 1987).

IV. Creation and Alteration of Surface Domains

The composition of the sperm surface is of obvious importance in controlling sperm function. A major feature of the sperm cell surface is its polarized distribution of surface molecules (for review see Eddy, 1988) and certain sperm cell surface functions (e.g., zona binding and fusion with the egg plasma membrane) are probably restricted to specific surface regions.

Because the sperm surface is separated into distinct domains, considerations of cell surface composition, from later stages of spermiogenesis on, may best be considered as the composition of these individual domains.

A. Expression of Surface Proteins in Localized Patterns

Not all proteins become localized at the same stage of sperm differentiation. Some proteins are localized to specific domains in the testis, but other proteins do not become localized until sperm reach the cauda epididymis. Patterns of protein distribution, once established, may be altered at later stages of sperm maturation. Therefore, to fully understand sperm cell surface functional maturation we need to address the questions of how these surface domains are created and altered. The related question of how these domains are maintained is discussed in section V.

1. SURFACE DOMAINS IN THE TESTIS

When spermatocytes have completed meiosis, the resulting spermatids begin the elaborate process of spermiogenesis during which two potential references for the development of membrane asymmetry are established. Internally, an asymmetry is established in the distribution of the cell's cytoplasmic structures that become associated with different regions of the sperm plasma membrane (Fawcett, 1975; Eddy, 1988). Externally, the Sertoli cell membrane shows a close association with the sperm plasma membrane in specific regions (Russell and Peterson, 1985).

Surface domains begin to develop during spermiogenesis, and on testicular sperm a clear separation of some surface proteins into head and tail regions can be demonstrated. The XT-1 antigen of mouse is detected by an mAb that binds to the entire surface of spermatocytes, but becomes restricted to the tail (apparently mainly to the anterior tail) and a small patch at the base of the head during spermatid elongation (Fig. 18–2A,B) (Bechtol, 1984). Antibodies that recognize identified surface proteins of guinea pig sperm (Primakoff and Myles, 1983) and show a whole head pattern on testicular sperm (Fig. 18–2C,D) are the mAbs AH-21, AH-50, WH-1, and WH-30 and the PH-30 pAb (Myles and Primakoff, 1983; Phelps et al., 1990).

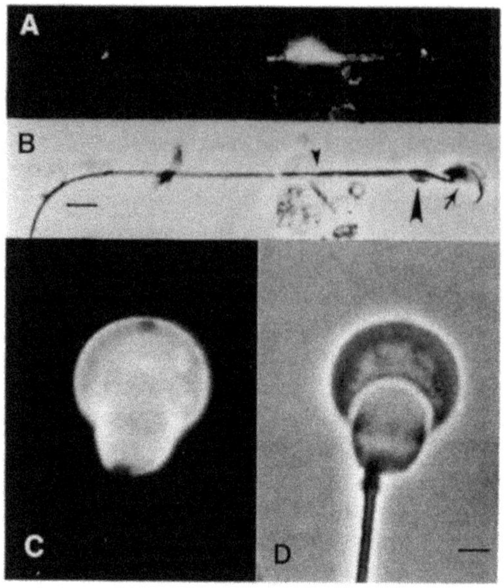

Figure 18–2. Testicular sperm of mouse (**A**,**B**) and guinea pig (**C**,**D**), stained by indirect immunofluorescence and corresponding phase contrast images. (**A**,**B**) The binding of the XT-1 mAb is strongest over the anterior tail (midpiece) and the cytoplasmic lobe, with a small area of staining at the base of the head. Bar, 5 μm. (**C**,**D**) Binding of the WH-1 mAb is localized to the whole head, but absent from the tail. Bar = 2 μm. (**A** and **B** are reproduced with permission from Bechtol, 1984.)

In some cases, subdomains of the head and tail may be already established on testicular sperm. The binding patterns of two mAbs, GHS-5 and GHS-7, are restricted to the anterior head region of hamster sperm from the testis (Koyama et al., 1988); the binding pattern of a pAb directed against GalTase is confined to the anterior half of the head of mouse testicular sperm (the surface region over the acrosome, Fig. 18–4G) (Scully et al., 1987), and a pAb raised against rat testis galactosyl receptor stains only the anterior part of the head of testicular sperm (Abdullah and Kierszenbaum, 1989). On the tail of guinea pig testicular sperm, some surface proteins have already partitioned into anterior tail and posterior tail domains. The PT-1 surface antigen is restricted to the posterior tail region of testicular sperm

(Myles and Primakoff, 1983), and two different mAbs bind to the anterior tail and residual body of guinea pig testicular sperm, indicating that an anterior tail–residual body domain may exist (Cowan and Myles, unpublished observation). Earlier work with polyclonal antibodies raised against spermatogonia and absorbed with somatic cells also demonstrated that there is one class of antigens that are sorted specifically into the residual body of mouse sperm (Millette and Bellve, 1980).

2. CHANGES IN THE PATTERNS OF SURFACE EXPRESSION DURING EPIDIDYMAL PASSAGE

During epididymal passage, localized surface patterns may either be retained, altered, or established. The staining pattern of the mAbs PT-1 (posterior tail) and WH-1 (whole head) are unaltered when sperm from the testis or cauda epididymis are compared (Myles and Primakoff, 1983; Phelps et al., 1990). Also in guinea pig, antibodies that bind to the whole head of testicular sperm may become restricted during epididymal maturation so that they ultimately bind to either the anterior head (mAbs, AH-21, and AH-40) or posterior head (pAb to PH-30 protein) (Myles and Primakoff, 1983; Phelps et al., 1990). The distribution of GalTase (determined with a pAb) is restricted from a location over the entire acrosome (see Fig. 18–4G) to the dorsal region (see Fig. 18–4H) of the head of mouse sperm (Scully et al., 1987). Other examples of antibody-binding patterns that are established during epididymal passage include the 1B3 mAb (recognizing a 28-kd protein), which binds uniformly to the surface of mouse sperm from the testis and caput, but is restricted in its binding to the anterior tip of sperm from the cauda epididymis and vas deferens (Gaunt, 1982). The distribution of PH-20 protein also goes from a whole cell surface distribution on sperm from testis (see Fig. 18–4A), caput, and corpus to being restricted to the posterior head surface of cauda sperm (see Fig. 18–4B) (Phelps and Myles, 1987; Phelps et al., 1990).

3. POSTEPIDIDYMAL CHANGES IN SURFACE DOMAINS

Testicular sperm surface proteins may continue to show altered patterns of localization up until fertilization. In a few cases, sperm surface proteins, known to originate in the testis, have been followed in their pattern of localization on postepididymal sperm. For example, during capacitation in vitro there is a change in the binding pattern of PT-1 on guinea pig sperm (Fig. 18–3A,D) (Myles and Primakoff, 1984) and 2B1, a protein of testicular origin (Gaunt et al., 1983), on rat sperm (Fig. 18–3E–H) (Jones et al., 1990). After the acrosome reaction there are changes in the patterns of localization of PH-20 protein on guinea pig sperm (Fig. 18–4B,C) (Myles and Primakoff, 1984), GalTase on mouse sperm (Lopez and Shur, 1987), and RSA on rabbit sperm (Fig. 18–4H,I) (Esaguy et al., 1988). In addition, the acrosome reaction results in a major loss of some testicular proteins found on the anterior head or whole head of acrosome-intact sperm when most of the anterior head plasma membrane is lost during the acrosome reaction (Myles and Primakoff, unpublished observations).

B. Mechanisms for the Changes in Protein Localization

During spermatogenesis, epididymal maturation, capacitation, and acrosome reaction, a variety of mechanisms may act initially to localize surface proteins or to alter subsequently their localizations. In some cases, localization of surface proteins to a domain may arise by the incorporation or adsorbtion of new proteins into localized regions. Protein localization may result from the redistribution to a domain of membrane proteins that are already on the surface, but initially are uniformly distributed. Furthermore, even after initial protein localization is established, the protein may redistribute to a new restricted domain.

1. CHANGES IN PROTEIN LOCALIZATION BY CHANGES IN SURFACE COMPOSITION

Any process that changes the composition of the surface (see section III) could act to change patterns of surface localization if the alterations are regionalized. This would include any modifications or losses of existing proteins, for example, the loss of surface proteins that results from the acrosome reaction (discussed previously). Patterns of surface expression of testicular proteins can also be altered by the

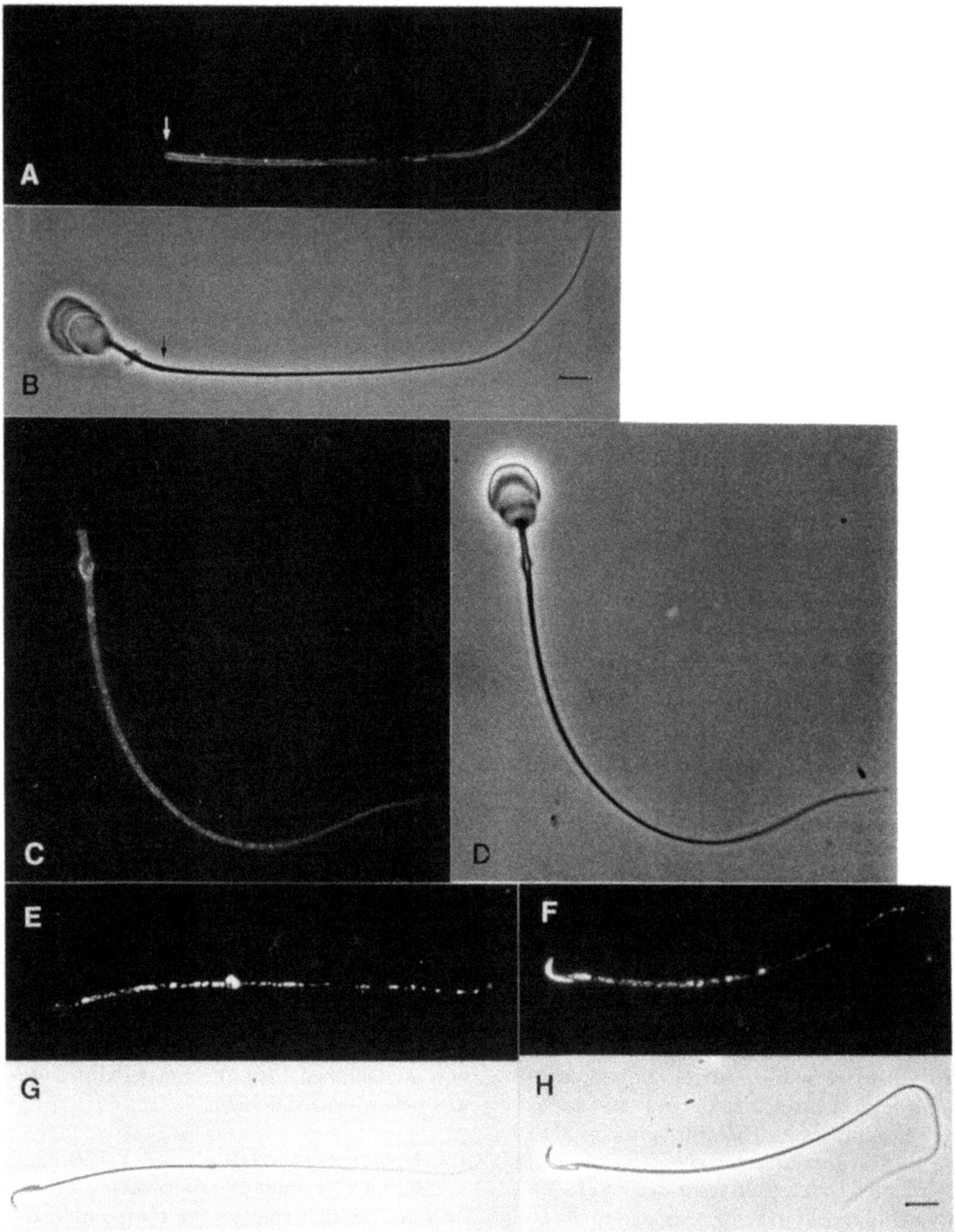

Figure 18–3. Cauda epididymal sperm of guinea pig (**A–D**) and rat (**E–H**), before and after capacitation, stained by indirect immunofluorescence and corresponding phase contrast images. (**A,B**) PT-1 mAb staining is restricted to the posterior tail region (principal piece and endpiece) before capacitation. *Arrow* indicates anterior tail–posterior tail junction. (**C,D**) After capacitation, the whole tail is stained. (**E,G**) 2B1 mAb binding is to the whole tail before capacitation. (**F,H**) After capacitation, staining patterns observed include uniform fluorescence over the acrosome region. **A–D** reprinted with permission from Myles and Primakoff (1984); **E–H** reprinted with permission from Jones et al. (1990). Bars = 5 μm.

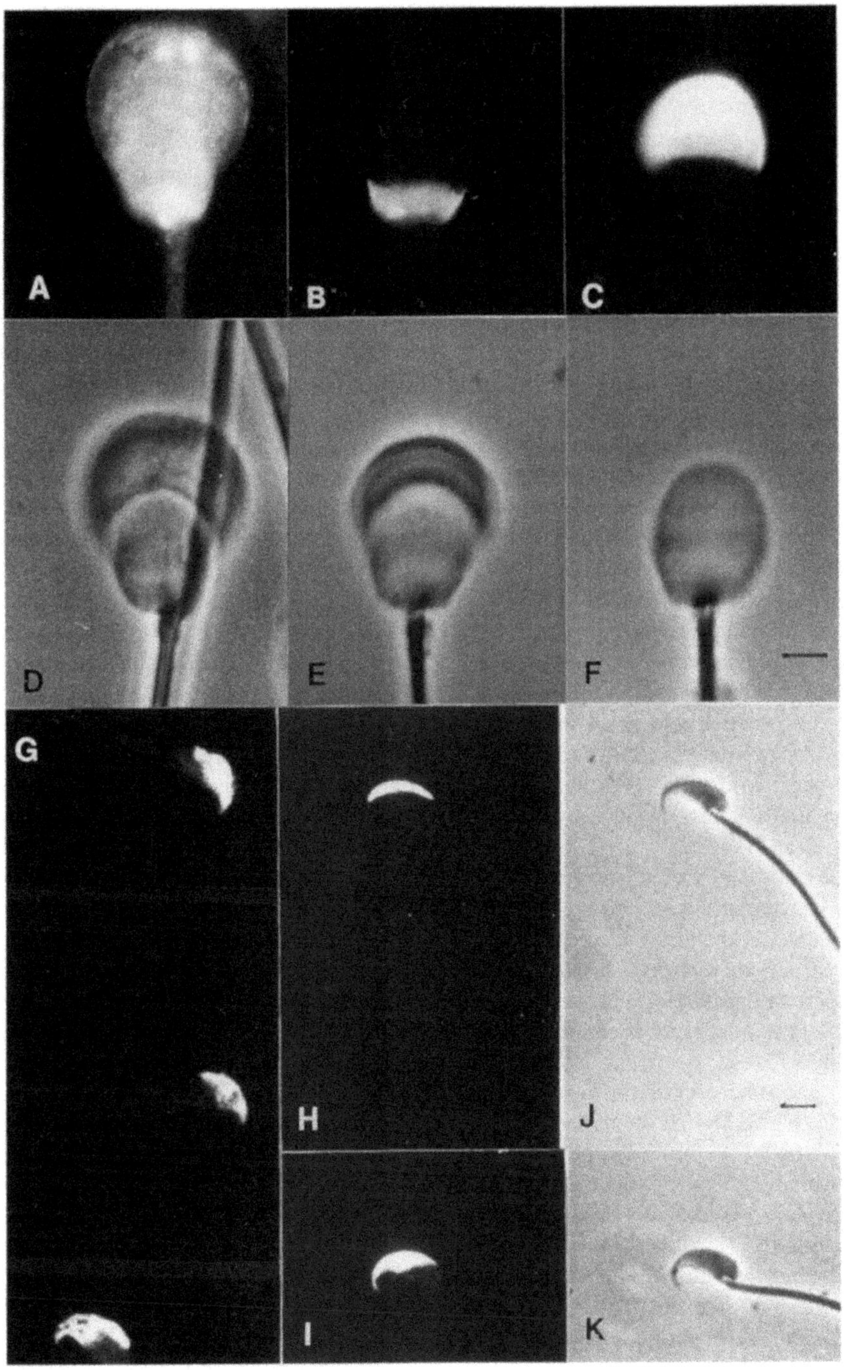

Figure 18–4. Comparison of antibody staining patterns, and corresponding phase contrast images from testicular and cauda epididymal (acrosome-intact and acrosome-reacted) sperm of guinea pig (**A–F**) and mouse (**G–K**). (**A,D**) PH-20 mAb stains the entire cell surface of testicular sperm; (**B,E**) PH-20 mAb binding is restricted to the posterior head region of acrosome-intact cauda sperm; (**C,F**) after the acrosome reaction, PH-20 staining is confined to the inner acrosome membrane. (**G**) Antibody binding to GalTase is restricted to the anterior head of testicular sperm; (**H,J**) GalTase becomes further restricted to the dorsal aspect of the anterior head of acrosome-intact cauda sperm; (**I,K**) after the acrosome reaction the GalTase staining pattern is shifted to the anterior half of the lateral sperm head (possibly the equatorial segment). (Figures are reprinted from (**A,D**) Phelps and Myles, 1987; (**B,C,E,F**) Cowan, et al., 1986; (**G**) Scully et al., 1987; (**H–K**) Lopez and Shur, 1987, with permission from Rockefeller University Press and Academic Press.) Bars = 2 μm.

addition (and then perhaps later loss) of molecules from the reproductive tract fluids, if the addition and/or loss alters surface expression of testicular proteins in localized regions.

2. CHANGES IN PROTEIN LOCALIZATION BY MIGRATION OF EXISTING SURFACE PROTEINS

Alternatively to changes in the composition of the membrane, changes in localization can result from the migration of surface molecules from an initial region into a new domain. Changes in patterns of intramembrane particles (IMPs) that occur post-testicularly could indicate rearrangements of integral membrane surface proteins (see, for example, Friend, 1984; Burgos, 1986; Suzuki and Yanagimachi, 1986; Suzuki, 1988; Aguas and Pinto da Silva, 1989). Demonstration that the proteins represented by the IMPs are exposed on the surface and are of testicular origin, however, would require identification of the protein associated with the IMP such as could be gained using techniques that can specifically identify proteins associated with the IMPs (Pinto da Silva and Kan, 1984; Esaguy et al., 1988).

The migration of individual testicular surface proteins has been tracked during epididymal passage, capacitation, and the acrosome reaction. As mentioned, the PH-20 proteins of guinea pig sperm migrate twice: once during epididymal passage and again after the acrosome reaction.

PH-20 is found on the entire surface of testicular sperm (Fig. 18–4A). It remains uniformly distributed on the sperm surface through the initial segments of the epididymis, and then becomes localized to the posterior head of the sperm in the cauda epididymis (Fig. 18–4B) (Phelps and Myles, 1987). Using an in vitro system of trypsin treatment that mimicks changes that occur during epididymal passage, it was possible to fluorescently label cells with antibody before protein localization and then quantitatively follow the change in PH-20 distribution (Phelps et al., 1990). A twofold increase in the amount of PH-20 associated with the posterior head region occurred during localization, indicating that there is an actual movement of PH-20 molecules from other surface regions into the posterior head domain.

Migration of two different tail antigens has been demonstrated during capacitation. In both cases proteins were prelabeled with fluorescent antibodies and then incubated in the presence of excess unlabeled antibody so that any stain appearing in a new region could be identified as prelabeled protein from another region. Both the PT-1 protein of guinea pig sperm (Myles and Primakoff, 1984) and the rat sperm protein 2B1 (Jones et al., 1990) migrate to new regions (anterior tail and anterior head, respectively) after incubation in capacitating media (Fig. 18–3).

Two head surface proteins have similarly been demonstrated to migrate after the acrosome reaction. The migration of PH-20 protein from the posterior head region of acrosome-intact sperm to the inner acrosomal membrane of acrosome-reacted sperm (Fig. 18–4B,C) was demonstrated by the same type of antibody prelabeling experiments (Myles and Primakoff, 1984). Migration of GalTase from the dorsal surface of the anterior sperm head overlying the acrosome of acrosome-intact sperm to the lateral surface of acrosome-reacted sperm (Fig. 18–4H–I) was demonstrated by the following observations: (1) the level of GalTase on the surface before and after the acrosome reaction is approximately the same, (2) no release of GalTase can be detected on membrane vesicles released after the acrosome reaction, (3) in contrast, nitrogen cavitation of the anterior head plasma membrane results in release of GalTase activity from the sperm (Lopez and Shur, 1987).

Although it is tempting to speculate about the function of these migrations, no function has yet been established. It is worth noting that three of the proteins, GalTase (Shur and Hall, 1982b; Lopez et al., 1985), the PH-20 protein (Primakoff et al., 1985), and 2B1 (Shalgi et al., 1990), have been reported to have a required role in sperm–zona binding. The migration of PH-20 into the inner acrosomal membrane allows PH-20 to join a covert population of PH-20 already localized to the inner acrosomal membrane (Cowan et al., 1986) and, therefore, concentrates the protein in the region where acrosome-reacted sperm binding to the zona has been observed (Huang and Yanagimachi, 1985; Myles et al., 1987a). Antibody inhibition studies indicate that PH-20 is required in binding acrosome-reacted sperm to the zona pellucida. Similarly the migration of 2B1 to the anterior head brings it into a region active in acrosome-intact sperm binding to the zona. The migration of GalTase results in its conservation on the sperm surface when it otherwise would have been lost with the acrosome reaction-released vesicles. It has been suggested that after the acrosome reaction GalTase may facilitate binding of acrosome-reacted sperm during the initial stages of zona penetration and/or attachment to the egg plasma membrane before membrane fusion (Shur, 1989).

Although it has been established that one way to alter patterns of surface localization is by the migration of surface proteins from one surface domain to another, how these migrations are executed is less clear. Studies of cells in the process of protein migration reveal that more than one mechanism may be responsible.

Migration of the PT-1 protein on guinea pig sperm from the posterior tail to the anterior tail during capacitation is energy and temperature dependent (Myles and Primakoff, 1984). Measurements of fluorescence redistribution after photobleaching (FRAP) of PT-1 on sperm where partial migration of the PT-1 protein had occurred produced data that were similar to computer simulations of PT-1 diffusion that was partially restricted at the anterior tail–posterior tail boundary. A substantial immobile population of PT-1 associated with the anterior tail region was also detected. These results are consistent with a model for PT-1 migration that would result from diffusion of PT-1 protein within the entire tail region and trapping in the anterior tail region (Koppel et al., 1986).

Analysis of PH-20 protein migration on the surface of sperm after the acrosome reaction indicates that an alternative mechanism to diffusion and trapping may be responsible for this migration. Using computer-assisted quantitative fluorescence microscopy, migration of PH-20 protein after the acrosome reaction was followed in single cells. During migration, the PH-20 protein moves against a concentration gradient. This result is not consistent with diffusion and trapping, but indicates that migration may be an active process (Cowan et al., 1991). Because PH-20 is a GPI-anchored protein (Phelps et al., 1988), any interactions responsible for its movement must occur in the ectodomain of the protein.

V. Maintenance of Membrane Domains

There are basically two ways in which an asymmetric distribution of protein in the membrane can be maintained: either through interactions with other molecules or by barriers in the membrane that prevent the diffusion of molecules across the domain boundaries. Maintenance of asymmetry by interactions implies that there would be either a tethering of the molecules or a transient interaction that would result in a restriction to lateral protein diffusion. Proteins that are maintained within a domain by some type of membrane barrier, on the other hand, could be freely diffusing within that domain. Therefore, the technique of fluorescence redistribution after photobleaching (FRAP) is useful because it can quantitate the diffusion characteristics of membrane proteins. Two characteristics are measured with this technique: (1) the rate of diffusion or diffusion coefficient (D) and (2) the percent mobile fraction or percent recovery (%R) (Koppel, 1979).

A. Molecular Interactions

A variety of interactions could maintain the localization of membrane proteins (Fig. 18–1). Interactions of membrane proteins could be with structures on either the cytoplasmic or extracellular side of the membrane or interactions within the membrane itself. During spermiogenesis, a submembranous cytoskeleton forms that is region specific (for discussion, see Eddy, 1988). On the external surface of the sperm is an extracellular coat or glycocalyx and the organization of the glycocalyx has been correlated with the arrangement of IMPS in the plasma membrane (Friend and Fawcett, 1974; Suzuki and Nagano, 1980). It is also possible that interactions of proteins with lipids that have partitioned into regions of the sperm plasma membrane would lead to a localized distribution of the protein (Wolf, 1986). An increasing population of immobilized lipid has been measured on sperm using lipid analogues and the technique of FRAP. This immobile population develops during spermiogenesis (Wolf et al., 1986c), and increases during epididymal maturation (Wolf and Voglmayr, 1984) and capacitation (Wolf et al., 1986a). Neither treatment of the extracellular surface with pronase, in an attempt to remove extracellular coat or glycocalyx proteins, or blebbing of the sperm membrane, to separate it from potential cytoplasmic interactions, had any effect on either D or %R of the lipid analogues. Therefore, it was concluded that lipid composition itself could be causing lateral phase separations and/or the nondiffusing lipid (Wolf et al., 1988). If proteins interact with lipids that are restricted in their surface distribution and lateral diffusion, the distribution and diffusion of the protein would also be restricted.

PH-20 diffusion has been shown to be restricted on guinea pig sperm at two different stages in development. On testicular sperm, when PH-20 is still uniformly distributed on the whole cell surface, diffusion is highly restricted (Phelps et al., 1988). It is also restricted on acrosome-intact sperm from the cauda epididymis when it is localized to the posterior head region; this restriction falls into an intermediate range and may reflect some type of transient interaction (Cowan et al., 1987). Measurements of diffusion of a lipid probe inserted into the plasma membrane of acrosome-intact guinea pig sperm did not reveal a substantial immobile lipid population (Cowan et al., 1987), indicating that PH-20 localization probably does not result from interactions with immobile lipid or lateral phase separations of lipids. Because at least 50 percent (and perhaps all) of the PH-20 protein is anchored in the membrane by GPI, the interactions of PH-20 that would lead to its restricted diffusion cannot be directly with cytoplasmic structures but must occur in the extracellular half of the membrane (Phelps et al., 1988). Therefore, restriction of PH-20 diffusion might be through its interaction with the extracellular coat or glycocalyx.

B. Barriers

In the absence of molecular interactions proteins may demonstrate free diffusion in the membrane. Although sometimes a "patchy" appearance of an antigen after binding to a divalent ligand (see Eddy, 1988, for review) has been used to indicate protein mobility in the membrane such a staining appearance does not necessarily correlate with ability to diffuse (Wolf, 1983). Furthermore, even transient interactions, which might slow but not immobilize a protein, could be responsible for maintenance of proteins within a localized domain (Cowan et al., 1987).

Two testicular-derived proteins that have been shown to be diffusing freely in the sperm membrane of cauda epididymal sperm by direct measurement using the FRAP technique are PT-1 and PH-20. PT-1 is freely diffusing in the posterior tail domain of sperm taken fresh from the cauda epididymis (Myles et al., 1984). After incubation in capacitating media, when part of the PT-1 population has migrated into the anterior tail region so that PT-1 is found in both anterior tail and posterior tail regions (see section IV.A.3.), PT-1 is freely diffusing in both the posterior tail and the anterior tail regions (Koppel et al., 1986). The PH-20 protein shows free diffusion when it is restricted to the inner acrosomal membrane after the acrosome reaction (Cowan et al., 1987). Free diffusion of a third protein, ESA 152, has been measured on ejaculated ram sperm, but this protein has been suggested to be inserted into the membrane during epididymal passage from the epididymal fluid (Wolf et al., 1986b).

The existence of freely diffusing proteins within strictly demarcated domains suggests that there are barriers in the membrane that are able to prevent the free diffusion of proteins from one domain to the adjacent domain. For the guinea pig sperm three such barriers might exist at (1) the equatorial region, (2) the head–tail junction, and (3) the junction between the anterior tail (midpiece) and the posterior tail (principal piece) (Fig. 18–5). In each of these membrane zones distinct structures have been observed in thin sections and freeze fracture replicas (Friend and Fawcett, 1974). The equatorial region contains a ring of fingerlike projections at its posterior margin that divides the head into anterior and posterior regions. In this region a narrow band of filaments (intermediate filaments?) have also been described that may participate in forming or in marking a barrier to membrane protein diffusion (Friend, 1989). At the base of the head is a specialized membrane region where nuclear envelope and plasma membrane are in close apposition ("posterior striated ring," Friend, 1989). At the anterior–posterior tail junction is an electron dense ring or annulus also closely apposed to a

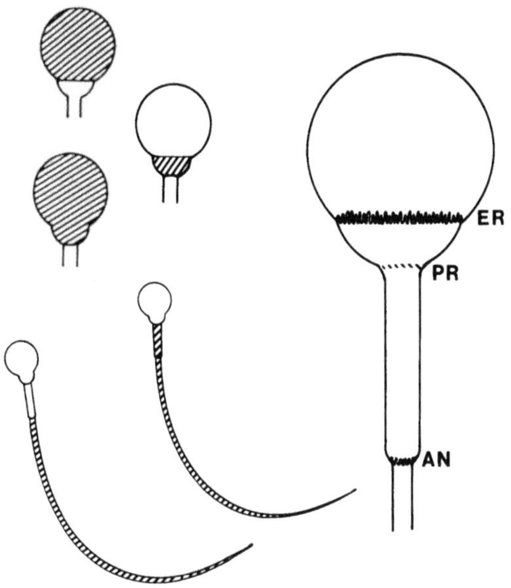

Figure 18–5. Five antibody binding patterns, previously described on the surface of cauda epididymal sperm of the guinea pig (Primakoff and Myles, 1983), are shown on the left. On the right, the locations of membrane specializations that may act as membrane barriers to localize surface proteins are indicated. ER, equatorial region; PR, posterior striated ring; AN, annulus. Structures associated with these regions are described in the text. (Reprinted from Myles and Primakoff, 1986, with permission from the Japanese Society of Electron Microscopists.)

specialized region of the membrane. Friend (1989) has suggested in all three cases that cytoskeletal elements are associated with these structures. It has been suggested previously that these images are morphological representations of barriers to membrane protein diffusion that maintain the localization of freely diffusing membrane proteins within a localized region (Myles and Primakoff, 1986; Myles et al., 1987b; Friend, 1989). The patterns observed on guinea pig sperm using monoclonal antibodies (Primakoff and Myles, 1983) are such that any of these patterns could be explained by the existence of these three membrane barriers (Myles and Primakoff, 1986).

VI. Summary and Conclusions

The male germ cell surface is a complex array of molecules that control the germ cell's interaction with other cells and are capable of receiving and responding to external signals. Unlike many mammalian cells, it does not spend the majority of its life in association with the same set of cells, responding repeatedly to the same set of signals. Rather it goes through a varied set of interactions both with the surrounding cells and with the fluid environments with which it comes into contact. It also changes in its ability to carry out specific functions.

Spermatogenic cells in the testis are apparently active in synthesizing new surface proteins, and differences in the composition of the surface have been detected at the three major subdivisions of spermatogenesis: mitotic proliferation, meiosis, and spermiogenesis. Once the sperm has finished spermiogenesis, any new protein synthesis is very limited. Rather, the sperm changes its surface composition, and presumably thereby regulates its surface functions by interaction with the surrounding environment (both cells and fluids). Although the sperm may acquire new surface molecules from this surrounding environment, it also modifies molecules that are associated with the sperm surface. Potential mechanisms for modification of testicular sperm proteins include elimination, rearrangements, changes in molecular interactions, and enzymatic modifications. Potential sources of enzymes include surrounding fluids and cells, as well as internal and surface enzymes of the sperm itself. This change in existing surface protein begins in the testes and continues until the sperm ultimately fuses with the egg.

ACKNOWLEDGMENTS

I would like to acknowledge Drs. Bruce Fenderson, E. M. Eddy, Barry Hinton, Barry Shur, and David Wolf for helpful conversations during the writing of the manuscript. I also appreciate the photographs supplied by Drs. Kathleen Bechtol, Roy Jones, and Barry Shur. Thanks to Dr. Paul Primakoff for reading the manuscript and for helpful suggestions and to Ms. Betty Burrows for help with the final preparation of the manuscript. I also appreciate the encouragement and useful advice of Dr. Larry Ewing.

The work from my laboratory cited in this review was supported by National Institutes of Health grant HD16580 and salary support was received from an American Cancer Society Faculty Research Award during the writing of this article.

REFERENCES

Abdullah, M., and Kierszenbaum, A. L. (1989). Identification of rat testis galactosyl receptor using antibodies to liver asialoglycoprotein receptor: Purification and localization on surfaces of spermatogenic cells and sperm. J. Cell Biol. 108:367–375.

Aguas, A. P., and Pinto da Silva, P. (1989). Biomodal redistribution of surface transmembrane glycoproteins during Ca^{2+}-dependent secretion (acrosome reaction) in boar spermatozoa. J. Cell Sci. 93:467–479.

Ailenberg, M., and Fritz, I. B. (1988). Control of levels of plasminogen activator activity secreted by Sertoli cells maintained in a two-chamber assembly. Endocrinology 122:2613–2618.

Bechtol, K. B. (1984). Characterization of a cell-surface differentiation antigen of mouse spermatogenesis: Timing and localization of expression by immunohistochemistry using a monoclonal antibody. J. Embryol. Exp. Morph. 81:93–104.

Bechtol, K. B., Brown, S. C., and Kennett, R. H. (1979). Recognition of differentiation antigens of spermatogenesis in the mouse by using antibodies from spleen cell-myeloma hybrids after syngeneic immunization. Proc. Natl. Acad. Sci. 76:363–367.

Blobel, C. P., Myles, D. G., Primakoff, P., and White, J. M. (1990). Proteolytic processing of a protein involved in sperm–egg fusion correlates with acquisition of fertilization competence. J. Cell Biol. 111:69–78.

Bragg, P. W., and Handel, M. A. (1979). Protein synthesis in mouse spermatozoa. Biol. Reprod. 20:333–337.

Burgos, M. H., Cavicchia, J. C., de Rosas, J. C., and Gutierrez, L. S. (1986). Freeze fracture of sperm membranes during epididymal maturation. J. Electron Microsc. (Suppl) 35:1881–1884.

Chapman, D. A., and Killian, G. J. (1984). Glycosidase activities in principal cells, basal cells, fibroblasts and spermatozoa isolated from the rat epididymis. Biol. Reprod. 31:627–636.

Cossu, G., and Boitani, C. (1984). Lactosaminoglycans synthesized by mouse male germ cells are fucosylated by an epididymal fucosyltranferase. Dev. Biol. 102:402–408.

Cowan, A. E., Myles, D. G., and Koppel, D. E. (1991). Migration of the guinea pig sperm membrane protein, PH-20, from one localized surface domain to another does not occur by a simple diffusion-trapping mechanism. Dev. Biol. 144:189–198.

Cowan, A. E., Myles, D. G., and Koppel, D. E. (1987). Lateral diffusion of the PH-20 protein on guinea pig sperm: Evidence that barriers to diffusion maintain plasma membrane domains in mammalian sperm. J. Cell Biol. 104:917–923.

Cowan, A. E., Primakoff, P., and Myles, D. G. (1986). Sperm exocytosis increases the amount of PH-20 antigen on the surface of guinea pig sperm. J. Cell Biol. 103:1289–1297.

D'Agostino, A., and Stefanini, M. (1990). An intercellular adhesion protein of rat testis: Isolation and preliminary characterization. Mol. Reprod. Dev. 25:172–176.

D'Agostino, A., and Stefanini, M. (1987). A rat spermatocyte surface protein is involved in adhesion of pachytene spermatocytes to Sertoli cells in vitro. Mol. Cell. Biol. 7:1250–1255.

Eddy, E. M. (1988). The spermatozoon, in *The Physiology of Reproduction* (Knobil, E., Neill, J., et al., eds.) pp. 27–68, Raven Press, Ltd., New York.

Eddy, E. M., Vernon, R. B., Muller, C. H., Hahnel, A. C., and Fenderson, B. A. (1985). Immunodissection of sperm surface modifications during epididymal maturation. Am. J. Anat. 174:225–237.

Enders, G. C., and Millette, C. F. (1988). Pachytene spermatocyte and round spermatid binding to Sertoli cells in vitro. J. Cell Sci. 90:105–114.

Esaguy, N., Welch, J. E., and O'Rand, M. G. (1988). Ultrastructural mapping of a sperm plasma membrane autoantigen before and after the acrosome reaction. Gam. Res. 19:387–399.

Fawcett, D. W. (1975). The mammalian spermatozoon. Dev. Biol. 44:394–436.

Fenderson, B. A., O'Brien, D. A., Millette, C. F., and Eddy, E. M. (1984). Stage-specific expression of three cell surface carbohydrate antigens during murine spermatogenesis detected with monoclonal antibodies. Dev. Biol. 103:117–128.

Ferguson, M.A.J., and Williams, A. F. (1988). Cell-surface anchoring of proteins via glycosyl-phosphatidylinositol structures. Ann. Rev. Biochem. 57:285–320.

Friend, D. S. (1989). Sperm maturation: Membrane domain boundaries. Ann. NY Acad. Sci. 567:208–221.

Friend, D. S. (1984). Membrane organization and differentiation in the guinea-pig spermatozoon, in *Ultrastructure of Reproduction* (Van Blerkom, J., and Motta, P. M., eds.) pp. 75–85, Martinus Nijhoff Publishers, Boston.

Friend, D. S. (1982). Plasma-membrane diversity in a highly polarized cell. J. Cell Biol. 93:243–249.

Friend, D. S., and Fawcett, D. W. (1974). Membrane differentiations in freeze-fractured mammalian sperm. J. Cell Biol. 63:641–664.

Gaunt, S. J. (1982). A 28k-dalton cell surface autoantigen of spermatogenesis: Characterization using a monoclonal antibody. Dev. Biol. 89:92–100.

Gaunt, S. J., Brown, C. R., and Jones, R. (1983). Identification of mobile and fixed antigens on the plasma membrane of rat spermatozoa using monoclonal antibodies. Exp. Cell Res. 144:275–284.

Gerton, G. L., and Millette, C. F. (1986). Stage-specific synthesis and fucosylation of plasma membrane proteins by mouse pachytene spermatocytes and round spermatids in culture. Biol. Reprod. 35:1025–1035.

Griswold, M. D. (1988). Protein secretions of Sertoli cells. Int. Rev. Cytol. 110:133–156.

Hall, J. C., and Killian, G. J. (1987). Changes in rat sperm membrane glycosidase activities and carbohydrate and protein contents associated with epididymal transit. Biol. Reprod. 36:709–718.

Hamilton, D. W. (1981). Evidence for a α-lactalbumin-like activity in reproductive tract fluids of the male rat. Biol. Reprod. 25:385–392.

Hamilton, D. W. (1980). UDP–galactose: *N*-acetylglucosamine galactosyltransferase in fluids from rat rete testis and epididymis. Biol. Reprod. 23:377–385.

Hamilton, D. W., Wenstrom, J. C., and Baker, J. B. (1986). Membrane glycoproteins from spermatozoa: Partial characterization of an integral $M_r = \sim 24,000$ molecule from rat spermatozoa that is glycosylated during epididymal maturation. Biol. Reprod. 34:925–936.

Hecht, N. B. (1988). Post-meitotic gene expression during spermatogenesis. Prog. Clin. Biol. Res. 267:291–313.

Hettle, J. A., Balekjian, E., Tung, P. S., and Fritz, I. B. (1988). Rat testicular peritubular cells in culture secrete an inhibitor of plasminogen activator activity. Biol. Reprod. 38:359–371.

Hohlbrugger, G., Schweisfurth, H., and Dahlheim, H. (1982). Angiotensin I converting enzyme in rat testis, epididymis and vas deferens under different conditions. J. Reprod. Fert. 65:97–103.

Huang, T.T.F. Jr., and Yanagimachi, R. (1985). Inner acrosomal membrane of mammalian spermatozoa: Its properties and possible functions in fertilization. Am. J. Anat. 174:249–268.

Huarte, J., Belin, D., Bosco, D., Sappino, A.-P., and Vassalli, J. D. (1987). Plasminogen activator and mouse spermatozoa: Urokinase synthesis in the male genital tract and binding of the enzyme to the sperm cell surface. J. Cell Biol. 104:1281–1289.

Jones, R., Shalgi, R., Hoyland, J., and Phillips, D. M. (1990). Topographical rearrangement of a plasma membrane antigen during capacitation of rat spermatozoa in vitro. Dev. Biol. 139:349–362.

Klint, M., Fridberger, A., Menge, A., Sallstrom, J., and Ploen, L. (1987). Boar sperm surface glycoproteins: Isolation, localization, and temporal expression during spermatogenesis. Gam. Res. 17:173–190.

Koppel, D. E. (1979). Fluorescence redistribution after photobleaching: A new multipoint analysis of membrane translational dynamics. Biophys. J. 28:281–291.

Koppel, D. E., Primakoff, P., and Myles, D. G. (1986). Fluorescence photobleaching analysis of cell surface regionalization, in *Applications of Fluorescence in the Biomedical Sciences* (Taylor, D. L., Waggoner, A. S., Lanni, F., Murphy, R. F., and Birge, R., eds.) pp. 477–497, Alan R. Liss, New York.

Koyama, K., Kameda, K., Ishida, Y., Shigeta, M., and Isojima, S. (1988). Analyses of epididymal sperm surface antigens by monoclonal antibodies against hamster spermatozoa. J. Exper. Zool. 245:294–303.

LaCroix, M., Parvinen, M., and Fritz, I. B. (1981). Localization of testicular plasminogen activator in discrete portions (stages VII and VIII) of the seminiferous tubule. Biol. Reprod. 25:143–146.

Lakoski, K., Williams, C., and Saling, P. (1989). Proteins of the acrosomal region in mouse sperm: Immunological probes reveal post-testicular modifications. Gam. Res. 23:21–37.

Lakoski, K. A., Carron, C. P., Cabot, C. L., and Saling, P. M. (1988). Epididymal maturation and the acrosome reaction in mouse sperm: Response to zona pellucida develops coincident with modification of M42 antigen. Biol. Reprod. 38:221–233.

Lopez, L. C., and Shur, B. D. (1987). Redistribution of mouse sperm surface galactosyltransferase after the acrosome reaction. J. Cell Biol. 105:1663–1670.

Lopez, L. C., Bayna, E. M., Litoff, D., Shaper, N. L., Shaper, J. H., and Shur, B. D. (1985). Receptor function of mouse sperm surface galactosyltransferase during fertilization. J. Cell Biol. 101:1501–1510.

Mahi-Brown, C. A., Yule, T. D., and Tung, K.S.K. (1988). Evidence for active immunological regulation in prevention of testicular autoimmune disease independent of the blood–testis barrier. Am. J. Reprod. Immunol. 16:165–170.

McLaughlin, J. D., and Shur, B. D. (1987). Binding of caput epididymal mouse sperm to the zona pellucida. Dev. Biol. 124:557–561.

Millette, C. F., Cardullo, R. A., Armant, D. R., and Gerton, G. L. (1987). Fucosylation events during mammalian spermatogenesis. Ann. NY Acad. Sci. 513:58–73.

Millette, C. F., and Scott, B. K. (1984). Identification of spermatogenic cell plasma membrane glycoproteins by two-dimensional electrophoresis and lectin blotting. J. Cell Sci. 65:233–248.

Millette, C. F., and Moulding, C. T. (1981). Cell surface marker proteins during mouse spermatogenesis: Two-dimensional electrophoretic analysis. J. Cell Sci. 48:367–382.

Millette, C. F., and Bellve, A. R. (1980). Selective partitioning of plasma membrane antigens during mouse spermatogenesis. Dev. Biol. 79:309–324.

Moore, A., White, T. W., Ensrud, K. M., and Hamilton, D. W. (1989). The major maturation glycoprotein found on rat cauda epididymal sperm surface is linked to the membrane via phosphatidylinositol. Biochem. Res. Comm. 160:460–468.

Myles, D. G., Hyatt, H., and Primakoff, P. (1987a). Binding of both acrosome-intact and acrosome-reacted guinea pig sperm to the zona pellucida during in vitro fertilization. Dev. Biol. 121:559–567.

Myles, D. G., Koppel, D. E., Cowan, A. E., Phelps, B. M., and Primakoff, P. (1987b). Rearrangement of sperm surface antigens prior to fertilization. Ann. NY Acad. Sci. 513:262–273.

Myles, D. G., and Primakoff, P. (1986). Topographical distribution of sperm surface proteins and sperm-egg interaction. J. Electron Microsc. 35(Suppl):1889–1892.

Myles, D. G., and Primakoff, P. (1984). Localized surface antigens of guinea pig sperm migrate to new regions prior to fertilization. J. Cell Biol. 99:1634–1641.

Myles, D. G., Primakoff, P., and Koppel, D. E. (1984). A localized surface protein of guinea pig sperm exhibits free diffusion in its domain. J. Cell Biol. 98:1905–1909.

Myles, D. G., and Primakoff, P. (1983). Establishment of sperm surface topography during development analyzed with monoclonal antibodies, in *The Sperm Cell* (J. Andre, ed.) pp. 127–130, Martinus Nijhoff, The Hague.

Neurath, H. (1989). Proteolytic processing and physiological regulation. Trends Biochem. Sci. 268–271.

O'Brien, D. A., and Millette, C. F. (1986). Immunochemical identification of multiple cell surface antigens appearing during specific stages of mouse spermatogenesis. Gam. Res. 13:199–211.

Olson, G. E., and Hamilton, D. W. (1978). Characterization of the surface glycoproteins of rat spermatozoa. Biol. Reprod. 19:26–35.

O'Rand, M. G., and Romrell, L. J. (1981). Localization of a single sperm membrane autoantigen (RSA-1) on spermatogenic cells and spermatozoa. Dev. Biol. 84:322–331.

O'Rand, M. G., and Romrell, L. J. (1980). Appearance of regional surface autoantigens during spermatogenesis: Comparison of anti-testis and anti-sperm autoantisera. Dev. Biol. 75:431–441.

Orgebin-Crist, M.-C., and Olson, G. E. (1984). Epididymal sperm maturation, in *The Male in Farm Animal Reproduction* (M. Courot, ed.) pp. 80–102, Martinus Nijhoff Publishing, The Hague.

Phelps, B. M., Koppel, D. E., Primakoff, P., and Myles, D. G. (1991). Evidence that proteolysis of the surface is an initial step in the mechanism of formation of sperm cell surface domains. J. Cell. Biol. 111:1839–1847.

Phelps, B. M., Primakoff, P., Koppel, D. E., Low, M. G., and Myles, D. G. (1988). Restricted lateral diffusion of PH-20, a PI-anchored sperm membrane protein. Science 240:1780–1782.

Phelps, B. M., and Myles, D. G. (1987). The guinea pig sperm plasma membrane protein, PH-20, reaches the surface via two transport pathways and becomes localized to a domain after an initial uniform distribution. Dev. Biol. 123:63–72.

Pinto da Silva, P., and Kan, F.W.K. (1984). Label-fracture: A method for high resolution labeling of cell surfaces. J. Cell Biol. 99:1156–1161.

Primakoff, P., Lathrop, W., Woolman, L., Cowan, A., and Myles, D. G. (1988). Fully effective contraception in male and female guinea pigs immunized with the sperm protein PH-20. Nature 335:543–546.

Primakoff, P., Hyatt, H., and Tredick-Kline, J. (1987). Identification and purification of a sperm surface protein with a potential role in sperm–egg membrane fusion. J. Cell Biol. 104:141–149.

Primakoff, P., Hyatt, H., and Myles, D. G. (1985). A role for the migrating sperm surface antigen PH-20 in guinea pig sperm binding to the egg zona pellucida. J. Cell Biol. 101:2239–2244.

Primakoff, P., and Myles, D. G. (1983). A map of the guinea pig sperm surface constructed with monoclonal antibodies. Dev. Biol. 98:417–428.

Ram, P. A., Cardullo, R. A., and Millette, C. F. (1989). Expression and topographical localization of cell surface fucosyltransferase activity during epididymal sperm maturation in the mouse. Gam. Res. 22:321–332.

Romrell, L. J., O'Rand, M. G., Sandow, P. R., and Porter, J. P. (1982). Identification of surface autoantigens which appear during spermatogenesis. Gam. Res. 5:35–48.

Russell, L. D., and Peterson, R. N. (1985). Sertoli cell junctions: Morphological and functional correlates. Int. Rev. Cytol. 94:177–211.

Scully, N. F., Shaper, J. H., and Shur, B. D. (1987). Spatial and temporal expression of cell surface galactoysyltransferase during mouse spermatogenesis and epididymal maturation. Dev. Biol. 124:111–124.

Shalgi, R., Matityahu, A., Gaunt, S. J., and Jones, R. (1990). Identification of antigens on rat spermatozoa with a potential role in fertilization. Mol. Reprod. Dev. 25:286–296.

Shur, B. D. (1989). Galactosyltransferase as a recognition molecule during fertilization and development, in *The Molecular Biology of Fertilization* (Schatten, H., and Schatten, G., eds.) pp. 37–71, Academic Press, San Diego.

Shur, B. D., and Hall, N. G. (1982a). Sperm surface galactosyltransferase activities during in vitro capacitation. J. Cell Biol. 95:567–573.

Shur, B. D., and Hall, N. G. (1982b). A role for mouse sperm surface galactosyltransferase in sperm binding to the egg zona pellucida. J. Cell Biol. 95:574–579.

Skudlarek, M. D., and Orgebin-Crist, M.-C. (1986). Glycosidases in cultured rat epididymal cells: Enzyme activity, synthesis and secretion. Biol. Reprod. 35:167–178.

Suzuki, F. (1988). Changes in the distribution of intramembranous particles and filipin-sterol complexes during epididymal maturation of golden hamster spermatozoa. J. Ultrastruc. Mol. Struct. Res. 100:39–54.

Suzuki, F., and Yanagimachi, R. (1986). Membrane changes in Chinese hamster spermatozoa during epididymal maturation. J. Ultrastruc. Mol. Struct. Res. 96:91–104.

Suzuki, F., and Nagano, T. (1980). Epididymal maturation of rat spermatozoa studied by thin sectioning and freeze-fracture. Biol. Reprod. 22:1219–1231.

Sylvester, S. R., Skinner, M. K., and Griswold, M. D. (1984). A sulfated glycoprotein synthesized by Sertoli cells and by epididymal cells is a component of the sperm membrane. Biol. Reprod. 31:1087–1101.

Talbot, P., and Chacon, R. (1981). Detection of modifications in the tail of capacitated guinea pig sperm using lectins. J. Exp. Zool. 216:435–444.

Thorley-Lawson, D. A., Swendeman, S. L., and Edson, C. M. (1986). Biochemical analysis suggests distinct functional roles for the Blast-1 and Blast-2 antigens. J. Immunol. 136:1745–1751.

Tung, S. K., Han, L.-P. B., and Evan, A. P. (1979). Differentiation autoantigen of testicular cells and spermatozoa in the guinea pig. Dev. Biol. 68:224–238.

Tung, P. S., and Fritz, I. B. (1978). Specific surface antigens on rat pachytene spermatocytes and successive classes of germinal cells. Dev. Biol. 64:297–315.

Vanha-Perttula, T., Mather, J. P., Bardin, C. W., Moss, S. B., and Bellve, A. R. (1985). Localization of the angiotensin-converting enzyme activity in testis and epididymis. Biol. Reprod. 33:870–877.

Wolf, D. E. (1986). Overcoming random diffusion in polarized cells—Corralling the drunken beggar. BioEssays 6:116–121.

Wolf, D. E. (1983). The plasma membrane in early embryogenesis, in *Development in Mammals,* Vol. 5 (Johnson, M. H., ed.) pp. 187–208, Elsevier/North Holland, Amsterdam.

Wolf, D. E., Lipscomb, A. C., and Maynard, V. M. (1988). Causes of nondiffusing lipid in the plasma membrane of mammalian spermatozoa. Biochemistry 27:860–865.

Wolf, D. E., Hagopian, S. S., and Ishijima, S. (1986a). Changes in sperm plasma membrane lipid diffusibility after hyperactivation during in vitro capacitation in the mouse. J. Cell Biol. 102:1372–1377.

Wolf, D. E., Hagopian, S. S., Lewis, R. G., Voglmayr, J. K., and Fairbanks, G. (1986b). Lateral regionalization and diffusion of a maturation-dependent antigen in the ram sperm plasma membrane. J. Cell Biol. 102:1826–1831.

Wolf, D. E., Scott, B. K., and Millette, C. F. (1986c). The development of regionalized lipid diffusibility in the germ cell plasma membrane during spermatogenesis in the mouse. J. Cell Biol. 103:1745–1750.

Wolf, D. E., and Voglmayr, J. K. (1984). Diffusion and regionalization in membranes of maturing ram spermatozoa. J. Cell Biol. 98:1678–1684.

Yanagimachi, R. (1988). Mammalian fertilization, in *Physiology of Reproduction* (Knobil, E., Neill, J., et al., eds.) pp. 135–185, Raven Press, New York.

Yule, T. D., Montoya, G. D., Russell, L. D., Williams, T. M., and Tung, K.S.K. (1988). Autoantigenic germ cells exist outside the blood–testis barrier. J. Immunol. 141:1161–1167.

Zeheb, R., and Orr, G. A. (1984). Characterization of a maturation-associated glycoprotein on the plasma membrane of rat caudal epididymal sperm. J. Biol. Chem. 259:839–848.

Ziparo, E., Geremia, R., Russo, M. A., and Stefanini, M. (1980). Surface interaction in vitro between Sertoli cells and germ cells at different stages of spermatogenesis. Am. J. Anat. 159:385–388.

CHAPTER NINETEEN

Regulation of Testicular Autoimmune Disease

Kenneth S. K. Tung

I. Introduction

At the beginning of this century, it was thought that the chief function of the immune system was the defense against microbes, and that immune reaction against self, or "horror autotoxicus," was not anticipated (Ehlich, 1905). Very soon after, however, autoimmunization against spermatozoa resulted in antisperm antibodies and this finding, among others, heralded the concept of autoimmunity and autoimmune disease (Metchnikoff, 1899).

Fifty years later, immunization of guinea pigs with homologous testis antigens in complete Freund's adjuvant (CFA) induced autoimmune disease of the testis, or experimental autoimmune orchitis (EAO) (Voisin et al., 1951; Freund et al., 1953). This experiment indicates that (1) the testis contains antigens immunogenic to the homologous (later shown for autologous as well) host, (2) the immune reactants, including antibody and/or sensitized T cells, can recognize antigens in the animal's testis, and (3) the immune reactions, amplified by humoral and cellular mediators, can result in an inflamed testis. These findings, in retrospect, already suggested that testis antigens relevant to autoimmune orchitis must be accessible to circulating antibodies and/or T cells—otherwise orchitis would not have occurred. The possibility was not discussed probably because complete antigen sequestration was then the most popular mechanism of defense against autoimmunity.

A few years later, the concept of complete sequestration of tissue autoantigens began to erode. Thyroglobulin, the antigen in autoimmume thyroiditis, was directly demonstrated in thyroid venous blood. Moreover, the theoretical and experimental framework for dynamic regulation against autoimmunity (or self-tolerance) began to emerge (Nossal, 1989). These early studies predicted (1) deletion in the thymus of clones of T cells that are potentially self-reactive (clonal deletion) (Burnet, 1957) and (2) the possible existence of T cells of which the effector function is to regulate or suppress an immune response (reviewed in Hodes, 1989). Most important, it was found that maintenance of tolerance requires the presence of circulating antigen (Weigle, 1973).

As the dynamic processes of tolerance to self-antigens gained acceptance as a seminal phenomenon in immunology, the situation of testis autoantigens has remained largely unchanged. Most recent texts still consider the testis and the eye as the vestigial examples of organs with completely sequestered autoantigens. This idea, however, is not without theoretical support (reviewed in Tung, 1980). In the first place, most testis autoantigens, found only in the haploid germ cells, do not develop until puberty—well past the neonatal period in mice (earlier in humans) when induction of tolerance is known to occur. Second, a strong anatomical and physiological blood–testis (BT) barrier has been defined around the haploid germ cells at the seminiferous tubules and in this region it would serve as an immunologic barrier (see Chapter 11). Third, antigens of haploid germ cells that develop in puberty are highly immunogenic to the autologous host when injected outside the testes.

Experimental results obtained in the past few years have significantly changed our concept of testicular autoantigens. They have shown unequivocally that testis autoantigens are not completely sequestered. Some autoantigenic germ cells are located outside the anatomical BT barrier, and the epithelial barrier downstream from the seminiferous tubules is permeable to germ cell antigens produced within the seminiferous tubules. In addition, there have been seminal studies that define critically the self-tolerance mechanism (reviewed in Schwartz, 1989) and they have significant implications for immunology of the testis. The two major topics that are covered in this chapter are (1) the testicular autoantigens and their relevance in the pathogenesis of autoimmune orchitis and (2) the possible local and systemic mechanisms that normally protect testicular antigens from autoimmune injury. This chapter begins with an outline of the experimental models of testis autoimmunity and ends with a brief account of the possible clinical relevance of testicular autoimmunity.

II. Experimental Models of Testicular Autoimmune Diseases

Experimental autoimmune orchitis has been for some time a model in search of clinical relevance. It is now clear, that as a model, autoimmune orchitis can provide insight on how tolerance to organ specific self-antigen is controlled and when defective control can cause autoimmunity. Autoimmune orchitis can be elicited by many experimental approaches (Table 19–1) and the attendant immunopathologies are similar. Each model, however, has certain unique features that offer an opportunity for studies that address specific questions.

Experimental autoimmune orchitis is elicited in genetically susceptible animals by injection of testis antigen in adjuvant. Studies of EAO are beginning to clarify the nature of target antigens in the testes. Moreover, study of passive EAO in mice for which immunological reagents are available has led to the elucidation of the pathogenetic mechanisms. Although study of EAO will not likely clarify disease etiology, it is relevant to studies of antifertility vaccine based on testicular antigens targeted for male subjects. Moreover, autoimmune response to testis/sperm antigens is a common sequela of vasoligation, and autoimmune orchitis develops in some species after vasectomy (Teuscher et al., 1981). Studies of naturally occurring models of autoimmune orchitis will more likely elucidate the etiologic basis of the disease. For example, the infertile dark mink has two demonstrable defects: testicular failure associated with hypogonadotrophic hypogonadism and autoimmune orchitis (Tung et al., 1984). This finding, therefore, suggests that organ-specific disease can occur when hormonal control of the organ is defective.

Table 19–1. Models of Autoimmune Orchitis

Experimental autoimmune orchitis (EAO)
Active EAO: Immunization of susceptible animals with orchitogenic testis antigen in adjuvant
Passive EAO: Disease elicited by adoptive transfer of lymphoid cells from animals with active EAO to naive syngeneic recipients
Postvasectomy autoimmune orchitis
Naturally occurring autoimmune orchitis
Infertile dark mink. Association of autoimmune orchitis and abnormal hypothalamo-pituitary-gonadal function
Autoimmune orchitis and autoimmune thyroiditis in beagle dogs. Strong genetic association in polyendocrine autoimmunity
Autoimmune orchitis in t locus mouse
Autoimmune orchitis elicited by deliberately perturbing the immune functions of normal mice
Neonatal thymectomy
Transfer of T cell subpopulation to athymic nu/nu mice

There are several approaches for genetic analysis of autoimmune orchitis. One of these is the finding that autoimmune orchitis develops in mice with a special t locus genotype. The t locus in mice is controlled by multiple genes critical for development, including the male reproductive system (Silver, 1985). The t loci are mapped centromeric to the major histocompatibility complex in chromosome 17 and specific gene products are being elucidated (Sarvetnick et al., 1989). Autoimmune orchitis can be a part of polyendocrinopathy syndrome for which firm immunologic basis has been established. It is, therefore, of interest that a closed colony of beagle dogs spontaneously develop autoimmune orchitis and autoimmune thyroiditis, both of which are under the control of limited ancestral genes (Fritz et al., 1976).

Autoimmune orchitis, as a component of the polyendocrinopathy syndrome, has also been elicited by deliberate perturbation of T cell development of the normal mouse. The manipulations include neonatal thymectomy (Taguchi and Nishizuka, 1981; Tung et al., 1987) and transfer of a subpopulation of normal T cells to athymic nu/nu mice (Sagaguchi et al., 1985); injection of adjuvants is not required. As studies of these models will likely elucidate the cellular mechanisms controlling tolerance and defects underlying autoimmunity, they should help to bring autoimmune orchitis to the mainstream of autoimmunity research.

The studies to be described emphasize the target antigens and immunoregulation of testicular autoimmunity. They are based largely on studies of murine EAO and the neonatal thymectomy model.

III. Testis-Specific Autoantigens and Testicular Autoimmunity

A. The Ontogeny of Testicular Autoantigens

During fetal life, the developing immune system acquires immunologic tolerance to antigens recognized as "self," so that at birth, humans are already immunologically endowed with the ability to differentiate nonself from self. In mice, the process is delayed so that the immune system may still acquire tolerance on the day of birth but not 1 to 2 days later. In view of this, the unique events beginning immediately before and at puberty in the male reproductive tract require special explanation. The testis harbors an array of endogenous antigens that develop late in ontogeny. In mice, preleptotene spermatocytes are detectable at 10 days of age and the more mature germ cells do not appear until puberty. Moreover, puberty is accompanied by hormonal changes that alter the function and perhaps the antigenic composition of testicular cells, including Sertoli cells and Leydig cells. Thus, many testis antigens are not present when tolerance to most self-antigens may be occurring.

Available evidence strongly indicates that when the immune system encounters testis-specific autoantigens, they are recognized as foreign despite their endogenous origin. Thus, antibodies to sperm surface and spermatid surface antigens, as well as antibodies to acrosomal antigens, are elicited in guinea pigs, rats, and mice that are immunized with autologous sperm or testicular cells (reviewed in Tung and Menge, 1985). Moreover, in humans and animals, the production of autoantibodies to sperm may result from testicular injury associated with biopsy, vasectomy, or an obstructed vas deferens (Teuscher et al., 1981).

For many years it was believed that the sole mechanism necessary to protect these antigens and the cells that bear them against immune destruction was their sequestration behind an impenetrable BT barrier. That concept has recently been challenged by evidence that the barrier is incomplete and at least some cells bearing the testis autoantigens are exposed to circulating antibodies and immune T cells under normal conditions. Indeed, recent studies indicate that active immunoregulatory mechanisms are also involved in preventing autoimmune responses to testis-specific antigens.

B. The Distribution of Testicular Autoantigens

1. MOST ANTIGENS ARE FOUND BEHIND THE BLOOD–TESTIS BARRIER

The BT barrier at the seminiferous tubules separates the germ cells in the basal compartment from those in the adluminal compartment

(Dym and Fawcett, 1970). Antibody and lymphoid cells are prevented by the barrier from reaching the adluminal compartment, as shown by electron-dense protein tracers that failed to penetrate the barrier, and normally, lymphocytes and macrophages are not detected in the adluminal side of the tubule.

The germ cells located behind the BT barrier include the pachytene spermatocytes, spermatids, and testicular spermatozoa. Essentially all of the numerous diploid-transcribed, haploid-expressed testis-specific antigens are located behind the BT barrier (reviewed in Tung and Menge, 1985). They include the LDHC4, cytochrome c, nuclear basic protein protamine, and testicular isoforms of other somatic molecules. Also present in this compartment are the known orchitogenic antigens: antigens with the capacity of eliciting active EAO. The orchitogenic antigens that have been isolated are first detected in pachytene spermatocytes. Thus, in normal testes most of the immunogenic autoantigens, including orchitogenic antigens, are sequestered behind the BT barrier at the level of the seminiferous tubule.

2. SOME AUTOANTIGENS ALSO EXIST OUTSIDE THE BLOOD–TESTIS BARRIER

Recent studies indicate that germ cell autoantigens relevant to the pathogenesis of EAO can be detected in two additional sites: (1) tissue spaces surrounding the straight tubules and rete testis, and (2) the basal compartment of the seminiferous tubules. Epithelia of the straight tubules, the rete testis, and the ductus efferentes do not possess the elaborate tight junction complexes that exist in the seminiferous tubules. Antibodies in the testis interstitium can enter the rete testis. Intravenously injected antisperm antibody is found on spermatozoa in the rete testis and the caput epididymis within 30 minutes (Tung et al., 1971). Moreover, ferritin particles injected into the lumen of the rete testis become bound to the apical surface of epithelial cells and they are endocytosed into vesicles that, after being transported to the basolateral and basal surface of the cells, are released into the interstitial space (Morales et al., 1984). Although this study demonstrates a mechanism for translocation of macromolecules from the lumen to the interstitial space, the possible leakage of luminal germ cell antigens at these sites is based on indirect evidence obtained in a study on passive EAO to be described later.

More direct evidence has been obtained for another group of testis autoantigens that exists outside the BT barrier. They include (1) germ cells in the basal compartment of the seminiferous tubules, (2) Sertoli cell surface antigens, and (3) structural antigens that form the basal lamina of the seminiferous tubules.

3. AUTOANTIGENIC GERM CELLS EXIST IN THE BASAL COMPARTMENT OF THE SEMINIFEROUS TUBULES

Previous studies on testis autoantigens have relied on their detection by polyclonal or monoclonal antibodies to testis germ cells produced in heterologous, homologous, or autologous animals. The immunized animals have intact testes, thus antigens existing outside the BT barrier can bind to antibodies and remove them from circulation. As a result, detection of testis antigens based on these serum antibodies may fail to localize antigens that exist outside the BT barrier.

Our initial experiment demonstrating the existence of the novel autoantigenic germ cells was made when testes of mice immunized with homologous testis in adjuvants was studied for in vivo binding of immunoglobulin G (IgG), presumed antibodies (Yule et al., 1988). Intense deposits of IgG on the edge of the seminiferous tubules were found (Fig. 19–1). In most cases, the IgG deposits appeared as small rings, but some were granular. Interestingly, testis IgG deposits were detected as early as 7 days after immunization and had peaked by day 10. This occurred 5 to 6 days before the onset of orchitis, thus it could not have been an epiphenomenon that followed inflammation.

Complement component C3 was not co-localized with IgG. The lack of complement activation was explained on the basis that only IgG1 and IgG3 were detected in the testis IgG deposits, and murine antibodies of these IgG subclasses are known to activate serum complement poorly (Perlmutter et al., 1978). The antibody response was in fact isotype restricted as IgG2, IgM, and IgA were absent. Thus, the antigens located on germ cells outside the BT bar-

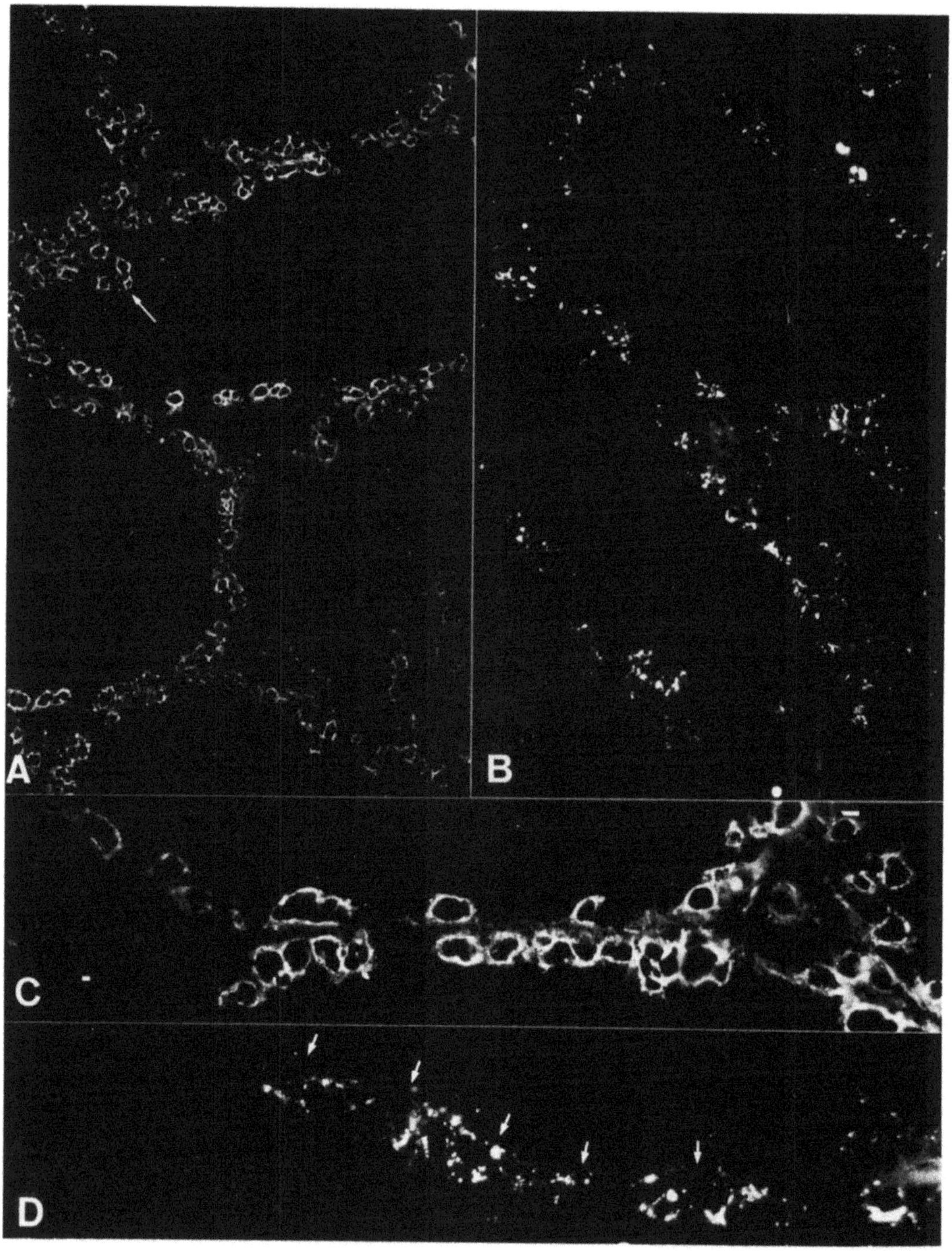

Figure 19–1. Testicular immunoglobulin G (IgG) deposits in mice immunized with testis in adjuvant. Ten days after immunization with testis in adjuvant, IgG can be detected by immunofluorescence along the perimeter of the seminiferous tubules. Some IgG appears as rings (**A**, high magnification in **C**) or granular (**B**, higher magnification in **D**). These testes have no evidence of orchitis. Similar IgG deposits are found in identical locations in normal recipients of serum from architectomized, testis-immunized donors.

rier have unique immunogenic properties; as a result, the lysis of antigen-bearing cells by complement-dependent pathway is unlikely to occur.

The induction of testis IgG deposits was antigen specific and was detected in many mouse strains. Testis IgG deposits were elicited by immunization with testis antigens with incomplete Freund's adjuvant, CFA, pertussis, or CFA and pertussis combined. The fact that it is inducible by immunization without bacterial adjuvant indicates that the antigens are highly immunogenic to the autologous host. The following studies help to confirm the antibody nature of the testis IgG deposits and to identify the autoantigenic germ cells.

Testis IgG deposits can be passively transferred to normal mice by serum from donors immunized with testis in adjuvant. This was possible, however, only if the donors were bilaterally orchiectomized. Presumably, the donors' antigenic germ cells removed most of the relevant antibodies from circulation in vivo. This possibility was supported by the experiment in which immune sera of orchiectomized mice were found to differ from immune sera of nonorchiectomized mice with respect to their reactivity with prepuberal germ cell antigens.

The reactions of these two antibodies with collagenase-dissociated testis cells from mice of different ages by an enzyme immunoassay were compared. The results showed that only antiserum produced by orchiectomized mice reacted with testis cells of puberal mice from day 15 to 17, which consisted mainly of diploid germ cells (located outside the BT barrier). The difference was less obvious when assayed against day 18 and 19 testis cells. In contrast, both sera reacted equally with adult testis cells and did not react with prepuberal (day 9 and 10) testis cells. The reactivity to adult testis cells was not unexpected as both groups of mice were immunized with adult testis. Their sera should, therefore, contain antibodies to haploid germ cells that form the bulk of dissociated adult testis cells.

The antibody nature of testis IgG deposits was also defined by quantitative elution of testis IgG. Testes with IgG deposits were homogenized in chilled phosphate-buffered saline (PBS), and washed repeatedly until essentially all contaminating serum IgG was eliminated. The tissue was acidified to dissociate antibody from antigen, and the gammaglobulins were precipitated in 50 percent ammonium sulfate at pH 3. The acid eluate equilibrated in PBS was compared with the pooled sera from the same mice for mouse IgG content and for antibody against dissociated puberal germ cells. Based on these parameters, the acid eluate was found to contain 24 to 40 times more antibody per unit of IgG than the animals' sera. This finding further confirms the conclusions that testis IgG represents antibody and that this antibody is directed to germ cells outside the BT barrier.

The next observation of interest was that the testis IgG deposits are spermatogenic cycle specific. In all of the experiments in which testis IgG deposits were found, the deposits were associated with only 30 to 45 percent of the seminiferous tubules. By immunoperoxidase study, IgG was found to be associated with 5 percent of the tubules at stages 1 to 6 of the cycle, 45 percent at stages 6 to 7 (when preleptotene spermatocytes are most prominent), and 95 percent at stages 8 to 12. Because preleptotene spermatocytes traverse the BT barrier as meiotic cell division is initiated, antibodies bound to these cells are most likely carried passively into the adluminal compartment of the tubules as the germ cells migrate from the basal to the adluminal compartment.

Finally, to determine the type of germ cells associated with IgG deposition, tritiated thymidine was injected intravenously into immunized mice. Two hours later, combined immunofluorescence and autoradiography showed that 25 percent of the radiolabeled cells had IgG on their surface. It has been well established that spermatogonia and preleptotene spermatocytes are the only germ cells that can be labeled with a 2-hour pulse of tritiated thymidine (Courot et al., 1970). Moreover, most of the co-labeled cells were in close clusters—resembling the location of preleptotene spermatocytes. Hence, it can be concluded that IgG was found in this class of spermatogenic cells.

4. AUTOANTIGENS OF THE SERTOLI CELL AND THE BASAL LAMINA

The simultaneous injection of two monoclonal antibodies, one to the Sertoli cell and the other to the basal lamina, has been reported to elicit aspermatogenesis in mouse testes (Ichinohasama et al., 1986). In addition, rats immunized with laminin develop autoimmune orchitis (Lustig et al., 1978).

C. *Target Antigens and the Pathogenesis of Murine Experimental Autoimmune Orchitis*

To elicit EAO, animals are immunized with testis antigen and adjuvant outside the testes (e.g., in the footpads). Antigens that drain to the regional lymph nodes stimulate antibody and T cell responses to testicular antigens that circulate to the testes, react with the target antigens in the testes, and initiate the inflammatory disease. The experimental approaches that investigate the role of antibody and T cells in EAO have included passive transfer of serum antibodies to normal animals or of lymphoid cells to syngeneic recipients and determination whether the recipients develop orchitis. In general, antibodies fail to transfer disease, whereas lymphoid cells can elicit orchitis (reviewed in Tung et al., 1980). These early studies, however, involved local transfer in which lymphoid cells were placed directly into the testes. Also the studies were made in guinea pigs, a species with few available inbred strains.

1. INDUCTION OF MURINE EXPERIMENTAL AUTOIMMUNE ORCHITIS

Unlike EAO in the guinea pig, the requirement for murine EAO induction is more stringent. The two critical factors are the use of genetically susceptible inbred mouse strains and pertussis toxin as adjuvant (Kohno et al., 1983). Histopathologic evidence of EAO includes orchitis, aspermatogenesis, and epididymovasitis, which are detectable in over 90 percent of animals 14 days after immunization. After orchitis develops, immune complexes in the basal lamina of seminiferous tubules become detectable. The integrity of the BT barrier is disrupted and antibodies enter the tubules to react with the intratubular germ cells. Similar peritubular immune complexes have been detected in autoimmune orchitis of the infertile dark mink, in vasoligated rabbits, and in rabbits with EAO (Bigazzi et al., 1976; Tung and Woodroffe, 1978; Tung et al., 1981). In these situations, the specificity of the IgG in the immune complexes has been determined to include antibody to the acrosomal antigen.

2. REQUIREMENT OF ANTIGEN-SPECIFIC $CD4^+$ T CELLS IS DETERMINED BY ADOPTIVE TRANSFER OF MURINE EXPERIMENTAL AUTOIMMUNE ORCHITIS

Lymphoid cells from mice with EAO adoptively transfer testis autoimmune disease to syngeneic recipients as long as they are first activated in vitro by testis antigens or by the T cell mitogen, concanavalin A (Mahi-Brown et al., 1987). Important controls show that disease transfer was antigen specific and that the T cells from the donor (and not from recipient) were sufficient (Mahi-Brown and Tung, 1989). By selectively deleting the CD4 or CD8 subset of T cells with antibody and complement or by adherence of respective T cells to solid phase antibody, it was determined that CD8 T cells were not required for the transfer of EAO. In contrast, CD4 T cells were essential.

A critical realization in current immunology is that CD4 and CD8 T cells do not accurately assign subpopulations of T cells with specific functions (reviewed in Sprent, 1989). Instead, they define more precisely the major histocompatibility complex (MHC) molecules required to present antigenic peptides to the T cells. Thus, CD4 T cells recognize antigenic peptides complexed to class II MHC (or Ia) molecules, and CD8 T cells recognize antigenic peptides complexed to class I MHC molecules. Moreover, it has been shown conclusively that T cells, unlike antibodies, do not recognize intact antigens (reviewed in Unanue, 1989). Instead, their receptors for antigens recognize the denatured, proteolytic peptide fragments of the antigens in association with class I or class II MHC molecules present on specialized antigen presenting cells (APC). Based on these new rules of immunology, and the finding that CD4 T cells are required for passive transfer of EAO, it was clear that APC with Ia and associated germ cell peptides must be accessible to circulating T cells in the normal testis for adoptive transfer of EAO to occur. Unlike class I MHC, Ia expression is normally restricted to B cells, monocyte/macrophages, and dendritic cells. These findings prompted the investigation of Ia positive APC in the testis.

3. DISTRIBUTION OF CLASS II MAJOR HISTOCOMPATIBILITY COMPLEX MOLECULE IN TESTES CORRELATES WITH THE DISTRIBUTION OF ORCHITIS IN PASSIVE AND ACTIVE EXPERIMENTAL AUTOIMMUNE ORCHITIS

Although the testes contain numerous macrophages in the interstitial space between seminiferous tubules detectable by the specific F4/80 monoclonal antibody, they rarely have detectable Ia (Tung et al., 1987). In the region of the straight tubules (or terminal segment of the seminiferous tubules), however, Ia^+ macrophages are readily detected around the epithelial cells. It is, therefore, of great interest that orchitis in passive EAO is invariably initiated in the region of the straight tubules and the rete testis (Fig. 19–2). Blockage of the sperm

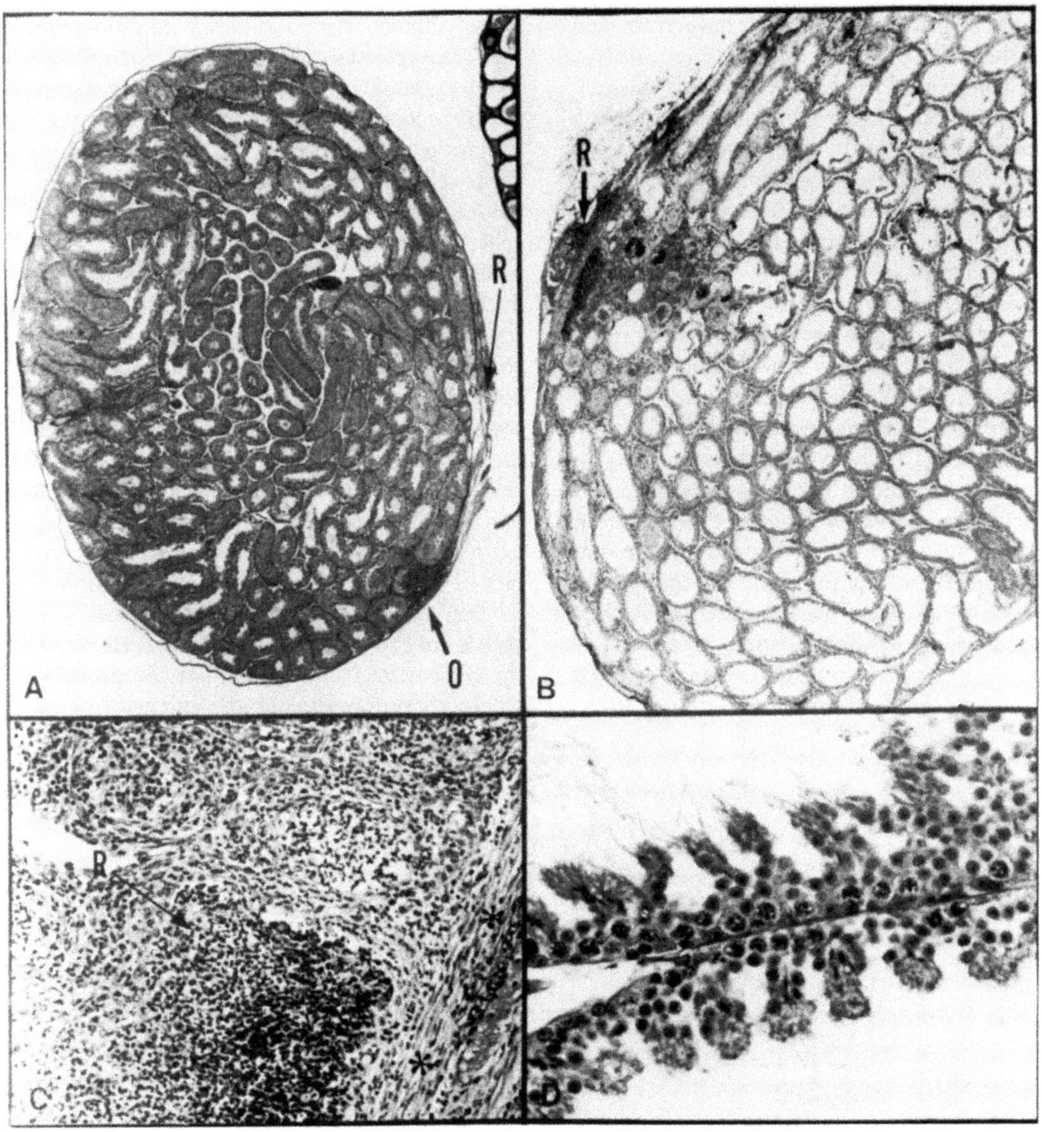

Figure 19–2. Histopathology of orchitis in recipients after adoptive transfer by lymphoid cells from donors immunized with testis antigens (**B**). Unlike early disease in active experimental autoimmune orchitis (EAO) (**A**), early disease in passive EAO almost invariably occurs in the region of the rete testis (R) (**B**,**C**) resulting in dilatation of seminiferous tubules (**B**,**D**). (Reprinted with permission from Tung et al., 1987.)

and fluid outflow leads to severe dilatation of seminiferous tubules and enlargement of the testes. As has already been mentioned, the straight tubules and rete testis do not possess the tight BT barrier found in the seminiferous tubules, thus bilateral diffusion of macromolecules may occur. For this reason, germ cell antigens most likely can reach the interstitial spaces surrounding the straight tubules and the rete testis, be processed by the unique Ia+ macrophages, and be presented to activated germ cell-specific T cells that have trafficked into the testis.

In contrast to passive EAO, orchitis associated with active EAO is distributed randomly throughout the testis, often under the testis capsule. This disease localization has been a puzzle in orchitis research as it coincides with a region with a formidable BT barrier deficient in Ia^+ macrophages.

It was, therefore, of considerable interest to find that, beginning 6 days after immunization, numerous macrophages surrounding the seminiferous tubules stained positively for Ia. The number of Ia^+ macrophages increased 20-fold. We do not yet know whether this change represents an increase in Ia expression of resident testicular macrophages or recent immigration of Ia^+ monocytes from circulation. Injection of adjuvants alone was sufficient to increase the number of Ia^+ macrophages in the testes and is dependent on circulating gamma interferon as it was completely abrogated in mice injected with monoclonal antibody to gamma interferon.

4. FORMULATION OF A WORKING HYPOTHESIS FOR THE PATHOGENESIS OF EXPERIMENTAL AUTOIMMUNE ORCHITIS

Activated CD4 T cells are pivotal in EAO induction and are directly responsible for initiating lesions in the straight tubules and the vas deferens (Fig. 19–3). Germ cell antigens that leak outside the straight tubules, having been processed by the abundant Ia^+ macrophages in the region, are presented to the activated CD4 T cells. Pathogenesis of disease in the seminiferous tubules has additional requirements. First, Ia expression on interstitial macrophages is upregulated by adjuvants by gamma interferon. Second, the target antigens for disease initiation at this site are likely derived from germ cells located outside the BT barrier, including the preleptotene spermatocytes, Sertoli cells, and antigens in the basal lamina. Antibodies to preleptotene spermatocytes are produced early in active EAO and are bound to testicular cells in vivo before the onset of orchitis. It is possible that these antigen/antibody complexes are shed from the cell surface, processed by the interstitial macrophages that then present the antigenic peptide to the CD4 T cells. This situation is similar to a model of experimental autoimmune nephritis, where the antigens have been found to shed from the surface of the target cells after binding to antibodies (Kerjaschki et al., 1987). In addition, antibodies are carried passively by germ cells across the BT barrier to the adluminal compartment where they may disrupt the integrity of the germ cell epithelium. After orchitis begins,

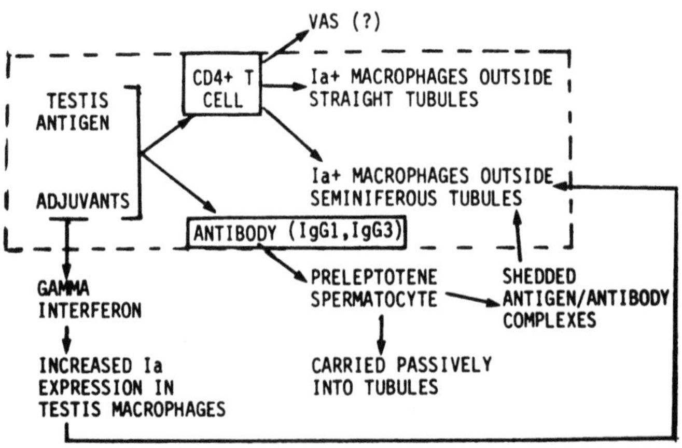

Figure 19–3. A model of pathogenesis and target autoantigens in murine EAO. See text for description.

there is evidence of disruption of the BT barrier and immune complexes are found on peritubular basal lamina. Thus, in EAO antibody may facilitate the T cell response by providing target antigens outside the seminiferous tubules, by potentiating the inflammatory response, and by assisting removal of the intratubular germ cells by macrophages.

IV. Mechanisms That Regulate Autoimmune Response to Testis Antigens

If germ cell autoantigens exist outside the BT barrier and they are immunogenic to the autologous host, what mechanisms may normally prevent autoimmune responses from occurring? This question addresses the fundamental issue of self and nonself discrimination by the immune system, and for testis autoantigens research it has been initiated very recently.

Despite their late ontogeny and their immunological behavior as foreign antigens, systemic tolerance mechanisms may operate for germ cell autoantigens. There is additional evidence that the interstitial space of the testis is an environment unfavorable for induction and/or expression of immune reaction. The magnitude of autoimmune response to testis-specific antigens and the development of autoimmune orchitis are also under genetic control.

A. Systemic Immunoregulation (Tolerance)

Immunologic tolerance is a state of unresponsiveness to an antigen resulting from prior exposure of the host to the same antigen (reviewed in Nossal, 1989). Thus, injection of a foreign protein into a neonatal mouse results in the long-lasting unresponsiveness to the foreign protein. The current hypothesis is that a similar phenomenon accounts for tolerance to most self-antigens that are present at birth. Importantly, tolerance can also be induced in the adult mouse by a larger dose of the foreign proteins. The state of tolerance in adult mice, however, is readily terminated unless the circulating antigen is sustained by repeated injections. Because most testis autoantigens are absent at birth, the state of tolerance they elicit is expected to be similar to tolerance induced in adult animals.

Evidence supporting systemic tolerance to testis antigens is derived from two recent studies. First, it has been observed that immune responses to testis-specific antigen LDHC4 are consistently lower in male than in female mice (Marsh et al., 1981). Conversely, the T cell responses of male mice to the female-specific ZP3 antigen are greater than that of female mice (Rhim and Tung, in preparation). Thus, the sex-specific difference in immune response to gonadal antigens is most likely related to the presence of endogenous gonadal antigens and is not a function of the difference in sex hormone levels. Second, recent studies on the model of autoimmune orchitis elicited by neonatal thymectomy has provided evidence for systemic tolerance (Smith et al., 1989). Moreover, these studies are beginning to elucidate the tolerance mechanisms.

1. MECHANISMS OF IMMUNOLOGIC TOLERANCE AND THE MODEL OF AUTOIMMUNE ORCHITIS IN NEONATALLY THYMECTOMIZED MICE

The thymus is the organ where T cells develop and where they first acquire their receptors for antigens. The specificity of the T cell receptor (TCR) is dictated in part by the variable (V) genes randomly selected from the many V genes in the genomic library for constructing the Vα and Vβ components of the TCR. Because the expression of TCR on T cells is clonal, the TCR expressed on any given T cell are identical. The immature thymocytes express a low level of TCR and in this population are cells with receptors for self-antigens, as well as cells with receptors for nonself antigens. As these thymocytes mature and gain a high density of TCR, the cells with TCR that recognize self-antigens are deleted by nonlymphoid cells in thymus and they do not leave the adult thymus (Kappler et al., 1987). The process of elimination of self-reactive T cells during their development, or clonal deletion, is considered to be a major mechanism of tolerance.

An interesting model of autoimmune orchitis is found in mice that have been thymectomized on day 3 (D3TX) after birth. The autoimmune disease develops spontaneously at age 5 to 7 weeks and no immunization is required (Taguchi and Nishizuka, 1981; Tung et al., 1987). Thymectomy has to be done within the period of day 1 to day 4; thymectomy on day

7 fails to elicit the disease. As a possible disease mechanism, D3TX may disrupt the process of T cell development and, as a result, the T cell repertoire is altered. Thus, if T cells responsible for orchitis are normally deleted in the *adult* thymus but not neonatal thymus, and they are enriched because the thymus is removed neonatally, they may cause testicular autoimmunity. A second possibility is that autoimmunity develops because of the elimination of a special subpopulation of T cells, the function of which is to regulate or suppress the self-reactive T cells that have escaped clonal deletion. Thus, if suppressor cells develop late in the thymus (e.g., 4 to 5 days after birth), they would be eliminated by D3TX. The two mechanisms are not mutually exclusive and they have been investigated in D3TX mice.

2. PERTURBED T CELL REPERTOIRE IN D3TX MICE

Experimental analysis of clonal deletion takes advantage of the available monoclonal antibodies to the products of several Vβ genes. The use of these antibodies in flow cytometric analysis of lymphocytes bearing the various Vβ proteins permits an accurate quantitation of the proportion of T cells that use each of the Vβ in their TCR. In addition, mice have been found to express certain *superantigens*. Although the function of these superantigens is unknown, they have been very useful to investigate tolerance mechanisms. Interestingly, each of these superantigens has been found to delete intrathymically T cells bearing a specific Vβ gene (reviewed in Schwartz, 1989). For example, in adult B6AF1 mice that express the superanti-

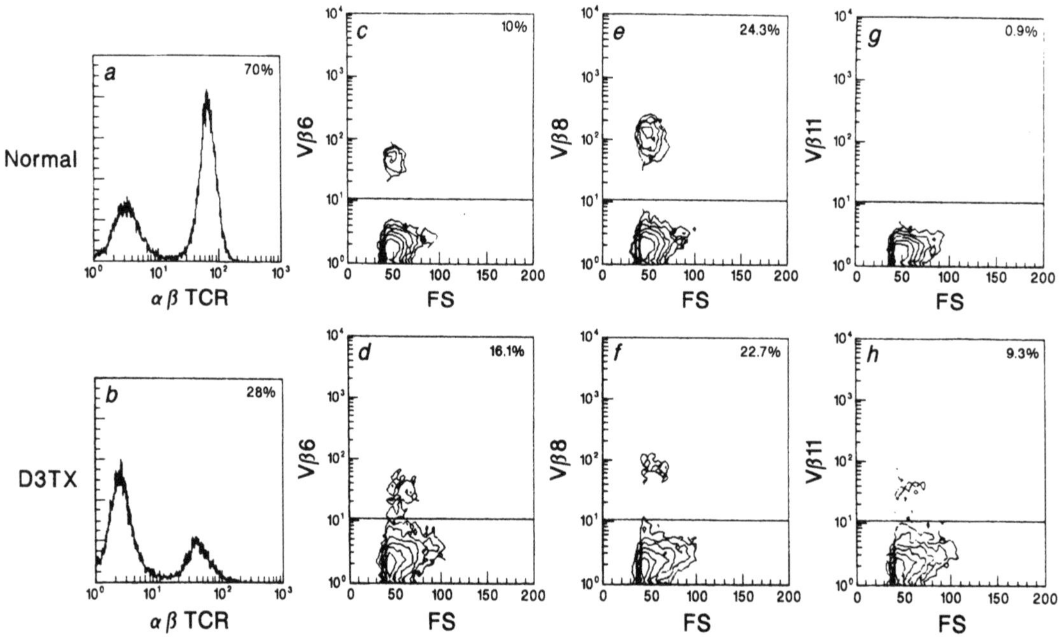

Figure 19–4. Quantitation of αβ TCRs and Vβ-specific TCRs on thymocytes and spleen cells of normal young B6AF1 mice. Thymocytes pooled from 3-day-old (**A** and **C**) and 14-day-old (**B** and **D**) female B6AF1 mice were analyzed by flow cytometry for $V\beta6^+$ (**A** and **B**) and $V\beta11^+$ (**C** and **D**) T cells. Comparable percentages of mature thymocytes with bright staining $V\beta6^+$ TCRs are detected in mice of both ages. In contrast, mature thymocytes with bright staining $V\beta11^+$ TCR are detected in 3-day-old but not in 14-day-old mice. Spleen cells from 3-day-old (**E** and **G**) and 7-day-old (**F** and **H**) mice were analyzed for $\alpha\beta$ TCR^+ (**E** and **F**) and for $V\beta11^+$ (**G** and **H**) T cells. In the 3-day-old spleen, 4.8 percent of the cells are $\alpha\beta$ TCR^+ (**D**) and 2.2 percent of the spleen cells are $V\beta11^+$ (**F**); thus 46 percent of $\alpha\beta$ TCR^+ cells are $V\beta11^+$. On day 7, the percentage of $\alpha\beta$ TCR^+ T cells has doubled (**G**) and the percentage of $V\beta11^+$ T cells is reduced to 0.9 percent (**H**); hence the $V\beta11^+$ T cells per $\alpha\beta$ TCR^+ T cells are reduced to 11 percent. Fluorescence intensity is analyzed in log scale. (FS, forward scatter analysis.) (Reprinted with permission from Smith et al., 1989.)

gens I-E and mls-2[a], all the thymocytes expressing Vβ11 and Vβ3 T cells, respectively, are deleted in the adult thymus. In fact, less than 1 percent of their lymph node T cells express these Vβ. In contrast, T cells that express the Vβ6 and Vβ8 genes constitute a majority of the peripheral T cells in B6AF1 mice because these mice lack the self-superantigens (e.g., mlsla for Vβ6) that would delete them in the thymus.

When Vβ-specific T cells were analyzed in D3TX mice, a very interesting picture emerged (Smith et al., 1989). Adult B6AF1 mice that were thymectomized on day 3 after birth had a 10-fold enrichment of Vβ3 and Vβ11 T cells over nonthymectomized mice (Fig. 19–4). In contrast, the proportions of Vβ6 and Vβ8 T cells between D3TX mice and normal mice were comparable. To determine the source of the enriched Vβ11 T cells, neonatal thymus and spleens were investigated for clonal deletion of the Vβ-specific T cells. This led to the unexpected discovery that clonal deletion for Vβ11 T cells was not occurring during the first neonatal week. As a result, Vβ11 T cells mature and leave the neonatal thymus and can be found in the neonatal spleen (Fig. 19–5). This is probably a general phenomenon as the lack of clonal deletion of Vβ6 T cells in neonatal mlsl[a] mice has also been reported (Schneider et al., 1989). The potentially self-reactive T cells found to be enriched in D3TX mice are, therefore, likely derived from the neonatal T cell repertoire. By removing the thymus—the source of T cells—D3TX has fixed effectively the T cell repertoire of the mice to that of the 3-day-old repertoire, which is rich in self-reactive T cells. In a recent study, we found that T cells obtained from 3-day-old mice were able to transfer autoimmune orchitis when they were transferred to syngeneic athymic nude mice (Smith, Lou, Lacy, and Tung, in preparation).

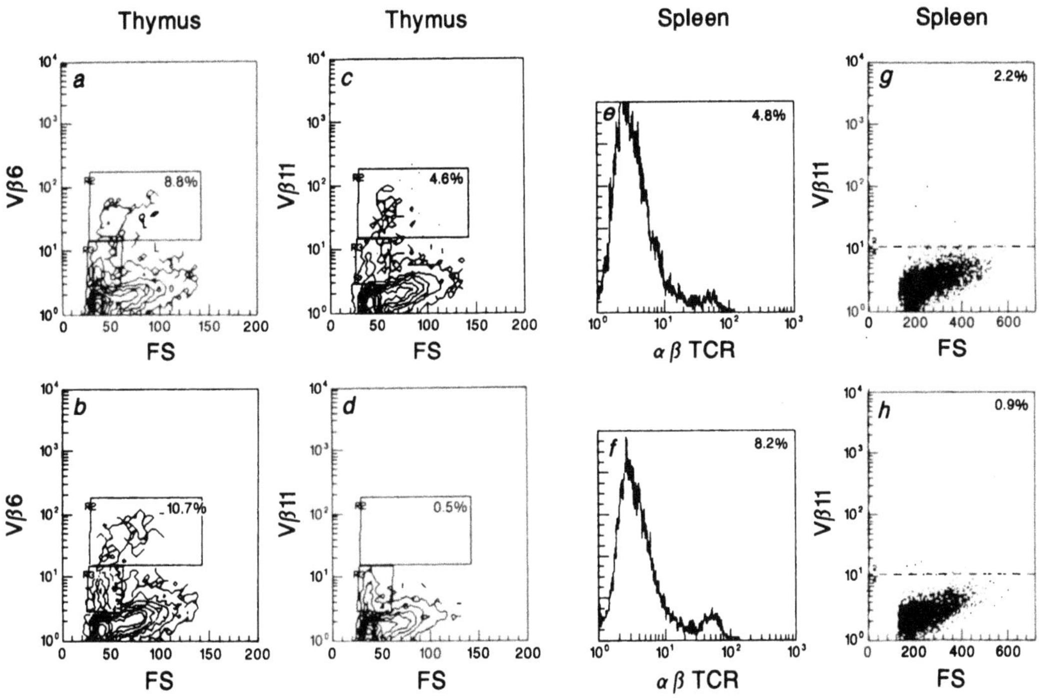

Figure 19–5. Quantitation of cell surface expression of αβ TCRs and Vβ-specific TCR-positive T cells in lymph nodes of 4-month-old D3TX and normal mice. (**a,c,e,g**) Data for normal mice; (**b,d,f,h**) data for D3TX mice. The total αβ TCR$^+$ T cells of D3TX mice are reduced to 40 percent of normal (**a** and **b**). The percentages of Vβ6$^+$ (**c** and **d**) and of Vβ8$^+$ (8.1 + 8.2) (**e** and **f**) T cells in D3TX and normal mice are similar. In contrast, D3TX mice have 10 times more Vβ11$^+$ T cells than normal mice (**g** and **h**). Fluorescence intensity is analyzed in log scale on the horizontal axis for (**a**) and (**b**) and on vertical axis for (**c**) to (**h**). (FS, forward scatter analysis.) (Reprinted with permission from Smith et al., 1989.)

The study has, therefore, shown that an altered T cell repertoire enriched in self-reactive T cells is associated with the development of autoimmune orchitis. As a corollary, it can be argued that the maintenance of the T cell repertoire is a normal physiologic function of the thymus critical for control against testis autoimmunity. Formal proof for this important conclusion will require further studies and these most likely will come from experiments based on transgenic mice endowed with the TCR gene of orchitogenic T cell clones.

3. TESTIS-ANTIGEN SPECIFIC SUPPRESSOR CELLS

Autoimmune orchitis induced by D3TX is prevented if the D3TX mice receive viable T cells from normal syngeneic mice before they reach 2 weeks of age (Taguchi and Nishizuka, 1981). The cells responsible for disease suppression are CD4 and are not detected in mice that have been thymectomized on day 3 or day 7 after birth (Smith et al., 1991). Moreover, for orchitis suppression, male spleen cells are more effective than spleen cells from female mice or male mice that had been orchiectomized at birth (Taguchi and Nishizuka, 1983). This observation suggests that endogenous testis antigens may be required for generation and maintainence of suppressor cells in the normal mice, and that orchitis suppression may be an antigen-specific phenomenon.

In summary, current research on immunologic tolerance to self-antigens is moving at a rapid pace. New information derived from these studies will help to clarify the mechanisms of tolerance to testis antigens. On the other hand, gonad-specific self-antigens provide a unique opportunity for tolerance investigation as these antigens exist in only one sex and their late ontogeny provides several windows of opportunity for critical dissection of tolerance mechanisms.

A. Local Immunoregulation

The testis, the anterior chamber of the eye, and the brain have been cited as immunologically privileged sites (reviewed in Barker and Billingham, 1977). Allografts, including parathyroid (Head et al., 1983) and pancreatic islet (Selawry et al., 1987), that are rapidly rejected under the renal capsule survive for a prolonged period when engrafted inside the testis (presumably between seminiferous tubules). The eye and the brain lack afferent lymphatics and, for these organs, ineffectual drainage of antigens to regional lymph nodes provides a possible explanation for their immunological status. Because the testis is not devoid of lymphatics other alternative methods must be invoked.

The effect of in vivo depletion or replacement of sex steroid hormones on rejection of allograft in testis has obtained conflicting results. In one study, estradiol treatment before transplantation led to rapid rejection of the parathyroid allograft in the testis (Head and Billingham, 1985). Hence, suppression of Leydig cell function and reduction of testosterone may be responsible for local immunosuppression. In contrast, other studies reported that hypophysectomy and steroid implants reduced serum levels of testosterone and progesterone, but had no effect on the survival of parathyroid allografts in the rat testes. Pancreatic islet allograft survival was also not affected by reduction of the serum progesterone and testosterone levels (Selawry and Whittington, 1988). On the other hand, there is general agreement that the cryptorchid testis offers a more profound immunosuppressive environment.

Recent studies have been directed to nonsteroidal testicular factors. Tissue fluid from the testis, and an increasing number of biologically active molecules from the testis have been found to modify lymphocyte growth (reviewed in Hedger et al., 1990). It is interesting that both stimulatory and inhibitory effects have been observed. For example, fluids drained from interstitial spaces of rat testes enhanced proliferation of lymphocyte or thymocyte in response to mitogens or alloantigens. Growth factors and IL-1-like factors produced by Sertoli cells may possibly account for this effect. On the other hand, the media of Sertoli cell culture inhibit mitogen-stimulated lymphocyte growth (Wyatt et al., 1988). Similarly, Leydig cells form rosettes with thymocytes and lymphocytes, and, in the presence of Leydig cells proliferation of lymphocytes in response to alloantigens and mitogens is reduced (Born and Wekerle, 1981, 1982).

Resident testicular cells may influence lymphoid cell function via a paracrine pathway. Several cytokines known to regulate lymphocyte response to mitogens have been identified in the testis. These include IL-1-like factor,

TGF-β, fibroblast growth factors, transglutaminase, and prostaglandin E2 (Hedger et al., 1990). More recently, inhibin and activin produced in the testes were shown to accentuate and to reduce, respectively, the proliferation of thymocytes in response to mitogen (Lee et al., 1989; Hedger et al., 1989).

Finally, proteins known to inhibit complement activation exist in testes where they may reduce complement-mediated inflammation and avert complement-dependent cytolysis (Tarter and Alexander, 1984). A major synthetic product of the Sertoli cell is the sulfated glycoprotein 2 (homologous to SGP2), which has been shown to inhibit cell lysis by C56789, the complement attack complex (Griswold et al., 1986; Jenne and Tschopp, 1989). The key step in activation of the complement system involves the cleavage of component C3 into C3b, an active serine esterase. C3b becomes covalently linked to the amide or carboxyl groups that exist on the target cells by a diester bond. Because the potentially cytolytic chemical reaction is thought to occur normally, albeit at a slow rate, several regulatory proteins exist on most cell surfaces to ensure that the C3b molecules generated on normal cells are rapidly degraded (Lublin and Atkinson, 1990). One of these is the membrane co-factor protein that has been detected recently on the spermatozoa and in association with spermatids in the testes (Anderson et al., 1989).

Although an increasing number of factors known to affect lymphocyte function have been discovered, their physiologic role remains to be determined. Future studies will need to examine their influence on testis antigen-specific T cell and B cell responses, including active and passive induction of EAO, and investigate relevant T cell functions such as cytokine production and expression of cytokine receptors in response to ligation of antigen receptors on T cells. As informative monoclonal antibodies to putative testicular factors become available, it will also be critical to evaluate their in vivo effects. Finally, the functions of most cytokines are pleiotropic and complex; thus, conclusions based on the study of a single cytokine should be interpreted with caution. Nevertheless, the ongoing research is exciting and the findings are consistent with the concept that cells of target organs and those of hemopoietic origin, including lymphocytes, communicate with one another through multiple paracrine-receptor pathways. In this way the effector cells in an autoimmune reaction may be modulated by the local physiologic and pathologic environments of the target organ, and vice versa.

C. Genetic Control of Autoimmune Orchitis

Inbred mouse strains vary greatly in their capacity to develop EAO. Studies of congenic mice that share the same non-MHC genes and differ only in the MHC genes show that MHC-linked genes influence disease susceptibility (Teuscher et al., 1985). Further analysis mapped an EAO susceptible gene to the H-2D end of the mouse MHC. In addition, non-MHC genes are also critical. Thus, in mice with the H-2^d haplotype, BALB/c is a responder, whereas DBA/2 is a nonresponder. In the case of DBA/2 mice, the nonresponsiveness was an autosomal dominant trait. Finally, BALB/c substrains also differed greatly in EAO susceptibility.

V. Clinical Testis Autoimmunity

A. Male Infertility

The common histopathologic findings in the various models of autoimmune orchitis encompass monocytic orchitis, aspermatogenic tubules, and epididymovasitis. Immune complexes containing sperm antigens and antisperm antibodies have also been detected in the tubular basement membrane of seminiferous tubules. Similar findings have been described in the testes of infertile men (reviewed in Tung and Mahi-Brown, 1990).

Human granulomatous orchitis, most resembling the histopathology of EAO, accounts for 5 percent of testes removed for testicular mass. Patients with this disease have not been adequately studied for testis autoimmune responses, however, and the etiology is uncertain. Another possible candidate is mumps orchitis, for which an autoimmune etiology has been postulated. In support of this is the finding that guinea pigs immunized with salivary glands in Freund's complete adjuvant develop aspermatogenesis. Moreover, mumps virus is isolated from the salivary gland but not from the testis, and orchitis is usually a complication of adults (20 percent) and is rarely found (1 percent) in prepuberal individuals suffering from mumps.

An immunologic basis for another group of patients whose infertility is associated with abnormal semen analysis and aspermatogenesis has been reported recently. In one study, using ultrastructural immunohistochemical analysis, 35 percent of the specimens had discrete electron dense deposits and clusters of membranovesicular particles in thickened basal lamina surrounding the seminiferous tubules (Salomon et al., 1982). Moreover, IgG and/or C3 were localized in the electron dense areas in some testes. In the other study, 1-μm-thick testis biopsy sections were studied by immunofluorescence and IgG, IgM, and C3 were detected in about 50 percent of the cases (Lehmann et al., 1987). Moreover, histopathology was found to correlate positively with immune deposits in the testis. By co-staining the tissue sections with the DNA-binding Hoechst stain, it was possible to identify the testis cells and their topographic relation to the immune deposits. Immunoglobulins were found around basal spermatogonia, intratubular germ cells, as linear deposits in tubular basement membranes, and in Leydig cells.

B. Contraceptive Vaccines

As an approach for population control, and as an alternative contraceptive, the idea of immunization of normal female or male subjects with reproductive hormones or gamete antigens has received serious consideration. Of the numerous sperm antigens, LDHC4, SP10, and the PH20 molecules are among the best defined. LDHC4 is the sperm-specific isoenzyme, an intracellular antigen that is detectable on the surface of viable spermatozoa. B cell epitopes of murine LDHC4 have been identified, and a peptide vaccine has been tested on female baboons with promising results (Goldberg, personal communication). The PH20 molecule has been described in detail by Myles (see Chapter 18). It is interesting that male guinea pigs immunized with PH20 develop reversible infertility (Primakoff et al., 1989), and the infertility is associated with a reversible form of autoimmune orchitis (Tung and Primakoff, unpublished data).

VI. Summary and Conclusions

The studies in the past few years based on several models of autoimmune orchitis have sharpened our understanding of the testis as an immunologic target organ. Results from immunologic analysis on murine autoimmune orchitis have allowed us to construct several pathogenetic pathways for the disease. Although CD4 T cells are most critical in disease pathogenesis, their function is assisted by antibody that targets the antigen at the level of the seminiferous tubules. It is clear that not all testis autoantigens are sequestered behind the blood–testis barrier. The recent localization of these antigens has suggested the nature of the effector mechanisms likely to operate and has also predicted the existence of more dynamic protective mechanisms against autoimmune responses to testis antigens.

In a relatively short time considerable progress has been made on the important issue of systemic and local regulations of testis autoimmune responses. Evidence for systemic tolerance for testis-specific antigens at the level of the thymus and in the form of immunoregulatory T cells has been forthcoming. In addition, testis-derived cytokines with the capacity to modulate lymphocyte functions have been identified. It is anticipated that future studies of testis autoimmunity will benefit from the rapid advances of current immunological research. It is likely that the converse will also be true.

Finally, the increasing clinical evidence for autoimmune diseases of the testis and the successful application of testicular and sperm antigens in contraceptive vaccine development will provide a new impetus to research in testicular autoimmunity.

ACKNOWLEDGMENTS

I am indebted to Cherrie Mahi-Brown, Terecita Yule, and Hedy Smith who contributed many ideas and completed much of the recent work in my laboratory. I also thank Mark Hedger for sending preprints and Julie Negrete for manuscript preparation. The studies are supported by NIH grants HD-14504 and HD-21953.

REFERENCES

Anderson, D. J., Michaelson, J. S., and Johnson, P. M. (1989). Trophoblast/leukocyte-common antigen is expressed by human testicular germ cells and appears on the surface of acrosome-reacted sperm. Biol. Reprod. 41:285–293.

Barker, C. F., and Billingham, R. E. (1977). Immunologically previleged sites. Adv. Immunol. 25:1–54.

Bigazzi, P. E., Kosuda, L. L., Hsu, K. C., and Andres, G. A. (1976). Immune complex orchitis in vasectomized rabbits. J. Exp. Med. 143:382–404.

Born, W., and Wekerle, H. (1981). Selective, immunologically nonspecific adherence of lymphoid and myeloid cells to Leydig cells. Eur. J. Cell Biol. 25:76–81.

Born, W., and Wekerle, H. (1982). Leydig cells nonspecifically suppress lymphoproliferation in vitro: Implications for the testis as an immunologically privileged site. Am. J. Reprod. Immunol. 2:291–295.

Burnet, F. M. (1957). A modification of Jerne's theory of antibody production using the concept of clonal selection. Aust. J. Sci. 20:67–69.

Courot, M., Hochereau-de Reviers, M-T., and Ortavant, R. (1970). Spermatogenesis, in *The Testis,* Vol. 1 (Johnson, A. D., Gomes, W. R., and Vandemark, N. L., eds.) pp. 339–432, Academic Press, New York.

Dym, M., and Fawcett, D. W. (1970). The blood–testis barrier in the rat and the physiological compartmentation of the seminiferous epithelium. Biol. Reprod. 3:308–326.

Ehlich, P. (1905). *Collected Studies in Immunity.* John Wiley, New York.

Freund, J., Lipton, M. M., and Thompson, G. E. (1953). Aspermatogenesis in the guinea pig induced by testicular tissue and adjuvant. J. Exp. Med. 97:711–725.

Fritz, T. E., Lombard, L. S., Tyler, S. A., and Norris, W. P. (1976). Pathology and familial incidence of orchitis and its relation to thyroiditis in a closed beagle colony. Exp. Mol. Pathol. 24:42–158.

Griswold, M. D., Robert, K., and Bishop, P. (1986). Purification and characterization of a sulfated glycoprotein secreted by Sertoli cells. Biochemistry 25:7265–7270.

Head, J. R., and Billingham, R. E. (1985). Immune privilege in the testis. I. Evaluation of potential local factors. Transplantation 40:269–275.

Head, J. R., Neaves, W. B., and Billingham, R. E. (1983). Immune privilege in the testis. I. Basic parameters of allograft survival. Transplantation 36:423–431.

Hedger, M. P., Qin, J-X., Robertson, D. M., and de Kretser, D. M. (1990). Intragonadal regulation of immune system functions. Reprod. Fertil. Dev., in press.

Hedger, M. P., Drummond, A. E., Robertson, D. M., Risbridger, G. P., and de Kretser, D. M. (1989). Inhibin and activin regulate [^{3}H]-thymidine uptake by rat thymocytes and 3T3 cells in vitro. Mol. Cell. Endocrinol. 61:133–138.

Hintz, M., and Goloberg, E. (1977). Immunohistochemical localization of LDH-X during spermatogenesis in mouse testis. Develop. Biol. 57:375–384.

Hodes, R. J. (1989). T-cell mediated regulation: Help and suppression, in *Fundamental Immunology,* 2nd ed. (Paul, W. E., ed.) pp. 587–620, Raven Press, New York.

Ichinohasama, R., Hirokawa, K., and Hatekeyama, S. (1986). Spermatogenic disturbance induced in mice by combined local injection of monoclonal antibodies to sertoli cell and to basal lamina of seminiferous tubule. Am. J. Reprod. Immunol. 10:158–165.

Jenne, D. E., and Tschopp, J. (1989). Molecular structure and functional characterization of a human complement cytolysis inhibitor found in blood and seminal plasma: Identity to sulfated glycoprotein 2, a constituent of rat testis fluid. Proc. Natl. Acad. Sci. 86:7123–7127.

Kappler, J., Roehm, N., and Marrack, P. (1987). T cell tolerance by clonal elimination in the thymus. Cell 49:273–280.

Kerjaschki, D., Miettinen, A., and Farquhar, M. G. (1987). Initial events in the formation of immune deposits in passive Heymann nephritis. Gp330-anti-gp330 immune complexes form in epithelial coated pits and rapidly become attached to the glomerular basement membrane. J. Exp. Med. 166:109–128.

Kohno, S., Munoz, J. A., Williams, T. S., Teuscher, C., Bernard, C.C.A., and Tung, K.S.K. (1983). Immunopathology of murine experimental allergic orchitis. J. Immunol. 130:2675–2682.

Lee, W., Mason, A. J., Schwall, R., Szonyi, E., and Mather, J. P. (1989). Secretion of activin by interstitial cells in the testis. Science 243:396–398.

Lehmann, D., Temminck, D., Da Rugna, D., Leibundgut, B., Sulmoni, A., and Muller, H. J. (1987). Role of immunological factors in male infertility: Immunohistochemical and serological evidence. Lab Invest 57:21–28.

Lublin, D. M., and Atkinson, J. P. (1990). Decay-accelerating factor and membrane cofactor protein. Cur. Topics Microbiol. Immunol., in press.

Lustig, L., Denduchis, B., Gonzalez, N. N., and Puig, R. P. (1978). Experimental orchitis induced in rats by passive transfer of an antiserum to seminiferous tubule basement membrane. Arch. Androl. 1:333–343.

Mahi-Brown, C. A., and Tung, K.S.K. (1989). Activation requirements of donor T cells and host T cell recruitment in adoptive transfer of murine experimental autoimmune orchitis (EAO). Cell Immunol. 124:368–379.

Mahi-Brown, C. A., Yule, T. D., and Tung, K.S.K. (1987). Adoptive transfer of murine autoimmune orchitis to naive recipients with immune lymphocytes. Cell Immunol. 106:408–419.

Marsh, J. A., O'Hern, P., and Goldberg, E. (1981). The role of an x-linked gene in the regulation of secondary humoral response kinetics to sperm-specific LDH-C_4 antigen. J. Immunol. 126:100–106.

Metchnikoff, E. (1899). Etudes sur la resorption des cellules. Ann. Inst. Pasteur. 13:734–769.

Morales, C., Hermo, C. M., and Clermont, Y. (1984). Endocytosis in epithelial cells lining the rete testis of the rat. Anat. Rec. 209:185–195.

Nossal, G.J.V. (1989). Immunologic tolerance, in *Fundamental Immunology,* 2nd ed. (Paul, W. E., ed.) pp. 571–586, Raven Press, New York.

Perlmutter, R. M., Hansburg, D., Briles, D. E., Nicolotti,

R. A., and Davie, J. M. (1978). Subclass restriction of murine anti-carbohydrate antibodies. J. Immunol. 121:566–572.

Primakoff, P., Lathrop, W., Woolman, L., Cowan, A., and Myles, D. (1988). Fully effective contraception in male and female guinea pigs immunized with the sperm protein PH-20. Nature 335:543–546.

Sagaguchi, S., Fukuma, K., Kuribayashi, K., and Masuda, T. (1985). Organ-specific autoimmune diseases induced in mice by elimination of T cell subset. I. Evidence for active participation of T cells in natural self-tolerance; deficit of a T cell subset as a possible cause of autoimmune disease. J. Exp. Med. 161:72–87.

Salomon, F., Saremaslani, P., Jakob, M., and Hedinger, C. E. (1982). Immune complex orchitis in infertile men. Lab Invest 47:555–567.

Sarvetnick, N., Tsai, J-Y, Fox, H., Pilder, S. H., and Silver, L. M. (1989). A mouse chromosome 17 gene encodes a testes-specific transcript with unusual properties. Immunogenetics 30:34–41.

Schneider, R., Lees, R. K., Pedrazzini, T., Zinkernagel, R. M., Hengartner, H., and McDonald, H. R. (1989). Postnatal disappearance of reef-reactive ($V\beta G^+$) cells from the thymus of Mls^a mice. Implication for T cell development and autoimmunity. J. Exp. Med. 169:2149–2158.

Schwartz, R. H. (1989). Acquisition of immunologic self-tolerance. Cell 57:1073–1081.

Selawry, H., Fojaco, R., and Whittington, K. (1987). Extended survival of MHC-compatible islet grafts from diabetes-resistant donors in spontaneously diabetic BB/W rat. Diabetes 36:1061–1067.

Selawry, H. P., and Whittington, K. B. (1988). Prolonged intratesticular islet allograft survival is not dependent on local steroidogenesis. Horm. Metab. Res. 20:562–565.

Silver, L. M. (1985). Mouse t haplotypes. Ann. Rev. Genet. 19:179–208.

Smith, H., Sakamoto, Y., Kasai, K., and Tung, K.S.K. (1991). Effector and regulatory cells in autoimmune oophoritis elicited by neonatal thymectomy. J. Immunol. 147:2928–2933.

Smith, H., Chen, I. M., Kubo, R., and Tung, K.S.K. (1989). Neonatal thymectomy results in a repertoire enriched in T cells deleted in adult thymus. Science 245:749–752.

Sprent, J. (1989). T lymphocytes and the thymus, in *Fundamental Immunology,* 2nd ed. (Paul, W. E., ed.) pp. 69–93, Raven Press, New York.

Taguchi, O., and Nishizuka, Y. (1981). Experimental autoimmune orchitis after neonatal thymectomy in the mouse. Clin. Exp. Immunol. 46:425–434.

Tarter, T. H., and Alexander, N. J. (1984). Complement-inhibiting activity of seminal plasma. Am. J. Reprod. Immunol. 6:28–32.

Teuscher, C., Smith, S. M., Goldberg, E. H., Shearer, G. M., and Tung, K.S.K. (1985). Experimental allergic orchitis in mice. I. Genetic control of susceptibility and resistance to induction of autoimmune orchitis. Immunogenetics 22:323–333.

Teuscher, C., Wild, G. C., Johnson, E., and Tung, K.S.K. (1981). Vasectomy (an experimental autoimmune disease state). La Ricerca Clin Lab 11:313–329.

Tung, K.S.K. (1980). Autoimmunity of the testis, in *Immunological Aspects of Infertility and Fertility Regulation* (Dhindsa, D. S., and Schumacher, G.F.B., eds.) pp. 33–91, Elsevier North Holland, New York.

Tung, K.S.K., and Mahi-Brown, C. A. (1990). Autoimmune orchitis and oophoritis. Immunol. Clin. N. Am. 10:1–20.

Tung, K.S.K., Yule, T. D., Mahi-Brown, C. A., and Listrom, M. B. (1987). Distribution of histopathology and Ia positive cells in actively induced and passively transferred experimental autoimmune orchitis. J. Immunol. 138:752–759.

Tung, K.S.K., Ellis, L. E., Childs, G. V., and Dufau, M. (1984). The dark mink: A model of male infertility. Endocrinology 114:922–929.

Tung, K.S.K., Ellis, L., Teuscher, C., Meng, A., Blaustein, J. C., Kohno, S., and Howell, R. (1981). The black mink *(Mustela vison).* A natural model of immunologic male infertility. J. Exp. Med. 154:1016–1032.

Tung, K.S.K., Teuscher, C., and Meng, A. L. (1980). Autoimmunity to spermatozoa and the testis. Immunol. Rev. 55:217–255.

Tung, K.S.K., and Woodroffe, A. J. (1978). Immunopathology of experimental allergic orchitis in the rabbit. J. Immunol. 120:320–328.

Tung, K.S.K., Unanue, E. R., and Dixon, F. J. (1971). Pathogenesis of experimental allergic orchitis. II. The role of antibody. J. Immunol. 106:1463–1472.

Unanue, E. R. (1989). Macrophages, antigen-presenting cells, and the phenomenon of antigen handling and presentation, in *Fundamental Immunology,* 2nd ed., pp. 95–115, Raven Press, New York.

Voisin, G. A., Delaunay, A., and Barber, M. (1951). Lesions testiculaires provoquées chez le cobaye par injection d'extrait de testicule homologue. C.R. Hedb. Seances. Acad. Sci. 232:48–63.

Weigle, W. O. (1973). Immunological unresponsiveness. Adv. Immunol. 16:61–122.

Wyatt, C. R., Law, L., Magnuson, J. A., Griswold, M. D., and Magnuson, N. S. (1988). Suppression of lymphocyte proliferation by proteins secreted by cultured Sertoli cells. J. Reprod. Immunol. 14:27–40.

Yule, T. D., Montoya, G. D., Russell, L. D., Williams, T. M., and Tung, K.S.K. (1988). Autoantigenic germ cells exist outside the blood-testis barrier. J. Immunol. 141:1161–1167.

Index

www.ingramcontent.com/pod-product-compliance
Ingram Content Group UK Ltd.
Pitfield, Milton Keynes, MK11 3LW, UK
UKHW051048150926
475595UK00005B/174

9 780195 062694